Fundamental Physical Constants

Quantity	Symbol	Value	Comments
Speed of light	c	2.99792458×10^8 m-s^{-1}	
Charge of an electron	q_e	$-1.602177 \times 10^{-19}$ C	
Gravitational constant	G	6.67259×10^{-11} N-m^2-kg^{-2}	
Gravitational acceleration	g	9.78049 m-s^{-2}	Sea level, at equator
Planck's constant	h	$6.6260755 \times 10^{-34}$ J-s	
Boltzmann's constant	k_B	1.38066×10^{-23} J-(K)$^{-1}$	
Avogadro's number	N_A	6.022×10^{23} (mole)$^{-1}$	Molecules per mole
Permittivity of free space	ϵ_0	8.854×10^{-12} F-m^{-1}	$\sim(36\pi)^{-1} \times 10^{-9}$ F-m^{-1}
Permeability of free space	μ_0	$4\pi \times 10^{-7}$ H-m^{-1}	
Rest mass of an electron	m_e	9.10939×10^{-31} kg	
Rest mass of a proton	m_p	1.67262×10^{-27} kg	$(m_p/m_e) \simeq 1836$
Rest mass of a neutron	m_n	1.67493×10^{-27} kg	
Bohr radius	a	0.529177×10^{-10} m	Radius of hydrogen atom
1 electron-volt	eV	1.60217×10^{-19} J	
Frequency of 1-eV photon	f	2.41796×10^{14} Hz	

Powers of Ten

Power	Symbol	Prefix
10^{18}	E	Exa
10^{15}	P	Peta
10^{12}	T	Tera
10^9	G	Giga
10^6	M	Mega
10^3	k	kilo
10^{-3}	m	milli
10^{-6}	μ	micro
10^{-9}	n	nano
10^{-12}	p	pico
10^{-15}	f	femto
10^{-18}	a	atto

ELECTROMAGNETIC
Waves

UMRAN S. INAN

Stanford University

AZIZ S. INAN

University of Portland

Prentice Hall
Upper Saddle River, New Jersey 07458

Library of Congress Cataloging-in-Publication Data

Inan, Umran S.   Electromagnetic waves / Umran S. Inan, Aziz S. Inan.
        p. cm.
    Includes bibliographical references and index.
    ISBN 0-201-36179-5
    1. Electromagnetic waves. I. Inan, Aziz S. II. Title.
    QC661.I63 1999
    539.2–dc21   99-28489
                                                                  CIP

Publisher: Tom Robbins
Editor-in-chief: Marcia Horton
Production editor: Publication Services
Executive managing editor: Vince O'Brien
Assistant managing editor: Eileen Clark
Art director: Jayne Conte
Cover design: Karl Miyajima
Manufacturing manager: Trudy Pisciotti
Assistant vice president of production and manufacturing: David W. Riccardi

©2000 by Prentice Hall
Prentice-Hall, Inc.
Upper Saddle River, NJ 07458

The author and publisher of this book have used their best efforts in preparing this book.
These efforts include the development, research, and testing of the theories to determine their
effectiveness. The author and publisher make no warranty of any kind, expressed or implied,
with regard to the documentation contained in this book.

Printed in the United States of America

10 9 8 7 6 5 4 3

ISBN 0-201-36179-5

    Prentice-Hall International (UK) Limited, *London*
    Prentice-Hall of Australia Pty. Limited, *Sydney*
    Prentice-Hall Canada, Inc., *Toronto*
    Prentice-Hall Hispanoamericana, S.A., *Mexico*
    Prentice-Hall of India Private Limited, *New Delhi*
    Prentice-Hall of Japan, Inc., *Tokyo*
    Prentice-Hall (Singapore) Pte. Ltd., *Singapore*
    Editora Prentice-Hall do Brazil, Ltda., *Rio de Janeiro*

To our parents, wives, and children

Contents

reflection and refraction of electromagnetic waves from simple planar boundaries, followed by (Chapter 4) the guiding of electromagnetic waves within planar metallic or dielectric structures. Guiding of electromagnetic waves within cylindrical structures and electromagnetic resonators are discussed next (Chapter 5), followed by coverage of the interaction of electromagnetic waves with matter and their generation by simple sources (Chapter 6).

Emphasis on Physical Understanding

Future engineers and scientists need a clear understanding and a firm grasp of the basic principles so that they can understand, formulate, and interpret the results of complex practical problems. Engineers and scientists nowadays do not (and should not) spend time on working out formulas and obtaining numerical results by substitution. As most of the number crunching and formula manipulations are left to computers and packaged application and design programs, a solid grasp of fundamentals is now more essential than ever before. We maintain a constant link with established as well as new and emerging applications throughout the text, so that the reader's interest remains perked up, while at the same time emphasizing fundamental physical insight and solid understanding of basic principles. We strive to empower the reader with more than just a working knowledge of the manipulation of a dry set of vector relations and formulas. We supplement rigorous analyses with extensive discussions of the physical nature of the electromagnetic fields and waves involved, often from alternative points of view. In emphasizing physical understanding, we attempt to distill the essentials of physically based treatments available in physics texts, while presenting them in the context of traditional and emerging engineering applications.

Detailed Examples and Abundant Illustrations

We present the material in a clear and simple yet precise and accurate manner, with interesting examples illustrating each new concept. Many examples emphasize selected applications of electromagnetic waves. A total of 74 illustrative examples are detailed over six chapters, with two of the chapters having more than 20 examples. Each example is presented with an abbreviated topical title, a clear problem statement, and a detailed solution. Recognizing the importance of visualization in the reader's understanding, especially in view of the three-dimensional nature of electromagnetic wave phenomena, over 180 diagrams, graphs, and illustrations appear throughout the book.

Numerous End-of-Chapter Problems

Each chapter concludes with a variety of practice problems to allow the students to test their understanding of the material covered in the chapter, with a total of over 230 exercise problems spread over five chapters. The topical content of each problem is clearly identified in an abbreviated title (e.g., "Infrared antireflection coating" or "GPS signal transmission through the ionosphere"). Many of the problems explore interesting applications, and many practical "real-life" problems are included in each chapter to motivate students.

Historical Notes and Abbreviated Biographies

The history of the development of electromagnetic waves is laden with outstanding examples of pioneering scientists and development of scientific thought. Throughout our text we maintain a constant link with the pioneering giants and their work in order to bring about a better appreciation of the complex physical concepts and how they were discovered, which helps to motivate readers to keep their interest. We provide abbreviated biographies of the pioneers, emphasizing their scientific work in electromagnetics as well as in other fields, such as optics, heat, chemistry, and astronomy.

Emphasis on Clarity without Sacrificing Rigor and Completeness

This textbook presents the material at a simple enough level to be readable by senior undergraduate and first-year graduate students, but it is also rigorous in providing references and footnotes for in-depth analyses of selected concepts and applications. We provide the students with a taste of the rigor and completeness of classical reference texts, combined with a level of physical insight that was exemplified so well in some very old texts, while still maintaining the necessary level of organization and presentation clarity required for a modern textbook. We also provide not just a superficial but a sufficiently rigorous and in-depth exposure to a diverse range of applications of electromagnetic waves, in the body of the text, in examples, and in end-of-chapter problems.

Hundreds of Footnotes

In view of its fundamental physical nature, and its wide-encompassing generality, electromagnetics is a subject that lends itself particularly well to alternative ways of thinking about physical and engineering problems and is also particularly rich in terms of available scientific literature and many outstanding textbooks. Almost every new concept encountered can be thought of in different ways, and its implications can be explored further by the interested reader. We encourage such scholarly pursuit of enhanced knowledge and understanding by providing tens to hundreds of footnotes in each chapter, providing further comments, qualifications of statements made in the text, and references for in-depth analyses of selected concepts and applications. A total of over 360 footnotes are spread over six chapters. These footnotes do not interrupt the flow of ideas and the development of the main topics, but they provide an unprecedented degree of completeness for a textbook at this level, with interesting and sometimes thought-provoking content to make the subject more appealing.

ELECTROMAGNETICS IN ENGINEERING

The organization as well as the philosophy of this textbook is motivated by our view of the current status of electromagnetic fundamentals and waves in engineering curricula. *Electromagnetic Waves* is designed specifically for what normally is the second fields and waves course in most schools, the first one being a course in electromagnetic fundamentals. This book is the second of a sequence of two books by the same authors, the first one, *Engineering Electromagnetics*, having been designed specifically for the one-semester first course.

Understanding electromagnetics and appreciating its applications require a generally higher level of abstraction than most other topics encountered by electrical engineering students. The first course in electromagnetics, which most students take after having had vector calculus, aims at the development and understanding of Maxwell's equations, which requires the utilization of the full three-dimensional vector form of the fields and their relationships. It is this very step that makes the subject of electromagnetics appear insurmountable to many students and turns off their interest, especially when coupled with a lack of presentation and discussion of important and stimulating applications and the physical (and experimental) bases of the fundamental laws of physics. In our first book, we attempt to overcome this difficulty by (i) using a modern chapter organization, starting with an initial exposure to transmission lines and transients on high-speed distributed circuits, to bridge electrical circuits and electromagnetics, and (ii) emphasizing a combination of physical understanding; practical applications; historical notes and biographies; clarity without sacrificing rigor; and abundant examples, illustrations, and end-of-chapter problems. A first course based on *Engineering Electromagnetics* provides the students with a working knowledge of transmission lines as well as a solid, physically based background and a firm understanding of Maxwell's equations and their experimental bases. At that point, the student is ready to study in detail the most important manifestation of Maxwell's equations: electromagnetic waves, which is the subject of the present book.

Since electromagnetics is a mature basic science, and the topics covered in introductory texts are well established, the various texts differ primarily in their organization as well as range and depth of coverage. In formulating our approach, we were cognizant of the many challenges and opportunities that lie ahead in teaching electromagnetic waves.[1] Challenges include (i) the need to return to fundamentals (rather than rely on derived concepts), especially in view of the many emerging new applications that exploit unusual properties of materials and that rely on unconventional device concepts, and (ii) the need to maintain student interest, in spite of the reputation of electromagnetics as a difficult and abstract subject. Opportunities are

---

[1] J. R. Whinnery, The Teaching of Electromagnetics, *IEEE Trans. Education,* 33(1), pp. 3–7, February 1990.

abundant, especially as the engineers working in electronics and computer science are now finding that as devices get smaller and faster, circuit theory is insufficient in describing system performance or facilitating design. The need for a basic understanding of electromagnetic waves and their guided propagation is underscored by the explosive expansion of the use of optical fibers and consideration of extremely high data rates (ranging to 10 Gb-s^{-1})[2] and the emerging use of high-performance, high-density cables for communication within systems that will soon be required to carry digital signals at Gb-s^{-1} rates over distances of a few meters.[3] The rapidly increasing demand for personal wireless communication systems similarly requires a thorough understanding of the electromagnetic propagation channel, varying from simple line of sight to one that is severely obstructed by buildings and foliage.[4] In addition, issues of electromagnetic interference (EMI) and electromagnetic compatibility (EMC) are beginning to limit the performance of system-, board-, and chip-level designs, and electrostatic discharge phenomena significantly impact the design and performance of integrated circuits.[5]

1.3   RECOMMENDED COURSE CONTENT

This book is specifically designed for a one-term course in electromagnetic waves, nowadays typically the second (required or optional) fields and waves course in most electrical engineering curricula. The recommended course content for a regular three-unit one-semester course (42 contact hours) is provided in Table 1. The sections under "Cover" are recommended for complete coverage, including illustrative examples, while those marked "Skim" are recommended to be covered lightly, although the material provided is complete in case individual students want to go into more detail. The sections marked with a superscript asterisk are intended to provide flexibility to the individual instructor. For example, depending on the desired emphasis, one may want to choose between covering oblique reflection from lossy interfaces (Sec. 3.8), wave propagation in plasmas (Sec. 6.2), or ferrites (Sec. 6.4).

Table 1 also shows a recommended course content for a three-unit one-quarter course (27 contact hours) identical to the "Electromagnetic Waves" course (entry

[2]R. Heidelmann, B. Wedding, and G. Veith, 10-Gb/s Transmission and Beyond, *Proc. IEEE*, 81(11), pp. 1558–1567, November 1993.

[3]Falk, H., Prolog to Electrical Characteristics of Interconnections for High-Performance Systems, *Proc. IEEE*, 86(2), pp. 313–314, February 1998.

[4]See Special Issue on Wireless Communications, *IEEE Trans. Antennas and Propagation*, 46(6), June 1998.

[5]Vinson, J. E., and J. J. Liou, Electrostatic Discharge in Semiconductor Devices: An Overview, *Proc. IEEE*, 86(2), pp. 399–418, February 1998.

TABLE 1: Suggested course content

Chapter	Quarter course (27 hours)			Semester course (42 hours)		
	Cover	Skim	Skip	Cover	Skim	Skip
1	All			All		
2	2.1–2.6	2.7	2.4.4	All		
3	3.1–3.7	3.1.1	3.8	3.1–3.7, 3.8*		
4	All			All		
5	5.1, 5.2.2	5.2.1	5.2.3	5.1, 5.2	5.2.3*	
	5.3.1	5.3.2, 5.3.3		5.3		
6	6.1, 6.2.1	6.5.1, 6.5.2	6.2.2, 6.2.3	6.1, 6.2*, 6.3*		
			6.3–6.5, 6.5.3	6.4*, 6.5*		

course for the fields and waves specialization for the BSEE degree) that one of us taught at Stanford for the past seven years. This topical coverage provides the students with a solid, physically based background and a working knowledge of electromagnetic wave phenomena and their applications.

INSTRUCTOR'S MANUAL

As educators with a good deal of experience, we firmly believe that practice is the key to learning and that homework and exams are all instruments of teaching—although they may not be regarded as such by the students at the time. In our own courses, we take pride in providing the students with detailed solutions of homework and exam problems, rather than cryptic and abbreviated answers. To aid the instructors who choose to use this text, we have thus taken it upon ourselves to prepare a well-laid-out solutions manual, describing the solution of *every* end-of-chapter problem, in the same step-by-step detailed manner as our illustrative examples within the chapters. The solution for each end-of-chapter problem has been typeset by the authors themselves, with special attention to pedagogical detail. This instructor's manual is available to instructors upon request from Prentice Hall.

As authors of this book, we are looking forward to interacting with its users, both students and instructors, to collect and respond to their comments, questions, and corrections. We can most easily be reached by electronic mail at inan@nova.stanford.edu (URL: http://nova.stanford.edu/~vlf/) and at ainan@egr.up.edu.

ACKNOWLEDGEMENTS

We gratefully acknowledge those who have made significant contributions to the successful completion of this text. We thank Professor J. W. Goodman of Stanford for his generous support of textbook writing by faculty throughout his term as department chair. We thank many students at both Stanford and the University of Portland who have identified errors and suggested clarifications. We owe special thanks to our reviewers for their valuable comments and suggestions, including J. Bredow of the University of Texas—Arlington, S. Castillo of New Mexico State University, R. J. Coleman of the University of North Carolina—Charlotte, A. Dienes of the University of California—Davis, J. Dunn of the University of Colorado, D. S. Elliott of Purdue University, R. A. Kinney of Louisiana State University, L. Rosenthal of Farleigh Dickinson University, E. Schamiloglu of the University of New Mexico, T. Shumpert of Auburn University, D. Stephenson of Iowa State University, E. Thomson of the University of Florida, J. Volakis of the University of Michigan, and A. Weisshaar of Oregon State University.

We greatly acknowledge the efforts of the Addison Wesley Longman staff, including Patti Myers, Anna Eberhard Friedlander, Kamila Storr, Kevin Berry, and our original editor Paul Becker, as well as our current Prentice Hall editor Tom Robbins, whose dedication and support were crucially important in completing this project. We also thank Kris Engberg and other staff of Publication Services, who did an outstanding job with the layout and production of this textbook.

We dedicate this book to our parents, Mustafa and Hayriye Inan, for their dedication to our education; to our wives Elif and Belgin, for their persistent support and understanding as this project expanded well beyond our initial expectations and consumed literally all of our available time for too many years; and to our children, Ayse, Ali, Baris, and Cem, for the joy they bring to our lives.

—Umran S. Inan
—Aziz S. Inan

Maxwell's Equations and Electromagnetic Waves

This book provides an introduction to electromagnetic waves, their propagation in empty space or material media, their reflection from boundaries, their guiding within metallic boundaries or other structures, their interaction with matter, and their generation by simple sources.

All classical electromagnetic phenomena are governed by a compact and elegant set of fundamental rules known as *Maxwell's equations.* These four coupled partial differential equations were put forth as the complete classical theory of electromagnetics by James Clerk Maxwell in a series of brilliant papers[1] between 1856 and 1865. Maxwell's work culminated in the classic paper[2] in which he provided a mathematical framework for Faraday's primarily experimental results, elucidated the different behavior of conductors and insulators under the influence of fields, introduced the concept of displacement current,[3] and inferred the electromagnetic nature of light. A fundamental prediction of this theoretical framework is the existence of electromagnetic waves, a conclusion at which Maxwell arrived in the absence of experimental evidence that such waves can exist and propagate through

---

[1] For an excellent account with passages quoted from Maxwell's papers, see Chapter 5 of R. S. Elliott, *Electromagnetics,* IEEE Press, New Jersey, 1993.

[2] J. C. Maxwell, A dynamical theory of the electromagnetic field, *Phil. Trans. Royal Soc.* (London), 155, p. 450, 1865.

[3] See Section 7.4 of U. S. Inan and A. S. Inan, *Engineering Electromagnetics,* Addison Wesley Longman, 1999.

empty space. His bold hypotheses were to be confirmed 23 years later (in 1887) by the experiments of Heinrich Hertz.[4]

When most of classical physics was fundamentally revised as a result of Einstein's introduction[5] of the special theory of relativity, Maxwell's equations remained intact.[6] To this day, they stand as the most general mathematical statement of the fundamental natural laws that govern all of classical electrodynamics. The basic justification and validity of Maxwell's equations lies in their consistency with physical experiments over the entire range of the experimentally observed electromagnetic spectrum (see Section 2.2.1 and Table 2.1), extending from cosmic rays at frequencies greater than 10^{22} Hz to the so-called micropulsations at frequencies of 10^{-3} Hz. The associated practical applications cover an equally wide range, from the use of gamma rays (10^{18}–10^{22} Hz) for cancer therapy to the use of waves at frequencies of a few Hz and below for geophysical prospecting. Electromagnetic wave theory as embodied in Maxwell's equations has provided the underpinning for the development of many vital practical tools of our technological society, including broadcast radio, radar, television, cellular phones, optical communications, the Global Positioning Systems (GPS), microwave heating and processing, and X-ray imaging.

Maxwell's equations are based on experimentally established facts, namely Coulomb's law,[7] Ampère's law,[8] Faraday's law,[9] and the principle of conservation

---

[4]H. Hertz, On the finite velocity of propagation of electromagnetic actions, *Sitzb. d. Berl. Akad. d. Wiss.,* Feb. 2, 1888; for a collected English translation of this and other papers by H. Hertz, see H. Hertz, *Electric Waves,* MacMillan, London, 1893.

[5]Einstein, A., *Annalen der Physik,* 1905. The English translation of this paper is remarkably readable and is available in a collection of original papers titled *The Principle of Relativity,* Dover Publications, New York.

[6]Maxwell's formulation was in fact one of the major motivating factors that led to the development of the theory of special relativity. The fact that Galilean relativity was consistent with classical mechanics but inconsistent with electromagnetic theory suggested that either Maxwell's equations were incorrect or that the laws of mechanics needed to be modified. For discussions of the relationship between electromagnetism and the special theory of relativity, see Section 15 of D. M. Cook, *The Theory of the Electromagnetic Field,* Prentice-Hall, Inc., 1975; Chapter 10 of D. J. Griffiths, *Introduction to Electrodynamics,* 2nd ed., Prentice-Hall, Inc., New Jersey, 1989; Chapter 2 of R. S. Elliott, *Electromagnetics,* IEEE Press, Piscataway, New Jersey, 1993; Chapter 11 of J. D. Jackson, *Classical Electrodynamics,* 2nd ed., Wiley, New York, 1975.

[7]Coulomb's law states that electric charges attract or repel one another in a manner inversely proportional to the square of the distance between them; C. A. de Coulomb, Première mémoire sur l'électricité et magnétisme (First memoir on electricity and magnetism), *Histoire de l'Académie Royale des Sciences,* p. 569, 1785.

[8]Ampère's law states that current-carrying wires create magnetic fields and exert forces on one another, with the amplitude of the magnetic field (and thus force) depending on the inverse square of the distance; A. M. Ampère, *Recueil d'observations électrodynamiques,* Crochard, Paris, 1820–1833.

[9]Faraday's law states that magnetic fields that vary with time induce electromotive force or electric field; M. Faraday, *Experimental Researches in Electricity,* Taylor, London, Vol.I, 1839, pp. 1–109.

of electric charge. Detailed discussion of the experimental bases of Maxwell's equations is available elsewhere.[10]

Our purpose in this book is to discuss the most important consequence of Maxwell's equations: electromagnetic waves, which can propagate from one point to another through intervening space, whether that be empty or filled with matter. After a brief introduction and review of Maxwell's equations and wave concepts in this chapter, the following five chapters are devoted to the consideration of the propagation of electromagnetic waves under different circumstances. In Chapter 2 we restrict our attention to electromagnetic waves in an *unbounded, simple,* and *source-free* medium. In Chapter 3 we remove the first condition and consider how waves reflect and refract in the presence of planar boundaries. In Chapters 4 and 5 we consider the case of guided wave propagation in regions bounded by conducting and dielectric structures. In Chapter 6 we return to wave propagation in an unbounded medium but now allow it not to be simple, with complicated material properties significantly affecting the nature of electromagnetic waves. We also provide a brief discussion of sources of electromagnetic waves (namely, antennas) and the principles of electromagnetic radiation.

Maxwell, James Clerk   *(Scottish mathematician and physicist, b. November 13, 1831, Edinburgh; d. November 5, 1879, Cambridge, England)   Maxwell, born of a well-known Scottish family, was an only son. His mother died of cancer when he was eight, but except for that, he had a happy childhood.*

Early in life, he showed signs of mathematical talent. He entered Cambridge in 1850 and graduated second in his class in mathematics. Maxwell was appointed to his first professorship at Aberdeen in 1856. In 1857 he made his major contribution to astronomy by showing mathematically that Saturn's rings must consist of numerous small solid particles in order to be dynamically stable. In 1860 Maxwell brought his mathematics to bear upon the problem of particles making up gases. He considered the molecules to move not only in all directions but at all velocities, and bouncing off each other and off the walls of the container with perfect elasticity. Along with Boltzmann [1844–1906], he worked out the Maxwell–Boltzmann kinetic theory of gases, bringing about an entirely new view of heat and temperature.

The crowning work of Maxwell's life took place between 1864 and 1873, when he placed into mathematical form the speculations of Faraday [1791–1867] concerning magnetic lines of force. (Maxwell resembled Faraday, by the way, in possessing deep religious convictions and in having a childless, but very happy, marriage.)

---

[10]Numerous books on fundamental electromagnetics discuss Coulomb's law, Ampère's law, Faraday's law, and Maxwell's equations. For a recent reference that provides a physical and experimentally based point of view, see Chapters 4 through 7 of U. S. Inan and A. S. Inan, *Engineering Electromagnetics,* Addison Wesley Longman, Inc., 1999.

In developing the concept of lines of force, Maxwell was able to work out a few simple equations that expressed all the varied phenomena of electricity and magnetism and bound them indissolubly together. Maxwell's theory showed that electricity and magnetism could not exist in isolation—where one exists, so does the other—so his work is usually referred to as the electromagnetic theory.

Maxwell showed that the oscillation of an electric charge produced an electromagnetic field that radiated outward from its source at a constant speed. This speed could be calculated by taking the ratio of certain units expressing magnetic phenomena to units expressing electrical phenomena. This ratio worked out to be just about 300,000 kilometers per second, or 186,300 miles per second, which is approximately the speed of light (for which the best available figure at present is 299,792.5 kilometers per second, or 186,282 miles per second).

To Maxwell, this seemed to be more than coincidence, and he suggested that light itself arose through an oscillating electric charge and was therefore an electromagnetic radiation. Furthermore, since charges could oscillate at any speed, it seemed to Maxwell that a whole family of electromagnetic radiations existed, of which visible light was only a small part.

Maxwell believed not only that the waves of electromagnetic radiation were carried by the ether, but that the magnetic lines of force were actually disturbances of the ether. In this way, he conceived that he had abolished the notion of action at a distance. It had seemed to some experimenters in electricity and magnetism (Ampère [1775–1836], for instance), that a magnet attracted iron without actually making contact with the iron. To Maxwell it seemed that the disturbances in the ether set up by the magnet touched the iron and that everything could be worked out as action on contact. Maxwell also rejected the notion that electricity was particulate in nature, even though that was so strongly suggested by Faraday's laws of electrolysis.

Maxwell died of cancer before the age of fifty. Had he lived out what would today be considered a normal life expectancy, he would have seen his prediction of a broad spectrum of electromagnetic radiation verified by Hertz [1857–1894]. However, he would also have seen the ether, which his theory had seemed to establish firmly, brought into serious question by the epoch-making experiment of Michelson [1852–1931] and Morley [1838–1923], and he would have seen electricity proved to consist of particles after all. His electromagnetic equations did not depend on his own interpretations of the ether, however, and he had wrought better than he knew. When Einstein's [1879–1955] theories, a generation after Maxwell's death, upset almost all of classical physics, Maxwell's equations remained untouched—as valid as ever. [Adapted with permission from I. Asimov, Biographical Encyclopedia of Science and Technology, *Doubleday, 1982.]*

1700   1831   1879   2000

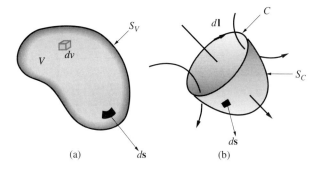

FIGURE 1.1.   Contour, surface, and volume. (a) A closed surface S_V enclosing a volume V. (b) A closed contour C enclosing a surface S_C.

1.1   REVIEW OF MAXWELL'S EQUATIONS

We start with a brief review of Maxwell's equations[11] and their foundations. These four fundamental equations of electromagnetics are based on three separate experimentally established facts: Coulomb's law, Ampère's law (or the Biot–Savart law), Faraday's law, and the principle of conservation of electric charge. The validity of Maxwell's equations is based on their consistency with all of our experimental knowledge to date concerning electromagnetic phenomena. The physical quantities that appear in Maxwell's equations are the electric field $\overline{\mathscr{E}}$, the magnetic field[12] $\overline{\mathscr{B}}$, the electric flux density $\overline{\mathscr{D}}$, the magnetic field intensity $\overline{\mathscr{H}}$, electric current density $\overline{\mathscr{J}}$, and electric charge density $\tilde{\rho}$. The physical meaning of the equations is better perceived in the context of their integral forms, which are listed below together with their differential counterparts:

1. Faraday's law is based on the experimental fact that time-changing magnetic flux induces electromotive force:

$$\oint_C \overline{\mathscr{E}} \cdot d\mathbf{l} = -\int_{S_C} \frac{\partial \overline{\mathscr{B}}}{\partial t} \cdot d\mathbf{s} \qquad \nabla \times \overline{\mathscr{E}} = -\frac{\partial \overline{\mathscr{B}}}{\partial t} \qquad [1.1]$$

where the contour C encloses the surface S_C, as shown in Figure 1.1b, and where the sense of the line integration over the contour C (i.e., the direction of $d\mathbf{l}$) must be consistent with the direction of the surface vector $d\mathbf{s}$ in accordance with the right-hand rule.

2. Gauss's law is a mathematical expression of the experimental fact that electric charges attract or repel one another with a force inversely proportional to the

[11]J. C. Maxwell, *A Treatise on Electricity and Magnetism,* Clarendon Press, Oxford, 1892, Vol. 2, pp. 247–262.

[12]The $\overline{\mathscr{B}}$ field is often referred to as the "magnetostatic induction field" or "magnetic flux density." However, both of these terms are in fact not appropriate and we choose to refer to $\overline{\mathscr{B}}$ simply as the "$\overline{\mathscr{B}}$ field." For a discussion of this issue see Sections 6.1 and 6.8 of U. S. Inan and A. S. Inan, *Engineering Electromagnetics,* Addison Wesley Longman, 1999.

square of the distance between them (i.e., Coulomb's law):

$$\oint_{S_V} \overline{\mathcal{D}} \cdot d\mathbf{s} = \int_V \tilde{\rho} \, dv \qquad \nabla \cdot \overline{\mathcal{D}} = \tilde{\rho} \qquad [1.2]$$

where the surface S_V encloses the volume V, as shown in Figure 1.1a. The volume charge density is represented with $\tilde{\rho}$ to distinguish it from the phasor form ρ used in the time-harmonic form of Maxwell's equations.

3. Maxwell's third equation is a generalization of Ampère's law, which states that the line integral of the magnetic field over any closed contour must equal the total current enclosed by that contour:

$$\oint_C \overline{\mathcal{H}} \cdot d\mathbf{l} = \int_{S_C} \overline{\mathcal{J}} \cdot d\mathbf{s} + \int_{S_C} \frac{\partial \overline{\mathcal{D}}}{\partial t} \cdot d\mathbf{s} \qquad \nabla \times \overline{\mathcal{H}} = \overline{\mathcal{J}} + \frac{\partial \overline{\mathcal{D}}}{\partial t} \qquad [1.3]$$

where the contour C encloses the surface S_C, as shown in Figure 1.1b. Maxwell's third equation expresses the fact that time-varying electric fields produce magnetic fields. This equation with only the first term on the right-hand side (also referred to as the conduction-current term) is Ampère's law, which is a mathematical statement of the experimental findings of Oersted, whereas the second term, known as the displacement-current term, was introduced theoretically by Maxwell in 1862 and verified experimentally many years later (1888) in Hertz's experiments.[13]

4. Maxwell's fourth equation is based on the fact that there are no magnetic charges (i.e., magnetic monopoles) and that, therefore, magnetic field lines always close on themselves:

$$\oint_{S_V} \overline{\mathcal{B}} \cdot d\mathbf{s} = 0 \qquad \nabla \cdot \overline{\mathcal{B}} = 0 \qquad [1.4]$$

where the surface S_V encloses the volume V, as shown in Figure 1.1a. This equation can actually be derived[14] from the Biot–Savart law, so it is not completely independent.[15]

---

[13]H. Hertz, On the finite velocity of propagation of electromagnetic actions, *Sitzb. d. Berl. Akad. d. Wiss.*, Feb. 2, 1888; for a collected English translation of this and other papers by H. Hertz, see H. Hertz, *Electric Waves,* MacMillan, London, 1893.

[14]See Sections 6.5 and 6.7 of U. S. Inan and A. S. Inan, *Engineering Electromagnetics,* Addison Wesley Longman, 1999.

[15]It is also interesting that [1.4] can be derived from [1.1] by taking the divergence of the latter and using the vector identity of $\nabla \cdot (\nabla \times \overline{\mathcal{G}}) \equiv 0$, which is true for any vector $\overline{\mathcal{G}}$. We find

$$\nabla \cdot (\nabla \times \overline{\mathcal{E}}) = -\nabla \cdot \left(\frac{\partial \overline{\mathcal{B}}}{\partial t} \right) \quad \rightarrow \quad 0 = -\frac{\partial (\nabla \cdot \overline{\mathcal{B}})}{\partial t} \quad \rightarrow \quad \text{const.} = \nabla \cdot \overline{\mathcal{B}}$$

The constant can then be shown to be zero by the following argument. If we suppose that the $\overline{\mathcal{B}}$ field was produced a finite time ago (that is, it has not always existed) then if we go back far enough in time, we have $\overline{\mathcal{B}} = 0$ and therefore $\nabla \cdot \overline{\mathcal{B}} = 0$. Hence it would appear that

$$\nabla \cdot \overline{\mathcal{B}} = 0 \quad \text{and} \quad \oint_S \overline{\mathcal{B}} \cdot d\mathbf{s} = 0$$

The two constitutive relations $\overline{\mathcal{D}} = \epsilon\overline{\mathcal{E}}$ and $\overline{\mathcal{B}} = \mu\overline{\mathcal{H}}$ (more properly expressed as $\overline{\mathcal{H}} = \mu^{-1}\overline{\mathcal{B}}$, because in the magnetostatic case $\overline{\mathcal{H}}$ is the medium-independent quantity[16]) govern the manner by which the electric and magnetic fields, $\overline{\mathcal{E}}$ and $\overline{\mathcal{B}}$, are related to the medium-independent quantities, $\overline{\mathcal{D}}$ and $\overline{\mathcal{H}}$, in material media, where, in general, $\epsilon \neq \epsilon_0$ and $\mu \neq \mu_0$. The current density $\overline{\mathcal{J}}$ is in general given by $\overline{\mathcal{J}} = \overline{\mathcal{J}}_{\text{source}} + \overline{\mathcal{J}}_{\text{c}}$, where $\overline{\mathcal{J}}_{\text{source}}$ represents the source currents from which magnetic fields originate and $\overline{\mathcal{J}}_{\text{c}} = \sigma\overline{\mathcal{E}}$ represents the conduction current, which flows in conducting media ($\sigma \neq 0$) whenever there is an electric field present. The volume charge density $\tilde{\rho}$ represents the sources from which electric fields originate.

Note that ϵ, μ, and σ are macroscopic parameters that describe the relationships among macroscopic field quantities, but they are based on the microscopic behavior of the atoms and molecules in response to the fields. These parameters are simple constants only for *simple* material media, which are linear, homogeneous, time-invariant, and isotropic. Otherwise, for complex material media that are non-linear, inhomogeneous, time-variant, or anisotropic, ϵ, μ, and σ may depend on the magnitudes of $\overline{\mathcal{E}}$ and $\overline{\mathcal{B}}$ (nonlinear), on spatial coordinates (x, y, z) (inhomogeneous), on time (time-variant), or on the orientations of $\overline{\mathcal{E}}$ and $\overline{\mathcal{H}}$ (anisotropic). For anisotropic media, ϵ, μ, or σ is generally expressed as a matrix (called a *tensor*) whose entries relate each component (e.g., the x, y, or z component) of $\overline{\mathcal{E}}$ (or $\overline{\mathcal{H}}$) to the other three components (e.g., the x, y, and z components) of $\overline{\mathcal{D}}$ or $\overline{\mathcal{J}}$ (or $\overline{\mathcal{B}}$). In ferromagnetic materials, the magnetic field $\overline{\mathcal{B}}$ is determined by the past history of the field $\overline{\mathcal{H}}$ rather than by its instantaneous value; such substances are said to exhibit *hysteresis*. Some hysteresis effects can also be seen in certain dielectric materials.

It is possible to eliminate the vectors $\overline{\mathcal{D}}$ and $\overline{\mathcal{H}}$ from Maxwell's equations by substituting for them as follows:

$$\overline{\mathcal{D}} = \epsilon_0\overline{\mathcal{E}} + \overline{\mathcal{P}} \qquad [1.5]$$

$$\overline{\mathcal{H}} = \frac{\overline{\mathcal{B}}}{\mu_0} - \overline{\mathcal{M}} \qquad [1.6]$$

where $\overline{\mathcal{P}}$ is the polarization vector in a dielectric in units of coulombs-m^{-2}, and $\overline{\mathcal{M}}$ is the magnetization vector in a magnetic medium in units of amperes-m^{-1}. These two quantities account for the presence of matter at the points considered. Maxwell's equations then take the following form:

$$\nabla \times \overline{\mathcal{E}} = -\frac{\partial\overline{\mathcal{B}}}{\partial t} \qquad [1.7a]$$

$$\nabla \cdot \overline{\mathcal{E}} = \frac{1}{\epsilon_0}(\tilde{\rho} - \nabla \cdot \overline{\mathcal{P}}) \qquad [1.7b]$$

---

[16] See Section 6.2.3 and footnote 71 in Section 6.8.2 of U. S. Inan and A. S. Inan, *Engineering Electromagnetics*, Addison Wesley Longman, 1999.

$$\nabla \times \overline{\mathscr{B}} = \epsilon_0 \mu_0 \frac{\partial \overline{\mathscr{E}}}{\partial t} + \mu_0 \left(\overline{\mathscr{J}} + \frac{\partial \overline{\mathscr{P}}}{\partial t} + \nabla \times \overline{\mathscr{M}} \right) \qquad [1.7c]$$

$$\nabla \cdot \overline{\mathscr{B}} = 0 \qquad [1.7d]$$

The preceding equations are completely general but are expressed in a way that stresses the contributions of the medium. Note that the presence of matter has the effect of adding the bound volume charge density $-\nabla \cdot \overline{\mathscr{P}}$, the polarization current density $\partial \overline{\mathscr{P}}/\partial t$, and the equivalent volume magnetization current density $\nabla \times \overline{\mathscr{M}}$. Note also that $\tilde{\rho}$ is free charge density, while $\overline{\mathscr{J}}$ could be a source current $\overline{\mathscr{J}}_{\text{source}}$, a conduction current given by $\overline{\mathscr{J}}_c = \sigma \overline{\mathscr{E}}$, or a superposition thereof.

The continuity equation, which expresses the principle of conservation of charge in differential form, is contained in Maxwell's equations and in fact can be readily derived[17] by taking the divergence of [1.3] and using [1.2]. For the sake of completeness, we give the integral and differential forms of the continuity equation:

$$-\oint_{S_V} \overline{\mathscr{J}} \cdot d\mathbf{s} = \frac{\partial}{\partial t} \int_V \tilde{\rho} \, dv \qquad \nabla \cdot \overline{\mathscr{J}} = -\frac{\partial \tilde{\rho}}{\partial t} \qquad [1.8]$$

where the surface S_V encloses the volume V, as shown in Figure 1.1a.

Note that for all of the equations [1.1] through [1.4] and equation [1.8], the differential forms can be derived from the integral forms (or vice versa) by using either Stokes's or the divergence theorem, both of which are valid for any arbitrary vector field $\overline{\mathscr{G}}$. These theorems are

$$\oint_C \overline{\mathscr{G}} \cdot d\mathbf{l} = \int_{S_C} (\nabla \times \overline{\mathscr{G}}) \cdot d\mathbf{s} \qquad \text{(Stokes's theorem)} \qquad [1.9]$$

where the contour C encloses the surface S_C, and

$$\oint_{S_V} \overline{\mathscr{G}} \cdot d\mathbf{s} = \int_V (\nabla \cdot \overline{\mathscr{G}}) \, dv \qquad \text{(divergence theorem)} \qquad [1.10]$$

where the surface S_V encloses volume V.

1.2 TIME-HARMONIC MAXWELL'S EQUATIONS

Numerous practical applications (e.g., broadcast radio and TV, radar, optical, and microwaves) involve transmitting sources that operate in such a narrow band of frequencies that the behavior of all the field components is very similar to that of the central single-frequency sinusoid (i.e., the carrier). In many applications the transients involved at the time the signal is switched on (or off) are not of concern, so

---

[17]That the continuity equation can be derived from equations [1.2] and [1.3] indicates that Maxwell's equations [1.2] and [1.3] are not entirely independent, if we accept conservation of electric charge as a fact.

the steady-state sinusoidal approximation is most suitable. For example, for an AM broadcast station operating at a carrier frequency of 1 MHz, any turn-on transients last only a few μs and are of little consequence in practice. For all practical purposes, the signal propagating from the transmitting antenna to the receivers can be treated as a sinusoid, with its amplitude modulated within a narrow bandwidth (e.g., ±5 kHz) around the carrier frequency. Since the characteristics of the propagation medium do not vary significantly over this bandwidth, we can describe the propagation behavior of the AM broadcast signal by studying a single sinusoidal carrier at a frequency of 1 MHz.

The time-harmonic (sinusoidal steady-state) forms of Maxwell's equations[18] are listed here together with their more general versions:

$$\nabla \times \overline{\mathcal{E}} = -\frac{\partial \overline{\mathcal{B}}}{\partial t} \qquad \nabla \times \mathbf{E} = -j\omega \mathbf{B} \qquad [1.11a]$$

$$\nabla \cdot \overline{\mathcal{D}} = \tilde{\rho} \qquad \nabla \cdot \mathbf{D} = \rho \qquad [1.11b]$$

$$\nabla \times \overline{\mathcal{H}} = \overline{\mathcal{J}} + \frac{\partial \overline{\mathcal{D}}}{\partial t} \qquad \nabla \times \mathbf{H} = \mathbf{J} + j\omega \mathbf{D} \qquad [1.11c]$$

$$\nabla \cdot \overline{\mathcal{B}} = 0 \qquad \nabla \cdot \mathbf{B} = 0 \qquad [1.11d]$$

Note that in [1.11a–d] the field vectors $\overline{\mathcal{E}}$, $\overline{\mathcal{D}}$, $\overline{\mathcal{H}}$, and $\overline{\mathcal{B}}$ are real (measurable) quantities that vary with time, whereas the vectors $\mathbf{E}$, $\mathbf{D}$, $\mathbf{H}$, and $\mathbf{B}$ are complex phasors that do not vary with time. In general, we can obtain the former from the latter by multiplying by $e^{j\omega t}$ and taking the real part. For example,

$$\overline{\mathcal{E}}(x, y, z, t) = \mathcal{R}e\{\mathbf{E}(x, y, z)e^{j\omega t}\}$$

Note that the same is true for all of the quantities. For example,

$$\tilde{\rho}(x, y, z, t) = \mathcal{R}e\{\rho(x, y, z)e^{j\omega t}\}$$

Example 1-1 illustrates the use of equations [1.11].

Example 1-1:   Aircraft VHF communication signal.   The electric field component of an electromagnetic wave in air used by an aircraft to communicate with the air traffic control tower can be represented by

$$\overline{\mathcal{E}}(z, t) = \hat{\mathbf{y}}0.02 \cos(7.5 \times 10^8 t - \beta z) \quad \text{V-m}^{-1}$$

Find the corresponding wave magnetic field $\overline{\mathcal{H}}(z, t)$ and the constant β.

---

[18]For example, the actual derivation of the time-harmonic form for [1.1] is as follows:

$$\nabla \times \overline{\mathcal{E}} = -\frac{\partial \overline{\mathcal{B}}}{\partial t} \quad \rightarrow \quad \nabla \times [\underbrace{\mathcal{R}e\{\mathbf{E}(x, y, z)e^{j\omega t}\}}_{\overline{\mathcal{E}}}] = -\frac{\partial}{\partial t}[\underbrace{\mathcal{R}e\{\mathbf{B}(x, y, z)e^{j\omega t}\}}_{\overline{\mathcal{B}}}]$$

$$\rightarrow \quad \mathcal{R}e\{e^{j\omega t}\nabla \times \mathbf{E}\} = \mathcal{R}e\{-j\omega e^{j\omega t}\mathbf{B}\} \quad \rightarrow \quad \nabla \times \mathbf{E} = -j\omega \mathbf{B}$$

Solution: In view of the sinusoidal nature of the electric field signal, it is appropriate to work with phasors. The phasor form of the electric field is

$$\mathbf{E}(z) = \hat{\mathbf{y}}E_y(z) = \hat{\mathbf{y}}0.02e^{-j\beta z} \quad \text{V-m}^{-1}$$

Using [1.11a], we can write

$$\mathbf{H}(z) = -\frac{1}{j\omega\mu_0}\nabla \times \mathbf{E}(z) = \hat{\mathbf{x}}\frac{1}{j\omega\mu_0}\frac{\partial E_y}{\partial z}$$

$$= -\hat{\mathbf{x}}\frac{\beta}{\omega\mu_0}E_y = -\hat{\mathbf{x}}\frac{0.02\beta}{\omega\mu_0}e^{-j\beta z}$$

where $\omega = 7.5 \times 10^8$ rad-s^{-1} and $\mu_0 = 4\pi \times 10^{-7}$ H-m^{-1}. Substituting the expression for $\mathbf{H}(z)$ into [1.11c], we find

$$\mathbf{E}(z) = \frac{1}{j\omega\epsilon_0}\nabla \times \mathbf{H}(z) = \hat{\mathbf{y}}\frac{1}{j\omega\epsilon_0}\frac{\partial H_x}{\partial z} = \hat{\mathbf{y}}\frac{0.02\beta^2}{\omega^2\mu_0\epsilon_0}e^{-j\beta z}$$

where $\epsilon_0 \simeq 8.85 \times 10^{-12}$ F-m^{-1}. But this expression for $\mathbf{E}(z)$ must be the same as the electric field phasor expression we started with. Thus, we must have

$$\frac{0.02\beta^2}{\omega^2\mu_0\epsilon_0} = 0.02 \quad \rightarrow \quad \beta = \omega\sqrt{\mu_0\epsilon_0} \simeq \frac{7.5 \times 10^8 \text{ rad-s}^{-1}}{3 \times 10^8 \text{ m-s}^{-1}} = 2.5 \text{ rad-m}^{-1}$$

where we have used the fact that $(\mu_0\epsilon_0)^{-1/2} = c$, the speed of light in free space. The corresponding magnetic field $\mathbf{H}$ is then

$$\mathbf{H}(z) \simeq -\hat{\mathbf{x}}\frac{(0.02)(2.5)}{(7.5 \times 10^8)(4\pi \times 10^{-7})}e^{-j2.5z} \simeq -\hat{\mathbf{x}}53.1 \times e^{-j2.5z} \quad \mu\text{A-m}^{-1}$$

and the instantaneous magnetic field $\overline{\mathcal{H}}(z, t)$ is

$$\overline{\mathcal{H}}(z, t) = \Re e\{\mathbf{H}(z)e^{j\omega t}\} = \Re e\{-\hat{\mathbf{x}}53.1e^{-j2.5z}e^{j\omega t}\}$$

$$= -\hat{\mathbf{x}}53.1\cos(7.5 \times 10^8 t - 2.5z) \quad \mu\text{A-m}^{-1}$$

1.3 ELECTROMAGNETIC BOUNDARY CONDITIONS

The integral forms of equations [1.1] through [1.4] can be used to derive the relationships between electric- and magnetic-field components on both sides of interfaces between two different materials (i.e., different μ, ϵ, and/or σ).

The electromagnetic boundary conditions can be summarized as follows:

1. It follows from applying [1.1] to a contour C, as shown in Figure 1.2b, that the tangential component of the electric field $\overline{\mathcal{E}}$ is continuous across any interface:

$$\hat{\mathbf{n}} \times [\overline{\mathcal{E}}_1 - \overline{\mathcal{E}}_2] = 0 \quad \rightarrow \quad \mathcal{E}_{1t} = \mathcal{E}_{2t} \quad \quad [1.12]$$

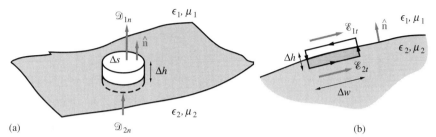

FIGURE 1.2. **Interfaces between two different materials.** The boundary conditions for the electromagnetic fields are derived by applying the surface integrals to the cylindrical surface as shown in (a) and the line integrals to the rectangular contour in (b).

where $\hat{\mathbf{n}}$ is the unit vector perpendicular to the interface and outward from medium 2 (shown as $\hat{\mathbf{n}}$ in Figure 1.2b).

2. It can be shown by applying [1.3] to the same contour C in Figure 1.2b that the tangential component of the magnetic field $\overline{\mathscr{H}}$ is continuous across any interface:

$$\hat{\mathbf{n}} \times [\overline{\mathscr{H}}_1 - \overline{\mathscr{H}}_2] = 0 \qquad \rightarrow \qquad \mathscr{H}_{1t} = \mathscr{H}_{2t} \qquad [1.13]$$

except where surface currents ($\overline{\mathscr{J}}_s$) may exist, such as at the surface of a perfect conductor (i.e., $\sigma = \infty$):

$$\hat{\mathbf{n}} \times \overline{\mathscr{H}}_1 = \overline{\mathscr{J}}_s \qquad [1.14]$$

noting that the field $\overline{\mathscr{H}}_2$ inside the perfect conductor is zero.[19]

3. It can be shown by applying [1.2] to the surface of the cylinder shown in Figure 1.2a that the normal component of electric flux density $\overline{\mathscr{D}}$ is continuous across interfaces, except where surface charge ($\tilde{\rho}_s$) may exist, such as at the surface of a metallic conductor or at the interface between two lossy dielectrics ($\sigma_1 \neq 0$, $\sigma_2 \neq 0$):

$$\hat{\mathbf{n}} \cdot (\overline{\mathscr{D}}_1 - \overline{\mathscr{D}}_2) = \tilde{\rho}_s \qquad \rightarrow \qquad \mathscr{D}_{1n} - \mathscr{D}_{2n} = \tilde{\rho}_s \qquad [1.15]$$

4. A consequence of applying [1.4] to the surface of the cylinder in Figure 1.2a is that the normal component of the magnetic field $\overline{\mathscr{B}}$ is continuous across interfaces:

$$\hat{\mathbf{n}} \cdot [\overline{\mathscr{B}}_1 - \overline{\mathscr{B}}_2] = 0 \qquad \rightarrow \qquad \mathscr{B}_{1n} = \mathscr{B}_{2n} \qquad [1.16]$$

5. It follows from applying [1.8] to the surface of the cylinder in Figure 1.2a that, at the interface between two lossy media (i.e., $\sigma_1 \neq 0$, $\sigma_2 \neq 0$), the normal component of the current density $\overline{\mathscr{J}}$ is continuous, except where time-varying surface charge may exist, such as at the surface of a perfect conductor or at the interface

---

[19]The interior of a solid perfect conductor is void of both static and dynamic magnetic fields; for a discussion at an appropriate level see Section 6.8.4 of U. S. Inan and A. S. Inan, *Engineering Electromagnetics,* Addison Wesley Longman, Inc., 1999.

between lossy dielectrics (the general case when $\epsilon_1 \neq \epsilon_2$ and $\sigma_1 \neq \sigma_2$):

$$\hat{\mathbf{n}} \cdot (\bar{\mathscr{J}}_1 - \bar{\mathscr{J}}_2) = -\frac{\partial \tilde{\rho}_s}{\partial t} \qquad \rightarrow \qquad J_{1n} - J_{2n} = -\frac{\partial \tilde{\rho}_s}{\partial t} \qquad [1.17]$$

Note that [1.17] is not completely independent of [1.12] through [1.16], since [1.8] is contained in [1.1] through [1.4], as mentioned previously. For the stationary case ($\partial/\partial t = 0$), [1.17] implies $\mathscr{J}_{1n} = \mathscr{J}_{2n}$, or $\sigma_1 \mathscr{E}_{1n} = \sigma_2 \mathscr{E}_{2n}$, which means that at the interface between lossy dielectrics (i.e., $\epsilon_1 \neq \epsilon_2$ and $\sigma_1 \neq \sigma_2$) there must in general be nonzero surface charge (i.e., $\tilde{\rho}_s \neq 0$), because otherwise [1.15] demands that $\epsilon_1 \mathscr{E}_{1n} = \epsilon_2 \mathscr{E}_{2n}$.

1.4   ELECTROMAGNETIC WAVES

Maxwell's equations embody all of the essential aspects of electromagnetics—the ideas that light is electromagnetic in nature, that electric fields that change in time create magnetic fields in the same way as time-varying voltages induce electric currents in wires, and that the source of electric and magnetic energy resides not only on the body that is electrified or magnetized but also, and to a far greater extent, in the surrounding medium. However, arguably the most important and far-reaching implication of Maxwell's equations is the idea that electric and magnetic effects can be transmitted from one point to another through the intervening space, whether that be empty or filled with matter.

Electromagnetic energy propagates, or travels from one point to another, as *waves*. The propagation of *electromagnetic waves* results in the phenomenon of *delayed action at a distance;* in other words, electromagnetic fields can exert forces, and hence can do work, at distances far away from the places where they are generated and at later times. Electromagnetic radiation is thus a means of transporting energy and momentum from one set of electric charges and currents (at the source end) to another (those at the receiving end). Since whatever can carry energy can also convey information, electromagnetic waves thus provide a means of transmitting energy and information over a distance.

To appreciate the concept of propagation of electromagnetic waves in empty space, it is useful to think of other wave phenomena that we may observe in nature. When a pebble is dropped into a body of water, the water particles in the vicinity of the pebble are immediately displaced from their equilibrium positions. The motion of these particles disturbs adjacent particles, causing them to move, and the process continues, creating a wave. Because of the finite velocity of the wave, a finite time elapses before the disturbance causes the water particles at distant points to move. Thus the initial disturbance produces, at distant points, effects that are *retarded* in time. The water wave consists of ripples that move along the surface away from the disturbance. Although the motion of any particular water particle is

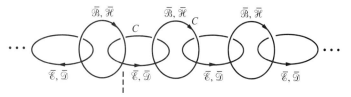

FIGURE 1.3.   **Plausibility of electromagnetic wave propagation as dictated by equations [1.1] and [1.3].** Starting with a time-varying magnetic field at the location of the dashed line, electric and magnetic fields are successively generated at surrounding regions.

essentially a small up-and-down movement, the cumulative effects of all the particles produce the wave, which moves radially outward from the point at which the pebble is dropped. Another excellent example of wave propagation is the motion of sound through a medium. In air this motion occurs through the to-and-fro movement of the air molecules, but these molecules do not actually move along with the wave.

Electromagnetic waves consist of time-varying electric and magnetic fields. Suppose an electrical disturbance, such as a change in the current through a conductor, occurs at a point in a region. The time-changing electric field resulting from the disturbance generates a time-changing magnetic field. The time-changing magnetic field, in turn, produces an electric field. These time-varying fields continue to generate one another in an ever-expanding region, and the resulting wave propagates away from the source. When electromagnetic waves propagate in a pure vacuum, there is no vibration of physical particles, as in the case of water and sound waves. The velocity of electromagnetic wave propagation in free space is finite (the speed of light), so the fields produced at distant points are *retarded* in time with respect to those near the source.

Careful examination of [1.1] to [1.4] provides a qualitative understanding of the plausibility of electromagnetic wave propagation in space. Consider, for example, [1.1] and [1.3] in a nonconducting medium (i.e., $\sigma = 0$) and in the absence of sources (i.e., $\tilde{\mathcal{J}}, \tilde{\rho} = 0$). According to [1.1], any magnetic field $\overline{\mathcal{B}}$ that varies with time generates an electric field along a contour C surrounding it, as shown in Figure 1.3. On the other hand, according to [1.3], this electric field $\overline{\mathcal{E}}$, which typically varies in time because $\overline{\mathcal{B}}$ is taken to be varying with time, in turn generates a magnetic field along a contour C surrounding itself. This process continues indefinitely, as shown in Figure 1.3. It thus appears that if we start with a magnetic field at one point in space and vary it with time, Maxwell's equations dictate that magnetic fields and electric fields are created at surrounding points—that is, that the disturbance initiated by the changing magnetic field propagates away from its point of origin, indicated by the vertical dashed line in Figure 1.3. Note that although Figure 1.3 shows propagation toward the left and right, the same sequence of events takes place in all directions.

1.5   SUMMARY

Maxwell's equations constitute the most general mathematical framework for classical electromagnetics, and they derive their validity from experimentally established laws of nature. These four partial differential equations embody a wide range of principles and concepts, including the propagation of electromagnetic waves through empty space or through matter. Propagation of electromagnetic waves in empty space or material media, their reflection from boundaries, their guiding within metallic boundaries or other structures, their interaction with matter, and their generation by simple sources are the subjects of the following chapters.

In Chapter 2 we restrict our attention to electromagnetic waves in an *unbounded, simple,* and *source-free* medium. In Chapter 3 we remove the first condition and consider the reflection and transmission of electromagnetic waves at planar boundaries between different material media. Many practical problems encountered in electromagnetics involve reflection of waves from interfaces between dielectrics and perfect conductors (e.g., air and copper) or between two different dielectrics (e.g., glass and air), and the treatment of such problems requires that we take into account the complicating effects of the boundary surfaces. In Chapters 4 and 5 we consider electromagnetic waves guided by metallic and dielectric boundaries and examine various guiding structures. Guiding of electromagnetic energy by confining it in two dimensions and allowing it to propagate in the third dimension forms the basis of a wide range of applications, including two-conductor transmission lines, hollow cylindrical waveguides, and dielectric waveguides or optical fibers. In Chapter 6, we return to considering an unbounded medium but now allow it not to be simple, but rather have material properties which significantly affect the nature of electromagnetic waves. We also provide a brief discussion of sources of electromagnetic waves, namely antennas.

Waves in an Unbounded Medium

Having reviewed the physical basis of the complete set of Maxwell's equations, we now proceed to discuss what is arguably their most important consequence: electromagnetic waves. Before we consider the space-time structure of electromagnetic waves and their propagation, reflection, transmission, and guiding, a few comments are in order on the general nature of waves.

We recognize a wave as some pattern of values in space that appears to move in time. The idea of a wave is one of the great unifying concepts of physics. Let us briefly discuss examples of different kinds of waves in order to place the characteristics of electromagnetic waves in context. The reader has no doubt observed the activity on a line of cars stopping or starting at a traffic light. The cars in the line do not all start at the same time when the light turns green; rather, the act of starting travels backwards through the line with a certain speed. This "start-up" wave is initiated by the first car in the line, and the speed of this wave depends on the reaction time of the drivers and the response characteristics of the cars (including the inertial mass and engine power). An important point to note is that no mass is transported by the wave; what travels is merely the "act" of start-up. In fact, in this particular case, the transport of mass (i.e., the motion of cars) is in the opposite direction to that of the start-up wave.

Similarly, when the end of a stretched rope is suddenly moved sideways, the event of sideways motion travels along the rope as a wave with a certain speed, which depends on the tension in the rope and its mass. The initial sideways displacement propagates away from the source, appearing at farther distances from the end at proportionally later times. Once again, no mass is transported by the wave; in fact, the mass elements move in the sideways direction. In a compressional seismic wave, the elastic displacement is in the same direction as that of the wave; however, clearly what travels is the "state of being compressed" rather than mass.

Consider water waves as perhaps the most familiar example. If we somehow (e.g., by dropping a pebble) initiate the vertical motion of a small volume of water in the middle of a lake, the surface of the water rises and falls in an oscillatory manner. However, it is not possible for the small volume of water to oscillate independently of the water surrounding it. Its periodic excesses and deficiencies of pressure as it rises and falls are transmitted to the surrounding water, which thereby receives energy and is, in turn, put into motion. In its resulting motion this body of water also transfers energy to its surrounding outer regions. By this process, energy is transmitted to the adjacent bodies of water, and a water wave propagates across the surface of the lake. The fundamental basis of the water wave is the fact that the motion of and the pressure exerted by a given volume of water are not independent of the motion and pressure of the water in its surrounding volume.

The fundamental reason for the existence of electromagnetic waves is very similar. Since a changing electric field produces a magnetic field, and that in turn produces an electric field, and so on, a series of energy transfers is initiated whenever any electric or magnetic disturbance takes place. A changing magnetic field induces an electric field, both in the region in which the magnetic field changes and also in the surrounding region; similarly, a changing electric field produces a magnetic field in the region in which the change occurs and also in the surrounding regions. This process was briefly discussed on a qualitative basis in Section 1.4, when we considered the plausibility of electromagnetic wave propagation. Consequently, when a disturbance of either the electric or the magnetic field takes place in a given region of space, it cannot be confined to that space. The changing fields within the region induce fields in surrounding regions also; these induced fields induce fields in further surrounding regions, and so forth; as a result, the electromagnetic energy travels in the surrounding region. When there is an excess of electromagnetic energy anywhere in unbounded space, it cannot stand still, any more than a mound of displaced water can be stationary on the surface of a lake! It cannot simply subside in a limited region of unbounded space. It can only travel as a wave until its energy is dissipated.

Although other types of waves travel only through material media (such as sound waves through air or water or seismic waves through earth), electromagnetic waves can also travel through free space (vacuum). Nineteenth-century physicists could not bring themselves to believe this fact and thought that the vacuum of space was permeated by an elusive substance, referred to as the *ether*, that allowed the transmission of light waves.[1]

It was mentioned above that waves do not necessarily involve transport of any mass. Like moving objects, however, waves carry *energy* from one point to another. For example, the electromagnetic energy that arrives at the earth from the sun can be converted into electrical energy via solar cells and into chemical energy by plants, which is subsequently released as we burn wood or coal. A radio or television transmitter broadcasts its programs in the form of electromagnetic signals at power levels of some tens of kilowatts. Each radio or TV receiver picks up a minute fraction of this power to reproduce the transmitted signal. Sound waves produced by thunder can shake a nearby house, while those produced by human voices dissipate to negligible levels within a few tens of feet. Although most waves carry energy, the amount of energy differs from case to case.

Like moving matter, waves also have *momentum*. When waves are absorbed or reflected by an object, they push the object in their direction of travel. For example, a steady sound wave imparts momentum on a membrane, in addition to causing it to vibrate. The momentum of a sound wave must be taken into account in understanding the behavior of solids. Ordinarily, however, the momentum of waves is less noticeable than their energy. Light and other types of electromagnetic waves also have momentum, although extremely high intensities are necessary to produce perceptible effects.

Another important property of waves is that they each have a *velocity*. The speed of wave propagation differs widely between different types of waves and the types of materials in which they propagate. For example, sound waves propagate at a velocity of ~ 340 m-s^{-1} in air but at ~ 1500 m-s^{-1} in water; heat in a concrete dam propagates at a few meters per second, whereas electromagnetic waves in air or in free space propagate at 300,000 kilometers per second.

We now proceed with the formal treatment of the propagation of electromagnetic waves. In this chapter, we consider the propagation in an *unbounded, simple,* and *source-free* medium of a special type of electromagnetic waves known as *uniform plane waves*. Uniform plane waves are waves in which the amplitude and phase of the electric field (and the magnetic field) at any instant of time are constants over infinite planes orthogonal to the direction of propagation. The characteristics of uniform plane waves are particularly simple, so their study constitutes an excellent starting point in understanding more complicated electromagnetic waves. Furthermore, many practically important electromagnetic waves can be approximated as uniform plane waves, so our study of such waves is also of significant practical importance. Uniform plane waves are often also referred to as transverse electromagnetic (or TEM) waves, since both the electric and magnetic fields of the wave are transverse to the propagation direction.

---

[1]E. Whittaker, *A History of the Theories of Aether and Electricity,* Thomas Nelson and Sons Ltd., London, 1951.

Many of the electromagnetic wave concepts that we study in this chapter correspond directly to the voltage and current wave concepts that are useful in modeling uniform two-conductor transmission lines.[2] The basic wave equations and their general solutions are identical regardless of whether they are written in terms of electric and magnetic fields or of voltages and currents, as are the concepts of propagation constant, wavelength, phase velocity, and attenuation constant. The similarity of voltage and current waves and uniform plane electromagnetic waves is, of course, not just a coincidence. The uniform two-conductor transmission lines are merely special cases of more general guiding structures that can efficiently transmit electromagnetic energy from one point to another. Thus, the voltage and current waves on uniform two-conductor transmission lines are in fact uniform plane electromagnetic waves and can be studied entirely in terms of electric and magnetic fields rather than voltages and currents. The relatively simple physical configuration of two-conductor transmission lines allows the analysis of the propagation of electromagnetic waves on them by means of voltages and currents, using Kirchhoff's voltage and current laws, rather than the more general Maxwell's equations.

Our coverage in this chapter of uniform plane waves in an unbounded, simple, and source-free medium starts a sequence of four chapters in each of which we consider the propagation of electromagnetic waves under different circumstances. In this chapter we restrict our attention to waves for which all three of these conditions hold true. In Chapter 3 we remove the first condition and consider how waves reflect and refract in the presence of planar boundaries. In Chapters 4 and 5 we consider the case of guided wave propagation in regions bounded by conducting or dielectric structures. In Chapter 6 we return to an unbounded medium, but now allow it not to be simple, but rather having material properties which in some cases significantly affect the nature of electromagnetic waves. We also provide a brief discussion of sources of electromagnetic waves, namely antennas.

We start our coverage of electromagnetic waves in an unbounded medium with a general discussion in Section 2.1 of the propagation of uniform plane waves in a simple, source-free, and lossless medium. Section 2.2 discusses the important practical case of time-harmonic uniform plane waves in a lossless medium, followed by coverage of wave propagation in a lossy medium in Section 2.3. Concepts of electromagnetic energy flow and the Poynting vector (which allows us to express power in terms of electric and magnetic fields) are discussed in Section 2.4. The different possible orientations and behavior of the electric field of uniform plane waves, described in terms of the polarization of the wave, are studied in Section 2.5. General expressions for uniform plane waves propagating in an arbitrary direction are developed in Section 2.6, followed by a brief introductory discussion of nonuniform and nonplanar waves, the detailed coverage of which is beyond the scope of this book.

---

[2]See Chapters 2, 3 and 8 of U. S. Inan and A. S. Inan, *Engineering Electromagnetics,* Addison Wesley Longman, 1999.

2.1   PLANE WAVES IN A SIMPLE, SOURCE-FREE, AND LOSSLESS MEDIUM

We mentioned in Chapter 1 that two of Maxwell's equations, namely the two curl equations [1.1] and [1.3], which represent the facts that changing magnetic fields produce electric fields and that changing electric fields produce magnetic fields, necessarily lead to propagation of electromagnetic waves. In this section we study the characteristics of such electromagnetic waves, as they propagate in unbounded, simple, source-free, and lossless media. Our starting point consists of Maxwell's equations, repeated below for convenience:

$$\nabla \times \overline{\mathscr{E}} = -\frac{\partial \overline{\mathscr{B}}}{\partial t} \qquad [2.1a]$$

$$\nabla \cdot \overline{\mathscr{D}} = \tilde{\rho} \qquad [2.1b]$$

$$\nabla \times \overline{\mathscr{H}} = \overline{\mathscr{J}} + \frac{\partial \overline{\mathscr{D}}}{\partial t} \qquad [2.1c]$$

$$\nabla \cdot \overline{\mathscr{B}} = 0 \qquad [2.1d]$$

The $\overline{\mathscr{J}}$ term in [2.1c] can in general be nonzero because of the presence of source currents $\overline{\mathscr{J}}_{\text{source}}$ (e.g., wires carrying current) and/or because of conduction current $\overline{\mathscr{J}}_{\text{c}} = \sigma \overline{\mathscr{E}}$, which flows in media with nonzero conductivity ($\sigma \neq 0$). The latter leads to loss of electromagnetic power, with volume density of power dissipation represented by $\overline{\mathscr{E}} \cdot \overline{\mathscr{J}}$. In this section we consider electromagnetic wave propagation in source-free (i.e., $\overline{\mathscr{J}}_{\text{source}}$, $\tilde{\rho} = 0$), simple,[3] and lossless (i.e., $\sigma = 0$, and thus $\overline{\mathscr{J}}_{\text{c}} = 0$) media, so that $\overline{\mathscr{J}} = \overline{\mathscr{J}}_{\text{source}} + \overline{\mathscr{J}}_{\text{c}} = 0$. Our goal is to describe the properties of different types of electromagnetic waves that can exist (i.e., can satisfy Maxwell's equations) in regions without any source currents and charges, regardless of where and when those fields may have originated—undoubtedly at faraway sources $\overline{\mathscr{J}}_{\text{source}}$ and/or $\tilde{\rho}$.

In general, the two coupled equations [2.1a] and [2.1c] can be combined to obtain equations in terms of only $\overline{\mathscr{E}}$ (or $\overline{\mathscr{H}}$). Taking the curl of [2.1a] and substituting into [2.1c] gives

$$\nabla \times \nabla \times \overline{\mathscr{E}} = -\mu \frac{\partial}{\partial t} \nabla \times \overline{\mathscr{H}}$$
$$= -\mu \frac{\partial^2 \overline{\mathscr{D}}}{\partial t^2}$$

---

[3]In this context, a "simple" medium is a material medium that is linear, time-invariant, isotropic, and homogeneous, so that ϵ, μ, and σ are simple constants.

where we have taken into account the fact that $\overline{\mathcal{J}} = 0$ and assumed that μ is a constant (i.e., the medium is magnetically linear, isotropic, homogeneous, and time-invariant) and that $\overline{\mathcal{H}}$ is a continuous function of time and space, so that spatial and temporal derivatives can be interchanged.

We now use the vector identity

$$\nabla \times \nabla \times \overline{\mathcal{E}} \equiv \nabla(\nabla \cdot \overline{\mathcal{E}}) - \nabla^2 \overline{\mathcal{E}}$$

and the fact that under source-free conditions ($\tilde{\rho} = 0$) we have

$$\nabla \cdot \overline{\mathcal{D}} = 0 \quad \rightarrow \quad \nabla \cdot (\epsilon \overline{\mathcal{E}}) = 0 \quad \rightarrow \quad \nabla \cdot \overline{\mathcal{E}} = 0$$

assuming that ϵ is constant (i.e., that the medium is electrically linear, isotropic, homogeneous, and time-invariant) to obtain

$$\boxed{\nabla^2 \overline{\mathcal{E}} - \mu\epsilon \frac{\partial^2 \overline{\mathcal{E}}}{\partial t^2} = 0} \qquad [2.2]$$

We can obtain a similar equation for the magnetic field $\overline{\mathcal{H}}$ by taking the curl of [2.1c] and making similar assumptions (i.e., a simple medium, being linear, isotropic, homogeneous, and time-invariant). We find

$$\boxed{\nabla^2 \overline{\mathcal{H}} - \mu\epsilon \frac{\partial^2 \overline{\mathcal{H}}}{\partial t^2} = 0} \qquad [2.3]$$

Equations of the type [2.2] and [2.3] are encountered in many branches of science and engineering and have natural solutions in the form of propagating waves; they are thus referred to as *wave equations*. The solutions of these equations describe the characteristics of the electromagnetic waves as dictated by Maxwell's equations and the properties (μ, ϵ) of lossless material media in regions without any source currents and charges. It is important to note that the equations [2.2] and [2.3] are not independent (since they were both obtained from [2.1a] and [2.1c]), and that either [2.2] or [2.3] (together with [2.1a] or [2.1c]) can be used to solve for both $\overline{\mathcal{E}}$ and $\overline{\mathcal{H}}$. We follow the usual convention of solving for $\overline{\mathcal{E}}$ from the electric field equation [2.2] and then determining $\overline{\mathcal{H}}$ from [2.1a].

We note that in the general case when $\overline{\mathcal{E}}$ has three nonzero components ($\mathcal{E}_x$, $\mathcal{E}_y$, and $\mathcal{E}_z$), which may vary with the three Cartesian spatial coordinates (x, y, and z), [2.2] is actually a set of three scalar equations:

$$\left[\frac{\partial^2}{\partial x^2} + \frac{\partial^2}{\partial y^2} + \frac{\partial^2}{\partial z^2}\right]\mathcal{E}_x - \mu\epsilon \frac{\partial^2 \mathcal{E}_x}{\partial t^2} = 0 \qquad [2.2a]$$

$$\left[\frac{\partial^2}{\partial x^2} + \frac{\partial^2}{\partial y^2} + \frac{\partial^2}{\partial z^2}\right]\mathcal{E}_y - \mu\epsilon \frac{\partial^2 \mathcal{E}_y}{\partial t^2} = 0 \qquad [2.2b]$$

$$\left[\frac{\partial^2}{\partial x^2} + \frac{\partial^2}{\partial y^2} + \frac{\partial^2}{\partial z^2} \right] \mathcal{E}_z - \mu\epsilon \frac{\partial^2 \mathcal{E}_z}{\partial t^2} = 0 \qquad [2.2c]$$

We do not need to consider the most general case in order to study the characteristics of propagating electromagnetic waves contained in the solution of [2.2]. For simplicity, we limit ourselves here to the special case in which the electric field $\mathcal{E}$ is independent of two dimensions (say x and y). Equation [2.2] then becomes

$$\frac{\partial^2 \overline{\mathcal{E}}}{\partial z^2} - \mu\epsilon \frac{\partial^2 \overline{\mathcal{E}}}{\partial t^2} = 0 \qquad [2.4]$$

which is equivalent to three scalar equations, one for each component of $\mathcal{E}$. With no loss of generality, we further restrict our attention to one of the components, say $\mathcal{E}_x$, the equation for which is then

$$\boxed{\frac{\partial^2 \mathcal{E}_x}{\partial z^2} - \mu\epsilon \frac{\partial^2 \mathcal{E}_x}{\partial t^2} = 0} \qquad [2.5]$$

Second-order partial differential equations of the type [2.5] commonly occur in many branches of engineering and science. For example, replacing $\mathcal{E}_x(z, t)$ with $\mathcal{V}(z, t)$ results in the wave equation that describes the voltage wave $\mathcal{V}(z, t)$ on a lossless transmission line,[4] whereas replacing $\mathcal{E}_x(z, t)$ with $u(z, t)$ gives the wave equation describing the variation of velocity $u(z, t)$ for acoustic waves in a fluid.

The general solution of [2.5] is of the form

$$\mathcal{E}_x(z, t) = p_1(z - v_p t) + p_2(z + v_p t) \qquad [2.6]$$

where $v_p = 1/\sqrt{(\mu\epsilon)}$ and p_1 and p_2 are arbitrary functions representing the shape (e.g., square pulse, sinusoid with a Gaussian envelope, exponentially decaying pulse) of the electric field excited by a remote source. Examples of functions[5] of $(z - v_p t)$ include $Ae^{-(z - v_p t)^2}$, $A\sqrt{z - v_p t}$, and $Ae^{-(z - v_p t)} \cos(z - v_p t)$. Note also that

---

[4]The corresponding equation is

$$\frac{\partial^2 \mathcal{V}}{\partial z} = LC \frac{\partial^2 \mathcal{V}}{\partial t^2}$$

where L and C are respectively the per-unit-length inductance and capacitance of the transmission line. For further details on the correspondence between uniform plane waves and voltage and current waves on transmission lines see Chapters 2, 3, 8 and Section 7.4 of U. S. Inan and A. S. Inan, *Engineering Electromagnetics,* Addison Wesley Longman, 1999.

[5]An important function of $(z - v_p t)$ that is often encountered and that we shall introduce in the next section and study in most of the rest of this book is the sinusoidal traveling-wave function, $A \cos[\omega(t - z/v_p)]$. Depending on the location of the observation point $z > 0$ along the z axis, this function replicates the sinusoidal variation $A \cos(\omega t)$ occurring at $z = 0$, except delayed by (z/v_p) seconds at the point z. Thus, (z/v_p) represents a time shift, or delay, which is a characteristic of the class of wave functions of the variable $(z - v_p t)$.

the functions p_1 and p_2 are not necessarily the same. The fact that [2.6] is a solution of [2.5] can be seen by simple substitution.[6]

That the solutions $p_1(z - v_p t)$ and $p_2(z + v_p t)$ represent waves propagating respectively in the $+z$ and $-z$ directions can be seen by observing their variation with z at different times t_i, as shown for one type of $p_1(z - v_p t)$ function in Figure 2.1. In this context, a wave is to be understood as some disturbance (e.g., an electric field variation) that occurs at one place at a given time and at other places at other times, with time delays proportional to the spatial separations from the first location. Since the wave travels with a velocity v_p, a time period of (z/v_p) elapses as the wave propagates from $z = 0$ to the position z. Thus, an observer at point $z > 0$ sees events that actually occurred (at $z = 0$) at an earlier time. For example, light waves from a supernova explosion may arrive at earth millions of years after their source has been extinguished. A wave is not necessarily a repetitive or oscillatory disturbance in time; a single pulse moving in space also constitutes a wave, such as when a transmission line is briefly connected to a battery or a tsunami is generated in an undersea earthquake.

Note that at any time, say t_1, the function $p_1(z - v_p t_1)$ is simply a function of z, because $v_p t_1$ is a constant. At another time t_2, the function is $p_1(z - v_p t_2)$, which has exactly the same type of dependence on z, but displaced to the right by an amount $v_p(t_2 - t_1)$. In other words, the disturbance (in this case the electric field) represented by $p_1(z - v_p t)$ has traveled in the positive z direction with a velocity v_p. Note that in free space, $v_p = 1/\sqrt{(\mu_0 \epsilon_0)} \simeq 3 \times 10^8$ m-s^{-1}, or the speed of light, a fact that led Maxwell to suggest that light is a form of electromagnetic radiation. The space-time dependence of a wave pulse such as $p_1(z - v_p t)$ is illustrated in Figure 2.1.

Note that the second term in [2.6], $p_2(z + v_p t)$, represents a wave traveling in the negative z direction. The general solution of [2.5] is thus a superposition of two waves: a forward wave traveling in the $+z$ direction (away from the source, if the source is assumed to be located far away in the $-z$ direction) and a reverse wave traveling in the $-z$ direction (back toward the source). The reverse wave term is nonzero

---

[6]Consider

$$\mathcal{E}_x(z, t) = f(z - v_p t) = f(\zeta)$$

To see that this function $f(\cdot)$ of the variable $\zeta = (z - v_p t)$ is a solution of [2.5], we can express the time and space derivatives of $\mathcal{E}_x(z, t)$ in terms of the derivatives of $f(\zeta)$ with respect to ζ:

$$\frac{\partial \mathcal{E}_x}{\partial t} = \frac{\partial f}{\partial \zeta}\frac{\partial \zeta}{\partial t} = -v_p \frac{\partial f}{\partial \zeta} \quad \text{and} \quad \frac{\partial^2 \mathcal{E}_x}{\partial t^2} = -v_p \frac{\partial^2 f}{\partial \zeta^2}\frac{\partial \zeta}{\partial t} = v_p^2 \frac{\partial^2 f}{\partial \zeta^2}$$

since $\partial \zeta/\partial t = -v_p$. Similarly, noting that $\partial \zeta/\partial z = 1$, we have

$$\frac{\partial \mathcal{E}_x}{\partial z} = \frac{\partial f}{\partial \zeta} \quad \text{and} \quad \frac{\partial^2 \mathcal{E}_x}{\partial z^2} = \frac{\partial^2 f}{\partial \zeta^2}$$

Substituting in [2.5] we find that the wave equation is indeed satisfied by any function $f(\cdot)$ of the variable $\zeta = (z - v_p t)$, where $v_p = 1/\sqrt{\mu \epsilon}$.

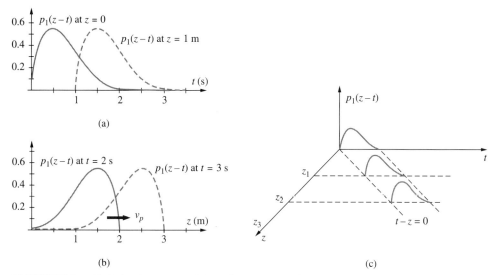

FIGURE 2.1.   Variation in space and time of an arbitrary function $p_1(z - v_p t)$**.** The function $p_1(\zeta)$ shown above is $p_1(\zeta) = \zeta^{1/2}(\sin \zeta/\zeta)^6 u(\zeta)$, where $u(\zeta)$ is the unit step function (i.e., $u(\zeta) = 0$ for $\zeta < 0$ and $u(\zeta) = 1$ for $\zeta > 0$). For the purpose of the plots on the left-hand side, the speed of propagation is taken to be $v_p = 1$ m-s^{-1}. (a) $p_1(z - t)$ as a function of t at two positions $z = 0$ and $z = 1$ m. (b) $p_1(z - t)$ as a function of z at two time instants, showing how the pulse travels in the z directions as time progresses. (c) $p_1(z - t)$ as a function of z and t.

only if there are discontinuities (surfaces with different ϵ and/or μ) that "reflect" some of the forward-traveling wave energy back to the source. The reflection and refraction of electromagnetic waves at planar interfaces are studied in Chapter 3.

Electromagnetic waves for which the field components have functional dependences as given by [2.6] belong to a class of waves known as *plane waves*. This term originates from the fact that the surfaces over which the argument of the function is a constant (e.g., $z - v_p t = $ constant) are planes.[7] Waves for which the field amplitudes do not vary with position over the planes of constant phase are known as *uniform plane waves*. Generalized expressions for uniform plane waves are given in Section 2.6. Note that although we studied the solution of [2.4] only for the x component of $\overline{\mathscr{E}}$, identical behavior would be expected for the y component $\mathscr{E}_y$, since the differential equation governing it is the y component of [2.4], which is identical in form to [2.5]. In general, both components may exist, in which case the total field is simply a linear superposition of the two separate solutions of the corresponding differential equations. Although it would seem that the $\mathscr{E}_z$ would also behave in a similar

---

[7]Such surfaces are known as *surfaces of constant phase, phase surfaces,* or *phase fronts*. Although the concept of phase is more commonly associated with sinusoidal steady-state or time-harmonic waves, in which case the function $p_1(z - v_p t)$ has the form $\cos[\omega(t - z/v_p)]$, it is nevertheless equally valid for nonsinusoidal functions. Time-harmonic waves are discussed in Section 2.2.

manner, a uniform plane wave propagating in the z direction in a simple medium cannot have a z component, as can be seen by examining equation [2.1b]. We have

$$\nabla \cdot \overline{\mathcal{D}} = 0 \longrightarrow \nabla \cdot \overline{\mathcal{E}} = 0 \longrightarrow \frac{\partial \mathcal{E}_x}{\partial x} + \frac{\partial \mathcal{E}_y}{\partial y} + \frac{\partial \mathcal{E}_z}{\partial z} = 0$$

The first two terms ($\partial \mathcal{E}_x / \partial x$ and $\partial \mathcal{E}_y / \partial y$) are zero (since the field components do not vary with x or y), which requires that $\partial \mathcal{E}_z / \partial z = 0$, which in turn means that $\mathcal{E}_z$ can at most be a constant in space. Substituting $\partial \mathcal{E}_z / \partial z = 0$ into the z component of [2.4], we find $\partial^2 \mathcal{E}_z / \partial t^2 = 0$, which means that $\mathcal{E}_z$ can at most be a linearly increasing function of time. Since a quantity that is constant in space and linearly increasing in time cannot contribute to wave motion, a uniform plane wave propagating in the z direction has no z component.

2.1.1   Relation between $\overline{\mathcal{E}}$ and $\overline{\mathcal{H}}$

Although we have just discussed the solution of the wave equation for the electric field, we could have just as easily solved [2.3] for the magnetic field $\overline{\mathcal{H}}$ by making the same assumption (i.e., uniform plane wave) and restricting our attention to a single component of the magnetic field. However, once we have described a solution for the electric field of a uniform plane wave (say the component $\mathcal{E}_x(z)$), we are no longer free to choose arbitrarily one of the two components of its magnetic field (i.e., $\mathcal{H}_x$ or $\mathcal{H}_y$) to be nonzero. Because the wave equation [2.4] for the electric field was obtained using the entire set of Maxwell's equations, the magnetic field is already determined once we specify the electric field. The proper means for finding the magnetic field of a uniform plane wave is thus to derive it from the electric field using Maxwell's equations, for example [2.1a].

Let us again assume that we have a uniform plane wave having an electric field with only an x component propagating in a simple medium in the $+z$ direction:

$$\mathcal{E}_x(z, t) = p_1(z - v_p t)$$

Since $\mathcal{E}_y, \mathcal{E}_z = 0$, $\mathcal{E}_x$ varies only with z, and any components of the magnetic field $\overline{\mathcal{H}}$ also vary only with z, we have from [2.1a]

$$\nabla \times \overline{\mathcal{E}} = -\frac{\partial \overline{\mathcal{B}}}{\partial t} \quad \rightarrow \quad \hat{\mathbf{y}} \frac{\partial \mathcal{E}_x}{\partial z} = -\mu \hat{\mathbf{x}} \frac{\partial \mathcal{H}_x}{\partial t} - \mu \hat{\mathbf{y}} \frac{\partial \mathcal{H}_y}{\partial t}$$

whereas from [2.1c] we have

$$\nabla \times \overline{\mathcal{H}} = \frac{\partial \overline{\mathcal{D}}}{\partial t} \quad \rightarrow \quad -\hat{\mathbf{x}} \frac{\partial \mathcal{H}_y}{\partial z} + \hat{\mathbf{y}} \frac{\partial \mathcal{H}_x}{\partial z} = \epsilon \hat{\mathbf{x}} \frac{\partial \mathcal{E}_x}{\partial t}$$

Equating the different components of the above two equations, we find

$$\frac{\partial \mathcal{E}_x}{\partial z} = -\mu \frac{\partial \mathcal{H}_y}{\partial t}; \qquad -\frac{\partial \mathcal{H}_y}{\partial z} = \epsilon \frac{\partial \mathcal{E}_x}{\partial t} \qquad [2.7]$$

It thus appears that the magnetic field of a uniform plane wave with an electric field of $\overline{\mathscr{E}} = \hat{\mathbf{x}}p_1(z - v_p t)$ has only a y component, which is related to $\mathscr{E}_x$ as described by [2.7]. For any given function $p_1(z - v_p t)$, $\mathscr{H}_y$ can be found using [2.7], as we will do so in the next section for sinusoidal uniform plane waves, for which case $p_1(z - v_p t) = A\cos(z - v_p t)$. By substituting $\mathscr{E}_x = p_1(z - v_p t)$ into [2.7], integrating with respect to time, and then taking the derivative with respect to z, it can be shown that

$$\mathscr{H}_y = \left(\sqrt{\frac{\epsilon}{\mu}}\right)p_1(z - v_p t) = \frac{1}{\eta}p_1(z - v_p t)$$

where $\eta \equiv \sqrt{\mu/\epsilon}$ is a quantity that has units of impedance (ohms) and is defined as the *intrinsic impedance* of the medium. Thus, for uniform plane waves in a simple lossless medium, the ratio of the electric and magnetic fields is η and is determined only by the material properties of the medium (i.e., μ, ϵ). The intrinsic impedance of free space[8] is $\eta = \sqrt{\mu_0/\epsilon_0} \simeq 120\pi \simeq 377\Omega$.

An important characteristic of uniform plane waves is that their electric and magnetic fields are perpendicular to one another. In simple and lossless media, the variation of $\overline{\mathscr{E}}$ and $\overline{\mathscr{H}}$ in space and time are identical [i.e., they are both proportional to $p_1(z - v_p t)$]. In other words, $\overline{\mathscr{E}}$ and $\overline{\mathscr{H}}$ propagate in unison along z, reaching their maxima and minima at the same points in space and at the same times. The orientation of $\overline{\mathscr{E}}$ and $\overline{\mathscr{H}}$ is such that the vector $\overline{\mathscr{E}} \times \overline{\mathscr{H}}$ is in the $+z$ direction, which is the direction of propagation of the wave. The orthogonality of the electric and magnetic fields and the propagation of a disturbance are illustrated in Figure 2.2.

Note that since the choice of the coordinate system cannot affect the physical relationship between $\overline{\mathscr{E}}$ and $\overline{\mathscr{H}}$, if we start with $\overline{\mathscr{E}}$ having only a y component (i.e., $\overline{\mathscr{E}} = \hat{\mathbf{y}}\mathscr{E}_y$), $\overline{\mathscr{H}}$ then comes out to be $\overline{\mathscr{H}} = -\hat{\mathbf{x}}\mathscr{H}_x$. The relationship between $\overline{\mathscr{E}} = \hat{\mathbf{y}}\mathscr{E}_y$ and $\overline{\mathscr{H}} = -\hat{\mathbf{x}}\mathscr{H}_x$ for a uniform plane wave propagating in the $+z$ direction is depicted in Figure 2.3.

The concept of uniform plane waves makes the solution of the wave equation [2.2] tractable, as we have seen. In a strict sense, uniform plane waves can be excited (created) only by sources infinite in extent (e.g., uniformly distributed over the entire xy plane to produce a uniform plane wave propagating in the $+z$ direction). However, uniform plane waves are in fact often excellent approximations in practice, especially when we observe (or receive) electromagnetic waves at large distances (compared to wavelength) from their sources (e.g., in broadcast radio and television applications). Electromagnetic fields emanating from a point source spread spherically, and since a very small portion of the surface of a large sphere is approximately planar, they can be considered plane waves at large distances from their sources. Uniform plane waves are also important because the electromagnetic fields of an arbitrary wave (i.e., nonuniform and/or nonplanar wave) can be expressed[9] as a

---

[8]Using $\mu_0 = 4\pi \times 10^{-7}$ H-m^{-1} and $\epsilon_0 = 1/(\mu_0 c^2) \simeq 10^{-9}/(36\pi)$ F-m^{-1}.

[9]P. C. Clemmow, *The Plane Wave Spectrum Representation of Electromagnetic Fields,* Pergamon Press, Oxford, 1966.

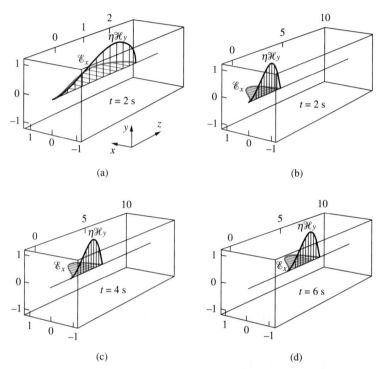

FIGURE 2.2. **The propagation of a uniform plane electromagnetic disturbance in the z direction.** The electric and magnetic fields are orthogonal at all times. For the purpose of this figure, the speed of propagation is taken to be $v_p = 1$ m-s^{-1}. (a) $\mathcal{E}_x(z, t)$ and $\eta\mathcal{H}_y(z, t)$ as a function of z at $t = 2$ s. (b) Same as (a) but shown on a compressed distance scale. (c) $\mathcal{E}_x(z, t)$ and $\eta\mathcal{H}_y(z, t)$ as a function of z at $t = 4$ s. (d) $\mathcal{E}_x(z, t)$ and $\eta\mathcal{H}_y(z, t)$ as a function of z at $t = 6$ s. The propagation of the pulse in the $+z$ direction is clearly evident. Note that the pulse shapes for the electric and magnetic fields are identical.

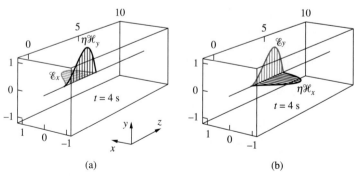

FIGURE 2.3. **Electric and magnetic fields of a uniform plane wave.** The relationship between $\overline{\mathcal{E}}$ and $\overline{\mathcal{H}}$ for a uniform plane wave is independent of the choice of a particular coordinate system. (a) $\mathcal{E}_x(z, t)$ and $\eta\mathcal{H}_y(z, t)$. (b) $\mathcal{E}_y(z, t)$ and $\eta\mathcal{H}_x(z, t)$.

superposition of component plane waves, much like a Fourier decomposition. Further discussion of uniform plane waves and nonuniform or nonplanar waves are provided in Section 2.7. We shall also see examples of nonuniform waves in Chapters 3 and 4 when we study reflection, refraction, and guiding of electromagnetic waves.

2.2 TIME-HARMONIC UNIFORM PLANE WAVES IN A LOSSLESS MEDIUM

We now study uniform plane waves for which the temporal behavior is harmonic, or sinusoidal. It was pointed out in Section 1.2 that a large number of practical applications (e.g., broadcast radio and TV, radar, optical and microwave applications) involve sources (transmitters) that operate in such a narrow band of frequencies that the behavior of all of the field components is very similar to that of the central single-frequency sinusoid (i.e., the carrier).

The AM radio broadcast signal was mentioned in Section 1.2 as an example of a signal with a narrow bandwidth centered around a carrier frequency. Other examples include FM radio broadcasts, which utilize carrier frequencies of 88 to 108 MHz with a bandwidth of ~200 kHz, or UHF television broadcasts, where the signal bandwidth is ~6 MHz for carrier frequencies in the range 470 to 890 MHz. For all practical purposes, the propagation of the signals from the transmitters to receivers can be described by studying single sinusoids at the carrier frequency, since the characteristics of the propagation medium do not vary significantly over the signal bandwidth. In other words, while at the transmitter the actual signal is constituted by a superposition of Fourier components at different frequencies within the signal bandwidth, each of the frequency components propagates to the receiver in a manner identical to the propagation of the carrier.

In other applications, such as in the case of guided propagation of electromagnetic waves, the propagation characteristics may vary significantly over the bandwidth of the signal (especially when terminated transmission lines or waveguides are concerned). However, sinusoidal (or time-harmonic) analysis is still useful, because any arbitrary waveform can be represented as a linear superposition of its Fourier components whose behavior is well represented by the time-harmonic (or sinusoidal steady-state) analysis.

To study the propagation of time-harmonic electromagnetic waves, we use the time-harmonic form of Maxwell's equations as given in [1.11]. Note that these are written in terms of the vector phasor quantities represented with the boldface symbols ($\mathbf{E}$, $\mathbf{H}$, etc.,) which are related to the real space-time fields in the usual manner previously described in Section 1.2; for example,

$$\overline{\mathscr{E}}(x, y, z, t) = \mathscr{R}e\left\{\mathbf{E}e^{j\omega t}\right\}$$

For lossless ($\sigma = 0$), source-free ($\mathbf{J}_{source}$, $\rho = 0$), linear [i.e., ϵ, $\mu \neq f(\mathbf{E}, \mathbf{H})$], homogeneous [i.e., ϵ, $\mu \neq f(x, y, z)$], isotropic [i.e., ϵ, $\mu \neq f(\text{direction})$], and time-invariant [i.e., ϵ, $\mu \neq f(t)$] medium, we can derive the wave equation for the electric field from the preceding equation by first taking the curl of [1.11a]:

$$\nabla \times \nabla \times \mathbf{E} = -j\omega\mu(\nabla \times \mathbf{H})$$

Note that μ must be independent of direction and spatial coordinates in order to write $\nabla \times \mathbf{B} = \mu\nabla \times \mathbf{H}$. We now use a vector identity to replace the left-hand side of the preceding equation:

$$\nabla \times \nabla \times \mathbf{E} \equiv \nabla(\nabla \cdot \mathbf{E}) - \nabla^2\mathbf{E}$$

Since ϵ is independent of direction and spatial coordinates, we have from [1.11b]

$$\nabla \cdot \mathbf{D} = 0 \longrightarrow \nabla \cdot (\epsilon\mathbf{E}) = \epsilon\nabla \cdot \mathbf{E} = 0 \longrightarrow \nabla \cdot \mathbf{E} = 0$$

Using [1.11c] further reduces the preceding equation to

$$-\nabla^2\mathbf{E} = -j\omega\mu(j\omega\epsilon\mathbf{E})$$

or

$$\boxed{\nabla^2\mathbf{E} + \beta^2\mathbf{E} = 0} \qquad [2.8]$$

where $\beta = \omega\sqrt{\mu\epsilon}$ is the phase constant, also often called the wave number or propagation constant, in units of radians-m^{-1}. Note that the propagation constant $\beta = \omega\sqrt{\mu\epsilon}$ for uniform plane waves in free space is in fact identical[10] to the phase constant for voltage waves on uniform lossless transmission lines, namely $\beta = \omega\sqrt{LC}$, where L and C are respectively the per-unit-length inductance and capacitance of the line. Equation [2.8] is known as the *vector wave equation* or *Helmholtz equation* and is actually a set of three equations:

$$\left[\frac{\partial^2}{\partial x^2} + \frac{\partial^2}{\partial y^2} + \frac{\partial^2}{\partial z^2}\right]E_x + \beta^2 E_x = 0 \qquad [2.8a]$$

$$\left[\frac{\partial^2}{\partial x^2} + \frac{\partial^2}{\partial y^2} + \frac{\partial^2}{\partial z^2}\right]E_y + \beta^2 E_y = 0 \qquad [2.8b]$$

$$\left[\frac{\partial^2}{\partial x^2} + \frac{\partial^2}{\partial y^2} + \frac{\partial^2}{\partial z^2}\right]E_z + \beta^2 E_z = 0 \qquad [2.8c]$$

We now limit our attention to uniform plane waves by considering solutions of [2.8] for which $\mathbf{E}$ has only an x component and is only a function of z. In other words, we let

$$\mathbf{E}(x, y, z) = \hat{\mathbf{x}}E_x(z)$$

---

[10]The reader is encouraged to check the various per-unit-length capacitance and inductance expressions for lossless uniform transmission lines, readily derivable from first principles (see U. S. Inan and A. S. Inan, *Engineering Electromagnetics,* Addison Wesley Longman, 1999) or available in tabulated form (e.g., see *Reference Data for Engineers,* 8th ed., Sams Prentice Hall Computer Publishing, Carmel, Indiana, 1993) to show that $LC = \mu\epsilon$, regardless of the particular type of transmission line.

The wave equation [2.8] then reduces to

$$\frac{d^2 E_x}{dz^2} + \beta^2 E_x = 0 \tag{2.9}$$

which is a second-order ordinary differential equation encountered in sinusoidal steady-state applications in many branches of physics and engineering. For example, replacing E_x with V gives the wave equation for the voltage phasor V on a lossless transmission line. The general solution of [2.9] is

$$E_x(z) = \underbrace{C_1 e^{-j\beta z}}_{E_x^+(z)} + \underbrace{C_2 e^{+j\beta z}}_{E_x^-(z)} \tag{2.10}$$

where C_1 and C_2 are constants to be determined by boundary conditions. Note that the quantity $E_x(z)$ is a complex phasor. The real (or instantaneous) electric field $\overline{\mathscr{E}}(z, t)$ can be found from $E_x(z)$ as

$$\mathscr{E}_x(z, t) = \mathscr{R}e\{(C_1 e^{-j\beta z} + C_2 e^{+j\beta z})e^{j\omega t}\} \tag{2.11}$$

or

$$\mathscr{E}_x(z, t) = \underbrace{C_1 \cos(\omega t - \beta z)}_{\mathscr{E}_x^+(z,t)} + \underbrace{C_2 \cos(\omega t + \beta z)}_{\mathscr{E}_x^-(z,t)} \tag{2.12}$$

assuming that C_1 and C_2 are real constants. The first term $\mathscr{E}_x^+(z, t)$ represents a wave traveling in the $+z$ direction, as can be seen from the successive snapshots in time plotted as a function of z in Figure 2.4a. The z axis is normalized to wavelength, which is defined as $\lambda = 2\pi/\beta$. Note that the electric field at any given point in space varies sinusoidally in time, as illustrated in Figure 2.4b.

By the same token, the second term $\mathscr{E}_x^-(z, t)$ in [2.12] represents a wave traveling in the $-z$ direction. Both terms represent waves traveling at a speed given by the phase velocity $v_p = \omega/\beta$, which is the velocity of travel of any point identified by a certain phase (i.e., argument) of the sinusoid. In other words, if we were to observe a fixed phase point on the wave, we have

$$\omega t - \beta z = \text{const.} \quad \longrightarrow \quad \frac{dz}{dt} = v_p = \frac{\omega}{\beta}$$

Note that $v_p = \omega/(\omega\sqrt{\mu\epsilon}) = 1/\sqrt{\mu\epsilon}$, which for free space ($\mu = \mu_0, \epsilon = \epsilon_0$) is the speed of light in free space, or $v_p = c$. Note also that substituting $\beta = 2\pi/\lambda$ in $v_p = \omega/\beta$ gives the familiar expression $v_p = f\lambda$, where $f = \omega/(2\pi)$. It is interesting to note that, for a simple lossless medium, the phase velocity is a function only of the medium parameters and is independent of frequency. This is an important property of uniform plane waves in a simple lossless medium, which does not hold true for waves propagating in a lossy medium or those that propagate in guiding structures, as we shall see later. However, it should be noted that ϵ and μ are in general functions

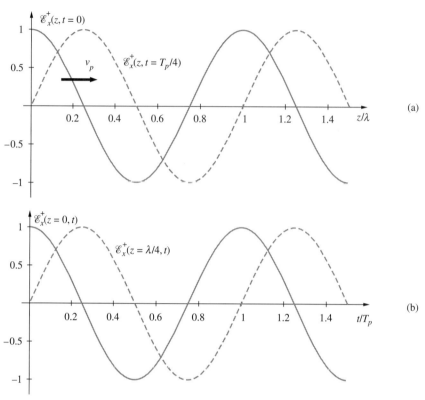

FIGURE 2.4. **Wave behavior in space and time.** (a) $\mathcal{E}_x^+(z,t) = \cos[(2\pi/T_p)t - (2\pi/\lambda)z]$ versus z/λ and for $t = 0$ and $t = T_p/4$. (b) $\mathcal{E}_x^+(z,t)$ versus t/T_p for $z = 0$ and $z = \lambda/4$.

of frequency even in a lossless medium and can be considered to be constant only over limited frequency ranges.

Although we have discussed the solution of the wave equation for the electric field, we could have just as easily derived a wave equation identical to [2.8] for the magnetic field **H**, and we could have solved it by making the same assumption (i.e., uniform plane wave) and restricting our attention to a single component of the magnetic field. However, once we have described a solution for the electric field of a uniform plane wave (say the component E_x), we are no longer free to choose one of the two components of its magnetic field (i.e., H_x or H_y) arbitrarily to be nonzero, because the wave equation for the electric field (i.e., [2.8]) was obtained using the entire set of equations [1.11] and the associated magnetic field is already determined once we specify its electric field. The proper means for finding the magnetic field of a uniform plane wave is thus to derive it from the electric field using [1.11a]. We have

$$\nabla \times \mathbf{E} = \hat{\mathbf{y}}\frac{dE_x(z)}{dz} = -j\omega\mu[\hat{\mathbf{x}}H_x + \hat{\mathbf{y}}H_y + \hat{\mathbf{z}}H_z]$$

It thus appears that the corresponding **H** has only a y component, which for $E_x(z)$ as given in [2.10] can be found as

$$H_y(z) = \frac{1}{-j\omega\mu}\frac{d}{dz}(C_1 e^{-j\beta z} + C_2 e^{j\beta z})$$

$$= \frac{\beta}{\omega\mu}(C_1 e^{-j\beta z} - C_2 e^{j\beta z}) = \frac{1}{\eta}(C_1 e^{-j\beta z} - C_2 e^{j\beta z})$$

where we have used $\eta = \sqrt{\mu/\epsilon} = \omega\mu/\beta$. Note that the intrinsic impedance η is a real number for a simple lossless medium, but, as we will see in Section 2.3, it is a complex number for a lossy medium.

The instantaneous magnetic field $\mathcal{H}(x, y, z, t)$ can be obtained from the phasor $\mathbf{H}(x, y, z)$ in the same way as in [2.11]. Considering only the first term in [2.12] and the associated magnetic field, the electric and magnetic fields of a time-harmonic uniform plane wave propagating in the $+z$ direction are

$$\boxed{\begin{aligned} \mathcal{E}_x(z, t) &= C_1 \cos(\omega t - \beta z) \\ \mathcal{H}_y(z, t) &= \frac{1}{\eta}C_1 \cos(\omega t - \beta z) \end{aligned}}$$

[2.13]

We find, as before (Section 2.1), that the electric and magnetic fields of a uniform plane wave are perpendicular to one another. The $\mathcal{E}$ and $\mathcal{H}$ fields are also both perpendicular to the direction of propagation, which is why uniform plane waves are also often referred to as *transverse electromagnetic,* or *TEM,* waves. In a simple lossless medium, the variation of $\mathcal{E}$ and $\mathcal{H}$ in space and time are identical (i.e., they are in phase); in other words, the $\mathcal{E}$ and $\mathcal{H}$ propagate in unison along z, reaching their maxima and minima at the same points in space and at the same times. The orientation of $\mathcal{E}$ and $\mathcal{H}$ is such that the vector $\mathcal{E} \times \mathcal{H}$ is in the $+z$ direction, which is the direction of propagation for the wave. Note that since the choice of the coordinate system cannot change the physical picture, if we start with $\mathcal{E}$ having only a y component (i.e., $\mathcal{E} = \hat{\mathbf{y}}\mathcal{E}_y$), then $\mathcal{H}$ comes out to be $\mathcal{H} = -\hat{\mathbf{x}}\mathcal{H}_x$, i.e., $\mathcal{E} \times \mathcal{H}$ would still be in the direction of propagation (i.e., the z direction). The relationship between $\mathcal{E}$ and $\mathcal{H}$ for a time-harmonic uniform plane wave is depicted in Figure 2.5. As time progresses, the electric and magnetic fields propagate in the $+z$ direction, staying in phase at all points and at all times.

Equations [2.13] and Figure 2.5 completely describe the properties of uniform plane waves in a simple lossless unbounded medium. As long as μ and ϵ are simple constants, an electromagnetic wave in free space differs from that in the material medium (with $\epsilon \neq \epsilon_0$ and/or $\mu \neq \mu_0$) primarily in terms of its wavelength, given by $\lambda = (f\sqrt{\mu\epsilon})^{-1}$. We shall see in the next section that the characteristics of uniform plane waves in a lossy medium are substantially more influenced by the material properties.

Examples 2-1 through 2-4 illustrate the use of [2.13] for different applications.

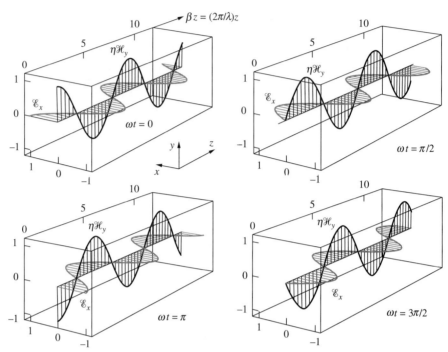

FIGURE 2.5. **Electric and magnetic fields of a uniform plane wave in a lossless medium.** Snapshots of $\mathscr{E}_x(z, t)$ and $\eta\mathscr{H}_y(z, t)$ for a sinusoidal uniform plane wave shown as a function of βz at $\omega t = 0, \pi/2, \pi$, and $3\pi/2$. The plots shown are for $C_1 = 1$.

Example 2-1: **AM broadcast signal.** The instantaneous expression for the electric field component of an AM broadcast signal propagating in air is given by

$$\overline{\mathscr{E}}(x, t) = \hat{z}10 \, \cos(1.5\pi \times 10^6 t + \beta x) \quad \text{V-m}^{-1}$$

(a) Determine the direction of propagation and frequency f. (b) Determine the phase constant β and the wavelength λ. (c) Write the instantaneous expression for the corresponding magnetic field $\overline{\mathscr{H}}(x, t)$.

Solution:

(a) The wave propagates in the $-x$ direction. From $\omega = 2\pi f = 1.5\pi \times 10^6$ rad-s^{-1}, $f = 750$ kHz.

(b) The phase constant is given by

$$\beta = \omega \sqrt{\mu_0 \epsilon_0} \simeq 1.5\pi \times 10^6/(3 \times 10^8) = 0.005\pi \text{ rad-m}^{-1}$$

The wavelength in air is equal to that in free space, or

$$\lambda = 2\pi/\beta \simeq 2\pi/0.005\pi = 400 \text{ m}$$

(c) The phasor electric field is given by

$$\mathbf{E}(x) = \hat{\mathbf{z}}10e^{+j\beta x} = \hat{\mathbf{z}}10e^{j0.005\pi x} \quad \text{V-m}^{-1}$$

To find the corresponding **H**, we use [1.11a] to find

$$-j\omega\mu_0\mathbf{H} = \nabla \times \mathbf{E} \quad \rightarrow \quad H_y(x) = -\frac{1}{j\omega\mu_0}\left[-\frac{\partial E_z(x)}{\partial x}\right] = \frac{1}{\eta}E_z(x)$$

$$\mathbf{H}(x) = \hat{\mathbf{y}}\frac{1}{\eta}E_z(x) \simeq \hat{\mathbf{y}}\frac{10 \text{ V-m}^{-1}}{377\Omega}e^{j0.005\pi x}$$

Therefore,

$$\overline{\overline{\mathscr{H}}}(x, t) \simeq \hat{\mathbf{y}}26.5 \times 10^{-3}\cos(1.5\pi \times 10^6 t + 0.005\pi x) \quad \text{A-m}^{-1}$$

Example 2-2:   FM broadcast signal.   An FM radio broadcast signal traveling in the y direction in air has a magnetic field given by the phasor

$$\mathbf{H}(y) = 2.92 \times 10^{-3}e^{-j0.68\pi y}(-\hat{\mathbf{x}} + \hat{\mathbf{z}}j) \quad \text{A-m}^{-1}$$

(a) Determine the frequency (in MHz) and wavelength (in m). (b) Find the corresponding **E**(y). (c) Write the instantaneous expression for $\overline{\overline{\mathscr{E}}}(y, t)$ and $\overline{\overline{\mathscr{H}}}(y, t)$.

Solution:
(a) We have

$$\beta = \omega\sqrt{\mu_0\epsilon_0} = 0.68\pi \text{ rad-m}^{-1}$$

from which we find

$$f = \frac{\omega}{2\pi} \simeq \frac{0.68\pi \text{ rad-m}^{-1} \times 3 \times 10^8 \text{ m-s}^{-1}}{2\pi} = 102 \text{ MHz}$$

(b) Using [1.11c]

$$\nabla \times \mathbf{H} = \hat{\mathbf{x}}\frac{\partial H_z}{\partial y} - \hat{\mathbf{z}}\frac{\partial H_x}{\partial y} = j\omega\epsilon_0\mathbf{E}$$

and performing the partial differentiation yields

$$\mathbf{E}(y) \simeq 1.1e^{-j0.68\pi y}(-\hat{\mathbf{x}}j - \hat{\mathbf{z}}) \quad \text{V-m}^{-1}$$

(c) The instantaneous expressions for $\overline{\overline{\mathscr{E}}}(y, t)$ and $\overline{\overline{\mathscr{H}}}(y, t)$ are given by

$$\overline{\overline{\mathscr{E}}}(y, t) = -\hat{\mathbf{x}}1.1\cos(2.04\pi \times 10^8 t - 0.68\pi y + \pi/2)$$

$$= -\hat{\mathbf{z}}1.1\cos(2.04\pi \times 10^8 t - 0.68\pi y) \quad \text{V-m}^{-1}$$

$$\overline{\overline{\mathscr{H}}}(y, t) = -\hat{\mathbf{x}}2.92 \times 10^{-3}\cos(2.04\pi \times 10^8 t - 0.68\pi y)$$

$$= +\hat{\mathbf{z}}2.92 \times 10^{-3}\cos(2.04\pi \times 10^8 t - 0.68\pi y + \pi/2) \quad \text{A-m}^{-1}$$

Example 2-3:   Uniform plane wave.   Consider a uniform plane wave travel-ing in the z direction in a simple lossless nonmagnetic medium (i.e., $\mu = \mu_0$) with a y-directed electric field of maximum amplitude of 60 V-m^{-1}. If the wavelength is 20 cm and the velocity of propagation is 10^8 m-s^{-1}, (a) determine the frequency of the wave and the relative permittivity of the medium; (b) write complete time-domain expressions for both the electric and magnetic field components of the wave.

Solution:

(a) We know that $f = v_p/\lambda = 10^8/0.2 = 500$ MHz. We also know that $v_p = 1/\sqrt{\mu\epsilon} = 1/\sqrt{\mu_0\epsilon_r\epsilon_0} \approx 3 \times 10^8/\sqrt{\epsilon_r} = 10^8$, from which we find the relative permittivity to be $\epsilon_r = 9$. The phase constant is $\beta = 2\pi/\lambda = 2\pi/0.2 = 10\pi$ rad-m^{-1}, and the intrinsic impedance is $\eta = \sqrt{\mu_0/\epsilon} = \sqrt{\mu_0/(\epsilon_r\epsilon_0)} \approx 120\pi/3 \approx 126\Omega$.

(b) Therefore, the instantaneous electric and magnetic fields are given by

$$\overline{\mathscr{E}}(z, t) = \hat{\mathbf{y}}60 \cos(10^9\pi t - 10\pi z) \quad \text{V-m}^{-1}$$

$$\overline{\mathscr{H}}(z, t) \approx -\hat{\mathbf{x}}0.477 \cos(10^9\pi t - 10\pi z) \quad \text{A-m}^{-1}$$

Example 2-4:   UHF cellular phone signal.   The magnetic field component of a UHF electromagnetic signal transmitted by a cellular phone base station is given by

$$\mathbf{H}(y) = \hat{\mathbf{x}}50e^{-j(17.3y-\pi/3)} \quad \mu\text{A-m}^{-1}$$

where the coordinate system is defined such that the z axis is in the vertical direction above a horizontal ground (xy plane or the $z = 0$ plane). (a) Determine the frequency f and the wavelength λ. (b) Write the corresponding expression for the electric field, $\mathbf{E}(y)$. (c) An observer located at $y = 0$ is using a vertical electric dipole antenna to measure the electric field as a function of time. Assuming that the observer can measure the vertical component of the electric field without any loss, what is the electric field at the time instants corresponding to $\omega t_1 = 0$, $\omega t_2 = \pi/2$, $\omega t_3 = \pi$, $\omega t_4 = 3\pi/2$, $\omega t_5 = 2\pi$ radians?

Solution:

(a) From $\beta = \omega/v_p = 2\pi f/c = 17.3$ rad-m^{-1}, we find $f \approx 826$ MHz. The corresponding wavelength is $\lambda = 2\pi/\beta \approx 36.3$ cm.

(b) From $\nabla \times \mathbf{H} = j\omega\epsilon\mathbf{E}$, we can solve for $\mathbf{E}$ as

$$\mathbf{E}(y) = \hat{\mathbf{z}}\eta H_x(y) \approx \hat{\mathbf{z}}377 \times 50 \times 10^{-6}e^{-j(17.3y-\pi/3)} = \hat{\mathbf{z}}18.85e^{-j(17.3y-\pi/3)} \quad \text{mV-m}^{-1}$$

(c) The instantaneous electric field is given by

$$\overline{\mathscr{E}}(y, t) = \mathscr{R}e\{\mathbf{E}(y)e^{j\omega t}\} \approx \hat{\mathbf{z}}18.85 \cos(5.19 \times 10^9 t - 17.3y + \pi/3) \quad \text{mV-m}^{-1}$$

Substituting $y = 0$ and $t_1 = 0$ yields $\overline{\mathscr{E}}(0, t_1) \approx \hat{\mathbf{z}}9.425$ mV-m^{-1}. Similarly, $\overline{\mathscr{E}}(0, t_2) \approx -\hat{\mathbf{z}}16.3$ mV-m^{-1}, $\overline{\mathscr{E}}(0, t_3) = -\hat{\mathbf{z}}9.425$ mV-m^{-1}, $\overline{\mathscr{E}}(0, t_4) \approx \hat{\mathbf{z}}16.3$ mV-m^{-1}, and $\overline{\mathscr{E}}(0, t_5) = \hat{\mathbf{z}}9.425$ mV-m^{-1}.

2.2.1   The Electromagnetic Spectrum

The properties of uniform plane waves, as defined by the electric and magnetic field expressions given in [2.13], are identical over all of the electromagnetic spectrum that has been investigated experimentally, frequencies ranging from millihertz to 10^{24} Hz. Regardless of their frequency, all uniform plane electromagnetic waves propagate in unbounded free space with the same velocity, namely $v_p = c$, but with different wavelengths, as determined by $\lambda = c/f$. The speed of propagation of electromagnetic waves is also independent of frequency in a simple lossless material medium. (However, the material properties themselves are often functions of frequency, so the assumption of a simple lossless material medium does not hold true over the entire electromagnetic spectrum for any material medium.) Table 2.1 lists the various designated frequency and wavelength ranges of the electromagnetic spectrum and selected applications for each range. Maxwell's equations and the results derived from them thus encompass a truly amazing range of physical phenomena and applications that affect nearly every aspect of human life and our physical environment. In this section, we briefly comment on a few of the many applications listed in Table 2.1.

At frequencies in the ultraviolet range and higher, physicists are more accustomed to thinking in terms of the associated energy level of the photon (a quantum of radiation), which is given by hf, where $h \simeq 6.63 \times 10^{-34}$ J-s is *Planck's constant*. Cosmic rays, consisting typically of photons at energies 10 MeV or greater, are constantly present in our universe; they ionize the highest reaches of Earth's atmosphere and help maintain the ionosphere at night, in the absence of solar radiation. Short-duration bursts of γ-rays, which bathe our solar system (and our galaxy) about three times a day, are believed to be produced in the most powerful explosions in the universe, releasing (within a few seconds or minutes) energies of 10^{51} ergs—more energy than our sun will produce in its entire ten billion years of existence.[11] Recently, brief flashes of γ-rays have been observed to be originating from Earth, believed to be generated at high altitudes above thunderstorms.[12]

It is also interesting to note that only a very narrow portion of the electromagnetic spectrum is perceptible to human vision, namely the visible range. We can also "feel" infrared as heat, and our bodies can be damaged by excessive amounts of microwave radiation, X-rays, and γ-rays. Applications of X-rays, ultraviolet, visible, and infrared light are far too numerous to be commented on here and include vision, lasers, optical fiber communications, and astronomy.

At present, a relatively unexplored part of the electromagnetic spectrum is the transition region between the low-frequency end of infrared and the high-frequency

---

[11]G. J. Fishman and D. H. Hartmann, Gamma-ray bursts, *Scientific American,* pp. 46–51, July 1997.

[12]G. J. Fishman, P. N. Bhat, R. Mallozzi, J. M. Horack, T. Koshut, C. Kouveliotou, G. N. Pendleton, C. A. Meegan, R. B. Wilson, W. S. Paciesas, S. J. Goodman, and H. J. Christian, Discovery of intense gamma-ray flashes of atmospheric origin, *Science,* 264, pp. 1313–1316, May 1994.

TABLE 2.1. The electromagnetic spectrum and related applications

Frequency	Designation	Selected applications	Wavelength (in free space)
$> 10^{22}$ Hz	Cosmic rays	Astrophysics	
10^{18}–10^{22} Hz	γ-rays	Cancer therapy, astrophysics	
10^{16}–10^{21} Hz	X-rays	Medical diagnosis	
10^{15}–10^{18} Hz	Ultraviolet	Sterilization	0.3–300 nm
3.95×10^{14}–7.7×10^{14} Hz	Visible light	Vision, astronomy, optical communications	390–760 nm
		Violet	390–455
		Blue	455–492
		Green	492–577
		Yellow	577–600
		Orange	600–625
		Red	625–760
10^{12}–10^{14} Hz	Infrared	Heating, night vision, optical communications	3–300 μm
0.3–1 THz	Millimeter	Astronomy, meteorology	0.3–1 mm
30–300 GHz	EHF	Radar, remote sensing	0.1–1 cm
80–100		W-band	
60–80		V-band	
40–60		U-band	
27–40		K_a-band	
3–30 GHz	SHF	Radar, satellite comm.	1–10 cm
18–27		K-band	
12–18		K_u-band	
8–12		X-band	
4–8		C-band	
0.3–3 GHz	UHF	Radar, TV, GPS, cellular phone	10–100 cm
2–4		S-band	
2.45		Microwave ovens	
1–2		L-band, GPS system	
470–890 MHz		TV Channels 14–83	
30–300 MHz	VHF	TV, FM, police,	1–10 m
174–216		TV Channels 7–13	
88–108		FM radio	
76–88		TV Channels 5–6	
54–72		TV Channels 2–4	
3–30 MHz	HF	Short-wave, citizens' band	10–100 m
0.3–3 MHz	MF	AM broadcasting	0.1–1 km
30–300 kHz	LF	Navigation, radio beacons	1–10 km
3–30 kHz	VLF	Navigation, positioning, naval communications	10–100 km
0.3–3 kHz	ULF	Telephone, audio	0.1–1 Mm
30–300 Hz	SLF	Power transmission, submarine communications	1–10 Mm
3–30 Hz	ELF	Detection of buried metals	10–100 Mm
< 3 Hz		Geophysical prospecting	> 100 Mm

end of the millimeter range. The frequency in this region is measured in terahertz (THz), or 10^{12} Hz, with the corresponding wavelengths being in the submillimeter range. Fabrication of antennas, transmission lines, and other receiver components usable in these frequencies requires the use of novel quasi-optical techniques.[13] At present, the terahertz region is primarily used in astronomy and remote sensing; however, industrial, medical, and other scientific applications are currently under investigation.

Each decade of the electromagnetic spectrum below the millimeter range of frequencies is divided[14] into designated ranges, with the acronyms indicated in Table 2.1: extremely high frequency (EHF); super high frequency (SHF); ultra high frequency (UHF); very high frequency (VHF); high frequency (HF); medium frequency (MF); low frequency (LF); very low frequency (VLF); ultra low frequency (ULF); super low frequency (SLF); extremely low frequency (ELF).

The microwave band of frequencies is vaguely defined as the range from 300 MHz up to 1 THz, including the millimeter range. It is extensively utilized for radar, remote sensing, and a host of other applications too numerous to cite here.[15] In radar work, the microwave band is further subdivided into bands with alphabetical designations, which are listed in Table 2.1.

The VLF range of frequencies is generally used for global navigation and naval communications, using transmitters that utilize huge radiating structures.[16] Although VLF transmissions are used for global communications with surface ships and submarines near the water surface, even lower frequencies are required for communication with deeply submerged submarines. The Sanguine system operated by the U.S. Navy utilizes two sets of orthogonal large (22.5 km length) horizontal antennas, located in Wisconsin and Michigan at a distance of 240 km apart, and operates[17] at 72 to 80 Hz.

The lowest frequencies of the experimentally investigated electromagnetic spectrum are commonly used to observe so-called micropulsations, which are electromagnetic waves at frequencies 0.001 to 10 Hz generated as a result of large-scale currents flowing in the earth's auroral regions and by the interaction between the earth's magnetic field and the energetic particles that stream out of the sun in the

---

[13] See Special Issue on Terahertz Technology of the *Proceedings of the IEEE,* 80 (11), November 1992.

[14] There is a certain arbitrariness in the designations of these frequency ranges. In geophysics and solar terrestrial physics, the designation ELF is used for the range 3 Hz to 3 kHz, while ULF is used to describe frequencies typically below 3 Hz.

[15] See J. Thuery, *Microwaves: Industrial, Scientific and Medical Applications,* Artech House, 1992. Also see O. P. Gandhi, editor, *Biological Effects and Medical Applications of Electromagnetic Energy,* Prentice Hall, 1990.

[16] See A. D. Watt, *VLF Radio Engineering,* Pergamon Press, New York, 1967; J. C. Kim and E. I. Muehldorf, *Naval Shipboard Communications Systems,* Prentice Hall, 1995.

[17] C. H. Richard, *Sub vs. Sub,* Orion Books, New York, 1988; M. F. Genge and R. D. Carlson, Project ELF Electromagnetic Compatibility Assurance Program, *IEEE Journal of Oceanic Engineering,* 9(3), pp. 143–153, July 1984.

form of the solar wind.[18] These natural signals are also often used for geophysical prospecting and magnetotelluric studies.[19] Another potentially very important practical use of this band may yet emerge, based on experimental evidence of detectable electromagnetic precursors produced many hours prior to major earthquakes.[20]

2.3 PLANE WAVES IN LOSSY MEDIA

Many of the more interesting electromagnetic applications involve the interactions between electric and magnetic fields and matter. The manner in which waves interact with matter is described in Chapter 6. For now, we represent the microscopic interactions of electromagnetic waves with matter in terms of the macroscopic parameters ϵ, μ, and σ. In general, most dielectric media exhibit small but nonzero conductivity or complex permittivity and can absorb electromagnetic energy, resulting in the attenuation of an electromagnetic wave as it propagates through the medium. The performance of most practical transmission lines, and of other devices that convey electromagnetic energy from one point to another, is limited by small losses in conductors or dielectrics. The inherently lossy nature of some media (e.g., seawater, animal tissue) determines the range of important applications (e.g., submarine communications and medical diagnostic implants). All media exhibit losses in some frequency ranges; for example, although air is largely transparent (lossless) over the radio and microwave ranges, it is a highly lossy medium at optical frequencies. The fact that upper atmospheric ozone is lossy at ultraviolet frequencies (i.e., it absorbs ultraviolet light) protects life on Earth from this harmful radiation.

When a material exhibits a nonzero conductivity σ, the electric field of a propagating wave causes a conduction current of $\mathbf{J}_c = \sigma\mathbf{E}$ to flow. This current, which is in phase with the wave electric field, leads to dissipation of some of the wave energy as heat within the material, with the power dissipated per unit volume being given by $\mathbf{E} \cdot \mathbf{J}$. This dissipation requires that the wave electric and magnetic fields attenuate with distance as they propagate in the lossy material, much like the attenuation of voltage and current waves propagating on a lossy transmission line. To determine the characteristics of uniform plane waves in lossy media, we start with Maxwell's equations and follow a procedure quite similar to that used in the previous section.

---

[18]J. A. Jacobs, *Geomagnetic Micropulsations,* Springer-Verlag, New York, Heidelberg, and Berlin, 1970; J. K. Hargreaves, *The Solar-Terrestrial Environment,* Cambridge University Press, 1992.

[19]M. N. Nabighian, editor, *Electromagnetic Methods in Applied Geophysics, Vol. 2: Application,* Parts A and B, Society of Exploration Geophysics, 1991.

[20]A. C. Fraser-Smith, A. Bernardi, P. R. McGill, M. E. Ladd, R. A. Helliwell, and O. G. Villard, Jr., Low-frequency magnetic field measurements near the epicenter of the M_s 7.1 Loma Prieta earthquake, *Geophys. Res. Lett.,* 17, pp. 1465–1468, 1990; also see B. Holmes, Radio hum may herald quakes, *New Scientist,* p. 15, 23/30 December 1995.

In a conducting (or lossy) medium the $\mathbf{J}$ term in [1.11c] is nonzero even in the absence of sources, since a current of $\mathbf{J}_c = \sigma\mathbf{E}$ flows in response to the wave's electric field. We then have

$$\nabla \times \mathbf{H} = \sigma\mathbf{E} + j\omega\epsilon\mathbf{E} \qquad [2.14]$$

and, taking the curl of [1.11a], we have

$$\nabla \times \nabla \times \mathbf{E} \equiv \nabla(\nabla \cdot \mathbf{E}) - \nabla^2\mathbf{E} = -j\omega\nabla \times \mathbf{B}$$

Using the fact that $\nabla \cdot \mathbf{E} = 0$ for a simple, source-free medium, and substituting from [1.11c], we find

$$\nabla^2\mathbf{E} - j\omega\mu(\sigma + j\omega\epsilon)\mathbf{E} = 0$$

or

$$\boxed{\nabla^2\mathbf{E} - \gamma^2\mathbf{E} = 0} \qquad [2.15]$$

where $\gamma = \sqrt{j\omega\mu(\sigma + j\omega\epsilon)}$ is known as the propagation constant (or wave number) and is in general a complex quantity, which can be expressed in terms of its real and imaginary parts as $\gamma = \alpha + j\beta$. Note that the complex propagation constant γ for uniform plane waves in an unbounded lossy medium is equivalent to the complex propagation constant for propagation of voltage waves on a lossy transmission line.[21]

Realizing that [2.15] represents three scalar equations similar to [2.8], we limit our attention once again to uniform plane waves propagating in the z direction (i.e., all field quantities varying only in z), with the electric field having only an x component, for which [2.15] becomes

$$\boxed{\frac{d^2E_x}{dz^2} - \gamma^2E_x = 0} \qquad [2.16]$$

which, when we replace $E_x(z)$ with $V(z)$, is identical to the equation describing the variation of the voltage phasor $V(z)$ on a lossy transmission line.[22] The general solution of [2.16] is

$$E_x(z) = C_1e^{-\gamma z} + C_2e^{+\gamma z} = \underbrace{C_1e^{-\alpha z}e^{-j\beta z}}_{E_x^+(z)} + \underbrace{C_2e^{+\alpha z}e^{+j\beta z}}_{E_x^-(z)} \qquad [2.17]$$

.

---

[21]For a uniform lossy transmission line with per-unit-length resistance, inductance, capacitance, and conductance of respectively R, L, C, and G, the complex propagation constant is given by

$$\gamma = \sqrt{(R + j\omega L)(G + j\omega C)} = \alpha + j\beta$$

For more details see Section 3.8 of U. S. Inan and A. S. Inan, *Engineering Electromagnetics,* Addison Wesley Longman, 1999.

[22]See Section 3.8 of U. S. Inan and A. S. Inan, *Engineering Electromagnetics,* Addison Wesley Longman, 1999.

For $\alpha, \beta > 0$, the two terms $E_x^+(z)$ and $E_x^-(z)$ represent waves propagating in the $+z$ and $-z$ directions, respectively. Note that the constants C_1 and C_2 are in general complex numbers to be determined by boundary conditions at material interfaces.

The instantaneous electric field can be found from [2.17] in the same way[23] as described in [2.11]. We find

$$\mathcal{E}_x(z,t) = \underbrace{C_1 e^{-\alpha z} \cos(\omega t - \beta z)}_{\mathcal{E}_x^+(z,t)} + \underbrace{C_2 e^{+\alpha z} \cos(\omega t + \beta z)}_{\mathcal{E}_x^-(z,t)} \qquad [2.18]$$

assuming C_1 and C_2 to be real. The nature of the waves described by the two terms of [2.18] is shown in Figure 2.6. We see, for example, that the wave propagating in the $+z$ direction has a decreasing amplitude with increasing distance at a fixed instant of time. In other words, the wave is attenuated as it propagates in the medium. The rate of this attenuation is given by the *attenuation constant α*.

To understand better the parameter α and its dimensions, consider the fact that the ratio of the magnitude of $E_x^+(z)$ corresponding to two different points separated by a distance d is a constant. In other words, the magnitude of the ratio of the electric field at position $(z + d)$ to that at position z is

$$\left| \frac{E_x^+(z)}{E_x^+(z+d)} \right| = \frac{C_1 e^{-\alpha z}}{C_1 e^{-\alpha(z+d)}} = e^{\alpha d}$$

Taking the natural logarithm of both sides we have

$$\alpha d = \ln\left[\left| \frac{E_x^+(z)}{E_x^+(z+d)} \right| \right]$$

Note that αd is a dimensionless number, since the units of α are in m^{-1} and d is in meters. However, to underscore the fact that αd is the natural, or Naperian, logarithm of the ratio of the fields, αd is commonly expressed as a number of *nepers*. In the context of this conventional usage, the unit[24] of the attenuation constant α is nepers-m^{-1}, or in abbreviated form, np-m^{-1}.

The quantity β (in rad-m^{-1}) determines the phase velocity and wavelength of the wave and is referred to as the *phase constant*. We shall see subsequently that the

---

[23] $\mathfrak{Re}\{C_1 e^{-\alpha z} e^{-j\beta z} e^{j\omega t}\} = C_1 e^{-\alpha z} \mathfrak{Re}\{e^{j(\omega t - \beta z)}\} = C_1 e^{-\alpha z} \cos(\omega t - \beta z)$, where C_1 is a real constant.

[24] In most engineering applications, a more common unit for attenuation is the *decibel*. The decibel is a unit derived from *bel*, which in turn was named after Alexander Graham Bell and was used in early work on telephone systems. Specifically, the decibel is defined as

$$\text{Attenuation in decibels (dB)} \equiv 20 \log_{10}\left[\left| \frac{E_x^+(z)}{E_x^+(z+d)} \right| \right]$$

The attenuation can be expressed in terms of either decibels or nepers. We have

$$\text{Attenuation in dB} = 20 \log_{10} e^{\alpha d} = (\alpha d) 20 \log_{10} e \simeq 8.686(\alpha d)$$
$$\simeq 8.686[\text{Attenuation in nepers}]$$

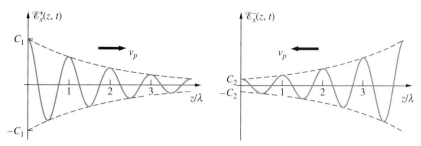

FIGURE 2.6.   Snapshots of waves in a lossy medium. The two terms of [2.18] plotted as a function of z at $\omega t = 0$

wavelength of a uniform plane wave in lossy materials can be substantially different from that in free space.

Note that the complex propagation constant γ is given by

$$\gamma = \alpha + j\beta = \sqrt{j\omega\mu(\sigma + j\omega\epsilon)}$$

Explicit expressions for α and β can be found[25] by squaring and equating the real and imaginary parts. We find

$$\alpha = \omega\sqrt{\frac{\mu\epsilon}{2}}\left[\sqrt{1 + \left(\frac{\sigma}{\omega\epsilon}\right)^2} - 1\right]^{1/2} \text{ np-m}^{-1} \qquad [2.19]$$

and

$$\beta = \omega\sqrt{\frac{\mu\epsilon}{2}}\left[\sqrt{1 + \left(\frac{\sigma}{\omega\epsilon}\right)^2} + 1\right]^{1/2} \text{ rad-m}^{-1} \qquad [2.20]$$

Consider now only the wave propagating in the $+z$ direction, the first term of [2.17]:

$$E_x^+(z) = C_1 e^{-\alpha z} e^{-j\beta z} \qquad [2.21]$$

---

[25] We have

$$\gamma^2 = (\alpha + j\beta)^2 = j\omega\mu(\sigma + j\omega\epsilon) \quad \rightarrow \quad \underbrace{\alpha^2 - \beta^2 = -\omega^2\mu\epsilon}_{\text{Real part}} \quad \text{and} \quad \underbrace{2\alpha\beta = \omega\mu\sigma}_{\text{Imaginary part}}$$

Squaring and adding the two equations and taking the square root, we have

$$(\alpha^2 - \beta^2)^2 + (2\alpha\beta)^2 = \omega^4\mu^2\epsilon^2 + (\omega\mu\sigma)^2 \rightarrow \alpha^4 + 2\alpha^2\beta^2 + \beta^4 = \omega^4\mu^2\epsilon^2 + \omega^2\mu^2\sigma^2$$

$$\rightarrow (\alpha^2 + \beta^2)^2 = \omega^4\mu^2\epsilon^2\left[1 + \left(\frac{\sigma}{\omega\epsilon}\right)^2\right] \rightarrow \alpha^2 + \beta^2 = \omega^2\mu\epsilon\sqrt{1 + \left(\frac{\sigma}{\omega\epsilon}\right)^2}$$

Equation [2.19] follows by adding the expressions for $(\alpha^2 + \beta^2)$ and $(\alpha^2 - \beta^2)$.

The wave magnetic field that accompanies this electric field can be found by substituting [2.21] into [1.11a], as we did for the lossless case, or by making an analogy with the lossless-media solution obtained in the previous section. Following the latter approach, we can rewrite [2.14] as

$$\nabla \times \mathbf{H} = \sigma \mathbf{E} + j\omega\epsilon\mathbf{E} = j\omega\epsilon_{\text{eff}}\mathbf{E}$$

where $\epsilon_{\text{eff}} = \epsilon - j\sigma/\omega$. It thus appears that the solutions we obtained in the previous section for a lossless medium can be used as long as we make the substitution $\epsilon \rightarrow \epsilon_{\text{eff}}$. On this basis, the intrinsic impedance of a conducting medium can be found as

$$\eta_c = |\eta_c|e^{j\phi_\eta} = \sqrt{\frac{\mu}{\epsilon_{\text{eff}}}} = \sqrt{\frac{\mu}{\epsilon - j\dfrac{\sigma}{\omega}}} = \frac{\sqrt{\dfrac{\mu}{\epsilon}}}{\left[1 + \left(\dfrac{\sigma}{\omega\epsilon}\right)^2\right]^{1/4}} e^{j(1/2)\tan^{-1}[\sigma/(\omega\epsilon)]} \qquad [2.22]$$

and the associated magnetic field $\mathbf{H}$ is

$$\mathbf{H} = \hat{\mathbf{y}}H_y^+(z) = \hat{\mathbf{y}}\frac{1}{\eta_c}C_1 e^{-\alpha z}e^{-j\beta z}$$

The instantaneous magnetic field $\overline{\mathcal{H}}(z, t)$ can be found as

$$\mathcal{H}_y^+(z, t) = \mathcal{R}e\{H_y^+(z)e^{j\omega t}\} = \mathcal{R}e\left\{\frac{1}{|\eta_c|e^{j\phi_\eta}}C_1 e^{-\alpha z}e^{-j\beta z}e^{j\omega t}\right\}$$

$$= \frac{1}{|\eta_c|}C_1 e^{-\alpha z}\mathcal{R}e\{e^{j(\omega t - \beta z)}e^{-j\phi_\eta}\} \qquad [2.23]$$

$$\mathcal{H}_y^+(z, t) = \frac{C_1}{|\eta_c|}e^{-\alpha z}\cos(\omega t - \beta z - \phi_\eta)$$

Thus we see that the electric and magnetic fields of a uniform plane wave in a conducting medium do not reach their maxima at the same time (at a fixed point) or at the same point (at a fixed time); in other words, $\overline{\mathcal{E}}(z, t)$ and $\overline{\mathcal{H}}(z, t)$ are *not in phase*. The magnetic field lags the electric field by a phase difference equal to the phase of the complex intrinsic impedance, ϕ_η. This relationship is depicted in Figure 2.7.

We can see from [2.19], [2.20], and [2.22] that the rate of attenuation α, the phase constant β, and the intrinsic impedance η_c depend sensitively on $\sigma/(\omega\epsilon)$. For $\sigma \ll (\omega\epsilon)$, the attenuation rate is small, and the propagation constant and intrinsic impedance are only slightly different from that for a lossless medium; namely, $\beta \simeq \omega\sqrt{\mu\epsilon}$ and $\eta_c \simeq \sqrt{\mu/\epsilon}$. For $\sigma \gg (\omega\epsilon)$, the attenuation rate α is large, causing the uniform plane wave to decay rapidly with distance, and the intrinsic impedance is very small, approaching zero as $\sigma \rightarrow \infty$, as it does for a perfect conductor. A perfect conductor, in other words, is a medium with zero intrinsic

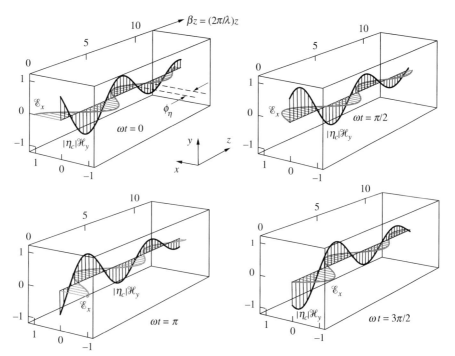

**FIGURE 2.7.   Time snapshots of the electric and magnetic fields in a conducting
medium.** The curves shown are for $C_1 = 1$, $\phi_\eta = 45°$, and $\alpha = 0.1\beta$, where $\beta = 2\pi/\lambda$.

impedance,[26] so the presence of an electric field of nonzero magnitude $C_1 \neq 0$ requires a magnetic field of an infinite magnitude. Thus, no time-harmonic electric or
magnetic fields can exist in a perfect conductor.[27]

The quantity $\tan \delta_c = \sigma/(\omega\epsilon)$ is called the *loss tangent* of the medium and is a
measure of the degree to which the medium conducts. A medium is considered to be
a good conductor if $\tan \delta_c \gg 1$; most metals are good conductors at frequencies of up
to 100 GHz or so. A medium is considered to be a poor conductor, or a good insulator,
if $\tan \delta_c \ll 1$. Example 2-5 illustrates the use of expressions [2.19], [2.20], and
[2.22] to determine the parameter values for uniform plane waves in muscle tissue.

---

[26] A perfect conductor is thus the ultimate lossy material. Although this may at first appear counterin-
tuitive, the behavior of a perfect conductor as a medium with zero impedance is analogous to a short
circuit. The voltage drop across a short circuit is always zero, while the current through it is determined
by the external circuits it is connected to. Similarly, we shall see in Chapter 3 that when a uniform plane
wave is incident on a perfect conductor, the wave is perfectly reflected, since no electromagnetic fields
can exist inside the conductor, while the current that flows on the surface of the perfect conductor is
determined by the magnetic field intensity of the incident uniform plane wave.

[27] Note that static electric fields inside metallic conductors must also be zero and that static magnetic
fields cannot exist in a perfect conductor because of the Meissner effect. See Sections 4.7 and 6.8.4 of
U. S. Inan and A. S. Inan, *Engineering Electromagnetics,* Addison Wesley Longman, 1999.

Example 2-5: Microwave exposure of muscle tissue. Find the complex propagation constant γ and the intrinsic impedance η_c of a microwave signal in muscle tissue[28] at 915 MHz ($\sigma = 1.6$ S-m^{-1}, $\epsilon_r = 51$).

Solution: The complex propagation constant is given by

$$\gamma = \alpha + j\beta$$

where α and β are given by [2.19] and [2.20]. The loss tangent of the muscle tissue at 915 MHz is

$$\tan \delta_c = \frac{\sigma}{\omega \epsilon} \simeq \frac{1.6 \text{ S-m}^{-1}}{2\pi \times 915 \times 10^6 \text{ rad-s}^{-1} \times 51 \times 8.85 \times 10^{-12} \text{ F-m}^{-1}} \simeq 0.617$$

So α and β are given by

$$\left\{ \begin{matrix} \alpha \\ \beta \end{matrix} \right\} \simeq (2\pi \times 915 \times 10^6 \text{ rad-s}^{-1}) \frac{\sqrt{51}}{\sqrt{2} \times 3 \times 10^8 \text{ m-s}^{-1}} \left[\sqrt{1 + (0.617)^2} \mp 1 \right]^{1/2}$$

$$\simeq \left\{ \begin{matrix} 40.7 \text{ np-m}^{-1} \\ 142.7 \text{ rad-m}^{-1} \end{matrix} \right\}$$

So the complex propagation constant is given by

$$\gamma \simeq 40.7 + j142.7$$

The complex intrinsic impedance can be found using [2.22] as

$$\eta_c \simeq \frac{377/\sqrt{51}}{[1 + (0.617)^2]^{1/4}} \exp \left[j \frac{1}{2} \tan^{-1}(0.617) \right] \simeq 48.7 e^{j15.8°} \, \Omega$$

At this point it is important to remember that the material properties themselves (i.e., σ and ϵ) may well be functions of frequency (as discussed further in Sections 2.3.1 and 6.3). For typical good conductors, both σ and ϵ are nearly independent of frequency, at frequencies below the optical range, but for lossy dielectrics the material constants σ and ϵ tend to be functions of frequency. For some materials, the loss tangent $\tan \delta_c = \sigma/(\omega \epsilon)$ tends to vary less over the frequency range of interest. Properties of dielectrics are usually given in terms of ϵ and $\tan \delta_c$. The ratio $\sigma/(\omega \epsilon)$ is plotted as a function of frequency in Figure 2.8 for some common materials, assuming the parameters ϵ and σ to be constants. The values above the microwave region (>10 GHz) are not likely to be accurate since σ and ϵ may vary significantly with frequency. Table 2.2 gives the values of σ and ϵ used in Figure 2.8. The most common cases of practical interest involving the propagation of electromagnetic waves in lossy media concern lossy dielectrics, which exhibit complex

---

[28]C. C. Johnson and A. W. Guy, Nonionizing electromagnetic wave effects in biological materials and systems, *Proceedings of the IEEE*, 60(6), pp. 692–718, June 1972.

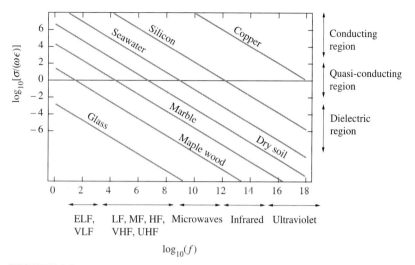

FIGURE 2.8.   **Loss tangent versus frequency.** The loss tangent $\tan \delta_c = \sigma/(\omega \epsilon)$ for selected materials plotted as a function of frequency, assuming that the material constants σ and ϵ are as given in Table 2.2 and that they do not vary with frequency.

TABLE 2.2.   Relative permittivity and conductivity of selected materials

Medium	Relative permittivity ϵ_r (dimensionless)	Conductivity σ, (S-m^{-1})
Copper	1	5.8×10^7
Seawater	81	4
Doped silicon	12	10^3
Marble	8	10^{-5}
Maple wood	2.1	3.3×10^{-9}
Dry soil	3.4	10^{-4} to 10^{-2}
Fresh water	81	$\sim 10^{-2}$
Mica	6	10^{-15}
Flint glass	10	10^{-12}

permittivity, and good conductors. We consider these two cases separately in the next two sections.

2.3.1   Uniform Plane Wave Propagation in Lossy Dielectrics

In most dielectrics that are good insulators, the direct conduction current (which is due to finite σ) is usually negligible. However, at high frequencies an alternating current that is in phase with the applied field is present, because the rapidly varying applied electric field alternately polarizes the bound electrons, thus doing work against molecular forces. As a result, materials that are good insulators at low frequencies can consume considerable energy when they are subjected to high-frequency fields.

The heat generated as a result of such radio-frequency heating is used in molding plastics; in microwave cooking, and in microwave drying of a wide variety of materials.[29]

The microphysical bases of such effects are different for solids, liquids, and gases and are too complex to be summarized here.[30] When an external time-varying field is applied to a material, bound charges are displaced, giving rise to volume polarization density **P**. At sinusoidal steady state, the polarization **P** varies at the same frequency as the applied field **E**. At low frequencies, **P** is also in phase with **E**, both quantities reaching their maxima and minima at the same points in the radio-frequency cycle. As the frequency is increased, however, the charged particles resist acceleration by the changing field, not just because of their mass but also because of the elastic and frictional forces that keep them attached to their molecules. This "inertia" tends to prevent the polarization **P** from keeping in phase with the applied field. The work that must be done against the frictional damping forces causes the applied field to lose power, and this power is deposited in the medium as heat. This condition of out-of-phase polarization that occurs at higher frequencies can be characterized by a complex electric susceptibility χ_e, and hence a complex permittivity ϵ_c. In such cases, both the frictional damping and any other ohmic losses (due to nonzero conductivity σ) are included in the imaginary part of the complex dielectric constant ϵ_c:

$$\epsilon_c = \epsilon' - j\epsilon''$$

We can analyze the resulting effects by substituting the preceding equation into [1.11c]:

$$\nabla \times \mathbf{H} = +j\omega\epsilon_c\mathbf{E} = (\omega\epsilon'')\mathbf{E} + j\omega\epsilon'\mathbf{E} \qquad [2.24]$$

Note that we have assumed $\mathbf{J} = \sigma\mathbf{E} = 0$, since the effects of any nonzero σ are included in ϵ''. It thus appears that the imaginary part of ϵ_c (namely, ϵ'') leads to a volume current density term that is in phase with the electric field, as if the material had an effective conductivity $\sigma_{\text{eff}} = \omega\epsilon''$. At low frequencies, $\omega\epsilon''$ is small, because ω is small and ϵ'' is itself small, so the losses are largely negligible. However, at high frequencies $\omega\epsilon''$ increases and produces the same macroscopic effect as if the dielectric had effective conductivity $\sigma_{\text{eff}} = \omega\epsilon''$. When a steady current $\mathbf{J} = \sigma\mathbf{E}$ flows in a conducting material in response to a constant applied field **E**, the electrical power per unit volume dissipated in the material is given by $\mathbf{E} \cdot \mathbf{J} = \sigma E^2$ in units of W-m^{-3}. Similarly, when a dielectric is excited at frequencies high enough for $\omega\epsilon''$ to be appreciable, an alternating current density of $\omega\epsilon''\mathbf{E}$ flows (in addition to the displacement current density given by $j\omega\epsilon'\mathbf{E}$) in response to an applied alternating field **E**, leading to an instantaneous power dissipation of $[(\omega\epsilon''\mathbf{E}) \cdot \mathbf{E}] = \omega\epsilon''|\mathbf{E}|^2$ in units

---

[29] J. Thuery, *Microwaves: Industrial, Scientific and Medical Applications,* Artech House, 1992.

[30] A brief introduction is provided in Section 6.3; for further discussion at an appropriate level, see C. Kittel, *Introduction to Solid State Physics,* 5th ed., Wiley, New York, 1976.

of W-m^{-3}. When the electric field is time-harmonic, i.e., $\mathscr{E}(t) = |\mathbf{E}|\cos(\omega t \pm \beta z)$, the time-average power dissipated in the material per unit volume at any arbitrary point (e.g., $z = 0$) is given by

$$P_{\mathrm{av}} = T_p^{-1}\int_0^{T_p} \omega\epsilon''|\mathscr{E}(t)|^2\, dt = T_p^{-1}\int_0^{T_p} \omega\epsilon''|\mathbf{E}|^2\cos^2(\omega t)\, dt = \tfrac{1}{2}\omega\epsilon''|\mathbf{E}|^2$$

This dissipated power density is the basis for microwave heating of dielectric materials. Since the values of ϵ'' are often determined by measurement, it is not necessary to distinguish between losses due to nonzero conductivity σ and the dielectric losses discussed here.

In general, both ϵ' and ϵ'' can depend on frequency in complicated ways, exhibiting several resonances over wide frequency ranges. The typical behavior in the vicinity of resonances is enhanced losses (i.e., large values of ϵ'') and a reduction of ϵ' to a new level.[31] Different materials exhibit a large variety of resonances, which originate in the basic energy-level structure of the atoms and molecules. For example, for water at room temperature, the resonance frequency is ~24 GHz, whereas for ice at $-20°C$ the relaxation frequency is ~1 kHz. Table 2.3 provides values of real and imaginary parts of the relative permittivity for selected materials.[32] Since these properties (especially ϵ'') are strongly dependent on frequency, and also to some degree on temperature, the frequency and temperature to which the given values correspond are specified. In some cases, the moisture content of the material is also noted, since the permittivity values also depend on this parameter.

It is clear from the preceding discussion that the degree to which losses in a dielectric are important depends on the ratio ϵ''/ϵ' and that the electrical properties of materials are functions of frequency. In practice, the electrical properties of a lossy

---

[31]The frequency response of a typical dielectric is shown in Figure 6.13 of Section 6.3, with contributions from the different types of polarizations (orientational, ionic, and electronic) identified in terms of the frequency ranges in which they are significant.

[32]The values in Table 2.3 were taken from the following sources: *Reference Data for Radio Engineers,* 8th ed., Sams Prentice Hall Computer Publishing, Carmel, Indiana, 1993; J. Thuery, *Microwaves: Industrial, Scientific and Medical Applications,* Artech House, Inc., 1992; D. I. C. Wang and A. Goldblith, Dielectric properties of food, *Tech. Rep. TR-76-27-FEL,* MIT, Dept. of Nutrition and Food Science, Cambridge, Massachusetts, November 1976; S. O. Nelson, Comments on "Permittivity measurements of granular food products suitable for computer simulations of microwave cooking processes," *IEEE Trans. Instrum. Meas.,* 40(6), pp. 1048–1049, December 1991; A. D. Green, Measurement of the dielectric properties of cheddar cheese, *J. Microwave Power Electromagn. Energy,* 32(1), pp. 16–27, 1997; S. Puranik, A. Kumbharkhane, and S. Mehrotra, Dielectric properties of honey-water mixtures between 10 MHz to 10 GHz using time domain technique, *J. Microwave Power Electromagn. Energy,* 26(4), pp. 196–201,1991; H. C. Rhim and O. Buyukozturk, Electromagnetic properties of concrete at microwave frequency range, *ACI Materials Journal,* 95(3), pp. 262–271, May–June 1998; J. M. Osepchuk, Sources and basic characteristics of microwave/RF radiation, *Bull. N. Y. Acad. Med.,* 55(11), pp. 976–998, December 1979; S. O. Nelson, Microwave dielectric properties of fresh onions, *Trans. Amer. Soc. Agr. Eng.,* 35(3), pp. 963–966, May–June 1992; R. F. Shiffmann, Understanding microwave reactions and interactions, *Food Product Design,* pp. 72–88, April 1993.

Preface

───────────────■───────────────

This book provides engineering students with a solid grasp of electromagnetic waves, by emphasizing physical understanding and practical applications. Starting with Maxwell's equations, the text provides a comprehensive treatment of the propagation of electromagnetic waves in empty space or material media, their reflection and refraction at planar boundaries, their guiding within metallic or dielectric boundaries, their interaction with matter, and their generation by simple sources.

Electromagnetic Waves is designed for senior and first-year graduate college and university engineering and physics students, for those who wish to learn the subject through self-study, and for practicing engineers who need an up-to-date reference text. The student using this text is assumed to have completed a typical (third- or fourth-year) undergraduate course in electromagnetic fundamentals.

KEY FEATURES

The key features of this textbook are

- Intuitive and progressive chapter organization
- Emphasis on physical understanding and practical applications
- Detailed examples, selected application examples, and abundant illustrations
- Numerous end-of-chapter problems, emphasizing selected practical applications
- Historical notes and abbreviated biographies of the great scientist pioneers
- Emphasis on clarity without sacrificing rigor and completeness
- Hundreds of footnotes providing physical insights and leads for further reading

Progressive Chapter Organization

We use a physical, intuitive, and progressive approach, covering electromagnetic wave phenomena in order of increasing complexity. Starting with Maxwell's equations (Chapter 1) as the experimentally based fundamental laws, we first (Chapter 2) study their most profound implications: the propagation of electromagnetic waves through empty space or unbounded simple media. We then (Chapter 3) discuss the

TABLE 2.3. Dielectric properties of selected materials

Material	f (GHz)	ϵ_r'	ϵ_r''	T(°C)
Aluminum oxide (AL$_2$O$_3$)	3.0	8.79	8.79×10^{-3}	25
Barium titanate (BaTiO$_3$)	3.0	600	180	26
Bread	2.45	4.6	1.20	
Bread dough	2.45	22.0	9.00	
Butter (salted)	2.45	4.6	0.60	20
Cheddar cheese	2.45	16.0	8.7	20
Concrete (dry)	2.45	4.5	0.05	25
Concrete (wet)	2.45	14.5	1.73	25
Corn (8% moisture)	2.45	2.2	0.2	24
Corn oil	2.45	2.5	0.14	25
Distilled water	2.45	78	12.5	20
Dry sandy soil	3.0	2.55	1.58×10^{-2}	25
Egg white	3.0	35.0	17.5	25
Frozen beef	2.45	4.4	0.528	−20
Honey (100% pure)	2.45	10.0	3.9	25
Ice (pure distilled)	3.0	3.2	2.88×10^{-3}	−12
Milk	3.0	51.0	30.1	20
Most plastics	2.45	2 to 4.5	0.002 to 0.09	20
Papers	2.45	2 to 3	0.1 to 0.3	20
Potato (78.9% moisture)	3.0	81.0	30.8	25
Polyethylene	3.0	2.26	7.01×10^{-4}	25
Polystyrene	3.0	2.55	8.42×10^{-4}	25
Polytetrafluoroethylene (Teflon)	3.0	2.1	3.15×10^{-4}	22
Raw beef	2.45	52.4	17.3	25
Snow (fresh fallen)	3.0	1.20	3.48×10^{-4}	−20
Snow (hard packed)	3.0	1.50	1.35×10^{-3}	−6
Some glasses (Pyrex)	2.45	~4.0	0.004 to 0.02	20
Smoked bacon	3.0	2.50	0.125	25
Soybean oil	3.0	2.51	0.151	25
Steak	3.0	4.0	12.0	25
White onion (78.7% moisture)	2.45	53.8	13.5	22
White rice (16% moisture)	2.45	3.8	0.8	24
Wood	2.45	1.2 to 5	0.01 to 0.5	25

dielectric are identified by specifying the real part of its dielectric constant (i.e., ϵ') and its loss tangent $\tan \delta_c$, given as

$$\tan \delta_c = \frac{\sigma_{\text{eff}}}{\omega \epsilon'} = \frac{\epsilon''}{\epsilon'}$$

If a given material has both nonzero conductivity ($\sigma \neq 0$) and complex permittivity, it can be treated as if it has an effective conductivity of $\sigma_{\text{eff}} = \sigma + \omega \epsilon''$ and a loss tangent $\tan \delta_c = (\sigma + \omega \epsilon'')/(\omega \epsilon')$. In practice, it is generally not necessary to distinguish between losses due to σ or $\omega \epsilon''$, since ϵ' and σ_{eff} (or $\tan \delta_c$) are often determined by measurement. The attenuation in a lossy dielectric is frequently expressed as the attenuation distance or penetration depth d over which the field intensity decreases by a factor of e^{-1}. Using [2.19] we can write

$$d \equiv \frac{1}{\alpha} = \frac{\sqrt{2}}{\omega\sqrt{\mu\epsilon'}[\sqrt{1 + \tan^2\delta_c} - 1]^{1/2}} = \frac{c\sqrt{2}}{\omega\sqrt{\mu_r\epsilon_r'}[\sqrt{1 + \tan^2\delta_c} - 1]^{1/2}} \quad [2.25]$$

where c is the speed of light in free space.

Examples 2-6 and 2-7 provide numerical values of penetration depth for two different lossy dielectric materials.

Example 2-6: Complex permittivity of distilled water. Consider distilled water at 25 GHz ($\epsilon_{cr} \simeq 34 - j9.01$). Calculate the attenuation constant α, propagation constant β, penetration depth d, and wavelength λ.

Solution: The loss tangent of distilled water at 25 GHz is $\tan\delta_c = \epsilon''/\epsilon' = 0.265$. From [2.19], the attenuation constant is then

$$\alpha \simeq \frac{(2\pi \times 25 \times 10^9 \text{ rad-m}^{-1})\sqrt{34}}{\sqrt{2} \times (3 \times 10^8 \text{ m-s}^{-1})}[\sqrt{1 + (0.265)^2} - 1]^{1/2} \simeq 401 \text{ np-m}^{-1}$$

In a similar manner, the phase constant β can be found from [2.20] to be $\beta \simeq 3079$ rad-m^{-1}. The penetration depth is $d = \alpha^{-1} \simeq 2.49$ mm, and the wavelength is $\lambda = 2\pi/\beta \simeq 2.04$ mm.

In other words, the depth of penetration into distilled water is of the order of one wavelength. Although the condition $\tan\delta_c \ll 1$ is not satisfied for this case, the value of β is nevertheless within a few percent of that for a lossless dielectric with $\epsilon_r = 34$.

Example 2-7: Complex permittivity of bread dough. The dielectric properties of commercially available bread dough are investigated in order to develop more efficient microwave systems and sensors for household and industrial baking.[33] The material properties depend not only on the microwave frequencies used, but also on the time of the baking process. Table 2.4 provides the dielectric constants of the bread baked for 10, 20 (not done yet), and 30 (baking complete) minutes, respectively, at two microwave frequencies, 600 MHz and 2.4 GHz. Calculate the depth of penetration for each case and compare the results.

Solution: At $f = 600$ MHz, after baking for 10 minutes, the loss tangent of bread dough is

$$\tan\delta_c = \frac{11.85}{23.1} \simeq 0.513$$

---

[33] J. Zuercher, L. Hoppie, R. Lade, S. Srinivasan, and D. Misra, Measurement of the complex permittivity of bread dough by an open-ended coaxial line method at ultra high frequencies, *J. Microwave Power Electromagn. Energy,* 25(3), pp. 161–167, 1990.

TABLE 2.4. Relative permittivity $\epsilon_{cr} = \epsilon_r' - j\epsilon_r''$ of bread dough versus frequency and baking time

	10 min	20 min	30 min
600 MHz	$23.1 - j11.85$	$16.78 - j6.66$	$8.64 - j2.51$
2.4 GHz	$12.17 - j4.54$	$9.53 - j3.12$	$4.54 - j1.22$

TABLE 2.5. Loss tangent $\tan \delta_c$ and depth of penetration d versus frequency and baking time

	10 min		20 min		30 min	
	$\tan \delta_c$	d	$\tan \delta_c$	d	$\tan \delta_c$	d
600 MHz	0.513	6.65 cm	0.397	9.97 cm	0.291	18.8 cm
2.4 GHz	0.373	3.11 cm	0.327	3.99 cm	0.269	7.01 cm

from which the depth of penetration can be calculated as

$$d \simeq \frac{3 \times 10^8 \text{ m-s}^{-1}}{2\pi \times 6 \times 10^8 \text{ rad-m}^{-1}} \left[\frac{2}{23.1(\sqrt{1 + (0.513)^2} - 1)} \right]^{1/2} \simeq 6.65 \text{ cm}$$

Similarly, we can calculate the other values. The results are tabulated in Table 2.5.

As we see from Table 2.5, the loss tangent decreases with both increasing baking time and increasing frequency. The depth of penetration increases with baking time but decreases with frequency.

In most applications we do not need to use the exact expressions for α, β, and η_c given, respectively, by [2.19], [2.20], and [2.22], since materials can be classified either as good conductors or low-loss dielectrics, depending respectively on whether their loss tangents are much larger or much smaller than unity.

A low-loss dielectric (or a good dielectric) is one for which $\tan \delta_c \ll 1$. In many applications, dielectric materials that are used are very nearly perfect but nevertheless do cause some nonzero amount of loss. A good example is the dielectric fillings of a coaxial line; the materials used for this purpose are typically nearly perfect insulators, and an approximate evaluation of the small amount of losses is entirely adequate for most applications.

When $\tan \delta_c \ll 1$ the expressions [2.19] and [2.20] for α and β can be simplified using the binomial approximation[34]

$$\gamma = \alpha + j\beta = j\omega \sqrt{\mu\epsilon'} \left(1 - j\frac{\sigma_{\text{eff}}}{\omega\epsilon'} \right)^{-1/2} \simeq j\omega \sqrt{\mu\epsilon'} \left[1 - j\frac{\sigma_{\text{eff}}}{2\omega\epsilon'} + \frac{1}{8} \left(\frac{\sigma_{\text{eff}}}{\omega\epsilon'} \right)^2 + \cdots \right]$$

---

[34] $(1 \pm x)^k = 1 \pm kx + \frac{k(k-1)}{2} x^2 \pm \cdots$

from which we can write

$$\alpha \simeq \frac{\sigma_{\text{eff}}}{2}\sqrt{\frac{\mu}{\epsilon'}} = \frac{\omega\epsilon''}{2}\sqrt{\frac{\mu}{\epsilon'}} \text{ np-m}^{-1}$$

as the attenuation constant and

$$\beta \simeq \omega\sqrt{\mu\epsilon'}\left[1 + \frac{1}{8}\left(\frac{\sigma_{\text{eff}}}{\omega\epsilon'}\right)^2\right] = \omega\sqrt{\mu\epsilon'}\left[1 + \frac{1}{8}\left(\frac{\epsilon''}{\epsilon'}\right)^2\right] \text{ rad-m}^{-1}$$

as the phase constant. Note that since $\tan\delta_c \ll 1$, the phase constant β (and therefore the wavelength $\lambda = 2\pi/\beta$) in a low-loss dielectric are very nearly equal to those in a lossless medium. The attenuation constant imposed on the wave by the medium is usually very small. For example, for dry earth (take $\sigma = 10^{-5}$ S-m^{-1} and $\epsilon = 3\epsilon_0$) at 10 MHz, $\tan\delta_c \simeq 0.006$ and $\alpha \simeq 10^{-3}$ np/m, so the signal is attenuated down to $1/e$ of its value in ~920 m. The wavelength ($2\pi/\beta$), on the other hand, is ~17.3 m, so the depth of penetration for a low-loss dielectric is typically many wavelengths. Such a picture is illustrated in Figure 2.9.

Note that a simplified expression for the phase velocity ($v_p = \omega/\beta$) can be obtained by using [2.20] and the binomial expansion for $(1 + x)^k$ and noting that the exponent k is negative:

$$v_p \simeq \frac{1}{\sqrt{\mu\epsilon'}}\left[1 - \frac{1}{8}\left(\frac{\sigma_{\text{eff}}}{\omega\epsilon'}\right)^2\right] = \frac{1}{\sqrt{\mu\epsilon'}}\left[1 - \frac{1}{8}\left(\frac{\epsilon''}{\epsilon'}\right)^2\right] \text{ m-s}^{-1}$$

Since $1/\sqrt{\mu\epsilon'}$ is the phase velocity in the lossless medium, we see that the small loss introduced by the low-loss dielectric also slightly reduces the velocity of the wave.

An approximate expression for the intrinsic impedance of a good dielectric can also be obtained, again using the binomial expansion. We find:

$$\eta_c = \sqrt{\frac{\mu}{\epsilon'}}\left(1 - j\frac{\sigma_{\text{eff}}}{\omega\epsilon'}\right)^{-1/2} \simeq \sqrt{\frac{\mu}{\epsilon'}}\left(1 + j\frac{\sigma_{\text{eff}}}{2\omega\epsilon'}\right) = \sqrt{\frac{\mu}{\epsilon'}}\left(1 + j\frac{\epsilon''}{2\epsilon'}\right) \quad \Omega$$

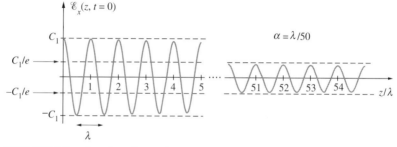

FIGURE 2.9.   Low-loss dielectric. The electric field component of a uniform plane wave propagating in a low-loss (i.e., good) dielectric, in which the depth of penetration is typically much larger than a wavelength.

We see that the small amount of loss also introduces a small reactive (in this case inductive) component to the intrinsic impedance of the medium. This will in turn result in a slight phase shift between the electric and magnetic fields, since ϕ_η in [2.22] is nonzero. We note, however, that the phase shift is usually quite small; for dry earth, for example, at 10 MHz, $\phi_\eta \simeq 0.172°$.

The use of the approximate expressions for α, β, and η_c for a low-loss dielectric is illustrated in Example 2-8.

Example 2-8:   Uniform plane wave in gallium arsenide. A uniform plane wave of frequency 10 GHz propagates in a sufficiently large sample of gallium arsenide (GaAs, $\epsilon_r \simeq 12.9$, $\mu_r = 1$, $\tan \delta_c \simeq 5 \times 10^{-4}$), which is a commonly used substrate material for high-speed solid-state circuits. Find (a) the attenuation constant α in np-m^{-1}, (b) phase velocity v_p in m-s^{-1}, and (c) intrinsic impedance η_c in Ω.

Solution: Since $\tan \delta_c = 5 \times 10^{-4} \ll 1$, we can use the approximate expressions for a good dielectric.

(a) We have

$$\alpha \simeq \frac{\sigma}{2}\sqrt{\frac{\mu}{\epsilon}} = \frac{\omega\epsilon \tan \delta_c}{2}\sqrt{\frac{\mu}{\epsilon}}$$

$$= \frac{2\pi \times 10^{10} \times 5 \times 10^{-4}}{2}\sqrt{\mu\epsilon}$$

$$= \frac{2\pi \times 10^{10} \times 5 \times 10^{-4}\sqrt{\mu_r\epsilon_r}\sqrt{\mu_0\epsilon_0}}{2}$$

$$= \frac{2\pi \times 10^{10} \times 5 \times 10^{-4}}{2 \times 3 \times 10^8}\sqrt{12.9} \simeq 0.188 \text{ np-m}^{-1}$$

(b) Since phase velocity $v_p = \omega/\beta$ where $\beta \simeq \omega\sqrt{\mu\epsilon}$, we have $v_p \simeq 1/\sqrt{\mu\epsilon} \simeq 3 \times 10^8/\sqrt{12.9} \simeq 8.35 \times 10^7$ m-s^{-1}. Note that the phase velocity is ~3.59 times slower than that in air.

(c) The intrinsic impedance $\eta_c \simeq \sqrt{\mu/\epsilon} \simeq 377/\sqrt{12.9} \simeq 105\Omega$. Note that the intrinsic impedance is ~3.59 times smaller than that in air.

Complex Permeability: Magnetic Relaxation   Although our discussion above dealt exclusively with lossy dielectrics, an analogous effect also occurs in magnetic materials, although it is of less importance for electromagnetic wave applications. If a magnetic field is suddenly applied to a paramagnetic[35] material, the magnetization

---

[35]Paramagnetic materials are those for which the permeability μ is slightly above μ_0. See Chapter 9 of M. A. Plonus, *Applied Electromagnetism,* McGraw-Hill, 1978 and Chapter 14 of P. S. Neelakanta, *Handbook of Electromagnetic Materials,* CRC Press, 1995.

exhibits some inertia and does not immediately reach its static value but instead approaches it gradually. Similar inertial delay also occurs when the field is turned off. The inertia exhibited is attributable to the energy exchanges between the spinning electrons and the lattice vibrations as well as energy exchanges between neighboring spins.[36] In analogy with dielectric relaxation, such effects can be described by introducing a complex permeability μ such that

$$\mu_c = \mu' - j\mu''$$

To represent the fact that this effect would lead to power dissipation in the medium, consider the time rate of change of magnetic energy density, or

$$\frac{\partial w_m}{\partial t} = \frac{\partial}{\partial t}\left(\frac{1}{2}\overline{\mathscr{H}} \cdot \mathscr{B}\right) \qquad \rightarrow \qquad \frac{1}{2}j\omega(\mathbf{H} \cdot \mathbf{B})$$

With $\mathbf{B} = (\mu' - j\mu'')\mathbf{H}$, we have

$$\frac{1}{2}j\omega(\mathbf{H} \cdot \mathbf{B}) = \frac{1}{2}(j\omega\mu')H^2 + \frac{1}{2}\omega\mu''H^2$$

where $H = |\mathbf{H}|$. For a time-harmonic magnetic field, the real part of this quantity represents the time-average power dissipated in a medium. We note that the real part of the first term is zero; that is, this term simply represents the volume density of energy stored in the magnetic field per unit time. The second term, on the other hand, is purely real and represents the time-average power density of the magnetization loss. The additional losses due to magnetization damping can be represented by an effective conductivity $\sigma_{\text{eff}} = \epsilon'\omega\mu''/\mu'$.

In principle, diamagnetic materials also exhibit time-varying relaxation behavior. Although the resultant effects are so small that one might think such phenomena are of little practical interest, resonant absorption in diamagnetic liquids is the cause of the phenomenon of nuclear magnetic resonance,[37] which in turn is the basis for magnetic resonance imaging (MRI) technology.[38]

2.3.2   Uniform Plane Wave Propagation in a Good Conductor

Another special case of wave propagation in lossy media that is of practical importance involves propagation in nearly perfect conductors. Many transmission lines and guiding systems are constructed using metallic conductors, such as copper, aluminum, or silver. In most cases these conductors are nearly perfect, so the losses are relatively small. Nevertheless, these small losses often determine the range of

---

[36]For further details, see Section 7.16 of R. S. Elliott, *Electromagnetics,* IEEE Press, 1993, and Chapter 16 of C. Kittel, *Introduction to Solid State Physics,* 5th ed., Wiley, New York, 1976.

[37]Resonant absorption occurs because of the highly frequency-dependent nature of the permeability μ, analogous to that shown in Figure 6.13 for ϵ_c. For further discussion and relevant references, see Chapter 16 of C. Kittel, *Introduction to Solid State Physics,* 5th ed., Wiley, New York, 1976.

[38]See M. A. Brown and R. C. Semelka, *MRI Basic Principles and Applications,* Wiley-Liss, 1995.

applicability of the systems utilized and need to be quantitatively determined. Fortunately, approximate determinations of the losses are perfectly adequate in most cases, using simplified formulas derived on the basis that the materials are good conductors; that is, that $\sigma \gg \omega\epsilon$.

For $\tan \delta_c \gg 1$, the propagation constant γ can be simplified as follows:

$$\gamma = \alpha + j\beta = \sqrt{(j\omega\mu\sigma)\left(1 + j\frac{\omega\epsilon}{\sigma}\right)} \simeq \sqrt{j\omega\mu\sigma} = \sqrt{\omega\mu\sigma}\,e^{j45°}$$

where we have used the fact that $\sqrt{j} = e^{j\pi/4}$. The real and imaginary parts (i.e., the attenuation and phase constants) of γ for a good conductor are thus equal:

$$\alpha = \beta \simeq \sqrt{\frac{\omega\mu\sigma}{2}}$$

The phase velocity and wavelength can be obtained from β, as $v_p = \sqrt{2\omega/(\mu\sigma)}$ and $\lambda = 2\sqrt{\pi/(f\mu\sigma)}$. Since both α and β are proportional to $\sigma^{1/2}$ and σ is large, it appears that uniform plane waves not only are attenuated heavily but also undergo a significant phase shift per unit length as they propagate in a good conductor. The phase velocity of the wave and the wavelength are both proportional to $\sigma^{-1/2}$ and are therefore significantly smaller than the corresponding values in free space.

For example, for copper ($\sigma = 5.8 \times 10^7$ S-m^{-1}) at 300 MHz, we have $v_p \simeq$ 7192 m-s^{-1}, and $\lambda \simeq 0.024$ mm, which are much smaller than the free-space values of $c \simeq 3 \times 10^8$ m-s^{-1} and $\lambda \simeq 1$ m at 300 MHz. At 60 Hz, the values for copper are even more dramatic, being $v_p \simeq 3.22$ m-s^{-1} and $\lambda \simeq 53.6$ mm, compared to the free-space wavelength of $\sim$5000 km. As an example of a nonmetallic conductor, for seawater ($\epsilon = 81\epsilon_0$, $\sigma = 4$ S-m^{-1}) at 10 kHz we have $v_p \simeq 1.58 \times 10^5$ m-s^{-1} and $\lambda \simeq 15.8$ m, compared to a free-space wavelength of 30 km. It is interesting to note here that in the context of undersea (submarine) communications, a half-wavelength antenna (i.e., a reasonably efficient radiator) operating in seawater at 10 kHz has to be only $\sim$7.91 m long, whereas a half-wavelength antenna for a ground-based very low frequency (VLF) transmitter radiating at 10 kHz in air would have to be $\sim$15 km long. In practice, VLF communication and navigation transmitters use electrically short (a few hundred meters high) vertical monopole antennas above ground planes and/or in top-loaded fashion.[39]

Using [2.22] we can express the intrinsic impedance for $\sigma \gg \omega\epsilon$ as

$$\eta_c = \sqrt{\frac{\mu}{\epsilon}}\frac{1}{\left(1 - j\dfrac{\sigma}{\omega\epsilon}\right)^{1/2}} \simeq \sqrt{\frac{j\omega\mu}{\sigma}} = \sqrt{\frac{\mu\omega}{\sigma}}\,e^{j45°}$$

[39] A. D. Watt, *VLF Radio Engineering,* Pergamon Press, New York, 1967.

The magnitude of η_c for good conductors is typically quite small (proportional to $\sigma^{-1/2}$), and the medium presents an inductive impedance (i.e., $\overline{\mathcal{H}}$ lags $\overline{\mathcal{E}}$). The phase of the intrinsic impedance (ϕ_η) is very nearly 45° for all good conductors. For copper at 300 MHz, the magnitude of the intrinsic impedance is ~$6.39 \times 10^{-3}\Omega$, compared to 377$\Omega$ in free space. Thus, for a given uniform plane-wave electric field intensity, the corresponding wave magnetic field in copper is ~6×10^4 times as large as that in free space. For seawater at 10 kHz, $|\eta_c| \simeq 0.14\Omega$, which is a factor of ~2700 times smaller than the free-space value of 377Ω.

A useful parameter used in assessing the degree to which a good conductor is lossy and the degree to which electromagnetic waves can penetrate into a good conductor is the so-called *skin depth* δ, defined as the depth in which the wave is attenuated to $1/e$ (or ~36.8%) of its original intensity (i e , its value immediately below the surface if a wave is incident on the material from above). The skin depth[40] is given by

$$\delta \equiv \frac{1}{\alpha} \simeq \sqrt{\frac{2}{\omega\mu\sigma}} = \frac{1}{\sqrt{\pi f \mu \sigma}}$$ [2.26]

For example, for copper at 300 MHz, $\delta \simeq 0.00382$ mm $= 3.82$ μm, which is a very small distance both in absolute terms and also as compared to the wavelength ($\lambda \simeq 0.024$ mm $= 24$ μm, see above). The nature of the wave propagation in a good conductor, as shown in Figure 2.10, is quite different from the case of that in a good dielectric shown in Figure 2.9. In a good conductor, since $\lambda = 2\pi\delta$, the wave attenuates rapidly with distance and reaches negligible amplitudes after traveling only a small fraction of a wavelength. Note that the skin depth in good conductors such as copper is very small even at audio or low radio frequencies (tens to hundreds of kHz). For example, at 30 kHz, we have $\delta \simeq 0.382$ mm, which means that any

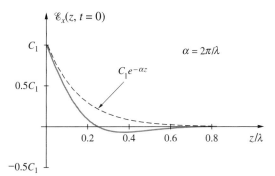

FIGURE 2.10.   **Good conductor.** The electric field component of a uniform plane wave in a good conductor.

---

[40]Note that the definition of the skin depth δ as being equal to $1/\alpha$ is identical to that of the penetration depth d in [2.25]. However, the term "skin depth" is used exclusively for good conductors and emphasizes that the penetration is confined to a thin region on the skin of the conductor.

time-varying fields and current densities exist only in a thin layer on the surface. We shall see in Section 2.3 that in typical transmission lines (e.g., a coaxial line), electromagnetic power propagates in the dielectric region between the conductors, typically at the velocity of light in free space (or at the phase velocity appropriate for the dielectric medium). The electromagnetic wave is guided by the conductor structure and penetrates into the conductor only to the degree that the conductor is imperfect. For example, in the case of a coaxial line, the propagation characteristics of the wave are not different in a solid inner metallic conductor versus one that consists of a thin cylindrical shell of metallic conductor that is hollow or filled with any other material (conductor or insulator).

Examples 2-9, 2-10, and 2-11 quantitatively illustrate skin depth values in different cases.

Example 2-9: Comparison of skin depths. Calculate the skin depth in each of the following nonmagnetic ($\mu = \mu_0$) media at 10 kHz: Seawater ($\epsilon_r = 81$, $\sigma = 4$ S-m^{-1}); wet earth ($\epsilon_r = 10$, $\sigma = 10^{-2}$ S-m^{-1}); and dry earth ($\epsilon_r = 3$, $\sigma = 10^{-4}$ S-m^{-1}).

Solution: Note that for any medium, the loss tangent is $\tan\delta_c = \sigma/(\omega\epsilon)$. For seawater we have

$$\tan\delta_c \simeq \frac{4 \text{ S-m}^{-1}}{(2\pi \times 10^4 \text{ rad-s}^{-1})(81 \times 8.85 \times 10^{-12} \text{ F-m}^{-1})} \simeq 8.88 \times 10^4$$

Similarly, for wet earth we find $\tan\delta_c \simeq 1.8 \times 10^3$, and for dry earth $\tan\delta_c \simeq 59.9$. Thus, at 10 kHz, all three media can be considered to be good conductors, and we can use the skin depth formula given by [2.26]. For seawater at 10 kHz we then have

$$\delta \simeq \frac{1}{\sqrt{(\pi \times 10^4 \text{ rad-s}^{-1})(4\pi \times 10^{-7} \text{ H-m}^{-1})(4 \text{ S-m}^{-1})}} \simeq 2.52 \text{ m}$$

Similarly, the skin depths for wet and dry earth can be found to be, respectively, $\sim$50.3 m and $\sim$503 m. The relatively large skin depths facilitate the use of VLF frequencies (3–30 kHz) for various applications such as undersea communication, search for buried objects, and geophysical prospecting. Note that the skin depths are even larger for the ELF, SLF, and ULF ranges of frequencies (see Table 2.1), in the range 3 Hz to 3 kHz. These relatively large penetration distances facilitate the use of these frequencies for geophysical prospecting.

Example 2-10: Propagation of radio waves through lake water versus seawater. Consider the propagation of a 10 MHz radio wave in lake water (assume $\sigma = 4 \times 10^{-3}$ S-m^{-1} and $\epsilon_r = 81$) versus seawater ($\sigma = 4$ S-m^{-1} and $\epsilon_r = 81$). Calculate the attenuation constant and the penetration depth in each medium, and comment on the difference. Note that both media are nonmagnetic.

Solution: For lake water, the loss tangent at 10 MHz is given by

$$\tan \delta_c = \frac{\sigma}{\omega \epsilon_r \epsilon_0} \simeq \frac{4 \times 10^{-3} \text{ S-m}^{-1}}{2\pi \times 10^7 \text{ rad-s}^{-1} \times 81 \times 8.85 \times 10^{-12} \text{ F-m}^{-1}} \simeq 8.88 \times 10^{-2} \ll 1$$

Therefore, at 10 MHz, lake water behaves like a low-loss dielectric. The penetration depth for lake water is

$$d = \frac{1}{\alpha} \simeq \frac{2}{\sigma} \sqrt{\frac{\epsilon_r \epsilon_0}{\mu_0}} \simeq \left[\frac{2}{4 \times 10^{-3} \text{ S-m}^{-1}} \right] \left[\frac{\sqrt{81}}{377\Omega} \right] \simeq 11.9 \text{ m} \simeq 12 \text{ m}$$

For seawater, the loss tangent at 10 MHz is given by

$$\tan \delta_c \simeq \frac{4 \text{ S-m}^{-1}}{2\pi \times 10^7 \text{ rad-s}^{-1} \times 81 \times 8.85 \times 10^{-12} \text{ F-m}^{-1}} \simeq 88.8 \gg 1$$

Therefore, at 10 MHz, seawater behaves like a good conductor. The skin depth for seawater is

$$\delta = \frac{1}{\alpha} \simeq \sqrt{\frac{2}{\omega \mu \sigma}} = \frac{1}{\sqrt{\pi \times 10^7 \text{ rad-s}^{-1} \times 4\pi \times 10^{-7} \text{ H-m}^{-1} \times 4 \text{ S-m}^{-1}}}$$

$$= \frac{1}{4\pi} \text{ m} \simeq 8 \text{ cm}$$

The results clearly indicate that communication through seawater at 10 MHz is not feasible, since the skin depth is only 8 cm. However, 10 MHz electromagnetic waves do penetrate lake water, making communication possible at relatively shallow depths of tens of meters.

Example 2-11: VLF waves in the ocean. The electric field component of a uniform plane VLF electromagnetic field propagating vertically down in the z direction in the ocean ($\sigma = 4$ S-m^{-1}, $\epsilon_r = 81$, $\mu_r = 1$) is approximately given by

$$\overline{\mathscr{E}}(z, t) = \hat{x} E_0 e^{-\alpha z} \cos(6\pi \times 10^3 t - \beta z) \quad \text{V-m}^{-1}$$

where E_0 is the electric field amplitude at $z = 0^+$ immediately below the air-ocean interface, at $z = 0$. Note that the frequency of the wave is 3 kHz. (a) Find the attenuation constant α and phase constant β. (b) Find the wavelength λ, phase velocity v_p, skin depth δ, and intrinsic impedance η_c, and compare them to their values in air. (c) Write the instantaneous expression for the corresponding magnetic field, $\overline{\mathscr{H}}(z, t)$. (d) A submarine located at a depth of 100 m has a receiver antenna capable of measuring electric fields with amplitudes of 1 μV-m^{-1} or greater. What is the minimum required electric field amplitude immediately below the ocean surface (i.e., E_0) in order to communicate with the submarine? What is the corresponding value for the amplitude of the magnetic field?

Solution: At $\omega = 6\pi \times 10^3$ rad-s^{-1}, the loss tangent of the ocean is given by

$$\tan \delta_c = \frac{\sigma}{\omega \epsilon_r \epsilon_0} \simeq \frac{4}{6\pi \times 10^3 \text{ rad-s}^{-1} \times 81 \times (8.85 \times 10^{-12} \text{ F-m}^{-1})} \simeq 2.96 \times 10^5 \gg 1$$

Therefore, the ocean acts as a good conductor at $f = 3$ kHz.

(a) Using the approximate expressions we have

$$\alpha = \beta \simeq \sqrt{\frac{\omega\mu\sigma}{2}} = \sqrt{\frac{(6\pi \times 10^3 \text{ rad-s}^{-1})(4\pi \times 10^{-7} \text{ H-m}^{-1})(4 \text{ S-m}^{-1})}{2}}$$

$$\simeq 0.218 \text{ np-m}^{-1} \text{ or rad-m}^{-1}$$

(b) The wavelength in the ocean is

$$\lambda = \frac{2\pi}{\beta} \simeq \frac{2\pi}{0.218} \simeq 28.9 \text{ m}$$

Since in air $\lambda \simeq 100$ km at 3 kHz, the wavelength in the ocean is approximately 3464 times smaller than that in air. The phase velocity is

$$v_p = f\lambda \simeq 3 \times 10^3 \times 28.9 \simeq 8.66 \times 10^4 \text{ m-s}^{-1}$$

Compared to $c \simeq 3 \times 10^8$ m-s^{-1} in air, the phase velocity in the ocean at 3 kHz is approximately ~3464 times smaller. The skin depth and the intrinsic impedance in the ocean are

$$\delta = 1/\alpha \simeq 4.59 \text{ m}$$

$$\eta_c = |\eta_c|e^{j\phi_\eta} \simeq \sqrt{\frac{\mu\omega}{\sigma}}e^{j45°}$$

$$= \sqrt{\frac{(4\pi \times 10^{-7} \text{ H-m}^{-1})(6\pi \times 10^3 \text{ rad-s}^{-1})}{4 \text{ S-m}^{-1}}}e^{j45°} \simeq 7.70 \times 10^{-2}e^{j45°}\Omega$$

Compared to $\eta \simeq 377\Omega$ in air, the magnitude of η_c in the ocean is approximately 4900 times smaller. In addition, η_c in the ocean is a complex number with a phase angle of ~45°.

(c) The corresponding magnetic field is given by

$$\overline{\mathcal{H}}(z, t) = \hat{\mathbf{y}}\frac{E_0}{|\eta_c|}e^{-\alpha z}\cos(6\pi \times 10^3 t - \beta z - \phi_\eta)$$

$$\simeq 13E_0 e^{-0.218z}\cos(6\pi \times 10^3 t - 0.218z - \pi/4) \text{ A-m}^{-1}$$

(d) At $z_1 = 100$ m, the amplitude of the electric field is given by

$$E_0 e^{-\alpha z_1} \simeq E_0 e^{-(0.218)(100)} \geq E_{min} = 1 \text{ } \mu\text{V-m}^{-1}$$

Therefore, the minimum value for E_0 to establish communication with the submarine located at a depth of 100 m is

$$E_0 \simeq 2.84 \text{ kV-m}^{-1}$$

The corresponding value for H_0 is

$$H_0 = \frac{E_0}{|\eta_c|} \simeq 13E_0 \simeq 36.9 \text{ kA-m}^{-1}$$

2.4 ELECTROMAGNETIC ENERGY FLOW AND THE POYNTING VECTOR

Electromagnetic waves carry power through space, transferring energy and momentum from one set of charges and currents (i.e., the sources that generated them) to another (those at the receiving points). Our goal in this section is to derive (directly from Maxwell's equations) a simple relation between the rate of this energy transfer and the electric and magnetic fields $\overline{\mathscr{E}}$ and $\overline{\mathscr{H}}$. We can in fact attribute definite amounts of energy and momentum to each elementary volume of space occupied by the electromagnetic fields. The energy and momentum exchange between waves and particles is described by the Lorentz force equation, which is discussed in detail in Section 6.1. This equation represents the forces exerted on a charged particle (having a charge q and velocity $\tilde{\mathbf{v}}$) by electric and magnetic fields: $\overline{\mathscr{F}}_{\text{elec}} = q\overline{\mathscr{E}}$ and $\overline{\mathscr{F}}_{\text{mag}} = q\tilde{\mathbf{v}} \times \overline{\mathscr{B}}$, where $\overline{\mathscr{B}} = \mu_0 \overline{\mathscr{H}}$. When fields exert forces on the charges, the charges respond to the force (i.e., they move) and gain energy, which must be at the expense of the energy lost by the fields. It is thus clear that electromagnetic fields store energy, which is available to charged particles when the fields exert forces (do work) on them. Since energy is transmitted to the particles at a specific rate, we can think of the fields as transmitting power.

2.4.1 Flow of Electromagnetic Energy and Poynting's Theorem

The densities of energy stored in static electric and magnetic fields in linear and isotropic media are, [41] respectively, $\frac{1}{2}\epsilon E^2$ and $\frac{1}{2}\mu H^2$. When these fields vary with time, the associated stored energy densities can also be assumed to vary with time. If we consider a given volume of space, electromagnetic energy can be transported into or out of it by electromagnetic waves, depending on whether the source of the waves is inside or outside the volume. In addition, electromagnetic energy can be stored in the volume in the form of electric and magnetic fields, and electromagnetic power can be dissipated in it in the form of Joule heating (i.e., as represented by $\overline{\mathscr{E}} \cdot \overline{\mathscr{J}}$). Poynting's theorem concerns the balance between power flow into or out of a given volume and the rate of change of stored energy and power dissipation within it. This theorem provides the physical framework by which we can express the flow of electromagnetic power in terms of the fields $\overline{\mathscr{H}}$ and $\overline{\mathscr{E}}$.

We now proceed to find an expression for electromagnetic power in terms of the field quantities $\overline{\mathscr{E}}$ and $\overline{\mathscr{H}}$. Since we expect the power flow to be related to the volume

---

[41] These stored electrostatic and magnetostatic energies represent the work that needs to be done to establish the fields, or the charge and distributions that maintain them. For a physical discussion of electric and magnetic energy densities, see Section 4.12 and 7.3 of U. S. Inan and A. S. Inan, *Engineering Electromagnetics,* Addison Wesley Longman, 1999.

density of power dissipation represented by $\overline{\mathscr{E}} \cdot \overline{\mathscr{J}}$, we can start with [2.1c]:

$$\overline{\mathscr{J}} = \nabla \times \overline{\mathscr{H}} - \frac{\partial \overline{\mathscr{D}}}{\partial t}$$

and take the dot product of both sides with $\overline{\mathscr{E}}$ to obtain $\overline{\mathscr{E}} \cdot \overline{\mathscr{J}}$ in units of W-m^{-3}:

$$\overline{\mathscr{E}} \cdot \overline{\mathscr{J}} = \overline{\mathscr{E}} \cdot (\nabla \times \overline{\mathscr{H}}) - \overline{\mathscr{E}} \cdot \frac{\partial \overline{\mathscr{D}}}{\partial t}$$

and use the vector identity

$$\nabla \cdot (\overline{\mathscr{E}} \times \overline{\mathscr{H}}) \equiv \overline{\mathscr{H}} \cdot (\nabla \times \overline{\mathscr{E}}) - \overline{\mathscr{E}} \cdot (\nabla \times \overline{\mathscr{H}})$$

to find

$$\overline{\mathscr{E}} \cdot \overline{\mathscr{J}} = \overline{\mathscr{H}} \cdot (\nabla \times \overline{\mathscr{E}}) - \nabla \cdot (\overline{\mathscr{E}} \times \overline{\mathscr{H}}) - \overline{\mathscr{E}} \cdot \frac{\partial \overline{\mathscr{D}}}{\partial t} \qquad [2.27]$$

We now use [2.1a],

$$\nabla \times \overline{\mathscr{E}} = -\frac{\partial \overline{\mathscr{B}}}{\partial t}$$

and substitute into [2.27] to obtain

$$\overline{\mathscr{E}} \cdot \overline{\mathscr{J}} = -\overline{\mathscr{H}} \cdot \frac{\partial \overline{\mathscr{B}}}{\partial t} - \overline{\mathscr{E}} \cdot \frac{\partial \overline{\mathscr{D}}}{\partial t} - \nabla \cdot (\overline{\mathscr{E}} \times \overline{\mathscr{H}}) \qquad [2.28]$$

We now consider the two terms with the time derivatives. For a simple medium[42] (i.e., ϵ and μ simple constants) we can write

$$\overline{\mathscr{H}} \cdot \frac{\partial \overline{\mathscr{B}}}{\partial t} = \overline{\mathscr{H}} \cdot \frac{\partial (\mu \overline{\mathscr{H}})}{\partial t} = \frac{1}{2} \frac{\partial (\mu \overline{\mathscr{H}} \cdot \overline{\mathscr{H}})}{\partial t} = \frac{\partial}{\partial t} \left(\frac{1}{2} \mu |\overline{\mathscr{H}}|^2 \right)$$

and

$$\overline{\mathscr{E}} \cdot \frac{\partial \overline{\mathscr{D}}}{\partial t} = \overline{\mathscr{E}} \cdot \frac{\partial (\epsilon \overline{\mathscr{E}})}{\partial t} = \frac{1}{2} \frac{\partial (\epsilon \overline{\mathscr{E}} \cdot \overline{\mathscr{E}})}{\partial t} = \frac{\partial}{\partial t} \left(\frac{1}{2} \epsilon |\overline{\mathscr{E}}|^2 \right)$$

respectively. With these substitutions, [2.28] can be rewritten as

$$\overline{\mathscr{E}} \cdot \overline{\mathscr{J}} = -\frac{\partial}{\partial t} \left(\frac{1}{2} \mu |\overline{\mathscr{H}}|^2 \right) - \frac{\partial}{\partial t} \left(\frac{1}{2} \epsilon |\overline{\mathscr{E}}|^2 \right) - \nabla \cdot (\overline{\mathscr{E}} \times \overline{\mathscr{H}}) \qquad [2.29]$$

---

[42]The more general form of Poynting's theorem, without these assumptions about the medium, is

$$\int_V \overline{\mathscr{E}} \cdot \overline{\mathscr{J}} \, dv = -\frac{\partial}{\partial t} \int_V \left(\overline{\mathscr{H}} \cdot \frac{\partial \overline{\mathscr{B}}}{\partial t} + \overline{\mathscr{E}} \cdot \frac{\partial \overline{\mathscr{D}}}{\partial t} \right) dv - \oint_S (\overline{\mathscr{E}} \times \overline{\mathscr{H}}) \cdot d\mathbf{s}$$

which is valid even for nonlinear media, as long as there are no hysteresis effects. For ferromagnetic materials, for which the relation between $\overline{\mathscr{B}}$ and $\overline{\mathscr{H}}$ is often multivalued due to hysteresis, an additional amount of energy is deposited within the material.

2.4   Electromagnetic Energy Flow and the Poynting Vector   ■   61

Integrating [2.29] over an arbitrary volume V, we have

$$\int_V \bar{\mathscr{E}} \cdot \bar{\mathscr{J}} \, dv = -\frac{\partial}{\partial t} \int_V \left(\frac{1}{2}\mu |\bar{\mathscr{H}}|^2 + \frac{1}{2}\epsilon |\bar{\mathscr{E}}|^2 \right) dv - \int_V \nabla \cdot (\bar{\mathscr{E}} \times \bar{\mathscr{H}}) \, dv \quad [2.30]$$

Using the divergence theorem on the last term in [2.30] we find

$$\boxed{\int_V \bar{\mathscr{E}} \cdot \bar{\mathscr{J}} \, dv = -\frac{\partial}{\partial t} \int_V \left(\frac{1}{2}\mu |\bar{\mathscr{H}}|^2 + \frac{1}{2}\epsilon |\bar{\mathscr{E}}|^2 \right) dv - \oint_S (\bar{\mathscr{E}} \times \bar{\mathscr{H}}) \cdot d\mathbf{s}} \quad [2.31]$$

where the surface S encloses the volume V.

We can now interpret the various terms in [2.31] physically. The left-hand term is the generalization of Joule's law and represents the instantaneous power dissipated in the volume V. If $\bar{\mathscr{E}}$ is the electric field that produces $\bar{\mathscr{J}}$ in a lossy medium, this term represents the ohmic (I^2R) power loss in the medium. Note that in a simple isotropic medium (i.e., σ a simple constant), $\bar{\mathscr{E}}$ and $\bar{\mathscr{J}}$ are in the same direction. However, in general this is not true. For example, in Earth's ionosphere (which is an anisotropic medium because of the presence of Earth's magnetic field, as discussed in Section 6.2.3), an applied electric field in one direction can cause current flow not only in that direction but also in other directions. Even in such cases, however, $\bar{\mathscr{E}} \cdot \bar{\mathscr{J}}$ still represents the power dissipated per unit volume, although $\bar{\mathscr{J}}$ and $\bar{\mathscr{E}}$ are not parallel. Alternatively, there could be an energy source within the volume V, such as an antenna carrying current, in which case $\bar{\mathscr{E}} \cdot \bar{\mathscr{J}}$ is negative and represents power flow out of that region.

The first term on the right-hand side represents the rate at which the electromagnetic energy stored in volume V decreases (negative sign), with the terms $\frac{1}{2}\mu |\bar{\mathscr{H}}|^2$ and $\frac{1}{2}\epsilon |\bar{\mathscr{E}}|^2$ representing, respectively, the magnetic and electric energy densities. Note that, strictly speaking, the quantities $W_e = \frac{1}{2}\epsilon |\bar{\mathscr{E}}|^2$ and $W_m = \frac{1}{2}\mu |\bar{\mathscr{H}}|^2$ are known to represent electric and magnetic energy densities for static fields. However, it is generally assumed that these quantities also represent stored energy densities in the case of time-varying fields.[43]

From conservation of energy, the last term in [2.31] must represent the flow of energy inward or outward through the surface S enclosing the volume V. Thus, the vector $\bar{\mathscr{S}} = \bar{\mathscr{E}} \times \bar{\mathscr{H}}$, which has dimensions of watts per square meter (W-m^{-2}), appears to be a measure of the rate of energy flow per unit area at any point on the surface S. The direction of power flow is perpendicular to both $\bar{\mathscr{E}}$ and $\bar{\mathscr{H}}$. In other

---

[43]Such an assumption is entirely reasonable, since the energy density is defined at a given point. From another point of view, we can consider Poynting's theorem or [2.31] to be the definition of energy density for time-varying fields. The correct amount of total electromagnetic energy is always obtained by assigning an amount $\frac{1}{2}(\bar{\mathscr{B}} \cdot \bar{\mathscr{H}} + \bar{\mathscr{D}} \cdot \bar{\mathscr{E}}) = \frac{1}{2}(\epsilon |\bar{\mathscr{E}}|^2 + \mu |\bar{\mathscr{H}}|^2)$ to each unit of volume. Other expressions for static energy densities, such as $\frac{1}{2}\rho\Phi$ for electrostatic fields, where Φ is the electrostatic potential, are not applicable for time-varying fields. See Chapter II-27 of R. P. Feynman, R. B. Leighton, and M. Sands, *The Feynman Lectures on Physics,* Addison-Wesley, 1964, and Section 2.19 of J. A. Stratton, *Electromagnetic Theory,* McGraw-Hill, 1941.

words, the power density in an electromagnetic wave is given by

$$\overline{\mathscr{S}} = \overline{\mathscr{E}} \times \overline{\mathscr{H}}$$

Equation [2.31] is known as *Poynting's theorem,* and the vector $\overline{\mathscr{S}}$ is known as the *Poynting vector.* In words, we have, "Electromagnetic power flow into a closed surface at any instant equals the sum of the time rates of increase of the stored electric and magnetic energies plus the ohmic power dissipated (or electric power generated, if the surface encloses a source) within the enclosed volume."

Poynting, John Henry   *(b. Sept 9, 1852; d. March 10, 1914) Poynting was a professor of physics at Mason Science College, later the University of Birmingham, from 1880 until his death. In papers published in 1884 and in 1885, he showed that the flow of energy at a point can be expressed by a simple formula in terms of the electric and magnetic forces at that point. This is Poynting's theorem. The value assigned to the rate of power flow of electrical energy is known as Poynting's vector. He also wrote papers on radiation and the pressure of light. After 12 years of experiments, he determined the mean density of the Earth in 1891, and in 1893 he determined the gravitational constant, a measure of the effect of gravity. He published his results in* The Mean Density of the Earth *(1894) and* The Earth: Its Shape, Size, Weight and Spin *(1893). [Adapted in part from* Encyclopedia Britannica, *14th ed., 1929, vol. 18, p. 396.]*

1700	*1852*	*1914*	*2000*

Note that the interpretation of $\overline{\mathscr{S}}$ as the local power flux vector does not follow from [2.31] with mathematical rigor. In principle, any vector $\overline{\mathscr{Y}}$ for which the integral over the closed surface S is zero (as is true for any vector that is the curl of another vector)[44] can be added to $\overline{\mathscr{E}} \times \overline{\mathscr{H}}$ without changing the result. Similar concerns may be raised about the interpretation of the terms $\frac{1}{2}\epsilon|\overline{\mathscr{E}}|^2$ and $\frac{1}{2}\mu|\overline{\mathscr{H}}|^2$ as energy densities. In general, Poynting's theorem makes physical sense in terms of its integral form, enabling us to describe the net flow of electromagnetic power through a closed surface. However, we run into difficulty when we try to describe where the energy resides.[45] This problem is reminiscent of the potential energy of a raised object, which is given by the weight times the height above the ground, but for which we also find it difficult to determine *where* the energy actually resides. The fact that we cannot pinpoint the "location" of electromagnetic energy is reminiscent of the fact that it is neither necessary nor possible to "determine" the location of electric and magnetic energies stored in static fields.[46]

---

[44] Let $\overline{\mathscr{Y}} = \nabla \times \overline{\mathscr{X}}$. Then $\oint_S \overline{\mathscr{Y}} \cdot d\mathbf{s} = \int_V \nabla \cdot (\nabla \times \overline{\mathscr{X}})\, dv \equiv 0$, since $\nabla \cdot \nabla \times \overline{\mathscr{X}} \equiv 0$ for any vector $\overline{\mathscr{X}}$.

[45] For a qualitative discussion see Section II-27-4 of R. P. Feynman, R. B. Leighton, and M. Sands, *The Feynman Lectures on Physics,* Addison-Wesley, 1964.

[46] See Sections 4.12 and 7.3 of U. S. Inan and A. S. Inan, *Engineering Electromagnetics,* Addison Wesley Longman, 1999.

The work done per unit time per unit volume by the fields (i.e., $\overline{\mathscr{J}} \cdot \overline{\mathscr{E}}$) is a conversion of electromagnetic energy into mechanical energy or heat. Since matter ultimately consists of charged particles (electrons and nuclei), we can think of this rate of conversion as a rate of increase of energy of the charged particles per unit volume. In this sense, Poynting's theorem for the microscopic fields can be interpreted as a statement of conservation of energy of the combined system of particles and fields.

Examples 2-12, 2-13, and 2-14 illustrate the application of Poynting's theorem to determine electromagnetic power flow in a wire carrying direct current, in a coaxial transmission line, and in a parallel-plate capacitor.

Example 2-12:   Wire carrying direct current.   Consider a long cylindrical conductor of conductivity σ and radius a, carrying a direct current I as shown in Figure 2.11. Find the power dissipated in a portion of the wire of length l, using (a) the left-hand side of [2.31]; (b) the right-hand side of [2.31]. Note that the cross-sectional area of the wire is $A = \pi a^2$.

Solution:

(a) The current density and electric field within the conductor are respectively $\overline{\mathscr{J}} = \hat{\mathbf{z}}(I/A)$ and $\overline{\mathscr{E}} = \hat{\mathbf{z}} I/(\sigma A)$. We can evaluate the left-hand side of [2.31] for a cylindrical volume V of arbitrary radius greater than a as shown in Figure 2.11. Noting that the current density is zero outside the wire, the limits of integration for r need extend only up to $r = a$, and we have

$$\int_V \overline{\mathscr{E}} \cdot \overline{\mathscr{J}} \, dv = \int_0^l \int_0^{2\pi} \int_0^a \left(\frac{I}{\sigma A} \hat{\mathbf{z}} \right) \cdot \left(\hat{\mathbf{z}} \frac{I}{A} \right) r \, dr \, d\phi \, dz$$

$$= \frac{I^2}{\sigma A^2} (\pi a^2 l) = \frac{I^2 l}{\sigma A} = I^2 R$$

where $R = l/(\sigma A)$ is the resistance of the cylindrical conductor. Thus, the left-hand side of [2.31] represents the total ohmic losses in the volume V.

(b) Since the current I and thus the associated electric and magnetic fields are static ($\partial/\partial t = 0$), the rate of increase of the electric and magnetic energies within the conductor is zero. Thus, the right-hand side of [2.31] reduces to

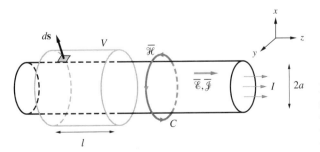

FIGURE 2.11.   Long straight wire. The power dissipated in a cylindrical conductor carrying a direct current I is given by $I^2 R$.

the surface integral of the $\overline{\mathscr{E}} \times \overline{\mathscr{H}}$ term. Using Ampère's law with a contour of radius $r \geq a$, the magnetic field surrounding the conductor is

$$\overline{\mathscr{H}} = \hat{\boldsymbol{\phi}} \frac{I}{2\pi r}$$

At the surface of the conductor ($r = a$), we thus have $\overline{\mathscr{H}} = \hat{\boldsymbol{\phi}} I/(2\pi a)$, so that the Poynting vector is

$$\overline{\mathscr{S}} = \overline{\mathscr{E}} \times \overline{\mathscr{H}} = \frac{I}{\sigma A}\hat{\mathbf{z}} \times \hat{\boldsymbol{\phi}} \frac{I}{2\pi a} = -\hat{\mathbf{r}} \frac{I^2}{2\pi a \sigma A}$$

which is directed radially inwards toward the surface of the wire. The right-hand side of [2.31] is simply the negative of the integral of $\overline{\mathscr{S}}$ over the closed surface of the cylindrical wire (note that there is no contribution from the ends of the cylinder, since $\overline{\mathscr{S}}$ is parallel to those surfaces):

$$-\oint_S \overline{\mathscr{S}} \cdot d\mathbf{s} = -\int_0^l \int_0^{2\pi} \frac{I^2}{2\pi a \sigma A}(-\hat{\mathbf{r}}) \cdot \hat{\mathbf{r}} a\, d\phi\, dz = \frac{I^2 l}{\sigma A} = I^2 R$$

which is equal to the left-hand side of [2.31] evaluated in part (a), thus verifying the Poynting theorem. Note here that if the wire were a perfect conductor ($\sigma = \infty$), then the electric field inside the wire (and thus just outside the wire) must be zero. In such a case, $\overline{\mathscr{E}} \times \overline{\mathscr{H}} = 0$ and there is no power flow into the wire, consistent with the fact that there is no power dissipation in the wire.

Example 2-13: Power flow in a coaxial line. Consider a coaxial line delivering power to a resistor as shown in Figure 2.12. Assume the wires to be perfect conductors, so that there is no power dissipation in the wires, and the electric field inside them is zero. The configurations of electric and magnetic field lines in the coaxial line are as shown, the electric field being due to the applied potential difference between the inner and outer conductors. Find the power delivered to the resistor R.

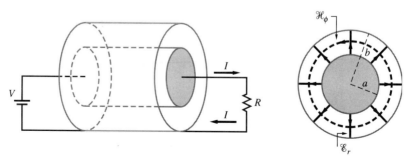

FIGURE 2.12. Power flow in a coaxial line. Coaxial wire delivering power to a resistor R.

Solution: From Ampère's law we have

$$\overline{\mathcal{H}} = \hat{\phi}\frac{I}{2\pi r} \qquad a \le r \le b$$

To find the electric field, we recall that the capacitance (per unit length) of a coaxial line is $C = (2\pi\epsilon)/\ln(b/a)$, so the applied voltage V induces a charge per unit length of $Q = CV$, from which we can find the $\mathcal{E}$ field using Gauss's law (i.e., the integral form of equation [1.2]) as

$$\overline{\mathcal{E}} = \hat{\mathbf{r}}\frac{Q}{2\pi\epsilon r} = \hat{\mathbf{r}}\frac{V}{r\ln(b/a)}$$

Therefore, the power delivered through the cross-sectional area of the coaxial line can be found by integrating the Poynting vector $\overline{\mathcal{S}} = \overline{\mathcal{E}} \times \overline{\mathcal{H}}$ over the area,

$$\oint_S \overline{\mathcal{E}} \times \overline{\mathcal{H}} \cdot d\mathbf{s} = \int_a^b \int_0^{2\pi} \frac{V}{r\ln(b/a)}\left(\frac{I}{2\pi r}\right)\hat{\mathbf{z}} \cdot (\hat{\mathbf{z}}r\,dr\,d\phi)$$

$$= \int_a^b \frac{V}{r\ln(b/a)}\left(\frac{I}{2\pi r}\right)2\pi r\,dr = \frac{VI}{\ln(b/a)}\int_a^b \frac{dr}{r} = VI$$

as expected on a circuit theory basis. We note that the electromagnetic power is carried entirely outside the conductors. Even if the inner and outer conductors of the coaxial line were imperfect conductors, this simply leads to a component of the Poynting vector radially inward, as we saw in the previous example. No part of the axially directed Poynting flux (i.e., the power flux flowing along the coaxial line in the z direction) is carried within the conductors.

Example 2-14: Energy flow in a capacitor. Figure 2.13 shows a parallel-plate capacitor of capacitance $C = \epsilon A/d$, being charged by a current I flowing in the connecting wires. When it is charged, an electric field $\mathcal{E}_z$ exists (taking the z axis to be along the current-carrying wire) between the capacitor plates, with $\frac{1}{2}\epsilon\mathcal{E}_z^2$ representing the stored energy. Investigate the manner in which the energy enters into the region between the capacitor plates.

Solution: Neglecting fringing fields, the electric field between the capacitor plates is uniform, so the total electrical energy stored between the plates is

$$W_e = \int_V \tfrac{1}{2}\epsilon\mathcal{E}_z^2\,dv = (\tfrac{1}{2}\epsilon\mathcal{E}_z^2)Ad$$

The time rate of change of this energy is given by

$$\frac{dW_e}{dt} = Ad\frac{d}{dt}\left(\frac{1}{2}\epsilon\mathcal{E}_z^2\right) = \epsilon Ad\mathcal{E}_z\frac{d\mathcal{E}_z}{dt}$$

which means that there must be a flow of energy into the volume between the plates from somewhere. Can it be arriving through the wire? Apparently not,

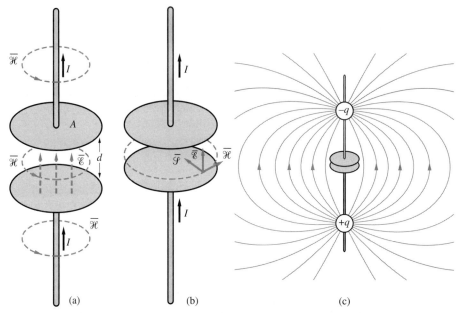

FIGURE 2.13. **A parallel-plate capacitor being charged.** (a) Magnetic fields encircling the conduction (outside the region between plates) or displacement (inside the region between the plates) currents. (b) Power evidently enters the storage region (i.e., the region between plates) through the sides via $\overline{\mathscr{E}} \times \overline{\mathscr{H}}$. (c) The intensification of the electric fields near the capacitor as two point charges come closer to the plates.

according to Poynting's theorem, since we note that $\overline{\mathscr{E}}$ is perpendicular to the plates, and thus $\overline{\mathscr{E}} \times \overline{\mathscr{H}}$ must necessarily be parallel to the plates.

A magnetic field $\overline{\mathscr{H}}$ encircles the z axis, supported by the conduction current along the wires and the displacement current in the region between the plates, as shown in Figure 2.13a. In the region between the plates, where $\mathbf{J} = 0$, the magnetic field at any radial distance r is given by [1.3] as

$$\oint_C \overline{\mathscr{H}} \cdot d\mathbf{l} = \int_S \frac{\partial(\epsilon\overline{\mathscr{E}})}{\partial t} \cdot d\mathbf{s} \quad \rightarrow \quad \mathscr{H}_\phi(2\pi r) = \epsilon(\pi r^2)\frac{\partial \mathscr{E}_z}{\partial t} \quad \rightarrow \quad \mathscr{H}_\phi = \frac{\epsilon r}{2}\frac{\partial \mathscr{E}_z}{\partial t}$$

Thus, Poynting's vector $\overline{\mathscr{P}} = \overline{\mathscr{E}} \times \overline{\mathscr{H}} = \mathscr{E}_z\hat{\mathbf{z}} \times \hat{\boldsymbol{\phi}}\mathscr{H}_\phi = \mathscr{E}_z\mathscr{H}_\phi(-\hat{\mathbf{r}})$ has only an r component, pointed inward, as shown in Figure 2.13b. The energy stored in the capacitor is apparently being supplied through the sides of the cylindrical region between the plates. The total power flow into the region can be found by integrating $\overline{\mathscr{P}}$ over the cylindrical side surface of the capacitor at $r = a$:

$$\oint_S \overline{\mathscr{P}} \cdot d\mathbf{s} = \int_0^{2\pi} \int_{z=0}^d \mathscr{E}_z\left(\frac{\epsilon a}{2}\frac{\partial \mathscr{E}_z}{\partial t}\right)a\,dz\,d\phi = \pi a^2\,d\frac{\partial}{\partial t}\left(\frac{1}{2}\epsilon\mathscr{E}_z^2\right)$$

which is identical to the dW_e/dt expression we found above, where we note that $A = \pi a^2$. Thus, we see that the power supplied into the region from the sides is precisely equal to the time rate of change of stored energy.

This result is quite contrary to our intuitive expectations of the power being supplied to the capacitor via the wires. One way to think about this result[47] is illustrated in Figure 2.13c. Consider two opposite charges above and below the capacitor plates, approaching the capacitor plates via the wire. When the charges are far away from the plates, the electric field between the plates is weak but spread out in the region surrounding the capacitor. As the charges approach the plates, the electric field between the capacitor plates becomes stronger (i.e., the field energy moves toward the capacitor) and in the limit is eventually completely confined to the region between the plates. Thus, as shown, the electromagnetic energy flows into the region between the capacitor plates through the openings on the sides.

2.4.2   Electromagnetic Power Carried by a Uniform Plane Wave in a Lossless Medium

Consider the uniform plane wave propagating in a lossless medium that was studied in Section 2.2:

$$\mathscr{E}_x = E_0 \cos(\omega t - \beta z)$$

$$\mathscr{H}_y = \frac{1}{\eta} E_0 \cos(\omega t - \beta z)$$

where $\beta = \omega\sqrt{\mu\epsilon}$ and $\eta = \sqrt{\mu/\epsilon}$. The electric and magnetic fields of the wave are as shown in Figure 2.14a. The Poynting vector for this wave is given by

$$\overline{\mathscr{S}} = \overline{\mathscr{E}} \times \overline{\mathscr{H}} = \hat{\mathbf{z}} E_0 \left(\frac{E_0}{\eta}\right) \cos^2(\omega t - \beta z) = \hat{\mathbf{z}} \frac{E_0^2}{2\eta}[1 + \cos 2(\omega t - \beta z)]$$

We see from Figure 2.14b that $\overline{\mathscr{S}}$ everywhere is directed in the $+z$ direction. The power carried by the wave has a constant component and a component varying at a frequency twice that of the fields, exhibiting two positive peaks per wavelength of the wave, as shown in Figure 2.14b. In most cases, the constant component, called the time-average Poynting vector $\mathbf{S}_{av}$, is the quantity of interest. The quantity $\mathbf{S}_{av}$ is defined as the average over one period ($T_p = 2\pi/\omega$) of the wave of $\overline{\mathscr{S}}$. For a uniform plane wave in a lossless medium, we have

$$\mathbf{S}_{av} = \frac{1}{T_p} \int_0^{T_p} \overline{\mathscr{S}}(z, t)\, dt$$

$$= \frac{1}{T_p} \int_0^{T_p} \hat{\mathbf{z}} \frac{E_0^2}{2\eta}[1 + \cos 2(\omega t - \beta z)]\, dt \quad \rightarrow \quad \boxed{\mathbf{S}_{av} = \hat{\mathbf{z}} \frac{E_0^2}{2\eta}} \quad [2.32]$$

---

[47] Taken from Section II-27-5 of R. P. Feynman, R. B. Leighton, and M. Sands, *The Feynman Lectures on Physics,* Addison-Wesley, 1964.

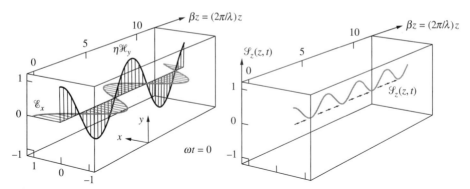

FIGURE 2.14.   **The Poynting vector of a uniform plane wave.** (a) The wave electric and magnetic fields at $\omega t = 0$. (b) The Poynting vector at $\omega t = 0$.

According to Poynting's theorem as expressed in [2.31], and since the medium is assumed to be lossless and source-free, so $\overline{\mathscr{J}} = 0$, the integral of the Poynting vector $\overline{\mathscr{S}}$ over any closed surface should be balanced by the time rate of change of the sum of stored electric and magnetic energies in the volume enclosed by that surface. The electric and magnetic energy densities for this uniform plane wave are

$$W_{\mathrm{e}} = \tfrac{1}{2}\epsilon|\mathscr{E}_x|^2 = \tfrac{1}{2}\epsilon E_0^2 \cos^2(\omega t - \beta z) \qquad [2.33a]$$

$$W_{\mathrm{m}} = \frac{1}{2}\mu|\mathscr{H}_y|^2$$

$$= \frac{1}{2}\frac{\mu}{(\sqrt{\mu/\epsilon})^2}E_0^2\cos^2(\omega t - \beta z) = \frac{1}{2}\epsilon E_0^2\cos^2(\omega t - \beta z) \qquad [2.33b]$$

We thus note that $W_{\mathrm{e}} = W_{\mathrm{m}}$, indicating that the instantaneous values of the electric and magnetic stored energies are equal at all points in space and at all times. The reader is encouraged to choose a suitable closed surface and show that the volume integral of the partial derivative with respect to time of $W_{\mathrm{m}} + W_{\mathrm{e}}$ is indeed equal to the surface integral of $\overline{\mathscr{S}}$. Alternatively, one can consider the differential form of Poynting's theorem, as given in [2.29], which for $\overline{\mathscr{E}} \cdot \overline{\mathscr{J}} = 0$ reduces to

$$\frac{\partial}{\partial t}\left(\frac{1}{2}\mu|\overline{\mathscr{H}}|^2\right) + \frac{\partial}{\partial t}\left(\frac{1}{2}\epsilon|\overline{\mathscr{E}}|^2\right) = -\nabla \cdot (\overline{\mathscr{E}} \times \overline{\mathscr{H}})$$

and show its validity by simply substituting [2.33a] and [2.33b] and carrying out the differentiation with respect to time and the curl operation. The manipulations are left as an exercise for the reader.

The instantaneous variations of the stored energies are often not of practical interest. Instead, one is often interested in the time averages of these quantities over one period ($T_p = 2\pi/\omega$). Denoting these time averages as $\overline{W}_{\mathrm{e}}$ and $\overline{W}_{\mathrm{m}}$, we have

$$\overline{W}_e = \overline{W}_m = \frac{1}{T_p}\int_0^{T_p}\frac{1}{2}\epsilon E_0^2 \cos^2(\omega t - \beta z)\, dt$$

$$= \frac{1}{2}\epsilon E_0^2\frac{1}{T_p}\int_0^{T_p}\left\{\frac{1}{2} + \frac{1}{2}\cos[2(\omega t - \beta z)]\right\}dt = \frac{1}{4}\epsilon E_0^2$$

[2.34]

We undertake a more general discussion of instantaneous and time-average electromagnetic power in the next section, applicable to uniform plane waves in lossy media. Examples 2-15 and 2-16 illustrate the use of [2.32] to determine the solar radiation flux on the earth and to determine the intensity at earth of a signal transmitted from a spaceship near the moon. Examples 2-17 and 2-18 concern the application of [2.32] to determine the power density of an FM and a VHF/UHF broadcast signal.

Example 2-15:   Solar radiation at Earth.   If the time-average power density arriving at the surface of the earth due to solar radiation is about 1400 W-m^{-2} (an average over a wide band of frequencies), find the peak electric and magnetic field values on the earth's surface due to solar radiation. The radii of Earth and the sun are ~6.37 Mm and ~700 Mm, respectively, and the radius of Earth's orbit around the sun is ~ 1.5×10^8 km.

Solution: Using [2.32] we can write

$$|\mathbf{S}_{av}| = \frac{1}{2}\frac{E_0^2}{\eta} = 1400 \text{ W-m}^{-2}$$

where $\eta \simeq 377\Omega$. Solving for E_0 we find $E_0 \simeq \sqrt{2 \times 377 \times 1400} \simeq 1027$ V-m^{-1}. The corresponding magnetic field H_0 is $H_0 = E_0/\eta \simeq 1027/377 \simeq 2.73$ A-m^{-1}.

Example 2-16:   A spaceship in lunar orbit.   A spaceship in lunar orbit (the Earth–Moon distance is ~380 Mm) transmits plane waves with an antenna operating at 1 GHz and radiating a total power of 1 MW isotropically (i.e., equally in all directions). Find (a) the time-average power density and (b) the peak electric field value on Earth's surface, and (c) the time it takes for these waves to travel from the spaceship to Earth.

Solution:
(a)   The time-average power density on Earth's surface a distance $R \simeq 380$ Mm away from the spaceship radiating a total power of $P_{tot} = 1$ MW isotropically is

$$|\mathbf{S}_{av}| = \frac{P_{tot}}{4\pi R^2} = \frac{10^6 \text{ W}}{4\pi(380 \times 10^6)^2 \text{ m}^2} \simeq 5.51 \times 10^{-13} \text{ W-m}^{-2}$$

(b) The corresponding peak electric field value E_0 can be found from [2.32]:

$$|\mathbf{S}_{av}| = \frac{1}{2}\frac{E_0^2}{\eta} \quad \rightarrow \quad E_0 \simeq \sqrt{2 \times 377 \times 5.51 \times 10^{-13}} \simeq 2.04 \times 10^{-5} \text{ V-m}^{-1}$$

(c) The time it takes for the waves to travel from the spaceship to the earth is given by

$$t = \frac{R}{c} \simeq \frac{380 \times 10^6 \text{ m}}{3 \times 10^8 \text{ m-s}^{-1}} \simeq 1.27 \text{ s}$$

Example 2-17:   FM broadcasting.   A transmitter antenna for an FM radio broadcasting station operates around 92.3 MHz (KGON station in Portland, Oregon) with an average radiated power of 100 kW, as limited by Federal Communications Commission (FCC) regulations to minimize interference problems with other broadcast stations outside its coverage area. For simplicity, assume isotropic radiation, i.e., radiation equally distributed in all directions, such as from a point source. (a) For a person standing 50 m away from the antenna, calculate the time-average power density of the radiated wave and compare it with the IEEE safety limit[48] for an uncontrolled environment at that frequency, namely 2 W-m^{-2} or 0.2 mW-(cm)$^{-2}$. (b) Find the minimum distance such that the time-average power density is equal to the IEEE safety standard value at that frequency. (c) If the antenna is mounted on a tower that is 200 m above the ground, repeat part (a) for a person standing 50 m away from the foot of the tower.

Solution: Assume the antenna to be a point source and neglect all effects from boundaries.

(a) The wavelength at 92.3 MHz is

$$\lambda \simeq \frac{3 \times 10^8}{92.3 \times 10^6} \cong 3.25 \text{ m}$$

Since 50 m $\gg$ 3.25 m, we can assume far field, as if the wave were a uniform plane wave. The total power of P_{tot} = 100 kW is radiated equally in all directions, so if we take a sphere of radius 50 m centered at the antenna, the average power density on the surface of the sphere is given by

$$|\mathbf{S}_{av}| = \frac{P_{tot}}{4\pi R^2} = \frac{100 \times 10^3}{4\pi(50)^2}$$

$$\simeq 3.18 \text{ W-m}^{-2} = 0.318 \text{ mW-(cm)}^{-2} > 0.2 \text{ mW-(cm)}^{-2}$$

Therefore, it is not safe to stand at a distance of 50 m from this FM broadcast antenna.

---

[48]IEEE c95.1-1991, *IEEE Standard for Safety Levels with Respect to Human Exposure to Radio Frequency Electromagnetic Fields,* 3 kHz to 300 GHz. For a related discussion, see M. Fischetti, The cellular phone scare, *IEEE Spectrum,* pp. 43–47, June 1993.

(b) From

$$\frac{P_{tot}}{4\pi R_{min}^2} = \frac{100 \times 10^3}{4\pi R_{min}^2} = 2 \text{ W-m}^{-2}$$

we find $R_{min} \simeq 63.1$ m.

(c) Considering a 200-m tower height and a person standing on the ground at a distance 50 m from the base of the tower, the distance between the antenna and the person is $\sqrt{(200)^2 + (50)^2}$, so that we have

$$|\mathbf{S}_{av}| = \frac{100 \times 10^3}{4\pi[(200)^2 + (50)^2]} \simeq 0.187 \text{ W-m}^{-2} < 2 \text{ W-m}^{-2}$$

Therefore, the radiation level on the ground is well below the IEEE safety standard.

Example 2-18:   VHF/UHF broadcast radiation.   A survey[49] conducted in the United States indicates that ~50% of the population is exposed to average power densities of approximately 0.005 μW-(cm)$^{-2}$ due to VHF and UHF broadcast radiation. Find the corresponding amplitudes of the electric and magnetic fields.

Solution: By direct application of [2.32] we have

$$|\mathbf{S}_{av}| = \frac{1}{2}\frac{E_0^2}{\eta} = 0.005 \text{ }\mu\text{W-(cm)}^{-2}$$

so

$$E_0 \simeq \sqrt{2 \times 377 \times 5 \times 10^{-9}/10^{-4}} \simeq 194 \text{ mV-m}^{-1}$$

$$H_0 = \frac{E_0}{\eta} \simeq \frac{194 \text{ mV-m}^{-1}}{377\Omega} \simeq 515 \text{ }\mu\text{A-m}^{-1}$$

where we have noted that $\eta \simeq 377\Omega$ in air.

2.4.3   Instantaneous and Time-Average Power and the Complex Poynting Theorem

In most applications we are interested in the Poynting vector averaged over time rather than in its instantaneous value. Consider the expressions for the instantaneous electric and magnetic fields of a time-harmonic uniform plane wave in the general case of a lossy medium. From [2.18] and [2.23] we have

$$\overline{\mathscr{E}}(z, t) = \hat{\mathbf{x}}C_1 e^{-\alpha z} \cos(\omega t - \beta z)$$

---

[49]R. A. Tell and E. D. Mantiply, Population exposure to VHF and UHF broadcast radiation in the United States, *Proc. IEEE,* 68(1), pp. 6–12, January 1980.

and

$$\overline{\mathcal{H}}(z, t) = \hat{\mathbf{y}} \frac{C_1}{|\eta_c|} e^{-\alpha z} \cos(\omega t - \beta z - \phi_\eta)$$

Noting that $\hat{\mathbf{x}} \times \hat{\mathbf{y}} = \hat{\mathbf{z}}$, the instantaneous Poynting vector given by the cross product $\overline{\mathcal{E}} \times \overline{\mathcal{H}}$ is in the $+z$ direction:

$$\overline{\mathcal{S}}(z, t) = \overline{\mathcal{E}}(z, t) \times \overline{\mathcal{H}}(z, t) = \hat{\mathbf{z}} \frac{C_1^2}{|\eta_c|} e^{-2\alpha z} \cos(\omega t - \beta z) \cos(\omega t - \beta z - \phi_\eta) \quad [2.35]$$

As expected, the electromagnetic power carried by a uniform plane wave propagating in the $+z$ direction flows in the $+z$ direction. We can simplify [2.35] to

$$\overline{\mathcal{S}}(z, t) = \overline{\mathcal{E}}(z, t) \times \overline{\mathcal{H}}(z, t) = \hat{\mathbf{z}} \frac{C_1^2}{2|\eta_c|} e^{-2\alpha z} [\cos(\phi_\eta) + \cos(2\omega t - 2\beta z - \phi_\eta)] \quad [2.36]$$

We see that the instantaneous Poynting vector has a component that does not change with time and a component that oscillates at twice the rate at which electric and magnetic fields change in time. In most applications the time-average value of the power transmitted by a wave is more significant than the fluctuating component. The relationship between the electric and magnetic fields and the Poynting vector $\overline{\mathcal{S}}(z, t)$ is illustrated in Figure 2.15. Note that the instantaneous Poynting vector itself can be negative in certain regions; however, the time-average value is always positive for a uniform plane wave and represents real power flow in the $+z$ direction.

The time-average value can be obtained from [2.36] by integrating over one period T_p of the sinusoidal variation. We find

$$\mathbf{S}_{\text{av}}(z) = \frac{1}{T_p} \int_0^{T_p} \overline{\mathcal{S}}(z, t) \, dt = \hat{\mathbf{z}} \frac{C_1^2}{2|\eta_c|} e^{-2\alpha z} \cos(\phi_\eta) \quad [2.37]$$

As an example, consider a uniform plane wave at 300 MHz in copper, for which $\eta_c \simeq 6.4 \times 10^{-3} e^{j45°} \, \Omega$. The time-average electromagnetic power carried by a wave with 1 V-m^{-1} field amplitude at $z = 0$ (i.e., $C_1 = 1$ V-m^{-1}) is ~55.3 W-m^{-2}, compared to ~0.0013 W-m^{-2} in free space for the same C_1.

Finding the time-average Poynting flux $\mathbf{S}_{\text{av}}$ by an integration over one period as in [2.37] is in general not necessary for sinusoidal signals, since $\mathbf{S}_{\text{av}}$ can be directly obtained from the phasors $\mathbf{E}$ and $\mathbf{H}$, as shown in the next section.

The Poynting theorem is expressed mathematically in [2.31] in terms of the instantaneous field quantities $\overline{\mathcal{E}}$ and $\overline{\mathcal{H}}$. The complex version of the theorem in terms of the phasors $\mathbf{E}$ and $\mathbf{H}$ cannot be obtained from [2.31] by substituting $\partial/\partial t \rightarrow j\omega$, since the terms in [2.31] involve vector products of field quantities. Thus, the complex Poynting theorem has to be derived from the phasor form of Maxwell's equations [1.11a] and [1.11c]:

$$\nabla \times \mathbf{E} = -j\omega \mathbf{B} \qquad \nabla \times \mathbf{H} = \mathbf{J} + j\omega \mathbf{D}$$

We start from the complex form of the vector identity

$$\nabla \cdot (\mathbf{E} \times \mathbf{H}^*) = \mathbf{H}^* \cdot (\nabla \times \mathbf{E}) - \mathbf{E} \cdot (\nabla \times \mathbf{H}^*)$$

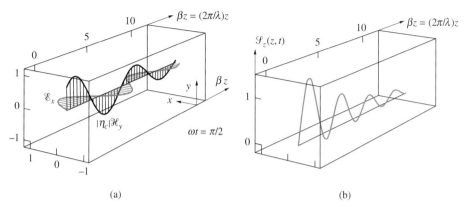

FIGURE 2.15. **The Poynting vector and fields for a uniform plane wave in a lossy medium.**
(a) The wave electric and magnetic fields at $\omega t = \pi/2$; (b) the Poynting vector $\overline{\mathcal{S}}(z, \omega t = \pi/2)$.

Substituting the two curl equations into this identity, we have

$$\nabla \cdot (\mathbf{E} \times \mathbf{H}^*) = \mathbf{H}^* \cdot (-j\omega \mathbf{B}) - \mathbf{E} \cdot (\mathbf{J}^* - j\omega \mathbf{D}^*) \qquad [2.38]$$

Integrating [2.38] over a volume V and using the divergence theorem,

$$\int_V \nabla \cdot (\mathbf{E} \times \mathbf{H}^*)\, dv = \oint_S (\mathbf{E} \times \mathbf{H}^*) \cdot d\mathbf{s} = \int_V -[\mathbf{E} \cdot \mathbf{J}^* + j\omega(\mathbf{H}^* \cdot \mathbf{B} - \mathbf{E} \cdot \mathbf{D}^*)]\, dv \quad [2.39]$$

If the medium is isotropic and if all the losses occur through conduction currents $\mathbf{J} = \sigma \mathbf{E}$, equation [2.39] becomes

$$\oint_S (\mathbf{E} \times \mathbf{H}^*) \cdot d\mathbf{s} = -\int_V \sigma \mathbf{E} \cdot \mathbf{E}^*\, dv - j\omega \int_V (\mu \mathbf{H} \cdot \mathbf{H}^* - \epsilon \mathbf{E} \cdot \mathbf{E}^*)\, dv \quad [2.40]$$

In the case of lossy dielectrics, for which losses occur through effective conduction currents $\overline{\mathbf{J}} = \sigma_{\mathrm{eff}} \overline{\mathbf{E}} = \omega\epsilon'' \overline{\mathbf{E}}$, equation [2.39] is valid when we replace σ with $\sigma_{\mathrm{eff}} = \omega\epsilon''$ and ϵ with ϵ'. Recognizing from [2.34] that $\overline{W}_e = \frac{1}{4}\epsilon \mathbf{E} \cdot \mathbf{E}^*$ and $\overline{W}_m = \frac{1}{4}\mu \mathbf{H} \cdot \mathbf{H}^*$ are the time averages of the energy densities $W_e = \frac{1}{2}\epsilon \mathcal{E} \cdot \mathcal{E}$ and $W_m = \frac{1}{2}\mu \mathcal{H} \cdot \mathcal{H}$ respectively, and noting that the volume density of time-average power dissipation in conduction currents is $P_c = \frac{1}{2}\sigma \mathbf{E} \cdot \mathbf{E}^*$, we can equate the real and imaginary parts of [2.40] as

$$\mathcal{R}e\left\{ \oint_S (\mathbf{E} \times \mathbf{H}^*) \cdot d\mathbf{s} \right\} = -2 \int_V P_c\, dv \qquad [2.41]$$

$$\mathcal{I}m\left\{ \oint_S (\mathbf{E} \times \mathbf{H}^*) \cdot d\mathbf{s} \right\} = -4\omega \int_V (\overline{W}_m - \overline{W}_e)\, dv \qquad [2.42]$$

The instantaneous Poynting vector $\overline{\mathcal{S}}(z, t)$ can be written in terms of the phasors $\mathbf{E}(z)$ and $\mathbf{H}(z)$, as follows:

$$\overline{\mathcal{S}}(z, t) = \mathcal{R}e\{\mathbf{E}(z)e^{j\omega t}\} \times \mathcal{R}e\{\mathbf{H}(z)e^{j\omega t}\} = \frac{1}{2}\mathcal{R}e\{\mathbf{E}(z) \times \mathbf{H}^*(z) + \mathbf{E}(z) \times \mathbf{H}(z)e^{j2\omega t}\}$$

where we have used the fact that in general, for two different complex vectors $\mathbf{G}$ and $\mathbf{F}$, and noting that $(\mathbf{G} \times \mathbf{F}^*)^* = \mathbf{G}^* \times \mathbf{F}$, we have

$$\mathscr{R}e\{\mathbf{G}\} \times \mathscr{R}e\{\mathbf{F}\} = \tfrac{1}{2}(\mathbf{G} + \mathbf{G}^*) \times \tfrac{1}{2}(\mathbf{F} + \mathbf{F}^*)$$

$$= \tfrac{1}{4}[(\mathbf{G} \times \mathbf{F}^* + \mathbf{G}^* \times \mathbf{F}) + (\mathbf{G} \times \mathbf{F} + \mathbf{G}^* \times \mathbf{F}^*)]$$

$$= \tfrac{1}{2}\mathscr{R}e\{\mathbf{G} \times \mathbf{F}^* + \mathbf{G} \times \mathbf{F}\}$$

since in general we have $\mathscr{R}e\{\mathbf{G}\} = \tfrac{1}{2}(\mathbf{G} + \mathbf{G}^*)$ for any complex vector $\mathbf{G}$. The time-average power density, which was denoted $\mathbf{S}_{av}$ in [2.37], can then be obtained by integrating $\overline{\mathscr{S}}(z, t)$ over one period $T_p = 2\pi/\omega$. Since the time-average of the $\mathbf{E}(z) \times \mathbf{H}(z)e^{j2\omega t}$ term vanishes, we find that the time-average Poynting vector is given by

$$\boxed{\mathbf{S}_{av} = \tfrac{1}{2}\mathscr{R}e\{\mathbf{E} \times \mathbf{H}^*\}} \qquad\qquad [2.43]$$

This result allows us to evaluate $\mathbf{S}_{av}$ conveniently from the phasor field quantities. Note that $\mathbf{S} = \mathbf{E} \times \mathbf{H}^*$ is referred to as the *complex Poynting vector.* The imaginary part of the integral of $\mathbf{S}$ as given in [2.42] is equal to the difference between the average energies stored in the magnetic and the electric fields and represents reactive power flowing back and forth to supply the instantaneous changes in the net stored energy in the volume.

Example 2-19 illustrates the application of the complex Poynting theorem to determine electromagnetic power flow in selected practical cases.

Example 2-19: VLF waves in the ocean. Consider VLF wave propagation in the ocean as discussed in Example 2-11. Find the time-average Poynting flux at the sea surface ($z = 0$) and at the depth of the submarine ($z = 100$ m).

Solution: The phasor expressions for the minimum electric and magnetic fields to communicate with the submarine at 100 m depth are given by

$$\mathbf{E}(z) \simeq \hat{\mathbf{x}}2.84e^{-0.218z}e^{-j0.218z} \text{ kV-m}^{-1}$$

$$\mathbf{H}(z) \simeq \hat{\mathbf{y}}36.9e^{-0.218z}e^{-j0.218z}e^{-j\pi/4} \text{ kA-m}^{-1}$$

Using [2.43], the time-average Poynting vector is given by

$$\mathbf{S}_{av}(z) = \tfrac{1}{2}\mathscr{R}e\{\mathbf{E} \times \mathbf{H}^*\}$$

$$\simeq \hat{\mathbf{z}}\tfrac{1}{2}(2.84)(36.9)e^{-2(0.218z)}\cos(\pi/4)$$

$$\simeq \hat{\mathbf{z}}36.9e^{-0.435z} \text{ MW-m}^{-2}$$

At the sea surface (i.e., $z = 0^+$), $\mathbf{S}_{av} \simeq \hat{\mathbf{z}}36.9$ MW-m^{-2}. At the location of the submarine (i.e., $z = 100$ m), $\mathbf{S}_{av} \simeq \hat{\mathbf{z}}4.59 \times 10^{-12}$ W-m^{-2}. The power density required at the sea surface is extremely high, indicating that it is not feasible to use 3 kHz signals to communicate with a submarine at 100 m depth.

2.4.4   Radiation Pressure and Electromagnetic Momentum

In addition to carrying energy as discussed above, an electromagnetic wave can also exert pressure (i.e., force per unit area) or transport linear momentum in the direction of propagation. In other words, pressure is exerted on an object by shining an electromagnetic wave (e.g., a beam of light) on it.[50] Clearly, such forces must ordinarily be quite small,[51] since we do not notice them in our ordinary environment. However, ultrahigh-intensity lasers[52] can nowadays have power densities of $\sim 10^{24}$ W-m^{-2}, corresponding to gigantic light pressures of up to tens of 10^9 bars (1 bar is defined as 10^5 N-m^{-2}; ~ 1.013 bar is equal to 1 atmospheric pressure (atm), i.e., the air pressure at sea level). A simple rotating device that uses radiation pressure of sunlight as its power source is *Crookes's radiometer*,[53] consisting of four vanes delicately mounted on a vertical axis. One side of the vanes is silvered, and the other is blackened, to reflect and absorb light, respectively, thus acquiring relative angular momentum, leading to rotational motion (also see Section 3.1.1). In astrophysics, consideration of radiation pressure is essential in understanding physical processes in stellar interiors and the mechanisms governing supernovae, nebulae, and black holes.[54]

The first suggestion that light could exert pressure dates back to Johannes Kepler, who first proposed in 1619 that it was the radiation pressure of sunlight that "blew" back a comet's tail so that it points away from the sun. Maxwell's new theories of electromagnetic waves and the wave nature of light allowed him in 1873 to predict formally the magnitude of momentum carried by electromagnetic waves and the radiation pressure exerted by them. Maxwell calculated the radiation pressure to be ordinarily quite small, but he ventured to say, "Such rays falling on a thin metallic disk, delicately suspended in vacuum, might perhaps produce an observable effect." The first experimental observation of radiation pressure

[50]G. E. Henry, Radiation pressure, *Scientific American,* pp. 99–108, June 1957; A. Ashkin, The pressure of laser light, *Scientific American,* pp. 63–71, February 1972.

[51]In 1905, J. H. Poynting, in his presidential address to the British Physical Society, said, "A very short experience in attempting to measure these light forces is sufficient to make one realize their extreme minuteness, which appears to put them beyond consideration in terrestrial affairs" [J. H. Poynting, Radiation pressure, *The London, Edinburgh, and Dublin Philosophical Magazine and Journal of Science,* 9(52), pp. 393–406, April 1905]. This view of electromagnetic light forces as being of no practical importance held true until the advent of the laser in 1960.

[52]G. A. Mourou, C. P. J. Barty, and M. D. Perry, Ultrahigh-intensity lasers: Physics of the extreme on a table top, *Physics Today,* pp. 22–28, January 1998.

[53]W. Crookes, *Phil. Trans.,* p. 501, 1873. Crookes apparently believed in 1873 that he had found the true radiation pressure with his newly invented radiometer and cautiously suggested that his experiments might have some bearing on the prevailing theory of the nature of light. He discovered certain gas pressures for which the combined gas and radiation forces neutralized, but since he did not discriminate between forces due to radiation and gas forces, his results were apparently capricious and his reasoning was somewhat confused. For further discussion, see E. F. Nichols and G. F. Hull, The pressure due to radiation, *Phys. Rev.,* 17(1), pp. 26–50, July 1903.

[54]See p. 252 of D. Goldsmith, *The Evolving Universe,* Benjamin Cummings, 1985, and Section 16.6 of A. J. McMahon, *Astrophysics and Space Science,* Prentice Hall, 1965.

was to be achieved about 30 years after Maxwell's predictions, by measuring with careful instrumentation the force exerted on an object by a light beam.[55]

The radiation pressure is by definition the force per unit area exerted by a wave upon a material on which it is incident. If we assume that the material completely absorbs the electromagnetic energy of a wave traveling at the speed of light c and carrying a time-average Poynting vector $\mathbf{S}_{av}$ normally incident on it for a time duration of Δt, the momentum $\Delta \overline{\mathcal{G}}$ delivered to an elemental surface area ΔA of the object is

$$\Delta \overline{\mathcal{G}} = \frac{\mathbf{S}_{av}\Delta A \Delta t}{c} \qquad \text{kg-m-s}^{-1} \text{ (or N-s)}$$

From Newton's second law, and assuming that the time-average power flux is steady during the time interval Δt, the time-average force exerted on the elemental surface ΔA of the object is given by

$$\mathbf{F}_{av} = \frac{\mathbf{P}_{av}}{\Delta t} = \frac{\mathbf{S}_{av}\Delta A}{c} \qquad \text{N}$$

so that the time-average force per unit area, or time-average pressure, is given by

$$\boxed{\mathbf{P}_{av} = \frac{\mathbf{S}_{av}}{c}} \qquad \text{N-m}^{-2} \qquad\qquad [2.44]$$

The quantity $\mathbf{P}_{av}$ is essentially a surface force density or stress. Considerations of this type constitute the subject of electromagnetic stress or the so-called *Maxwell stress tensor*, a topic beyond the scope of this book.[56]

The pressure of electromagnetic radiation encountered in our daily environment is extremely small, as mentioned above. For example, the power density of all electromagnetic radiation from the sun at the Earth's surface is $|\mathbf{S}_{av}| \simeq 1400$ W-m^{-2}, so that the pressure is only $(|\mathbf{S}_{av}|/c) \simeq 4.67 \times 10^{-6}$ N-m^{-2}, exceedingly small indeed.

---

[55]The first such measurement was realized independently in both the United States and Russia [E. F. Nichols and G. F. Hull, *Phys. Rev.* 13, p. 307, October 1901; P. Lebedev, *Ann. Physik,* 6, p. 433, October 1901; E. F. Nichols and G. F. Hull, The pressure due to radiation, *Phys. Rev.* 17(1), p. 26–50, July 1903]. One of the experiments, as described in E. F. Nichols and G. F. Hull, Pressure due to radiation, *Amer. Acad. Proc.,* 38(20), pp. 559–599, April 1903, used a torsion balance in a carefully undertaken experiment to measure a pressure of $\sim 7 \times 10^{-6}$ N-m^{-2}, or a force on the ~ 1 cm^2 area of their silvered glass vanes of $\sim 7 \times 10^{-10}$ N. The Poynting flux in the light beam they used was $|\mathbf{S}_{av}| \simeq 0.1$ W-(cm)$^{-2}$. In comparison, typical laboratory laser beams in use today may have a total power of ~ 5 mW and a width of ~ 0.5 mm (or $|\mathbf{S}_{av}| \simeq 2$ W-(cm)$^{-2}$).

[56]Faraday was the first to speak of lines of force as elastic bands that transmit tension and compression. Maxwell also spent considerable time on this concept and placed Faraday's notions into clear mathematical focus. For brief discussions of the concept of electromagnetic stress see C. C. Johnson, *Field and Wave Electrodynamics,* Section 1.19, McGraw-Hill, 1965; W. B. Cheston, *Elementary Theory of Electric and Magnetic Fields,* Section 10.7–10.8, Wiley, 1964. For more complete coverage, see L. M. Landau and E. M. Lifshitz, *Electrodynamics of Continuous Media,* Addison-Wesley, 1960; J. A. Stratton, *Electromagnetic Theory,* McGraw-Hill, 1941.

The radiation pressure varies as the inverse square of the distance from the source, as does the Poynting vector $\mathbf{S}_{av}$. Thus, we can expect the radiation pressure to be much higher at the surface of the sun. Using a mean Earth–sun distance of $\sim 1.5 \times 10^{11}$ m, we find the radiation pressure on the surface of the sun to be only ~ 0.21 N-m^{-2}, equivalent to a weight (on Earth) of $\sim 2.2 \times 10^{-3}$ gm-(cm)$^{-2}$. Radiation pressure is unimportant even in the interior of the sun, but it is important in the dynamics of brighter stars.

Much higher electromagnetic power densities and thus radiation pressures can be attained in waveguides, where power densities of up to 10^9 W-m^{-2} are possible (see Chapter 5). With propagation at velocities very close to c, 10^9 W-m^{-2} corresponds to a radiation pressure of ~ 3.33 N-m^{-2}, equivalent to a weight of 0.034 gm-(cm)$^{-2}$. For a powerful laser, which may utilize narrow (~ 0.1 mm radius) pulsed beams of a thousand gigawatts (i.e., $|\mathbf{S}_{av}| = 10^{12}/[\pi(0.1 \times 10^{-3})^2] \simeq 3.2 \times 10^{19}$ W-m^{-2}), the force exerted on an absorbing material across the size of the beam would be ~ 31 newtons, or as much as the weight of a ~ 3 kg object.

The radiation pressure of a wave acts in the direction of wave propagation, and the force of the wave is transferred to an object upon which the wave is incident. If the incident wave energy is completely absorbed by the object (no reflection), a force of $\mathbf{S}_{av}\Delta A/c$ acts on it. If the incident energy is entirely reflected, as it would be if the object was a perfect conductor, the force acting on the body is twice[57] that for absorption, namely $2\mathbf{S}_{av}\Delta A/c$. We shall revisit this subject when we study reflection of plane waves from conducting surfaces in Chapter 3.

The question of the momentum of an electromagnetic wave can be more formally studied by using the Lorentz force equation (see Section 6.1). The force on an object is the rate at which momentum is transferred to that object. The force per unit volume on the charges and currents in a given region is the rate of transfer per unit volume. Using a development very similar to that used in the derivation of Poynting's theorem (see above), it can be shown[58] that the total rate of transfer is equal to the rate of decrease of momentum residing in the field. One finds that each unit of volume in the electromagnetic field appears to carry a momentum of

$$\bar{g} = \overline{\mathscr{D}} \times \overline{\mathscr{B}} = \epsilon\mu(\overline{\mathscr{E}} \times \overline{\mathscr{H}}) \qquad [2.45]$$

which is dimensionally a *momentum per unit volume*. The total momentum of the electromagnetic field contained in a volume V is

$$\overline{\mathscr{G}} = \int_V \bar{g}\, dv \qquad \text{N-s}$$

---

[57]Note that this is similar to the fact that twice as much momentum is delivered to an object when a perfectly elastic ball is bounced from it as when it is struck with a perfectly inelastic ball. We can also think of this circumstance as the conductor first absorbing the energy and then recoiling as it re-emits it.

[58]See W. T. Scott, *The Physics of Electricity and Magnetism,* Section 10.4, Wiley, 1966. The procedure basically involves starting with the Lorentz force equation describing the electromagnetic force $\overline{\mathscr{F}}$ on a material volume (see Section 6.5.1), namely $\overline{\mathscr{F}} = \tilde{\rho}\overline{\mathscr{E}} + \overline{\mathscr{J}} \times \overline{\mathscr{B}}$, and eliminating $\tilde{\rho}$ and $\overline{\mathscr{J}}$ using Maxwell's equations.

An interesting implication of the fact that electromagnetic fields carry momentum is that Newton's third law and the principle of conservation of momentum are strictly valid only when the momentum of an electromagnetic field is taken into account along with that of the matter with which it interacts.[59] A related issue is the fact that the magnetostatic force law (i.e., $q\bar{v} \times \overline{\mathscr{B}}$) is not by itself consistent with Newton's third law, in that if one considers two moving charges $q_1\bar{v}_1$ and $q_2\bar{v}_2$, the vector sum of the force exerted by one on the other and vice versa is not zero[60] (i.e., $\mathbf{F}_{12} + \mathbf{F}_{21} \neq 0$).

The fact that an electromagnetic wave carries momentum of amount $\mathbf{S}_{av}/c$ is also consistent with atomic physics. According to quantum theory, electromagnetic energy is transported in units of hf, where h is Planck's constant ($h \simeq 6.63 \times 10^{-34}$ joule-s) and f is the frequency. Thus, each photon of energy must carry a momentum of $hf/c = h/\lambda$, as has been verified in experiments, such as the Compton effect.[61]

The above discussion concerns *linear* momentum. Electromagnetic waves can also possess *angular* momentum.[62] The angular momentum about a given origin carried by an electromagnetic radiation is $\overline{\mathscr{L}} = \mathbf{r} \times \overline{\mathscr{G}}$, where $\mathbf{r}$ is the vector from the origin to the point at which the momentum is measured. When a beam of circularly polarized (see Section 2.5) light with Poynting flux $\mathbf{S}_{av}$ is incident on a completely absorbing object of area ΔA for a time Δt, the angular momentum transferred to the object is given by $\overline{\mathscr{L}} = (\mathbf{S}_{av}\Delta A\Delta t)/(2\pi f)$, where f is the frequency of light. The first experimental verification of the angular momentum of electromagnetic waves was the fact that a doubly refracting slab that produces circularly polarized light experiences a reaction torque.[63]

2.5 POLARIZATION OF ELECTROMAGNETIC WAVES

In previous sections we noted that the electric field vector of a uniform plane wave lies in the plane perpendicular to the direction of propagation, but we did not concern ourselves with the particular direction in which the electric field oscillates. In a practical communication environment, the electric field vector of a monochromatic (i.e., single-frequency) uniform plane wave propagating in the z direction can have both an x and a y component, which can oscillate independently, depending on the

[59] See Section 2.5 of J. A. Stratton, *Electromagnetic Theory,* McGraw-Hill, 1941.

[60] For further discussion, see Sections 6.1, 6.2, and 10.8 of W. B. Cheston, *Elementary Theory of Electric and Magnetic Fields,* Wiley, 1962.

[61] A. H. Compton, Quantum Theory of the Scattering of X-Ray by Light Elements, *Phys. Rev.,* 21, pp. 483–502, 1923. A. H. Compton earned a Nobel Prize for this work in 1927. For a historical account, see A. H. Compton, The Scattering of X-Rays as Particles, *Am. J. Phys.,* pp. 817–820, 1961.

[62] A. M. Portis, *Electromagnetic Fields: Sources and Media,* Sec. 11–18, Wiley, 1978; R. P. Feynman, R. B. Leighton, and M. Sands, *The Feynman Lectures in Physics,* Sec. II-27-6, Addison-Wesley, 1964.

[63] R. A. Beth, Mechanical Detection and Measurement of the Angular Momentum of Light, *Phys. Rev.,* 50, p.115–125, 1936.

manner in which the wave was generated at its source. The resultant effect produced by two independent oscillations at right angles to one another can result in a variety of different behavior of the total electric field vector. The orientation and the behavior of the total electric field vector is determined by the *polarization* of the wave. Polarized electromagnetic waves and light are commonly encountered in our everyday environment and are also put to very good use in practice, as discussed in Section 2.5.5.

The polarization of a uniform plane wave refers to the behavior of its electric field vector as a function of time at a fixed point in space. More specifically, polarization describes the manner in which the shape and orientation of the locus of the tip of the electric field vector varies with time at a fixed plane in space. Note that polarization refers to the wave electric field by convention; the wave magnetic field can be obtained from the wave electric field and should thus behave in a corresponding manner.[64] For uniform plane waves, the wave electric field is confined to the plane perpendicular to the direction of propagation, so that the polarization of the wave necessarily describes the behavior of the electric field vector in this plane. For nonuniform plane waves, of the type we shall encounter in Chapters 3 and 4, the wave electric field vector may well have components in the direction of propagation. In such cases, polarization of the wave can be separately described in different planes, including that which is perpendicular to the direction of propagation, as well those which are parallel to it.

2.5.1   Linear Polarization

A uniform plane wave propagating in the z direction is said to be *linearly* polarized (LP) when it has only one component ($\mathscr{E}_x$ or $\mathscr{E}_y$) or when its two transverse

---

[64]The choice of the wave electric field vector (rather than the wave magnetic field vector) as the one to which we refer in delineating the orientation of the wave with respect to the plane of reference is somewhat arbitrary. After all, the electric field vector of a propagating wave cannot exist without the magnetic field vector and vice versa. However, when we pay attention to the physical effect of these vectors, we might think that $\overline{\mathscr{E}}$ is more qualified to represent the electromagnetic field than $\overline{\mathscr{H}}$ or $\overline{\mathscr{B}}$. To see this, consider the effect of the electromagnetic field on a small particle of charge q moving with a velocity $\tilde{\mathbf{v}}$. The total force experienced by the particle is

$$\overline{\mathscr{F}} = q(\overline{\mathscr{E}} + \tilde{\mathbf{v}} \times \overline{\mathscr{B}})$$

which is the Lorentz force discussed in detail in Section 6.1. For a uniform plane wave propagating in the z direction, say with only $\mathscr{E}_x$ and $\mathscr{B}_y$ components, the magnitudes of the fields are related via by Faraday's Law, or $\beta\mathscr{E}_x = \omega\mathscr{B}_y \longrightarrow \mathscr{B}_y = \sqrt{\mu\epsilon}\mathscr{E}_x = \mathscr{E}_x/v_p$, so that we have $|\overline{\mathscr{B}}| = |\overline{\mathscr{E}}|/v_p$. Thus, the Lorentz force acting on the particle is

$$\overline{\mathscr{F}} = q\left[\overline{\mathscr{E}} + \left(\frac{|\overline{\mathscr{E}}|}{|\tilde{\mathbf{v}}|}\right)\frac{\tilde{\mathbf{v}} \times \overline{\mathscr{B}}}{|\overline{\mathscr{B}}|}\right]$$

The electric field vector can thus act on the particle even when the particle is at rest, whereas the magnetic force is comparatively smaller in magnitude since typically $|\tilde{\mathbf{v}}| \ll v_p$.

components ($\mathscr{E}_x$ and $\mathscr{E}_y$) are *in phase*. In other words, we have

$$
\boxed{
\begin{aligned}
\mathscr{E}_x(z, t) &= C_{1x} \cos(\omega t - \beta z + \zeta) \\
\mathscr{E}_y(z, t) &= C_{1y} \cos(\omega t - \beta z + \zeta)
\end{aligned}
}
\qquad \text{Linear polarization} \qquad [2.46]
$$

where C_{1x} and C_{1y} are constants, either of which may be zero, or in general both can be nonzero. When viewed at a fixed location (i.e., fixed z) in the plane perpendicular to the propagation direction (i.e., the xy plane), the tip of the total electric field vector of a linearly polarized wave vibrates in time along a straight line. The orientation of the straight line to which the electric field vector is confined is shown in Figure 2.16 for different combinations of values of C_{1x} and C_{1y}.

2.5.2 Circular Polarization

A uniform plane wave is *circularly* polarized when its two transverse components are out of phase by 90° (or $\pi/2$ radians) but have equal amplitudes. In other words:

$$
\boxed{
\begin{aligned}
\mathscr{E}_x(z, t) &= C_1 \cos(\omega t - \beta z) \\
\mathscr{E}_y(z, t) &= C_1 \cos\left(\omega t - \beta z \pm \frac{\pi}{2}\right)
\end{aligned}
}
\qquad \text{Circular polarization} \qquad [2.47]
$$

As shown in Figure 2.17, the tip of the total electric field vector as observed at a fixed point in the xy plane (chosen to be at the origin in this case) moves along a fixed circle as time progresses. For the case shown in Figure 2.17a, with the motion of the tip of the total electric field vector being counterclockwise when viewed looking toward the $-z$ direction, the wave is said to be *right-hand* circularly polarized (RHCP). An easy way to remember this convention is to use your right hand with the thumb pointing in the direction of propagation (in this case the $+z$ direction); if the electric field moves in the direction of your other fingers, then the wave is polarized in the right-handed sense.[65] If the total electric field vector moves in the opposite sense, the wave is said to be *left-hand* circularly polarized (LHCP). Note that for a left-hand polarized wave, $\mathscr{E}_y(z, t)$ would lead $\mathscr{E}_x(z, t)$ by 90° (i.e., $+\pi/2$ instead of $-\pi/2$ radians), as shown in Figure 2.17b.

---

[65]The right-hand rule described here is the IEEE convention. Amazingly enough, there is considerable disagreement on how to define the sense of polarization of a circularly polarized wave. Most physicists, as well as scientists and engineers specializing in optics, prefer to have the thumb point to *where the wave is coming from,* exactly opposite of the IEEE convention. A further source of confusion is the preference of physicists to use $e^{-j\omega t}$ instead of $e^{j\omega t}$, which of course reverses the sense of rotation. In view of these ambiguities, it is wise in any given case to examine the actual sense of rotation of the total instantaneous electric field $\mathscr{E}$ carefully by examining the field vector orientation at a fixed point in space at two specific times (separated by less than $T_p/2$, where $T_p = 2\pi/\omega$), such as $\omega t = 0$ and $\omega t = \pi/2$, and compare with Figure 2.17.

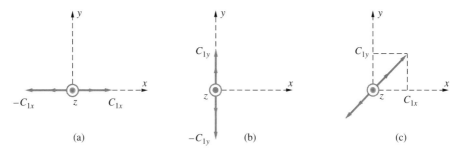

FIGURE 2.16.   **The electric field vector of a linearly polarized wave.** (a) $C_{1x} \neq 0$ and $C_{1y} = 0$. (b) $C_{1x} = 0$ and $C_{1y} \neq 0$. (c) $C_{1x} \neq 0$ and $C_{1y} \neq 0$.

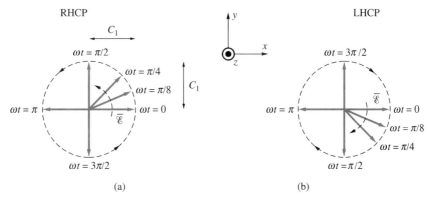

FIGURE 2.17.   **The electric field vectors of a right-hand and left-hand circularly polarized waves.** (a) RCHP. (b) LHCP. In both cases, the positions of the total electric field vector are shown at different times as indicated. Note that the wave is assumed to be propagating in the $+z$ direction.

That the locus of the total electric field vector traces a circle as time progresses can be seen by examining $\mathscr{E}_x(z, t)$ and $\mathscr{E}_y(z, t)$ given by [2.47] at any fixed observation point, say at $z = 0$, as in Figure 2.17. We then have

$$\mathscr{E}_x = C_1 \cos(\omega t) \qquad \mathscr{E}_y = C_1 \cos\left(\omega t - \frac{\pi}{2}\right) = C_1 \sin(\omega t)$$

for the right-hand circularly polarized wave. Thus, the magnitude of the total electric field vector can be written as

$$|\mathscr{E}|^2 = \mathscr{E}_x^2 + \mathscr{E}_y^2 = C_1^2$$

which is the equation for a circle.

As observed at a fixed point, the magnitude of the total electric field vector $|\mathscr{E}|^2$ remains constant in time but changes its direction. At a fixed instant of time, the orientation of the total electric field vector at different points in space can be found by vector addition of the $\mathscr{E}_x$ and $\mathscr{E}_y$ components, as illustrated in Figure 2.18.

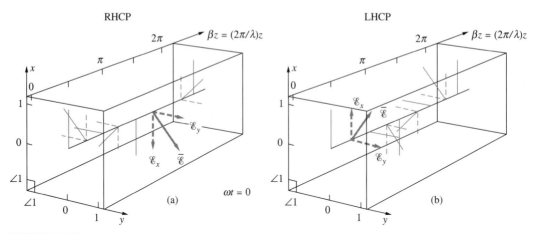

FIGURE 2.18. **Circularly polarized wave in space.** The total electric field of the wave at any given point in space and at any time is given by the vector addition of the $\mathcal{E}_x$ and $\mathcal{E}_y$ components. The direction of propagation is the $+z$ direction. (a) Right-hand circularly polarized (RHCP) wave. (b) Left-hand circularly polarized (LHCP) wave.

For a right-hand circularly polarized wave, the locus of the tip of the total electric field describes a helix in space, which advances in the propagation direction (i.e. the $+z$ direction) as time progresses in the same way as a right-handed screw, as shown in Figure 2.19. The corresponding picture for a left-hand circularly polarized wave is shown in Figure 2.20.

Note that we can alternately express the electric field components of the circularly polarized wave in phasor form. We have

$$\text{RHCP:} \quad E_x(z) = C_1 e^{-j\beta z} \quad E_y(z) = C_1 e^{-j\beta z} e^{-j\frac{\pi}{2}} = -jC_1 e^{-j\beta z} \quad [2.48a]$$

$$\text{LHCP:} \quad E_x(z) = C_1 e^{-j\beta z} \quad E_y(z) = C_1 e^{-j\beta z} e^{j\frac{\pi}{2}} = jC_1 e^{-j\beta z} \quad [2.48b]$$

Polarization of the Wave Magnetic Field As noted before, our discussion of wave polarization concerns only the behavior of the total wave electric field vector, essentially as a matter of convention. However, each of the total wave electric field components must necessarily be accompanied by an associated magnetic field component, and the $\overline{\mathcal{H}}$-components exhibit time variations (as observed at a fixed point in space) consistent with the $\overline{\mathcal{E}}$-components. For a right-hand circularly polarized (RHCP) wave propagating in the z direction we have

$$\overline{\mathcal{E}}(z,t) = \hat{\mathbf{x}}\mathcal{E}_x + \hat{\mathbf{y}}\mathcal{E}_y = C_1\left[\hat{\mathbf{x}}\cos(\omega t - \beta z) + \hat{\mathbf{y}}\cos\left(\omega t - \beta z - \frac{\pi}{2}\right)\right] \quad [2.49a]$$

$$\overline{\mathcal{H}}(z,t) = \hat{\mathbf{x}}\mathcal{H}_x + \hat{\mathbf{y}}\mathcal{H}_y = \frac{C_1}{\eta}\left[-\hat{\mathbf{x}}\cos\left(\omega t - \beta z - \frac{\pi}{2}\right) + \hat{\mathbf{y}}\cos(\omega t - \beta z)\right] \quad [2.49b]$$

Note that $\mathcal{E}_x$ is associated with $\mathcal{H}_y$ and $\mathcal{E}_y$ with $\mathcal{H}_x$. If we were to observe the tip of the total wave magnetic field vector as a function of time, looking towards the

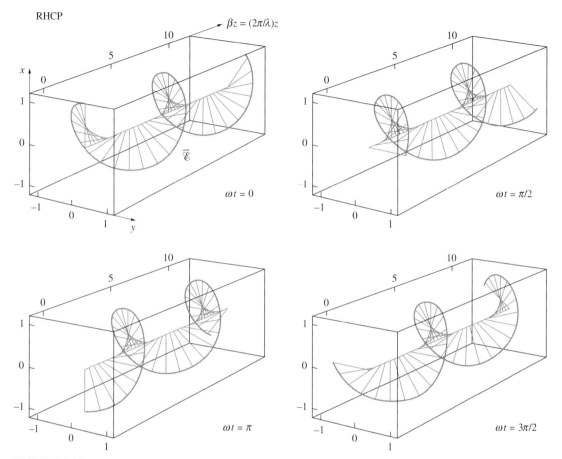

FIGURE 2.19.   **Helical loci of RHCP wave.** The total electric field vector of a right-hand circularly polarized wave propagating in the $+z$ direction shown at different points in space at $t = 0$.

propagation direction, we would see it rotate clockwise together with the total electric field vector, always remaining perpendicular to it and oriented such that $\overline{\mathscr{E}} \times \overline{\mathscr{H}}$ (i.e., the direction of electromagnetic power flow) is always in the $+z$ direction.

Example 2-20:   Geopositioning satellite.   A low earth orbit (LEO) geopositioning satellite orbiting at an altitude of 1000 km transmits a total power of $P_{\text{tot}} = 40$ kW isotropically at a downlink frequency of about 137.5 MHz, as shown in Figure 2.21a. The electric field of the received wave on the earth at a point immediately below the satellite is given by

$$\mathbf{E}(z) = C_1(-j\hat{\mathbf{x}} + \hat{\mathbf{y}})e^{-j\beta z}$$

(a) Find the polarization of this wave. (b) Find the values of β and C_1. (c) Find the corresponding magnetic field, $\mathbf{H}(z)$.

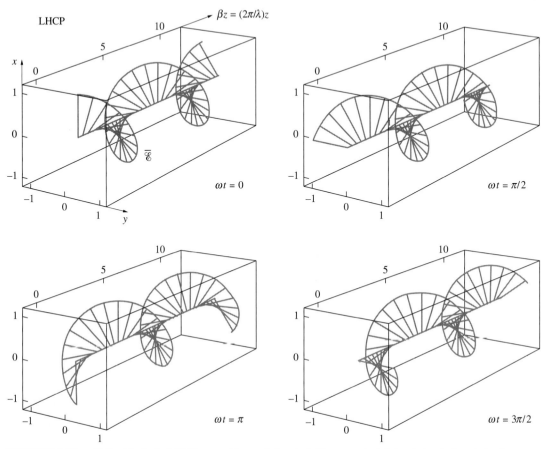

FIGURE 2.20. Helical loci of a LHCP wave. The total electric field vector of a left-hand circularly polarized wave wave propagating in the $+z$ direction shown at different points in space at $t = 0$.

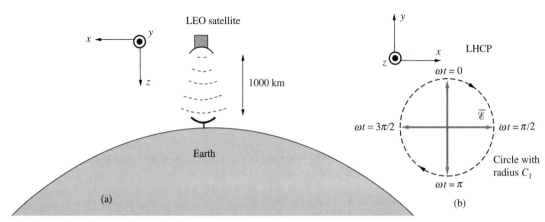

FIGURE 2.21. A satellite at low earth orbit and the polarization of the received signal. The direction of propagation is the $+z$ direction.

Solution:

(a) The instantaneous expression for the total electric field is given by

$$\overline{\mathscr{E}}(z, t) = \mathcal{R}e\{\mathbf{E}(z)e^{j\omega t}\}$$

$$= \mathcal{R}e\{\hat{\mathbf{x}}C_1 e^{-j\pi/2}e^{-\beta z}e^{j\omega t} + \hat{\mathbf{y}}C_1 e^{-j\beta z}e^{j\omega t}\}$$

$$= \hat{\mathbf{x}}C_1 \cos(\omega t - \beta z - \pi/2) + \hat{\mathbf{y}}C_1 \cos(\omega t - \beta z)$$

$$= \hat{\mathbf{x}}C_1 \sin(\omega t - \beta z) + \hat{\mathbf{y}}C_1 \cos(\omega t - \beta z)$$

We see that the two components of the wave electric field are equal in magnitude but 90° out of phase, meaning that this wave is circularly polarized. Further, if we sketch the total electric field vector as a function of time at $z = 0$, as seen in Figure 2.21b, we find that the wave is left-hand circularly polarized (i.e., LHCP).

(b) $\beta = 2\pi f/c \simeq 2\pi(137.5 \times 10^6)/(3 \times 10^8) \simeq 2.88$ rad-m^{-1}. Neglecting all losses and effects from boundaries, we have

$$|\mathbf{S}_{av}| = \frac{P_{tot}}{4\pi R^2} = \left(\frac{1}{2}\frac{C_1^2}{\eta}\right)$$

so

$$C_1 \simeq \sqrt{\frac{2 \times 377 \times 40 \times 10^3}{4\pi(10^6)^2}} \simeq 1.55 \times 10^{-3} \text{ V-m}^{-1} = 1.55 \text{ mV-m}^{-1}$$

(c) The phasor expression for the corresponding wave magnetic field can written as

$$\mathbf{H}(z) = \frac{C_1}{\eta}(-\hat{\mathbf{x}} - j\hat{\mathbf{y}})e^{-j\beta z} \simeq 4.11(-\hat{\mathbf{x}} - j\hat{\mathbf{y}})e^{-j2.88z} \quad \mu\text{A-m}^{-1}$$

2.5.3   Elliptical Polarization

The general case of elliptical polarization occurs when $\mathscr{E}_x$ and $\mathscr{E}_y$ have different magnitudes or if the phase difference between them is different from 90°. The case of 90° phase difference with different amplitudes is shown in Figure 2.22, for which we have the two transverse wave electric field components expressed in phasor form as

$$E_x(z) = C_{1x}e^{-j\beta z} \qquad [2.50a]$$

$$E_y(z) = C_{1y}e^{-j\beta z}e^{j\zeta} \qquad [2.50b]$$

where $\zeta = \pm\pi/2$, with the + and − representing respectively left- and right-hand polarization senses.

Even if the magnitudes of the two field components were equal, the wave is elliptically polarized if $0 < \zeta < 90°$, i.e., $\zeta \neq 0$ or $\zeta \neq 90°$. In the general case we can write the two transverse components of the wave electric field in the time domain as

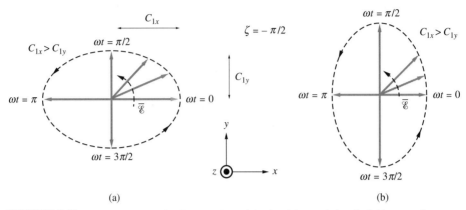

FIGURE 2.22. **Elliptical polarization.** The loci of the tip of the total electric field vector shown at the origin for an elliptically polarized wave propagating in the $+z$ direction represented by [2.51] with $\zeta = -\pi/2$. (a) The major axis of the ellipse is along the x axis when $C_{1x} > C_{1y}$. (b) The major axis of the ellipse is along the y axis when $C_{1x} < C_{1y}$. As in the case of circular polarization, it can be shown using the right-hand rule that this wave is right-hand elliptically polarized (RHEP).

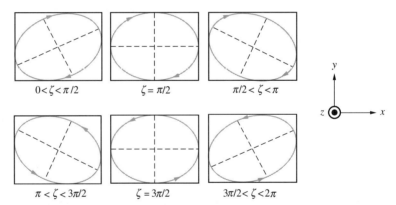

FIGURE 2.23. **Elliptical polarization ellipses.** The loci of the tip of the total electric field vector for an elliptically polarized wave as given by [2.51] for different ranges of values of ζ. Top panels are LHEP whereas the bottom panels are RHEP waves. The straight-line loci for the cases of $\zeta = 0$ and π, corresponding to linear polarization, are not shown.

$$\begin{aligned} \mathcal{E}_x(z,t) &= C_{1x}\cos(\omega t - \beta z) \\ \mathcal{E}_y(z,t) &= C_{1y}\cos(\omega t - \beta z + \zeta) \end{aligned}$$ Elliptical polarization   [2.51]

The corresponding shape of the polarization ellipses for different values of ζ are given in Figure 2.23. The top panels represent left-hand elliptical polarization (LHEP), whereas the bottom panels show right-hand elliptical polarization (RHEP).

For the case when $C_{1x} = C_{1y}$, the full range of possible polarizations can be realized by simply varying the phase angle ζ. As illustrated in Figure 2.24, when the phase difference ζ varies from $\zeta = 0$ to $\zeta = \pi$, the polarization of a wave with

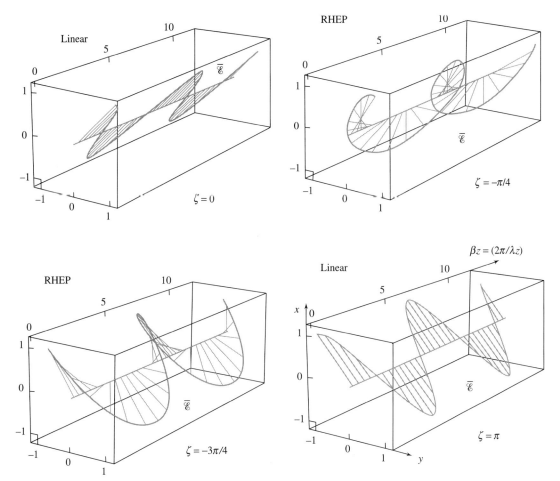

FIGURE 2.24.   Polarization as a function of phase angle ζ. The electric field vectors at $t = 0$ and at different points in space for $C_{1x} = C_{1y}$ and for $\zeta = 0, -\pi/4, -3\pi/4$, and π.

$|\mathscr{E}_x| = |\mathscr{E}_y|$ varies from linear ($\zeta = 0$) tilted at 45° with respect to the x-axis to elliptical ($0 < \zeta < \pi/2$) to circular ($\zeta = \pi/2$; see Figure 2.19) to elliptical ($\pi/2 < \zeta < \pi$) and back to linear tilted at $-45°$ ($\zeta = \pi$).

We can show that the tip of the total electric field vector does indeed trace an ellipse by eliminating ωt between the two equations of [2.51] and rewriting them in the form[66]

$$\left(\frac{\mathscr{E}_x}{C_{1x}}\right)^2 + \left(\frac{\mathscr{E}_y}{C_{1y}}\right)^2 - 2\frac{\mathscr{E}_x}{C_{1x}}\frac{\mathscr{E}_y}{C_{1y}}\cos\zeta = \sin^2\zeta \qquad [2.52]$$

This equation represents a tilted polarization ellipse, as shown in Figure 2.25.

---

[66]See *Principles of Optics,* M. Born and E. Wolf, 5th ed., by Pergamon Press, 1975, pp. 24–30.

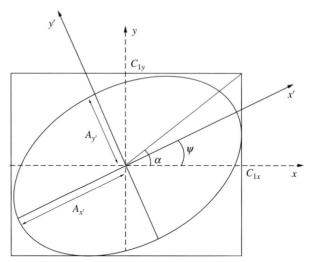

FIGURE 2.25.   The tilted polarization ellipse.

The major and minor axes of this ellipse do not coincide with the x and y axes of the rectangular coordinate system; rather, they coincide with the x' and y' axes obtained by rotating the x and y axes by a tilt angle ψ, as shown in Figure 2.25, given by

$$\psi = \tfrac{1}{2}\tan^{-1}[\tan(2\alpha)\cos\zeta] \qquad [2.53]$$

where $\tan\alpha = C_{1y}/C_{1x}$. The principal semiaxes $A_{x'}$ and $A_{y'}$ (in other words, the half-lengths of the principal axes) can be written in terms of C_{1x}, C_{1y}, and ψ as

$$A_{x'} = C_{1x}\cos\psi + C_{1y}\sin\psi$$

$$A_{y'} = -C_{1x}\sin\psi + C_{1y}\cos\psi$$

where $A_{x'}$, $A_{y'}$, C_{1x}, and C_{1y} are related by

$$A_{x'}^2 + A_{y'}^2 = C_{1x}^2 + C_{1y}^2$$

One can rewrite the equation of the polarization ellipse given by [2.51] with respect to the rotated axes x' and y' as

$$\left(\frac{\mathscr{E}_{x'}}{A_{x'}}\right)^2 + \left(\frac{\mathscr{E}_{y'}}{A_{y'}}\right)^2 = 1$$

where $\mathscr{E}_{x'}$ and $\mathscr{E}_{y'}$ are the components of the wave electric field in the rotated coordinate system.

Also of interest is the so-called axial ratio, AR, which is defined as the ratio of the major to the minor axes. If $A_{x'} > A_{y'}$, then $AR = A_{x'}/A_{y'}$, whereas if $A_{x'} < A_{y'}$, then $AR = (A_{x'}/A_{y'})^{-1}$, the value of which satisfies $1 \le AR \le \infty$.

The determination of the polarization ellipse of an elliptically polarized wave is illustrated in Example 2-21.

Example 2-21:   Elliptical polarization.   An electromagnetic wave travels in the x direction with its magnetic field given by

$$\overline{\mathcal{H}}(x, t) = \hat{\mathbf{y}} 8 \cos(\omega t - \beta x) + \hat{\mathbf{z}} 12 \cos(\omega t - \beta x + 70°)$$

in mA-m^{-1}. Find the tilt angle, axial ratio, and rotation sense of the polarization ellipse.

Solution: From geometry, we have

$$\tan \alpha = \frac{C_{1y}}{C_{1z}} = \frac{8}{12} = \frac{2}{3}$$

from which

$$\tan(2\alpha) = \frac{2 \tan \alpha}{1 - \tan^2 \alpha} = 2.4$$

So, the tilt angle measured from the z axis is given by

$$\psi = \tfrac{1}{2} \tan^{-1}[\tan(2\alpha) \cos \zeta] = \tfrac{1}{2} \tan^{-1}[(2.4) \cos 70°] \simeq 19.7°$$

The principal semiaxes $A_{x'}$ and $A_{y'}$ are found as

$$A_{x'} \simeq 8 \sin(19.7°) + 12 \cos(19.7°) \simeq 14$$
$$A_{y'} \simeq -8 \cos(19.7°) + 12 \sin(19.7°) \simeq -3.49$$

So the axial ratio is

$$\text{AR} \simeq \frac{14}{3.49} \simeq 4.01$$

Figure 2.26 shows the polarization ellipse and the sense of rotation of the magnetic field vector as a function of time at the origin. Since the magnetic field vector rotates around the ellipse in the clockwise direction as shown, this is a left-hand elliptically polarized (LHEP) wave.

Sometimes it is helpful to express an elliptically polarized wave as a sum of two oppositely rotating circularly polarized waves, as shown in Figure 2.27. To see this decomposition analytically, note that the total electric field phasor for a left-hand circularly polarized wave can be written as

$$\mathbf{E}_1 = \hat{\mathbf{x}} C_1 e^{-j\beta z} + \hat{\mathbf{y}} j C_1 e^{-j\beta z}$$

whereas that for a right-hand circularly polarized wave is

$$\mathbf{E}_2 = \hat{\mathbf{x}} C_2 e^{-j\beta z} - \hat{\mathbf{y}} j C_2 e^{-j\beta z}$$

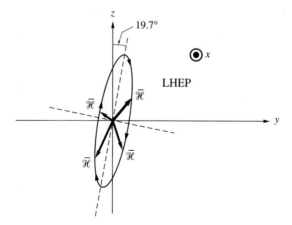

FIGURE 2.26. Polarization ellipse of Example 2-21.

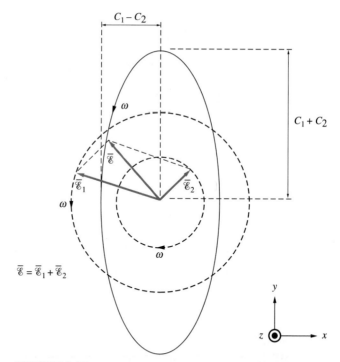

FIGURE 2.27. **The decomposition of an elliptically polarized wave into two circularly polarized waves.** Any elliptically polarized wave ($\overline{\mathscr{E}}$) can be decomposed into two counter-rotating circularly polarized components $\overline{\mathscr{E}}_1$ and $\overline{\mathscr{E}}_2$. In the limiting case when $\overline{\mathscr{E}}$ is linearly polarized (i.e., $\zeta = 0$), $|\overline{\mathscr{E}}_1| = |\overline{\mathscr{E}}_2|$. Otherwise, the sense of rotation of $\overline{\mathscr{E}}$ is in the direction of the larger of $\overline{\mathscr{E}}_1$ and $\overline{\mathscr{E}}_2$.

the sum of the electric fields for these two waves is then

$$\mathbf{E} = \hat{\mathbf{x}}(C_1 + C_2)e^{-j\beta z} + \hat{\mathbf{y}}\,j(C_1 - C_2)e^{-j\beta z}$$

which is an elliptically polarized wave (since $\mathscr{E}_x$ and $\mathscr{E}_y$ have different magnitudes and are 90° out of phase). Note that for elliptically polarized waves in which the major and minor axes are not aligned with the x (or y) and y (or x) axes, a similar type of decomposition can be used after rotating the coordinate system to align the polarization axes with the x (or y) and y (or x) axes.

Example 2-22 illustrates the construction of an elliptically polarized wave by superposition of two circularly polarized waves.

Example 2-22:   Two linearly polarized waves.   Two linearly polarized waves propagating in the same direction at the same frequency are given by

$$\mathbf{E}_1(y) = \hat{\mathbf{x}}C_1 e^{-j\beta y}$$

$$\mathbf{E}_2(y) = \hat{\mathbf{z}}C_2 e^{-j\beta y}e^{j\theta}$$

where C_1, C_2, and θ are constants and $\beta = \omega\sqrt{\mu\epsilon}$. Find the polarization of the sum of these waves for the following cases: (a) $\theta = 0°$; (b) $\theta = \pi/2$; (c) $\theta = \pi/2$ and $C_1 = C_2$; (d) $\theta = \pi$.

Solution: The real-time expression for the sum of these two waves can be found as

$$\overline{\mathscr{E}}(y, t) = \mathscr{R}e\{[\mathbf{E}_1(y) + \mathbf{E}_2(y)]e^{j\omega t}\}$$

$$= \mathscr{R}e[\hat{\mathbf{x}}C_1 e^{-j\beta y}e^{j\omega t} + \hat{\mathbf{z}}C_2 e^{-j\beta y}e^{j\theta}e^{j\omega t}]$$

$$= \hat{\mathbf{x}}C_1 \cos(\omega t - \beta y) + \hat{\mathbf{z}}C_2 \cos(\omega t - \beta y + \theta)$$

(a) When $\theta = 0°$, the two components of the total electric field are in phase, therefore, resulting in a linearly polarized (LP) wave oscillating in the direction as shown in Figure 2.28a.

(b) When $\theta = \pi/2$, the two components are 90° out of phase. Since the amplitudes are in general different, the total wave is elliptically polarized. Figure 2.28b shows the variation of the total electric field at $y = 0$ as a function of time (for $C_2 > C_1$) from which we can conclude that the wave is right-hand elliptically polarized (RHEP).

(c) When $\theta = \pi/2$ and $C_1 = C_2$, the wave is circularly polarized (i.e., RHCP), the time variation of which is shown at $y = 0$ in Figure 2.28c.

(d) When $\theta = \pi$, the two components are 180° out of phase. Rewriting the total electric field yields

$$\overline{\mathscr{E}}(y, t) = \hat{\mathbf{x}}C_1 \cos(\omega t - \beta y) + \hat{\mathbf{z}}C_2 \cos(\omega t - \beta y + \pi)$$

$$= \hat{\mathbf{x}}C_1 \cos(\omega t - \beta y) - \hat{\mathbf{z}}C_2 \cos(\omega t - \beta y)$$

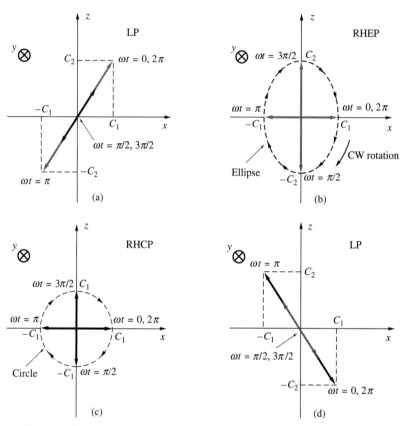

FIGURE 2.28.   Two linearly polarized waves. The variation of the field vector at $y = 0$ is shown as a function of time at five different time instants, namely $\omega t = 0, \pi/2,$ $\pi, 3\pi/2$ and 2π, for the cases when (a) $\theta = 0°$; (b) $\theta = \pi/2$; (c) $\theta = \pi/2$ and $C_1 = C_2$; and (d) $\theta = \pi$. The direction of propagation is the y direction. Also note that when $C_1 \neq C_2$, it is assumed that $C_1 < C_2$.

At any time instant t, the direction of the total electric field is given by the unit vector as

$$\hat{u}_\epsilon = \frac{\overline{\mathscr{E}}(y, t)}{|\overline{\mathscr{E}}(y, t)|}$$

$$= \hat{\mathbf{x}} \frac{C_1}{\sqrt{C_1^2 + C_2^2}} - \hat{\mathbf{z}} \frac{C_2}{\sqrt{C_1^2 + C_2^2}}$$

which does not change with time. Therefore, the wave is linearly polarized, as shown in Figure 2.28d.

Example 2-23:   Polarization of a nonuniform plane wave.   At large distances, the electric field of a vertical electric dipole of length l and current

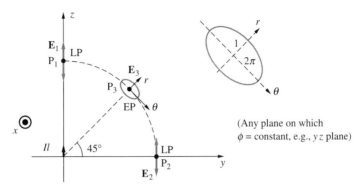

FIGURE 2.29.   Vertical electric dipole. Polarization of the electric field at points $P_1(r = 2\lambda, \phi, \theta = 0°)$, $P_2(r = 2\lambda, \phi, \theta = 90°)$, and $P_3(r = 2\lambda, \phi, \theta = 45°)$. Note that at $\theta = 45°$ (i.e., P_3), the major and minor axes of the ellipse are normalized by $E_0/(2\sqrt{2\lambda^2})$.

$I_0 \cos(\omega t)$ located at the origin and oriented along the z axis in air is given by[67]

$$\mathbf{E}(r, \theta) = E_0 e^{-j\beta r}\left(\hat{\mathbf{r}}\frac{2\cos\theta}{r^2} + \hat{\boldsymbol{\theta}}\frac{j\beta\sin\theta}{r}\right)$$

where $E_0 = \eta_0 I_0 l/(4\pi)$, $\eta = \sqrt{\mu_0/\epsilon_0}$, and $\beta = \omega/c = 2\pi/\lambda$. Find the polarization of the electric field at three equidistant field points from the dipole, each with $r = 2\lambda$ and θ given as (a) 0° at P_1; (b) 90° at P_2; and (c) 45° at P_3.

Solution: Note that the problem has azimuthal symmetry.
(a) At $P_1(r = 2\lambda, \phi, \theta = 0°)$, the expression for the electric field reduces to

$$\mathbf{E}_1 = \hat{\mathbf{r}}\frac{E_0 e^{-j4\pi}}{2\lambda^2} = \hat{\mathbf{z}}\frac{E_0}{2\lambda^2}$$

which is linearly polarized along the z axis.
(b) At $P_2(r = 2\lambda, \phi, \theta = 90°)$, the electric field simplifies to

$$\mathbf{E}_2 = \hat{\boldsymbol{\theta}}\frac{j\pi E_0}{\lambda^2} = -\hat{\mathbf{z}}\frac{j\pi E_0}{\lambda^2}$$

which is also linearly polarized along the z axis.
(c) At $P_3(r = 2\lambda, \phi, \theta = 45°)$, the electric field is given by

$$\mathbf{E}_3 = \frac{\sqrt{2}E_0}{4\lambda^2}[\hat{\mathbf{r}} + \hat{\boldsymbol{\theta}}j(2\pi)] = \frac{E_0}{2\sqrt{2\lambda^2}}[\hat{\mathbf{r}} + \hat{\boldsymbol{\theta}}(2\pi)e^{j\pi/2}]$$

which is elliptically polarized on any $\phi = $ const. plane. Figure 2.29 shows the polarization of the electric field of an electric dipole at points P_1, P_2, and P_3.

[67] E. C. Jordan and K. G. Balmain, *Electromagnetic Waves and Radiating Systems,* Prentice Hall, 1968, p. 323.

2.5.4 Poynting Vector for Elliptically Polarized Waves

Consider the electric field phasor for an elliptically polarized wave given by

$$\mathbf{E}(z) = \mathbf{E}_1(z) + \mathbf{E}_2(z) = \hat{\mathbf{x}} C_{1x} e^{-j\beta z} + \hat{\mathbf{y}} C_{1y} e^{-j(\beta z - \zeta)}$$

with its accompanying magnetic field, which follows as

$$\mathbf{H}(z) = \mathbf{H}_1(z) + \mathbf{H}_2(z) = -\hat{\mathbf{x}} \frac{C_{1y}}{\eta} e^{-j(\beta z - \zeta)} + \hat{\mathbf{y}} \frac{C_{1x}}{\eta} e^{-j\beta z}$$

where $\mathbf{E}_1 = \hat{\mathbf{x}}(\cdots)$ is associated with $\mathbf{H}_1 = \hat{\mathbf{y}}(\cdots)$, whereas $\mathbf{E}_2 = \hat{\mathbf{y}}(\cdots)$ is associated with $\mathbf{H}_2 = -\hat{\mathbf{x}}(\cdots)$. The time-average Poynting vector for this wave is given by

$$\mathbf{S}_{av} = \frac{1}{2} \mathcal{R}e\{\mathbf{E} \times \mathbf{H}^*\}$$

and can be shown to be equal to

$$\mathbf{S}_{av} = \hat{\mathbf{z}} \frac{1}{2\eta} \left\{ C_{1x}^2 + C_{1y}^2 \right\} \tag{2.54}$$

It should be noted here that $\mathbf{S}_{av}$ is independent of ζ. For a lossy medium the time-average Poynting vector of an elliptically polarized wave can be shown to be

$$\mathbf{S}_{av} = \hat{\mathbf{z}} \frac{1}{2|\eta_c|} \left[C_{1x}^2 + C_{1y}^2 \right] \cos \phi_\eta \tag{2.55}$$

Example 2-24: Elliptical polarization. Calculate the total time-average power carried by the electromagnetic wave given in Example 2-21.

Solution: The total time-average power per unit area is equal to the time-average Poynting vector for this wave given by

$$\mathbf{S}_{av} = \frac{1}{2} \eta [C_{1y}^2 + C_{1z}^2]$$

$$\simeq \frac{1}{2}(377)[8^2 + (12)^2] \simeq 39.2 \ \text{mW-m}^{-2}$$

2.5.5 Polarization in Practice

Polarization of electromagnetic waves and thus light can be observed in our everyday environment and is also put to very good use in practice. Direct sunlight is virtually unpolarized, but the light in the rainbow exhibits distinct polarization, as does the blue sky light itself.[68] Polarization studies were crucial to the early investigations of the nature of light[69] and to the determination that X rays were electromagnetic in nature.[70] Much useful information about the structure of atoms and nuclei is gained from polarization studies of the emitted electromagnetic radiation. Polarized light also has many practical applications in industry and in engineering science. In the following, we describe some of the more common engineering applications at radio frequencies.

Linearly polarized waves are commonly utilized in radio and TV broadcast applications. AM broadcast stations operate at relatively lower frequencies and utilize large antenna towers and generate *vertically* polarized signals, with the wave electric field perpendicular to the ground. Although the primary polarization of the AM signal is vertical, there is typically also a component of the electric field in the direction of propagation, due to ground-losses. For maximum reception of an AM radio signal, the receiver antenna should be parallel to the electric field, or perpendicular to the ground (i.e., vertical). Most AM radio receivers use ferrite-rod antennas to detect the wave magnetic field, which is horizontal, instead of using wire antennas to detect the wave electric field. For maximum reception, the ferrite-rod antenna should be parallel to the magnetic field. In North America, TV broadcast signals are typically horizontally polarized, which is why typical rooftop antennas are horizontal. Most FM radio broadcast stations in the United States utilize circularly polarized waves, so that the orientation of an FM receiving antenna is not critical as long as the antenna lies in a plane orthogonal to the direction from which the signal arrives.

Polarization of an electromagnetic signal is initially determined by the source antenna that launches the signal.[71] However, as a signal propagates through some medium, its polarization may change. A good example of this occurs in propagation of waves through Earth's ionosphere, when a rotation of the plane of polarization occurs, known as Faraday rotation (see Section 6.2.3). Circularly polarized waves are utilized in some communication and radar systems, primarily to relax the requirement for the receiving antenna to be carefully aligned with the wave electric

---

[68]The color of the sky is blue because of the scattering of sunlight from tiny molecules of air, a process which leads to polarized light. Sometimes, small clouds, hardly visible in air, can be seen more clearly in a reflection in water (e.g., a lake), because the light from the clouds is not polarized and is thus more efficiently reflected by the water than the blue sky light, which is polarized. Reflection and refraction of electromagnetic waves will be studied in Chapter 3. For more on the color of the sky and many other interesting phenomena, see M. Minnaert, *The Nature of Light & Colour in the Open Air,* Dover, 1954.

[69]Polarization was discovered by E. L. Malus [*Nouveau Bulletin des Sciences, par la Soc. Philomatique,* p. 266, 1809]; also see C. Huygens, *Traité de la Lumière,* 1690.

[70]C. G. Barkla, Polarization in Secondary Röntgen Radiation, *Roy. Soc. Proc. Ser. A,* 77, pp. 247–255, 1906.

[71]A very interesting antenna that can selectively generate right-hand or left-hand circularly polarized signals is the *helical* antenna, discovered by J. D. Kraus. See J. D. Kraus, The Helical Antenna, *Proc. IRE,* 37(3), pp. 263–272, 1949. For detailed discussion of the helical antenna and other antennas see J. D. Kraus, *Antennas,* McGraw-Hill, 1988.

vector. Many international communication satellites also utilize circularly polarized signals. In radar applications, circular polarization is used because such a wave reflects from a metal target with the opposite sense (i.e., RHCP becomes LHCP), providing a method to distinguish metal targets (i.e., aircraft) from clouds and clutter.

Use of polarized transmissions also provides a means of multiple simultaneous usage of the same frequency band. Most North American communication satellites use linear polarization, alternating between horizontal and vertical polarization between adjacent satellites.[72]

2.6 ARBITRARILY DIRECTED UNIFORM PLANE WAVES

In previous sections we used expressions for uniform plane wave electric and magnetic fields for which the direction of propagation was chosen to coincide with one of the coordinate axes, namely, the z axis. This choice obviously results in no loss of generality in our considerations of waves in an unbounded medium in this chapter. In Chapter 3, however, we consider electromagnetic wave propagation in the presence of planar boundaries, resulting in their reflection and refraction. When uniform plane waves are normally incident on an infinite planar boundary, the direction of propagation of the wave can be chosen to be along one of the principal axes without any loss of generality. However, the general case of reflection and refraction of waves at infinite planar boundaries involves waves that are incident on the boundaries at an arbitrary angle, in which case we shall need to work with uniform plane waves propagating in arbitrary directions. In this section, we derive general expressions for such waves and introduce the notation to be used. We also briefly discuss nonuniform plane waves, because such waves are encountered in later chapters.

2.6.1 Uniform Plane Waves in an Arbitrary Direction

Before we proceed with deriving some general relationships concerning uniform plane waves, it is useful to note that the physical nature of the waves we consider here is still the same as those studied in Sections 1.1 through 2.5. In other words, the waves considered here are uniform plane waves in which the field vectors are confined to an infinite plane (which is the surface of constant phase) and exhibit spatial variations only in the direction perpendicular to that plane.

Consider the vector wave equation [2.8] for time-harmonic fields in a lossless medium, repeated here for convenience:

$$\nabla^2\mathbf{E} + \beta^2\mathbf{E} = 0 \qquad\qquad [2.56]$$

---

[72]W. Sinnema, *Electronic Transmission Technology,* Prentice-Hall, 1988.

where $\beta = \omega\sqrt{\mu\epsilon}$. We know that one solution of [2.56] is

$$\mathbf{E}(z) = \mathbf{E}_0 e^{-j\beta z} = [\hat{\mathbf{x}}E_{0x} + \hat{\mathbf{y}}E_{0y}]e^{-j\beta z}$$

A more general solution representing a wave propagating in an arbitrary direction $\mathbf{r} = \hat{\mathbf{x}}x + \hat{\mathbf{y}}y + \hat{\mathbf{z}}z$ is

$$\mathbf{E}(x, y, z) = \mathbf{E}_0 e^{-j\beta_x x - j\beta_y y - j\beta_z z} = [\hat{\mathbf{x}}E_{0x} + \hat{\mathbf{y}}E_{0y} + \hat{\mathbf{z}}E_{0z}]e^{-j\beta z} \qquad [2.57]$$

It can be shown by direct substitution in [2.56] that [2.57] satisfies the wave equation as long as we have

$$\beta^2 = \beta_x^2 + \beta_y^2 + \beta_z^2 = \omega^2 \mu\epsilon$$

We can rewrite [2.57] in a more compact form by using vector notation for the exponent:

$$\boxed{\mathbf{E}(\mathbf{r}) = \mathbf{E}_0 e^{-j\beta \hat{\mathbf{k}}\cdot\mathbf{r}}} \qquad [2.58]$$

where $\hat{\mathbf{k}}$ is the unit vector in the direction of propagation, as shown in Figure 2.30. Note that $\hat{\mathbf{k}}$ is given by

$$\hat{\mathbf{k}} = \frac{\hat{\mathbf{x}}\beta_x + \hat{\mathbf{y}}\beta_y + \hat{\mathbf{z}}\beta_z}{[\beta_x^2 + \beta_y^2 + \beta_z^2]^{1/2}} = \frac{\hat{\mathbf{x}}\beta_x + \hat{\mathbf{y}}\beta_y + \hat{\mathbf{z}}\beta_z}{\omega\sqrt{\mu\epsilon}}$$

The vector $\hat{\mathbf{k}}\beta$ is the wavenumber vector, commonly referred in other texts as $\mathbf{k}$. For a uniform plane wave propagating in the z direction as considered in previous sections,

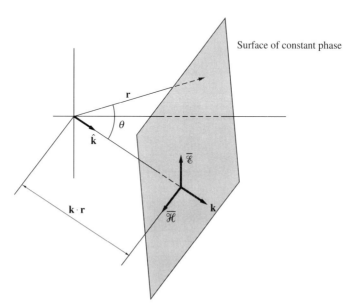

FIGURE 2.30.   Constant-phase surface. The planar surface of constant phase, characteristic of a plane wave, and the orientation of the field vectors.

the phase constant β represents the space rate of change of wave phase (in units of rad-m^{-1}) along the z axis. In the general case of a uniform plane wave propagating in an arbitrary direction, the component phase constants β_x, β_y and β_z represent the space rates of change of the wave phase as measured along the respective axes. These component phase constants and the corresponding component wavelengths are further discussed below in connection with Figure 2.31.

We note that $\hat{\mathbf{k}} \cdot \mathbf{r}$ = constant represents the equation for the planes over which the wave phase is a constant, referred to as the planes of constant phase, or phase fronts (Figure 2.30). By substituting [2.58] into [1.11b] for a source-free medium (i.e., $\nabla \cdot \mathbf{E} = 0$), we find

$$\nabla \cdot \mathbf{E}_0 e^{-j\beta\hat{\mathbf{k}}\cdot\mathbf{r}} = 0$$

$$\mathbf{E}_0 \cdot \nabla(e^{-j\beta\hat{\mathbf{k}}\cdot\mathbf{r}}) = 0$$

$$\mathbf{E}_0 \cdot \left(\hat{\mathbf{x}}\frac{\partial}{\partial x} + \hat{\mathbf{y}}\frac{\partial}{\partial y} + \hat{\mathbf{z}}\frac{\partial}{\partial z}\right) e^{-j(\beta_x x + \beta_y y + \beta_z z)} = 0$$

$$\mathbf{E}_0 \cdot [-j(\hat{\mathbf{x}}\beta_x + \hat{\mathbf{y}}\beta_y + \hat{\mathbf{z}}\beta_z) e^{-j(\beta_x x + \beta_y y + \beta_z z)}] = 0$$

$$\mathbf{E}_0 \cdot (-j\beta\hat{\mathbf{k}} e^{-j\beta\hat{\mathbf{k}}\cdot\mathbf{r}}) = 0$$

$$-j\beta(\mathbf{E}_0 \cdot \hat{\mathbf{k}}) e^{-j\beta\hat{\mathbf{k}}\cdot\mathbf{r}} = 0$$

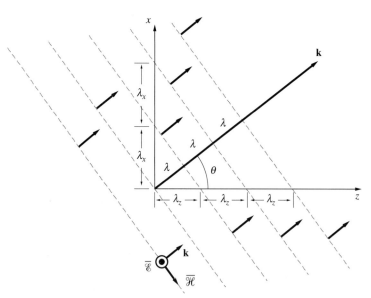

FIGURE 2.31. **A uniform plane wave propagating at an angle θ to the z axis.** Although the wavelength in the propagation direction is λ, the projections of the constant phase fronts on the z and x axes are separated by λ_z and λ_x respectively.

which yields

$$\boxed{\hat{\mathbf{k}} \cdot \mathbf{E}_0 = 0}$$ [2.59]

Thus, $\mathbf{E}_0$ must be transverse to the propagation direction for [1.11b] ($\nabla \cdot \mathbf{E} = 0$) to be satisfied. (This result is the generalization of what we saw earlier for the simplest uniform plane wave propagating in the z direction, where we had to have $E_z = 0$ for the field to have zero divergence.)

To find the magnetic field $\mathbf{H}$ in terms of the electric field $\mathbf{E}$, we use Faraday's law (equation [1.11a]):

$$\mathbf{H}(\mathbf{r}) = \frac{1}{-j\omega\mu}\nabla \times \mathbf{E}(\mathbf{r})$$ [2.60]

We can further manipulate [2.60] as follows

$$\mathbf{H} = \frac{1}{-j\omega\mu}\nabla \times \mathbf{E}_0 e^{-j\beta\hat{\mathbf{k}}\cdot\mathbf{r}}$$

$$= \frac{-1}{-j\omega\mu}\mathbf{E}_0 \times \nabla e^{-j\beta\hat{\mathbf{k}}\cdot\mathbf{r}} \qquad \text{since } \mathbf{E}_0 \text{ is a constant vector}$$

$$= \frac{-1}{-j\omega\mu}\mathbf{E}_0 \times (-j\beta\hat{\mathbf{k}})e^{-j\beta\hat{\mathbf{k}}\cdot\mathbf{r}} \qquad \text{see derivation of [2.59] above.}$$

$$= \frac{\beta}{\omega\mu}\hat{\mathbf{k}} \times \mathbf{E}_0 e^{-j\beta\hat{\mathbf{k}}\cdot\mathbf{r}}$$

$$= \frac{1}{\eta}\hat{\mathbf{k}} \times \underbrace{\mathbf{E}_0 e^{-j\beta\hat{\mathbf{k}}\cdot\mathbf{r}}}_{\mathbf{E}(\mathbf{r})}$$

or

$$\boxed{\mathbf{H}(\mathbf{r}) = \frac{1}{\eta}\hat{\mathbf{k}} \times \mathbf{E}(\mathbf{r})}$$ [2.61]

The magnetic field vector $\mathbf{H}$ is thus perpendicular to both the electric field vector $\mathbf{E}$ and $\hat{\mathbf{k}}$, as shown in Figure 2.30.

Equations [2.58] and [2.61] constitute general expressions for transverse electromagnetic (TEM) waves propagating in an arbitrary direction $\hat{\mathbf{k}}$. These expressions will be useful in the discussion of reflection and refraction of uniform plane waves obliquely incident on dielectric or conductor interfaces, which is the topic of the next chapter.

As an example, we consider a uniform plane wave propagating in the xz plane (i.e., in a direction perpendicular to the y axis) but in a direction oriented at an angle θ with respect to the z axis, as shown in Figure 2.31. Assuming that the wave electric field is in the y direction, and noting that we have $\hat{\mathbf{k}} = \hat{\mathbf{x}}\sin\theta + \hat{\mathbf{z}}\cos\theta$, the expression for the wave electric field phasor can be written as

$$\mathbf{E}(\mathbf{r}) = \hat{\mathbf{y}}E_0 e^{-j\beta\hat{\mathbf{k}}\cdot\mathbf{r}} = \hat{\mathbf{y}}E_0 e^{-j\beta(x\sin\theta + z\cos\theta)}$$

whereas the wave magnetic field can be found using [2.61] as

$$\mathbf{H}(\mathbf{r}) = \frac{1}{\eta}\hat{\mathbf{k}} \times [\hat{\mathbf{y}}E_0 e^{-j\beta(x\sin\theta + z\cos\theta)}]$$

$$= \frac{1}{\eta}(\hat{\mathbf{x}}\sin\theta + \hat{\mathbf{z}}\cos\theta) \times [\hat{\mathbf{y}}E_0 e^{-j\beta(x\sin\theta + z\cos\theta)}]$$

$$= \frac{E_0}{\eta}(-\hat{\mathbf{x}}\cos\theta + \hat{\mathbf{z}}\sin\theta)e^{-j\beta(x\sin\theta + z\cos\theta)}$$

$$\mathbf{H}(\mathbf{r}) = \frac{E_0}{\eta}(-\hat{\mathbf{x}}\cos\theta + \hat{\mathbf{z}}\sin\theta)e^{-j(\beta_x x + \beta_z z)}$$

where $\beta_x = \beta\sin\theta$ and $\beta_y = \beta\cos\theta$. Note from Figure 2.31 that the expression for $\mathbf{H}(\mathbf{r})$ could also have been written by inspection, since $\mathbf{E} \times \mathbf{H}$ must be in the direction of $\hat{\mathbf{k}}$.

Further consideration of Figure 2.31 allows us to better understand the properties of a uniform plane wave. It was mentioned above that the component phase constants $\beta_{x,y,z}$ represent the space rate of change of the phase of the wave along the principal axes. Instead of considering the rate of change of wave phase, we can consider the distances between successive equiphase surfaces on which the wave phase differs by exactly 2π. These distances are *apparent wavelengths,* because the phase shift is linearly proportional to distance in any direction. Along the propagation direction, we have the conventional wavelength λ over which the wave phase changes by 2π. Measured along any other direction, the distances between equiphase surfaces are greater, as is apparent from Figure 2.31. The apparent wavelengths along the x and z axes are respectively λ_x and λ_z, which are both greater than λ. In fact, it is easy to see from Figure 2.31 that we have

$$\frac{1}{\lambda_x^2} + \frac{1}{\lambda_z^2} = \frac{1}{\lambda^2}$$

which is a direct consequence of

$$\beta^2 = \beta_x^2 + \beta_z^2$$

Similarly, because a particular planar phase front moves uniformly (i.e., with same speed at all points on the plane) with time along the propagation direction at the phase velocity $v_p = \omega/\beta$, its intercept along any other direction also moves uniformly in time but with a *greater* speed. In the time of one full cycle of the wave (i.e., $T_p = 1/f$), a phase front moves a distance along the $\hat{\mathbf{k}}$ direction of $\lambda = 2\pi/\beta$ and a distance along the x direction of $\lambda_x = \lambda/\sin\theta = 2\pi/(\beta\sin\theta) = 2\pi/\beta_x$. Thus, the various *apparent phase velocities* along the principal axes are in general given by

$$v_{x,y,\text{ or }z} = \frac{\omega}{\beta_{x,y,\text{ or }z}}$$

and, in the case shown in Figure 2.31, are related to one another as

$$\frac{1}{v_x^2} + \frac{1}{v_z^2} = \frac{1}{v_p^2} = \epsilon\mu$$

Component phase velocities greater than the speed of propagation in the medium can be observed in water waves approaching at an angle to a breakwater or a shoreline. Visualize water waves striking a beach obliquely. The distance between successive crests (i.e., maxima), for example, is large along the beach, especially if the waves are only slightly off normal incidence. To keep up with a given crest, one has to run much faster along the beach than the speed with which the waves progress in their own directions of propagation.

2.6.2   Nonuniform Plane Waves

Nonuniform plane waves are those electromagnetic waves for which the amplitudes of the electric and magnetic fields of the wave are not constant over the planes of constant phase. An example of a *nonuniform* plane wave is

$$\overline{\mathcal{H}}(z, t) = \hat{\mathbf{x}} \underbrace{C_1 \cos(\pi y)}_{H_0(y)} \cos(\omega t - \beta z)$$

which is a plane wave, because the constant phase surfaces are panes of $z =$ constant (i.e., planes parallel to the xy plane, but it is nonuniform, because the amplitude of the field is a function of y and hence varies as a function of position along the planes of constant phase. We shall encounter nonuniform plane waves when we discuss reflection of uniform plane waves from obliquely oriented boundaries in Chapter 3, and also when we study guiding of electromagnetic waves by metallic and dielectric boundaries in Chapters 4 and 5.

2.7   NONPLANAR ELECTROMAGNETIC WAVES

We have seen that uniform plane waves are natural solutions of Maxwell's equations and the wave equations derived from them. The general character of these uniform plane waves is apparent in the functional form of any one of the electric or magnetic field components, namely

$$\overline{\mathcal{E}}(\mathbf{r}, t) = \hat{\mathbf{k}} E_0 \cos(\omega t - \beta \hat{\mathbf{k}} \cdot \mathbf{r})$$

As discussed before, such waves[73] are *plane* waves because the surfaces of constant phase, i.e., the surfaces over which $\overline{\mathcal{E}}(\mathbf{r}, t)$ is constant, are planes. These planar surfaces are defined by $\hat{\mathbf{k}} \cdot \mathbf{r} =$ constant. The waves are *uniform* because the field $\overline{\mathcal{E}}(\mathbf{r}, t)$ does not vary as a function of position over the surfaces of constant phase; in other words, E_0 is not a function of spatial coordinates.

---

[73]Note that the functional form of the quantity does not need to be a cosinusoid for it to be a uniform plane wave. In the general case, an electromagnetic field is considered to be a uniform plane wave if the field quantities vary with space and time as $A f(\omega t - \beta \hat{\mathbf{k}} \cdot \mathbf{r})$, where $f(\cdot)$ is any function and A is a constant.

Examples of *nonplanar* waves are those electromagnetic fields for which the variation of the field quantities exhibits a more complicated dependence on spatial coordinates than is implied by $(\omega t - \beta \hat{\mathbf{k}} \cdot \mathbf{r})$. In general, the electric field vector of a nonplanar time-harmonic wave can be expressed as

$$\overline{\mathscr{E}}(\mathbf{r}, t) = \overline{\mathscr{E}}_0 \cos[\omega t - g(\mathbf{r})]$$

where $g(\cdot)$ is some arbitrary function. The surfaces of constant phase are now given by

$$g(\mathbf{r}) = \text{constant}$$

which could be any arbitrary surface, depending on the function $g(\cdot)$. For example, if $g(\mathbf{r}) = g(x, y, z) = xy$, then the surfaces of constant phase are hyperbolic cylinders, with infinite extent in z (i.e., no variation in z) and with the functional form in planes parallel to the xy plane given as $xy = $ constant. Many of the properties of plane waves that we may take for granted may not be applicable when the wave fronts are nonplanar. For example, the speed of wave propagation, or the phase velocity v_p, may not be constant in space and may instead be a function of position.[74]

2.7.1 Uniform Cylindrical Waves

A *uniform cylindrical* wave is one for which the equiphase surfaces form a family of coaxial cylinders (i.e., cylindrical) and the amplitude of which is the same at all points on such cylinders (i.e., uniform). The equations governing uniform cylindrical waves can be obtained from Maxwell's equations, following a procedure very similar to that followed in Section 2.2 for uniform plane waves. First, we choose the axis of cylindrical symmetry for such waves to be the z axis of our cylindrical coordinate system (r, ϕ, z). Second, we assume that the field quantities vary only with one of the coordinates, namely the r direction, and have components only in the directions perpendicular to r. In other words, we assume that $\partial/\partial\phi = 0$, $\partial/\partial z = 0$ for all field components and that $E_r = 0$ and $H_r = 0$. Under these conditions, the governing equations reduce to the following:

Cylindrical Waves	Planar Waves
$\dfrac{dE_z}{dr} = j\omega\mu H_\phi$	$\dfrac{dE_x}{dz} = -j\omega\mu H_y$
$\dfrac{d(rE_\phi)}{dr} = -j\omega\mu(rH_z)$	$\dfrac{dH_y}{dz} = -j\omega\epsilon E_x$
$\dfrac{d(rH_\phi)}{dr} = j\omega\epsilon(rE_z)$	$\dfrac{dE_y}{dz} = +j\omega\mu H_x$
$\dfrac{dH_z}{dr} = -j\omega\epsilon E_\phi$	$\dfrac{dH_x}{dz} = +j\omega\epsilon E_y$

---

[74]See M. Born and E. Wolf, *Principles of Optics,* 5th ed., Section 1.3, Pergamon Press, 1975.

Note that these equations are very similar to the corresponding rectangular coordinate equations given on the right. Based on the above, uniform cylindrical waves are transverse electromagnetic (TEM). Eliminating H_ϕ from the first and the third equations on the left, we find

$$\frac{d^2 E_z}{dr^2} + \frac{1}{r}\frac{dE_z}{dr} + \beta^2 E_z = 0$$

where $\beta = \omega\sqrt{\mu\epsilon}$.

The above equation is the simplest form (i.e., zeroth order) Bessel equation, and its general solution can be written as a linear combination of two independent solutions, namely $J_0(\beta r)$ and $Y_0(\beta r)$, where $J_0(\cdot)$ and $Y_0(\cdot)$ are the Bessel functions of the zeroth order and respectively of the first and second kind.[75] To obtain functions that are propagating waves, we can construct a linear combination[76] of the two solutions of the form

$$E_z(r) = A\underbrace{[J_0(\beta r) - jY_0(\beta r)]}_{\text{Hankel function}}$$

where A is a constant. At large distances, $r \gg 1$, we can use the asymptotic forms of the Bessel functions to show[77] that

$$E_z(r) \simeq \frac{A}{(\omega^2\mu\epsilon)^{1/4}}\sqrt{\frac{2}{\pi r}}e^{-j(\beta r - \frac{\pi}{4})}$$

which represents an outward-bound wave, as can be seen by finding the corresponding space-time function

$$\mathcal{E}_z(r, t) = \frac{A}{(\omega^2\mu\epsilon)^{1/4}}\sqrt{\frac{2}{\pi r}}\cos\left(\omega t - \beta r + \frac{\pi}{4}\right)$$

Note that for large values of r the uniform cylindrical wave is very similar to a uniform plane wave, except that its amplitude steadily decreases as $r^{-1/2}$. For small deviations around any faraway point, the cylindrical phase fronts are indistinguishable from planar wave fronts, as can be seen from Figure 2.32. Further discussion on cylindrical waves can be found elsewhere.[78]

---

[75] See M. Abramowitz and I. A. Stegun (Eds.), *Handbook of Mathematical Functions*, Dover, New York, 1964.

[76] This is analogous to constructing $e^{-j\beta z} = \cos(\beta z) - j\sin(\beta z)$ type of solutions as linear combinations of the basic $\cos(\beta z)$ and $\sin(\beta z)$ solutions of the wave equation for rectangular coordinates.

[77] See p. 192 and 195 of H. B. Dwight, *Table of Integrals and Other Mathematical Data*, Macmillan, New York, 1961.

[78] J. A. Stratton, *Electromagnetic Theory*, Chapter 6, McGraw-Hill, 1941; S. A. Schelkunoff, *Electromagnetic Waves*, Chapter 8, Van Nostrand Company, Boston, 1943.

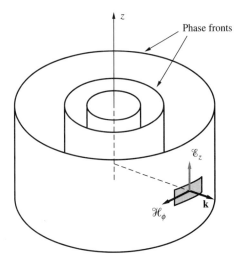

FIGURE 2.32. **Constant phase surfaces of a uniform cylindrical wave.** At large distances from the source (i.e., the z axis) portions of the cylindrical phase fronts (e.g., the shaded region) look planar.

2.7.2 Uniform Spherical Waves

Similar to uniform waves in planar and cylindrical geometries, uniform spherical waves are those for which the field quantities vary only as a function of radial distance r, and for which the electric and magnetic fields are confined entirely to the direction transverse to the radial direction. In other words, we have $\mathcal{E}_r = 0$ and $\mathcal{H}_r = 0$. Analysis of these waves can be carried out in a manner similar to that of planar waves. In general, one finds that $r\mathcal{E}_\theta$ and $r\mathcal{H}_\phi$ vary respectively in the same manner as $\mathcal{E}_x$ and $\mathcal{H}_y$, while $r\mathcal{E}_\phi$ and $r\mathcal{H}_\theta$ behave as $\mathcal{E}_y$ and $\mathcal{H}_x$. The general functional form of the field quantities is

$$\mathcal{E}_\theta(r, \theta, \phi, t) = T(\theta, \phi)\cos(\omega t - \beta r)$$

where $T(\theta, \phi)$ is the solution of the spherical harmonic equations.[79] If we require in spherical coordinates that $\mathcal{E}_\theta$ be a function only of r and θ, we find that, off the $\theta = 0$ axis, the wave equation allows solutions of the type

$$\mathcal{E}_\theta = C_1\frac{e^{-j\beta r}}{r\sin\theta} + C_2\frac{e^{+j\beta r}}{r\sin\theta}$$

with $\mathcal{H}_\phi$ perpendicular to $\mathcal{E}_\theta$ and to the radial direction and with $|\mathcal{H}_\phi| = |\mathcal{E}_\theta|/\eta$, where $\eta = \sqrt{\mu/\epsilon}$. The wave propagates along the radial direction, and the field components are transverse to the direction of propagation. The surfaces of constant phase are concentric spheres.

---

[79]J. A. Stratton, *Electromagnetic Theory,* Chapter 5, McGraw-Hill, 1941; S. A. Schelkunoff, *Electromagnetic Waves,* Section 8.12, Van Nostrand Company, Boston, 1943.

2.8   SUMMARY

This chapter discussed the following topics:

- **Uniform plane waves.** The characteristics of electromagnetic waves in source-free and simple media are governed by the wave equation derived from Maxwell's equations:

$$\nabla^2 \mathscr{E} - \mu\epsilon \frac{\partial^2 \mathscr{E}}{\partial t^2} = 0$$

Uniform plane waves are the simplest type of solution of the wave equation, with the electric and magnetic fields both lying in the direction transverse to the direction of propagation (i.e., z direction):

$$\mathscr{E}_x(z, t) = p_1(z - v_p t)$$

$$\mathscr{H}_y(z, t) = \frac{1}{\eta} p_1(z - v_p t)$$

where $v_p = 1/\sqrt{\mu\epsilon}$ is the phase velocity, $\eta = \sqrt{\mu/\epsilon}$ is the intrinsic impedance of the medium, and p_1 is an arbitrary function.

- **Time-harmonic waves in a lossless medium.** The electric and magnetic fields of a time-harmonic uniform plane wave propagating in the z direction in a simple lossless medium are

$$\mathbf{E}(z) = \hat{\mathbf{x}} C_1 e^{-j\beta z} \qquad \overline{\mathscr{E}}(z, t) = \hat{\mathbf{x}} C_1 \cos(\omega t - \beta z)$$

and

$$\mathbf{H}(z) = \hat{\mathbf{y}} \frac{C_1}{\eta} e^{-j\beta z} \qquad \overline{\mathscr{H}}(z, t) = \hat{\mathbf{y}} \frac{C_1}{\eta} \cos(\omega t - \beta z)$$

where $\mathbf{E}$ and $\mathbf{H}$ are the phasor quantities, and $\beta = \omega\sqrt{\mu\epsilon}$ is the propagation constant.

- **Uniform plane waves in a lossy medium.** The electric and magnetic fields of a uniform plane wave propagating in the z direction in a lossy medium are

$$\mathbf{E}(z) = \hat{\mathbf{x}} C_1 e^{-\gamma z} \qquad \overline{\mathscr{E}}(z, t) = \hat{\mathbf{x}} C_1 e^{-\alpha z} \cos(\omega t - \beta z)$$

and

$$\mathbf{H}(z) = \hat{\mathbf{y}} \frac{C_1}{\eta_c} e^{-\gamma z} \qquad \overline{\mathscr{H}}(z, t) = \hat{\mathbf{y}} \frac{C_1 e^{-\alpha z}}{|\eta_c|} \cos(\omega t - \beta z - \phi_\eta)$$

where

$$\gamma = \sqrt{j\omega\mu(\sigma + j\omega\epsilon)} = \alpha + j\beta$$

is the complex propagation constant and

$$\eta_c = \sqrt{\mu/[\epsilon - j(\sigma/\omega)]} = |\eta_c|e^{j\phi_\eta}$$

is the complex intrinsic impedance of the medium.

The characteristics of uniform plane waves in a lossy medium are largely determined by the loss tangent $\tan\delta_c = \sigma/(\omega\epsilon)$. When $\tan\delta_c \ll 1$, the material is classified as a low-loss dielectric, and the propagation constant β is only negligibly different from that in a lossless medium. The attenuation and propagation constants α, β, and the intrinsic impedance for a low-loss dielectric are given by

$$\alpha \simeq \frac{\omega\epsilon''}{2}\sqrt{\frac{\mu}{\epsilon'}} \qquad \beta \simeq \omega\sqrt{\mu\epsilon'}\left[1 + \frac{1}{8}\left(\frac{\epsilon''}{\epsilon'}\right)^2\right] \qquad \eta_c \simeq \sqrt{\frac{\mu}{\epsilon'}}\left(1 + j\frac{\epsilon''}{2\epsilon'}\right)$$

Material media for which $\tan\delta_c \gg 1$ are classified as good conductors. The attenuation and propagation constants α and β and the intrinsic impedance for a good conductor are given by

$$\alpha = \beta \simeq \sqrt{\frac{\omega\mu\sigma}{2}} \qquad \eta_c \simeq \sqrt{\frac{\mu\omega}{\sigma}}e^{j45°}$$

Another important parameter for good conductors is the skin depth δ, given by

$$\delta = \frac{1}{\alpha} \simeq \frac{1}{\sqrt{\pi f\mu\sigma}}$$

The skin depth for metallic conductors is typically extremely small, being $\sim 3.82\ \mu$m for copper at 300 MHz.

■ **Electromagnetic power flow and the Poynting vector.** Poynting's theorem states that electromagnetic power flow entering into a given volume through the surface enclosing it equals the sum of the time rates of increase of the stored electric and magnetic energies and the ohmic power dissipated within the volume. The instantaneous power density of the electromagnetic wave is identified as

$$\overline{\mathscr{S}}(z, t) = \overline{\mathscr{E}}(z, t) \times \overline{\mathscr{H}}(z, t)$$

although in most cases the quantity of interest is the time-average power, which can be found either from $\overline{\mathscr{S}}(z, t)$ or directly from the phasor fields $\mathbf{E}$ and $\mathbf{H}$ as

$$\mathbf{S}_{av}(z) = \frac{1}{T_p}\int_0^{T_p}\overline{\mathscr{S}}(z, t)\,dt \qquad \mathbf{S}_{av}(z) = \frac{1}{2}\mathscr{R}e\{\mathbf{E} \times \mathbf{H}^*\}$$

The time-average Poynting flux for a uniform plane wave propagating in the z direction in an unbounded medium is

$$\mathbf{S}_{av}(z) = \hat{\mathbf{z}}\frac{C_1^2}{2|\eta_c|}e^{-2\alpha z}\cos(\phi_\eta)$$

■ **Wave polarization.** The polarization of an electromagnetic wave describes the behavior of its electric field vector as a function of time, at a fixed point in space. In the general case of an elliptically polarized wave, the two components of the electric field vector of a wave propagating in the z direction are given by

$$\mathscr{E}_x(z, t) = C_{1x} \cos(\omega t - \beta z)$$

$$\mathscr{E}_y(z, t) = C_{1y} \cos(\omega t - \beta z + \zeta)$$

A wave is linearly polarized if $\zeta = 0$ or π, and is circularly polarized if $C_{1x} = C_{1y}$ and $\zeta = \pm\pi/2$, with the negative sign corresponding to a right-hand circularly polarized wave. For any other value of ζ, the wave is elliptically polarized, in the left-hand sense for $0 < \zeta < \pi$, and right-hand sense for $\pi < \zeta < 2\pi$.

■ **Uniform plane wave propagating in arbitrary direction.** The electric and magnetic field phasors of a uniform plane wave propagating in an arbitrary direction $\hat{\mathbf{k}}$ in a simple lossless medium are given as

$$\mathbf{E}(\mathbf{r}) = \mathbf{E}_0 e^{-j\beta\hat{\mathbf{k}}\cdot\mathbf{r}}$$

$$\mathbf{H}(\mathbf{r}) = \frac{1}{\eta}\hat{\mathbf{k}} \times \mathbf{E}(\mathbf{r})$$

2.9   PROBLEMS

2-1. Uniform plane wave. The electric field of a uniform plane wave in air is given by

$$\overline{\mathscr{E}}(x, t) = \hat{\mathbf{z}}15 \times 10^{-2} \cos(1.7\pi \times 10^9 t - \beta x) \quad \text{V-m}^{-1}$$

(a) Find the phase constant β and the wavelength λ. (b) Sketch $\mathscr{E}_z(x, t)$ as a function of t at $x = 0$ and $x = \lambda/4$. (c) Sketch $\mathscr{E}_z(x, t)$ as a function of x at $t = 0$ and $t = \pi/(\omega)$.

2-2. Uniform plane wave. A uniform plane wave is traveling in the x direction in air with its magnetic field oriented in the z direction. At the instant $t = 0$, the wave magnetic field has two adjacent zero values, observed at locations $x = 2.5$ cm and $x = 7.5$ cm, with a maximum value of 70 mA-m^{-1} at $x = 5$ cm. (a) Find the wave magnetic field $\mathscr{H}(x, t)$ and its phasor $\mathbf{H}(x)$. (b) Find the corresponding wave electric field $\overline{\mathscr{E}}(x, t)$ and its phasor $\mathbf{E}(x)$.

2-3. TV broadcast signal. The magnetic field of a TV broadcast signal propagating in air is given as

$$\overline{\mathscr{H}}(z, t) = \hat{\mathbf{x}}0.1 \sin(\omega t - 9.3z) \text{ mA-m}^{-1}$$

(a) Find the wave frequency $f = \omega/(2\pi)$. (b) Find the corresponding $\overline{\mathscr{E}}(z, t)$.

2-4. Uniform plane wave. The electric field phasor of an 18 GHz uniform plane wave propagating in free space is given by

$$\mathbf{E}(y) = \hat{\mathbf{x}}5e^{j\beta y} \text{ V-m}^{-1}$$

(a) Find the phase constant, β, and wavelength, λ. (b) Find the corresponding magnetic field phasor $\mathbf{H}(y)$.

2-5. Lossless nonmagnetic medium. The magnetic field component of a uniform plane wave propagating in a lossless simple nonmagnetic medium ($\mu = \mu_0$) is given by

$$\overline{\mathscr{B}}(x, t) = \hat{\mathbf{y}}0.25 \sin[2\pi(10^8 t + 0.5x - 0.125)] \, \mu\text{T}$$

(a) Find the frequency, wavelength, and the phase velocity. (b) Find the relative permittivity, ϵ_r, and the intrinsic impedance, η, of the medium. (c) Find the corresponding $\overline{\mathscr{E}}$. (d) Find the time-average power density carried by this wave.

2-6. Short electric dipole antenna. Consider a short electric dipole antenna of length l, carrying a current $I_0 \cos(\omega t)$, located at the origin and oriented along the z axis in free space. The electric field component of the wave at distances r very much greater than the wavelength is approximately given by

$$\mathbf{E}(r, \theta) = \hat{\boldsymbol{\theta}} j \frac{30\beta I_0 l}{r} \sin\theta \, e^{-j\beta r}$$

where $\beta = \omega\sqrt{\mu_0\epsilon_0}$. (a) Find the corresponding magnetic field, $\mathbf{H}(r, \theta)$. (b) Find the time-averaged Poynting vector, $\mathbf{S}_{\text{av}}$. (c) Find the total power radiated by the dipole source by integrating the radial Poynting vector over a spherical surface centered at the dipole.

2-7. Maxwell's Equations. The electric field component of an electromagnetic wave in free space is given by

$$\overline{\mathscr{E}}(y, z, t) = \hat{\mathbf{x}}E_0 \cos(ay)\cos(\omega t - bz)$$

(a) Find the corresponding magnetic field $\overline{\mathscr{H}}(y, z, t)$. (b) Find the relationship between the constants a, b, and ω such that all of Maxwell's equations are satisfied. (c) Assuming that this wave may be regarded as a sum of two uniform plane waves, determine the direction of propagation of the two component waves.

2-8. Cellular phones. The electric field component of a uniform plane wave in air emitted by a mobile communication system is given by

$$\overline{\mathscr{E}}(x, y, t) = \hat{\mathbf{z}}100 \cos(\omega t + 4.8\pi x - 3.6\pi y + \theta) \, \text{mV-m}^{-1}$$

(a) Find the frequency f and wavelength λ. (b) Find θ if $\overline{\mathscr{E}}(x, y, t) \simeq \hat{\mathbf{z}}80.9 \, \text{mV-m}^{-1}$ at $t = 0$ and at $y = 2x = 0.2$ m. (c) Find the corresponding magnetic field $\overline{\mathscr{H}}(x, y, t)$. (d) Find the time-average power density carried by this wave.

2-9. Superposition of two waves. The sum of the electric fields of two time-harmonic (sinusoidal) electromagnetic waves propagating in opposite directions in air is given as

$$\overline{\mathscr{E}}(z, t) = \hat{\mathbf{x}}95 \sin(\beta z)\sin(21 \times 10^9 \pi t) \, \text{mV-m}^{-1}$$

(a) Find the constant β. (b) Find the corresponding $\overline{\mathscr{H}}$. (c) Assuming that this wave may be regarded as a sum of two uniform plane waves, determine the direction of propagation of the two component waves.

2-10. Unknown material. The intrinsic impedance and the wavelength of a uniform plane wave traveling in an unknown dielectric at 3 GHz are measured to be $\sim$194Ω and $\sim$5.14 cm respectively. Determine the constitutive parameters (i.e., ϵ_r and μ_r) of the material.

2-11. Uniform plane wave. A 10 GHz uniform plane wave with maximum electric field of 100 V-m^{-1} propagates in air in the direction of a unit vector given by $\hat{\mathbf{u}} = 0.6\hat{\mathbf{x}} - 0.8\hat{\mathbf{y}}$. The wave magnetic field has only an z component, which is approximately equal to 100 mA-m^{-1} at $x = y = t = 0$. Find $\overline{\mathscr{E}}$ and $\overline{\mathscr{H}}$.

2-12. Standing waves. The magnetic field phasor of an electromagnetic wave in air is given by

$$\mathbf{B}(z) = \hat{x}0.8 \sin(200\pi z) \quad \text{mT}$$

(a) Find the wavelength, λ. (b) Find the corresponding electric field $\mathbf{E}(z)$. (c) Is this a traveling wave?

2-13. Propagation through wet versus dry earth. Assume the conductivities of wet and dry earth to be $\sigma_{\text{wet}} = 10^{-2}$ S-m^{-1} and $\sigma_{\text{dry}} = 10^{-4}$ S-m^{-1}, respectively, and the corresponding permittivities to be $\epsilon_{\text{wet}} = 10\epsilon_0$ and $\epsilon_{\text{dry}} = 3\epsilon_0$. Both media are known to be nonmagnetic (i.e., $\mu_r = 1$). Determine the attenuation constant, phase constant, wavelength, phase velocity, penetration depth, and intrinsic impedance for a uniform plane wave of 20 MHz propagating in (a) wet earth, (b) dry earth. Use approximate expressions whenever possible.

2-14. Propagation in lossy media. (a) Show that the penetration depth (i.e., the depth at which the field amplitude drops to $1/e$ of its value at the surface) in a lossy medium with $\mu = \mu_0$ is approximately given by

$$d \simeq \frac{0.225(c/f)}{\sqrt{\epsilon_r'}\sqrt{\sqrt{1 + \tan^2 \delta_c} - 1}}$$

where $\tan \delta_c$ is the loss tangent.[80] (b) For $\tan \delta_c \ll 1$, show that the above equation can be further approximated as $d \simeq [0.318(c/f)]/[\tan \delta_c \sqrt{\epsilon_r'}]$. (c) Assuming the properties of fat tissue at 2.45 GHz to be $\sigma = 0.12$ S-m^{-1}, $\epsilon_r = 5.5$, and $\mu_r = 1$, find the penetration depth of a 2.45 GHz plane wave in fat tissue using both expressions, and compare the results.

2-15. Glacier ice. A glacier is a huge mass of ice that, unlike sea ice, sits over land. Glaciers are formed in the cold polar regions and in high mountains. They spawn billions of tons of icebergs each year from tongues that reach the sea. The icebergs drift over an area of 70,000,000 km^2, which is more than 20% of the ocean area, and pose a serious threat to navigation and offshore activity in many areas of the world. Glacier ice is a low-loss dielectric material that permits significant microwave penetration.[81] The depth of penetration of an electromagnetic wave into glacier ice with loss tangent of $\tan \delta_c \simeq 0.001$ at X-band (assume 3 cm air wavelength) is found to be ~5.41 m. (a) Calculate the dielectric constant and the effective conductivity of the glacier ice. (Note that $\mu_r = 1$.) (b) Calculate the attenuation of the signal measured in dB-m^{-1}.

2-16. Good dielectric. Alumina (Al_2O_3) is a low-loss ceramic material that is commonly used as a substrate for printed circuit boards. At 10 GHz, the relative permittivity and loss tangent of alumina are approximately equal to $\epsilon_r = 9.7$ and $\tan \delta_c = 2 \times 10^{-4}$. Assume $\mu_r = 1$. For a 10 GHz uniform plane wave propagating in a sufficiently large sample of alumina, determine the following: (a) attenuation constant, α, in np-m^{-1}; (b) penetration depth, d; (c) total attenuation in dB over thicknesses of 1 cm and 1 m.

---

[80]J. M. Osepchuk, Sources and basic characteristics of microwave radiation, *Bull. N. Y. Acad. Med.,* 55(11), pp. 976–998, December 1979.

[81]E. O. Lewis, B. W. Currie, and S. Haykin, *Detection and Classification of Ice,* Chapter 3, John Wiley & Sons, New York, 1987.

2-17. Concrete wall. The effective complex dielectric constant of walls in buildings are investigated for wireless communication applications.[82] The relative dielectric constant of the reinforced concrete wall of a building is found to be $\epsilon_r = 6.7 - j1.2$ at 900 MHz and $\epsilon_r = 6.2 - j0.69$ at 1.8 GHz, respectively. (a) Find the appropriate thickness of the concrete wall to cause a 10 dB attenuation in the field strength of the 900 MHz signal traveling over its thickness. Neglect the reflections from the surfaces of the wall. (b) Repeat the same calculations at 1.8 GHz.

2-18. Unknown medium. The magnetic field phasor of a 100 MHz uniform plane wave in a nonmagnetic medium is given by

$$\mathbf{H}(z) = \hat{\mathbf{x}}3\,e^{j(17-j10)z} \ \text{A-m}^{-1}$$

(a) Find the conductivity σ and relative permittivity ϵ_r of the medium. (b) Find the corresponding time-domain electric field $\overline{\mathscr{E}}(z, t)$.

2-19. Unknown biological tissue: The electric field component of a uniform plane wave propagating in a biological tissue with relative dielectric constant $\epsilon_r = \epsilon_r' - j\epsilon_r''$ is given by

$$\overline{\mathscr{E}}(y, t) = \hat{\mathbf{x}}0.5e^{-39y}\cos(1.83\pi \times 10^9 t - 141y) \quad \text{V-m}^{-1}$$

(a) Find ϵ_r' and ϵ_r''. Assume $\sigma = 0$ and $\mu_r = 1$. (b) Write the corresponding expression for the wave magnetic field. (c) Write the mathematical expression for the time-average Poynting vector $\mathbf{S}_{av}$ and sketch its magnitude from $y = 0$ to 5 cm.

2-20. Propagation in seawater. Transmission of electromagnetic energy through the ocean is practically impossible at high frequencies because of the high attenuation rates encountered. For seawater, take $\sigma = 4$ S-m^{-1}, $\epsilon_r = 81$, and $\mu_r = 1$, respectively. (a) Show that seawater is a good conductor for frequencies much less than ~890 MHz. (b) For frequencies less than 100 MHz, calculate, as a function of frequency (in Hz), the approximate distance over which the amplitude of the electric field is reduced by a factor of 10.

2-21. Wavelength in seawater. Find and sketch the wavelength in seawater as a function of frequency. Calculate λ_{sw} at the following frequencies: 1 Hz, 1 kHz, 1 MHz, and 1 GHz. Sketch log λ_{sw} vs. log f. Use the following properties for seawater: $\sigma = 4$ S-m^{-1}, $\epsilon_r = 81$, and $\mu_r = 1$.

2-22. Communication in seawater. ELF communication signals ($f \leq 3$ kHz) can more effectively penetrate seawater than VLF signals (3 kHz $\leq f \leq$ 30 kHz). In practice, an ELF signal used for communication can penetrate and be received at a depth of up to 80 m below the ocean surface.[83] (a) Find the ELF frequency at which the skin depth in seawater is equal to 80 m. For seawater, use $\sigma = 4$ S-m^{-1}, $\epsilon_r = 81$, and $\mu_r = 1$. (b) Find the ELF frequency at which the skin depth is equal to half of 80 m. (c) At 100 Hz, find the depth at which the peak value of the electric field propagating vertically downward in seawater is 40 dB less than its value immediately below the

---

[82]C. F. Yang, C. J. Ko, and B. C. Wu, A free space approach for extracting the equivalent dielectric constants of the walls of the buildings, *IEEE AP-S Int. Symp. URSI Radio Sci. Meet.*, pp. 1036-1039, Baltimore, MD, July 1996.

[83]J. C. Kim and E. I. Muehldorf, *Naval Shipboard Communications Systems*, Prentice Hall, 1995, p. 111.

surface of the sea. (d) A surface vehicle–based transmitter operating at 1 kHz generates an electromagnetic signal of peak value 1 V-m^{-1} immediately below the sea surface. If the antenna and the receiver system of the submerged vehicle can measure a signal with a peak value of as low as 1 μV-m^{-1}, calculate the maximum depth beyond which the two vehicles cannot communicate.

2-23. Submarine communication near a river delta. A submarine submerged in the sea ($\sigma = 4$ S-m^{-1}, $\epsilon_r = 81$, $\mu_r = 1$) wants to receive the signal from a VLF transmitter operating at 20 kHz. (a) How close must the submarine be to the surface in order to receive 0.1% of the signal amplitude immediately below the sea surface? (b) Repeat part (a) if the submarine is submerged near a river delta, where the average conductivity of seawater is ten times smaller.

2-24. Human brain tissue. Consider a 1.9 GHz electromagnetic wave produced by a wireless communication telephone inside a human brain tissue[84] ($\epsilon_r = 43.2$ and $\sigma = 1.29$ S-m^{-1}) such that the peak electric field magnitude at the point of entry ($z = 0$) inside the tissue is about 100 V-m^{-1}. Assuming plane wave approximation, do the following: (a) Calculate the electric field magnitudes at points $z = 1$ cm, 2 cm, 3 cm, 4 cm, and 5 cm inside the brain tissue and sketch it with respect to z. (b) Calculate the time-average power density at the same points and sketch it with respect to z. (c) Calculate the time-average power absorbed in the first 1 cm thickness of a tissue sample having a cross-sectional area of 1 cm^2.

2-25. Dispersion in sea water. A uniform plane electromagnetic wave in free space propagates with the speed of light, namely, $c \simeq 3 \times 10^8$ m-s^{-1}. In a conducting medium, however, the velocity of propagation of a uniform plane wave depends on the signal frequency, leading to the "dispersion" of a signal consisting of a band of frequencies. (a) For sea water ($\sigma = 4$ S-m^{-1}, $\epsilon_r = 81$, and $\mu_r = 1$), show that for frequencies much less than ~890 MHz, the velocity of propagation is approximately given by $v_p \simeq k_1 \sqrt{f}$, where k_1 is a constant. What is the value of k_1? (b) Consider two different frequency components of a signal, one at 1 kHz, the other at 2 kHz. If these two signals propagate in the same direction in seawater and are in phase at $z = 0$, what is the phase delay (in degrees) between them (e.g., between their peak values) at a position 100 m away?

2-26. Electromagnetic earthquake precursor. A group of Stanford scientists measured[85] mysterious electromagnetic waves varying with ultralow frequencies in the range of 0.01 to 10 Hz during two different earthquakes which occurred in Santa Cruz, California, in 1989 and in Parkfield, California, in 1994. A member of the group speculates that these waves may result from a local disturbance in the earth's magnetic field caused by charged particles carried by water streams that flow along the fault lines deep in the earth's crust as a result of the shifts that led to the quake. These low-frequency waves can penetrate rock much more easily than those of higher frequencies but can still travel only about 15 km through the ground. Since this low-frequency electromagnetic

[84]Gandhi, O. P., G. Lazzi, and C. M. Furse, Electromagnetic absorption in the human head and neck for mobile telephones at 835 and 1900 MHz, *IEEE Transactions on Microwave Theory and Techniques,* 44(10), pp. 1884–1897, October 1996.

[85]A. C. Fraser-Smith, A. Bernardi, P. R. McGill, M. E. Ladd, R. A. Helliwell, and O. G. Villard, Jr., Low-frequency magnetic field measurements near the epicenter of the M_s 7.1 Loma Prieta earthquake, *Geophys. Res. Lett.,* 17, pp. 1465–1468, 1990; also see B. Holmes, Radio hum may herald quakes, *New Scientist,* p. 15, 23/30 December 1995.

activity was recorded close to a month before the quake and lasted about a month after, this phenomenon has a potential use as an earthquake predictor. Consider three plane waves of equal amplitudes with frequencies of 0.1 Hz, 1 Hz, and 10 Hz, all produced at a depth of 15 km below the earth's surface during an earthquake. Assuming each of these waves to be propagating vertically up toward the surface of the earth, (a) calculate the percentage time-average power of each wave that reaches the surface of the earth, and (b) using the results of part (a), comment on which one of the three signals is more likely to be picked up by a receiver located on the earth's surface, based on their signal strengths. For simplicity, assume the earth's crust to be homogeneous, isotropic, and nonmagnetic with properties $\sigma = 10^{-3}$ S-m^{-1} and $\epsilon_r = 10$ respectively.

2-27. Phantom muscle tissue. In order to develop radiofrequency (RF) heating techniques for treating tumors at various locations and depths in patients, it is necessary to carry out experiments to determine the energy absorbed by an object exposed to electromagnetic fields over a wide range of RF frequencies. An artificial muscle tissue ("muscle phantom") is designed to be used in these experiments to simulate actual muscle tissue for applications in the frequency range used for RF hyperthermia.[86] (a) Given the relative dielectric constant and the conductivity of the muscle phantom at 915 MHz and 22°C to be $\epsilon_r \simeq 51.1$ and $\sigma \simeq 1.27$ S-m^{-1}, calculate the depth of penetration in the phantom. Note that $\mu_r = 1$. (b) Repeat part (a) at 2.45 GHz when $\epsilon_r \simeq 47.4$ and $\sigma \simeq 2.17$ S-m^{-1}. Which frequency can penetrate deeper into the muscle phantom? (c) Calculate the total dB attenuation over a muscle phantom of 1.5 cm thickness at both frequencies.

2-28. Unknown medium. The skin depth and the loss tangent of a nonmagnetic conducting medium at 21.4 kHz are approximately equal to 1.72 m and 4.15×10^4, respectively. (a) Find the conductivity σ and the relative dielectric constant ϵ_r of the medium. What medium is this? (b) Write the mathematical expressions for the electric and magnetic field components of a 21.4 kHz uniform plane wave propagating in this medium, assuming the maximum peak value of the electric field to be 10 V-m^{-1}. (c) Repeat part (b) at 2.14 MHz. Assume the properties of the medium to be the same at both frequencies.

2-29. Unknown medium. A uniform plane wave propagates in the x direction in a certain type of material with unknown properties. At $t = 0$, the wave electric field is measured to vary with x as shown in Figure 2.33a. At $x = 40$ m, the temporal variation of the wave electric field is measured to be in the form shown in Figure 2.33b. Using the data in these two figures, find (a) σ and ϵ_r (assume nonmagnetic case), (b) the depth of penetration and the attenuation in dB-m^{-1}, and (c) the total dB attenuation and the phase shift over a distance of 100 m through this medium.

2-30. Lossy material: In general, for both an electrically and magnetically lossy material, where $\epsilon_c = \epsilon' - j\epsilon''$ and $\mu_c = \mu' - j\mu''$, a loss tangent is defined[87] as

$$\tan \delta_c = \frac{\delta_r''}{\delta_r'}$$

---

[86]C. K. Chou, G. W. Chen, A. W. Guy, and K. H. Luk, Formulas for preparing phantom muscle tissue at various radio frequencies, *Bioelectromagnetics*, 5, pp. 435–441, 1984.

[87]W. B. Weir, Automatic measurement of complex dielectric constant and permeability at microwave frequencies, *Proc. IEEE*, 62(1), pp. 33–36, January 1974.

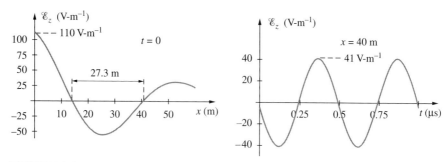

FIGURE 2.33.   **Unknown medium.** Problem 2-29.

where

$$\delta'_r = \mu'_r\epsilon'_r - \mu''_r\epsilon''_r \qquad \text{and} \qquad \delta''_r = \mu'_r\epsilon''_r + \mu''_r\epsilon'_r$$

(a) Verify the above expression for loss tangent. (b) Show that for this case, the attenuation constant is given by

$$\alpha = \frac{\pi f \sqrt{2\delta'_r}}{c}\left[\sqrt{1 + \tan^2\delta_c} - 1\right]^{1/2}$$

where c is the speed of light in free space. (c) Consider a large sample of radar-absorbing (RAM) magnetic coating material at 1 GHz with its properties $\epsilon_r = 15$ and $\mu_r = 5 - j10$ respectively.[88] Calculate the values of $\tan\delta_c$ and α.

2-31. A ferrite absorber. Ferrite tiles are magnetic materials that are used commercially as electromagnetic absorbers in anechoic chambers for emissions and immunity testing of electronic systems because they have an intrinsic impedance that is very close to the intrinsic impedance of free space over a limited range of frequencies.[89] An electromagnetic wave incident from air on the surface of a ferrite material that exhibits this property penetrates into the material with very little of its energy reflecting back into the air. Furthermore, because the ferrite material is magnetically lossy, the wave energy is absorbed as it travels in the ferrite. Consider a large sample of a commercially available ferrite material whose permittivity and permeability at 500 MHz are stated by its manufacturers to be $\epsilon_r = 11.03 - j0.02$ and $\mu_r = 0.47 - j15.66$, respectively. Calculate the loss tangent $\tan\delta_c$, the intrinsic impedance η_c, and the attenuation constant α in dB-(cm)$^{-1}$ of this material at 500 MHz. *Hint:* Use the expressions given in Problem 2-30.

2-32. Thickness of beef products. Microwave heating is generally uniform over the entire body of the product being heated if the thickness of the product does not exceed about

---

[88]E. Michielssen, J. M. Sajer, S. Ranjithan, and R. Mittra, Design of lightweight, broad-band microwave absorbers using genetic algorithms, *IEEE Trans. on Microwave Theory and Techniques,* 41(6/7), pp. 1024–1031, June/July 1993.

[89]C. L. Holloway, R. R. DeLyser, R. F. German, P. McKenna, and M. Kanda, Comparison of electromagnetic absorbers used in anechoic and semi-anechoic chambers for emissions and immunity testing of digital devices, *IEEE Trans. on Electromagnetic Compatibility,* 39(1), pp. 33–47, February 1997.

1 to 1.5 times its penetration depth.[90] (a) Consider a beef product to be heated in a microwave oven operating at 2.45 GHz. The dielectric properties of raw beef at 2.45 GHz and 25°C are $\epsilon_r' = 52.4$ and $\tan \delta_c = 0.33$.[91] What is the maximum thickness of this beef product for it to be heated uniformly? (b) Microwave ovens operating at 915 MHz are evidently more appropriate for cooking products with large cross sections and high dielectric loss factors. The dielectric properties of raw beef at 915 MHz and 25°C are $\epsilon_r' = 54.5$ and $\tan \delta_c = 0.411$. Find the maximum thickness of the beef product at 915 MHz and compare it with the results of part (a).

2-33. Beef versus bacon. The dielectric properties of cooked beef and smoked bacon at 25°C are given by $\epsilon_r \simeq 31.1 - j10.3$ at 2.45 GHz and $\epsilon_r \simeq 2.5 - j0.125$ at 3 GHz, respectively (see the references in the preceding problem). Calculate the loss tangent and the penetration depth for each and explain the differences.

2-34. Aluminum foil. A sheet of aluminum foil of thickness ~25 μm is used to shield an electronic instrument at 100 MHz. Find the dB attenuation of a plane wave that travels from one side to the other side of the aluminum foil. (Neglect the effects from the boundaries.) For aluminum, $\sigma = 3.54 \times 10^7$ S-m^{-1} and $\epsilon_r = \mu_r = 1$.

2-35. Unknown material. Using the results of a reflection measurement technique, the intrinsic impedance of a material at 200 MHz is found to be approximately given by

$$\eta \simeq 22.5 e^{j37°} \; \Omega$$

Assuming that the material is nonmagnetic, determine its conductivity σ and the relative dielectric constant ϵ_r'.

2-36. Poynting flux. The electric and magnetic field expressions for a uniform plane wave propagating in a lossy medium are as follows:

$$\mathscr{E}_x(z, t) = 2e^{-4z} \cos(\omega t - \beta z) \quad \text{V-m}^{-1}$$

$$\mathscr{H}_y(z, t) = H_0 e^{-4z} \cos(\omega t - \beta z - \zeta) \quad \text{A-m}^{-1}$$

The frequency of operation is $f = 10^8$ Hz, and the electrical parameters of the medium are $\epsilon = 18.5\epsilon_0$, μ_0, and σ. (a) Find the time-average electromagnetic power density *entering* a rectangular box-shaped surface like that shown in Figure 2.34, assuming

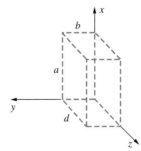

FIGURE 2.34. **Poynting flux.** Problem 2-36.

[90] J. Thuery, *Microwaves: Industrial, Scientific and Medical Applications*, Artech House, 1992.

[91] D. I. C. Wang and A. Goldblith, Dielectric properties of food, *Tech. Rep. TR-76-27-FEL*, MIT, Dept. of Nutrition and Food Science, Cambridge, Massachusetts, November, 1976.

$a = d = 1$ m and $b = 0.5$ m. (b) Determine the power density *exiting* this region and compare with (a). (c) The difference between your results in (a) and (b) should represent electromagnetic power lost in the region enclosed by the square-box region. Can you calculate this dissipated power by any other method (i.e., without using the Poynting vector)? If yes, carry out this calculation. *Hint:* You may first need to find σ.

2-37. **Laser beams.** The electric field component of a laser beam propagating in the z direction is approximated by

$$\mathscr{E}_r = E_0 e^{-r^2/a^2} \cos(\omega t - \beta z)$$

where E_0 is the amplitude on the axis and ω is the effective beam radius, where the electric field amplitude is a factor of e^{-1} lower than E_0. (a) Find the corresponding expression for the magnetic field $\mathscr{H}$. (b) Show that the time-average power density at the center of the laser beam is given by

$$|\mathbf{S}_{av}| = \frac{E_0^2}{2\eta} e^{-2r^2/a^2} \text{ W-m}^{-2}$$

where $\eta = 377\Omega$. (c) Find the total power of the laser beam. Consider a typical laboratory helium-neon laser with a total power of 5 mW and an effective radius of $a = 400$ μm. What is the power density at the center of the beam? (d) The power density of solar electromagnetic radiation at the Earth's surface is 1400 W-m^{-2}. At what distance from the Sun would its power density be equal to that for the helium-neon laser in part (c)? (e) One of the highest-power lasers built for fusion experiments operates at $\lambda = 1.6$ μm, produces 10.2 kJ for 0.2 ns, and is designed for focusing on targets of 0.5 mm diameter. Estimate the electric field strength at the center of the beam. Is the field large enough to break down air? What is the radiation pressure of the laser beam? How much weight can be lifted with the pressure of this beam?

2-38. **Maxwell's equations.** Consider a parallel-plate transmission line with perfectly conducting plates of large extent, separated by a distance of d meters. As shown in Figure 2.35, an alternating surface current density J_s in the z direction flows on the conductor surface:

$$\overline{\mathscr{J}}_s(z, t) = \hat{\mathbf{z}} J_0 \cos\left[\omega\left(t - \frac{z}{c}\right)\right] \text{ A-m}^{-1}$$

(a) Find an expression for the electric field, and determine the voltage between the plates, for $d = 0.1$ m and $J_0 = 1$ A-m^{-1}. (b) Use the continuity equation to find an expression for the surface charge density $\rho_s(z, t)$.

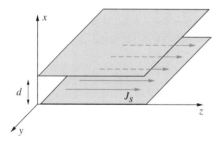

FIGURE 2.35. **Surface current.**
Problem 2-38.

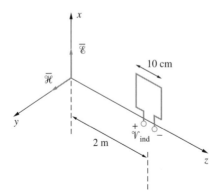

FIGURE 2.36. **Uniform plane wave.** Problem 2-39.

2-39. Uniform plane wave. A uniform plane electromagnetic wave propagates in free space with electric and magnetic field components as shown in Figure 2.36:

$$\overline{\mathscr{E}} = \hat{\mathbf{x}} E_0 \cos(\omega t - \beta z)$$

$$\overline{\mathscr{H}} = \hat{\mathbf{y}} \frac{E_0}{\eta} \cos(\omega t - \beta z)$$

The wave frequency is 300 MHz and the electric field amplitude $E_0 = 1$ V-m^{-1}. A square loop antenna with side length $a = 10$ cm is placed at $z = 2$ m as shown. (a) Find the voltage $\mathscr{V}_{\text{ind}}(t)$ induced at the terminals of the loop. (b) Repeat (a) for the loop located at a distance of $d - 3$ m from the x axis instead of 2 m as shown. Compare your answers in (a) and (b).

2-40. FM radio. An FM radio station operating at 100 MHz radiates a circularly polarized plane wave with a total isotropically radiated power of 200 kW. The transmitter antenna is located on a tower 500 m above the ground. (a) Find the rms value of the electric field 1 km away from the base of the antenna tower. Neglect the effects of reflections from the ground and other boundaries. (b) If the primary coverage radius of this station is ~100 km, find the approximate time-average power density of the FM wave at this distance.

2-41. Mobile phones. Cellular phone antennas installed on cars have a maximum output power of 3 W, set by the Federal Communications Commission (FCC) standards. The incident electromagnetic energy to which the passengers in the car are exposed does not pose any health threats, both because they are some distance away from the antenna and also because the body of the car and glass window shield them from much of the radiation.[92] (a) For a car with a synthetic roof, the maximum localized power density in the passenger seat is about 0.3 mW-(cm)$^{-2}$. For cars with metal roofs, this value reduces to 0.02 mW-(cm)$^{-2}$ or less. Antennas mounted on the trunk or in the glass of the rear windshield deliver power densities of about 0.35 to 0.07 mW-(cm)$^{-2}$ to passengers in the back seat. Compare these values with the IEEE safety limit (IEEE Standard C95.1-1991) in the cellular phone frequency range (which is typically 800 to 900 MHz) and comment on the safety of the passengers. Note that from 300 MHz to 15 GHz, IEEE

[92]M. Fischetti, The cellular phone scare, *IEEE Spectrum,* pp. 43–47, June 1993.

safety limits[93] specify a maximum allowable power density that increases linearly with frequency as $|\mathbf{S}_{av}|_{max} = f/1500$ in mW-(cm)$^{-2}$, where the frequency f is in MHz. For reference, the maximum allowable power density is 1 mW-(cm)$^{-2}$ at 1.5 GHz. (b) Calculate the maximum output power of a cellular phone antenna installed on the metal roof of a car such that the localized power density in the passenger compartment is equal to the IEEE safety limit at 850 MHz.

2-42. Radar aboard a Navy ship. Some shipboard personnel work daily in an environment where the radio frequency (RF) power density only a few feet above their heads may exceed safe levels. Some areas on the deck of the ship are not even allowed to personnel due to high power densities. Pilots of aircrafts routinely fly through the ship's radar beams during takeoff and landing operations. On one of the Navy aircraft carriers, the average power density along the axis of the main beam (where the field intensity is the greatest) 100 ft away from a 6 kW missile control radar operating in the C-band is measured[94] to be about 300 mW-(cm)$^{-2}$. (a) If it is assumed that human exposures in such environments are limited to less than 10 mW-(cm)$^{-2}$, calculate the approximate distance along the beam axis that can be considered as the hazardous zone. (b) Assuming the operation frequency of the C-band radar to be 5 GHz, recalculate the hazardous zone along the radar's main beam based on the IEEE safety limit[95] for the average power density given by $|\mathbf{S}_{av}|_{max} = f/1500$ in mW-(cm)$^{-2}$ (valid over the frequency range 300 MHz to 15 GHz), where f is in MHz.

2-43. Radio frequency exposure time. In 1965, the U.S. Army and Air Force amended their use of the prevailing 10 mW-(cm)$^{-2}$ exposure guideline to include a time limit for exposures, given by the formula

$$t_{max} = \frac{6000}{|\mathbf{S}_{av}|^2}$$

where $|\mathbf{S}_{av}|$ is the average power density [in mW-(cm)$^{-2}$] of exposure and t_{max} is the maximum recommended exposure duration, in minutes.[96] Consider a person standing on the deck of a Navy ship where the peak rms electric field strength due to a microwave radar transmitter is measured to be 140 V-m^{-1}. Using the above formula, calculate the maximum exposure time allowed (in hours) for this person to stay at that location.

2-44. Microwave cataracts in humans. Over 50 cases of human cataract induction have been attributed to microwave exposures, primarily encountered in occupational situations involving acute exposure to presumably relatively high-intensity fields.[97] The following are three reported incidents of cataracts caused by microwave radiation: (1) A 22-year-old technician exposed approximately five times to 3 GHz radiation at an

---

[93] American National Standards Institute—Safety levels with respect to exposure to radio frequency electromagnetic fields, 3 kHz to 300 GHz, ANSI/IEEE C95.1, 1992.

[94] Z. R. Glaser and G. M. Heimer, Determination and elimination of hazardous microwave fields aboard naval ships, *IEEE Transactions on Microwave Theory and Techniques,* 19(2), pp. 232-238, February 1971.

[95] American National Standards Institute—Safety levels with respect to exposure to radio frequency electromagnetic fields, 3 kHz to 300 GHz, ANSI/IEEE C95.1, 1992.

[96] W. F. Hammett, *Radio Frequency Radiation, Issues and Standards,* McGraw-Hill, 1997.

[97] S. F. Cleary, Microwave cataractogenesis, *Proc. IEEE,* 68(1), pp. 49–55, January 1980.

estimated average power density of 300 mW-(cm)$^{-2}$ for 3 min/exposure developed bilateral cataracts. (2) A person was exposed to microwaves for durations of approximately 50 hr/month over a 4-year period at average power densities of less than 10 mW-(cm)$^{-2}$ in most instances, but with a period of 6 months or more during which the average power density was approximately 1 W-(cm)$^{-2}$. (3) A 50-year-old woman was intermittently exposed to leakage radiation from a 2.45 GHz microwave oven of approximately 1 mW-(cm)$^{-2}$ during oven operation, with levels of up to 90 mW-(cm)$^{-2}$ when the oven door was open, presumably over a period of approximately 6 years prior to developing cataract. For each of these above cases, compare the power densities with the IEEE standards (see Problem 2-41) and comment.

2-45. Radiation pressure of the sun. In Example 2-15, the sun's time-averaged solar radiation power density at the earth's surface is quoted to be about 1400 W-m^{-2}, a value that is averaged over a wide band of frequencies. Calculate the radiation pressure exerted by the sun on a flat object located at the earth's surface if the object is (a) a perfect absorber and (b) a perfect reflector.

2-46. High-power microwave beam. The power density of a high-power microwave beam cannot be increased indefinitely but has an upper bound set by the dielectric breakdown of the medium supporting the beam.[98] For altitudes below 10 km, a microwave beam of power density greater than 10^5 W-(cm)$^{-2}$ can be sustained. Calculate the radiation pressure exerted by such a beam on a perfectly conducting object having a flat surface.

2-47. Ultrahigh-intensity lasers. Calculate the radiation pressure (in bars where 1 bar = 10^5 N-m^{-2}) exerted by an ultrahigh-intensity laser[99] with a time-average power density of 10^{20} W-(cm)$^{-2}$.

2-48. VHF TV signal. The magnetic field component of a 10 μW-m^{-2}, 200 MHz TV signal in air is given by

$$\overline{\mathcal{H}}(x, y, t) = \hat{\mathbf{z}}H_0 \sin(\omega t - ax - ay + \pi/3)$$

(a) What are the values of H_0 and a? (b) Find the corresponding electric field $\overline{\mathcal{E}}(x, y, t)$. What is the polarization of the wave? (c) An observer at $z = 0$ is equipped with a wire antenna capable of detecting the component of the electric field along its length. Find the maximum value of the measured electric field if the antenna wire is oriented along the (i) x direction, (ii) y direction, (iii) 45° line between the x and y directions.

2-49. FM polarization. Find the type (linear, circular, elliptical) and sense (right- or left-handed) of the polarization of the FM broadcast signal given in Example 2-2.

2-50. Unknown wave polarization. The magnetic field component of a uniform plane wave in air is given by

$$\mathbf{H}(x) = 2e^{j10\pi x}[\hat{\mathbf{y}}e^{-j\pi/4} - \hat{\mathbf{z}}e^{ja\pi/4}] \quad \text{mA-m}^{-1}$$

where a is a real constant. (a) Find the wavelength λ and frequency f. (b) Find the total time-average power carried by this wave. (c) Determine the type (linear, circular,

[98]C. D. Taylor and D. V. Giri, *High-Power Microwave Systems and Effects,* Chapter 4, Francis & Taylor, Washington, D.C., 1994.

[99]G. A. Mourou, C. P. J. Barty, and M. D. Perry, Ultrahigh-intensity lasers: Physics of the extreme on a tabletop, *Physics Today,* pp. 22–28, January 1998.

elliptical) and sense (right- or left-handed) of the polarization of this wave when $a = 1$.
(d) Repeat part (c) when $a = 3$.

2-51. Linear and circularly polarized waves. Two electromagnetic waves operating at the same frequency and propagating in the same direction (y direction) in air are such that one of them is linearly polarized in the x direction whereas the other is left-hand circularly polarized (LHCP). However, the electric and magnetic field components of the two waves appear identical at one instant within every 50 ps time interval. The linearly polarized wave carries a time-average power density of 1.4 W-m^{-2}. An observer located at $y = 0$ uses a receiving antenna to measure the x component of the total electric field only and records a maximum field magnitude of about 65 V-m^{-1} over every time interval of 0.5 ns. (a) Write the mathematical expressions for the electric field components of each wave, using numerical values of various quantities whenever possible. (b) Find the ratio of the time-average power densities of the LHCP and the linearly polarized waves.

2-52. Two circularly polarized waves. Consider two circularly polarized waves traveling in the same direction transmitted by two different satellites operating at the same frequency given by

$$\mathbf{E}_1 = E_{01}(\hat{\mathbf{x}} + \hat{\mathbf{z}}e^{j\pi/2})e^{j\beta y}$$

$$\mathbf{E}_2 = E_{02}(\hat{\mathbf{x}} + \hat{\mathbf{z}}e^{j3\pi/2})e^{j\beta y}$$

where E_{01} and E_{02} are real constants. (a) If the time-average power densities of these two waves are equal, find the polarization of the total wave. (b) Repeat part (a) for the case when the time-average power of the first wave is 4 times the time-average power of the second wave.

2-53. Wave polarization. Consider the following complex phasor expression for a time-harmonic electric field in free space:

$$\mathbf{E} = [3\hat{\mathbf{x}} + 4\hat{\mathbf{y}} + j5\hat{\mathbf{z}}]e^{-j(8x-6y)\pi} \quad \text{mV-m}^{-1}$$

(a) Is this a uniform plane wave? What is its frequency? (b) What is the direction of propagation and the state of polarization (specify both the type and sense of polarization) of this electromagnetic field? (c) Find the associated magnetic field and the time-average power density in the direction of propagation.

2-54. Wave polarization. The electric field component of a communication satellite signal traveling in free space is given by

$$\mathbf{E}(z) = [\hat{\mathbf{x}} - \hat{\mathbf{y}}(1 + j)]12e^{j50\pi z} \quad \text{V-m}^{-1}$$

(a) Find the corresponding $\mathbf{H}(z)$. (b) Find the total time-average power carried by this wave. (c) Determine the polarization (both type and sense) of the wave.

2-55. Superposition of two waves. The electric field components of two electromagnetic waves at the same frequency and propagating in free space are represented by

$$\mathbf{E}_1 = 10(j\hat{\mathbf{x}} - \hat{\mathbf{y}})e^{j28\pi z} \quad \text{V-m}^{-1}$$

$$\mathbf{E}_2 = \hat{\mathbf{y}}E_0 e^{-j28\pi x}$$

Find and sketch the locus of the total electric field measured at the origin ($x = y = z = 0$) if E_0 is equal to (a) 10 V-m^{-1}, (b) 20 V-m^{-1}, (c) 40 V-m^{-1}, respectively.

CHAPTER 3

Reflection, Transmission, and Refraction of Waves at Planar Interfaces

---

Up to now, we have considered the relatively simple propagation of uniform plane waves in homogeneous and unbounded media. However, practical problems usually involve waves propagating in bounded regions, in which different media may be present, and require that we take account of the complicating effects at boundary surfaces. Typical boundary surfaces may lie between regions of different permittivity (e.g., glass and air) or between regions of different conductivity (e.g., air and copper). Boundary surfaces between regions of different permeability (e.g., air and iron) are also interesting but of less practical importance for electromagnetic wave phenomena.

When a wave encounters the boundary between two different homogeneous media, it is split into two waves: a *reflected* wave that propagates back to the first medium and a *transmitted* (or *refracted*) wave that proceeds into the second medium. Reflection of waves is very much a part of our everyday experience. When a guitar string is plucked, a wave is generated that runs back and forth between the ends of the string, reflecting repeatedly at each end. When we shout toward a cliff, we hear the reflection of the sound waves we generated as an echo. The basis for operation of radars, sonars, and lidars is the reflection of radio, acoustic, or light waves from various objects. When sound waves of sufficient intensity hit an object (such as an eardrum), they transmit energy into it by making it vibrate. Seismic waves transfer their energy quite efficiently through boundaries between different layers of the earth, as do ultrasound waves through tissues. Earthquake waves generated deep under the ocean floors transmit wave energy into the ocean, generating tsunamis.

When we look at the way ocean waves hit a beach or go around a ship standing still, we realize that the reflection of waves from boundaries can be very complicated indeed. The shape of a wave can be greatly altered on reflection, and a wave striking a rough (nonsmooth) surface can be reflected in all directions. We also know from our experiences with physical optics (e.g., refraction of light by lenses or reflection by mirrors) that the direction and magnitude of the waves reflected and refracted at a boundary depend on the angle of arrival of the waves at the boundary. In this book we confine our attention to planar waves of infinite extent incident on planar boundaries. When uniform plane waves encounter planar interfaces, the reflected and transmitted (refracted) waves are also planar, and expressions for their directions of propagation and their amplitudes and phases can be derived with relative ease. Restricting our attention to the reflection and refraction of uniform plane waves from planar boundaries is also appropriate because many practical situations can be modeled in this way to a high degree of accuracy. Optical beams usually have diameters much larger than their free-space wavelength, so their encounter with planar boundaries is quite accurately represented by treating them as uniform plane waves. Furthermore, nonuniform waves, such as those that propagate in hollow cylindrical waveguides, can be represented as a superposition of uniform plane waves reflecting from the conducting waveguide boundaries.

In this chapter, we discuss the simplest cases of propagation of electromagnetic waves in two or more different homogeneous media separated by plane boundaries of infinite extent. In general, when a uniform plane wave propagating in one medium is *incident* on a boundary, part of its energy reflects back into the first medium and is carried away from the boundary by a reflected wave, and the other part propagates into the second medium, carried by a transmitted wave. The existence of these two waves (reflected and transmitted) is a direct result of the boundary conditions imposed by the properties of the media, which in general cannot be satisfied without postulating the presence of both of these waves. Using Maxwell's equations with the appropriate boundary conditions, we can find mathematical expressions for the reflected and transmitted waves in terms of the properties of the incident wave and the two media. We limit our coverage to interfaces that are infinitely sharp, with the electromagnetic properties (e.g., dielectric constant) changing suddenly (within a distance very small compared to the wavelength) from that of one medium to that of the other. For example, for optical applications we require the medium properties near the interface to change from one medium to another over distances much smaller than the free-space wavelength of visible light (390 to 760 nm). This is why the surfaces of optical coatings have to be very smooth, since any thin layer of impurities will change the reflection properties.

The reflection and transmission of uniform plane waves from planar interfaces is in many ways analogous to the reflection and transmission of voltage and current waves at junctions between different transmission lines or at the terminations of transmission lines at a load.[1] When appropriate, we make references to this analogy as we study uniform plane wave reflection and transmission, and in some cases we utilize it for more efficient solution of interface problems.

We now proceed with our coverage of reflection and refraction of electromagnetic waves from planar boundaries. We consider different cases in order of increasing complexity. We start in Section 3.1 with normal (or perpendicular) incidence of uniform plane waves on perfect conductors; in that case all of the incident wave energy is reflected, so there is no refracted wave. We then consider normal incidence on single dielectric interfaces in Section 3.2 and multiple dielectric interfaces in Section 3.3, with the associated important applications of coating. We continue with oblique incidence on planar interfaces, first on perfect conductors in Section 3.4 and then on lossless dielectrics in Section 3.5, and consider selected associated applications. The important phenomenon of total internal reflection is introduced and discussed in Section 3.6, followed by a discussion in Section 3.7 of reflection and transmission of waves normally incident on lossy media. We conclude the chapter with a discussion in Section 3.8 of oblique incidence of uniform plane waves on lossy media—a topic that is particularly important in the view of the rapid expansion of wireless personal personal communication applications.

3.1 NORMAL INCIDENCE ON A PERFECT CONDUCTOR

We first consider the case of a uniform plane wave propagating in the $+z$ direction in a simple lossless ($\sigma_1 = 0$) medium occupying the half-space $z < 0$ and normally incident from the left on a perfectly conducting ($\sigma_2 = \infty$) medium that occupies the $z > 0$ half-space, as shown in Figure 3.1. The interface between the two media is the entire xy plane (i.e., infinite in transverse extent). We arbitrarily[2] assume that the electric field of the incident wave is oriented in the x direction and that therefore its magnetic field is in the y direction. The phasor field expressions for the incident wave (presumably originating at $z = -\infty$) and the reflected wave are given as

Incident wave:	Reflected wave:
$\mathbf{E}_i(z) = \hat{\mathbf{x}} E_{i0} e^{-j\beta_1 z}$	$\mathbf{E}_r(z) = \hat{\mathbf{x}} E_{r0} e^{+j\beta_1 z}$
$\mathbf{H}_i(z) = \hat{\mathbf{y}} \dfrac{E_{i0}}{\eta_1} e^{-j\beta_1 z}$	$\mathbf{H}_r(z) = -\hat{\mathbf{y}} \dfrac{E_{r0}}{\eta_1} e^{+j\beta_1 z}$

---

[1]For a detailed discussion of voltage and current waves on transmission lines, see Chapters 2 and 3 of U. S. Inan and A. S. Inan, *Engineering Electromagnetics*, Addison Wesley Longman, 1999.

[2]Note that for normal incidence on a planar boundary of infinite extent in two dimensions, we lose no generality by taking the direction in which the electric field of the incident wave vibrates to be along one of the principal axes.

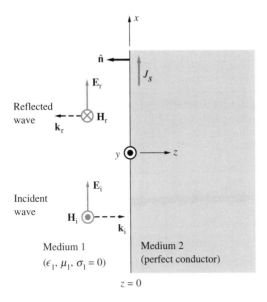

FIGURE 3.1.   Uniform plane wave incident normally on a plane, perfectly conducting boundary. The unit vectors $\hat{\mathbf{k}}_i$ and $\hat{\mathbf{k}}_r$ represent the propagation directions of the incident and reflected waves, respectively.

where $\beta_1 = \omega\sqrt{\mu_1\epsilon_1}$ and $\eta_1 = \sqrt{\mu_1/\epsilon_1}$ are, respectively, the propagation constant and the intrinsic impedance for medium 1. The presence of the boundary requires that a reflected wave propagating in the $-z$ direction exist in medium 1, with phasor field expressions as given above. Note that the reflected wave propagates in the $-z$ direction, so its magnetic field is oriented in the $-y$ direction for an x-directed electric field.

The boundary condition on the surface of the conductor requires the total tangential electric field to vanish for $z = 0$ for *all x and y,* since the electric field inside medium 2 (perfect conductor) must be zero. This condition requires that the electric field of the reflected wave is also confined to the x direction. The total electric and magnetic fields in medium 1 are

$$\mathbf{E}_1(z) = \mathbf{E}_i(z) + \mathbf{E}_r(z) = \hat{\mathbf{x}}[E_{i0}e^{-j\beta_1 z} + E_{r0}e^{+j\beta_1 z}]$$

$$\mathbf{H}_1(z) = \mathbf{H}_i(z) + \mathbf{H}_r(z) = \hat{\mathbf{y}}\left[\frac{E_{i0}}{\eta_1}e^{-j\beta_1 z} - \frac{E_{r0}}{\eta_1}e^{+j\beta_1 z}\right]$$

Since $\mathbf{E}_1(z)$ is tangential to the boundary, application of the boundary condition at $z = 0$, namely that $\mathbf{E}_1(z = 0) = 0$, gives

$$\mathbf{E}_1(z = 0) = \hat{\mathbf{x}}[E_{i0}e^{-j\beta_1 z} + E_{r0}e^{+j\beta_1 z}]_{z=0} = \hat{\mathbf{x}}[E_{i0} + E_{r0}] = 0$$

from which

$$\boxed{E_{r0} = -E_{i0}}$$

so that

$$\mathbf{E}_1(z) = \hat{\mathbf{x}}E_{i0}\underbrace{[e^{-j\beta_1 z} - e^{+j\beta_1 z}]}_{-j2\sin(\beta_1 z)} = -\hat{\mathbf{x}}E_{i0}\,j2\sin(\beta_1 z)$$

With the constant E_{r0} determined in terms of E_{i0} (i.e., $E_{r0} = -E_{i0}$), we can also write the total magnetic field in medium 1 as

$$\mathbf{H}_1(z) = \mathbf{H}_i(z) + \mathbf{H}_r(z) = \hat{\mathbf{y}}\frac{1}{\eta_1}[E_{i0}e^{-j\beta_1 z} + E_{i0}e^{+j\beta_1 z}] = \hat{\mathbf{y}}\frac{2E_{i0}}{\eta_1}\cos(\beta_1 z)$$

The corresponding space-time functions $\overline{\mathscr{E}}_1(z, t)$ and $\overline{\mathscr{H}}_1(z, t)$ can be found from the phasors $\mathbf{E}_1(z)$ and $\mathbf{H}_1(z)$ as

$$\overline{\mathscr{E}}_1(z, t) = \mathscr{R}e\{\mathbf{E}_1(z)e^{j\omega t}\} = \mathscr{R}e\{\hat{\mathbf{x}}2E_{i0}\sin(\beta_1 z)e^{j\omega t}\underbrace{e^{-j\pi/2}}_{-j}\}$$

$$\overline{\mathscr{H}}_1(z, t) = \mathscr{R}e\{\mathbf{H}_1(z)e^{j\omega t}\} = \mathscr{R}e\left\{\hat{\mathbf{y}}\frac{2E_{i0}}{\eta_1}\cos(\beta_1 z)e^{j\omega t}\right\}$$

which gives

$$\boxed{\begin{aligned} \overline{\mathscr{E}}_1(z, t) &= \hat{\mathbf{x}}2E_{i0}\sin(\beta_1 z)\sin(\omega t) \\ \overline{\mathscr{H}}_1(z, t) &= \hat{\mathbf{y}}2\frac{E_{i0}}{\eta_1}\cos(\beta_1 z)\cos(\omega t) \end{aligned}} \qquad [3.1]$$

assuming that E_{i0} is a real number.[3]

The fields described by [3.1] are plotted in Figure 3.2 as a function of distance z at different time instants. Note that the $\overline{\mathscr{E}}$ and $\overline{\mathscr{H}}$ fields do not represent a propagating wave, because as time advances, the peaks (or nulls) always occur at the same points in space. Such a wave is termed to be a *pure standing wave;* it consists of a superposition of two waves traveling in opposite directions. The maxima and minima *stand* at the same location as time advances. This standing wave, which is established when a uniform plane wave is incident on a perfect conductor, is analogous in all respects to the one that occurs in the case of sinusoidal excitation of a lossless transmission line terminated in a short circuit.[4] This analogy is illustrated in Figure 3.3.

A pure standing wave such as the one just described does not carry electromagnetic energy, as expected on physical grounds, since the perfect conductor boundary reflects all of the incident energy. To verify this notion, we can consider the Poynting vector for the standing wave just described. Using the phasor forms of the fields and

---

[3]If, instead, E_{i0} was a complex number, $E_{i0} = |E_{i0}|e^{j\zeta}$, the expressions for $\overline{\mathscr{E}}_1$ and $\overline{\mathscr{H}}_1$ would be

$$\overline{\mathscr{E}}_1(z, t) = \hat{\mathbf{x}}2|E_{i0}|\sin(\beta_1 z)\sin(\omega t + \zeta)$$

$$\overline{\mathscr{H}}_1(z, t) = \hat{\mathbf{y}}2\frac{|E_{i0}|}{\eta_1}\cos(\beta_1 z)\cos(\omega t + \zeta)$$

amounting to a simple shift of the time origin.

[4]See Section 3.2 of U. S. Inan and A. S. Inan, *Engineering Electromagnetics*, Addison Wesley Longman, 1999.

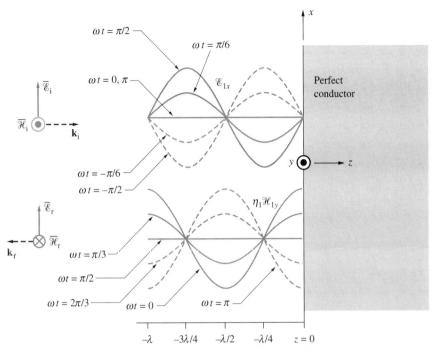

FIGURE 3.2. **Instantaneous electric and magnetic field waveforms.** Total electric and magnetic field waveforms in medium 1 are shown at selected instants of time. The "standing" nature of the waves is apparent as the peaks and nulls remain at the same point in space as time progresses. Note that medium 1 is a perfect dielectric, while medium 2 is a perfect conductor.

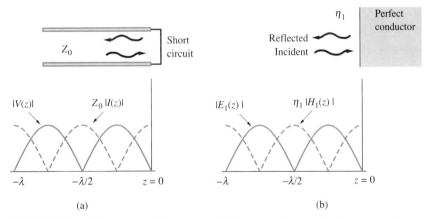

FIGURE 3.3. **Transmission line analogy.** (a) The voltage/current standing-wave patterns on a short-circuited lossless transmission line. Here $V(z)$ and $I(z)$ are the total voltage and current phasors along the lossless transmission line, respectively, and Z_0 is the characteristic impedance of the line, given by $Z_0 = \sqrt{L/C}$, with L and C respectively being the per-unit-length inductance and capacitance of the line. (b) The electric and magnetic fields for normal incidence of a uniform plane wave from a lossless medium on a perfect conductor. In this context, note that the intrinsic impedance η_1 is analogous to Z_0, while the electric field $\mathbf{E}_1$ and magnetic field $\mathbf{H}_1$ are analogous to the voltage $V(z)$ and current $I(z)$ respectively, although $\mathbf{E}_1$ and $\mathbf{H}_1$ are of course vector quantities while $V(z)$ and $I(z)$ are scalars.

125

noting that E_{i0} is in general complex, we have

$$(\mathbf{S}_{av})_1 = \frac{1}{2}\mathcal{R}e\{\mathbf{E}_1 \times \mathbf{H}_1^*\} = \frac{1}{2}\mathcal{R}e\left\{\hat{\mathbf{x}}[-2jE_{i0}\sin(\beta_1 z)] \times \hat{\mathbf{y}}\left[\frac{2E_{i0}^*}{\eta_1}\cos(\beta_1 z)\right]\right\}$$

$$= \frac{1}{2}\mathcal{R}e\left\{\hat{\mathbf{z}}\left[-j\frac{2|E_{i0}|^2}{\eta_1}\sin(2\beta_1 z)\right]\right\} = 0$$

The instantaneous Poynting vector can also be obtained from $\overline{\mathscr{E}}_1$ and $\overline{\mathscr{H}}_1$ as

$$\overline{\mathscr{S}}_1(z, t) = \overline{\mathscr{E}}_1(z, t) \times \overline{\mathscr{H}}_1(z, t) = \hat{\mathbf{z}}\frac{4|E_{i0}|^2}{\eta_1}\sin(\beta_1 z)\cos(\beta_1 z)\sin(\omega t + \zeta)\cos(\omega t + \zeta)$$

$$= \hat{\mathbf{z}}\frac{|E_{i0}|^2}{\eta_1}\sin(2\beta_1 z)\sin[2(\omega t + \zeta)]$$

where ζ is the arbitrary phase angle of E_{i0}; that is, we assume that $E_{i0} = |E_{i0}|e^{j\zeta}$. The time average value of $\overline{\mathscr{S}}$ is clearly zero. Although electromagnetic energy does not flow, it surges back and forth. In other words, at any given point z, the Poynting vector fluctuates between positive and negative values at a rate of 2ω.

With respect to the discussion of electromagnetic power flow in Section 2.4, we note from [2.29] that since there is no dissipation term (i.e., $\overline{\mathscr{J}}_1 = 0$, and thus $\overline{\mathscr{E}}_1 \cdot \overline{\mathscr{J}}_1 = 0$), the fluctuating Poynting vector $\mathscr{S}_1(z, t)$ represents the time variation of the stored electric and magnetic energy densities, w_e and w_m, which are given by

$$w_e(z, t) = \tfrac{1}{2}\epsilon_1|\overline{\mathscr{E}}_1(z, t)|^2 = 2\epsilon_1|E_{i0}|^2\sin^2(\beta_1 z)\sin^2(\omega t + \zeta) \quad \text{J-m}^{-3}$$

$$w_m(z, t) = \tfrac{1}{2}\mu_1|\overline{\mathscr{H}}_1(z, t)|^2 = 2\epsilon_1|E_{i0}|^2\cos^2(\beta_1 z)\cos^2(\omega t + \zeta) \quad \text{J-m}^{-3}$$

The variations of w_m and w_e with z at different times are shown in Figure 3.4. We can see that the stored energy at point z alternates between fully magnetic energy ($\omega t + \zeta = 0$) and fully electric energy ($\omega t + \zeta = \pi/2$), much like the fluctuation of stored energy between the inductance and capacitance in a resonant LC circuit.

Note that the total wave magnetic field phasor exhibits a maximum at the conductor surface ($z = 0$). This magnetic field is supported by a surface current whose density can be obtained from the boundary condition

$$\mathbf{J}_s = \hat{\mathbf{n}} \times \mathbf{H}_1(0) = (-\hat{\mathbf{z}} \times \hat{\mathbf{y}})\frac{2E_{i0}}{\eta_1} = \hat{\mathbf{x}}\frac{2E_{i0}}{\eta_1} \quad \text{A-m}^{-1}$$

where $\hat{\mathbf{n}}$ is the outward normal to the first medium as shown in Figure 3.1. Note that the direction of the surface current $\mathbf{J}_s$ is related to that of $\mathbf{H}_1$ by Ampère's law, with its sense determined by the right-hand rule. If the conductor is not perfect, this current flow leads to dissipation of electromagnetic power, as discussed in Section 3.7.

Examples 3-1 and 3-2 quantitatively illustrate the properties of the reflected wave for linearly and circularly polarized uniform plane waves normally incident on a perfect conductor.

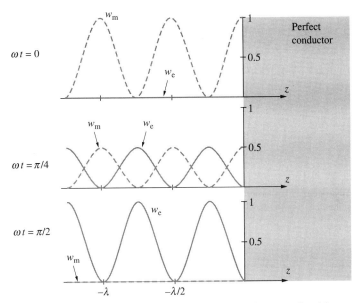

FIGURE 3.4.   Instantaneous electric and magnetic energy densities.
Oscillation of the total wave energy between electric and magnetic fields in a
standing wave. For the purpose of these plots, we have taken E_{i0} to be real and
equal to $1/\sqrt{2\epsilon_1}$, so that $\zeta = 0$.

**Example 3-1:   Linearly polarized wave incident on an air–conductor
interface.**   A z-polarized uniform plane wave having a electric field with peak
value 10 V-m^{-1} and operating at 1.5 GHz is normally incident from air on a perfectly
conducting surface located at $y = 0$, as shown in Figure 3.5. (a) Write the phasor
expressions for the total electric and magnetic fields in air. (b) Determine the nearest

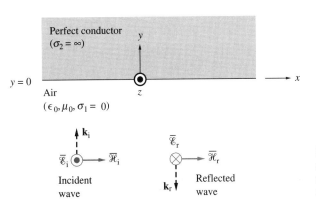

**FIGURE 3.5.   Air–perfect
conductor interface.** A
uniform plane wave incident
from air on a perfect conductor.

location to the reflecting surface where the total electric field is zero at all times. (c) Determine the nearest location to the reflecting surface where the total magnetic field is zero at all times.

Solution:

(a) The electric and magnetic fields of the incident wave are

$$\mathbf{E}_i(y) = \hat{\mathbf{z}}10e^{-j\beta_1 y} \quad \text{V-m}^{-1}$$

$$\mathbf{H}_i(y) = \hat{\mathbf{x}}\frac{10}{\eta_1}e^{-j\beta_1 y} \quad \text{A-m}^{-1}$$

where $\beta_1 \simeq [2\pi(1.5 \times 10^9) \text{ rad-s}^{-1}]/(3 \times 10^8 \text{ m-s}^{-1}) = 10\pi \text{ rad-m}^{-1}$ and $\eta_1 \simeq 377\Omega$. The electric and magnetic fields of the reflected wave are

$$\mathbf{E}_r(y) = -\hat{\mathbf{z}}10e^{+j\beta_1 y} \quad \text{V-m}^{-1}$$

$$\mathbf{H}_r(y) = \hat{\mathbf{x}}\frac{10}{\eta_1}e^{+j\beta_1 y} \quad \text{A-m}^{-1}$$

The total fields in air are

$$\mathbf{E}_1(y) = \mathbf{E}_i(y) + \mathbf{E}_r(y) \simeq -\hat{\mathbf{z}}j20\sin(10\pi y) \quad \text{V-m}^{-1}$$

$$\mathbf{H}_1(y) = \mathbf{H}_i(y) + \mathbf{H}_r(y) \simeq \hat{\mathbf{x}}\frac{20}{377}\cos(10\pi y) \simeq \hat{\mathbf{x}}(0.0531)\cos(10\pi y) \quad \text{A-m}^{-1}$$

The corresponding time-domain expressions for the total fields in air are

$$\overline{\mathscr{E}}_1(y, t) \simeq \hat{\mathbf{z}}20\sin(10\pi y)\sin(3\pi \times 10^9 t) \quad \text{V-m}^{-1}$$

$$\overline{\mathscr{H}}_1(y, t) \simeq \hat{\mathbf{x}}53.1\cos(10\pi y)\cos(3\pi \times 10^9 t) \quad \text{mA-m}^{-1}$$

(b) Therefore, the nearest location to $y = 0$ boundary where the total electric field is zero at all times can be found from

$$10\pi y = -\pi \quad \longrightarrow \quad y = -0.1 \text{ m} = -10 \text{ cm}$$

(c) Similarly, the nearest location to $y = 0$ where the total magnetic field is zero at all times can be found from

$$10\pi y = -\pi/2 \quad \longrightarrow \quad y = -5 \text{ cm}$$

Note that the two answers correspond to 10 cm $= \lambda/2$ and 5 cm $= \lambda/4$, as expected on the basis of Figure 3.2.

Example 3-2: Circularly polarized wave normally incident on an air–conductor interface. Show that circularly polarized light normally incident on the plane surface of a perfectly conducting mirror changes its sense of polarization (i.e., left-handed becomes right-handed, and vice versa) upon reflection from the surface.

Solution: Assume the surface of the mirror to be the $z = 0$ plane. The electric field of a circularly polarized wave propagating in air (half-space $z < 0$) can be

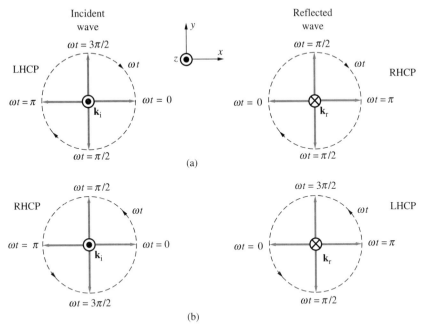

FIGURE 3.6.   Circularly polarized wave incident on an air-conductor interface. An LHCP wave normally incident on a perfect conductor boundary results in a reflected wave which is a RHCP wave as shown in (a) and vice versa as shown in (b). In examining the motion of the electric field vectors as a function of time, it is useful to remember to use the right-hand rule for defining the sense of polarization as discussed in Chapter 2. With the thumb oriented *in the direction of propagation*, the electric field vector should rotate in the same (opposite) direction as the fingers of the right hand for a RHCP (LHCP) wave.

written as

$$\mathbf{E}_i(z) = C_1(\hat{\mathbf{x}} \pm \hat{\mathbf{y}}j)e^{-j\beta_1 z}$$

where the plus sign represents an LHCP wave and the minus sign represents a RHCP wave. The corresponding time-domain expression is

$$\overline{\mathscr{E}}_i(z, t) = \hat{\mathbf{x}}C_1 \cos(\omega t - \beta_1 z) \pm \hat{\mathbf{y}}C_1 \underbrace{\cos(\omega t - \beta_1 z + \pi/2)}_{-\sin(\omega t - \beta_1 z)}$$

Since the tangential component of the total electric field at the surface is zero, the reflected wave can be written as

$$\mathbf{E}_r(z) = -C_1(\hat{\mathbf{x}} \pm \hat{\mathbf{y}}j)e^{+j\beta_1 z}$$

The corresponding time-domain expression $\overline{\mathscr{E}}_r(z, t)$, written together with $\overline{\mathscr{E}}_i(z, t)$ for easy comparison, is

$$\overline{\mathscr{E}}_i(z, t) = \hat{\mathbf{x}}C_1 \cos(\omega t - \beta_1 z) \pm \hat{\mathbf{y}}C_1 \cos(\omega t - \beta_1 z + \pi/2)$$

$$\overline{\mathscr{E}}_r(z, t) = -\hat{\mathbf{x}}C_1 \cos(\omega t + \beta_1 z) \mp \hat{\mathbf{y}}C_1 \cos(\omega t + \beta_1 z + \pi/2)$$

The orientation of the electric field vectors of the incident and reflected waves as obtained from the above expressions are shown in Figure 3.6 at $z = 0$ and at different time instants. Noting the directions of propagation of the waves ($\hat{\mathbf{k}}_i$, $\hat{\mathbf{k}}_r$) as indicated, it is clear from these that the reflected wave has the opposite sense of polarization with respect to the incident wave.

3.1.1   More on the Reflection Process and Radiation Pressure

Consideration of the current flow on the conductor surface provides an alternative way of thinking about the reflection of the incident wave at the boundary. Allowing for the fact that the conductor may not be perfect, the current can be thought to result from the electromotive force induced in the conductor by the time-varying magnetic field (via Faraday's law), which, due to the very low resistivity (i.e., high conductivity), causes a large current flow. This current reradiates[5] a field of its own. In the ideal case of zero resistivity (perfect conductor), the reradiated field must exactly cancel the incident field within the conductor, since any finite field would produce an infinite current. Therefore, the polarity of the electric and magnetic fields of the reradiated field propagating in the $+z$ direction within the conductor are opposite to those of the incident wave, as shown in Figure 3.7. Thus, the electric field of the reradiated wave propagating in the $-z$ direction in medium 1 is $\overline{\mathscr{E}}_{\text{rad}}(z, t) = -\hat{\mathbf{x}} E_{i0} \cos(\omega t + \beta_1 z)$. The magnetic field of the reradiated wave can be found from $\overline{\mathscr{J}}_s$ using the right-hand rule[6] and is in the y direction in medium 1 for an x-directed current. It should be noted that thinking about the reflection process in terms of a reradiated field is simply an alternative method that produces the same result previously arrived at via the resolution of boundary conditions. Both ways of looking at this problem are based on the steady-state (time-harmonic) solution of Maxwell's equations. Note also that, in considering the reradiated field, it is not appropriate to think in terms of a time-sequence of events (e.g., incident wave sets up the current, which *then* reradiates), since our discussion above is limited to sinusoidal steady-state and that therefore the

---

[5]An infinite planar sheet of alternating surface current $\overline{\mathscr{J}}_s = \hat{\mathbf{x}} J_0 \cos(\omega t)$ located at $z = 0$ in free space radiates uniform plane waves propagating away from the sheet in the $+z$ and $-z$ directions, with electric fields of $\overline{\mathscr{E}}(z, t) = \pm \hat{\mathbf{x}}(1/2)\mu_0 c J_0 \cos(\omega t \mp \beta z)$. To see this, one can first find the retarded vector magnetic potential (see Section 6.5) $\overline{\mathscr{A}}$ at an observation point z given by

$$\overline{\mathscr{A}}(z, t) = \hat{\mathbf{x}} \frac{\mu_0}{4\pi} J_0 \int_{\text{entire plane}} \frac{\mathscr{R}e\{e^{j(\omega t - \frac{\omega}{c} z)}\}}{z} \, ds$$

and then determine the fields from $\overline{\mathscr{B}} = \nabla \times \overline{\mathscr{A}}$ and $\partial(\epsilon_0 \overline{\mathscr{E}})/\partial t = \nabla \times (\mu_0^{-1}\overline{\mathscr{B}})$, or from $\overline{\mathscr{E}} = -\hat{\mathbf{x}}\partial\overline{\mathscr{A}}/\partial t$. In order to avoid infinite values of the vector potential, it is necessary to evaluate $\nabla \times \overline{\mathscr{A}}$ before carrying out the surface integral. For further information, see Section 10.7 of W. T. Scott, *Physics of Electricity and Magnetism*, 2nd ed., Wiley, New York, 1966.

[6]Which, by the way, is in essence what the boundary condition $\overline{\mathscr{J}}_s = \hat{\mathbf{n}} \times \overline{\mathscr{H}}$ represents.

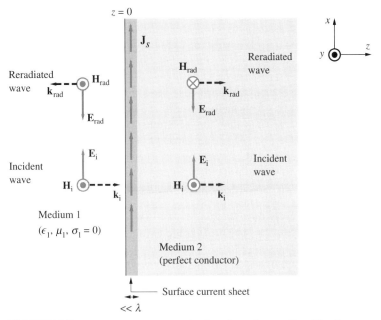

FIGURE 3.7.   A reflected wave can be thought to be reradiated by the surface current in the perfect conductor. Note that the direction of the wave magnetic field is out of the page (into the page) on the left side (right side) of the current sheet, consistent with the right-hand rule as required by Ampère's law. The wave electric field is pointed in the $-x$ direction, as required for the Poynting vector of the two reradiated waves to be respectively in the $-z$ and $+z$ directions.

surface current and the fields are different components of the same time-harmonic solution that have to coexist in order for Maxwell's equations to be satisfied.

Based on our discussions in Section 2.4.4, the uniform plane wave incident at the boundary carries a momentum per unit volume of $(\mu_1\epsilon_1)(\mathcal{E}_1 \times \mathcal{H}_1)$. Since the wave travels at a speed of $(\mu_1\epsilon_1)^{-1/2}$, a time-average momentum per unit area and per unit time (i.e., radiation pressure) of $(\mu_1\epsilon_1)^{1/2}E_{i0}^2/(2\eta_1)$ is delivered to the surface. The reradiated wave carries away the same amount of momentum in the opposite direction, thus recoiling the conductor (via Newton's third law), so the total radiation pressure on the perfectly reflecting surface is

$$\mathbf{P}_{av} = \frac{2\mathbf{S}_{av}}{v_{p1}} = \hat{\mathbf{z}}\frac{(\mu_1\epsilon_1)^{1/2}E_{i0}^2}{\eta_1} = \hat{\mathbf{z}}\epsilon_1 E_{i0}^2$$

The magnitude of this pressure is of course very small, as discussed with specific examples in Section 2.4.4. Nevertheless, the device called Crookes's radiometer operates[7] by taking advantage of the fact that $\mathbf{P}_{av}$ on a perfectly reflecting surface is twice that on a perfectly absorbing surface. As mentioned in Section 2.4.4, this device

---

[7]See Section 6.1 of J. R. Meyer-Arendt, *Introduction to Classical and Modern Optics*, Prentice-Hall, 1972.

consists of four vanes delicately mounted on a vertical axis, one side of each of the vanes being silvered, the other blackened. When light falls on the vanes, they start rotating, the sense of rotation depending on whether the system is in vacuum or whether there is residual gas left in the enclosure. In a near-perfect vacuum, the radiation pressure on the silvered (i.e., reflecting) surfaces is twice that on the absorbing (i.e., black) surfaces, moving the vanes in one way. However, if some residual gas is left in the enclosure, the black surfaces absorb more radiant energy and become warmer than the silvered surfaces. As a result, molecules that collide with the blackened surface rebound with higher velocities than those that strike the silvered surface, thus imparting more momentum to the blackened surface, causing the device to rotate in the opposite direction.

3.2   NORMAL INCIDENCE ON A LOSSLESS DIELECTRIC

In the previous section we considered the special case of normal incidence on a plane boundary in which the second medium was a perfect conductor, within which no electromagnetic field can exist. We now consider the more general case in which the second medium is a lossless dielectric, within which a nonzero electromagnetic field can propagate. When a uniform plane wave propagating in medium 1 is normally incident on an interface with a second medium with a different dielectric constant as shown in Figure 3.8, some of the incident wave energy is transmitted into medium 2 and continues to propagate to the right ($+z$ direction). In the following discussion, we assume both media to be lossless dielectrics (i.e., $\sigma_1, \sigma_2 = 0$). Once again we assume, with no loss of generality under conditions of normal incidence on a planar boundary, that the incident electric field is oriented in the x direction. We also assume that the amplitude E_{i0} of the incident wave is real—once again, with no loss of generality, since this basically amounts to the choice of the time origin, as shown in the previous section. The phasor fields for the incident, reflected, and transmitted waves are given as

$$\mathbf{E}_i(z) = \hat{\mathbf{x}}E_{i0}e^{-j\beta_1 z}$$

$$\mathbf{H}_i(z) = \hat{\mathbf{y}}\frac{E_{i0}}{\eta_1}e^{-j\beta_1 z}$$

Incident wave

$$\mathbf{E}_r(z) = \hat{\mathbf{x}}E_{r0}e^{+j\beta_1 z}$$

$$\mathbf{H}_r(z) = -\hat{\mathbf{y}}\frac{E_{r0}}{\eta_1}e^{+j\beta_1 z}$$

Reflected wave

$$\mathbf{E}_t(z) = \hat{\mathbf{x}}E_{t0}e^{-j\beta_2 z}$$

$$\mathbf{H}_t(z) = \hat{\mathbf{y}}\frac{E_{t0}}{\eta_2}e^{-j\beta_2 z}$$

Transmitted wave

where $\beta_2 = \omega\sqrt{\mu_2\epsilon_2}$ and $\eta_2 = \sqrt{\mu_2/\epsilon_2}$ are, respectively, the phase constant and the intrinsic impedance for medium 2. Note that E_{t0} is the amplitude (yet to be

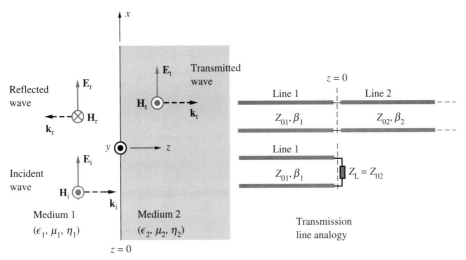

FIGURE 3.8.   Uniform plane wave normally incident on a lossless dielectric boundary. Also shown on the right is the transmission line equivalent. Note that the second medium can either be thought of as an infinitely long line with characteristic impedance Z_{02} or as a load with load impedance $Z_L = Z_{02}$.

determined) of the transmitted wave at $z = 0$. In Figure 3.8 we have defined the polarities of $\mathbf{E}_i$ and $\mathbf{E}_r$ to be the same, and taken (as before in Figure 3.1) $\mathbf{H}_r$ to be in the $-y$ direction so that $\mathbf{E}_r \times \mathbf{H}_r$ is in the $-z$ direction. Note that, at this point, the selected orientations of $\mathbf{E}$ and $\mathbf{H}$ for the different waves (incident, reflected, transmitted) are simply convenient choices.[8] The boundary conditions will determine whether the phasor fields at the boundary are positive or negative according to these assumed conventions.

3.2.1   Reflection and Transmission Coefficients

We now proceed by taking the incident wave as given and determine the properties of the reflected and transmitted waves so that the fundamental boundary conditions for electromagnetic fields are satisfied at the interface, where all three waves can be related to one another. We have two unknown quantities E_{r0} and E_{t0} to be determined in terms of the incident field amplitude E_{i0}. We will use two boundary conditions to determine them. These two conditions are the continuity of the tangential components of both the electric and magnetic fields across the interface. We thus have

$$\mathbf{E}_i(z = 0) + \mathbf{E}_r(z = 0) = \mathbf{E}_t(z = 0) \quad \longrightarrow \quad E_{i0} + E_{r0} = E_{t0}$$

$$\mathbf{H}_i(z = 0) + \mathbf{H}_r(z = 0) = \mathbf{H}_t(z = 0) \quad \longrightarrow \quad \left(\frac{E_{i0}}{\eta_1} - \frac{E_{r0}}{\eta_1}\right) = \frac{E_{t0}}{\eta_2}$$

---

[8]The fact that selected orientations are simply convenient choices is important, especially since a different convention may be adopted in other texts.

The solution of these two equations yields

$$E_{r0} = \frac{\eta_2 - \eta_1}{\eta_2 + \eta_1} E_{i0} \qquad E_{t0} = \frac{2\eta_2}{\eta_2 + \eta_1} E_{i0}$$

We now define[9] the reflection and transmission coefficients as follows:

$$\Gamma = \frac{E_{r0}}{E_{i0}} = \frac{\eta_2 - \eta_1}{\eta_2 + \eta_1} \qquad \text{Reflection coefficient} \qquad [3.2]$$

$$\mathcal{T} = \frac{E_{t0}}{E_{i0}} = \frac{2\eta_2}{\eta_2 + \eta_1} \qquad \text{Transmission coefficient} \qquad [3.3]$$

where we note that $(1 + \Gamma) = \mathcal{T}$. Although we limited our discussions here to loss-less dielectrics, the above relations are fully applicable when the media are dissipa-tive, as long as the proper complex values of η_1 and η_2 are used. Note that, physi-cally, the above coefficients are derived from application of the boundary conditions, which are valid for all media[10] in general. Complex reflection and transmission coef-ficients may result when η_2 and/or η_1 are complex (i.e., one or both of the media are lossy), meaning that in addition to the differences in amplitudes, phase shifts are also introduced between the incident, reflected, and transmitted fields at the interface.

The reflection coefficient expression also could have been obtained on the basis of the transmission line analogy depicted in Figure 3.8. For a transmission line of characteristic impedance Z_{01} terminated in a load impedance Z_L, the load reflection coefficient Γ_L is given[11] by

$$\Gamma_L = \frac{Z_L - Z_{01}}{Z_L + Z_{01}}$$

which, upon substitution of $Z_L \rightarrow \eta_2$ and $Z_{01} \rightarrow \eta_1$, is identical to [3.2]. Note that the load could just as well be an infinitely long transmission line with characteristic impedance $Z_{02} \neq Z_{01}$, presenting an impedance $Z_L = Z_{02}$ at the junction, as depicted in Figure 3.8. The analogy of the junction between two transmission lines of infinite extent and the interface between two different dielectric media of infinite extent is thus quite clear.

For most dielectrics and insulators, the magnetic permeability does not differ appreciably from its free-space value, so the expressions derived for Γ and $\mathcal{T}$ can also be simply rewritten in terms of ϵ_1 and ϵ_2. So, when $\mu_1 = \mu_2 = \mu_0$, we have

$$\Gamma = \frac{\sqrt{\epsilon_1} - \sqrt{\epsilon_2}}{\sqrt{\epsilon_1} + \sqrt{\epsilon_2}} \qquad \mathcal{T} = \frac{2\sqrt{\epsilon_1}}{\sqrt{\epsilon_1} + \sqrt{\epsilon_2}}$$

---

[9]Note that definition of Γ as the ratio of the reflected to the incident *electric* fields (rather than magnetic fields) is simply a matter of convention.

[10]Note that when medium 2 is a perfect conductor, for which we have $\eta_2 = 0$, the expressions reduce to $\Gamma = -1$ and $\mathcal{T} = 0$, as expected.

[11]See Chapter 3 of U. S. Inan and A. S. Inan, *Engineering Electromagnetics*, Addison Wesley Longman, 1999.

In most optical applications, the electromagnetic properties of the dielectric materials are expressed in terms of the *refractive index,* defined as $n \equiv c/v_p = \beta c/\omega = \sqrt{\mu_r \epsilon_r}$, where v_p is the wave phase velocity. For the nonmagnetic case ($\mu_1 = \mu_2 = \mu_0$), the reflection and transmission coefficients can also be expressed in terms of the refractive indices as

$$\Gamma = \frac{n_1 - n_2}{n_1 + n_2} \qquad \mathcal{T} = \frac{2n_1}{n_1 + n_2}$$

Examples 3-3 and 3-4 illustrate the application of [3.2] and [3.3] to simple dielectric interfaces.

Example 3-3:   Air–water interface.   Calculate the reflection and transmission coefficients for a radio-frequency uniform plane wave traveling in air incident normally upon a calm lake. Assume the water in the lake to be lossless with a relative dielectric constant of $\epsilon_r \simeq 81$. Note that for water, we have $\mu_r = 1$.

Solution: The reflection coefficient is given by

$$\Gamma = \frac{\sqrt{\epsilon_{1r}} - \sqrt{\epsilon_{2r}}}{\sqrt{\epsilon_{1r}} + \sqrt{\epsilon_{2r}}} = \frac{1 - \sqrt{81}}{1 + \sqrt{81}} = -0.8$$

and the transmission coefficient is

$$\mathcal{T} = \frac{2\sqrt{\epsilon_{1r}}}{\sqrt{\epsilon_{1r}} + \sqrt{\epsilon_{2r}}} = \frac{2 \times 1}{1 + \sqrt{81}} = 0.2$$

both of which satisfy the relationship $1 + \Gamma = \mathcal{T}$.

Example 3-4:   Air–germanium interface.   Germanium (Ge) is a popular material used in infrared optical system designs in either the 3- to 5-μm or 8- to 12-μm spectral bands. Germanium has an index of refraction of approximately 4.0 at these wavelengths. Calculate the reflection and transmission coefficients for an uncoated air–germanium surface.

Solution: The reflection coefficient is given by

$$\Gamma = \frac{n_1 - n_2}{n_1 + n_2} = \frac{1 - 4}{1 + 4} = -0.6$$

and the transmission coefficient is

$$\mathcal{T} = \frac{2n_1}{n_1 + n_2} = \frac{2 \times 1}{1 + 4} = 0.4$$

Once again, the relationship $1 + \Gamma = \mathcal{T}$ is satisfied.

3.2.2   Propagating and Standing Waves

In Section 3.1 we determined that for a uniform plane wave normally incident on a perfect conductor, the total electric field in medium 1 consisted of a purely standing wave. In the case of normal incidence on a lossless dielectric, we expect at least a portion of the total wave in medium 1 to be propagating in the z direction in order to supply the electromagnetic power taken away from the interface by the transmitted wave in medium 2. To determine the nature of the wave in medium 1, we now examine the total electric field in medium 1. We have

$$\mathbf{E}_1(z) = \mathbf{E}_i(z) + \mathbf{E}_r(z) = \hat{\mathbf{x}} E_{i0}(e^{-j\beta_1 z} + \Gamma e^{+j\beta_1 z})$$

$$= \hat{\mathbf{x}} E_{i0}[(1 + \Gamma)e^{-j\beta_1 z} - \Gamma e^{-j\beta_1 z} + \Gamma e^{+j\beta_1 z}]$$

$$\mathbf{E}_1(z) = \hat{\mathbf{x}} E_{i0}[(1 + \Gamma)e^{-j\beta_1 z} + \Gamma j2 \sin(\beta_1 z)]$$

The corresponding space-time field is

$$\overline{\mathscr{E}}_1(z, t) = \hat{\mathbf{x}} E_{i0}[\ \underbrace{\mathscr{T} \cos(\omega t - \beta_1 z)}_{\text{Propagating wave}} + \underbrace{(-2\Gamma) \sin(\beta_1 z) \sin(\omega t)}_{\text{Standing wave}}\ ]$$

where we note once again that E_{i0} was assumed to be real. Note that the propagating wave in medium 1 sustains the transmitted wave in medium 2, whereas the standing wave is produced by the sum of the reflected wave and a portion of the incident wave. We can also express the total electric field phasor in medium 1 as

$$\mathbf{E}_1(z) = \hat{\mathbf{x}} E_{i0} e^{-j\beta_1 z}(1 + \Gamma e^{j2\beta_1 z}) \qquad\qquad [3.4]$$

The associated total magnetic field phasor in medium 1 is

$$\mathbf{H}_1(z) = \hat{\mathbf{y}} \frac{E_{i0}}{\eta_1}(e^{-j\beta_1 z} - \Gamma e^{j\beta_1 z})$$

$$= \hat{\mathbf{y}} \frac{E_{i0}}{\eta_1}[(1 + \Gamma)e^{-j\beta_1 z} - \Gamma e^{-j\beta_1 z} - \Gamma e^{+j\beta_1 z}]$$

$$= \hat{\mathbf{y}} \frac{E_{i0}}{\eta_1}[(1 + \Gamma)e^{-j\beta_1 z} - 2\Gamma \cos(\beta_1 z)]$$

The corresponding space-time function is

$$\overline{\mathscr{H}}_1(z, t) = \hat{\mathbf{y}} \frac{E_{i0}}{\eta_1}[\ \underbrace{\mathscr{T} \cos(\omega t - \beta_1 z)}_{\text{Propagating wave}} + \underbrace{(-2\Gamma) \cos(\beta_1 z) \cos(\omega t)}_{\text{Standing wave}}\ ]$$

The total magnetic field phasor in medium 1 can also be expressed in a compact form as

$$\mathbf{H}_1(z) = \hat{\mathbf{y}} \frac{E_{i0}}{\eta_1} e^{-j\beta_1 z}(1 - \Gamma e^{j2\beta_1 z})$$

In medium 2, we have only one wave propagating in the $+z$ direction, represented as

$$\mathbf{E}_2(z) = \mathbf{E}_t(z) = \hat{\mathbf{x}}\mathcal{T}E_{i0}e^{-j\beta_2 z}$$

$$\mathbf{H}_2(z) = \mathbf{H}_t(z) = \hat{\mathbf{y}}\frac{\mathcal{T}}{\eta_2}E_{i0}e^{-j\beta_2 z}$$

with the corresponding space-time field expressions of

$$\overline{\mathcal{E}}_t(z, t) = \hat{\mathbf{z}}\mathcal{T}E_{i0}\cos(\omega t - \beta_2 z)$$

$$\overline{\mathcal{H}}_t(z, t) = \hat{\mathbf{y}}\frac{\mathcal{T}}{\eta_2}E_{i0}\cos(\omega t - \beta_2 z)$$

To understand better the relationships between the incident, reflected, and transmitted waves, and the matching of the fields at the boundary, we can examine the instantaneous electric field waveforms shown in Figure 3.9a. These displays are similar to the waveforms in Figure 3.2, except that the incident and reflected waves in medium 1 are shown separately and only at one instant of time, namely, at $\omega t = 0$. The waveforms are shown for the two different cases of $\epsilon_1 < \epsilon_2$ and $\epsilon_1 > \epsilon_2$ assuming nonmagnetic media, or $\mu_1 = \mu_2 = \mu_0$. The difference in wavelength ($\lambda = 2\pi/\beta$) in the two media is readily apparent; for the cases shown, ϵ_1 and ϵ_2 differ by a factor of 4, which means that wavelengths differ by a factor of 2.

For $\epsilon_1 < \epsilon_2$, once again assuming $\mu_1 = \mu_2$, we note that the amplitude of the transmitted wave is smaller than that of the incident wave (i.e., $E_{t0} < E_{i0}$, or $\mathcal{T} < 1$) and the reflection coefficient Γ is negative. On the other hand, for $\epsilon_1 > \epsilon_2$, the transmitted wave amplitude is actually larger than that of the incident wave ($E_{t0} > E_{i0}$), and the reflected wave electric field is in the same direction as that of the incident wave (i.e., $\Gamma > 0$) as expected from [3.2]. Note, however, that $E_{t0} > E_{i0}$ does not pose any problems with the conservation of wave energy, since the Poynting vector of the transmitted wave is given by $E_{t0}^2/(2\eta_2)$, and power conservation holds true, as we shall see below.

Also shown, in Figure 3.9b, are the standing-wave patterns: the magnitude of $\mathbf{E}_1(z)$ and $\mathbf{E}_2(z)$ for the two different cases of $\epsilon_1 < \epsilon_2$ and $\epsilon_1 > \epsilon_2$. Note that $\mathbf{E}_1(z)$ is given by [3.4], so its magnitude varies between a maximum value of $|\mathbf{E}_1(z)|_{\text{max}} = 1 + |\Gamma|$ and a minimum value of $|\mathbf{E}_1(z)|_{\text{min}} = |-|\Gamma||$ while $|\mathbf{E}_2(z)| = \mathcal{T}E_{i0}$. These patterns are entirely analogous to standing-wave patterns for transmission lines.[12] In medium 2, there is only one wave, and the standing-wave ratio is thus unity. In medium 1, however, the interference between the reflected and incident waves

---

[12]For transmission lines, standing-wave patterns are important in practice because, although the rapid temporal variations of the line voltages and currents are not easily accessible, the locations of the voltage minima and maxima and the ratio of the voltage maxima to minima are often readily measurable. A key parameter used to describe a standing-wave pattern is the standing-wave ratio, or S, defined as $S = |V(z)|_{\text{max}}/|V(z)|_{\text{min}}$.

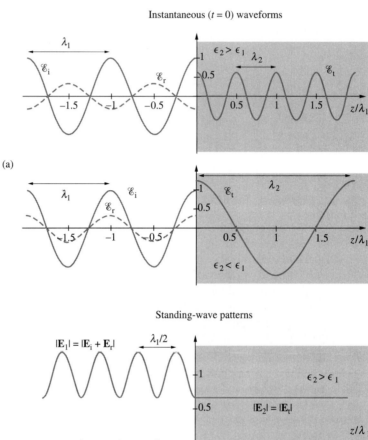

FIGURE 3.9. Electric field waveforms and standing-wave patterns on both sides of the interface. (a) Electric field waveforms are given separately for $\epsilon_1 < \epsilon_2$ (in the example shown, $\epsilon_{2r} = 4\epsilon_{1r}$) and for $\epsilon_1 > \epsilon_2$ ($\epsilon_{1r} = 4\epsilon_{2r}$). The electric field waveforms are shown as a function of position (z) at time instant $\omega t = 0$, with E_{i0} taken to be unity. (b) Also shown are the standing-wave patterns for the same two cases, $\epsilon_{2r} = 4\epsilon_{1r}$ and $\epsilon_{1r} = 4\epsilon_{2r}$. Note that the standing-wave patterns are identical in form to those for voltage waves on a transmission line.

produces a standing-wave ratio of 2, since we have chosen $\epsilon_{2r} = 4\epsilon_{1r}$ or $\epsilon_{1r} = 4\epsilon_{2r}$ for the two cases shown. Note that in this context, the standing-wave ratio S_1 is defined as

$$S_1 = \frac{|\mathbf{E}_1(z)|_{\max}}{|\mathbf{E}_1(z)|_{\min}} \equiv \frac{1 + |\Gamma|}{1 - |\Gamma|} = \frac{1 + 1/3}{1 - 1/3} = 2$$

3.2.3   Electromagnetic Power Flow

We can now examine the electromagnetic power flow in both medium 1 and medium 2. Noting that using the complex Poynting vector and the phasor expressions for the fields is most convenient, we have in medium 2

$$(\mathbf{S}_{av})_2 = \frac{1}{2}\mathcal{R}e\{\mathbf{E}_t(z) \times \mathbf{H}_t^*(z)\} = \hat{\mathbf{z}}\frac{E_{i0}^2}{2\eta_2}\mathcal{T}^2$$

which follows from Section 2.4 assuming E_{i0} is real. In medium 1 we have

$$(\mathbf{S}_{av})_1 = \tfrac{1}{2}\mathcal{R}e\{\mathbf{E}_1 \times \mathbf{H}_1^*\}$$

$$(\mathbf{S}_{av})_1 = \hat{\mathbf{z}}\frac{E_{i0}^2}{2\eta_1}\mathcal{R}e\{e^{-j\beta_1 z}(1 + \Gamma e^{j2\beta_1 z})e^{+j\beta_1 z}(1 - \Gamma e^{-j2\beta_1 z})\}$$

$$= \hat{\mathbf{z}}\frac{E_{i0}^2}{2\eta_1}\mathcal{R}e\{(1 - \Gamma^2) + \Gamma(e^{j2\beta_1 z} - e^{-j2\beta_1 z})\}$$

$$= \hat{\mathbf{z}}\frac{E_{i0}^2}{2\eta_1}\mathcal{R}e\{(1 - \Gamma^2) + j2\Gamma \sin(2\beta_1 z)\}$$

$$(\mathbf{S}_{av})_1 = \hat{\mathbf{z}}\frac{E_{i0}^2}{2\eta_1}(1 - \Gamma^2)$$

Note that the time-average power in medium 1 propagates in the $+z$ direction and basically supplies the power carried by the transmitted wave in medium 2. The average power flows in both media must be equal, since no energy is stored or dissipated at the interface between the two media. We thus have

$$(\mathbf{S}_{av})_1 = (\mathbf{S}_{av})_2 \quad \longrightarrow \quad \boxed{1 - \Gamma^2 = \frac{\eta_1}{\eta_2}\mathcal{T}^2}$$

which can also be independently verified using the expressions [3.2] and [3.3] for Γ and $\mathcal{T}$.

Example 3-5 applies the foregoing expressions for electromagnetic power flow to an air–glass interface.

Example 3-5:   Reflectance of glass.   A beam of light is incident normally from air on a BK-7 glass interface. Calculate the reflection coefficient and percent of incident energy reflected if the BK-7 glass has an index of refraction of $n = 1.52$.

Solution: Assuming $\mu_1 = \mu_2$, the reflection coefficient is given by

$$\Gamma = \frac{n_1 - n_2}{n_1 + n_2} = \frac{1 - 1.52}{1 + 1.52} \simeq -0.206$$

The power density of the incident wave is given by $|(\mathbf{S}_{av})_i| = \frac{1}{2}C_1^2/\eta_1$, where C_1 is the magnitude of the electric field of the incident wave. The power density of the reflected wave is $|(\mathbf{S}_{av})_r| = \frac{1}{2}(\Gamma C_1)^2/\eta_1 = \Gamma^2 |(\mathbf{S}_{av})_i|$. Therefore, the percent of incident energy reflected back is given by

$$\frac{|(\mathbf{S}_{av})_r|}{|(\mathbf{S}_{av})_i|} \times 100 \simeq \Gamma^2 \times 100 \simeq (0.206)^2 \times 100 \simeq 4.26\%$$

Thus, only ~4% of the incident power is reflected by the glass interface. In some applications, this loss may be considered as a significant loss. For example, a camera lens often consists of three or more separate lenses, representing six or more air–glass interfaces. If ~4% of the incoming energy reflects back every time light passes through one of these interfaces, up to ~22% of the original energy is lost during each traverse of light through the lens. It is possible to reduce these losses significantly by introducing antireflection coating on the glass surface, as discussed in the next section.

3.3 MULTIPLE DIELECTRIC INTERFACES

Many practical applications involve the reflection and refraction of electromagnetic waves from dielectric or metallic surfaces that are coated with another dielectric material to reduce reflections and to improve the coupling of the wave energy. The underlying principle in such cases is identical to that in the case of quarter-wave transformer matching of transmission lines.[13] Some of the applications include antireflection coatings to improve light transmission of a lens, thin-film coatings on optical components to reduce losses selectively over narrow wavelength ranges, metal mirror coatings, flexible metal foils to provide EMI shielding, radar domes (radomes), and stealth technology in which aircraft are specially coated to reduce radar reflectivity.[14]

---

[13]See Section 3.5 of U. S. Inan and A. S. Inan, *Engineering Electromagnetics*, Addison Wesley Longman, 1999.

[14]An example of a coating material that reduces reflections is ferrite-titanate, for which $\mu_r \simeq \epsilon_r \simeq 60(2 - j1)$ at 100 MHz, so that its characteristic impedance, $\eta = \sqrt{\mu/\epsilon}$, is approximately equal to that of free space (377Ω). See Sections 12–16 of J. D. Kraus, *Electromagnetics,* 4th ed., McGraw-Hill, 1992. For tabulated properties of a commercially used ferrite material as a function of frequency, see R. R. Delyser, C. L. Holloway, R. T. Johnk, A. R. Ondrejka, and M. Kauda, Figure of merit for low frequency analysis chambers based on absorber reflection coefficients, *IEEE Trans. on Electromagnetic Compatibility*, 38(4), pp. 576–584, November 1996. To provide radar invisibility over a range of frequencies, multilayered coatings are used. See J. A. Adam, How to design an "invisible" aircraft, *IEEE Spectrum*, p. 30, April 1988.

3.3.1 Application of Boundary Conditions

A fundamentally based determination of the reflected and transmitted waves requires the use of the general solutions of the wave equation with two undetermined constants (the amplitudes of the component waves propagating in the $+z$ and $-z$ directions) in each medium, and application of the boundary conditions to determine these unknown constants. An alternative method is to rely on a transmission line analogy as implied in Figure 3.10, which reduces the problem to one of impedance transformation. The latter approach is valid because *within* each medium the wave propagation is governed by equations identical to the transmission line equations (see footnote 4 of Chapter 2) and because the boundary conditions of the continuity of the tangential electric and magnetic fields are precisely equivalent to the boundary conditions for the continuity of voltage and current at a line termination. We proceed with the general solution of the wave equation, but we also take advantage of the transmission line analogy in interpreting and generalizing our results. The impedance transformation method is discussed in the next subsection.

The expressions for the total wave fields in the three media shown in Figure 3.10 can be written in their most general form as

$$\mathbf{E}_1(z) = \mathbf{E}_{1i} + \mathbf{E}_{1r} = \hat{\mathbf{x}} E_{i0}[e^{-j\beta_1(z+d)} + \Gamma_{\text{eff}} e^{+j\beta_1(z+d)}]$$

$$z < -d \quad \text{Medium 1}$$

$$\mathbf{H}_1(z) = \mathbf{H}_{1i} + \mathbf{H}_{1r} = \hat{\mathbf{y}} \frac{E_{i0}}{\eta_1}[e^{-j\beta_1(z+d)} \quad \Gamma_{\text{eff}} e^{+j\beta_1(z+d)}]$$

$$\mathbf{E}_2(z) = \mathbf{E}_{2f} + \mathbf{E}_{2r} = \hat{\mathbf{x}} E_{20}[e^{-j\beta_2 z} + \Gamma_{23} e^{+j\beta_2 z}]$$

$$-d < z < 0 \quad \text{Medium 2}$$

$$\mathbf{H}_2(z) = \mathbf{H}_{2f} + \mathbf{H}_{2r} = \hat{\mathbf{y}} \frac{E_{20}}{\eta_2}[e^{-j\beta_2 z} - \Gamma_{23} e^{+j\beta_2 z}]$$

$$\mathbf{E}_3(z) = \mathbf{E}_{3f} = \hat{\mathbf{x}} \mathcal{T}_{\text{eff}} E_{i0} e^{-j\beta_3 z}$$

$$z > 0 \quad \text{Medium 3}$$

$$\mathbf{H}_3(z) = \mathbf{H}_{3f} = \hat{\mathbf{y}} \frac{\mathcal{T}_{\text{eff}} E_{i0}}{\eta_3} e^{-j\beta_3 z}$$

where the wave components $\mathbf{E}_{2f}$ and $\mathbf{E}_{2r}$ in medium 2 are, respectively, the forward- and reverse-propagating waves that constitute the general solution of the wave equation, and where we recognize that there is no reverse-propagating wave in medium 3. Note that Γ_{23} is the reflection coefficient at the interface $z = 0$, whereas Γ_{eff} is an "effective" reflection coefficient[15] at the interface $z = -d$ that accounts for the

---

[15] In this context, by an "effective" reflection coefficient we mean one that is defined as the ratio of the reflected to the incident electric field phasors at the boundary at $z = -d$, so that Γ_{eff} determines the amplitude and phase of the reflected wave in medium 1 in terms of that of the incident wave. Note, however, that $\Gamma_{\text{eff}} \neq (\eta_2 - \eta_1)/(\eta_2 + \eta_1)$, since Γ_{eff} must also include the effect of the second boundary at $z = 0$.

effects of the presence of the third medium (and the second interface at $z = 0$). Similarly, $\mathcal{T}_{\text{eff}}$ is an "effective" transmission coefficient. However, due to the presence of the second boundary, Γ_{eff} is *not* simply the single-layer reflection coefficient $(\eta_2 - \eta_1)/(\eta_2 + \eta_1)$. The solution written above for medium 1 simply recognizes that the amplitudes of the wave components traveling in the $+z$ and $-z$ directions are represented by E_{i0} and $\Gamma_{\text{eff}} E_{i0}$, respectively. However, viewing Γ_{eff} as a reflection coefficient is useful, since in practice (e.g., dielectric coating of glass) the purpose of the second layer is typically to make $\Gamma_{\text{eff}} = 0$ in order to eliminate reflections from the entire multiple-dielectric interface system. Note that although the single-boundary reflection coefficient Γ was necessarily real (i.e., $\Gamma = \rho e^{j(0 \text{ or } \pi)}$) for lossless media, the effective reflection coefficient $\Gamma_{\text{eff}} = \rho_{\text{eff}} e^{j\phi_\Gamma}$ is in general complex even in the lossless case. That this should be so can be easily understood by considering the transmission line analogy shown in Figure 3.10; the dielectric layer of thickness d simply acts as an impedance transformer, which takes the intrinsic impedance η_3 of medium 3 and presents it at the $z = -d$ interface as another impedance, which in general is complex. Similarly, the effective transmission coefficient $\mathcal{T}_{\text{eff}} = |\mathcal{T}_{\text{eff}}| e^{j\phi_{\mathcal{T}}}$ is in general also complex.

We can now apply the electromagnetic boundary conditions, namely, the requirement that the tangential components of the electric and magnetic fields be continuous at the two interfaces. At $z = -d$, we find

$$E_{i0}(1 + \Gamma_{\text{eff}}) = E_{20}(e^{+j\beta_2 d} + \Gamma_{23} e^{-j\beta_2 d}) \qquad [3.5a]$$

$$\frac{E_{i0}}{\eta_1}(1 - \Gamma_{\text{eff}}) = \frac{E_{20}}{\eta_2}(e^{+j\beta_2 d} - \Gamma_{23} e^{-j\beta_2 d}) \qquad [3.5b]$$

Similarly, at $z = 0$ we find

$$E_{20}(1 + \Gamma_{23}) = \mathcal{T}_{\text{eff}} E_{i0} \qquad [3.6a]$$

$$\frac{E_{20}}{\eta_2}(1 - \Gamma_{23}) = \frac{\mathcal{T}_{\text{eff}} E_{i0}}{\eta_3} \qquad [3.6b]$$

Note that we have four simultaneous equations in terms of the four unknowns E_{20}, $\mathcal{T}_{\text{eff}}$, Γ_{eff}, and Γ_{23}, which can be determined in terms of the incident field amplitude E_{i0} and the parameters of the media, η_1, η_2, and η_3.

To solve, we start by multiplying [3.6b] by η_3 and subtracting from [3.6a] to find

$$E_{20}\left(1 - \frac{\eta_3}{\eta_2}\right) + \Gamma_{23} E_{20}\left(1 + \frac{\eta_3}{\eta_2}\right) = 0$$

or

$$\Gamma_{23} = \frac{\dfrac{\eta_3}{\eta_2} - 1}{\dfrac{\eta_3}{\eta_2} + 1} = \frac{\eta_3 - \eta_2}{\eta_3 + \eta_2}$$

which is as expected, since there is only one wave in medium 3, and therefore the matching of the boundary conditions at the $z = 0$ interface is identical to that of a two-medium, single-interface problem as studied in the previous section. Accordingly, the expression for Γ_{23} is simply [3.2] applied to the interface between media 2 and 3.

We can now rewrite equations [3.5a] and [3.5b] using the above expression for Γ_{23}, divide them by one another, and manipulate further to find

$$\Gamma_{\text{eff}} = \frac{(\eta_2 - \eta_1)(\eta_3 + \eta_2) + (\eta_2 + \eta_1)(\eta_3 - \eta_2)e^{-2j\beta_2 d}}{(\eta_2 + \eta_1)(\eta_3 + \eta_2) + (\eta_2 - \eta_1)(\eta_3 - \eta_2)e^{-2j\beta_2 d}} \qquad [3.7]$$

The effective transmission coefficient $\mathcal{T}_{\text{eff}}$ can be also determined by similar manipulations of equations [3.5a] and [3.5b]; we find

$$\mathcal{T}_{\text{eff}} = \frac{4\eta_2\eta_3 e^{-j\beta_2 d}}{(\eta_2 + \eta_1)(\eta_3 + \eta_2) + (\eta_2 - \eta_1)(\eta_3 - \eta_2)e^{-j2\beta_2 d}} \qquad [3.8]$$

Note that the foregoing expressions for Γ_{eff} and $\mathcal{T}_{\text{eff}}$ are completely general, since they were derived by applying the fundamental electromagnetic boundary conditions. Although our treatment in this section is limited to the cases of interfaces involving lossless dielectrics, [3.7] and [3.8] are also fully applicable when one or more of the three media is lossy, in which case η_1, η_2, and η_3 may be complex and $j\beta_2$ must be replaced by $\gamma_2 = \alpha_2 + j\beta_2$. We discuss multiple interfaces involving lossy media in Section 3.7, using the generalized forms of [3.7] and [3.8].

Based on power arguments similar to those used in Section 3.2 for single interfaces, the net time-average electromagnetic power propagating in the $+z$ direction in medium 1 must be equal to that carried in the same direction by the transmitted wave in medium 3, because medium 2 is lossless and therefore cannot dissipate power. Accordingly, the following relation between Γ_{eff} and $\mathcal{T}_{\text{eff}}$ must hold true:

$$1 - |\Gamma_{\text{eff}}|^2 = \frac{\eta_1}{\eta_3}|\mathcal{T}_{\text{eff}}|^2$$

To illustrate the dependence of the reflection coefficient on the thickness d of the dielectric slab, we study the practically useful case of nonmagnetic media with $\epsilon_1 < \epsilon_2 < \epsilon_3$ with $\epsilon_1 = \epsilon_0$, and for two different example media 3, namely, water ($\epsilon_3 = 81\epsilon_0$) and glass ($\epsilon_3 = 2.25\epsilon_0$). Plots of the magnitudes and phases of Γ_{eff} for the two cases are shown in Figure 3.11. For each case, we show results for three different values of ϵ_2.

We note from Figure 3.11 that Γ_{eff} varies significantly as a function of d. In each case, the minimum value of $|\Gamma_{\text{eff}}|$ (and hence the maximum value of $|\mathcal{T}_{\text{eff}}|$) occurs at $d = \lambda_2/4$, and further, $|\Gamma_{\text{eff}}| = 0$ when $\epsilon_2 = \sqrt{\epsilon_1\epsilon_3}$. Under these conditions, all the energy is transmitted from medium 1 into medium 3. We note that with $\epsilon_2 = \sqrt{\epsilon_1\epsilon_3}$

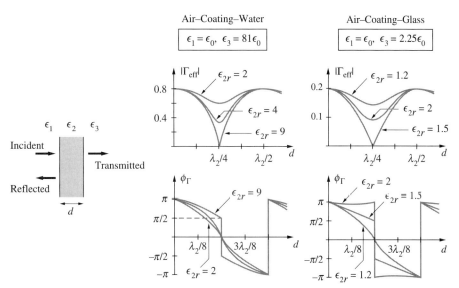

FIGURE 3.11. **Reflection and transmission from a dielectric slab.** The reflection coefficients of dielectric slabs are shown as functions of the slab thickness d (given in terms of the wavelength λ_2 in the slab) and dielectric constant ϵ_{2r}. Results are given for two examples, namely, the air–slab–water and air–slab–glass interfaces.

(or $\eta_2^2 = \eta_1 \eta_3$) and $d = \lambda_2/4$ we have from [3.8]

$$\mathcal{T}_{eff} = \sqrt{\frac{\eta_3}{\eta_1}} e^{-j\pi/2}$$

so that the condition

$$1 - |\Gamma_{eff}|^2 = \frac{\eta_1}{\eta_3} |\mathcal{T}_{eff}|^2$$

is satisfied with $\Gamma_{eff} = 0$ for $d = \lambda_2/4$. The $e^{-j\pi/2}$ factor simply represents the phase difference over a distance $d = \lambda_2/4$, or the fact that, for example at $\omega t = 0$, the peak of the incident wave is at $z = -d = -\lambda_2/4$, whereas the peak of the transmitted wave at $z = 0$ as measured occurs at $\omega t = \pi/2$ rather that at $\omega t = 0$.

When the slab thickness is $d = \lambda_2/2$, we have from [3.7]

$$\Gamma_{eff} = \frac{\eta_3 - \eta_1}{\eta_3 + \eta_1}$$

so that the system behaves as if the slab were not present. In the air–slab–water example we have $\Gamma_{13} = (1 - \sqrt{81})/(1 + \sqrt{81}) = -0.8 = 0.8e^{j\pi}$, as can be also seen from Figure 3.11. These results are reminiscent of the quarter-wavelength or half-wavelength transmission line impedance transformers and in fact can be arrived at by using the transmission line analogy implied in Figure 3.10, as we show in the next subsection.

3.3.2 Impedance Transformation and Transmission Line Analogy

The transmission line analogy implied in Figure 3.10 provides a useful and systematic method of solution of the problem of multiple dielectric interfaces, especially those involving more than three dielectric regions. Such a solution method, based on successive impedance transformations on transmission lines, can follow along lines very similar to methods used in transmission line analysis, specifically in determining the input impedance of cascaded transmission lines.[16] For this purpose, we define the concept of *wave impedance* of the total electromagnetic field at any position z in a given medium as

$$Z(z) \equiv \frac{[E_x(z)]_{\text{total}}}{[H_y(z)]_{\text{total}}}$$

where we implicitly assume a wave with an x-directed electric field propagating in the z direction. Note that, in this context, $E_x(z)$ and $H_y(z)$ are scalar quantities entirely analogous to the total line voltage $V(z)$ and the total line current $I(z)$ on a transmission line. To appreciate this analogy better, consider the expressions for the total fields $E_{2x}(z)$ and $H_{2y}(z)$ in medium 2 and the total line voltage and current expressions on the analogous line 2 (see Figure 3.10):

$$E_{2x}(z) = E_{20}[e^{-j\beta_2 z} + \Gamma_{23}e^{+j\beta_2 z}] \qquad V(z) = V_{20}[e^{-j\beta_2 z} + \Gamma_L e^{+j\beta_2 z}]$$

$$H_{2y}(z) = \frac{E_{20}}{\eta_2}[e^{-j\beta_2 z} - \Gamma_{23}e^{+j\beta_2 z}] \qquad I(z) = \frac{V_{20}}{Z_{02}}[e^{-j\beta_2 z} - \Gamma_L e^{+j\beta_2 z}]$$

where V_{20} is a constant and $\Gamma_L = (Z_L - Z_0)/(Z_L + Z_0)$. Note that the line impedance $Z(z)$ (analogous to the wave impedance) at any point on line 2 is similarly defined as $Z(z) = V(z)/I(z)$.

Referring back to the three-media problem depicted in Figure 3.10, the wave impedance of the total field in medium 3 is

$$Z_3(z) = \frac{E_{3x}(z)}{H_{3y}(z)} = \eta_3$$

Note that since the field in medium 3 consists only of a single wave traveling in the $+z$ direction, $Z_3(z) = \eta_3$ is independent of position z. This result is similar to the line impedance on an infinitely long or match-terminated transmission line, i.e., $Z(z) = Z_{03}$, and η_3 is thus entirely analogous to the characteristic impedance of this line.

In media 2 and 1, on the other hand, the total fields are constituted by two waves traveling in the $+z$ and $-z$ directions, so the wave impedances are not simply η_2 or η_1. In medium 2, for example, the wave impedance has to take into account medium

---

[16]See Sections 3.4 and 3.5 and Examples 3-14 and 3-19 of U. S Inan and A. S. Inan, *Engineering Electromagnetics*, Addison Wesley Longman, 1999.

3, which lies beyond the $z = 0$ interface and acts as a "load" (of impedance $Z_L = \eta_3$) on the second "transmission line." To find the ratio of E_{2x} and H_{2y}, we need to know the relative electric field amplitudes of the two traveling waves within the layer; that is, Γ_{23}. We already know that the reverse-propagating wave in medium 2 (i.e., $E_{20}\Gamma_{23}e^{+j\beta_2 z}$) is related to the forward-propagating wave in medium 3. In other words, the "reflection coefficient" that we "see" looking toward the load from a line with characteristic impedance Z_{02} at the $z = 0$ interface is

$$\Gamma_{23} = \frac{Z_L - Z_{02}}{Z_L + Z_{02}} \quad \rightarrow \quad \Gamma_{23} = \frac{\eta_3 - \eta_2}{\eta_3 + \eta_2}$$

This result is also expected because the field structure at the last interface is exactly the three-wave problem we studied in the case of a single dielectric interface, with an incident wave within the layer (here of unknown amplitude E_{20} rather than E_{i0}) and a reflected wave (of amplitude $\Gamma_{23}E_{20}$ rather than ΓE_{i0}). The above result for Γ_{23} then follows directly from expression [3.2] for Γ.

The wave impedance in medium 2 is then given by

$$Z_2(z) = \frac{E_{2x}(z)}{H_{2y}(z)} = \eta_2 \frac{e^{-j\beta_2 z} + \Gamma_{23}e^{+j\beta_2 z}}{e^{-j\beta_2 z} - \Gamma_{23}e^{+j\beta_2 z}}$$

which in general results in a complex impedance. At $z = 0$ this expression reduces to

$$Z_2(z = 0) = \eta_2 \frac{1 + \Gamma_{23}}{1 - \Gamma_{23}} = \eta_2 \frac{1 + \dfrac{\eta_3 - \eta_2}{\eta_3 + \eta_2}}{1 - \dfrac{\eta_3 - \eta_2}{\eta_3 + \eta_2}} = \eta_2 \frac{2\eta_3}{2\eta_2} = \eta_3$$

which is as expected. The wave impedance in medium 2 at $z = -d$ is given by

$$\boxed{Z_2(-d) = \eta_2 \frac{e^{+j\beta_2 d} + \Gamma_{23}e^{-j\beta_2 d}}{e^{+j\beta_2 d} - \Gamma_{23}e^{-j\beta_2 d}} = \eta_2 \frac{\eta_3 + j\eta_2 \tan(\beta_2 d)}{\eta_2 + j\eta_3 \tan(\beta_2 d)} = \eta_{23}} \quad [3.9]$$

Using the expressions for the total electric and magnetic fields in medium 1, we can write the wave impedance in medium 1 as

$$Z_1(z) = \frac{E_{1x}(z)}{H_{1y}(z)} = \eta_1 \frac{e^{-j\beta_1(z+d)} + \Gamma_{\text{eff}}e^{+j\beta_1(z+d)}}{e^{-j\beta_1(z+d)} - \Gamma_{\text{eff}}e^{+j\beta_1(z+d)}}$$

which at $z = -d$ reduces to

$$Z_1(z = -d) = \eta_1 \frac{1 + \Gamma_{\text{eff}}}{1 - \Gamma_{\text{eff}}}$$

Since the electromagnetic boundary conditions require the continuity of the tangential electric (E_x) and magnetic (H_y) field components, and since wave impedance is defined as the ratio of these two quantities, the wave impedances on both sides of any of the interfaces must be equal. In other words, we must have

$$Z_1(z = -d) = \eta_1 \frac{1 + \Gamma_{eff}}{1 - \Gamma_{eff}} = Z_2(z = -d) \quad \rightarrow \quad \Gamma_{eff} = \frac{Z_2(-d) - \eta_1}{Z_2(-d) + \eta_1} = \frac{\eta_{23} - \eta_1}{\eta_{23} + \eta_1}$$

where $Z_2(-d) = \eta_{23}$ is given by [3.9].

In summary, a dielectric slab of thickness d (i.e., medium 2) inserted between two media (i.e., media 1 and 3) essentially works like a transmission line impedance transformer, transforming the impedance of medium 3 from its intrinsic value of η_3 to η_{23}. The impedance η_{23} is what is "seen" looking toward the "load" at the $z = -d$ interface. Without the slab, the reflection coefficient at the interface of media 1 and 3 is given by

$$\Gamma_{13} = \frac{\eta_3 - \eta_1}{\eta_3 + \eta_1}$$

With the slab present, the effective reflection coefficient Γ_{eff} can be found by simply substituting η_{23} for η_3, namely,

$$\boxed{\Gamma_{eff} = \rho_{eff} e^{j\phi_\Gamma} = \frac{Z_2(-d) - \eta_1}{Z_2(-d) + \eta_1} = \frac{\eta_{23} - \eta_1}{\eta_{23} + \eta_1}} \quad [3.10]$$

As an example, consider the case when medium 1 and medium 3 are the same (i.e., $\eta_1 = \eta_3$) and the thickness of the medium 2 is an integer multiple of a half-wavelength in that medium. Since $\beta_2 d = (2\pi d)/\lambda_2$, where $d = (n\lambda_2)/2$, $n = 0, 1, 2, \ldots$, we have $\tan(\beta_2 d) = 0$, yielding

$$Z_2(-d) = \eta_{23} = \eta_3$$

In addition, since $\eta_1 = \eta_3$, we find $\Gamma_{eff} = 0$, resulting in total transmission from medium 1 into medium 3. Note that this is true for any dielectric slab (i.e., a dielectric material with any ϵ_2) as long as its thickness is an integer multiple of a half-wavelength in that dielectric at the frequency of operation.

Another case of interest is that in which the intrinsic impedance of medium 2 is the geometric mean of the intrinsic impedances of media 1 and 3 (i.e., $\eta_2 = \sqrt{\eta_1 \eta_3}$), and the thickness of medium 2 is an odd integer multiple of a quarter-wavelength in that medium. Since $\beta_2 d = (2\pi d)/\lambda_2$, where $d = [(2n + 1)\lambda_2]/4$, $n = 0, 1, 2, \ldots$, we have $\tan(\beta_2 d) = \pm\infty$, yielding

$$\lim_{\tan(\beta_2 d) \to \pm\infty} [\eta_{23}] = \lim_{\tan(\beta_2 d) \to \pm\infty} \left[\eta_2 \frac{\eta_3 + j\eta_2 \tan(\beta_2 d)}{\eta_2 + j\eta_3 \tan(\beta_2 d)} \right] = \frac{\eta_2^2}{\eta_3}$$

Because $\eta_2^2 = \eta_1 \eta_3$, however, we also have $\eta_{23} = \eta_1$, resulting in total transmission from medium 1 into medium 3.

The results displayed in Figure 3.11 can now be better understood as simply a manifestation of the quarter-wave type of transmission line matching. The variation of $|\Gamma_{eff}|$ with normalized slab thickness d/λ_2 provides a measure of the frequency bandwidth over which the matching is effective, since for a given value of d the curves show the variation of $|\Gamma_{eff}|$ with wavelength or frequency.

The transmission line analogy can easily be extended to problems involving more than two dielectric boundaries, which can be modeled in the same manner by using a series of cascaded transmission lines. Multiple-layer coatings to reduce reflections may be utilized where single-layer coating materials with the desired refractive index (i.e., $n_2 = \sqrt{\epsilon_{2r}}$) are not convenient to use (e.g., they are not structurally self-supporting at the thicknesses desired) or are not available. Also, multiple-layer coatings are generally more desirable in order to achieve the "matching" (i.e., no reflection) condition over a broader range of wavelengths,[17] in a manner quite analogous to multiple-stage quarter-wavelength transmission line systems utilized for wideband matching.[18]

Examples 3-6, 3-7, and 3-8 illustrate selected applications of reflection from multiple dielectric interfaces.

Example 3-6:   Radome design.   A radar dome, or *radome,* is a protective dielectric enclosure for a microwave antenna. A ground-based C-band microwave landing system used to help airplanes to land is to be protected from weather by enclosing it in a radome. The center frequency of the operating frequency band is 5 GHz. Thermoplastic PEI material (assume lossless, nonmagnetic, with $\epsilon_{2r} \approx 3$) is chosen for the design. (a) Assuming a flat planar radome as shown in Figure 3.12a, determine the minimum thickness of the radome that will give no reflections at 5 GHz. (b) If the frequency is changed to 4 GHz and the thickness of the radome remains as in part (a), what percentage of the incident power is reflected? (c) Repeat part (b) for 6 GHz.

Solution:

(a) At 5 GHz, the wavelength in thermoplastic PEI material is

$$\lambda_2 = \frac{\lambda_1}{\sqrt{\epsilon_{2r}}} = \frac{c}{f\sqrt{\epsilon_{2r}}} \approx \frac{3 \times 10^8 \text{ m-s}^{-1}}{(5 \times 10^9 \text{ Hz})\sqrt{3}} \approx 3.46 \times 10^{-2} \text{ m} = 3.46 \text{ cm}$$

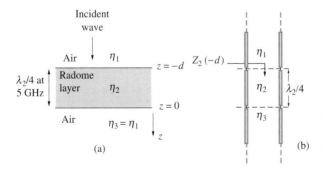

FIGURE 3.12.   **Radome design.** (a) Geometry of the radome layer. (b) Transmission line analog.

---

[17]For extensive discussion of multiple-layer coatings for optical applications see J. D. Rancourt, *Optical Thin Films User's Handbook,* McGraw-Hill, 1987.

[18]See Example 3-19 of U. S. Inan and A. S. Inan, *Engineering Electromagnetics*, Addison Wesley Longman, 1999.

In order not to affect the operation of the microwave landing system, the thickness d of the radome layer should be an integer multiple of $\lambda_2/2$, where $\lambda_2/2 \simeq 1.73$ cm, which is also the minimum thickness required.

(b) Using the transmission line method (see Figure 3.12b), we start by evaluating $Z_2(z = -d) = \eta_{23}$. Assuming that the radome is designed with the minimum thickness required (i.e., $d = d_{min} \simeq 1.73$ cm) and noting that at $f = 4$ GHz, $\lambda_1 = (3 \times 10^8)/(4 \times 10^9) = 0.075$ m or 7.5 cm, $\lambda_2 = \lambda_1/\sqrt{\epsilon_{2r}} = 7.5/\sqrt{3} \simeq 4.33$ cm, and $\tan(\beta_2 d) = \tan[(2\pi/\lambda_2)d] \simeq \tan[(2\pi/4.33)1.73] \simeq -0.727$, and $\eta_2 = \eta_1/\sqrt{\epsilon_{2r}} \simeq 377/\sqrt{3} \simeq 218\Omega$, we have from [3.9]

$$Z_2(z = -d) = \eta_{23} = \eta_2\frac{\eta_3 + j\eta_2\tan(\beta_2 d)}{\eta_2 + j\eta_3\tan(\beta_2 d)}$$

$$\simeq 218\frac{377 + j218(-0.727)}{218 + j377(-0.727)}$$

$$\simeq 218\frac{409e^{-j22.8°}}{350e^{-j51.5°}} \simeq 254e^{j28.8°}\,\Omega$$

Therefore, using [3.10], the effective reflection coefficient is

$$\Gamma_{eff} = \frac{\eta_{23} - \eta_1}{\eta_{23} + \eta_1} \simeq \frac{254e^{j28.8°} - 377}{254e^{j28.8°} + 377}$$

$$\simeq \frac{-154 + j122}{600 + j122} \simeq \frac{197e^{j142°}}{612e^{j11.5°}} \simeq 0.321e^{j130°}$$

Hence, the percentage of the incident power that reflects back can be calculated as

$$\frac{|(\mathbf{S}_{av})_r|}{|(\mathbf{S}_{av})_i|} \times 100 = |\Gamma_{eff}|^2 \times 100 \simeq (0.321)^2 \times 100 \simeq 10.3\%$$

(c) Following a similar procedure, at $f = 6$ GHz, we have $\lambda_1 = 5$ cm, $\lambda_2 \simeq 2.89$ cm, $\tan(\beta_2 d) = \tan(216°) \simeq 0.727$, $\eta_2 \simeq 218\Omega$, and [3.9] yields

$$Z_2(z = -d) = \eta_{23} \simeq 218\frac{377 + j218(0.727)}{218 + j377(0.727)} \simeq 218\frac{409e^{j22.8°}}{350e^{j51.5°}} \simeq 254e^{-j28.8°}\,\Omega$$

Using [3.10], the effective reflection coefficient is

$$\Gamma_{eff} = \frac{254e^{-j28.8°} - 377}{254e^{-j28.8°} + 377} \simeq \frac{-154 - j122}{600 - j122} \simeq 0.321e^{-j130°}$$

Hence, the percentage of the incident power that reflects back is $(0.321)^2 \times 100 \simeq 10.3\%$. Thus we see that the effective reflection coefficient varies quite symmetrically around the design frequency of 5 GHz, being down by the same amount in magnitude at 4 GHz as at 6 GHz.

Example 3-7:   Coated glass surface.   Consider Example 3-5 on the re-flectance of glass. Determine the refractive index and minimum thickness of a film to be deposited on the glass surface ($n_3 = 1.52$) such that no normally incident visible light of free-space wavelength 550 nm (i.e., ~545 THz) is reflected.

Solution: Since the permittivity of air is equal to ϵ_0, the wavelength in air (medium 1) is $\lambda_1 = 550$ nm. Since we have nonmagnetic media, the require-ment of $\eta_2 = \sqrt{\eta_1 \eta_3}$ is equivalent to $n_2 = \sqrt{n_1 n_3}$, and we can find the refrac-tive index of the film as $n_2 = \sqrt{1 \times 1.52} \simeq 1.23$. The minimum thickness of the film is

$$d = d_{\min} = \frac{\lambda_2}{4} = \frac{\lambda_1}{4 n_2} = \frac{550 \times 10^{-9}}{4 \times 1.23} \simeq 0.112 \ \mu\text{m}$$

Note that once this film is deposited on glass, it will eliminate reflections com-pletely only at 550 nm. Although there would still be reflections at other wave-lengths in the vicinity of 550 nm, the percentage of the reflected power will be less than the ~4% found in Example 3-5.

In practice, it is typically not possible to manufacture antireflection coating materials that have precisely the desired refractive index. A practical coating material that is commonly used is magnesium fluoride (MgF$_2$, $n_2 = 1.38$; see Figure 3.13a). For visible light with free-space wavelength of 550 nm, the wave-length in MgF$_2$ is $\lambda_2 = 550/1.38 \simeq 399$ nm, so the thickness of the quarter-wavelength MgF$_2$ layer is $d = \lambda_2/4 \simeq 99.6$ nm.

Because we know the matching is not perfect (i.e., $n_{\text{MgF}_2} \neq 1.23$), we can calculate the reflection coefficient for $\lambda_1 = 550$ nm. For this purpose, we use the transmission line method, which can also be applied to multiple coating layers, as discussed in the next example. We start by evaluating $Z_2(z = -d) = \eta_{23}$. Noting that $d = \lambda_2/4$, so that we have $\tan(\beta_2 d) = \tan[(2\pi/\lambda_2)(\lambda_2/4)] = \infty$, and that $\eta_1 \simeq 377\Omega$, $\eta_2 = \eta_1/1.38$, and $\eta_3 = \eta_1/1.52$, we have from [3.9]

$$Z_2(z = -d) = \eta_{23} = \eta_2 \frac{\eta_3 + j \eta_2 \tan(\beta_2 d)}{\eta_2 + j \eta_3 \tan(\beta_2 d)} = \frac{\eta_2^2}{\eta_3} \simeq \frac{(377)(1.52)}{(1.38)^2} \simeq 301\Omega$$

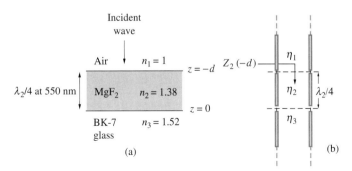

FIGURE 3.13.   **Air–glass interface with single MgF$_2$ coating layer.**
(a) Geometry of the antireflection coating layer. (b) Transmission line analog.

Therefore, using [3.10] we have

$$\Gamma_{\text{eff}} = \frac{Z_2(-d) - \eta_1}{Z_2(-d) + \eta_1} \simeq \frac{301 - 377}{301 + 377} \simeq -0.112$$

so the fraction of the incident power reflected at 550 nm is $\rho_{\text{eff}}^2 \simeq (0.112)^2 \simeq 0.0126$, or 1.26%, substantially less than the ~4% reflection that occurs without any coating. To assess the effectiveness of the MgF_2 coating layer in reducing reflections over the visible spectrum, we can calculate the reflection coefficient at other wavelengths. For example, at 400 nm (violet light) we have

$$\tan(\beta_2 d) = \tan\left(\frac{2\pi}{\lambda_2} \cdot \frac{\lambda_2}{4} \cdot \frac{550}{400}\right) = \tan(123.75°) \simeq -1.50$$

so we have

$$Z_2(z = -d) = \eta_{23} \simeq \left(\frac{377}{1.38}\right)\frac{(377/1.52) + j(377/1.38)(-1.50)}{(377/1.38) + j(377/1.52)(-1.50)}$$

$$= \left(\frac{377}{1.38}\right)\frac{1.38 - j(1.50)(1.52)}{1.52 - j(1.50)(1.38)} \simeq 283.45e^{-j5.11°}$$

and therefore from [3.10]

$$\Gamma_{\text{eff}} = \frac{Z_2(-d) - \eta_1}{Z_2(-d) + \eta_1} \simeq \frac{283e^{-j5.11°} - 377}{283e^{-j5.11°} + 377} \simeq \frac{-94.47 - j25.24}{659.3 - j25.24} \simeq 0.1485e^{-j163°}$$

so the percentage of reflected power in terms of the incident power is

$$\Gamma_{\text{eff}}^2 \times 100 = (0.149)^2 \times 100 \simeq 2.21\%$$

At the other end of the visible light spectrum, for red light at 750 nm, we have

$$\tan(\beta_2 d) = \tan\left(\frac{2\pi}{\lambda_2} \cdot \frac{\lambda_2}{4} \cdot \frac{550}{750}\right) \simeq 2.246$$

and using [3.9], we have

$$Z_2(-d) = \eta_{23} \simeq \left(\frac{377}{1.38}\right)\frac{(1.52)^{-1} + j(1.38)^{-1}(2.246)}{(1.38)^{-1} + j(1.52)^{-1}(2.246)} \simeq 290.65 + j20.90$$

which from [3.10] gives

$$\Gamma_{\text{eff}} = \frac{Z_2(-d) - \eta_1}{Z_2(-d) + \eta_1} \simeq 0.13e^{j164.6°}$$

so the percentage of incident power that reflects back is ~1.7%, still substantially lower than the ~4% we have without any coating. The foregoing results for MgF_2 coating are consistent with the plot of $|\Gamma_{\text{eff}}|$ versus d given in Figure 3.11 for $\epsilon_{3r} = 2.25$ (glass with $n_3 = 1.5$) and $\epsilon_{2r} = 2$ (close to $\epsilon_{MgF_2} \simeq 1.9$ or $n_{MgF_2} \simeq 1.38$).

Example 3-8:   Multiple-layered coatings.   Antireflection coatings are required in most optical applications in order to reduce unwanted reflections at the surfaces of optical elements. In most cases, it is desirable to reduce the surface reflectivity (or simply the reflection coefficient) to an extremely low value over an extended spectral region so as to maintain proper color balance and provide optimum efficiency. Single-layer coatings of magnesium fluoride generally do not satisfy these requirements because, as seen in Example 3-7, they can reduce reflectivity only from ~4% down to ~2.2%. A commonly used technique to achieve the desired requirements is triple-layer coatings.[19] To illustrate the improvements brought about by the use of multiple-layer coating, consider a wideband nonmagnetic antireflection coating system[20] for an air–glass interface, involving three layers of coating materials, as shown in Figure 3.14a. Analyze this system by extending the transmission line analogy method and plot $|\Gamma_{\text{eff}}|$ over the wavelength range 400 to 750 nm.

Solution: We take the design wavelength for this system to be green light, with $\lambda_1 = 550$ nm. With the refractive indices of the layers being $n_2 = 1.38$,

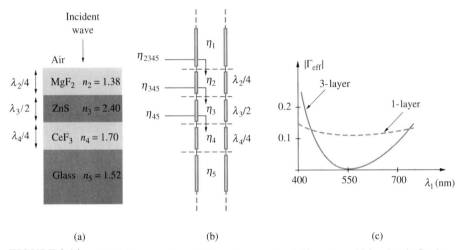

(a)	(b)	(c)

FIGURE 3.14.   **Triple-layer coating.** (a) A modern quarter–half–quarter wideband antireflection triple–layer coating system for air-glass interface. The outermost layer is MgF_2, the middle layer is ZnS, and the bottom layer is CeF_3. (b) The transmission line analog is also shown for reference. (c) Plot of $|\Gamma_{\text{eff}}|$ versus free space wavelength for single- and three-layer coatings.

---

[19]First put forth by J. T. Cox, G. Hass, and A. Theelen, Triple-layer antireflection coatings on glass for the visible and near infrared, *J. Opt. Soc. Am.,* 52, p. 965, 1962; also see J. D. Rancourt, *Optical Thin Film User's Handbook,* McGraw-Hill, 1987.

[20]J. T. Cox, G. Hass, and A. Theelen, Triple-layer antireflection coatings on glass for the visible and near infrared, *J. Opt. Soc. Am.,* 529, pp. 965–969, 1962.

$n_3 = 2.4$, and $n_4 = 1.70$, the thicknesses of the coating layers are

$$d_2 = \frac{\lambda_2}{4} = \frac{\lambda_1}{4n_2} \simeq 99.4 \text{ nm}$$

$$d_3 = \frac{\lambda_3}{2} = \frac{\lambda_1}{2n_3} \simeq 115 \text{ nm}$$

$$d_4 = \frac{\lambda_4}{4} = \frac{\lambda_1}{4n_4} \simeq 80.9 \text{ nm}$$

The phase constants in the three coating materials are

$$\beta_2 = \frac{2\pi}{\lambda_2} = \frac{2\pi}{\lambda_1/n_2} = \frac{2\pi(1.38)}{\lambda_1}$$

$$\beta_3 = \frac{2\pi}{\lambda_3} = \frac{2\pi}{\lambda_1/n_3} = \frac{2\pi(2.4)}{\lambda_1}$$

$$\beta_4 = \frac{2\pi}{\lambda_4} = \frac{2\pi}{\lambda_1/n_4} = \frac{2\pi(1.70)}{\lambda_1}$$

With reference to Figure 3.14b, we start at the glass (i.e., the "load") end and simply transform impedances toward air (i.e., the "source" end in the transmission line context). We have

$$\eta_{45} = \eta_4 \frac{\eta_5 + j\eta_4 \tan(\beta_4 d_4)}{\eta_4 + j\eta_5 \tan(\beta_4 d_4)}$$

where $\eta_4 = \eta_1/n_4 \simeq 377/1.7 \simeq 222\Omega$, $\eta_5 = \eta_1/n_5 \simeq 377/1.52 = 248\Omega$, and the subscript "45" indicates that η_{45} is the intrinsic impedance η_5 transformed by layer (or line) 4. The combined equivalent impedance η_{45} of materials 4 and 5 is now transformed over the next line segment into η_{345}:

$$\eta_{345} = \eta_3 \frac{\eta_{45} + j\eta_3 \tan(\beta_3 d_3)}{\eta_3 + j\eta_{45} \tan(\beta_3 d_3)}$$

where $\eta_3 = \eta_1/n_3 \simeq 377/2.4 \simeq 157\Omega$. Finally we have

$$\eta_{2345} = \eta_2 \frac{\eta_{345} + j\eta_2 \tan(\beta_2 d_2)}{\eta_2 + j\eta_{345} \tan(\beta_2 d_2)}$$

where $\eta_2 = \eta_1/n_2 \simeq 377/1.38 \simeq 273\Omega$. The effective reflection coefficient at the air interface is now given as

$$\Gamma_{\text{eff}} = \frac{\eta_{2345} - \eta_1}{\eta_{2345} + \eta_1}$$

where $\eta_1 \simeq 377\Omega$. The magnitude of Γ_{eff} is plotted in Figure 3.14c over the wavelength range 400–750 nm of the visible spectrum. The wideband performance of the triple-layer system is evident, with extremely low (<0.01%) value of $|\Gamma_{\text{eff}}|$ over a broad frequency range. For comparison, $|\Gamma_{\text{eff}}|$ for a single MgF$_2$

layer (Example 3-7) is also shown. Note that although the latter is also smooth across the wavelength band of interest, the value of reflectivity is much larger, with $|\Gamma_{\text{eff}}| > 0.1$.

3.4   OBLIQUE INCIDENCE UPON A PERFECT CONDUCTOR

Up to now, we have exclusively considered reflection and transmission of uniform plane waves from interfaces upon which they are *normally* incident. Many interesting applications, especially in optics but also at microwave and radio frequencies, involve oblique incidence of uniform plane waves upon planar interfaces. The optical mirror arrangements in a camera involve reflections from slanted surfaces, and radio wave propagation at distances beyond line of sight involve oblique reflections from the ionized region (see Section 6.2) of the Earth's upper atmosphere known as the ionosphere. We have all observed the reflection of ocean waves from a beachfront upon which they are obliquely incident; careful observation of a particular isolated incident wavefront under relatively calm conditions clearly reveals the production of a reflected wave, propagating away at an angle from the normal that is the mirror image of that of the incident wave, and a "surface" wave, propagating rapidly along the beach. The reflection of electromagnetic waves occurs in a very similar manner, except that such waves are somewhat more complicated than water waves, because they are characterized by two vectors[21] (**E** and **H**) perpendicular to the direction of propagation.

We consider a uniform plane wave propagating in a lossless dielectric, obliquely incident upon a perfect conductor occupying the $z > 0$ half-space, with the interface between the two media being the entire xy plane. As can be seen from Figure 3.15, it is necessary to distinguish between two distinctly different cases:

1. *Perpendicular polarization,*[22] in which the wave electric field is perpendicular to the plane of incidence, defined by the normal to the interface and the $\mathbf{k} = \beta\hat{\mathbf{k}}$ vector in the direction of the propagation (i.e., the xz plane in Figure 3.15). The

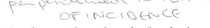

---

[21]Rather than a single vector (i.e., velocity) in the direction of propagation and a scalar (i.e., density or pressure), as is the case for acoustic waves in a fluid.

[22]This terminology may be somewhat confusing until one realizes that the "perpendicular" or "parallel" labels are with respect to the plane of incidence and not the interface plane (Figure 3.15). Other ways of distinguishing between the two different cases can be encountered in various texts, including TE versus TM, horizontal versus vertical, E- versus H-, or s- versus p-waves. We believe that the perpendicular versus parallel polarization terminology is quite appropriate, since it actually refers to the polarization of the wave by identifying the orientation of the wave electric field, in accordance with the conventional understanding of wave polarization as discussed in Chapter 2. In this context, and taking, for example, the plane of incidence as the horizontal ground, perpendicular polarization is akin to a vertically polarized wave, whereas parallel polarization corresponds to a horizontally polarized one.

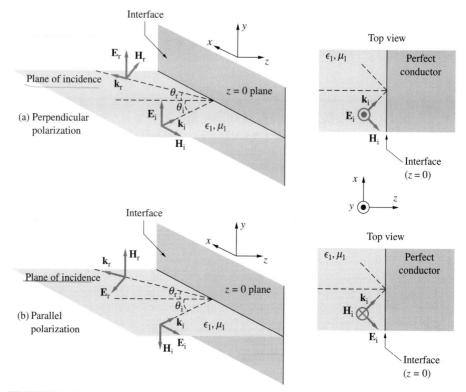

FIGURE 3.15. **Oblique incidence on a perfect conductor.** (a) Plane wave impinging on a perfect conductor with the electric field vector perpendicular to the plane of incidence (perpendicular polarization). (b) Electric field vector parallel to the plane of incidence (parallel polarization).

general geometry of oblique incidence and the identification of the plane of incidence and the interface plane are shown in Figure 3.15.

2. *Parallel polarization*, in which the electric field of the incident wave is parallel to the plane of incidence.

That the two cases lead to different results is evident when one considers the physical boundary conditions. For example, the continuity of the electric field at the boundary affects the entire electric field vector in the case of perpendicular polarization, whereas only one component of the wave electric field is affected in the parallel polarization case. In terms of the boundary condition on the tangential magnetic field, the surface currents[23] flowing on the conductor surface are in the x or y directions respectively, depending on whether the incident wave is parallel or perpendicularly polarized. We first consider the perpendicular polarization case.

---

[23]Note that $\mathbf{J}_s = \hat{\mathbf{n}} \times \mathbf{H}$, so that the surface current must be orthogonal to the tangential component of $\mathbf{H}$ (i.e., H_t) *and* the outward normal to the interface, i.e., $\hat{\mathbf{n}}$.

3.4.1   Perpendicular Polarization

We first consider the perpendicular polarization case as depicted in Figure 3.15a. In the context of the coordinate system shown in Figure 3.16, the incident plane wave can be described in phasor form as

$$\mathbf{E}_i(\mathbf{r}) = \hat{\mathbf{y}} E_{i0} e^{-j\beta_1 \hat{\mathbf{k}}_i \cdot \mathbf{r}}$$

where $\hat{\mathbf{k}}_i = \hat{\mathbf{x}} \sin \theta_i + \hat{\mathbf{z}} \cos \theta_i$ is the unit vector in the propagation direction of the incident wave, as shown in Figure 3.16, and $\mathbf{r}$ is the position vector $\mathbf{r} = \hat{\mathbf{x}} x + \hat{\mathbf{y}} y + \hat{\mathbf{z}} z$. Thus we have

$$\mathbf{E}_i(x, z) = \hat{\mathbf{y}} E_{i0} e^{-j\beta_1 (x \sin \theta_i + z \cos \theta_i)} \tag{3.11}$$

and the wave magnetic field is

$$\mathbf{H}_i(x, z) = \frac{1}{\eta_1} [\hat{\mathbf{k}}_i \times \mathbf{E}_i] = \frac{1}{\eta_1} [\hat{\mathbf{x}} \sin \theta_i + \hat{\mathbf{z}} \cos \theta_i] \times [\hat{\mathbf{y}} E_{i0} e^{-j\beta_1 (\hat{\mathbf{x}} \sin \theta_i + \hat{\mathbf{z}} \cos \theta_i) \cdot (\hat{\mathbf{x}} x + \hat{\mathbf{y}} y + \hat{\mathbf{z}} z)}]$$

$$= \frac{E_{i0}}{\eta_1} \underbrace{(-\hat{\mathbf{x}} \cos \theta_i + \hat{\mathbf{z}} \sin \theta_i)}_{\hat{\mathbf{h}}_i} e^{-j\beta_1 (x \sin \theta_i + z \cos \theta_i)} \tag{3.12}$$

where $\hat{\mathbf{h}}_i = -\hat{\mathbf{x}} \cos \theta_i + \hat{\mathbf{z}} \sin \theta_i$ is the unit vector in the direction of the magnetic field vector of the incident wave. Note that the orientation of $\mathbf{H}_i$ makes sense in that $\mathbf{E}_i \times \mathbf{H}_i$ is in the $\hat{\mathbf{k}}_i$ direction. To see this consider

$$\hat{\mathbf{e}}_i \times \hat{\mathbf{h}}_i = \hat{\mathbf{y}} \times (-\hat{\mathbf{x}} \cos \theta_i + \hat{\mathbf{z}} \sin \theta_i) = \hat{\mathbf{z}} \cos \theta_i + \hat{\mathbf{x}} \sin \theta_i = \hat{\mathbf{k}}_i$$

where $\hat{\mathbf{e}}_i$ is the unit vector in the direction of the electric field vector of the incident wave.

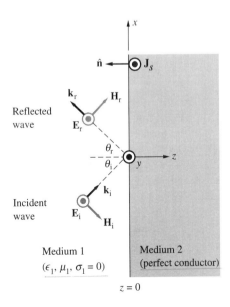

FIGURE 3.16.   Perpendicularly polarized wave at a perfect conductor boundary. Uniform plane wave having perpendicular polarization obliquely incident on a perfectly conducting boundary.

The reflected wave can be similarly described, except that the propagation direction vector $\hat{\mathbf{k}}_r$ is given[24] by

$$\hat{\mathbf{k}}_r = \hat{\mathbf{x}} \sin \theta_r - \hat{\mathbf{z}} \cos \theta_r$$

Thus we have

$$\mathbf{E}_r(x, z) = \hat{\mathbf{y}} E_{r0} e^{-j\beta_1 \hat{\mathbf{k}}_r \cdot \mathbf{r}} = \hat{\mathbf{y}} E_{r0} e^{-j\beta_1 (x \sin \theta_r - z \cos \theta_r)} \qquad [3.13]$$

and

$$\mathbf{H}_r(x, z) = \frac{1}{\eta_1}(\hat{\mathbf{k}}_r \times \mathbf{E}_r) = \frac{E_{r0}}{\eta_1}(+\hat{\mathbf{x}} \cos \theta_r + \hat{\mathbf{z}} \sin \theta_r) e^{-j\beta_1(x \sin \theta_r - z \cos \theta_r)} \qquad [3.14]$$

Note that the direction of the magnetic field vector $\hat{\mathbf{h}}_r = \hat{\mathbf{x}} \cos \theta_r + \hat{\mathbf{z}} \sin \theta_r$ in [3.14] can be deduced by inspection since $\mathbf{E}_r \times \mathbf{H}_r$ must be along $\hat{\mathbf{k}}_r$. In expressions [3.11] through [3.14], we have two unknowns E_{r0} and θ_r, which we wish to find in terms of E_{i0} and θ_i. At the boundary surface, $z = 0$, the total electric field must vanish, since the total electric field is tangential to the conductor and the electric field inside the conductor is zero. We thus have

$$\mathbf{E}_1(x, 0) = \mathbf{E}_i(x, 0) + \mathbf{E}_r(x, 0) = 0 \qquad [3.15]$$

where $\mathbf{E}_1$ is the total electric field in medium 1. Substituting [3.11] and [3.13] in [3.15] we find

$$\mathbf{E}_1(x, 0) = \hat{\mathbf{y}}[E_{i0} e^{-j\beta_1 x \sin \theta_i} + E_{r0} e^{-j\beta_1 x \sin \theta_r}] = 0 \qquad [3.16]$$

which must hold true at *all* values of x along the interface, which implies that

$$E_{r0} = -E_{i0}; \qquad \theta_i = \theta_r \qquad [3.17]$$

The equality of the incident and reflected angles is known as Snell's law of reflection. Using [3.17], we can write the total electric field intensity in medium 1 as

$$\mathbf{E}_1(x, z) = \hat{\mathbf{y}} E_{i0}(e^{-j\beta_1 z \cos \theta_i} - e^{+j\beta_1 z \cos \theta_i}) e^{-j\beta_1 x \sin \theta_i}$$

$$= -\hat{\mathbf{y}} j2 E_{i0} \sin(\beta_1 z \cos \theta_i) e^{-j\beta_1 x \sin \theta_i}$$

Similarly by substituting [3.17] in [3.12] and [3.14] we can find an expression for the total magnetic field $\mathbf{H}_1(x, z)$ in medium 1. We have

$$\mathbf{H}_1(x, z) = -\frac{2E_{i0}}{\eta_1}[\hat{\mathbf{x}} \cos \theta_i \cos(\beta_1 z \cos \theta_i) + \hat{\mathbf{z}} j \sin \theta_i \sin(\beta_1 z \cos \theta_i)] e^{-j\beta_1 x \sin \theta_i}$$

---

[24]Note that the propagation direction of the reflected wave must lie in the xz plane, because the presence of a y component of $\hat{\mathbf{k}}_r$ would require (since $\hat{\mathbf{k}}_r \perp \mathbf{E}_r$ for a uniform plane wave) the electric field of the reflected wave to have a component at an angle to the plane of incidence, in which case the boundary condition at the interface cannot be satisfied (since the incident wave electric field is entirely perpendicular to the plane of the interface).

The time-averaged Poynting vector associated with the total fields is given by

$$\mathbf{S}_{av} = \frac{1}{2}\mathcal{R}e\{\mathbf{E}_1(x, z) \times \mathbf{H}_1^*(x, z)\}$$

$$= \frac{1}{2}\mathcal{R}e\left\{-\hat{\mathbf{z}}j\frac{2E_{i0}^2}{\eta_1}\cos\theta_i\sin(2\beta_1 z\cos\theta_i) + \hat{\mathbf{x}}\frac{4E_{i0}^2}{\eta_1}\sin\theta_i\sin^2(\beta_1 z\cos\theta_i)\right\}$$

$$= \hat{\mathbf{x}}\frac{2E_{i0}^2}{\eta_1}\sin\theta_i\sin^2(\beta_1 z\cos\theta_i)$$

Note that the z component of the Poynting vector does not have a real part, because $E_{1y}(x, z)$ and $H_{1x}(x, z)$ are $90°$ out of phase in time. In other words, since $E_{1y}(x, z)$ is multiplied by j while $H_{1x}(x, z)$ is not, the $j = e^{j\pi/2}$ term appears as a $\pi/2$ phase factor in the expression for $\mathcal{E}_{1y}(x, z, t)$. This result is to be expected because no average power is absorbed by the perfect conductor. In the z direction, $E_{1y}(x, z)$ and $H_{1x}(x, z)$ thus represent standing-wave patterns described by $\sin(\beta_1 z\cos\theta_i)$ and $\cos(\beta_1 z\cos\theta_i)$ respectively, where $\beta_{1z} = \beta_1\cos\theta_i$ is the projection on the z axis of the wave phase constant, as was discussed in Section 2.6. On the other hand, the x component of the Poynting vector is real and positive, indicating nonzero average power flow in the x direction, parallel to the boundary. Note that the magnitude of $\mathbf{S}_{av}$ varies as a function of z.

We also observe that E_{1y} and H_{1z} are *in phase* in both space and time and represent a propagating wave with a velocity[25]

$$v_{1x} = \frac{\omega}{\beta_{1x}} = \frac{\omega}{\beta_1\sin\theta_i} = \frac{v_{p1}}{\sin\theta_i}$$

where $v_{p1} = \omega/\beta_1$ is the phase velocity of a uniform plane wave in medium 1 (i.e., the phase velocity of the incident wave along its own propagation direction). The x component of the Poynting vector (resulting from $\frac{1}{2}\mathcal{R}e\{E_yH_z^*\}$) is real and positive, indicating nonzero average power flow in the x direction, parallel to the interface. The wavelength of the wave propagating in the x direction is

$$\lambda_{1x} = \frac{2\pi}{\beta_{1x}} = \frac{\lambda_1}{\sin\theta_i}$$

where $\lambda_1 = 2\pi/\beta_1$ is the wavelength of the incident wave. We note that the wave propagating in the x direction is a *nonuniform* plane wave, since its amplitude varies[26] with z.

Physically, we can think of the wave field in the vicinity of the conductor as a pure standing wave extending along $-z$ that "slides" bodily (as a whole) along x.

---

[25]Note that since $v_{p1} = c$ when medium 1 is free space, v_{1x} can be larger than c. As discussed in Section 2.6, this circumstance does not pose any particular dilemma, because no energy or information travels at a speed of v_{1x}.

[26]Note that, based on our discussions in Section 2.7, the wave is still a *plane* wave because the phase fronts are planar, i.e., $x\sin\theta_i = $ constant, describes planes parallel to the yz plane. However, the wave is *nonuniform*, because the wave electric field varies as a function of position (namely z) over any given plane of constant phase.

That there should be power flow in the x direction is intuitively obvious, because the incident wave carries power in both the x and z directions.

We further note that the total electric field in medium 1 is identically zero at a discrete set of planes parallel to the xy plane, namely at

$$z = -\frac{m\lambda_1}{2\cos\theta_i} \qquad m = 1, 2, 3, \ldots$$

It is thus possible to place a thin conducting sheet at any one of these positions at which the electric field is nulled without affecting[27] the field pattern to the left of the sheet, as shown in Figure 3.17. Note that when such a thin conducting sheet is placed at a distance of $\lambda_1/(2\cos\theta_i)$ from the interface, the region between the sheet and the planar conducting boundary constitutes a parallel-plate waveguide. It appears, therefore, that the propagation of electromagnetic waves in such a guide may be interpreted as a superposition of multiple plane waves reflecting back and forth between the waveguide walls. We shall study parallel-plate and other waveguides extensively in Chapters 4 and 5, and refer back to Figure 3.17 when appropriate.

Current and Charge Induced in the Conductor As was the case for normal incidence of a uniform plane wave on a perfectly conducting boundary, the nonzero tangential component of the magnetic field (i.e., H_x) at $z = 0$ means that a current has to flow at the surface of the conductor to support this magnetic field. Furthermore, if the normal component of the wave electric field is nonzero at the surface of the conductor, surface charge has to be induced in accordance with the boundary condition $\rho_s = \hat{\mathbf{n}} \cdot (\epsilon \mathbf{E})$. The induced current can be estimated from the boundary condition $\mathbf{J}_s = \hat{\mathbf{n}} \times \mathbf{H}$, where $\hat{\mathbf{n}} = -\hat{\mathbf{z}}$ is the outward normal to the conductor surface. At $z = 0$ we have

$$\mathbf{H}_1(x, 0) = -\hat{\mathbf{x}}\frac{2E_{i0}}{\eta_1}\cos\theta_i e^{-j\beta_1 x \sin\theta_i}$$

so that

$$\mathbf{J}_s(x) = \hat{\mathbf{n}} \times \mathbf{H}_1(x, 0) = [(-\hat{\mathbf{z}}) \times (-\hat{\mathbf{x}})]\frac{2E_{i0}}{\eta_1}\cos\theta_i e^{-j\beta_1 x \sin\theta_i}$$

$$= \hat{\mathbf{y}}\frac{2E_{i0}}{\eta_1}\cos\theta_i e^{-j\beta_1 x \sin\theta_i}$$

---

[27]The only requirement such a thin conducting sheet imposes is to have the tangential electric field to be zero at its two surfaces. Since the electric field is in the y direction and is thus tangential to the sheet, and since it is already zero because of the interference between the incident and reflected waves, the conducting sheet can be placed without influencing the field. Note, however, that the tangential component of the wave magnetic field is nonzero on the conducting sheet, so surface currents flow on the sheet. To the degree that the thin conducting sheet consists of perfectly conducting material, these surface currents flow in an infinitesimally thin (macroscopically speaking) layer and do not lead to any power loss. If, however, the conductor is imperfect, there are losses and the current is confined to a layer of thickness of the order of one skin depth, as discussed further in Section 3.6.

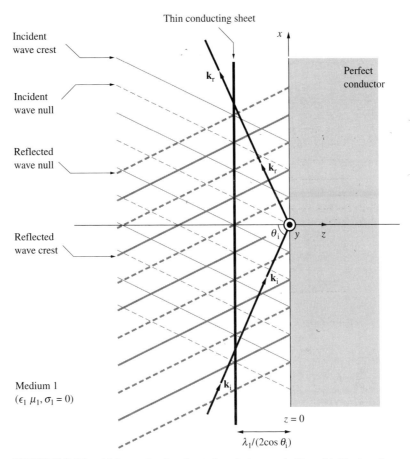

FIGURE 3.17. **Thin conducting sheet placed at** $z = -\lambda_1/(2\cos\theta_i)$**.** The interference between the incident and reflected wave fronts generates a null at $z = -\lambda_1/(2\cos\theta_i)$ and its other integer multiples. The crests (nulls) of the incident wave are shown with light solid (dashed) lines. The thicker solid (dashed) lines represent the crests (nulls) of the reflected wave.

Note that $J_s(x)$ is a surface current density in units of A-m^{-1}. For this case of perpendicular polarization, the induced current flows in the direction perpendicular to the plane of incidence. The corresponding space-time expression for this surface current is

$$\overline{\mathcal{J}}_s(x, t) = \hat{\mathbf{y}}\,\frac{2E_{i0}}{\eta_1}\cos\theta_i\cos(\omega t - \beta_1 x\sin\theta_i) \qquad [3.18]$$

It was mentioned earlier that this induced current, which flows on the surface of the conductor, is necessary to support the wave; alternatively, we can think of the reflected wave as having been reradiated by this planar surface current, as was discussed in connection with the case of normal incidence on a perfect conductor. If the conductor is imperfect, power is dissipated in the conductor by this current flow.

In Section 3.6 and in Chapters 4 and 5, we shall study methods of evaluating the magnitude of such losses, which occur on the walls of hollow metallic waveguides.

Because the wave electric field is entirely tangential to the conductor boundary (i.e., $E_z = 0$), no surface charge is induced in the conductor. This fact in turn means that the surface current must be divergence free,[28] which is true because $\bar{\mathcal{J}}$ has only a y component and does not depend on y.

3.4.2 Parallel Polarization

Now consider the case where the electric field vector $\mathbf{E}_i$ of the incident wave lies in the plane of incidence as shown in Figure 3.18 and in Figure 3.15b. The electric field of the incident wave is given by

$$\mathbf{E}_i(\mathbf{r}) = E_{i0}[\hat{\mathbf{x}} \cos \theta_i - \hat{\mathbf{z}} \sin \theta_i]e^{-j\beta_1 \hat{\mathbf{k}}_i \cdot \mathbf{r}}$$

where, as before, $\hat{\mathbf{k}}_i = \hat{\mathbf{x}} \sin \theta_i + \hat{\mathbf{z}} \cos \theta_i$ is the unit vector in the propagation direction of the incident wave as shown and $\mathbf{r}$ is the position vector $\mathbf{r} = \hat{\mathbf{x}}x + \hat{\mathbf{y}}y + \hat{\mathbf{z}}z$.

We follow the same procedure as before; in this case $\mathbf{E}_i$ and $\mathbf{E}_r$ have components in the x and z directions, but $\mathbf{H}_i$ and $\mathbf{H}_r$ have only y components. Thus we have

$$\begin{cases} \mathbf{E}_i(x, z) = E_{i0}[\hat{\mathbf{x}} \cos \theta_i - \hat{\mathbf{z}} \sin \theta_i]e^{-j\beta_1(x \sin \theta_i + z \cos \theta_i)} \\[2mm] \mathbf{H}_i(x, z) = \hat{\mathbf{y}} \dfrac{E_{i0}}{\eta_1} e^{-j\beta_1(x \sin \theta_i + z \cos \theta_i)} \end{cases}$$

Note that the orientations of $\mathbf{H}_i$ and $\mathbf{E}_i$ make sense in that $\mathbf{E}_i \times \mathbf{H}_i$ is in the $\hat{\mathbf{k}}_i$ direction.

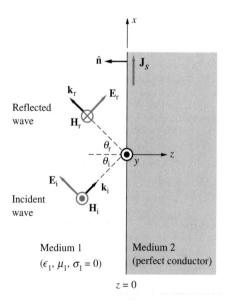

Medium 1
$(\epsilon_1, \mu_1, \sigma_1 = 0)$

Medium 2
(perfect conductor)

$z = 0$

FIGURE 3.18. Parallel polarized wave at a perfectly conducting boundary. Uniform plane wave with parallel polarization obliquely incident on a perfectly conducting boundary.

---

[28]Since $\nabla \cdot \bar{\mathcal{J}} = -\partial \bar{\rho}/\partial t$ implies that $\nabla \cdot \bar{\mathcal{J}}_s = -\partial \tilde{\rho}_s \partial t$, and with $\tilde{\rho}_s = 0$, we must have $\nabla \cdot \bar{\mathcal{J}}_s = 0$.

The reflected wave can be similarly described noting that

$$\hat{\mathbf{k}}_r = \hat{\mathbf{x}} \sin \theta_r - \hat{\mathbf{z}} \cos \theta_r$$

Thus we have

$$\mathbf{E}_r(x, z) = E_{r0}[\underbrace{\hat{\mathbf{x}} \cos \theta_r + \hat{\mathbf{z}} \sin \theta_r}_{\hat{\mathbf{e}}_r}]e^{-j\beta_1(x \sin \theta_r - z \cos \theta_r)}$$

$$\mathbf{H}_r(x, z) = -\hat{\mathbf{y}}\frac{E_{r0}}{\eta_1}e^{-j\beta_1(x \sin \theta_r - z \cos \theta_r)}$$

Note that the closer orientations for $\mathbf{H}_r$ and $\mathbf{E}_r$ make sense since $\hat{\mathbf{k}}_r \times \hat{\mathbf{e}}_r = -\hat{\mathbf{y}}$.

At the boundary surface, $z = 0$, the parallel component (i.e., x component) of the total electric field $\mathbf{E}_1(x, z)$ in medium 1 must vanish, since the field inside the conductor is zero. We thus have

$$E_{1x}(x, 0) = E_{ix}(x, 0) + E_{rx}(x, 0) = 0$$

$$(E_{i0} \cos \theta_i)e^{-j\beta_1 x \sin \theta_i} + (E_{r0} \cos \theta_r)e^{-j\beta_1 x \sin \theta_r} = 0$$

which requires $E_{r0} = -E_{i0}$ and $\theta_i = \theta_r$. The total electric field phasor in medium 1 can then be written as

$$\mathbf{E}_1(x, z) = \mathbf{E}_i(x, z) + \mathbf{E}_r(x, z)$$

$$= \hat{\mathbf{x}}E_{i0} \cos \theta_i(e^{-j\beta_1 z \cos \theta_i} - e^{+j\beta_1 z \cos \theta_i})e^{-j\beta_1 x \sin \theta_i}$$

$$- \hat{\mathbf{z}}E_{i0} \sin \theta_i(e^{-j\beta_1 z \cos \theta_i} + e^{+j\beta_1 z \cos \theta_i})e^{-j\beta_1 x \sin \theta_i}$$

$$= -2E_{i0}[\hat{\mathbf{x}}j \cos \theta_i \sin(\beta_1 z \cos \theta_i) + \hat{\mathbf{z}} \sin \theta_i \cos(\beta_1 z \cos \theta_i)]e^{-j\beta_1 x \sin \theta_i}$$

The associated total magnetic field in medium 1 is given by

$$\mathbf{H}_1(x, z) = \mathbf{H}_i(x, z) + \mathbf{H}_r(x, z)$$

$$= \hat{\mathbf{y}}\frac{2E_{i0}}{\eta_1} \cos(\beta_1 z \cos \theta_i)e^{-j\beta_1 x \sin \theta_i}$$

We can make similar observations as for the case of perpendicular polarization. For example, a conducting plane can be placed at points $z = -m\lambda_1/(2 \cos \theta_i)$ without affecting the fields to the left of the sheet. In the z direction, E_{1x} and H_{1y} maintain standing wave patterns with spacing of minima being $\lambda_{1z}/2$, where $\lambda_{1z} = \lambda_1 \cos \theta_i$. No average power is transmitted in the z direction, because E_{1x} and H_{1y} are 90° out of time phase. The time-average Poynting vector is given by

$$\mathbf{S}_{av} = \hat{\mathbf{x}}\frac{2E_{i0}^2}{\eta_1} \sin \theta_i \cos^2(\beta_1 z \cos \theta_i)$$

In the direction along the boundary (i.e., x direction), E_{1z} and H_{1y} are in time and space phase and represent a propagating *nonuniform* plane wave with a phase velocity of $v_{1x} = v_{p1}/\sin \theta_i$.

Current and Charge Induced in the Conductor As in the case of perpendicular polarization, we can find the current induced on the surface of the conductor using

$$\mathbf{J}_s(x) = \hat{\mathbf{n}} \times \mathbf{H}_1(x, 0)$$

where $\hat{\mathbf{n}} = -\hat{\mathbf{z}}$ and

$$\mathbf{H}_1(x, 0) = \hat{\mathbf{y}} \frac{2E_{i0}}{\eta_1} e^{-j\beta_1 x \sin \theta_i}$$

so that we find

$$\mathbf{J}_s(x) = \hat{\mathbf{x}} \frac{2E_{i0}}{\eta_1} e^{-j\beta_1 x \sin \theta_i}$$

which has dimensions of A-m^{-1}. In contrast to perpendicular polarization, the surface current for parallel polarization flows parallel to the plane of incidence. The space-time expression for the surface current is $\quad$ *m x dire lin*

$$\overline{\mathscr{J}}_s(x, t) = \hat{\mathbf{x}} \frac{2E_{i0}}{\eta_1} \cos(\omega t - \beta_1 x \sin \theta_i)$$

In the parallel polarization case, the wave electric field has a component that is oriented normally to the conductor surface (i.e., $E_z \neq 0$). Accordingly, through Gauss's law, or the boundary condition derived from it, we must have induced electric charge on the conductor surface given by

$$\rho_s(x) = \hat{\mathbf{n}} \cdot \mathbf{D} = \hat{\mathbf{n}} \cdot \epsilon_1 \mathbf{E} = -\epsilon_1 E_{1z}(x, 0)$$

since $\hat{\mathbf{n}}$ is the outward normal from the conductor surface and is thus given by $\hat{\mathbf{n}} = -\hat{\mathbf{z}}$. We thus have

$$\tilde{\rho}_s(x, t) = 2\epsilon_1 E_{i0} \sin \theta_i \cos(\omega t - \beta_1 x \sin \theta_i)$$

We encourage the reader to check whether charge conservation is satisfied in the conductor, namely whether $\nabla \cdot \overline{\mathscr{J}}_s = -\partial \tilde{\rho}_s / \partial t$ holds.

Example 3-9 illustrates the case of reflection of an unpolarized wave obliquely incident on a perfect conductor.

Example 3-9: Unpolarized wave at a perfect conductor boundary. A 10 W-m^{-2}, 200 MHz uniform plane wave propagating in a nonmagnetic lossless medium with its electric field vector given by

$$\mathbf{E}_i(x, y) = (\hat{\mathbf{x}} - \hat{\mathbf{y}}) \frac{E_0}{\sqrt{2}} e^{-j\sqrt{2}\pi(x+y)} + \hat{\mathbf{z}} j E_0 e^{-j\sqrt{2}\pi(x+y)} \quad \text{V-m}^{-1}$$

is obliquely incident upon a perfectly conducting surface located at the xz plane, as shown in Figure 3.19a. Find (a) the angle of incidence θ_i and the relative dielectric constant of the lossless medium, (b) E_0, (c) the expression for $\mathbf{E}_r(x, y)$ of the reflected wave, (d) the polarization of the incident and reflected waves.

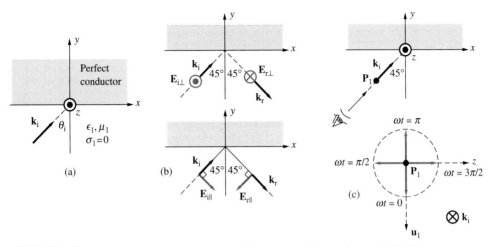

FIGURE 3.19.   **Unpolarized wave obliquely incident on a perfect conductor.** (a) Incidence geometry. (b) Separation of the incident wave into parallel and perpendicularly polarized component waves. (c) The polarization of the field at point P_1 as viewed by an observer looking toward the propagation direction.

Solution:

(a) From

$$\beta_1 \hat{\mathbf{k}}_i \cdot \mathbf{r} = (\hat{\mathbf{x}}\beta_{1x} + \hat{\mathbf{y}}\beta_{1y} + \hat{\mathbf{z}}\beta_{1z}) \cdot (\hat{\mathbf{x}}x + \hat{\mathbf{y}}y + \hat{\mathbf{z}}z) = \sqrt{2}\pi(x + y)$$

we have

$$\beta_{1x} = \beta_1 \sin\theta_i = \sqrt{2}\pi$$

$$\beta_{1y} = \beta_1 \cos\theta_i = \sqrt{2}\pi$$

$$\beta_{1z} = 0$$

Taking the ratio β_{1x}/β_{1y} yields

$$\tan\theta_i = 1 \quad \rightarrow \quad \underline{\theta_i = 45°}$$

Substituting back,

$$\beta_1 \sin\theta_i = \beta_1 \sin 45° = \sqrt{2}\pi \rightarrow \beta_1 = \omega\sqrt{\mu_0\epsilon_1} = 2\pi$$

$$\frac{2\pi \times 2 \times 10^8}{3 \times 10^8}\sqrt{\epsilon_{1r}} \simeq 2\pi \rightarrow \epsilon_{1r} \simeq (1.5)^2 = 2.25$$

(b) $$|\mathbf{S}_{av}| = \underbrace{\frac{(E_0/\sqrt{2})^2}{2\eta_1}}_{|E_x|^2/(2\eta_1)} + \underbrace{\frac{(E_0/\sqrt{2})^2}{2\eta_1}}_{|E_y|^2/(2\eta_1)} + \underbrace{\frac{E_0^2}{2\eta_1}}_{|E_z|^2/(2\eta_1)} = \frac{E_0^2}{\eta_1} = 10 \text{ W-m}^{-2}$$

$$E_0 \simeq \sqrt{\frac{10 \times 377}{\sqrt{2.25}}} \simeq 50.1 \text{ V-m}^{-1}$$

(c) Let us decompose the incident wave into its perpendicular and parallel components as

$$\mathbf{E}_{i\perp} = \hat{\mathbf{z}}jE_0e^{-j\sqrt{2}\pi(x+y)}$$

$$\mathbf{E}_{i\parallel} = (\hat{\mathbf{x}} - \hat{\mathbf{y}})\frac{E_0}{\sqrt{2}}e^{-j\sqrt{2}\pi(x+y)}$$

as shown in Figure 3.19b. Using the boundary condition (the tangential component of electric field at $y = 0$ must be continuous) we can write

$$\mathbf{E}_{r\perp} = -\hat{\mathbf{z}}jE_0e^{-j\sqrt{2}\pi(x-y)}$$

$$\mathbf{E}_{r\parallel} = (-\hat{\mathbf{x}} - \hat{\mathbf{y}})\frac{E_0}{\sqrt{2}}e^{-j\sqrt{2}\pi(x-y)}$$

or

$$\mathbf{E}_r(x, y) = \mathbf{E}_{r\perp} + \mathbf{E}_{r\parallel} = -(\hat{\mathbf{x}} + \hat{\mathbf{y}} + \hat{\mathbf{z}}j\sqrt{2})\frac{E_0}{\sqrt{2}}e^{-j\sqrt{2}\pi(x-y)}$$

(d) $\mathbf{E}_{i\perp}$ and $\mathbf{E}_{i\parallel}$ are both linearly polarized by themselves. $\mathscr{E}_{i\perp}$ oscillates along the z direction as a function of time as

$$\mathscr{E}_{i\perp}(x, y, z) = -\hat{\mathbf{z}}E_0 \sin[4\pi \times 10^8t - \sqrt{2}\pi(x + y)]$$

$\mathscr{E}_{i\parallel}$ oscillates along the $y = x$ direction (along a 45° line between the x and y axes) as a function of time as

$$\mathscr{E}_{i\parallel}(x, y, t) = \hat{\mathbf{x}}\frac{E_0}{\sqrt{2}} \cos[4\pi \times 10^8t - \sqrt{2}\pi(x + y)]$$

$$- \hat{\mathbf{y}}\frac{E_0}{\sqrt{2}} \cos[4\pi \times 10^8t - \sqrt{2}\pi(x + y)]$$

$$= \hat{\mathbf{u}}_1 E_0 \cos[4\pi \times 10^8t - \sqrt{2}\pi(x + y)]$$

where $\hat{\mathbf{u}}_1$ is a unit vector given by

$$\hat{\mathbf{u}}_1 = \frac{1}{\sqrt{2}}(\hat{\mathbf{x}} - \hat{\mathbf{y}})$$

which is perpendicular to $\mathscr{E}_{i\perp}(x, y, t)$. Since $\mathscr{E}_{i\parallel}$ and $\mathscr{E}_{i\perp}$ have the same magnitude but are out of phase by 90°, the total field $\mathscr{E}_i = \mathscr{E}_{i\parallel} + \mathscr{E}_{i\perp}$ given by

$$\mathscr{E}_i(x, y, t) = \hat{\mathbf{u}}_1 E_0 \cos[4\pi \times 10^8t - \sqrt{2}\pi(x + y)]$$

$$- \hat{\mathbf{z}}E_0 \sin[4\pi \times 10^8t - \sqrt{2}\pi(x + y)]$$

is circularly polarized. To find the sense of polarization, let us look at $\mathscr{E}_i(x, y, t)$ at an arbitrary fixed point, such as $P_1(0, 0, 0)$ (i.e., the origin):

$$\mathscr{E}_i(0, 0, 0) = \hat{\mathbf{u}}_1 E_0 \cos(4\pi \times 10^8t) - \hat{\mathbf{z}}E_0 \sin(4\pi \times 10^8t)$$

and examine its rotation as a function of time. The rotation of the electric field vector with respect to time is shown as seen when viewed in the direction of propagation, which clearly indicates that this is a right-hand circularly polarized (RHCP) wave.

Similarly, the time-domain expression for the electric field of the reflected wave can be written as

$$\overline{\mathscr{E}}_r(x, y, t) = \hat{\mathbf{u}}_2 E_0 \cos[4\pi \times 10^8 t - \sqrt{2}\pi(x - y)]$$
$$+ \hat{\mathbf{z}} E_0 \sin[4\pi \times 10^8 t - \sqrt{2}\pi(x - y)]$$

where $\hat{\mathbf{u}}_2 = \frac{1}{\sqrt{2}}(-\hat{\mathbf{x}} - \hat{\mathbf{y}})$. It can easily be shown that the reflected wave is left-hand circularly polarized (LHCP).

3.5   OBLIQUE INCIDENCE AT A DIELECTRIC BOUNDARY

We now analyze the more general case of oblique incidence of a uniform plane wave on an interface between two different lossless dielectric media. Many of our everyday optical experiences[29] involve refraction of light at dielectric interfaces, the most famous one being the rainbow. Reflection of waves from dielectric boundaries is also the basis of many useful applications, including polarizing filters for sunglasses, optical resonators, interferometers, and optical fibers.[30]

Consider a uniform plane wave obliquely incident at the boundary between two lossless dielectrics, as shown in Figure 3.20. We rely on established concepts of uniform plane wave fronts (see Section 2.6) to derive the relationships between the incident, reflected, and refracted waves. It is useful to remember that all three waves considered are uniform plane waves, with phase surfaces of infinite extent in the direction orthogonal to their propagation direction. If, for example, the wave fronts shown in Figure 3.20 (i.e., the dashed lines AC, EB, and BD) are taken to represent the position of the wave crests, the successive crests are separated by the wavelength in each medium, being λ_1 in medium 1 and λ_2 in medium 2. On a purely geometric basis, the advancing wave fronts of the three waves must be "in step" with one another[31] along the interface. In other words, the phase velocities of the three waves as measured along the x axis must have the same magnitude and sign.

Starting with the incident wave front (AC), the incident ray travels a distance CB in medium 1, the transmitted ray travels a distance AD, and the reflected wave travels a distance of AE to their respective phase fronts. Noting that the phase velocities of

---

[29] M. Minnaert, *The Nature of Light & Colour in the Open Air*, Dover Publications, Inc., 1954.

[30] See, for example, H. A. Haus, *Waves and Fields in Optoelectronics*, Prentice-Hall, Inc., 1984.

[31] The "in step" terminology is adopted from R. B. Adler, L. J. Chu, and R. M. Fano, *Electromagnetic Energy Transmission and Radiation*, Chapter 7, Wiley, New York, 1960.

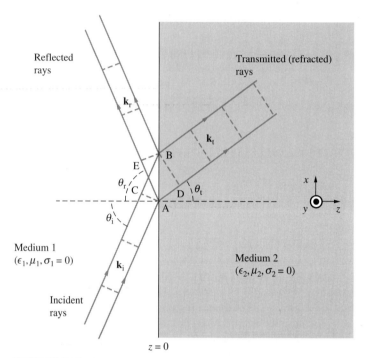

FIGURE 3.20. **Oblique incidence at a dielectric boundary.** Uniform plane wave obliquely incident at a dielectric boundary located at the $z = 0$ plane. Dashed lines AC, BD, and EB are the wavefronts (planar surfaces of constant phase).

uniform plane waves in media 1 and 2 are respectively v_{p1} and v_{p2}, in order for each wave to have a uniform phase we must have

$$\frac{CB}{AD} = \frac{v_{p1}}{v_{p2}}$$

We note, however, that $CB = AB \sin \theta_i$ and $AD = AB \sin \theta_t$, so

$$\boxed{\frac{\sin \theta_i}{\sin \theta_t} = \frac{v_{p1}}{v_{p2}} = \sqrt{\frac{\epsilon_2 \mu_2}{\epsilon_1 \mu_1}}} \qquad [3.19]$$

which is the well-known relationship between the angle of incidence and angle of refraction, commonly referred to as *Snell's law of refraction*. Also, since $CB = AE$, we have

$$\underbrace{AB \sin \theta_i}_{CB} = \underbrace{AB \sin \theta_r}_{AE}$$

which results in *Snell's law of reflection*, given by

$$\sin\theta_i = \sin\theta_r \quad \longrightarrow \quad \boxed{\theta_i = \theta_r}$$

Equation [3.19] is often expressed in terms of refractive indices of the media, defined as $n \equiv c/v_p = \sqrt{(\mu\epsilon)/(\mu_0\epsilon_0)} = \sqrt{\mu_r\epsilon_r}$

$$\frac{\sin\theta_i}{\sin\theta_t} = \frac{n_2}{n_1} \quad \text{or} \quad n_1\sin\theta_i = n_2\sin\theta_t$$

or, just in terms of the relative permittivities ϵ_{1r} and ϵ_{2r} for nonmagnetic media ($\mu_1 = \mu_2 = \mu_0$),

$$\boxed{\frac{\sin\theta_i}{\sin\theta_t} = \sqrt{\frac{\epsilon_{2r}}{\epsilon_{1r}}}} \quad \text{or} \quad \sqrt{\epsilon_{1r}}\sin\theta_i = \sqrt{\epsilon_{2r}}\sin\theta_t \qquad [3.20]$$

NON MAGNETIC

It is important to note that Snell's laws are independent of the polarization of the incident wave, because they are purely based on the three waves being "in step" with one another in terms of their planar phase surfaces. This is further illustrated in Figure 3.21. Furthermore, we note that the functional forms of the fields have not

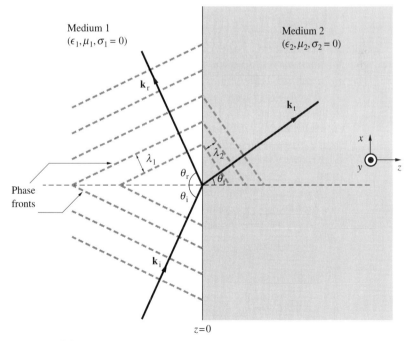

FIGURE 3.21.   Planar phase fronts of the incident, reflected and transmitted waves. Illustration of the 'in step' condition for the planar phase fronts of the three waves. Note that Snell's law, derived from this condition, holds for any polarization of the incident wave.

entered into our derivation, so the wave fields of the uniform plane wave do not need to be time-harmonic (i.e., sinusoidal) for Snell's laws to be valid. If the incident wave fields involve any arbitrary functional form, for example $f(t - \beta \hat{\mathbf{k}} \cdot \mathbf{r}) = f(t - \beta_x x - \beta_z z)$, both the refracted and the reflected waves must have the same functional form at the boundary in order for the boundary conditions at $z = 0$ to be satisfied at all values of x and y (i.e., all along the boundary).

Snell, Willebrord van Roijen   *(Dutch mathematician, b. 1580, Leiden; d. October 30, 1626, Leiden) Snell received his master's degree in 1608 and succeeded his father as professor of mathematics at the University of Leiden in 1613. In 1615, Snell developed the method of triangulation, starting with his house and the spires of nearby churches as reference points. He used a large quadrant over 2 meters long to determine angles, and, by building up a network of triangles, was able to obtain a value for the distance between two towns on the same meridian. From this, Snell made an accurate determination of the radius of the Earth.*

Snell is best known for his discovery in 1621 that when a ray of light passes obliquely from a rarer into a denser medium (as from air into water or glass) it is bent toward the vertical. The phenomenon (refraction of light) was known as long ago as the time of Ptolemy [∼100 to ∼170], but Ptolemy thought that as the angle to the vertical made by the light ray in air was changed, it maintained a constant relationship to the angle from the vertical made by the light ray in water or glass. Snell showed this was not so. It was the ratio of the sines *of the angles that bore the constant relationship. It was only because at small angles the sines are almost proportional to the angles themselves that Ptolemy was deluded.*

This key discovery in optics was not well publicized until 1638, when Descartes [1596–1650] published it—without giving credit to Snell. Descartes expressed the law differently from Snell but could have easily derived it from Snell's original formulation. It is not known whether Descartes knew of Snell's work or discovered the law independently. In 1617 Snell had also developed the method of determining distances by trigonometric triangulation and thus founded the modern art of mapmaking. [Partly adapted with permission from I. Asimov, Biographical Encyclopedia of Science and Technology, Doubleday, 1982].

1500	*1580*	*1626*	*2000*

To derive the reflection and transmission coefficients, we can follow a procedure similar to that in previous sections, which involves matching boundary conditions and solving for the various unknown field amplitudes. However, for the case in hand, we can derive the reflection and transmission coefficients more easily by considering the conservation of energy flow. For each of the three waves (incident, reflected, and transmitted), the power transmitted per square meter is equal to $E^2/(2\eta)$ and is in the direction of propagation [i.e, $\mathbf{S}_{\text{av}} = \hat{\mathbf{k}} E^2/(2\eta)$], where $\eta = \sqrt{\mu/\epsilon}$. We can consider the powers striking the surface AB as shown in Figure 3.20 carried by each of the waves as follows:

$$\text{Incident wave:} \quad |\mathbf{S}_{av}|_i \cos\theta_i = \frac{1}{2\eta_1} E_{i0}^2 \cos\theta_i$$

$$\text{Reflected wave:} \quad |\mathbf{S}_{av}|_r \cos\theta_r = \frac{1}{2\eta_1} E_{r0}^2 \cos\theta_r$$

$$\text{Transmitted wave:} \quad |\mathbf{S}_{av}|_t \cos\theta_t = \frac{1}{2\eta_2} E_{t0}^2 \cos\theta_t$$

where $\eta_1 = \sqrt{\mu_1/\epsilon_1}$ and $\eta_2 = \sqrt{\mu_2/\epsilon_2}$ are the intrinsic impedances of medium 1 and 2 respectively. Note that each of these can be obtained by taking the projection of the respective Poynting vectors on the area of the surface AB in Figure 3.20. For example, for the incident wave, we have

$$P_{AB} = (\mathbf{S}_{av})_i \cdot \hat{\mathbf{z}} = \frac{E_{i0}^2}{2\eta_1} \hat{\mathbf{k}}_i \cdot \hat{\mathbf{z}} = \frac{E_{i0}^2}{2\eta_1} \cos\theta_i$$

On the basis of conservation of power we must have[32]

$$|\mathbf{S}_{av}|_i \cos\theta_i = |\mathbf{S}_{av}|_r \cos\theta_r + |\mathbf{S}_{av}|_t \cos\theta_t$$

$$\rightarrow \quad \frac{1}{2\eta_1} E_{i0}^2 \cos\theta_i = \frac{1}{2\eta_1} E_{r0}^2 \cos\theta_r + \frac{1}{2\eta_2} E_{t0}^2 \cos\theta_t \qquad [3.21]$$

or

$$\boxed{\frac{E_{r0}^2}{E_{i0}^2} = 1 - \frac{\eta_1 E_{t0}^2 \cos\theta_t}{\eta_2 E_{i0}^2 \cos\theta_i}} \qquad [3.22]$$

We now separately consider the perpendicular and parallel polarizations. It is clear from Figure 3.22 that the boundary conditions will differently affect the electric and magnetic fields in the two cases.

3.5.1   Perpendicular Polarization

The expressions for the wave electric and magnetic field phasors of the incident, reflected, and refracted (transmitted) waves shown in Figure 3.23 can be expressed as

$$\mathbf{E}_i(x, z) = \hat{\mathbf{y}} E_{i0} e^{-j\beta_1(x\sin\theta_i + z\cos\theta_i)}$$

$$\mathbf{H}_i(x, z) = \frac{E_{i0}}{\eta_1}(-\hat{\mathbf{x}}\cos\theta_i + \hat{\mathbf{z}}\sin\theta_i)e^{-j\beta_1(x\sin\theta_i + z\cos\theta_i)}$$

$$\mathbf{E}_r(x, z) = \hat{\mathbf{y}} E_{r0} e^{-j\beta_1(x\sin\theta_r - z\cos\theta_r)}$$

---

[32]Note that this condition is required simply so that there is no indefinite accumulation of electromagnetic energy at the interface, which would otherwise occur under the assumed steady-state conditions.

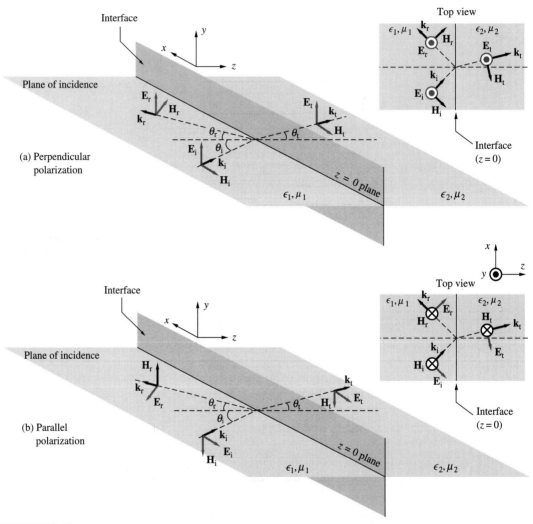

FIGURE 3.22. **Oblique incidence on a dielectric boundary.** (a) Plane wave impinging on a perfect dielectric with the wave electric field vector perpendicular to the plane of incidence (perpendicular polarization). (b) Wave electric field vector parallel to the plane of incidence (parallel polarization).

$$
\begin{cases}
\mathbf{H}_r(x, z) = \dfrac{E_{r0}}{\eta_1}(\hat{\mathbf{x}}\cos\theta_r + \hat{\mathbf{z}}\sin\theta_r)e^{-j\beta_1(x\sin\theta_r - z\cos\theta_r)} \\[2mm]
\mathbf{E}_t(x, z) = \hat{\mathbf{y}}E_{t0}e^{-j\beta_2(x\sin\theta_t + z\cos\theta_t)} \\[2mm]
\mathbf{H}_t(x, z) = \dfrac{E_{t0}}{\eta_2}(-\hat{\mathbf{x}}\cos\theta_t + \hat{\mathbf{z}}\sin\theta_t)e^{-j\beta_2(x\sin\theta_t + z\cos\theta_t)}
\end{cases}
$$

To determine the amplitudes of the reflected and transmitted wave fields in terms of the incident field amplitude E_{i0}, we can apply the boundary condition concerning

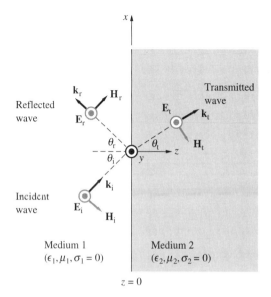

FIGURE 3.23.   A perpendicularly polarized wave incident on a dielectric boundary. A perpendicularly polarized wave incident on a dielectric boundary at $z = 0$ at an incidence angle of θ_i.

the continuity of the tangential component of the wave electric field across the interface. Considering the field orientations as defined in Figure 3.23, we have at $z = 0$,

$$E_{i0}e^{-j\beta_1 x \sin\theta_i} + E_{r0}e^{-j\beta_1 x \sin\theta_r} = E_{t0}e^{-j\beta_2 x \sin\theta_t}$$

Since this condition has to be satisfied at *all* values of x, all three exponents must be equal. Thus,

$$\beta_1 x \sin\theta_i = \beta_1 x \sin\theta_r = \beta_2 x \sin\theta_t$$

which leads to Snell's law,[33] as expressed in [3.19] and with $\theta_r = \theta_i$. We rewrite the boundary condition at any given value of x, say at $x = 0$:

$$E_{i0} + E_{r0} = E_{t0} \qquad \longrightarrow \qquad \frac{E_{t0}}{E_{i0}} = 1 + \frac{E_{r0}}{E_{i0}} \qquad [3.23]$$

Substituting[34] [3.23] in [3.22] and manipulating to solve for E_{r0}/E_{i0} (by eliminating E_{t0}) we find

$$\boxed{\Gamma_\perp \equiv \frac{E_{r0}}{E_{i0}} = \frac{\eta_2 \cos\theta_i - \eta_1 \cos\theta_t}{\eta_2 \cos\theta_i + \eta_1 \cos\theta_t}} \qquad [3.24]$$

[33]Note that matching of the exponents is essentially equivalent to the phase-matching or "in-step" condition that we used on the basis of geometric arguments in deriving Snell's law using Figure 3.20.

[34]An alternative method would have been to use the second boundary condition, i.e., the continuity of the tangential magnetic fields. However, since we have already derived [3.21] and [3.22] by matching the projections of the wave power densities at the boundary, we have in essence already utilized this second boundary condition. The same could be said about the boundary condition of the normal components of **D** and **B**; use of these conditions would not bring any additional information, because our analysis already yields complete information about the state of the reflected and refracted waves.

where $\Gamma_\perp$ is defined as the reflection coefficient for perpendicular polarization. Note that when the second medium is a perfect conductor, i.e., $\eta_2 = 0$, [3.24] reduces to $\Gamma_\perp = -1$, consistent with [3.17]. Note that although we assumed media 1 and 2 to be lossless dielectrics, the expression [3.24] is valid for lossy media also, as long as the proper complex values of η_1 and η_2 are used. Plots of the magnitude $(\rho_\perp = |\Gamma_\perp|)$ of $\Gamma_\perp$ for selected cases are given in Figure 3.30 in Section 3.5.4.

For magnetically identical media $(\mu_1 = \mu_2)$, and also using [3.20], we can find alternative expressions for $\Gamma_\perp$, such as

$$\Gamma_\perp \equiv \frac{E_{r0}}{E_{i0}} = \frac{\cos\theta_i - \sqrt{\epsilon_{2r}/\epsilon_{1r}}\cos\theta_t}{\cos\theta_i + \sqrt{\epsilon_{2r}/\epsilon_{1r}}\cos\theta_t} = \frac{\cos\theta_i - \sqrt{(\epsilon_{2r}/\epsilon_{1r}) - \sin^2\theta_i}}{\cos\theta_i + \sqrt{(\epsilon_{2r}/\epsilon_{1r}) - \sin^2\theta_i}}$$

that can be useful in some cases.

The transmission coefficient $\mathcal{T}_\perp$ can be found from [3.22] and [3.24] as

TRANSMISSION COEFFICIENT

$$\mathcal{T}_\perp \equiv \frac{E_{t0}}{E_{i0}} = \frac{2\eta_2\cos\theta_i}{\eta_2\cos\theta_i + \eta_1\cos\theta_t} \longrightarrow \mathcal{T}_\perp = \underbrace{\frac{2\cos\theta_i}{\cos\theta_i + \sqrt{\epsilon_{2r}/\epsilon_{1r}}\cos\theta_t}}_{\mu_1 = \mu_2}$$

$$\longrightarrow \quad \mathcal{T}_\perp = \frac{2\cos\theta_i}{\cos\theta_i + \sqrt{(\epsilon_{2r}/\epsilon_{1r}) - \sin^2\theta_i}} \qquad [3.25]$$

We note that

$$\boxed{1 + \Gamma_\perp = \mathcal{T}_\perp}$$

similar to the result obtained in Section 3.2.1 for normal incidence on a dielectric interface, where it was found that $1 + \Gamma = \mathcal{T}$.

Example 3-10: A perpendicularly polarized wave. A uniform plane wave traveling in air having an electric field given by

$$\mathbf{E}_i(y, z) = \hat{\mathbf{x}}E_0 e^{-j\beta_1(y\cos\theta_i - z\sin\theta_i)}$$

is obliquely incident on the $y = 0$ interface between air and polystyrene, as shown in Figure 3.24. If the time-average power, the frequency, and the incidence angle of this wave are given to be 1.4 W-m^{-2}, 3 GHz, and $\sim 58°$ respectively, find (a) E_0 (assume real) and β_1; (b) $\mathbf{H}_i(y, z)$; (c) $\mathbf{E}_r(y, z)$ and $\mathbf{H}_r(y, z)$; (d) the time-average power of the reflected and transmitted waves. For polystyrene, take $\sigma_2 = 0$, $\epsilon_{2r} \approx 2.56$, $\mu_{2r} = 1$ at 3 GHz.

Solution:

(a) Using the time-average power of the incident wave, the amplitude of the electric field can be calculated as

$$|\mathbf{S}_{av}|_i = \frac{1}{2}\frac{E_0^2}{\eta_1} = 1.4\ \text{W-m}^{-2} \rightarrow E_0 \approx \sqrt{2 \times 377 \times 1.4} \approx 32.5\ \text{V-m}^{-1}$$

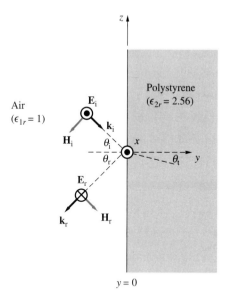

FIGURE 3.24.   Air–polystyrene interface. Incident, reflected, and refracted waves for Example 3-10.

The phase constant is given by

$$\beta_1 = \omega \sqrt{\mu_1 \epsilon_1} = \frac{\omega}{c} \simeq \frac{2\pi \times 3 \times 10^9}{3 \times 10^8} = 20\pi \quad \text{rad-m}^{-1}$$

(b) From Figure 3.24, the magnetic field of the incident wave can be found as

$$\mathbf{H}_i(y, z) = \frac{1}{\eta_1} \hat{\mathbf{k}}_i \times \mathbf{E}_i(y, z)$$

$$= \frac{1}{\eta_1} [\hat{\mathbf{y}} \cos \theta_i - \hat{\mathbf{z}} \sin \theta_i] \times [\hat{\mathbf{x}} E_0 e^{-j\beta_1(y \cos \theta_i - z \sin \theta_i)}]$$

$$= \frac{-E_0}{\eta_1} [\hat{\mathbf{z}} \cos \theta_i + \hat{\mathbf{y}} \sin \theta_i] e^{-j\beta_1(y \cos \theta_i - z \sin \theta_i)}$$

Substituting the numerical values, we have

$$\mathbf{H}_i(y, z) \simeq -86.2[\hat{\mathbf{y}}0.848 + \hat{\mathbf{z}}0.530]e^{-j20\pi(0.530y - 0.848z)} \quad \text{mA-m}^{-1}$$

(c) Using Snell's law of refraction as given by [3.20], we can find the transmission angle as

$$\sin 58° = \sqrt{2.56} \sin \theta_t \quad \rightarrow \quad \theta_t \simeq 32°$$

Since the incident wave is perpendicularly polarized, we can find the reflection coefficient using [3.24] as

$$\Gamma_\perp \simeq \frac{\cos 58° - \sqrt{2.56} \cos 32°}{\cos 58° + \sqrt{2.56} \cos 32°} \simeq -0.438$$

The electric field of the reflected wave can be written as

$$\mathbf{E}_r(y, z) = \hat{\mathbf{x}} \Gamma_\perp E_0 e^{j\beta_1(y\cos\theta_r + z\sin\theta_r)} \simeq -\hat{\mathbf{x}} 14.2 e^{j20\pi(0.530y + 0.848z)} \text{ V-m}^{-1}$$

The magnetic field of the reflected wave is given by

$$\mathbf{H}_r(y, z) = \frac{1}{\eta_1} \hat{\mathbf{k}}_r \times \mathbf{E}_r = \frac{\Gamma_\perp E_0}{\eta_1} [\hat{\mathbf{y}} \sin\theta_r - \hat{\mathbf{z}} \cos\theta_r] e^{j\beta_1(y\cos\theta_r + z\sin\theta_r)}$$

$$\simeq -37.8[\hat{\mathbf{y}} 0.848 - \hat{\mathbf{z}} 0.530] e^{j20\pi(0.530y + 0.848z)} \text{ mA-m}^{-1}$$

(d) The time-average power of the reflected wave is given by

$$|\mathbf{S}_{av}|_r = \frac{1}{2} \frac{(\Gamma_\perp E_0)^2}{\eta_1} \simeq \frac{1}{2} \frac{(14.2)^2}{377} \simeq 0.269 \text{ W-m}^{-2}$$

Using the principle of conservation of power ([3.21]), the time-average power of the transmitted wave can be found from

$$1.4 \cos 58° \simeq 0.269 \cos 58° + |\mathbf{S}_{av}|_t \cos 32°$$

yielding

$$|\mathbf{S}_{av}|_t \simeq 0.707 \text{ W-m}^{-2}$$

Note that one can also find this value using

$$|\mathbf{S}_{av}|_t = \frac{1}{2} \frac{(\mathcal{T}_\perp E_0)^2}{\eta_2}$$

where

$$\mathcal{T}_\perp \simeq \frac{2\cos 58°}{\cos 58° + \sqrt{2.56} \cos 32°} \simeq 0.562$$

and $\eta_2 \simeq 377/\sqrt{2.56} \simeq 236\Omega$. Substituting,

$$|\mathbf{S}_{av}|_t \simeq \frac{1}{2} \frac{(0.562 \times 32.5)^2}{236} \simeq 0.707 \text{ W-m}^{-2}$$

Note that $|\mathbf{S}_{av}|_i \neq |\mathbf{S}_{av}|_r + |\mathbf{S}_{av}|_t$. Although this result may at first appear counterintuitive, there is in fact no reason to expect the sum of power densities of the reflected and transmitted waves to be equal to that of the incident wave. The Poynting flux of a wave is a directional quantity (i.e., a vector), and the summation of the magnitude of the vectors pointing in different directions has no physical significance. Projection of the three Poynting vectors on any area must satisfy the conservation of power, a fact used ([3.21]) in deriving the expressions for $\Gamma_\perp$ and $\Gamma_\parallel$.

3.5.2   Parallel Polarization

The expressions for the wave electric and magnetic field phasors of the incident, reflected, and refracted (transmitted) waves illustrated in Figure 3.25 are given by

$$\mathbf{E}_i(x, z) = E_{i0}(\hat{\mathbf{x}} \cos \theta_i - \hat{\mathbf{z}} \sin \theta_i)e^{-j\beta_1(x \sin \theta_i + z \cos \theta_i)}$$

$$\mathbf{H}_i(x, z) = \hat{\mathbf{y}}\frac{E_{i0}}{\eta_1}e^{-j\beta_1(x \sin \theta_i + z \cos \theta_i)}$$

$$\mathbf{E}_r(x, z) = E_{r0}(\hat{\mathbf{x}} \cos \theta_r + \hat{\mathbf{z}} \sin \theta_r)e^{-j\beta_1(x \sin \theta_r - z \cos \theta_r)}$$

$$\mathbf{H}_r(x, z) = -\hat{\mathbf{y}}\frac{E_{r0}}{\eta_1}e^{-j\beta_1(x \sin \theta_r - z \cos \theta_r)}$$

$$\mathbf{E}_t(x, z) = E_{t0}(\hat{\mathbf{x}} \cos \theta_t - \hat{\mathbf{z}} \sin \theta_t)e^{-j\beta_2(x \sin \theta_t + z \cos \theta_t)}$$

$$\mathbf{H}_t(x, z) = \hat{\mathbf{y}}\frac{E_{t0}}{\eta_2}e^{-j\beta_2(x \sin \theta_t + z \cos \theta_t)}$$

where $\eta_1 = \sqrt{\mu_1/\epsilon_1}$ and $\eta_2 = \sqrt{\mu_2/\epsilon_2}$ are respectively the intrinsic impedances of media 1 and 2.

We follow a procedure similar to that used for the perpendicular polarization case and set out to find the amplitudes of the reflected and transmitted waves in terms of E_{i0}. For this purpose, we apply the boundary condition concerning the continuity of the tangential component of the wave electric field across the interface. Therefore,

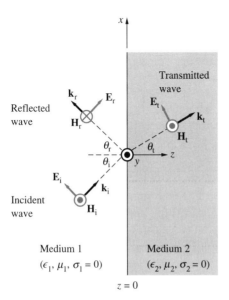

Reflected wave

Incident wave

Medium 1
$(\epsilon_1, \mu_1, \sigma_1 = 0)$

Medium 2
$(\epsilon_2, \mu_2, \sigma_2 = 0)$

$z = 0$

FIGURE 3.25.   **A parallel polarized wave at a dielectric boundary.** A parallel polarized wave incident on a dielectric boundary at $z = 0$ at an incidence angle of θ_i.

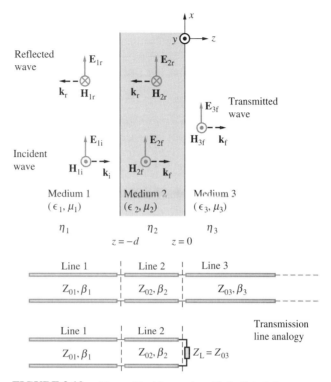

FIGURE 3.10.   Normal incidence at multiple dielectric interfaces. Normal incidence of a uniform plane on a multiple dielectric interface is directly analogous to a transmission line with characteristic impedance Z_{01} (medium 1) connected to a load Z_L (or a third transmission line with characteristic impedance Z_{03}) via another transmission line (characteristic impedance Z_{02}) segment of length d, as shown above.

We formulate the multiple dielectric interface problem as shown in Figure 3.10, in terms of three different dielectric media characterized by intrinsic impedances η_1, η_2, and η_3 and separated by infinite planar boundaries located at $z = -d$ and $z = 0$, so that medium 2 is a layer of thickness d.

We assume a uniform plane wave propagating in medium 1 to be normally incident at the boundary located at $z = -d$. In attempting to determine the amount of wave energy reflected, we might at first be tempted to treat the problem in terms of a series of reflections in which a portion of the energy of the incident wave is transmitted into medium 2, which then propagates to the right and encounters the boundary at $z = 0$, where a portion of it is reflected; this reflected energy propagates to the left in medium 2, encountering the boundary at $z = -d$ from the right, where some of it is reflected back and some transmitted into medium 1, and so on. Even though it might lead to the "correct" answer in some cases, such thinking, which implies a "sequence of events" type of scenario, is inconsistent with our initial steady-state (or time-harmonic) assumption.

at $z = 0$, we have

$$(E_{i0} + E_{r0}) \cos \theta_i = E_{t0} \cos \theta_t$$

or

$$\frac{E_{t0}}{E_{i0}} = \left(1 + \frac{E_{r0}}{E_{i0}}\right) \frac{\cos \theta_i}{\cos \theta_t}$$

Substituting in [3.22] and manipulating to solve for E_{r0}/E_{i0} (by eliminating E_{t0}) we find

REFLECTION
COEFFICIENT
PARALLEL
POLARISATION

$$\boxed{\Gamma_\| \equiv \frac{E_{r0}}{E_{i0}} = \frac{-\eta_1 \cos \theta_i + \eta_2 \cos \theta_t}{\eta_1 \cos \theta_i + \eta_2 \cos \theta_t}} \longrightarrow \Gamma_\| = \underbrace{\frac{\cos \theta_t - \sqrt{\epsilon_2/\epsilon_1} \cos \theta_i}{\cos \theta_t + \sqrt{\epsilon_2/\epsilon_1} \cos \theta_i}}_{\mu_1 = \mu_2}$$

$$\longrightarrow \quad \Gamma_\| = \frac{-\cos \theta_i + (\epsilon_{1r}/\epsilon_{2r})\sqrt{(\epsilon_{2r}/\epsilon_{1r}) - \sin^2 \theta_i}}{\cos \theta_i + (\epsilon_{1r}/\epsilon_{2r})\sqrt{(\epsilon_{2r}/\epsilon_{1r}) - \sin^2 \theta_i}} \qquad [3.26]$$

as the reflection coefficient for parallel polarization and eliminating E_{r0} we find

TRANSMISSION
COEFFICIENT

$$\boxed{\mathcal{T}_\| = \frac{E_{t0}}{E_{i0}} = \frac{2\eta_2 \cos \theta_i}{\eta_1 \cos \theta_i + \eta_2 \cos \theta_t}} \longrightarrow \mathcal{T}_\| = \underbrace{\frac{2 \cos \theta_i}{\cos \theta_t + \sqrt{\epsilon_2/\epsilon_{1r}} \cos \theta_i}}_{\mu_1 = \mu_2}$$

$$\longrightarrow \quad \mathcal{T}_\| = \frac{2\sqrt{\epsilon_{1r}/\epsilon_{2r}} \cos \theta_i}{\cos \theta_i + \sqrt{\epsilon_{1r}/\epsilon_{2r}}\sqrt{1 - (\epsilon_{1r}/\epsilon_{2r}) \sin^2 \theta_i}} \qquad [3.27]$$

as the transmission coefficient for parallel polarization. Plots of the magnitude $\rho_\|$ and the phase $\phi_\|$ of $\Gamma_\|$ (i.e., $\Gamma_\| = \rho_\| e^{j\phi_\|}$) are presented in Figure 3.30 in Section 3.5.4. From [3.26] and [3.27] we note that

$$\boxed{1 + \Gamma_\| = \mathcal{T}_\| \left(\frac{\cos \theta_t}{\cos \theta_i}\right)}$$

Example 3-11 provides numerical values for reflection and transmission coefficients for oblique incidence on an air–glass interface.

Example 3-11: Oblique incidence of light at air–glass interface. A light ray traveling in air is obliquely incident at an incidence angle of 40° on the air–glass ($n = 1.52$) plane boundary as shown in Figure 3.26. (a) Find the reflection and transmission coefficients for a perpendicularly polarized (s-polarized) light. (b) Find the reflection and transmission coefficients for a parallel polarized (p-polarized) light.

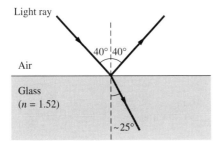

FIGURE 3.26.   Air–glass boundary. Figure for Example 3-11.

Solution:

(a) For the perpendicularly polarized case, using [3.24] and [3.25] we have

$$\Gamma_\perp = \frac{\cos 40° - \sqrt{(1.52)^2 - \sin^2 40°}}{\cos 40° + \sqrt{(1.52)^2 - \sin^2 40°}} \simeq -0.285$$

$$\mathcal{T}_\perp = \frac{2\cos 40°}{\cos 40° + \sqrt{(1.52)^2 - \sin^2 40°}} \simeq 0.715$$

and $1 + \Gamma_\perp = \mathcal{T}_\perp$ as expected.

(b) For the parallel polarized case we have from [3.26] and [3.27],

$$\Gamma_\parallel = \frac{-\cos 40° + (1.52)^{-2}\sqrt{(1.52)^2 - \sin^2 40°}}{\cos 40° + (1.52)^{-2}\sqrt{(1.52)^2 - \sin^2 40°}} \simeq -0.125$$

$$\mathcal{T}_\parallel = \frac{2(1.52)^{-1}\cos 40°}{\cos 40° + (1.52)^{-1}\sqrt{1 - (1.52)^{-2}\sin^2 40°}} \simeq 0.74$$

From Snell's law of refraction given by [3.20], the transmission angle can be found as

$$\sin 40° = 1.52 \sin \theta_t \rightarrow \theta_t \simeq 25°$$

Substituting, we find that $1 + \Gamma_\parallel = \mathcal{T}_\parallel(\cos \theta_t / \cos \theta_i)$, as expected.

3.5.3   Brewster's Angle

Examination of [3.26] indicates the possibility that for some parameter values we may have $\Gamma_\parallel = 0$, in which case no reflection (i.e., total transmission) occurs. This condition is satisfied when

$$\Gamma_\parallel = 0 \longrightarrow -\eta_1 \cos \theta_i + \eta_2 \cos \theta_t = 0 \longrightarrow \cos \theta_i = \frac{\eta_2}{\eta_1} \cos \theta_t$$

From Snell's law of refraction, we have

$$\frac{\sin \theta_t}{\sin \theta_i} = \sqrt{\frac{\epsilon_{1r}\mu_{1r}}{\epsilon_{2r}\mu_{2r}}} \quad \rightarrow \quad \sin \theta_t = \sqrt{\frac{\epsilon_{1r}\mu_{1r}}{\epsilon_{2r}\mu_{2r}}} \sin \theta_i$$

and, using $\cos\theta_t = \sqrt{1 - \sin\theta_t^2}$, we can rewrite the special incidence angle θ_i at which $\Gamma_{\parallel} = 0$ as

$$\sin\theta_i = \sqrt{\frac{1 - \mu_{2r}\epsilon_{1r}/(\mu_{1r}\epsilon_{2r})}{1 - (\epsilon_{1r}/\epsilon_{2r})^2}}$$

This special incidence angle θ_i that satisfies the $\Gamma_{\parallel} = 0$ condition is known as the *Brewster angle,* denoted as θ_{iB}. For nonmagnetic materials $(\mu_1 = \mu_2 = \mu_0)$ the Brewster angle can be simplified as

$$\theta_{iB} = \sin^{-1}\left(1 + \frac{\epsilon_{1r}}{\epsilon_{2r}}\right)^{-1/2} = \tan^{-1}\left[\sqrt{\frac{\epsilon_{2r}}{\epsilon_{1r}}}\right] = \tan^{-1}\frac{n_2}{n_1} \qquad [3.28]$$

where n_1 and n_2 are the refractive indices of the two media. Note that if the the wave is incident at this angle, then there is no reflected wave when the incident wave is parallel polarized.

Brewster, Sir David *(Scottish physicist, b. Dec. 11, 1781, Jedburgh, Roxburghshire; d. Feb. 10, 1868, Allerly, Roxburghshire) The son of a schoolmaster, Brewster entered the University of Edinburgh at the age of 12 and was educated for the ministry, becoming a licensed preacher at the age of 23. However, he hated to preach, and extreme nervousness and a fear of failure induced him to give up his clerical career to pursue his extensive scientific and literary interests. At age 27, he became editor of the* Edinburgh Encyclopedia, *a post he held for 22 years. He contributed to the 7th and 8th editions of the* Encyclopedia Britannica *and wrote more than 425 scientific papers, articles, and books (mostly on optics). In 1815 he found that a beam of light could be split into a reflected portion and a refracted portion, at right angles to each other, and that both would then be completely polarized. This is still called Brewster's law and earned him the Rumford medal in 1819. The law can be neatly explained by supposing light to consist of transverse waves. Neither the longitudinal wave theory nor the particle wave theory of light could explain it.*

Nevertheless, Brewster remained an ardent adherent to the particle theory all his life, refusing to accept the ether. This seemed at the time to be an example of scientific ultraconservatism, but a half century after his death he was to be vindicated by Einstein [1879–1955].

Brewster's name is permanently associated with impedance matching in not only optical but also a wide range of electromagnetic applications. When Hertz [1857–1894] showed that radio waves were quasi-optical, VHF and microwave investigators of the period (1888–1900) investigated the reflection phenomena and repeatedly demonstrated the "polarizing" and Brewster angle. When Marconi invented long-wave radio communication, Brewster's angle continued to play a role. Radiated waves impinging on lossy ground (i.e., wet earth) exhibit a reflection minimum (not zero, however) at the "pseudo Brewster angle," and navigational and radio aids, radars, and VHF communications take notice of this for system analysis and design.

Brewster was also the inventor of the kaleidoscope in 1816, a scientific toy that has never ceased to amuse the young—and the old. He patented it, and although thousands were sold in a few days, it was so easy to pirate that he earned virtually nothing from it. He also invented the stereoscope, through which one views two slightly different pictures, one with each eye, giving the illusion of three-dimensionality.

Brewster wrote a biography of Newton [1642–1727] and helped found the British Association for the Advancement of Science in 1831. He was knighted in 1832. He was successively the principal of the Scottish Universities of St. Andrews and Edinburgh. During his lifetime, he met savants galore, including Laplace, Poisson, Arago, Gay-Lussac, de la Rive, and the elite of the U. K. scientific community. At the age of 75, he married a second time and had a daughter some years later. He died at age 87 in Melrose in Scotland. [Partly adapted with permission from Biographical Encyclopedia of Science and Technology, I. Asimov, Doubleday, 1982].

1700	*1781*	*1868*	*2000*

As a useful way of thinking about the Brewster condition we consider Figure 3.27. An interesting property of the Brewster condition is that the reflected and

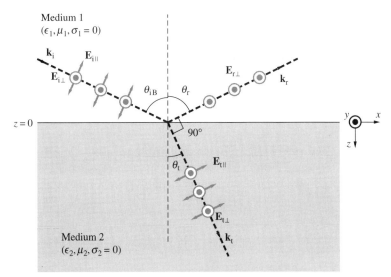

FIGURE 3.27.   Unpolarized light at a dielectric boundary. An unpolarized light wave is incident at the dielectric boundary at the Brewster angle, i.e., $\theta_i = \theta_{iB}$, produces a polarized reflected wave and a partially polarized transmitted wave. Note that for $\theta_i = \theta_{iB}$, the angle between the reflected and refracted rays is 90°.

refracted rays are at 90° with respect to one another, or that $\theta_{iB} + \theta_t = \pi/2$. To see this, we can simply use the Snell's law of refraction and the Brewster condition:

Brewster angle: $\quad \tan \theta_{iB} = \dfrac{\sin \theta_{iB}}{\cos \theta_{iB}} = \sqrt{\dfrac{\epsilon_{2r}}{\epsilon_{1r}}} \quad \rightarrow \quad \sin \theta_{iB} = \cos \theta_{iB} \sqrt{\dfrac{\epsilon_{2r}}{\epsilon_{1r}}}$

Snell's law $\quad \sin \theta_t = \sin \theta_i \sqrt{\dfrac{\epsilon_{1r}}{\epsilon_{2r}}} \quad \rightarrow \quad \sin \theta_t = \left[\cos \theta_{iB} \sqrt{\dfrac{\epsilon_{2r}}{\epsilon_{1r}}} \right] \sqrt{\dfrac{\epsilon_{1r}}{\epsilon_{2r}}}$

$$= \cos \theta_{iB}$$

and therefore θ_{iB} and its corresponding θ_t are complementary angles. Physically, the incident field gives rise to vibrations of electrons in the atoms of the second medium. These vibrations are in the direction of the electric vector of the transmitted wave. The vibrating electrons essentially radiate a new wave in the direction transverse to their motion[35] and give rise to the reflected wave, which propagates back into the first medium. When the reflected and transmitted rays are at right angles to each other, the reflected ray does not receive any energy for oscillations in the plane of incidence.[36]

It follows from Figure 3.27 that polarized light may be produced from unpolarized light. A randomly polarized wave incident on a dielectric at the Brewster angle becomes linearly polarized upon reflection.

Note that a Brewster angle may also be defined for perpendicular polarization. We find from [3.24] by setting $\Gamma_\perp = 0$ that

$$\sin^2 \theta_{iB\perp} = \frac{1 - \mu_{1r}\epsilon_{2r}/(\mu_{2r}\epsilon_{1r})}{1 - (\mu_{1r}/\mu_{2r})^2}$$

For nonmagnetic media ($\mu_1 = \mu_2 = \mu_0$) the right-hand side becomes infinite, so that $\theta_{iB\perp}$ does not exist. Thus, the case of a Brewster angle for perpendicular polarization is not of practical interest.

Examples 3-12 and 3-13 illustrate the calculation and use of the Brewster angle.

Example 3-12:   Brewster angle at air–silicon interface.   Find the Brewster angle for an air–silicon ($\epsilon_r = 11.7$, assume lossless) interface (Figure 3.28) for a parallel polarized wave incident (a) from air into silicon, and (b) from silicon into air.

---

[35]Radiation fields of oscillating dipoles are discussed in Section 6.5.

[36]This particular interpretation of the lack of a reflected wave at $\theta_i = \theta_{iB}$ is often put forth and is undoubtedly useful. However, the argument is at the same time highly qualitative and somewhat superficial; for example, there is a Brewster angle when medium 2 is free space (i.e., no electrons) with permittivity ϵ_0. Also, at interfaces of magnetic media ($\mu_1 \neq \mu_2$), there can also exist a Brewster angle for perpendicular polarization [P. Lorrain, D. R. Corson, and F. Lorrain, *Electromagnetic Fields and Waves,* 3rd ed., Freeman, New York, p. 575, 1988].

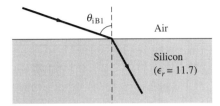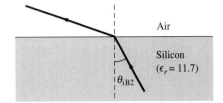

FIGURE 3.28.   Example 3-12.

Solution:

(a) For the air–silicon interface we have

$$\theta_{iB1} = \tan^{-1}\sqrt{\epsilon_r} = \tan^{-1}\sqrt{11.7} \simeq 73.7°$$

(b) For the silicon–air interface, we have

$$\theta_{iB2} = \tan^{-1}\left(\frac{1}{\sqrt{\epsilon_r}}\right) = \tan^{-1}\left(\frac{1}{\sqrt{11.7}}\right) \simeq 16.3°$$

(Note that θ_{iB1} and θ_{iB2} are complementary angles, as expected.)

Example 3-13:   A parallel polarized wave.   Consider the uniform plane wave incident on an air–polystyrene interface at an angle of $\sim 58°$ as considered in Example 3-10. Assuming that the wave is taken to be parallel polarized as shown in Figure 3.29, with its magnetic field represented by

$$\mathbf{H}_i(y, z) = \hat{\mathbf{x}}H_0 e^{-j\beta_1(\cos\theta_i y - \sin\theta_i z)}$$

and keeping the total time-average power, the frequency, and the incidence angle of the wave to be the same as in Example 3-10 (i.e., 1.4 W-m^{-2}, 3 GHz, and $\sim 58°$ respectively), find the following: (a) H_0 (assume real) and β_1, (b) $\mathbf{E}_i(y, z)$, (c) $\mathbf{E}_r(y, z)$ and $\mathbf{H}_r(y, z)$, and (d) $\mathbf{E}_t(y, z)$ and $\mathbf{H}_t(y, z)$.

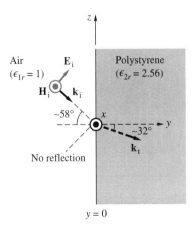

FIGURE 3.29.   Air–polystyrene interface. Incident and refracted waves for Example 3-13.

Solution:

(a) We have

$$|\mathbf{S}_{av}|_i = \frac{1}{2}\eta_1 H_0^2 = 1.4 \text{ W-m}^{-2} \quad \rightarrow \quad H_0 \simeq \sqrt{2 \times 1.4/377} \simeq 0.0862 \text{ A-m}^{-1}$$

and

$$\beta_1 = \omega\sqrt{\mu_1\epsilon_1} = \frac{\omega}{c} \simeq \frac{2\pi \times 3 \times 10^9}{3 \times 10^8} = 20\pi \text{ rad-m}^{-1}$$

Substituting the numerical values, we have

$$\mathbf{H}_i(y, z) \simeq \hat{\mathbf{x}}86.2e^{-j20\pi(0.53y-0.848z)} \text{ mA-m}^{-1}$$

(b) The electric field phasor of the incident wave can be written as

$$\mathbf{E}_i(y, z) = \eta_1 H_0(\hat{\mathbf{y}}\sin\theta_i - \hat{\mathbf{z}}\cos\theta_i)e^{-j\beta_1(\cos\theta_i y-\sin\theta_i z)}$$

$$\simeq 32.5(\hat{\mathbf{y}}0.848 - \hat{\mathbf{z}}0.53)e^{-j20\pi(0.53\hat{\mathbf{y}}-0.848\hat{\mathbf{z}})} \text{ V-m}^{-1}$$

(c) Since the incident wave is parallel-polarized, let us check the Brewster angle:

$$\theta_{iB} = \tan^{-1}\sqrt{\frac{\epsilon_{2r}}{\epsilon_{1r}}} = \tan^{-1}(1.6) \simeq 58°$$

So the incidence angle is the Brewster angle, which corresponds to the case of total transmission. As a result, we have

$$\Gamma_\parallel = 0 \quad \rightarrow \quad \mathbf{E}_r(y, z) = \mathbf{H}_r(y, z) = 0$$

Note that if the incidence were not at the Brewster angle, we could have simply evaluated $\Gamma_\parallel$ from [3.26] to determine E_{r0}.

(d) From the conservation-of-power principle, we have

$$|\mathbf{S}_{av}|_i = |\mathbf{S}_{av}|_t \rightarrow \frac{1}{2}\frac{E_0^2}{\eta_1}\cos\theta_i = \frac{1}{2}\frac{E_{t0}^2}{\eta_2}\cos\theta_t$$

So

$$E_{t0} \simeq \sqrt{\frac{(32.5)^2}{1.6}}\left[\frac{\cos 58°}{\cos 32°}\right] \simeq 20.3 \text{ V-m}^{-1}$$

From Figure 3.29, the electric field phasor of the transmitted wave can be written as

$$\mathbf{E}_t(y, z) \simeq 20.3[\hat{\mathbf{y}}0.53 + \hat{\mathbf{z}}0.848]e^{-j32\pi(0.848y-0.53z)} \text{ V-m}^{-1}$$

The corresponding magnetic field phasor is

$$\mathbf{H}_t(y, z) \simeq \hat{\mathbf{x}}53.9e^{-j32\pi(0.848y-0.53z)} \text{ mA-m}^{-1}$$

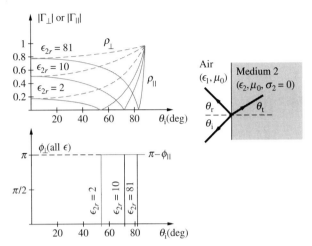

FIGURE 3.30. **Reflection coefficient versus the angle of incidence.** (a) Magnitude and phase of reflection coefficient for perpendicular ($\Gamma_\perp$) and parallel ($\Gamma_\parallel$) polarization versus angle of incidence θ_i for distilled water ($\epsilon_{2r} = 81$), flint glass ($\epsilon_{2r} = 10$) and paraffin ($\epsilon_{2r} = 2$), all assumed to be lossless. For clarity, the complement of the phase angle $\phi_\parallel$ (i.e., $\pi - \phi_\parallel$) is sketched, rather than $\phi_\parallel$ itself. In all cases, $\phi_\parallel = \pi$ for $\theta_i < \theta_{iB}$, and $\phi_\parallel = 0$ for $\theta_i > \theta_{iB}$, while $\phi_\perp = \pi$ for all θ_i.

3.5.4 Fresnel's Formulas

The reflection coefficient equations [3.24] and [3.26] are known as Fresnel's equations, or respectively Fresnel's sine law and Fresnel's tangent law,[37] after Augustin Fresnel, who put forth these and many other aspects of reflection, refraction, and diffraction of light waves during his short life.[38]

The behavior of the reflection coefficients can be investigated by plotting their magnitudes against incidence angle for different values of ϵ_1, ϵ_2. A plot of the magnitude and phases of $\Gamma_\perp$ and $\Gamma_\parallel$ for three different values of ϵ_2 (with $\epsilon_1 = \epsilon_0$) is given in Figure 3.30. Note that as θ_i is varied from 0 to 90°, $\Gamma_\parallel$ changes polarity at $\theta_i = \theta_{iB}$, where it is equal to zero. For the case of nonmagnetic media ($\mu_1 = \mu_2 = \mu_0$) and for $\epsilon_2 > \epsilon_1$, since $\theta_i > \theta_t$, both $\Gamma_\perp$ and $\Gamma_\parallel$ are negative for $\theta_i < \theta_{iB}$. For $\theta_i > \theta_{iB}$, $\Gamma_\perp$ remains negative, but $\Gamma_\parallel$ becomes positive. Note that the polarity of $\Gamma_{\perp,\parallel}$ is determined by our choice of the orientation of the incident and reflected electric field vectors in Figures 3.25 and 3.23.

---

[37]The reason for this terminology is the fact that Fresnel's version of $\rho_\perp$ and $\rho_\parallel$ were

$$\Gamma_\perp = \frac{\sin(\theta_t - \theta_i)}{\sin(\theta_t + \theta_i)} \quad \text{and} \quad \Gamma_\parallel = \frac{\tan(\theta_t - \theta_i)}{\tan(\theta_t + \theta_i)}$$

It is clear from the above that when $\theta_t + \theta_i = \pi/2$, we have $\tan(\theta_t + \theta_i) = \infty$, which means $\Gamma_\parallel = 0$, i.e., the Brewster condition. The reader is encouraged to determine that the above expressions for $\Gamma_\perp$ and $\Gamma_\parallel$ are indeed consistent with [3.24] and [3.26].

[38]For a very interesting discussion of Fresnel's scientific contributions, see Chapter IV of E. Whittaker, *History of the Theories of Aether and Electricity,* Thomas Nelson and Sons Lim., New York, 1951.

Consider first the perpendicular polarization case. With the sense of orientation defined as the same for $\mathbf{E}_i$ and $\mathbf{E}_r$, it is clear from the boundary condition that if the second medium is very dense, i.e., $\epsilon_2 \gg \epsilon_1$ or $\eta_2 \ll \eta_1$, then most of the wave energy would be reflected and the magnitude of E_t would be very small. Accordingly, to satisfy the boundary condition, the electric field should switch polarity upon reflection, or $\Gamma_\perp \simeq -1$ for $\theta_i = 0$.

For parallel polarization, and with the electric field polarities as defined in Figure 3.25, $\Gamma_\parallel < 0$ at low values of θ_i simply means that the tangential component of the transmitted wave is smaller than that of the incident wave, so the electric field of the reflected wave has to be negative to cancel some of the incident wave field in medium 1. At $\theta_i = \theta_{iB}$, $\Gamma_\parallel$ changes sign, as can be seen from Figure 3.30. Thus, at large angles $\theta_i > \theta_{iB}$, the transmitted wave field is larger than the incident one, and the reflected wave needs to be positive to add to the incident wave field.

Wave Polarization upon Reflection and Refraction

In all the above analyses we have implicitly assumed that the propagation directions of the reflected and refracted waves (i.e., $\hat{\mathbf{k}}_r$ and $\hat{\mathbf{k}}_t$) lie in the plane of incidence. For uniform plane waves, the incident fields are by definition independent of y, assuming the xz plane to be the plane of incidence. Since the layout of the boundary is also independent of y, the reflected and refracted fields must also be independent of y. We further note that there is complete symmetry in the $+y$ versus $-y$ directions and that there is no preferred direction for the waves to bend upon reflection or refraction. Another argument against a y dependence of the reflected and refracted fields is the fact that any solution of Maxwell's equations satisfying a given set of boundary conditions must be unique; thus, because we can find a solution for the fields by assuming $\hat{\mathbf{k}}_r$ and $\hat{\mathbf{k}}_t$ to be in the plane of incidence, these solutions must be the only possible solutions.

In the light of the above argument, an incident wave that is linearly polarized with $\mathbf{E}_i$ perpendicular to the plane of incidence is partially reflected, with the same polarization, and partially transmitted, again with the same polarization. When the incident wave is linearly polarized with $\mathbf{E}_i$ in the plane of incidence, the wave is again reflected and refracted with no change in polarization. This situation is analogous to dropping a uniform rigid rod on a smooth plane surface. If the rod is dropped with its axis either normal or parallel to the surface, then its orientation will not change as it bounces back from the surface (i.e., reflection). Otherwise, however, the rod rotates upon reflection.

For an incident wave that is linearly polarized, the polarizations of the reflected and refracted waves differs from that of the incident wave only if $\mathbf{E}_i$ is not entirely normal or parallel to the plane of incidence. That this is so is clear from Figure 3.30, which shows that, in general, the magnitudes of $\Gamma_\parallel$ and $\Gamma_\perp$ are different. If the incident wave is linearly polarized with its electric field $\mathbf{E}_i$ oriented at an angle to the plane of incidence, then we can decompose the field into two components, $\mathbf{E}_\perp$ and $\mathbf{E}_\parallel$, and separately calculate the reflection for each component, using $\Gamma_\perp$ and $\Gamma_\parallel$.

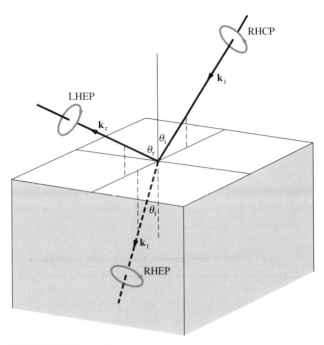

FIGURE 3.31. **Circularly polarized wave at a dielectric interface.** Right-hand circularly polarized (RHCP) wave is obliquely incident from air upon a dielectric. For $\theta_i < \theta_{iB}$, the reflected wave is in general left-hand elliptically polarized (LHEP) whereas the transmitted wave is also elliptically polarized but with the same sense of rotation as that of the incident wave (i.e., RHEP).

Since $|\Gamma_\perp| \neq |\Gamma_\parallel|$, the reflected wave, while still linearly polarized, now has an electric field oriented at a different angle with respect to the plane of incidence. If the second medium is lossy, then the reflection coefficients are complex (as we shall see in Section 3.7) with different phase shifts for the parallel and perpendicular polarizations, in which case the reflected wave is elliptically polarized. A similar situation occurs under conditions of total reflection, as discussed in the next section.

When the incident wave is circularly polarized, the relative signs of the two reflection coefficients determine the polarization sense of the reflected wave, as illustrated in Figure 3.31. For $\theta_i < \theta_{iB}$, the reflected wave is elliptically polarized with the opposite sense of rotation to that of the incident wave. Both $\Gamma_\perp$ and $\Gamma_\parallel$ are negative in this case, so the sense of polarization changes (i.e., from RHEP to LHEP or vice versa), because the $\hat{\mathbf{k}}$ vector of the wave has a component in the $+z$ direction, while $\hat{\mathbf{k}}_i$ has a component in the $+z$ direction. At $\theta_i = \theta_{iB}$, there is no reflection for the parallel-polarized component, so the reflected wave would be linearly polarized, with electric field perpendicular to the plane of incidence. For $\theta_i > \theta_{iB}$, $\Gamma_\perp$ remains negative while $\Gamma_\parallel$ is now positive, and the reflected wave is elliptically polarized with the same sense of rotation as that of the incident wave.

Fresnel, Augustin Jean   *(French engineer and physicist, b. May 10, 1788, Brogile, Eure, Normandy; d. July 14, 1827, Ville-d'Avray, near Paris) Fresnel, the son of an architect, was the very reverse of an infant prodigy. He was eight before he could read. Nevertheless his intelligence shone out with the passing years, and he became a civil engineer, working for the government for most of his professional life. There was a short break in 1814, when Fresnel opposed the return of Napoleon from exile in Elba, was taken prisoner, and as a result lost his post. However, Napoleon's return lasted only a hundred days and ended with Waterloo; it was then Fresnel's turn to return.*

About 1814 Fresnel grew interested in the problem of light and independently conducted some of the experiments that Young [1773–1829] had conducted a decade before. Arago [1786–1853] read Fresnel's reports and was converted to the wave theory. He called Fresnel's attention to Young's work and Fresnel's similar work was accelerated. The Frenchman began to construct a thorough mathematical basis for the wave theory.

Huygens [1629–1695] had constructed part of such a mathematical basis a century and a half before, but Fresnel went beyond him. For one thing, Huygens and all the wave theories after his time (except for Hooke [1635–1703], whose freewheeling conjectures hit the mark a number of times) had felt that light waves, if they existed, were longitudinal, with oscillations taking place along the line of propagation, as in sound waves. Young eventually suggested that light waves might be transverse, with oscillations at right angles to the line of propagation, as in water waves. Fresnel adopted the transverse wave view with alacrity and built up the necessary theoretical basis for it.

The greatest victory of the transverse wave theory was the explanation of the phenomenon of double refraction through Iceland spar, discovered by Bartholin [1625–1698]. Neither the particle theory nor the longitudinal wave theory could explain it. The transverse wave theory, however, could, and Fresnel showed that light could be refracted through two different angles because one ray would consist of waves oscillating in a particular plane, while the other ray consisted of waves oscillating in a plane perpendicular to the first plane. The two rays would therefore be expected to have different properties under certain conditions and to be refracted differently by certain solids.

Ordinary light, according to Fresnel's views, consisted of waves oscillating equally in all possible planes at right angles to the line of propagation, but light with oscillations unequally distributed among the planes was polarized light, a rather poor term introduced by Malus [1775–1812]. When the oscillations were restricted to a single plane, as in the case of the light rays passing through Iceland spar, the light was said to be plane-polarized.

Fresnel used his new view of light to design lenses for lighthouses, and they were more efficient than the mirrors they replaced. An understanding of polarized light, moreover, came to have an important application to organic chemistry, through the work of Pasteur [1822–1895] a generation later.

*Arago, after a period of collaboration with Fresnel, backed out nervously
when the transverse waves were adopted by the latter. Later he came round, but
Fresnel had published his work alone and got credit alone.*

*The difficulty that frightened Arago was this: If light consisted of waves,
something must be waving. Early wave theorists postulated an "ether" filling
space and all transparent substances. Light consisted of waves in this ether,
which thus carried light even through an apparent vacuum and could be called a
luminiferous ("light-carrying") ether. (The word "ether" is taken from Aristotle's
[384 B.C.–322 B.C.] name for the fifth element that he considered to make up the
heavens.) If light waves were longitudinal, the ether could be looked on as a very
fine gaslike substance, indetectable to ordinary instruments, and there would have
been no difficulty in accepting that, or at least no more difficulty than there was in
accepting Dalton's [1766–1844] indetectable atoms. However, transverse waves
can be transmitted through solids only, and if light waves were transverse, the
ether would have to be viewed as a solid, and a very rigid one at that, considering
the velocity of those waves. In that case how was it planets could move through
the ether without any detectable interference? Men such as Brewster [1781–1868]
refused to accept the wave theory if it meant accepting such an ether, and it was
only the work of Cauchy [1789–1857] that enabled others to swallow it at all.*

*But Fresnel's work was accepted by physicists generally, and Melloni [1798–
1854] carried it beyond the visible spectrum.*

*A half-year before his death Fresnel received the Rumford medal from
the Royal Society for his work. [Adopted with permission from* Biographical
Encyclopedia of Science and Technology, *I. Asimov, Doubleday, 1982].*

1700   *1788*   *1827*   *2000*

3.6   TOTAL INTERNAL REFLECTION

The phenomenon of *total internal reflection* has a broad range of applications, including reflections from prisms and the guiding of light waves in dielectric waveguides
or optical fibers. Total internal reflection occurs when an electromagnetic wave is
incident at a highly oblique angle from a dense medium onto a less dense medium.

We have so far implicitly assumed the incidence of plane waves from a less
dense medium to a denser one (i.e., $\epsilon_1 < \epsilon_2$), as is evident from the various diagrams
(e.g., Figures 3.23, 3.25, and 3.27) that illustrate the refracted wave to be "bent"
toward the normal (i.e., $\theta_t < \theta_i$). Also, when we showed the magnitudes and phases
of the Fresnel coefficients in Figure 3.30, we considered only cases with $\epsilon_2 > \epsilon_1$. In
this section, we consider the incidence of plane waves from a denser medium into a
less dense one, namely the cases in which $\eta_2 > \eta_1$ (or $\epsilon_2 < \epsilon_1$ when $\mu_1 = \mu_2$).

We start once again with Snell's law (Figure 3.20), which is valid regardless of
the polarization of the incident wave. For nonmagnetic media ($\mu_1 = \mu_2 = \mu_0$), the

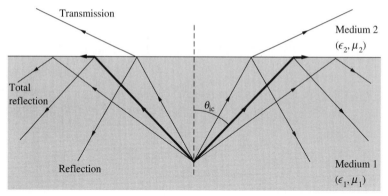

FIGURE 3.32. **Critical angle of incidence and total internal reflection.**
Reflection and refraction characteristics of a plane wave passing from a relatively dense dielectric medium into a less dense medium (i.e., $\epsilon_1 > \epsilon_2$ for $\mu_1 = \mu_2$).

sine of the angle of the transmitted (or refracted) ray is given by

$$\sin \theta_t = \sqrt{\frac{\epsilon_1}{\epsilon_2}} \sin \theta_i$$

and whenever $\epsilon_1 > \epsilon_2$, we have $\theta_t > \theta_i$. Thus, there exist large angles of incidence θ_i for which $\sin \theta_t$ exceeds unity, corresponding to total reflection of the incident wave (i.e., no transmitted wave), as illustrated in Figure 3.32. Specifically, $\sin \theta_t > 1$ for all angles $\theta_i > \theta_{ic}$, where the so-called critical angle θ_{ic}, corresponding to $\theta_t = 90°$, is given by

$$\sin \theta_{ic} = \sqrt{\frac{\epsilon_2}{\epsilon_1}} = \frac{n_2}{n_1} \qquad [3.29]$$

Thus, an observer near the interface in medium 2 is able to receive waves from medium 1 only over a range of incident angles of $2\theta_{ic}$. In other words, there is no transmission into medium 2 for $\theta_i \geq \theta_{ic}$; that is, there is total internal reflection.

3.6.1 Reflection Coefficients

At low angles of incidence, $\theta_i < \theta_{ic}$, all of our previous discussion for $\epsilon_1 < \epsilon_2$ (assuming $\mu_1 = \mu_2$) is also valid[39] for $\epsilon_1 > \epsilon_2$. The Brewster condition is still realized at $\theta_{iB} = \tan^{-1}(n_2/n_1) < \theta_{ic} = \sin^{-1}(n_2/n_1)$ (see Figure 3.33).

At larger angles of incidence $\theta_i > \theta_{ic}$, we have total internal reflection and

$$\sin \theta_t = \sqrt{\frac{\epsilon_1}{\epsilon_2}} \sin \theta_i > 1 \qquad \longrightarrow \qquad \theta_t \text{ imaginary} \qquad [3.30]$$

---

[39]Except that the phases of the reflected components (i.e., $\phi_\perp$ and $\phi_\parallel$) are reversed compared to the $\epsilon_1 < \epsilon_2$ case, as can be seen from [3.24] and [3.26]. In other words, while for $\epsilon_2 > \epsilon_1$ we had $\Gamma_\perp$ and $\Gamma_\parallel$ negative for $\theta_i < \theta_{iB}$, for $\epsilon_1 > \epsilon_2$ the reflection coefficients $\Gamma_\perp$ and $\Gamma_\parallel$ are positive for $\theta_i < \theta_{iB}$.

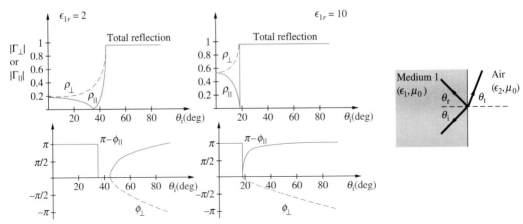

FIGURE 3.33. **Reflection coefficient versus the angle of incidence.** Magnitude and phase of the reflection coefficients for perpendicular ($\Gamma_\perp$) and parallel ($\Gamma_\parallel$) polarization versus angle of incidence for $\epsilon_1/\epsilon_2 = 2$ and $\epsilon_1/\epsilon_2 = 10$. For clarity of presentation, $(\pi - \phi_\parallel)$ is plotted instead of $\phi_\parallel$. The phase angle $\phi_\perp$ is shown as a dashed line, while $(\pi - \phi_\parallel)$ is plotted as a solid line.

Note that the condition $\sin\theta_t > 1$ cannot be satisfied if θ_t is a real angle. Although an imaginary angle does not pose any particular dilemma,[40] especially because we do not need to evaluate θ_t explicitly, since what is needed to determine the expressions for the fields is $\cos\theta_t$, which can be found from $\sin\theta_t$. For $\sin\theta_t > 1$, $\cos\theta_t$ is purely imaginary and can be found using [3.30], we can find $\cos\theta_t$ as

$$\cos\theta_t = \sqrt{1 - \sin^2\theta_t} = \pm j\sqrt{\frac{\epsilon_1}{\epsilon_2}\sin^2\theta_i - 1} \qquad [3.31]$$

where we choose the solution with the negative sign to ensure that the refracted electromagnetic field in medium 2 (see Section 3.6.2) attenuates with increasing distance from the boundary. Substituting [3.31] in [3.24] and [3.26] shows that the reflection coefficients are now complex, given by

$$\Gamma_\perp = \frac{\cos\theta_i + j\sqrt{\sin^2\theta_i - \epsilon_{21}}}{\cos\theta_i - j\sqrt{\sin^2\theta_i - \epsilon_{21}}} = 1e^{j\phi_\perp} \qquad [3.32]$$

$$\Gamma_\parallel = -\frac{\epsilon_{21}\cos\theta_i + j\sqrt{\sin^2\theta_i - \epsilon_{21}}}{\epsilon_{21}\cos\theta_i - j\sqrt{\sin^2\theta_i - \epsilon_{21}}} = 1e^{j\phi_\parallel} \qquad [3.33]$$

---

[40]For an arbitrary complex number $C = A + jB$, we can evaluate $\sin C$ as

$$\sin C = \frac{e^{jC} - e^{-jC}}{2j} = \frac{e^{jA}e^{-B} - e^{-jA}e^{+B}}{2j} = \sin A\cosh(B) + j\cos A\sinh(B)$$

In the case of $\theta_t = \mathcal{R}e\{\theta_t\} + j\mathcal{I}m\{\theta_t\}$, we note that $\sin\theta_t$ is a real number given by [3.30], so that we must have $\cos(\mathcal{R}e\{\theta_t\}) = 0$, or $\mathcal{R}e\{\theta_t\} = \pi/2$, so that

$$\theta_t = \frac{\pi}{2} + j\mathcal{I}m\{\theta_t\}$$

where we have introduced the permittivity ratio parameter $\epsilon_{21} = \epsilon_2/\epsilon_1$ for convenience.

Plots of the magnitude and phase of $\Gamma_\parallel$ and $\Gamma_\perp$ for nonmagnetic media ($\mu_2 = \mu_1 = \mu_0$) and for $\epsilon_1/\epsilon_2 = 2$ and 10 are shown as a function of incidence angle θ_i in Figure 3.33. The Brewster condition, occurring at an incident angle somewhat below that of total reflection, is apparent for the parallel polarization case. Note that for $\theta_i > \theta_{ic}$, the reflection coefficients have unity amplitude but different phase angles, as is apparent from [3.32] and [3.33]. Thus, upon total internal reflection an incident plane wave acquires a phase shift. By manipulating [3.32] and [3.33], explicit expressions for the phase shifts can be obtained. We find

$$\tan\frac{\phi_\perp}{2} = \frac{\sqrt{\sin^2\theta_i - \epsilon_{21}}}{\cos\theta_i}; \qquad \tan\frac{\phi_\parallel}{2} = \frac{\sqrt{\sin^2\theta_i - \epsilon_{21}}}{\epsilon_{21}\cos\theta_i}$$

The phase change that takes place in total reflection may be used to produce a circularly or elliptically polarized wave from one that is linearly polarized. For this purpose, the incident wave must be polarized in a direction that is neither parallel nor normal to the plane of incidence. As an example, we consider the incident wave to be polarized at an angle of 45° with respect to the plane of incidence, so that it possesses two components of equal magnitude (i.e., $|E_\parallel| = |E_\perp|$). Upon total internal reflection, the two components acquire different phases, so the phase difference between them is given by

$$\Delta\phi = \phi_\perp - \phi_\parallel \quad \rightarrow \quad \tan\frac{\Delta\phi}{2} = \frac{\tan(\phi_\perp/2) - \tan(\phi_\parallel/2)}{1 + \tan(\phi_\perp/2)\tan(\phi_\parallel/2)} = \frac{\cos\theta_i\sqrt{\sin^2\theta_i - \epsilon_{21}}}{\sin^2\theta_i}$$
$$[3.34]$$

A plot of $\Delta\phi$ as a function of θ_i is given in Figure 3.34 for $\epsilon_1/\epsilon_2 = 2, 10, 81$; we note that the relative phase is zero at grazing incidence ($\theta_i = \pi/2$) and at $\theta_i = \theta_{ic}$ and reaches a maximum in between. By differentiating the above expression for $\Delta\phi$ with respect to θ_i it can be shown that the incidence angle θ_{imax} at which the maximum occurs and the maximum value of the phase difference $(\Delta\phi)_{max}$ are given by

$$\sin^2\theta_{imax} = \frac{2\epsilon_{21}}{1 + \epsilon_{21}}; \qquad \tan\frac{(\Delta\phi)_{max}}{2} = \frac{1 - \epsilon_{21}}{2\sqrt{\epsilon_{21}}} \qquad [3.35]$$

It is evident from the foregoing discussion that circularly polarized light can be obtained from linearly polarized light by choosing the angle of incidence and the

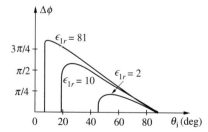

FIGURE 3.34. Phase difference between the two polarizations upon total internal reflection. Plots of $\Delta\phi = \phi_\perp - \phi_\parallel$ are shown for $\epsilon_{1r} = 2, 10, 81$, assuming $\epsilon_2 = \epsilon_0$. Note that $\Delta\phi = 0$ when $\theta_i = \theta_{ic}$ or $\theta_i = 90°$. The critical angles θ_{ic} for the cases of $\epsilon_{1r} = 2, 10, 81$ (with $\epsilon_2 = \epsilon_0$) are respectively $\theta_{ic} = 45°, \sim 18°,$ and $\sim 6.4°$.

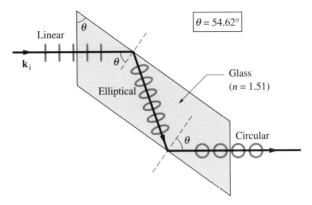

FIGURE 3.35. **Fresnel's rhomb.** Linearly polarized light incident on glass (let $n = 1.51$) with its electric field at 45° to the plane of incidence acquires a phase difference between its two components ($E_\perp$ and $E_\parallel$) of 45° at each total internal reflection. As a result, the wave exiting the rhomb is circularly polarized.

permittivity ratios in such a way that $\phi = 90°$. To obtain this in a single reflection, the refractive index of the first medium must be $n_1 \simeq 2.41$ (assuming the second medium is air, with $n_2 = 1$). In his first demonstration of this concept, Fresnel was not able to find any transparent material with such a high refractive index and instead made use of two total reflections on glass ($n_1 = 1.51$), as shown in Figure 3.35. For glass, we can show from [3.35] that the maximum relative phase difference of $(\Delta\phi)_{max} \simeq 45.93°$ occurs for $\theta_{imax} \simeq 51.33°$, and a $\Delta\phi = 45°$ phase difference can be obtained with either $\theta_i \simeq 48.62°$ or $\theta_i \simeq 54.62°$.

Example 3-14:   Water–air interface.   Consider a linearly polarized electromagnetic wave incident from water ($\epsilon_{1r} = 81$) on a water–air interface. The electric field vector of the wave is at an angle of 45° with respect to the plane of incidence. Find the incidence angle θ_i for which the reflected wave is circularly polarized. Neglect losses.

Solution: We have $\epsilon_1/\epsilon_2 = 81$ or $n_1/n_2 = 9$. From [3.35], we find $\phi_{max} \simeq 77.4°$, which occurs for $\theta \simeq 8.98°$. By trial and error, or by using a graphical solution (see Figure 3.34), we find that $\phi = 90°$ occurs for $\theta_i \simeq 6.5°$ and $\theta_i \simeq 44.6°$. The first solution is very close to the critical angle (see Figure 3.34), which for the water–air interface is $\theta_{ic} = \sin^{-1}(n_2/n_1) \simeq 6.38°$. Furthermore the phase angle ϕ in the vicinity of 6.5° depends highly sensitively on the incidence angle θ_i. The second solution is comfortably in the total internal reflection region and can be realized with relative ease.

3.6.2 The Electromagnetic Field in Medium 2: The Refracted "Wave"

Although the incident wave is *totally* reflected, there does exist a finite electromagnetic field in medium 2, and this field exhibits a number of surprisingly complicated properties. That there should be a field in the second medium is evident on the basis of the fact that the boundary conditions, as worked out in previous sections, must still be satisfied, and that $\mathbf{E}_i + \mathbf{E}_r = \mathbf{E}_i + \Gamma\mathbf{E}_i \neq 0$. Regardless of the fact that $\sin\theta_t > 1$, all of the field expressions that were derived earlier using the physical boundary conditions are still valid. Accordingly, the electric field of the wave in medium 2 for the perpendicular polarization[41] case is given by

$$\mathbf{E}_t(x, z) = \hat{\mathbf{y}}E_{t0}e^{-j\beta_2\hat{\mathbf{k}}_2\cdot\mathbf{r}} = \hat{\mathbf{y}}E_{t0}e^{-j\beta_2(x\sin\theta_t + z\cos\theta_t)} \qquad [3.36]$$

where $\beta_2 = \omega\sqrt{\epsilon_2\mu_0}$, assuming nonmagnetic media. Using expression [3.31] for $\cos\theta_t$ in [3.36] we find

$$\mathbf{E}_t(x, z) = \hat{\mathbf{y}}E_{t0}e^{-\alpha_t z}e^{-j\beta_t x}$$

where

$$\alpha_t = \pm\beta_2\sqrt{\frac{\epsilon_1}{\epsilon_2}\sin^2\theta_i - 1} = \frac{2\pi}{\lambda_1}\sqrt{\sin^2\theta_i - \epsilon_{21}} \qquad [3.37]$$

$$\beta_t = \beta_2\sqrt{\frac{\epsilon_1}{\epsilon_2}}\sin\theta_i = \underbrace{\omega\sqrt{\mu_0\epsilon_1}}_{\beta_1}\sin\theta_i - \frac{2\pi}{\lambda_1}\sin\theta_i \qquad [3.38]$$

Note that $\alpha_t = j\beta_2\cos\theta_t$ must be positive in order for the refracted wave amplitude to decay (rather than grow) with the distance into the second medium, meaning that $\cos\theta_t$ must be imaginary and negative; i.e., we must choose the negative sign in [3.31]. Thus, for $\theta_i > \theta_{ic}$ a wave exists along the interface, propagating in the x direction (i.e., parallel to the surface) with a phase velocity $v_{pt} = (\omega/\beta_t) = \omega/(\beta_1\sin\theta_i) = v_{p1}/\sin\theta_i$. The wavelength of this propagation is $\lambda_t = 2\pi/\beta_t = \lambda_1/\sin\theta_i > \lambda_1$, as is clear from Figure 3.36. However, this wave is exponentially attenuated very rapidly in the z direction.[42] Because it is typically tightly bound to the surface,[43] such a wave is known as a *surface wave*. This nonuniform wave in

---

[41]Note that although [3.36] is written for perpendicular polarization, the spatial variation of both components of the electric field for the parallel polarization case is also proportional to $e^{-j\beta_2\hat{\mathbf{k}}_t\cdot\mathbf{r}}$. Thus, the properties of the refracted wave as discussed here are also exhibited by the refracted wave in the parallel polarization case.

[42]Naturally only the positive sign in α_t corresponds to a physical situation.

[43]The value of α_t is typically such that the depth of penetration d, defined as the distance $d = 1/\alpha_t$ at which the amplitude drops to e^{-1} of its value at the interface, is of order λ_1 or smaller. For example, for a glass–air interface with $\epsilon_1/\epsilon_2 = 2.31$, we have $\alpha_t \simeq 2.47/\lambda_1$ (i.e., $d \simeq 0.406\lambda_1$) at $\theta_i = 50°$, just a few degrees above the critical angle $\theta_{ic} \simeq 41°$. At this rate of decay, the amplitude is reduced to $e^{-2.47} \simeq 0.085$ or $\sim 8.5\%$ of its value at the interface over a distance of λ_1. Note, however, that α_t becomes smaller for as $\theta_i \to \theta_{ic}$, becoming zero at $\theta_i = \theta_{ic}$, in which case the attenuation in medium 2 disappears and the transmitted wave travels as a uniform plane surface wave parallel to the boundary.

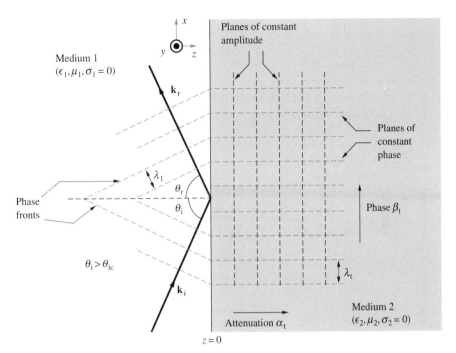

FIGURE 3.36. **Total internal reflection: Constant amplitude and phase fronts.** Oblique incidence beyond the critical angle $\theta_i > \theta_{ic}$ on a dielectric interface with $\mu_1 = \mu_2 = \mu_0$ and $\epsilon_1 > \epsilon_2$. The planes of constant amplitude are parallel to the interface, whereas the planes of constant phase are normal to it. This picture is valid for both perpendicular and parallel polarization.

medium 2 is also a *slow wave*, since its phase velocity $v_{pt} = \omega/\beta_t$ is less than the velocity of light in medium 2 (i.e., $v_{p2} = \omega/\beta_2 = 1/\sqrt{\mu_0\epsilon_2}$). To see this, we can use expression [3.38] for β and the fact under the total reflection condition we have $\sin\theta_t = \sqrt{\epsilon_1/\epsilon_2}\sin\theta_i > 1$. We have

$$v_{pt} = \frac{\omega}{\beta_t} = \frac{\omega}{\beta_2\sqrt{\dfrac{\epsilon_1}{\epsilon_2}}\sin\theta_i} = \frac{\omega}{\beta_2\sin\theta_t} < \frac{\omega}{\beta_2}$$

Figure 3.36 illustrates the structure of the phase fronts for the wave in medium 2. Since the planes of constant amplitude are parallel to the interface, whereas the planes of constant phase are orthogonal to it, the refracted wave is an inhomogeneous (or nonuniform) plane wave. In other words, the amplitude of the wave varies as a function of position (z) on the planes of constant phase (planes defined by $x =$ constant).

We can find the magnitude E_{t0} of the refracted field by substituting $\cos\theta_t$ as given by [3.31] into the transmission coefficient expression [3.25] of Section 3.5.

We have

$$\mathcal{T}_\perp = \frac{E_{t0}}{E_{i0}} = \frac{2\cos\theta_i}{\cos\theta_i + j\sqrt{\sin^2\theta_i - \epsilon_{21}}} = \left(\frac{2\cos\theta_i}{\sqrt{1 - \epsilon_{21}}}\right)e^{j(\phi_\perp/2)}$$

where $\phi_\perp$ is the phase of the reflection coefficient $\Gamma_\perp = 1e^{j\phi_\perp}$, plotted for two example cases in Figure 3.33.

The magnetic field of the refracted wave can be obtained from the electric field given in [3.36] (written below in terms of α_t and β_t) and Maxwell's equation [1.11a]. We have

$$\nabla \times \mathbf{E}_t = -j\omega\mu_0\mathbf{H}_t$$

$$\mathbf{E}_t(x, z) = \hat{\mathbf{y}}E_{t0}e^{-\alpha_t z}e^{-j\beta_t x}$$

$\longrightarrow$

$$-\frac{\partial E_{ty}}{\partial z} = -j\omega\mu_0 H_{tx}$$

$$\frac{\partial E_{ty}}{\partial x} = -j\omega\mu_0 H_{tz}$$

which upon manipulation gives

$$\mathbf{H}_t(x, z) = \frac{E_{t0}}{\omega\mu_0}\left(+\hat{\mathbf{x}}j\alpha_t + \hat{\mathbf{z}}\beta_t\right)e^{-\alpha_t z}e^{-j\beta_t x}$$

The spatial variation of the magnetic field is identical to that of the electric field in that the field components propagate in the x direction with a phase constant β_t and attenuate in the z direction with an attenuation constant α_t. Furthermore, the two components of the magnetic field are 90° out of phase with one another (i.e., the amplitude of one is multiplied by j while the other is not), which means that the wave magnetic field vector, which is confined to the plane of incidence, rotates in time [as observed at a fixed point (x, z)] somewhat similar to an elliptically polarized wave.[44] The refracted wave that exists in the second medium to support the total internal reflection is indeed rather complex.

Although there do exist electric and magnetic fields in medium 2, it can be shown that no energy flows across the boundary. More precisely, although the component of the Poynting vector in the direction normal to the boundary is in general finite, its time-average vanishes; this implies that the energy flows to and fro, but that there is no lasting flow into the second medium. To see this, it suffices to note that the electric field (in the y direction) and the x component of the magnetic field are 90° out of phase, so that the z component of $\mathbf{E} \times \mathbf{H}^*$ is purely imaginary. Such a wave that does not carry power and attenuates with distance is sometimes referred to as an *evanescent* wave.

The time-average of the component of the Poynting vector in the x direction is finite in medium 2, indicating that real electromagnetic power flows parallel to the

---

[44]Note, however, that the plane of this rotation is not transverse to the propagation direction, so the similarity with an elliptically polarized wave does not go beyond superficial.

interface. This is clear from the expressions for $\mathbf{E}_t(x, z)$ and $\mathbf{H}_t(x, z)$, from which we see that E_{ty} and H_{tz} are in phase, indicating that the time-average power in the x direction is nonzero. We can find an expression for $\mathbf{S}_{av}$ after straightforward manipulation; for the perpendicular polarization case we have

$$(\mathbf{S}_{av})_{t\perp} = \frac{1}{2}\mathcal{R}e\{\mathbf{E}_t \times \mathbf{H}_t^*\} = \hat{\mathbf{x}}\frac{\beta_t}{\omega\epsilon_0}\frac{E_{i0}^2 2\cos^2\theta_i}{1 - \epsilon_{21}}e^{-2\alpha_t z}$$

This result, namely that real power flows unattenuated parallel to the interface, appears problematic when we consider an incident wave of finite cross section (e.g., a beam of light). Does the wave power continue to flow and extend beyond the illuminated region? This clearly cannot be the case, since the refracted wave exists simply to satisfy the boundary conditions. However, our existing formulation cannot resolve this dilemma, since it is based on using uniform plane waves of infinite transverse extent.[45]

The very existence of an electromagnetic field in the second medium seems somewhat paradoxical when one considers the fact that all of the incoming wave energy is reflected. Since the electric and magnetic fields of the refracted wave are clearly different from zero, the average energy densities $w_e = (1/2)\epsilon E^2$ and $w_m = (1/2)\mu H^2$ must be different from zero, and it seems strange that there can be energy in the second medium when all of the incoming energy is reflected. In other words, how was the energy transmitted into medium 2? The explanation lies in the fact that our analysis has been based on assumptions of steady-state, time-harmonic solutions. At the time the incident wave first strikes the surface, a small amount of energy penetrates into the second medium and establishes the field. Once established, this energy cannot escape the second medium, since the steady-state solution does not allow any energy transfer between the two media. More generally, the field in the second medium can be thought to arise from the fact that the incident field is bounded in both time and space.

The existence of an electric field in medium 2 under total internal reflection conditions is experimentally verified by the fact in optics that, if a prism is causing an internal total reflection, and another one is pressed closely against one of the sides at which the total reflection is occurring (see Figure 3.37), most of the light goes into the second prism instead of being totally reflected. This means that there must have been some field *just* outside the original prism surface; otherwise, how could bringing up the second prism cause any change? Such coupling through the evanescent wave is illustrated in the diagram; however, note that for effective coupling, the second prism has to be within a few wavelengths, which means that at optical frequencies ($\lambda \simeq 10^{-5}$ cm), it must be physically pressed on to the first one. The refracted electromagnetic field that occurs under total internal reflection has

---

[45]A simplified treatment of the total reflection of a collimated beam of light is given in Sec 3.6 of D. L. Lee, *Electromagnetic Principles of Integrated Optics*, Wiley, 1986.

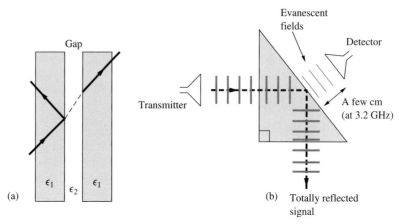

FIGURE 3.37. **Coupling through an air gap.**

also been measured in several experiments at microwave frequencies.[46] One should note, however, that any arrangement used to measure the refracted wave disturbs the boundary conditions, so that the experimental verifications noted can be considered only approximate confirmation of the existence of the wave in medium 2.

An interesting use of the refracted wave that exists during total internal reflection is as a primary diagnostic tool for evaluating the propagation characteristics of dielectric thin-film waveguides using a *prism coupler*. As depicted in Figure 3.38, the coupler consists of a high-refractive-index prism placed in close proximity to a

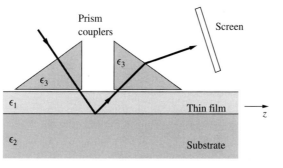

FIGURE 3.38.   The prism coupler. The high-refractive-index prisms are in fact tightly clamped on the film, with dust particles serving as the air gap.

---

[46]One set of experiments was carried out by G. F. Hull, Jr., Experiments with UHF Wave Guides, *Am. J. Phys.*, 13(6), pp. 384–389, December 1945. In Appendix J of J. Strong, *Concepts of Physical Optics*, Freeman, 1958, G. W. Hull describes his experiments carried out at wavelength of 3.2 cm using sizable prisms constructed out of paraffin ($\epsilon_r = 2.1$). The field behind the paraffin prism was found to decrease at a rate of ~5 dB-(cm)$^{-1}$ (comparing well with the calculated value of α_t, which for $\lambda = 3.2$ cm and $\epsilon_{2r} = 2.1$ is ~5.2 dB-(cm)$^{-1}$), being measurable out to a distance of ~7 cm. In another experiment, 1.25-cm waves were used with a metal plate behind the totally reflecting boundary; see W. Culshaw and D. S. Jones, Effect of Metal Plate on Total Reflection, *Proc. Phys. Soc.*, B, 66, 859, 1954.

slab dielectric waveguide. When an optical beam passing through the prism is incident upon its bottom at angle $\theta_i > \theta_{ic}$, the evanescent fields that extend below the prism base penetrate into the waveguide and couple electromagnetic power into a single chosen (as determined by θ_i; see Section 4.2.3) waveguide mode. A second prism is used to decouple the energy at some distance away. The measurement of the output beam can be used to determine the refractive index and thickness of the slab.[47]

Total internal reflection is widely utilized in a host of optical applications, including optical waveguides and binocular optics.

Examples 3-15 and 3-16 illustrate the application of total internal reflection to the refraction of light by prisms.

Example 3-15: **Refraction of light by prisms.** Consider a light ray normally incident on one side of each of three different right angled glass ($n = 1.52$) prisms, as shown in Figure 3.39. Determine the side of the prism at which the beam of light first exits the prism and the exit transmission angle θ_{te} (measured with respect to the normal to that side) in all three cases.

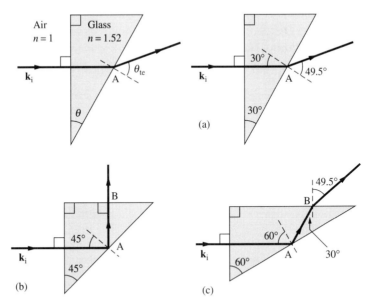

FIGURE 3.39. **Three different right-angle prisms.** Figure for Example 3-15.

---

[47]For a detailed discussion of the principles and operation of the prism coupler, see Chapter 6 of D. L. Lee, *Electromagnetic Principles of Integrated Optics,* Wiley, 1986.

Solution:

(a) Note that the critical angle for a glass–air interface is $\theta_{ic} = \sin^{-1}[(1.52)^{-1}] \simeq$ 41.14°. Thus, there is no total internal reflection in this case, and, based on the ray geometry shown, light exits the prism at point A. Using Snell's law of refraction, we have

$$(1.52) \sin 30° = \sin \theta_t \quad \rightarrow \quad \theta_t = \theta_{te} \simeq 49.5°$$

(b) In this case, light is incident on the hypotenuse of the prism at an incidence angle $\theta_i = 45° > \theta_{ic}$. Accordingly, total internal reflection occurs at point A, and the light ray first exits the prism at point B. Since the incidence to the glass–air interface at B is normal to this side of the prism, the exit angle is $\theta_{te} = 0$.

(c) In this case, once again, total internal reflection occurs at point A. The light ray first exits the prism at point B, at an angle of $\theta_{te} \simeq 49.5°$. The incidence on the glass–air interface at point B in this case is identical to the incidence at point A in case (a).

Example 3-16: Effect of manufacturing errors. Consider the 45°–90°–45° right-angled prism analyzed in part (b) of Example 3-15. Assume that due to manufacturing error, one of the angles of the prism is $45 + \zeta$, where $\zeta < 1°$, and the other is the complementary angle is $45 - \zeta$, as shown in Figure 3.40. Find the effect of this error on the exit transmission angle at point B.

Solution: Considering the worst-case error ($\zeta = 1°$), we have at point A

$$1.52 \sin 46° = \sin \theta_{tA} \quad \rightarrow \quad \sin \theta_{tA} \simeq 1.09 > 1$$

so total reflection occurs at point A. The ray is now incident on the glass–air interface at point B at an angle 2ζ. We apply Snell's law of refraction at point B at an incidence angle of 2ζ:

$$1.52 \sin(2\zeta) \simeq (1.52)2\zeta = \sin \theta_{tB} \quad \rightarrow \quad \theta_{tB} = \sin^{-1}(3.04\zeta) \simeq 3.04\zeta$$

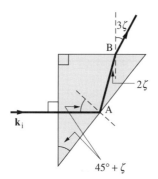

FIGURE 3.40. A 45° right-angled prism with a manufacturing error. Prism for Example 3-16.

We note that small manufacturing errors can lead to relatively large deviations of the ray at its eventual exit point.[48]

3.7   NORMAL INCIDENCE ON A LOSSY MEDIUM

Up to now, we have studied reflection and refraction of uniform plane waves at interfaces between different lossless media or between a lossless dielectric and a perfect conductor. In practice, all media have some loss, leading to absorption of the transmitted energy as it propagates through the lossy medium. In some cases such attenuation may be undesirable but unavoidable; in other cases the heat produced by the attenuation of the wave in the lossy material may constitute the primary application. In this section we consider two important cases involving reflection from an imperfect conducting plane and multiple lossy interfaces.

3.7.1   Reflection from an Imperfect Conducting Plane

We now consider the special case of incidence of a uniform plane wave on a "good" conductor with finite conductivity σ. We will show that the total current flowing within the conducting material is essentially independent of the conductivity. As the conductivity approaches infinity, the total current is squeezed into a narrower and narrower layer, until in the limit a true surface current (as discussed in Section 3.1 for the case of a perfectly conducting boundary) is obtained. We will further show that the conductor can be characterized as a boundary exhibiting a surface impedance of $Z_s = (\sigma\delta)^{-1}(1 + j)$, where δ is the skin depth for the conductor, being $\delta = \sqrt{2/(\omega\mu\sigma)}$, as given in [2.26].

Let a uniform plane wave be normally incident on a conducting interface located at $z = 0$ (i.e., the half-space $z \geq 0$ is filled with a conducting medium), as shown in Figure 3.41. The phasor fields for the incident, reflected, and transmitted waves are

$$\mathbf{E}_i(z) = \hat{\mathbf{x}}E_{i0}e^{-j\beta_1 z}$$

$$\mathbf{H}_i(z) = \hat{\mathbf{y}}\frac{E_{i0}}{\eta_1}e^{-j\beta_1 z}$$

$$\mathbf{E}_r(z) = \hat{\mathbf{x}}E_{r0}e^{+j\beta_1 z}$$

$$\mathbf{H}_r(z) = -\hat{\mathbf{y}}\frac{E_{r0}}{\eta_1}e^{+j\beta_1 z}$$

$$\mathbf{E}_t(z) = \hat{\mathbf{x}}E_{t0}e^{-\gamma_2 z}$$

$$\mathbf{H}_t(z) = \hat{\mathbf{y}}\frac{E_{t0}}{\eta_c}e^{-\gamma_2 z}$$

---

[48]See section 4.17 of W. J. Smith, *Modern Optical Engineering: The Design of Optical Systems*, McGraw-Hill, 1966.

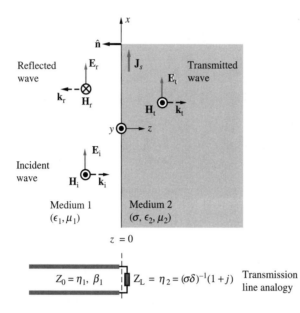

FIGURE 3.41. Uniform plane wave normally incident on the boundary between a dielectric and an imperfect conductor.

where

$$\beta_1 = \omega\sqrt{\mu_1\epsilon_1} \qquad\qquad \eta_1 = \sqrt{\mu_1/\epsilon_1} \simeq 377\Omega$$

$$\gamma_2 = \sqrt{j\omega\mu_2(\sigma_2 + j\omega\epsilon_2)} \qquad \eta_c = \sqrt{j\omega\mu_2/(\sigma_2 + j\omega\epsilon_2)}$$

In a good conducting medium (i.e., $\sigma \gg \omega\epsilon$) we have $\nabla \times \mathbf{H} = (j\omega\epsilon + \sigma)\mathbf{E} \simeq \sigma\mathbf{E}$, since the conduction current is much greater than the displacement current. Rewriting this equation as $\nabla \times \mathbf{H} \simeq j\omega[\sigma/(j\omega)]\mathbf{E}$, we recall from Section 2.3 that $\epsilon_{\text{eff}} \simeq \sigma/(j\omega)$ may be considered as the permittivity ϵ in Maxwell's equations in a lossless medium. Thus, propagation constant and intrinsic impedance of the lossy medium can be obtained from that of a lossless medium by the substitution $\epsilon_{\text{eff}} \longrightarrow \epsilon$. We thus have the propagation constant for medium 2 given by

$$\gamma_2 = j\omega\sqrt{\mu\epsilon_{\text{eff}}} \simeq j\omega\left(\frac{\mu_2\sigma_2}{j\omega}\right)^{1/2} = (j\omega\mu_2\sigma_2)^{1/2} = \delta^{-1}(1 + j) \qquad [3.39]$$

and the complex intrinsic impedance given by

$$\eta_c \equiv Z_s = R_s + jX_s = \sqrt{\frac{\mu_2}{\epsilon_{\text{eff}}}} \simeq \left(\frac{j\omega\mu_2}{\sigma_2}\right)^{1/2} = \frac{\gamma_2}{\sigma_2} = (\sigma_2\delta)^{-1}(1 + j) \qquad [3.40]$$

where $\delta = (\pi f \mu_2\sigma_2)^{-1/2}$ is the skin depth for medium 2. Note that for medium 1 we have the propagation constant and intrinsic impedance of a lossless medium, namely, $\beta_1 = \omega\sqrt{\mu_1\epsilon_1}$ and $\eta_1 = \sqrt{\mu_1/\epsilon_1}$.

The conductor presents an impedance $Z_s = \eta_c$ to the electromagnetic wave with equal inductive and resistive parts, defined above as R_s and jX_s, respectively. The resistance part is simply the resistance of a sheet of metal of 1 meter square and of

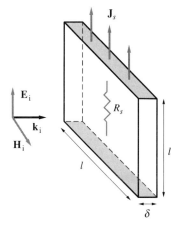

FIGURE 3.42. **Surface resistance concept.** The resistance between the shaded faces is independent of the linear dimension l and hence can be thought of as a surface resistance.

thickness δ, as illustrated in Figure 3.42; actually, the resistance is independent of the area (l^2) of the square plate. Thus, the resistance between the two shaded faces (perpendicular to the current flow) is given by

$$R_s = \frac{l}{l\delta\sigma_2} = \frac{1}{\delta\sigma_2} \qquad [3.41]$$

Since the resistance is independent of the linear dimension l, it is called a *surface resistance,* and the complex intrinsic impedance η_c can be thought of as the *surface impedance* of the conductor.

To find the reflection and transmission coefficients, we follow the same procedure as before and apply the boundary conditions:

$$E_{i0} + E_{r0} = E_{t0} \quad \text{and} \quad \frac{1}{\eta_1}(E_{i0} - E_{r0}) = \frac{E_{t0}}{Z_s}$$

Solving for E_{r0}/E_{i0} and E_{t0}/E_{i0}, we find

$$\Gamma = \frac{E_{r0}}{E_{i0}} = \frac{Z_s - \eta_1}{Z_s + \eta_1} = \rho e^{j\phi_\Gamma} \qquad \mathscr{T} = \frac{E_{t0}}{E_{i0}} = \frac{2Z_s}{Z_s + \eta_1} = \tau e^{j\phi_{\mathscr{T}}}$$

Note that we use the more general notation $\Gamma = \rho e^{j\phi_\Gamma}$ and $\mathscr{T} = \tau e^{j\phi_{\mathscr{T}}}$, since the reflection and transmission coefficients are in general complex.

For any reasonably good conductor, Z_s is very small compared to η_1 for free space (i.e., 377Ω), as was mentioned in Section 2.3. For example, for copper ($\sigma_2 = 5.8 \times 10^7$ S-m^{-1}) at 1 MHz, $\delta \simeq 66.1$ μm and $R_s = 2.61 \times 10^{-8}\Omega$. The reflection and transmission coefficients, respectively, are $\Gamma \simeq 0.9999986 e^{j179.999921°}$ and $\mathscr{T} \simeq 2 \times 10^{-6} e^{j45°}$. We thus note that for all practical purposes the field in front of the conductor ($z < 0$) is the same as exists for a perfect conductor, since $\rho = |\Gamma|$ is very close to unity. For the same reason, only a very small amount of power is transmitted into the conductor; that is, $\tau = |\mathscr{T}|$ is small. Nevertheless, this small amount of power that penetrates into the walls can cause significant attenuation of waves propagating in waveguides and coaxial lines, especially when these types of transmission lines are relatively long. Next we discuss a method for calculating the finite amount of power dissipated in the conductor.

In the case when medium 2 is a good conductor (i.e., $\sigma_2 \gg \omega\epsilon_2$), we can find approximate expressions for Γ and $\mathcal{T}$ as follows:

$$\Gamma \simeq \frac{\delta - \lambda_1(1-j)}{\delta + \lambda_1(1-j)} \quad \text{and} \quad \mathcal{T} \simeq \frac{2\delta}{\lambda_1(1-j)}$$

where $\lambda_1 = 2\pi/\beta_1$ is the wavelength in the first medium. Note that since $\lambda_1 \gg \delta$, we have $\Gamma \simeq 1e^{j\pi}$ and $\tau = |\mathcal{T}| \ll 1$. Other useful approximate expressions can be obtained for Γ by using power series expansions. For example,

$$\Gamma \simeq \left(-1 + \sqrt{\frac{2\omega\epsilon_1}{\sigma_2}}\right) + j\sqrt{\frac{2\omega\epsilon_1}{\sigma_2}}$$

Based on these expressions, the fraction of the incident power reflected is approximately given by

$$|\Gamma|^2 \simeq 1 - 4\sqrt{\frac{\omega\epsilon_1}{2\sigma_2}}$$

and the fraction transmitted into the conductor is

$$1 - |\Gamma|^2 \simeq 4\sqrt{\frac{\omega\epsilon_1}{2\sigma_2}}$$

The current density $\mathbf{J}$ in the conductor is $\mathbf{J} = \sigma\mathbf{E}_t = \hat{\mathbf{x}}\sigma_2\mathcal{T}E_{i0}e^{-\gamma_2 z}$. The total current per unit width of the conductor is

$$J_s = \sigma_2\mathcal{T}E_{i0}\int_0^{\infty} e^{-\gamma_2 z}\,dz = \frac{\sigma_2\mathcal{T}E_{i0}}{\gamma_2} \qquad [3.42]$$

Note that while $\mathbf{J}$ is in units of A-m^{-2}, the current per unit width is a surface current in units of A-m^{-1}. We can relate this surface current to the magnetic field at the surface of the conductor, namely, the total magnetic field $\mathbf{H}_1$ in medium 1. Note that the continuity of the tangential magnetic field at the interface requires that $\mathbf{H}_1(z = 0) = \mathbf{H}_t(z = 0)$. Since the magnetic field of the transmitted wave is $\mathbf{H}_t(z) = \hat{\mathbf{y}}(E_{t0}/\eta_c)e^{-\gamma z}$, then substituting $\eta_c = Z_s$ and $\gamma = \delta^{-1}(1 + j)$, we have

$$\mathbf{H}_1(0) = \mathbf{H}_t(0) \quad \rightarrow \quad \mathbf{H}_1(0) = \hat{\mathbf{y}}\frac{\mathcal{T}E_{i0}}{Z_s} \quad \rightarrow \quad \mathcal{T}E_{i0} = Z_s H_{1y}(0)$$

Substituting in [3.42] and using [3.39] and [3.40] we have

$$J_s = \frac{\sigma_2 Z_s H_{1y}(0)}{\gamma} = \frac{\sigma_2(\sigma_2\delta)^{-1}(1+j)}{\delta^{-1}(1+J)}H_{1y}(0) = H_{1y}(0)$$

It is thus apparent that if we let $\sigma_2 \to \infty$, we have $\delta \to 0$, $\rho \to -1$, but the total current J_s does not vanish; it remains equal to H_{1y}. More generally, we have $\mathbf{J}_s = \hat{\mathbf{n}} \times \mathbf{H}_1(0)$, as was found in Section 3.1 for the case of normal incidence of a uniform plane wave on a perfect conductor. However, since $\delta \to 0$ as $\sigma_2 \to \infty$, the current is squeezed into a narrower and narrower layer and in the limit becomes a true surface current. When σ_2 is finite, the current density $\mathbf{J}$ inside the conductor varies with z in the same manner as $\mathbf{E}_t(z)$ and is confined to a region of thickness δ, with the total current per unit width of the conductor being $\mathbf{J}_s$.

The time-average power loss per unit area in the xy plane may be evaluated from the complex Poynting vector at the surface. We have

$$|\mathbf{S}_{av}| = \left|\frac{1}{2}\mathcal{R}e\{\mathbf{E} \times \mathbf{H}^*\}\right| = \frac{1}{2}\mathcal{R}e\{E_{1x}H^*_{1y}\} = \frac{1}{2}\mathcal{R}e\left\{\mathcal{T}E_{i0}\left(\frac{\mathcal{T}^*E^*_{i0}}{Z^*_s}\right)\right\}$$

$$= \frac{1}{2}|\mathcal{T}E_{i0}|^2\mathcal{R}e\left\{\frac{1}{Z^*_s}\right\} = \frac{1}{2}|\mathcal{T}E_{i0}|^2\mathcal{R}e\left\{\frac{1}{R_s + jX_s}\left(\frac{R_s - jX_s}{R_s - jX_s}\right)\right\}$$

$$= \frac{1}{2}|\mathcal{T}E_{i0}|^2\mathcal{R}e\left\{\frac{R_s - jX_s}{R^2_s + X^2_s}\right\} = \frac{1}{2}|\mathcal{T}E_{i0}|^2\frac{R_s}{R^2_s + X^2_s} = \frac{1}{4}|\mathcal{T}E_{i0}|^2\sigma_2\delta$$

where we have used the fact that $R_s = X_s = (\sigma\delta)^{-1}$. This Poynting flux represents the total average power per unit area entering the conductor. All of this power must be dissipated in the conductor due to $\mathbf{E} \cdot \mathbf{J}$ losses, which can be evaluated by integrating over a volume with 1 m^2 cross section in the xy plane. In other words,

$$P_{\text{loss}} = \frac{1}{2}\int_V \mathbf{E} \cdot \mathbf{J}^*dv = \frac{1}{2}\int_0^1\int_0^1\int_0^\infty (E_x)(\sigma E^*_x)\,dx\,dy\,dz$$

$$= \frac{\sigma}{2}|E_{i0}\mathcal{T}|^2\int_0^\infty e^{-2z/\delta}dz$$

$$= \frac{1}{4}|\mathcal{T}E_{i0}|^2\sigma_2\delta$$

Thus we find that, as expected, the total power entering the conductor through a unit area on its surface is equal to the total power dissipated in the volume of the conductor behind the unit area.

We can express P_{loss} in a more useful form by replacing $|E_{i0}\mathcal{T}|$ by $|J_s\gamma_2/\sigma_2|$ using [3.42]. We find

$$P_{\text{loss}} = \frac{1}{4}|J_s\gamma|^2\frac{\delta}{\sigma_2} = \frac{1}{2}|J_s|^2R_s \qquad\qquad [3.43]$$

which underscores the term *surface resistance* for R_s.

In practice, an approximate method is generally used to evaluate power loss per unit area in conducting walls of waveguides and coaxial lines. The electric and magnetic field configurations are found using the assumption that the conductors are perfect (i.e., $\sigma = \infty$). The surface current density is then determined from the boundary condition

$$\mathbf{J}_s = \hat{\mathbf{n}} \times \mathbf{H} \quad \to \quad J_s = H_t$$

where $\hat{\mathbf{n}}$ is the outward normal to the conductor surface and H_t is the tangential field at the surface, evaluated for $\sigma = \infty$. Once $\mathbf{J}_s$ is found, we can use the surface resistance $R_s = (\sigma \delta)^{-1}$ to calculate the power loss per unit area using [3.43].

Note that the value of H_t calculated assuming $\sigma = \infty$ is a very good approximation of the actual field, since $\eta_1 \gg |Z_s|$. In the case of infinite conductivity the tangential electric field at the conductor surface is zero. However, for finite σ there has to be a finite value of tangential electric field to support a component of the Poynting vector directed into the conductor. It can be shown that this tangential electric field at the surface is $\mathbf{E}_t = \mathbf{J}_s Z_s$ and is thus normally quite small, since Z_s is small.

Example 3-17 illustrates the absorption of incident electromagnetic power at an air–copper interface, while Example 3-18 shows the effectiveness of a thin foil of aluminum in shielding electromagnetic fields.

Example 3-17: Air–copper interface. Consider a uniform plane wave propagating in air incident normally on a large copper block. Find the percentage time-average power absorbed by the copper block at 1, 10, and 100 MHz and at 1 GHz.

Solution: The amount of power absorbed by the copper block is the power transmitted into medium 2, which is equal to the Poynting flux moving in the $+z$ direction in medium 1:

$$(\mathbf{S}_{av})_1 = \hat{\mathbf{z}} \frac{E_{i0}^2}{2\eta_1}(1 - |\Gamma|^2)$$

as derived in Section 3.2.3. Since the power of the incident wave is $E_{i0}^2/(2\eta_1)$, the fraction of the incident power transmitted into medium 2 is $(1 - |\Gamma|^2)$. Thus, we need to determine the reflection coefficient Γ. The intrinsic impedance of copper is given by

$$\eta_c = \sqrt{\frac{j\omega \mu_0}{\sigma_2}} = \sqrt{\frac{\omega \mu_0}{\sigma_2}} e^{j45°} = \frac{2\pi \times 10^{-7}\sqrt{f}}{\sqrt{5.8}}(1 + j)$$

$$\simeq 2.61 \times 10^{-7}\sqrt{f}(1 + j)\,\Omega$$

where we have used $\mu_2 = \mu_0$ and $\sigma_2 = 5.8 \times 10^7$ S-m^{-1} for copper and $\sqrt{j} = e^{j\pi/4}$. The reflection coefficient at the air–copper interface is

$$\Gamma = \frac{\eta_c - \eta_0}{\eta_c + \eta_0} \simeq \frac{2.61 \times 10^{-7}\sqrt{f}(1 + j) - 377}{2.61 \times 10^{-7}\sqrt{f}(1 + j) + 377}$$

At $f = 1$ MHz, we have

$$\Gamma \simeq \frac{2.61 \times 10^{-4}\sqrt{f}(1 + j) - 377}{2.61 \times 10^{-4}\sqrt{f}(1 + j) + 377} \simeq 0.9999986 e^{j179.99992°}$$

so that the percentage of incident power absorbed by the copper block is

$$P_{Cu} \simeq 1 - |\Gamma|^2 \times 100 \simeq 0.000277\%$$

Similar calculations give the following results:

$$\begin{aligned}
&\text{10 MHz} &&\Gamma \simeq 0.9999956 e^{j179.99975°} &&P_{Cu} \simeq 0.000875\% \\
&\text{100 MHz} &&\Gamma \simeq 0.9999862 e^{j179.99921°} &&P_{Cu} \simeq 0.00277\% \\
&\text{1000 MHz} &&\Gamma \simeq 0.9999562 e^{j179.99749°} &&P_{Cu} \simeq 0.00875\%
\end{aligned}$$

We see that the percentage of the incident power absorbed by the copper block increases with the frequency of the incident wave. However, note that as the frequency is increased, more of the total absorbed power is dissipated in a narrower region near the surface (i.e., the skin depth in copper, proportional to $f^{-1/2}$).

Example 3-18:   Transmission through a metal foil: RF shielding.   Consider an x-polarized uniform plane radio-frequency (RF) wave propagating in air, incident normally on a metal foil of thickness d at $z = 0$, as shown in Figure 3.43. (a) Find a relationship between the electric field of the transmitted wave (i.e., $E_{t0} = |\mathbf{E}_t|$) and that of the incident wave (i.e., $E_{i0} = |\mathbf{E}_i|$). Assume the foil to be thick enough that multiple reflections can be neglected. (b) Consider an ordinary aluminum foil, which is approximately 0.025 mm thick. If a 100-MHz plane wave is normally incident from one side of the foil, find the percentage of the incident power transmitted to the other side. For aluminum, take $\sigma_2 = 3.54 \times 10^7$ S-m^{-1}, $\epsilon_2 = \epsilon_0$, and $\mu_2 = \mu_0$.

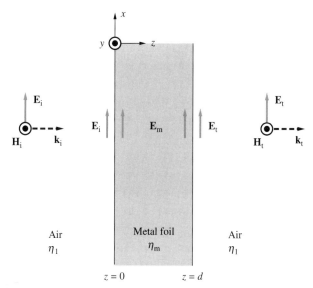

FIGURE 3.43.   RF shielding. A metal foil of thickness d designed to shield RF energy. Note that typically we have $|\mathbf{E}_t| \ll |\mathbf{E}_i|$.

Solution:

(a) Neglecting multiple reflections allows us to treat each boundary of the metal foil as a separate interface. Thus, the amplitude of the wave transmitted into the metal foil due to the incident wave at the first boundary is given by

$$|\mathbf{E}_m(z = 0)| = \tau_1 |\mathbf{E}_i(z = 0)|$$

where $\tau_1 = |\mathcal{T}_1|$ is the magnitude of the transmission coefficient at the air–metal interface. Note that we have

$$\mathcal{T}_1 = \frac{2\eta_m}{\eta_1 + \eta_m}$$

where $\eta_1 \simeq 377\,\Omega$ and $\eta_m \simeq \sqrt{j\omega\mu_0/\sigma_2} = \sqrt{\omega\mu_0/(2\sigma_2)}(1+j)$. This transmitted wave attenuates exponentially as it propagates into the metal foil, so its amplitude at $z = d$ is

$$|\mathbf{E}_m(z = d)| = |\mathbf{E}_m(z = 0)|e^{-d/\delta}$$

where $\delta = (\pi f \mu_0 \sigma_2)^{-1/2}$ is the skin depth in the metal. The attenuated transmitted wave incident on the metal–air boundary at $z = d$ transmits wave energy into medium 3 (air). The amplitude of the wave transmitted into medium 3 can be written as

$$|\mathbf{E}_t(z = d)| = \tau_2 |\mathbf{E}_m(z = d)|$$

where τ_2 is the amplitude of the transmission coefficient $\mathcal{T}_2$ at the metal–air boundary, given by

$$\mathcal{T}_2 = \frac{2\eta_1}{\eta_c + \eta_1}$$

The incident and transmitted electric field magnitudes are related by

$$|\mathbf{E}_t| = \tau_1 e^{-d/\delta} \tau_2 |\mathbf{E}_i|$$

(b) We can first calculate the skin depth of aluminum at 100 MHz:

$$\delta = \frac{1}{\sqrt{\pi f \mu_0 \sigma_2}} = \frac{1}{\sqrt{\pi(10^8)(4\pi \times 10^7)(3.54 \times 10^7)}}$$
$$\simeq 8.46 \times 10^{-6}\ \text{m} = 8.46\ \mu\text{m}$$

Noting that the thickness of the aluminum foil is $d = 0.025$ mm $\simeq 2.96\delta$, our assumption of neglecting multiple reflections is easily justified. In this connection, note that the reflected wave will be attenuated by an additional exponential factor of e^{-3} in the course of its propagation back through the metal to the first boundary. Thus, using the result found in (a) we have

$$\frac{|\mathbf{E}_t|}{|\mathbf{E}_i|} = \tau_1 e^{-d/\delta} \tau_2$$

where

$$|\tau_1 \tau_2| = \left| \frac{2\eta_m}{\eta_m + \eta_1} \right| \left| \frac{2\eta_1}{\eta_1 + \eta_m} \right| = \left| \frac{4\eta_m \eta_1}{(\eta_1 + \eta_m)^2} \right|$$

and $e^{-d/\delta} \simeq e^{-2.96} \simeq 0.0521$. Noting that $\eta_1 \simeq 377\Omega$ and

$$\eta_m = \sqrt{\frac{j\omega\mu_0}{\sigma_2}} \simeq \frac{2\pi \times 10^{-7}\sqrt{f}}{\sqrt{3.54}}(1 + j) \simeq 3.34 \times 10^{-3}(1 + j)\Omega$$

Since $\eta_1 \gg \eta_m$, we have

$$\tau_1 \tau_2 \simeq \left| \frac{4\eta_1 \eta_m}{\eta_1^2} \right| = \frac{4|\eta_m|}{\eta_1} \simeq \frac{4\sqrt{2}(3.34 \times 10^{-3})}{377} \simeq 5.01 \times 10^{-5}$$

so $|\mathbf{E}_t|/|\mathbf{E}_i| \simeq (0.0521)(5.01 \times 10^{-5}) \simeq 2.61 \times 10^{-6}$. Therefore, the percentage of the incident power that transmits to the other side of the foil is

$$\frac{|\mathbf{E}_t|^2/(2\eta_1)}{|\mathbf{E}_i|^2/(2\eta_1)} \times 100\% = \frac{|\mathbf{E}_t|^2}{|\mathbf{E}_i|^2} \times 100\% \simeq 6.80 \times 10^{-10}\% !$$

Thus, the thin aluminum foil works quite well indeed as a shield for RF fields at 100 MHz.

3.7.2   Reflection from Multiple Lossy Interfaces

In the case of multiple interfaces involving conducting (lossy) media, the various expressions for the effective reflection and transmission coefficients that we derived in Section 3.3 apply if we substitute the complex propagation constant $\gamma_2 = \alpha_2 + j\beta_2$ instead of $j\beta_2$ and allow the various intrinsic impedances (η) to be complex. In other words, we have from [3.7]

$$\Gamma_{eff} = \rho e^{j\phi_\Gamma} = \frac{(\eta_2 - \eta_1)(\eta_3 + \eta_2) + (\eta_2 + \eta_1)(\eta_3 - \eta_2)e^{-2\gamma_2 d}}{(\eta_2 + \eta_1)(\eta_3 + \eta_2) + (\eta_2 - \eta_1)(\eta_3 - \eta_2)e^{-2\gamma_2 d}}$$

The effective transmission coefficient $\mathcal{T}_{eff}$ can also be determined in a similar manner from [3.8] as

$$\mathcal{T}_{eff} = \tau e^{j\phi_\mathcal{T}} \frac{4\eta_2 \eta_3 e^{-\gamma_2 d}}{(\eta_2 + \eta_1)(\eta_3 + \eta_2) + (\eta_2 - \eta_1)(\eta_3 - \eta_2)e^{-2\gamma_2 d}}$$

We note, however, that when the second medium is lossy, we have

$$1 - |\Gamma_{eff}|^2 \neq \frac{\eta_1}{|\eta_3|}|\mathcal{T}_{eff}|^2$$

since some of the incident power is absorbed in medium 2.

Reflection and penetration (transmission) of electromagnetic signals at biological tissue interfaces constitute an interesting and important application of multiple lossy interfaces and are briefly covered in Examples 3-19 through 3-21. In order

TABLE 3.1. ϵ_r and σ for biological tissues

f (MHz)	Muscle, skin, and tissues with high water content		Fat, bone, and tissues with low water content	
	$(\epsilon_m)_r$	σ_m (S-m^{-1})	$(\epsilon_f)_r$	σ_f (mS-m^{-1})
100	71.7	0.889	7.45	19.1–75.9
300	54	1.37	5.7	31.6–107
750	52	1.54	5.6	49.8–138
915	51	1.60	5.6	55.6–147
1,500	49	1.77	5.6	70.8–171
2,450	47	2.21	5.5	96.4–213
5,000	44	3.92	5.5	162–309
10,000	39.9	10.3	4.5	324–549

*Note that σ_m is in S-m^{-1}, whereas σ_f is in mS-m^{-1}.

to appreciate these applications better, it should be noted that the dielectric properties of different biological tissues are different and have been studied over a very broad frequency range. The dielectric constant of tissues decreases gradually over many orders of magnitude as the frequency is varied from a few Hz to tens of GHz. The effective conductivity, on the other hand, rises with frequency, initially very slowly, and then more rapidly above 1 GHz. The permittivity of tissues depends on the tissue type and widely varies between different tissues. Tissues of higher water content, such as muscle, brain, kidney, heart, liver, and pancreas, have larger dielectric constant and conductivity than low-water-content tissues such as bone, fat, and lung. Table 3.1 lists the relative dielectric constant and conductivity of biological tissues with high water content versus those with low water content at discrete source frequencies over the radio-frequency spectrum.[49]

Examples 3-19 through 3-21 provide quantitative values for the reflection and transmission coefficients at planar biological interfaces.

Example 3-19: Air–muscle interface. Consider a planar interface between air and muscle tissue. If a plane wave is normally incident at this boundary, find the percentage of incident power absorbed by the muscle tissue at (a) 100 MHz; (b) 300 MHz; (c) 915 MHz; and (d) 2.45 GHz.

Solution:

(a) The reflection coefficient is given by

$$\Gamma = \rho e^{j\phi} = \frac{\eta_m - \eta_1}{\eta_m + \eta_1}$$

[49]C. C. Johnson and A. W. Guy, Nonionizing electromagnetic wave effects in biological materials and systems, *Proc. IEEE,* 60(6), pp. 692–718, June 1972.

where $\eta_1 \simeq 377\Omega$ and the intrinsic impedance of muscle tissue η_m is given by [2.22]. So, at 100 MHz, the loss tangent of muscle is given by

$$\tan \delta_m = \frac{\sigma_m}{\omega \epsilon_m} \simeq \frac{0.889 \text{ S-m}^{-1}}{2\pi \times 10^8 \text{ rad-s}^{-1} \times 71.7 \times 8.85 \times 10^{-12} \text{ F-m}^{-1}} \simeq 2.23$$

and the intrinsic impedance is

$$\eta_m \simeq \frac{377/\sqrt{71.7}}{[1 + (2.23)^2]^{1/4}} e^{j(1/2)\tan^{-1}(2.23)} \simeq 28.5 e^{j32.9°} \Omega$$

Substituting, we find

$$\Gamma \simeq \frac{28.5 e^{j32.9°} - 377}{28.5 e^{j32.9°} + 377}$$

$$\simeq \frac{23.9 + j15.5 - 377}{23.9 + j15.5 + 377} \simeq \frac{353 e^{j177°}}{401 e^{j2.21°}} \simeq 0.881 e^{j175°}$$

So the percentage of incident power absorbed by the muscle tissue can be calculated as

$$\frac{|(\mathbf{S}_{av})_t|}{|(\mathbf{S}_{av})_i|} \times 100\% = (1 - \rho^2) \times 100\% \simeq [1 - (0.881)^2] \times 100\% \simeq 22.4\%$$

(b) Similarly, at 300 MHz, we have $\tan \delta_m \simeq 1.52$, $\eta_m \simeq 38 e^{j28.3°}\Omega$, and the reflection coefficient is

$$\Gamma \simeq \frac{38 e^{j28.3°} - 377}{38 e^{j28.3°} + 377} \simeq 0.837 e^{j174°}$$

so that the percentage absorbed power is $\sim[1 - (0.837)^2] \times 100 \simeq 29.9\%$.

(c) At 915 MHz, $\tan \delta_m \simeq 0.617$, $\eta_m \simeq 48.7 e^{j15.8°}\Omega$, and

$$\Gamma \simeq \frac{48.7 e^{j15.8°} - 377}{48.7 e^{j15.8°} + 377} \simeq 0.779 e^{j176°}$$

so that the percentage absorbed power is $\sim[1 - (0.779)^2] \times 100 \simeq 39.3\%$.

(d) At 2.45 GHz, $\tan \delta_m \simeq 0.345$, $\eta_m \simeq 53.5 e^{j9.52°}\Omega$, and

$$\Gamma \simeq \frac{53.5 e^{j9.52°} - 377}{53.5 e^{j9.52°} + 377} \simeq 0.755 e^{j177°}$$

So the percentage of incident power absorbed is

$$(1 - \rho^2) \times 100 \simeq [1 - (0.755)^2] \times 100 \simeq 43\%$$

Example 3-20:   Muscle–fat interface.   Consider a planar interface between muscle and fat tissues. If a 1-mW-(cm)$^{-2}$ plane wave in muscle is normally incident at this boundary at 2.45 GHz, find the power density of the wave transmitted into the fat tissue. For fat tissue, take $\sigma_f = 155$ mS-m^{-1}.

Solution: At 2.45 GHz, the intrinsic impedances of muscle and fat tissues, η_m and η_f, are given by

$$\eta_m \simeq 53.5 e^{j9.52°} \,\Omega \text{ (from previous example)}$$

$$\eta_f \simeq \frac{377/\sqrt{5.5}}{[1 + (0.207)^2]^{1/4}} \, \exp\left[j\frac{1}{2}\tan^{-1}(0.207)\right] \simeq 159 e^{j5.84°} \,\Omega$$

The reflection coefficient can be calculated as

$$\Gamma = \frac{\eta_m - \eta_f}{\eta_m + \eta_f} \simeq \frac{53.5 e^{j9.52°} - 159 e^{j5.84°}}{53.5 e^{j9.52°} + 159 e^{j5.84°}}$$

$$\simeq \frac{52.7 + j8.84 - 158 - j16.2}{52.7 + j8.84 + 158 + j16.2} \simeq 0.498 e^{j177°}$$

Therefore the power transmitted into the fat tissue is given by

$$|(S_{av})_t| = (1 - \rho^2)|(S_{av})_i| \simeq [1 - (0.498)^2](1 \text{ mW-(cm)}^{-2}) \simeq 0.752 \text{ mW-(cm)}^{-2}$$

Example 3-21: Microwave treatment of hypothermia in newborn piglets. Newly born piglets are very vulnerable to cold temperatures, and many of them die because of hypothermia. At the moment, hypothermia is treated by placing the piglets under infrared lamps, which are not very effective and are very costly. It has been proposed[50] that microwaves can be used to treat hypothermia. Compared to the infrared lamp, a microwave heater is more expensive to build, but is more effective and consumes less power.

Consider a plane wave normally incident at the surface of the body of a pig (which to first order can be assumed to be a plane boundary). The body of the pig can be approximately modeled as a layer of fat tissue of a certain thickness followed by muscle tissue (which is assumed to be infinite in extent), as shown in Figure 3.44. For

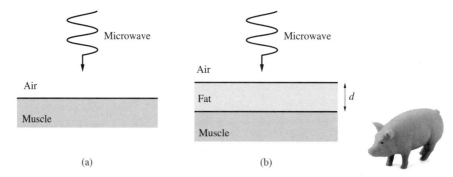

FIGURE 3.44.   Microwave warming of pigs. (a) Air–piglet interface, modeled as a single air–muscle interface. (b) Air–adult pig interface, modeled as an air–fat–muscle interface.

[50]M. Allen, Pinky perks up when popped in the microwave, *New Scientist,* p. 19, January 23, 1993.

a newly born piglet, the fat layer is so thin (less than a mm) that it can be neglected, so the problem reduces to that of a single boundary. However, for a developed pig, the thickness of the fat layer can vary anywhere from 2 to 5 cm (taken below to be 4 cm for a mature pig), so it must be taken into account. Calculate the percentage of the incident microwave power reflected back into air, the percentage of power dissipated in the fat layer, the percentage of power transmitted into the muscle tissue, and the depth of penetration into the muscle tissue of the newly born piglet at (a) 915 MHz; (b) 2.45 GHz. (c) Repeat (a) and (b) for a mature pig. Use the same parameters for tissues as given in Table 3.1.

Solution:

(a) For the newly born piglet, using the results of Example 3-20, for 915 MHz, the percentage of incident power reflected is

$$\frac{|(\mathbf{S}_{av})_r|}{|(\mathbf{S}_{av})_i|} \times 100 = \rho^2 \times 100 \simeq (0.779)^2 \times 100 \simeq 60.7\%$$

and the percentage of incident power transmitted is

$$\frac{|(\mathbf{S}_{av})_t|}{|(\mathbf{S}_{av})_i|} \times 100 = (1 - \rho^2) \times 100 \simeq 39.3\%$$

No significant power is dissipated in the fat layer, since its thickness is negligible. The depth of penetration into the muscle layer at 915 MHz can be calculated from [2.25] as

$$d = \frac{c}{\omega} \left[\frac{2}{\epsilon_r(\sqrt{1 + \tan^2 \delta} - 1)} \right]^{1/2}$$

$$\simeq \frac{(3 \times 10^{10} \text{ cm-s}^{-1})}{2\pi(915 \times 10^6 \text{ Hz})} \left[\frac{2}{51(\sqrt{1 + (0.617)^2} - 1)} \right]^{1/2} \simeq 2.47 \text{ cm}$$

(b) Similarly, for 2.45 GHz, the percentage of the incident power reflected is $(0.755)^2 \times 100 \simeq 57\%$, and the percentage of incident power transmitted is $(1 - 0.755^2) \times 100 \simeq 43\%$. The depth of penetration at 2.45 GHz is

$$d \simeq \frac{(3 \times 10^{10} \text{ cm-s}^{-1})}{2\pi(2.45 \times 10^9 \text{ Hz})} \left[\frac{2}{47(\sqrt{1 + (0.345)^2} - 1)} \right]^{1/2} \simeq 1.67 \text{ cm}$$

As expected, the depth of penetration at 2.45 GHz is less than that at 915 MHz. Thus, it may be more desirable to use 915 MHz rather than the more typical commercial microwave oven frequency of 2.45 GHz. At the lower frequency the microwave energy can penetrate deeper into the piglet's body and thus provide more effective heating.

(c) For the mature pig we have a two-boundary problem. To calculate the percentages of incident power reflected back into air and transmitted into the muscle tissue, we need to use the effective reflection (Γ_{eff}) and transmission ($\mathcal{T}_{eff}$) coefficient expressions. First, we need to find the intrinsic impedances of fat and muscle tissues at 915 MHz and 2.45 GHz. At 915 MHz, for fat tissue (assume $\sigma_f = 0.1$ S-m^{-1}) we have

$$\tan \delta_f \simeq \frac{0.1 \text{ S-m}^{-1}}{(2\pi \times 915 \times 10^6 \text{ rad-s}^{-1}) \times (5.6 \times 8.85 \times 10^{-12} \text{ F-m}^{-1})}$$
$$\simeq 0.351$$

so that from [2.22], the intrinsic impedance is

$$\eta_f \simeq \frac{377/\sqrt{5.6}}{[1 + (0.351)^2]^{1/4}} \exp\left[j\frac{1}{2} \tan^{-1}(0.351)\right] \simeq 155 e^{j9.67°} \, \Omega$$

and the propagation constant is

$$\gamma_f = \alpha_f + j\beta_f$$

where from [2.19] and [2.20], we have

$$\begin{Bmatrix} \alpha_f \\ \beta_f \end{Bmatrix} \simeq \frac{2\pi \times 915 \times 10^6}{3 \times 10^8} \sqrt{\frac{5.6}{2}} [\sqrt{1 + (0.351)^2} \mp 1]^{1/2}$$
$$\simeq \begin{cases} 7.84 & \text{np-m}^{-1} \\ 46.0 & \text{rad-m}^{-1} \end{cases}$$

At 915 MHz, for muscle tissue, we have from the previous example,

$$\eta_m \simeq 48.7 e^{j15.8°} \, \Omega$$

so the effective reflection and transmission coefficients are given by

$$\Gamma_{\text{eff}} = \frac{(\eta_f - \eta_1)(\eta_m + \eta_f) + (\eta_f + \eta_1)(\eta_m - \eta_f)e^{-2\gamma_f d}}{(\eta_f + \eta_1)(\eta_m + \eta_f) + (\eta_f - \eta_1)(\eta_m - \eta_f)e^{-2\gamma_f d}}$$
$$\simeq 0.231 e^{-j159°}$$

and

$$\mathcal{T}_{\text{eff}} = \frac{4\eta_f \eta_m e^{-\gamma_f d}}{(\eta_f + \eta_1)(\eta_m + \eta_f) + (\eta_f - \eta_1)(\eta_m - \eta_f)e^{-2\gamma_f d}}$$
$$\simeq 0.222 e^{-j99.2°}$$

Therefore, the percentages of the power reflected back to air and power transmitted into the muscle tissue are $\rho_{\text{eff}}^2 \times 100 \simeq (0.231)^2 \times 100 \simeq 5.33\%$ and $[\eta_1 \tau_{\text{eff}}^2/|\eta_m|] \simeq [377(0.222)^2/48.7] \times 100 \simeq 38.2\%$, respectively, from which the percentage of power absorbed by the fat layer can be found as ~56.4%. These results show that the mature pig absorbs, overall, a much higher percentage of the incident power than the newborn piglet does (~95% versus ~39%), because the fat layer acts as an impedance transformer and reduces the amount of reflections significantly. The calculations at 2.45 GHz are left as an exercise for the reader.

3.8   OBLIQUE INCIDENCE ON A LOSSY MEDIUM

Reflection and refraction of uniform plane waves obliquely incident on a lossy medium is a topic of significant practical importance. In the context of HF, VHF, and UHF radio wave propagation, reflections from imperfectly conducting ground interfere with the direct signal between a transmitter and receiver (see Figure 3.45a), causing constructive or destructive interference at the receiver.[51] Accurate modeling of the propagation channel requires the knowledge of the reflection coefficient, which in general is complex, depending in a more complicated manner on the incidence angle θ_i. The recent explosive expansion of mobile radio, cellular telephone, and other wireless communications systems has brought renewed attention to the topic, because the propagation channel, which may be severely obstructed by buildings, mountains, and foliage, places fundamental limitations on the performance of such systems.[52] The problem of radio wave reflection from ground has been under extensive study for some time,[53] but characterization of the mobile radio environment

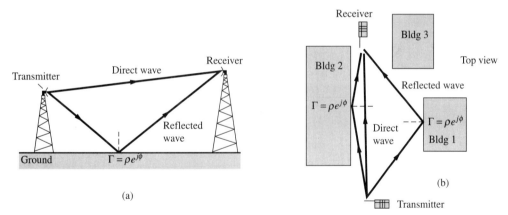

FIGURE 3.45.   Illustration of direct and reflected waves. (a) Radio wave (e.g., HF, VHF, or UHF) or microwave link between a transmitter and receiver, with the ground (or seawater or a lake surface) behaving as a lossy reflector. (b) A mobile radio environment (shown in top view) where surrounding buildings or structures act as lossy reflecting surfaces.

[51] See Chapter 6 of R. E. Collin, *Antennas and Radiowave Propagation,* McGraw-Hill, 1985.

[52] For a relatively simple discussion, see T. S. Rappaport, *Wireless Communications,* Prentice-Hall, 1996. For a recent discussion of current problems, see the Special Issue on Wireless Communications of *IEEE Transactions on Antennas and Propagation,* 46(6), June 1998.

[53] See Chapter 16 of E. C. Jordan and K. G. Balmain, *Electromagnetic Waves and Radiating Systems,* 2nd ed., Prentice-Hall, 1968.

requires new measurements and modeling[54] of reflection from boundaries such as buildings and concrete structures (see Figure 3.45b) at frequencies of a few GHz and higher. Given the enormous growth and demand for personal communication systems, and the anticipated expansion of these systems to nearly every aspect of our daily lives, a thorough understanding of the electromagnetic wave propagation channel in the presence of multiple lossy reflecting surfaces is of essential importance. Better understanding of reflecting properties of lossy surfaces is also important in other recently emerging consumer applications, such as millimeter-wave (30–100 GHz) systems used in car collision avoidance radar systems.[55]

The general treatment of this seemingly simple problem of a plane wave incident on a lossy plane surface is amazingly complex, because the refracted wave in the second medium is highly nonuniform and propagates in a direction determined by an effective refractive index whose magnitude is dependent on the incidence angle θ_i. While numerical results can be obtained for any general case, the physical nature of the reflected and refracted waves can be physically understood only in special cases (e.g., medium 2 being a good conductor, so that $\sigma_2 \gg \omega\epsilon_2$).

Before we proceed, it is useful to review the basis for Snell's law for the lossless case. The governing Maxwell's equations for a source-free and lossless medium are repeated here for convenience:

$$\nabla \times \mathbf{E} = -j\omega\mu\mathbf{H} \qquad [3.44a]$$

$$\nabla \times \mathbf{H} = j\omega\epsilon\mathbf{E} \qquad [3.44b]$$

Equations [3.44] were the basis for uniform plane wave propagation in each of the dielectric media with propagation constant $\beta = \omega\sqrt{\mu\epsilon}$, leading to wave solutions of the type

$$(\cdots)\,e^{-j\beta\hat{\mathbf{k}}\cdot\mathbf{r}}$$

and ultimately (on application of the boundary conditions) to Snell's law (equation [3.19]), repeated here as

$$\frac{\sin\theta_t}{\sin\theta_i} = \left(\frac{\mu_1\epsilon_1}{\mu_2\epsilon_2}\right)^{1/2} = \frac{\beta_1}{\beta_2} \qquad [3.45]$$

Now consider a plane wave obliquely incident on a lossy dielectric material, characterized by the parameters σ_2, ϵ_2 and μ_2. The governing equation corresponding to [3.44b] for the transmitted wave in medium 2 is now the complete Maxwell's equation [1.11c]:

$$\nabla \times \mathbf{H}_2 = \sigma_2\mathbf{E}_2 + j\omega\epsilon_2\mathbf{E}_2 = j\omega\left(\epsilon_2 + \frac{\sigma_2}{j\omega}\right)\mathbf{E}_2 = j\omega\epsilon_{\text{eff}}\mathbf{E}_2 \qquad [3.46]$$

---

[54]O. Landron, M. J. Feuerstein, and T. S. Rappaport, A comparison of theoretical and empirical reflection coefficients for typical exterior wall surfaces in a mobile radio environment, *IEEE Trans. on Antennas and Propagation*, 44(3), March 1996.

[55]K. Sato, T. Manabe, J. Polivka, T. Ihara, Y. Kasashima, and K. Yamaki, Measurement of the complex refractive index of concrete at 57.5 GHz, *IEEE Trans. on Antennas and Propagation*, 44(1), January 1996.

where $\epsilon_{\text{eff}} = [\epsilon_2 + \sigma_2/(j\omega)]$. The corresponding wave solutions can be written down by analogy with the lossless case and have the form

$$(\cdots)\, e^{-\gamma_2 \hat{\mathbf{k}} \cdot \mathbf{r}} \qquad\qquad [3.47]$$

where the propagation constant γ_2 in the lossy medium 2 is given by

$$\gamma_2 \equiv \alpha_2 + j\beta_2 = j\omega \sqrt{\mu_2 \epsilon_{\text{eff}}} = j\omega \left[\mu_2 \left(\epsilon_2 + \frac{\sigma_2}{j\omega} \right) \right]^{1/2} \qquad [3.48]$$

Again by analogy with [3.45], we can write Snell's law for this case as

$$\frac{\sin \theta_t}{\sin \theta_i} = \frac{j\beta_1}{\gamma_2} = \frac{j\beta_1}{\alpha_2 + j\beta_2} = \left[\frac{\mu_1 \epsilon_1}{\mu_2 \left(\epsilon_2 + \dfrac{\sigma_2}{j\omega} \right)} \right]^{1/2} \qquad [3.49]$$

from which we realize that θ_t must in general be a complex number. Note that the sine of a complex number does not pose a mathematical problem, since for a general complex number $\theta_t = \theta_{tr} + j\theta_{ti}$ we can evaluate $\sin(\theta_t)$ as

$$\sin \theta_t = \frac{e^{j\theta_t} - e^{-j\theta_t}}{2j} = \sin(\theta_{tr}) \cosh(\theta_{ti}) + j \cos(\theta_{tr}) \sinh(\theta_{ti}) \qquad [3.50]$$

so that $\sin \theta_t$ is simply another complex number. It is possible to proceed at this point by expressing θ_t in terms of its real and imaginary parts and rewriting Snell's law in terms of two separate equations (obtained, respectively, from the real and imaginary parts of [3.50]).[56] However, better physical insight into the nature of the refracted wave in medium 2 is obtained by expanding[57] (into its real and imaginary parts) the quantity $\gamma_2 \cos \theta_t$. We proceed by considering the refracted electromagnetic field in medium 2.

3.8.1   The Refracted Electromagnetic Field in Medium 2

With reference to Figure 3.46, the electric field phasor of the refracted wave in medium 2, for the case of perpendicular polarization[58] is given by

$$\mathbf{E}_t(x, z) = \hat{\mathbf{y}}\, E_{t0}\, e^{-\gamma_2 \hat{\mathbf{k}}_t \cdot \mathbf{r}} = \hat{\mathbf{y}}\, E_{t0}\, e^{-\gamma_2 (x \sin \theta_t + z \cos \theta_t)} \qquad [3.51]$$

---

[56] For a compact treatment along such lines, see Section 14.4 of R. W. P. King and S. Prasad, *Fundamental Electromagnetic Theory and Applications,* Prentice-Hall, 1986.

[57] An expanded version of the brief discussion provided here can be found in Sections 9.8 and 9.10 of J. A. Stratton, *Electromagnetic Theory,* McGraw-Hill, 1941.

[58] Note that although [3.51] is written for perpendicular polarization, the spatial variation of both components of the electric field for the parallel polarization case is also proportional to $e^{-\gamma_2 \hat{\mathbf{k}}_t \cdot \mathbf{r}}$. Thus, the properties of the refracted wave as discussed here are also exhibited by the refracted wave in the parallel polarization case.

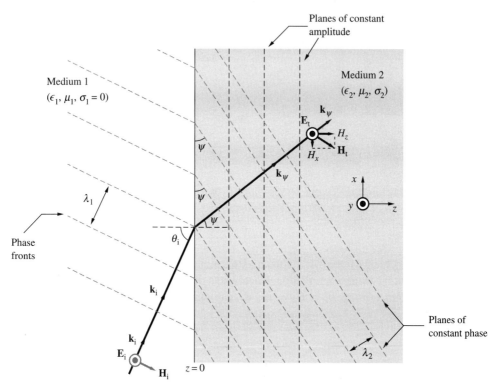

FIGURE 3.46. **Oblique incidence on a lossy medium: Constant amplitude and phase fronts.** The planes of constant amplitude are parallel to the interface, while the planes of constant phase are perpendicular to $\mathbf{k}_\psi$, which is at an angle ψ from the vertical. This "true" angle ψ is given by [3.59] and is also equal to the real part of the complex angle θ_t defined by [3.55].

which is identical to the expression provided in Section 3.5.1 for $\mathbf{E}_t(x, z)$ except for the fact that $j\beta_2$ is now replaced by γ_2. We now have to rewrite this expression by taking into account of the fact that γ_2, $\sin\theta_t$ and $\cos\theta_t$ are all complex numbers.

Noting that the total electric field phasor in medium 1 consists of the sum of the incident and reflected fields, we have

$$\mathbf{E}_1(x, z) = \hat{\mathbf{y}}\, E_{i0}\, e^{-j\beta_1 (x \sin\theta_i + z \cos\theta_i)} + \hat{\mathbf{y}}\, E_{r0}\, e^{-j\beta_1 (x \sin\theta_r - z \cos\theta_r)} \qquad [3.52]$$

The boundary condition on the continuity of the tangential electric field at the interface ($z = 0$) requires that $\mathbf{E}_1(x, 0) = \mathbf{E}_t(x, 0)$ or

$$\hat{\mathbf{y}}\, E_{i0}\, e^{-j\beta_1 x \sin\theta_i} + \hat{\mathbf{y}}\, E_{r0}\, e^{-j\beta_1 x \sin\theta_r} = \hat{\mathbf{y}} E_{t0}\, e^{-\gamma_2 x \sin\theta_t} \qquad [3.53]$$

For [3.53] to hold at all values of x, we must have three conditions:

$$\theta_i = \theta_r \quad \text{and} \quad E_{i0} + E_{r0} = E_{t0} \qquad [3.54]$$

and

$$\boxed{\gamma_2 \sin\theta_t = j\beta_1 \sin\theta_i} \qquad [3.55]$$

where [3.55] is the generalized form of Snell's law of refraction derived in Section 3.5 for an interface between two lossless dielectrics and also written in [3.49] by analogy with [3.45]. We thus see that although γ_2 and $\sin \theta_t$ are both complex, their product is purely imaginary, as is evident from [3.55]. On the other hand, the product $\gamma_2 \cos \theta_t$ is a complex number and we define its real and imaginary parts as

$$\gamma_2 \cos \theta_t = \gamma_2 \sqrt{1 - \sin^2 \theta_t} = \sqrt{\gamma_2^2 + \beta_1^2 \sin^2 \theta_i} \equiv p + jq \qquad [3.56a]$$

$$p \equiv \mathfrak{Re}\left\{ \sqrt{\gamma_2^2 + \beta_1^2 \sin^2 \theta_i} \right\} \quad \text{and} \quad q \equiv \mathfrak{Im}\left\{ \sqrt{\gamma_2^2 + \beta_1^2 \sin^2 \theta_i} \right\} \quad [3.56b]$$

By substituting [3.55] and [3.56a] into [3.51], we can rewrite the refracted field in medium 2 as

$$\boxed{\mathbf{E}_t(x, z) = \hat{\mathbf{y}} E_{t0} \, e^{-pz} e^{-j(x\beta_1 \sin \theta_i + qz)}} \qquad [3.57]$$

Examination of [3.57] indicates that the surfaces of constant amplitude (i.e., $pz = $ constant) of the refracted wave are parallel to the interface. The planes of constant phase are given by

$$qz + \beta_1 x \sin \theta_i = \text{constant}$$

$$\mathbf{k}_\psi \cdot \mathbf{r} = \text{constant}$$

$$|\mathbf{k}_\psi| (\hat{\mathbf{k}}_\psi \cdot \mathbf{r}) = \text{constant} \qquad [3.58]$$

where $\mathbf{k}_\psi = \hat{\mathbf{z}} q + \hat{\mathbf{x}} \beta_1 \sin \theta_i$ and $\hat{\mathbf{k}}_\psi = \hat{\mathbf{x}} \sin \psi + \hat{\mathbf{z}} \cos \psi$ is the unit vector in the direction of $\mathbf{k}_\psi$. Note that $|\mathbf{k}_\psi| = \sqrt{q^2 + \beta_1^2 \sin^2 \theta_i}$. We see from [3.58] that the refracted wave propagates at the "true" angle of refraction[59] ψ, which is defined by

$$\tan \psi = \frac{\beta_1 \sin \theta_i}{q} \qquad [3.59]$$

It is clear that, as in the case of total internal reflection, the refracted wave in medium 2 is a nonuniform wave. The planes of constant amplitude are parallel to the boundary, as in the case of total internal reflection. However, in contrast to the case of total internal reflection, the planes of constant phase are not perpendicular to the boundary

---

[59]This true angle of refraction is in fact the real part θ_{tr} of the imaginary angle θ_t. By expanding $\sin \theta_t$ using [3.50], Snell's law [3.55] can be written as two equations

$$\beta_2 \sin(\theta_{tr})\cosh(\theta_{ti}) - \alpha_2 \cos(\theta_{tr})\sinh(\theta_{ti}) = \beta_1 \sin(\theta_i)$$

$$\alpha_2 \sin(\theta_{tr})\cosh(\theta_{ti}) + \beta_2 \cos(\theta_{tr})\sinh(\theta_{ti}) = 0$$

Solving these equations we find

$$\tanh(\theta_{ti}) = -\frac{\alpha_2}{\beta_2} \tan(\theta_{tr}) \quad \text{and} \quad \sin(\theta_{tr}) \left\{ \frac{\beta_2^2 + \alpha_2^2}{[\beta_2^2 + \alpha_2^2 \tan^2(\theta_{tr})]^{1/2}} \right\} = \beta_1 \sin(\theta_i)$$

the latter of which is Snell's law in terms of real angles, essentially equivalent to [3.61]. For further details of an analysis in terms of θ_{tr} and θ_{ti} see Section 14.4 of R. W. P. King and S. Prasad, *Fundamental Electromagnetic Theory and Applications,* Prentice-Hall, 1986.

but are instead inclined at an angle ψ from the normal to the interface. This behavior is depicted in Figure 3.46.

Note that we have

$$\sin \psi = \frac{\mathbf{k}_\psi \cdot \hat{\mathbf{x}}}{|\mathbf{k}_\psi|} = \frac{\beta_1 \sin \theta_i}{\sqrt{q^2 + \beta_1^2 \sin^2 \theta_i}} \qquad [3.60]$$

so that a geometrical refractive index for the lossy material can be defined which relates this true angle ψ to the incident angle θ_i via a modified Snell's law of refraction in terms of real angles as

$$\frac{\sin \theta_i}{\sin \psi} = \frac{\sqrt{q^2 + \beta_1^2 \sin^2 \theta_i}}{\beta_1} \equiv n(\theta_i) \qquad [3.61]$$

Note that this geometrical index of refraction describing the angle ψ at which the refracted wave propagates is a function of the incident angle θ_i. The phase velocity in the lossy material, defined as the velocity of propagation of the planes of constant phase, is

$$v_{p2}(\theta_i) = \frac{\omega}{|\mathbf{k}_\psi|} = \frac{\omega}{\sqrt{q^2 + \beta_1^2 \sin^2 \theta_i}} = \frac{\omega}{\beta_1 n(\theta_i)} = \frac{v_{p1}}{n(\theta_i)} \qquad [3.62]$$

Note that the phase velocity depends on the incident angle and can exceed the velocity of light in medium 2. The wavelength in medium 2 can similarly be defined in terms of the refractive index $n(\theta_i)$ as

$$\lambda_2 = \frac{\lambda_1}{n(\theta_i)} = \frac{2\pi}{\sqrt{q^2 + \beta_1^2 \sin^2 \theta_i}} \qquad [3.63]$$

The quantities $n(\theta_i)$, p, and q can be written out explicitly in terms of the phase constant β_2 and the attenuation constant α_2 of medium 2. Useful relations between these quantities are

$$\beta_1^2\, n(\theta_i) = \beta_2^2 - \alpha_2^2; \qquad p(\theta_i)\, q(\theta_i) = \alpha_2 \beta_2; \qquad \beta_1\, p(\theta_i)\, n(\theta_i) = \frac{\alpha_2 \beta_2}{\cos \psi} \quad [3.64]$$

where we note that p and q are also functions of the incidence angle θ_i. In principle, the incidence angle dependent refractive index $n(\theta_i)$ can be optically measured (assuming that medium 2 is sufficiently transparent) and [3.64] can be used to determine the properties of medium 2 (i.e., α_2 and β_2).

As in the case of the refracted wave for total internal reflection (see Section 3.6.2), the refracted wave described by [3.57] has field components in its propagation direction $\hat{\mathbf{k}}_\psi$. To see this, we note that with the electric field taken to be in the y direction, i.e., $\mathbf{E}_t(x, z) = \hat{\mathbf{y}}\, E_{ty}(x, z)$, we have

$$\nabla \times \mathbf{E}_t = -j\omega \mu_0 \mathbf{H}_t$$

$$\mathbf{E}_t(x, z) = \hat{\mathbf{y}} E_{t0} e^{-pz} e^{-j(x\beta_1 \sin \theta_i + qz)}$$

$\longrightarrow$

$$-\frac{\partial E_{ty}}{\partial z} = -j\omega \mu_0 H_{tx}$$

$$\frac{\partial E_{ty}}{\partial x} = -j\omega \mu_0 H_{tz}$$

which upon manipulation yields

$$\mathbf{H}_t(x, z) = \frac{E_{t0}}{\omega \mu_0} \left[\hat{\mathbf{x}} \, j \, (p + jq) + \hat{\mathbf{z}} \, \beta_1 \sin \theta_i \right] e^{-pz} e^{-j(x\beta_1 \sin \theta_i + qz)}$$

Since in general $\mathbf{k}_\psi \cdot \mathbf{H}_t \neq 0$, we see that $\mathbf{H}_t$ in general has a component along the propagation direction $\mathbf{k}_\psi$. As in the case of total internal reflection, the two components of this magnetic field are not in phase; consequently, the magnetic field in medium 2 rotates in an elliptical path about the electric field $\mathbf{E}_t$ (i.e., in the xz plane).

Refraction into a Good Conductor As an important limiting case, we now consider oblique incidence of uniform plane waves on a good conductor, for which $\sigma_2 \gg \omega \epsilon_2$. In this case, we note from Section 2.3.2 that

$$\alpha_2 \simeq \beta_2 \simeq \sqrt{\frac{\omega \mu_2 \sigma_2}{2}} \qquad [3.65]$$

so that we have from [3.49] (also assuming $\mu_1 = \mu_2$)

$$\frac{\sin \theta_t}{\sin \theta_i} = \frac{j\beta_1}{\alpha_2 + j\beta_2} = \frac{j\omega \sqrt{\mu_1 \epsilon_1}}{\sqrt{\frac{\omega \mu_2 \sigma_2}{2}} + j\sqrt{\frac{\omega \mu_2 \sigma_2}{2}}} = \sqrt{\frac{\omega \epsilon_1}{\sigma_2}} \, e^{j\pi/4} \qquad [3.66]$$

indicating that $\sin \theta_t \to 0$ for large σ_2, which in turn means that $\cos \theta_t \to 1$, so that from [3.56a] we have

$$\gamma_2 \cos \theta_t \simeq \gamma_2 = \alpha_2 + j\beta_2 \simeq p + jq$$

$$\sqrt{\frac{\omega \mu_2 \sigma_2}{2}} + j\sqrt{\frac{\omega \mu_2 \sigma_2}{2}} \simeq p + jq$$

$$\rightarrow \quad p \simeq \alpha_2 \simeq \sqrt{\frac{\omega \mu_2 \sigma_2}{2}} \qquad \text{and} \qquad q \simeq \beta_2 \simeq \sqrt{\frac{\omega \mu_2 \sigma_2}{2}} \qquad [3.67]$$

and therefore from [3.59] we find

$$\psi = \tan^{-1}\left[\frac{\beta_1 \sin \theta_i}{q}\right] \simeq \tan^{-1}\left[\frac{\omega \sqrt{\mu_1 \epsilon_1} \sin \theta_i}{\sqrt{\frac{\omega \mu_2 \sigma_2}{2}}}\right] \rightarrow 0 \quad \text{for large } \sigma_2 \quad [3.68]$$

Thus we see that when medium 2 is a good conductor with $\sigma_2 \gg \omega \epsilon_2$, the true angle of refraction ψ tends to zero, and the planes of constant phase are oriented parallel to the reflecting plane boundary and to the planes of constant amplitude (which, based on [3.57], are generally parallel to the interface for any σ_2), as shown in Figure 3.47. The fact that the angle ψ is indeed extremely small in practice can be seen by assuming nonmagnetic materials ($\mu_1 = \mu_0$ and $\mu_2 = \mu_0$) and evaluating [3.68] at the highest end of microwave frequencies (e.g., 100 GHz) for a typical conductor (e.g., copper, so that $\sigma_2 \simeq 5.8 \times 10^7$ S-m^{-1} and $\epsilon_1 \simeq \epsilon_0$) and assuming

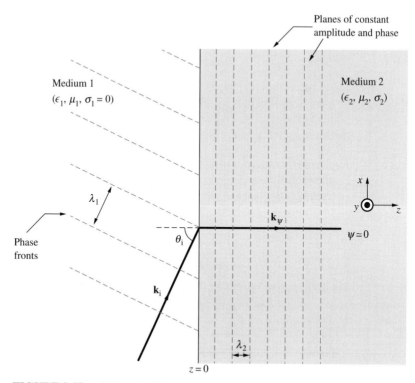

FIGURE 3.47. **Oblique incidence on a good conductor** $(\sigma_2 \gg \omega\epsilon_2)$. The planes of constant amplitude and constant phase are both parallel to the interface, the former being exactly parallel as in the general case, while the planes of constant phase are very nearly so, with ψ differing from zero by an imperceptible amount.

the angle of incidence to be as large as possible (i.e., $\sin\theta_i = 1$). We find

$$\psi \simeq \tan^{-1}\left[\frac{\omega\sqrt{\mu_0\epsilon_1}\sin\theta_i}{\sqrt{\dfrac{\omega\mu_0\sigma_2}{2}}}\right] \simeq \tan^{-1}\left[\sqrt{\frac{2\omega\epsilon_1}{\sigma_2}}\right] \simeq 0.025°$$

which means that $\mathbf{k}_\psi$ is indistinguishably close to being perpendicular to the interface. It is important to note that the angle ψ is very close to zero even for materials having much lower conductivity than metals. As another example, for seawater[60] at 1 MHz ($\sigma_2 \simeq 4\,\text{s-m}^{-1}$ and $\epsilon_2 \simeq 81\epsilon_0$), for $\sin\theta_i \simeq 1$ we find $\psi < 3°$. Using the approximate expressions [3.67] for p and q in [3.57], we can write the electric field

---

[60]The fact that the transmitted wave penetrates the ocean vertically, independent of θ_i, is important for communication with deeply submerged submarines using extremely low-frequency (ELF) signals guided by the Earth–ionosphere waveguide (see Example 4-2).

in medium 2 as

$$\mathbf{E}_t(x, z) \simeq \hat{\mathbf{y}} E_{t0} \, e^{-\alpha_2 z} e^{-j(x\beta_1 \sin\theta_i + \beta_2 z)} \qquad [3.69]$$

which indicates that the skin depth as defined in Section 2.3.2, namely $\delta \simeq \alpha_2^{-1} = \sqrt{2/(\omega\mu_2\sigma_2)}$, is indeed a valid measure of effective penetration of an electromagnetic wave into a conducting medium, regardless of the angle θ_i at which the wave is incident on the medium. The concept of skin depth thus applies to an electromagnetic wave incident at any angle on a good conductor. Independent of the angle of incidence θ_i, the wave transmitted (or refracted) into medium 2 can be considered to be a plane wave propagating normal to the surface, which is heavily attenuated as it propagates, as expected for an electromagnetic wave in a good conductor.

The fact that the wave transmitted into a good conductor propagates nearly normal to the surface regardless of the incident angle θ_i greatly broadens the applicability of the surface resistance method of characterizing power loss in a conductor, discussed in Section 3.7. Thus, the procedure described in Section 3.7 for the specific case of normal incidence is also valid for oblique incidence. In most cases, the calculation of the power loss using $\frac{1}{2}|H_t|^2 R_s$ (equation [3.43]) is entirely valid for arbitrary conducting surfaces having any incident electromagnetic fields.[61]

Note that the approximate expression [3.67] for q when $\sigma_2 \gg \omega\epsilon_2$ can be used in [3.61] to write an expression for the refractive index $n(\theta_i)$. We find

$$n(\theta_i) = \frac{\sqrt{q^2 + \beta_1^2 \sin^2\theta_i}}{\beta_1} = \sqrt{\frac{q^2}{\beta_1^2} + \sin^2\theta_i} \simeq \sqrt{\frac{\mu_2\sigma_2}{2\omega\mu_1\epsilon_1} + \sin^2\theta_i} \simeq \sqrt{\frac{\mu_2\sigma_2}{2\omega\mu_1\epsilon_1}}$$

$$[3.70]$$

since typically $\sigma_2 \gg \omega\epsilon_1$ and $\mu_1 = \mu_2 = \mu_0$, while $\sin\theta_i < 1$. We thus see that the refractive index for a good conductor is independent of the incident angle θ_i. Substituting numerical values for the previously discussed cases of copper at 100 GHz and seawater at 1 MHz yield very large refractive index values of respectively $n \simeq 2280$ and $n \simeq 21$, which in turn indicate from [3.63] that the wavelength λ_2 in a good conducting medium 2 is typically much smaller than it is in medium 1.

3.8.2   Oblique Incidence Reflection from a Lossy Medium

We have so far focused our attention on the refracted wave in the second medium. More important in practice—especially in applications involving reflection of radio waves from the ground and seawater or reflection of cellular phone signals from buildings and walls—are the characteristics of the reflected signals. The problem of reflection in oblique incidence from a lossy medium is generally quite involved, because the reflection coefficients are both complex and also dependent on the

---

[61] This method is not applicable if the conductor surface is curved with a radius of curvature comparable to the skin depth δ. Given the extremely small skin depths for most good conductors (see Section 2.3.2), conductors with such small radii of curvature are not encountered in practice.

polarization of the incident wave.[62] Fortunately, however, the numerical evaluation of the reflection coefficients is quite straightforward, since the reflection coefficients given by equations [3.24] (for perpendicular polarization) and [3.26] (for parallel polarization) are fully valid as long as we use the complex values of $\cos\theta_t$ and η_2. Using $\sin\theta_t$ as given in [3.55], adopting $\epsilon_{\text{eff}} = [\epsilon_2 + \sigma_2/(j\omega)]$ as defined in [3.46] to be the effective permittivity of medium 2, and noting that $\gamma_2 = j\omega\sqrt{\mu_2\epsilon_{\text{eff}}}$ as given in [3.48], we can write

$$\cos\theta_t = \sqrt{1 - \sin^2\theta_t} = \sqrt{1 + \frac{\beta_1^2\sin^2\theta_i}{\gamma_2^2}}$$

$$= \sqrt{1 + \frac{\mu_1\epsilon_1\sin^2\theta_i}{\mu_2\epsilon_{\text{eff}}}} = \sqrt{1 + \frac{j\omega\mu_1\epsilon_1\sin^2\theta_i}{\mu_2(\sigma_2 + j\omega\epsilon_2)}} \qquad [3.71a]$$

$$\eta_2 = \sqrt{\frac{\mu_2}{\epsilon_{\text{eff}}}} = \sqrt{\frac{j\omega\mu_2}{\sigma_2 + j\omega\epsilon_2}} \qquad [3.71b]$$

With $\cos\theta_t$ and η_2 as given in [3.71], the reflection coefficients for perpendicular and parallel polarization cases are given by [3.24] and [3.26] as

$$\Gamma_\perp = \frac{\eta_2\cos\theta_i - \eta_1\cos\theta_t}{\eta_2\cos\theta_i + \eta_1\cos\theta_t} = \rho_\perp e^{j\phi_\perp} \qquad [3.72a]$$

$$\Gamma_\parallel = \frac{-\eta_1\cos\theta_i + \eta_2\cos\theta_t}{\eta_1\cos\theta_i + \eta_2\cos\theta_t} = \rho_\parallel e^{j\phi_\parallel} \qquad [3.72b]$$

It is clear from [3.72] that both $\Gamma_\perp$ and $\Gamma_\parallel$ are in general complex, and that furthermore, in general we have $\phi_\perp \neq \phi_\parallel$. This in turn means that a linearly polarized incident wave, polarized such that its electric field is neither parallel nor perpendicular to the interface, in general becomes elliptically polarized upon reflection by the lossy medium.[63] Under the right circumstances (e.g., when the incident wave is polarized at an angle of 45° with respect to the plane of incidence, $\Delta\phi = \phi_\perp - \phi_\parallel = 90°$, $\rho_\perp = \rho_\parallel$), the reflected wave is circularly polarized.

Numerical evaluation of the reflection coefficients as given in [3.72] is straightforward, using $\cos\theta_t$ and η_2 as given in [3.71a] and [3.71b] respectively, for any given set of medium parameter values ϵ_1, μ_1, ϵ_2, μ_2, and σ_2. If medium 2 is a lossy dielectric, we may have $\sigma_2 = 0$, but instead have a complex permittivity

---

[62]The rich variety of resulting polarization effects have been well utilized to determine the optical constants of metals from measurements of reflected light. Note that while it is in general not possible to observe the refracted wave in a highly absorbing (i.e., lossy) medium, the reflected wave can be well measured and the dielectric properties of the reflecting medium can be deduced from polarization properties of the reflected wave. See Section 13.2 of M. Born and E. Wolf, *Principles of Optics*, 5th ed., Pergamon Press, 1975. Also see Section II-20 of A. R. Von Hippel, *Dielectrics and Waves*, Wiley, 1954.

[63]Note that this aspect of the problem of oblique reflection from a lossy medium is similar to the case of total internal reflection discussed in Section 3.6, except for the fact that in the present case $\rho_\perp$ and $\rho_\parallel$ are not unity.

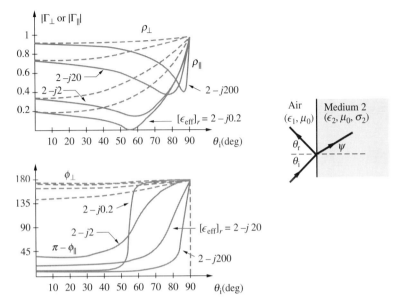

FIGURE 3.48.   Reflection coefficients for oblique incidence on a lossy medium. (a) Magnitude and phase of reflection coefficient for perpendicular ($\Gamma_\perp$) and parallel ($\Gamma_\parallel$) polarization versus angle of incidence θ_i for $[\epsilon_{\mathrm{eff}}]_r = (\epsilon_{\mathrm{eff}}/\epsilon_0) = 2 - j0.2$, $2 - j2, 2 - j20$, and $2 - j200$. For clarity, the complement of the phase angle $\phi_\parallel$ (i.e., $\pi - \phi_\parallel$) is sketched, rather than $\phi_\parallel$ itself.

$\epsilon_{2c} = \epsilon_2' - j\epsilon_2''$. In such a case, [3.71] can still be used by making the substitution $\epsilon_{\mathrm{eff}} \leftrightarrow \epsilon_{2c}$. Results for $\epsilon_{2r}' = 2$ and different values of ϵ_{2r}'' ranging from 0.2 to 200 are shown in Figure 3.48.

The results shown in Figure 3.48 can be compared with those for the lossless case given in Figure 3.30. We note that the variations with incident angle θ_i of the reflection coefficients are generally similar between the lossless case and the case of relatively low loss, i.e., $[\epsilon_{\mathrm{eff}}]_r = (\epsilon_{\mathrm{eff}}/\epsilon_0) = 2 - j0.2$. As the losses increase, the shapes of the curves for $\rho_\perp$ and $\phi_\perp$ remain generally the same. However, for parallel polarization we see that there is no Brewster angle at which $\rho_\parallel = 0$; instead we note that a pseudo-Brewster angle is quite evident, exhibiting a deep minimum in $\rho_\parallel$ at which $\phi_\parallel = 90°$. Results for $\epsilon_{2r}' = 10$ and $\epsilon_{2r}' = 80$ are shown respectively in Figure 3.49 and Figure 3.50, basically exhibiting similar behavior, with a pseudo-Brewster angle for parallel polarization, which occurs at increasingly higher θ_i with increasing losses. The depth of the minimum in $\rho_\parallel$ at first decreases with increasing losses, but then remains nearly constant in the range between $\rho_\parallel \simeq 0.4$ and $\rho_\parallel \simeq 0.5$ with increasing losses beyond $\epsilon_2''/\epsilon_2' > 10$.

Oblique Reflection from a Good Conductor   When medium 2 is a good conductor, so that $\sigma_2 \gg \omega\epsilon_2$, the angle $\theta_t \to 0$ and $\cos\theta_t \to 1$, as noted in the previous section and as is also evident from [3.71a]. Furthermore, from [3.40] we have

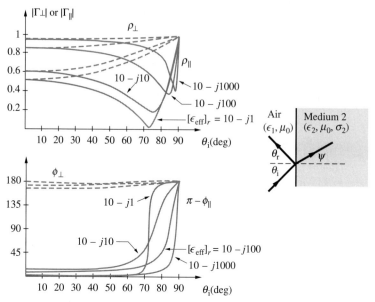

FIGURE 3.49. **Reflection coefficients for oblique incidence on a lossy medium.** (a) Magnitude and phase of reflection coefficient for perpendicular ($\Gamma_\perp$) and parallel ($\Gamma_\parallel$) polarization versus angle of incidence θ_i for $[\epsilon_{\text{eff}}]_r = (\epsilon_{\text{eff}}/\epsilon_0) = 10 - j1, 10 - j10, 10 - j100$, and $10 - j1000$. For clarity, the complement of the phase angle $\phi_\parallel$ (i.e., $\pi - \phi_\parallel$) is sketched, rather than $\phi_\parallel$ itself.

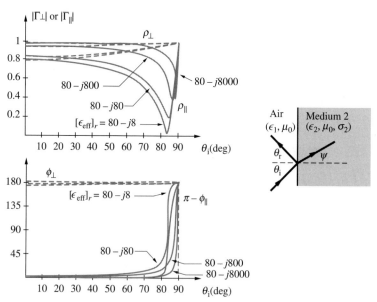

FIGURE 3.50. **Reflection coefficients for oblique incidence on a lossy medium.** (a) Magnitude and phase of reflection coefficient for perpendicular ($\Gamma_\perp$) and parallel ($\Gamma_\parallel$) polarization versus angle of incidence θ_i for $[\epsilon_{\text{eff}}]_r = (\epsilon_{\text{eff}}/\epsilon_0) = 80 - j8, 80 - j80, 80 - j800$ and $80 - j8000$. For clarity, the complement of the phase angle $\phi_\parallel$ (i.e., $\pi - \phi_\parallel$) is sketched, rather than $\phi_\parallel$ itself.

$\eta_2 \simeq \sqrt{j\omega\mu_2/\sigma_2}$, so that

$$\frac{\eta_1}{\eta_2} \simeq \frac{\sqrt{\mu_1/\epsilon_1}}{\sqrt{j\omega\mu_2/\sigma_2}} \simeq \sqrt{\frac{\sigma_2}{j\omega\epsilon_1}} \gg 1 \qquad [3.73]$$

once again assuming nonmagnetic media so that $\mu_1 = \mu_2$. The reflection coefficients as given in [3.72] can then be simplified as

$$\Gamma_\perp = \frac{\cos\theta_i - (\eta_1/\eta_2)}{\cos\theta_i + (\eta_1/\eta_2)} \simeq -1 = 1e^{j\pi} \qquad [3.74a]$$

$$\Gamma_\parallel = \frac{-(\eta_1/\eta_2)\cos\theta_i + 1}{(\eta_1/\eta_2)\cos\theta_i + 1} \simeq -1 = 1e^{j\pi} \qquad [3.74b]$$

since $\cos\theta_i$ is bounded by unity. This result indicates that the magnitude of the reflection coefficient is very nearly equal to unity for reflection from a good conductor, regardless of the angle of incidence. Note that this behavior is also evident from Figures 3.48, 3.49, and 3.50, where for higher values of ϵ_2'' (i.e., highly lossy material) both $\rho_\perp$ and $\rho_\parallel$ stay closer to unity regardless of the value of θ_i. Note that the parallel polarization case still exhibits a pseudo-Brewster angle, which itself approaches $90°$ as ϵ_2'' increases. Also note that for good conductors, such as metals, we have σ_2 of order $\sim 10^7$ s-m^{-1}, so that $\epsilon_{\text{eff}} \simeq [\sigma_2/(j\omega)]$, and $[\epsilon_{\text{eff}}]_r = \epsilon_{\text{eff}}/\epsilon_0 \simeq [\sigma_2/(j\omega\epsilon_0)]$, which is a number typically much larger than the ϵ_2'' values considered in Figures 3.48, 3.49, and 3.50. If we were to plot $\rho_\perp$ and $\rho_\parallel$ for a metal versus θ_i on the same scale as in Figure 3.48, the resulting curve would be indistinguishable from a parallel straight line at $\rho_\perp = \rho_\parallel \simeq 1$.

Example 3-22 illustrates the numerical evaluation of reflection coefficients for a wireless communication signal obliquely incident on a glass wall.

Example 3-22:   Wireless communication signal incident on a glass wall.
Consider a wireless communication signal propagating in air represented by

$$\mathbf{E}_i = \mathbf{E}_{i\perp} + \mathbf{E}_{i\parallel} = \left[C_1\hat{\mathbf{y}} + C_2\left(\hat{\mathbf{x}}\frac{1}{2} - \hat{\mathbf{z}}\frac{\sqrt{3}}{2}\right)\right] e^{-j[x(40\pi/\sqrt{3})+z(40\pi/3)]}$$

which is obliquely incident on a particular type of glass wall[64] ($\sigma = 2.5$ S-m^{-1}, $\epsilon_{2r} = 5$, and $\mu_{2r} = 1$) located at $z = 0$. (a) Find the frequency of operation f and the incidence angle θ_i. (b) Find the propagation constant γ_2 and the true angle of refraction ψ. (c) Find the reflected electric field $\mathbf{E}_r$.

---

[64]O. Landron, M. J. Feuerstein, and T. S. Rappaport, A comparison of theoretical and empirical reflection coefficients for typical exterior wall surfaces in a mobile radio environment, *IEEE Trans. on Antennas and Propagation*, 44(3), March 1996.

Solution:

(a) Comparing the given $\mathbf{E}_{i\perp}$ with [3.11], we have

$$\begin{matrix} \beta_1 \sin \theta_i = \dfrac{40\pi}{\sqrt{3}} \\[2mm] \beta_1 \cos \theta_i = \dfrac{40\pi}{3} \end{matrix} \quad \rightarrow \quad \tan \theta_i = \dfrac{40\pi/\sqrt{3}}{40\pi/3} = \sqrt{3} \quad \rightarrow \quad \theta_i = 60°$$

and

$$\beta_1 = \frac{40\pi/3}{\cos \theta_i} = \frac{80\pi}{3} = \frac{\omega}{c} \quad \rightarrow \quad f \simeq 4 \text{ GHz}$$

(b) The propagation constant γ_2 can be calculated from [3.48] as

$$\gamma_2 \simeq j(2\pi)(4 \times 10^9) \left[(4\pi \times 10^{-7}) \left(5 \times 8.85 \times 10^{-12} - \frac{j\,2.5}{2\pi \times 4 \times 10^9} \right) \right]^{1/2}$$

$$\simeq j\,(2\pi)(4 \times 10^9)\sqrt{4\pi \times 10^{-7}}(10^{-5})\underbrace{\sqrt{0.443 - j0.995}}_{\sim 1.089e^{-j66.0°}}$$

$$\simeq j\,(281.7)(1.043e^{-j33.0°}) \simeq e^{j90°}\,294e^{-j33.0°}$$

$$\simeq 160.1 + j\,246.5 \quad \rightarrow \quad \alpha_2 \simeq 160.1 \text{ np-m}^{-1} \quad \text{and} \quad \beta_2 \simeq 246.5 \text{ rad-m}^{-1}$$

Substituting this value of γ_2 in [3.55] we find

$$(160.1 + j246.5) \sin \theta_t \simeq j\left(\frac{80\pi}{3}\right) \sin 60° \quad \rightarrow \quad \sin \theta_t \simeq 0.207 + j\,0.134$$

which can be substituted in [3.56] yielding

$$\gamma_2 \cos \theta_t \simeq (160.1 + j246.5)\sqrt{1 - (0.207 + j\,0.134)^2} \simeq 165.1 + j239.0$$

$$\rightarrow \quad p \simeq 165.1 \text{ np-m}^{-1} \quad \text{and} \quad q \simeq 239.0 \text{ rad-m}^{-1}$$

the true angle of refraction can then be found from [3.59] as

$$\psi \simeq \tan^{-1}\left[\frac{(80\pi/3)\sin 60°}{239}\right] \simeq 16.88°$$

(c) Using [3.72], the reflection coefficients for the parallel and perpendicular components can be calculated. We first note from [3.71b] that

$$\Gamma_2 = \sqrt{j\omega\mu_2/(\sigma_2 + j\omega\epsilon_2)} \simeq 90.09 + j58.53$$

so that

$$\Gamma_\perp \simeq \frac{(90.09 + j58.53)\cos 60° - (377)(0.987 - j0.0281)}{(90.09 + j58.53)\cos 60° + (377)(0.987 - j0.0281)} \simeq 0.789e^{j170°}$$

$$\Gamma_\parallel \simeq \frac{-(377)\cos 60° + (90.09 + j58.53)(0.987 - j0.028)}{(377)\cos 60° + (90.09 + j58.53)(0.987 - j0.0281)} \simeq 0.395e^{j139°}$$

The reflected electric field follows as

$$\mathbf{E}_r = \Gamma_\perp C_1 \hat{\mathbf{y}} + \Gamma_\parallel C_2 \left(\hat{\mathbf{x}}\frac{1}{2} - \hat{\mathbf{z}}\frac{\sqrt{3}}{2}\right) e^{-j[x(40\pi/\sqrt{3})-z(40\pi/3)]}$$

It is also instructive to determine the effective permittivity of medium 2 for this case. We have

$$[\epsilon_{\text{eff}}]_r = \frac{\epsilon_{\text{eff}}}{\epsilon_0} = \epsilon_{2r} + \frac{\sigma_2}{j\omega\epsilon_0}$$

$$\simeq 5 - j\frac{2.5}{2\pi(4\times10^9)(8.85\times10^{-12})} \simeq 5 - j\,11.24$$

The behavior of the reflection coefficients with incident angle for this case falls somewhere between the $[\epsilon_{\text{eff}}] = 10 - j10$ case in Figure 3.49 and the $[\epsilon_{\text{eff}}] = 2 - j20$ case in Figure 3.48. Note that the pseudo-Brewster angle must therefore be between 70° to 80°, not too different from the $\theta_i = 60°$ at which the wave is incident on the glass wall.

3.9   SUMMARY

This chapter discussed the following topics:

■ **Normal incidence on a perfect conductor.** When a uniform plane wave is normally incident from a dielectric (medium 1) onto a perfect conductor, a standing wave pattern is produced with the total electric and magnetic fields in medium 1 given by

$$\overline{\mathscr{E}}_1(z, t) = \hat{\mathbf{x}}2E_{i0}\sin(\beta_1 z)\sin(\omega t)$$

$$\overline{\mathscr{H}}_1(z, t) = \hat{\mathbf{y}}2\frac{E_{i0}}{\eta_1}\cos(\beta_1 z)\cos(\omega t)$$

The standing wave is entirely analogous to the standing wave that is produced on a short-circuited lossless uniform transmission line and represents purely reactive power, with the time-average power being identically zero.

■ **Normal incidence on a lossless dielectric.** The reflection and transmission coefficients for the case of normal incidence of a uniform plane wave on a lossless dielectric are respectively given by

$$\Gamma = \frac{E_{r0}}{E_{i0}} = \frac{\eta_2 - \eta_1}{\eta_2 + \eta_1} \qquad \mathscr{T} = \frac{E_{t0}}{E_{i0}} = \frac{2\eta_2}{\eta_2 + \eta_1}$$

where η_1 and η_2 are the intrinsic impedances of the two dielectric media. Note that $(1+\Gamma) = \mathscr{T}$. The total electric and magnetic fields in medium 1 each consist of a component propagating in the z direction and another that is a standing wave, and can be expressed in phasor form as

$$\mathbf{E}_1(z) = \hat{\mathbf{x}} E_{i0} e^{-j\beta_1 z}(1 + \Gamma e^{j2\beta_1 z})$$

$$\mathbf{H}_1(z) = \hat{\mathbf{y}} \frac{E_{i0}}{\eta_1} e^{-j\beta_1 z}(1 - \Gamma e^{j2\beta_1 z})$$

The time-average Poynting flux in medium 1 is equal to that in medium 2 and is given as

$$(\mathbf{S}_{av})_1 = \hat{\mathbf{z}} \frac{E_{i0}^2}{2\eta_1}(1 - \Gamma^2)$$

so the fraction of the incident power transmitted into medium 1 is determined by $(1 - \Gamma^2)$.

■ **Multiple dielectric interfaces.** The effective reflection and transmission co-efficients for a uniform plane wave incident on an interface consisting of two dielectrics are given as

$$\Gamma_{\text{eff}} = \frac{(\eta_2 - \eta_1)(\eta_3 + \eta_2) + (\eta_2 + \eta_1)(\eta_3 - \eta_2)e^{-2j\beta_2 d}}{(\eta_2 + \eta_1)(\eta_3 + \eta_2) + (\eta_2 - \eta_1)(\eta_3 - \eta_2)e^{-2j\beta_2 d}}$$

$$\mathcal{T}_{\text{eff}} = \frac{4\eta_2\eta_3 e^{-j\beta_2 d}}{(\eta_2 + \eta_1)(\eta_3 + \eta_2) + (\eta_2 - \eta_1)(\eta_3 - \eta_2)e^{-j2\beta_2 d}}$$

These expressions are valid even for lossy media, as long as the proper complex intrinsic impedances are used and $j\beta_2$ is replaced by $\gamma_2 = \alpha_2 + j\beta_2$. In general, interfaces involving multiple dielectrics can be analyzed by relying on the transmission line analogy and treating the problem as one involving impedance transformation by a set of cascaded transmission lines.

■ **Oblique incidence on a perfect conductor.** When a uniform plane wave is obliquely incident at an angle θ_i on a perfect conductor, a reflected wave propagating away from the conductor at an angle $\theta_r = \theta_i$ is produced. The combination of the incident and reflected waves forms a standing wave pattern in the direction orthogonal to the boundary, with the electric field being identically zero at a discrete set of planes defined by

$$z = -\frac{m\lambda_1}{2\cos\theta_i} \qquad m = 1, 2, 3, \dots$$

The orientation of the total electric and magnetic fields, and the associated surface currents that flow within the conductor, are different depending on whether the incident wave is parallel or perpendicularly polarized.

■ **Oblique incidence at a dielectric boundary.** The reflection and refraction of a uniform plane wave obliquely incident at an angle θ_i to an interface between two dielectrics occurs in accordance with Snell's law, namely:

$$\frac{\sin\theta_i}{\sin\theta_t} = \frac{v_{p1}}{v_{p2}} = \sqrt{\frac{\epsilon_{2r}\mu_{2r}}{\epsilon_{1r}\mu_{1r}}} \quad \text{and} \quad \theta_r = \theta_i$$

where θ_r and θ_t are the angles from the vertical of respectively the reflected and refracted (transmitted) rays. The reflection and transmission coefficients for perpendicular polarization are

$$\Gamma_\perp = \frac{\eta_2 \cos\theta_i - \eta_1 \cos\theta_t}{\eta_2 \cos\theta_i + \eta_1 \cos\theta_t}$$

$$\mathcal{T}_\perp = \frac{2\eta_2 \cos\theta_i}{\eta_2 \cos\theta_i + \eta_1 \cos\theta_t}$$

with $(1 + \Gamma_\perp) = \mathcal{T}_\perp$. For parallel polarization we have

$$\Gamma_\parallel = \frac{-\eta_1 \cos\theta_i + \eta_2 \cos\theta_t}{\eta_1 \cos\theta_i + \eta_2 \cos\theta_t}$$

$$\mathcal{T}_\parallel = \frac{2\eta_2 \cos\theta_i}{\eta_1 \cos\theta_i + \eta_2 \cos\theta_t}$$

with $(1 + \Gamma_\parallel) = \mathcal{T}_\parallel(\cos\theta_t / \cos\theta_i)$. An interesting case with many applications occurs for the case of parallel polarization when the angle of incidence θ_i is equal to the Brewster angle θ_{iB}, given by

$$\theta_{iB} = \tan^{-1}\left[\sqrt{\frac{\epsilon_{2r}}{\epsilon_{1r}}}\right]$$

in which case $\Gamma_\parallel = 0$ and there is no reflected wave.

■ **Total internal reflection.** When a uniform plane wave is incident at an angle θ_i from a denser medium into a less dense medium (i.e., $\epsilon_1 > \epsilon_2$), total internal reflection occurs for $\theta_i \geq \theta_{ic}$, where θ_{ic} is given as

$$\sin\theta_{ic} = \sqrt{\frac{\epsilon_{2r}}{\epsilon_{1r}}}$$

Under conditions of total internal reflection, the reflection coefficients for parallel and perpendicular polarizations are both complex, with unity magnitude but, in general, different phase angles. When an incident wave is totally reflected, a nonuniform plane electromagnetic wave does exist in medium 2, but is heavily attenuated with distance into the medium, being confined to a spatial region of size comparable to the wavelength.

■ **Normal incidence on a lossy medium.** When a uniform plane wave is normally incident on an imperfect conductor, the surface current that flows within the conductor leads to power dissipation, although the magnitude of the reflection coefficient is very nearly unity. The amount of power lost in the conductor is given by

$$P_{\text{loss}} = \tfrac{1}{2}|J_s|^2 R_s$$

where the surface resistance $R_s = (\delta\sigma)^{-1}$, with δ being the skin depth of the imperfect conductor, and $|J_s| = |\mathbf{H}_1(0)|$ with $\mathbf{H}_1(0)$ being the total tangential magnetic field on the conductor surface.

- **Oblique incidence on a lossy medium.** When a uniform plane wave is obliquely incident on a lossy material having a propagation constant $\gamma_2 = \alpha_2 + j\beta_2$, the generalized form of Snell's law is

$$\gamma_2 \sin \theta_t = j\beta_1 \sin \theta_i$$

indicating that the transmitted angle θ_t is a complex number. A modified version of Snell's law of refraction can be derived as follows

$$\frac{\sin \theta_i}{\sin \psi} = \frac{\sqrt{q^2 + \beta_1^2 \sin^2 \theta_i}}{\beta_1} \qquad \begin{aligned} q &\equiv \mathcal{I}m\{\gamma_2 \cos \theta_t\} \\ &= \mathcal{I}m\{\sqrt{\gamma_2^2 + \omega^2 \mu_1 \epsilon_1 \sin^2 \theta_i}\} \end{aligned}$$

where ψ is the "true" angle of refraction, $\beta_1 = \omega\sqrt{\mu_1 \epsilon_1}$, and q is the imaginary part of $\gamma_2 \cos \theta_t$. The refracted wave is a nonuniform plane wave, with planes of constant amplitude parallel to the interface, whereas the planes of constant phase are at an angle ψ to the interface. When medium 2 is a good conductor, the true angle $\psi \simeq 0$, so that the planes of constant phase are nearly parallel to the interface, regardless of the incident angle θ_i. The reflection coefficients for oblique incidence on a lossy medium can be simply calculated using the general expressions [3.72] and the complex values of $\cos \theta_t$ and η_2. In general, a pseudo-Brewster angle exists for parallel polarization, at which $\rho_\parallel$ exhibits a deep minimum.

3.10 PROBLEMS

3-1. Air–perfect conductor interface. A uniform plane wave traveling in air with its electric field given by

$$\overline{\mathcal{E}}_i(y, t) = \hat{z} E_0 \cos(\omega t - \beta y)$$

is normally incident on a perfect conductor boundary located at $y = 0$. If the measured distance between any two successive zeros of the total electric field in air is 6 cm, and the maximum value of the electric field measured at $y = -75$ cm is 3 V-m^{-1}, determine the following: (a) the frequency (in GHz) and the power density (in μW-(cm)$^{-2}$) of the incident wave; (b) the instantaneous expression for the total electric field, $\overline{\mathcal{E}}_1(y, t)$, in air; (c) the instantaneous expression for the total magnetic field, $\overline{\mathcal{H}}_1(y, t)$, in air; (d) the maximum value of the total magnetic field measured at $y = -75$ cm.

3-2. Air–perfect conductor interface. A uniform plane wave of time-average power density 10 mW-cm^{-2} in air is normally incident on the surface of a perfect conductor located at $z = 0$. The total magnetic field phasor in air is given by

$$\mathbf{H}_1(z) = \hat{y} H_0 \cos(2\pi z)$$

(a) What is H_0? (b) What is the frequency, f? (c) At $z = -3.5$ m, what is the total electric field $\overline{\mathcal{E}}(z, t)$?

3-3. Air–perfect conductor interface. A uniform plane wave propagating in air given by

$$\mathbf{E}_i(x) = 60 e^{-j40\pi x}(\hat{y} - j\hat{z}) \quad \text{V-m}^{-1}$$

is normally incident on a perfectly conducting plane located at $x = 0$. (a) Find the frequency and the wavelength of this wave. (b) Find the corresponding magnetic field $\mathbf{H}_i(x)$. (c) Find the electric and magnetic field vectors of the reflected wave [i.e., $\mathbf{E}_r(x)$ and $\mathbf{H}_r(x)$]. (d) Find the total electric field in air [i.e., $\mathbf{E}_1(x)$], and plot the magnitude of each of its components as a function of x.

3-4. Air–perfect conductor interface. A uniform plane wave of frequency 10 GHz traveling in free space having an electric field given by

$$\overline{\mathscr{E}}_i(z, t) = \hat{\mathbf{x}}3.3 \sin(\omega t + \beta z) - \hat{\mathbf{y}}5 \cos(\omega t + \beta z) \qquad \text{V-m}^{-1}$$

is normally incident on a perfect conductor boundary located at $z = 0$. (a) Find the real-time expression for the reflected wave, $\overline{\mathscr{E}}_r(z, t)$. (b) Compare the polarizations of the incident and the reflected waves. Is there any difference? (c) Find the maximum value of the total electric field at $z = 0, 0.75, 1.5, 2.25$, and 3 cm, respectively.

3-5. Unknown material. When a uniform plane wave in air is normally incident onto a planar lossless material medium, the reflection coefficient is measured to be -0.25, and the phase velocity of the wave is reduced by a factor of 3. Find the relative permittivity and permeability of the unknown material.

3-6. Dielectric–dielectric interface. A uniform plane wave propagates from one dielectric into another at normal incidence. Find the ratio of the dielectric constants such that the magnitudes of the reflection and transmission coefficients are both equal to 0.5. Assume lossless nonmagnetic materials.

3-7. Air–GaAs interface. A uniform plane wave having a magnetic field given by

$$\mathbf{H}_i(z) = \hat{\mathbf{x}}10e^{j210z} \text{ mA-m}^{-1}$$

is normally incident from air onto a plane air–gallium arsenide (GaAs) interface located at $z = 0$. Assume GaAs to be a perfect dielectric with $\epsilon_r \approx 13$. (a) Find the reflected ($\mathbf{H}_r(z)$) and the transmitted ($\mathbf{H}_t(z)$) fields. (b) Calculate the power density of the incident, reflected, and transmitted waves independently and verify the conservation of energy principle. (c) Find the expression for the total magnetic field ($\mathbf{H}_1(z)$) in air and sketch its magnitude as a function of z between $z = 0$ and $z = 3$ cm.

3-8. Air–salty lake interface. Consider a uniform plane wave traveling in air with its electric field given by

$$\overline{\mathscr{E}}_i(y, t) = \hat{\mathbf{y}}100 \sin(2\pi \times 10^9 t - \beta z) \text{ V-m}^{-1}$$

normally incident on the surface ($z = 0$) of a salty lake ($\sigma = 0.88$ S-m^{-1}, $\epsilon_r = 78.8$, and $\mu_r = 1$ at 1 GHz). (a) Assuming the lake to be perfectly flat and lossless (i.e., assume $\sigma = 0$), find the electric fields of the reflected and transmitted waves (i.e., $\overline{\mathscr{E}}_r$ and $\overline{\mathscr{E}}_t$). (b) Modify the $\overline{\mathscr{E}}_t$ expression found in part (a) by introducing in it the exponential attenuation term due to the nonzero conductivity of lake water, given in part (a), and justify this approximation. (c) Using this expression, find the thickness of the lake over which 90% of the power of the transmitted wave is dissipated. What percentage of the power of the incident wave corresponds to this amount?

3-9. Aircraft–submarine communication. A submarine submerged in the ocean is trying to communicate with a Navy airplane, equipped with a VLF transmitter operating at 20 kHz, approximately 10 km immediately overhead from the location of the submarine. If the output power of the VLF transmitter is 200 kW and the receiver sensitivity of the submarine is 1 μV-m^{-1}, calculate the maximum depth of the submarine from the surface of the ocean for it to be able to communicate with the transmitter.

Assume the transmitter is radiating its power isotropically, and assume normal incidence at the air–ocean boundary. Use $\sigma = 4$ S-m^{-1}, $\epsilon_r = 81$, and $\mu_r = 1$ for the properties of the ocean.

3-10. Air–fat interface. Consider a planar interface between air and fat tissue (assume it to be of semi-infinite extent). If a plane wave is normally incident from air at this boundary, find the percentage of the power absorbed by the fat tissue at (a) 100 MHz; (b) 300 MHz; (c) 915 MHz; and (d) 2.45 GHz, and compare your results with the results of Example 3-19. Use Table 3-1 for the parameters of the fat tissue.

3-11. Air–concrete interface. A uniform plane wave operating at 1 GHz is normally incident from air onto the air–concrete interface. At 1 GHz, the complex relative dielectric constants of wet and dry concrete are measured as $\epsilon_{wr} \simeq 14.8 - j1.73$ and $\epsilon_{dr} \simeq 4.5 - j0.03$ respectively.[65] For each case, calculate (a) the percentage of the incident power reflected, and (b) the penetration depth in the concrete. Assume concrete to be semi-infinite in extent.

3-12. Shielding with a copper foil. A 1-GHz, 1-kW-m^{-2} microwave beam is incident upon a sheet of copper foil of 10 μm thickness. Consider neglecting multiple reflections, if justified. (a) Find the power density of the reflected wave. (b) Find the power density transmitted into the foil. (c) Find the power density of the wave that emerges from the other side of the foil. Comment on the shielding effectiveness of this thin copper foil.

3-13. Absorbing material. Consider a commercial absorber slab[66] made of EHP-48 material of 1 m thickness backed by a perfectly conducting metal plate. A 100-MHz uniform plane wave is normally incident from the air side at the air–absorber–metal interface. Find the percentage of incident power lost in the absorber material. For EHP-48, use $\epsilon_r = 6.93 - j8.29$ at 100 MHz.

3-14. Absorbing material. The relative permittivity of the commercially made graphite-impregnated pyramid cone foam absorber EHP-24 from Rantech (see footnote to Problem 3-13) is measured to be $\epsilon_r = 2.6 - j2.6$ at 300 MHz. A 300-MHz uniform plane wave propagating in air is normally incident at the surface of an EHP-24 slab. The absorber can be assumed to be nonmagnetic. Calculate the minimum thickness of the absorber slab such that the amplitude of the electric field reflected from this material coated on a perfect conductor (i.e., air–EHP-24–metal interface) would be attenuated by at least 100 dB. Can you neglect multiple reflections?

3-15. Radome design. A common material in dielectric radomes for aeronautical applications is fiberglass. For L-band (1–2 GHz), fiberglass has a typical relative dielectric constant of approximately $\epsilon_r \simeq 4.6$.(a) Assuming a flat-plane radome, determine the minimum thickness of fiberglass that causes no reflections at the center of the L-band. (b) Using the thickness found in part (a), find the percentage of the incident power which transmits to the other side of the radome at each end of the L-band (i.e., 1 GHz and 2 GHz).

3-16. Radome design. A radome is to be designed for the nose of an aircraft to protect an X-band weather radar operating between 8.5 and 10.3 GHz. A new type of foam

[65]H. C. Rhim and O. Buyukozturk, Electromagnetic properties of concrete at microwave frequency range, *ACI Materials Journal*, 95(3), pp. 262–271, May–June 1998.

[66]C. L. Holloway and E. F. Kuester, A low frequency model for wedge or pyramid absorber arrays—II: Computed and measured results, *IEEE Trans. on Electromagnetic Compatibility*, 36(4), pp. 307–313, November 1994.

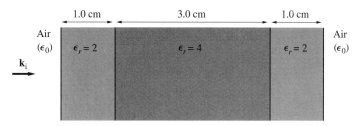

FIGURE 3.51.   Multilayered dielectric. Problem 3-17.

material with $\epsilon_r = 2$ (assume lossless) is chosen for the design. (a) Assuming a flat planar radome, determine the minimum thickness of the foam that will give no reflections at the center frequency of the band. (b) Using the thickness found in part (a), what percentage of the incident power is reflected at each end of the operating frequency band? (c) A thin layer of a different material ($\epsilon_r = 4.1$, $\tan \delta_c = 0.04$, thickness 0.25 mm) is added on one side of the radome designed in part (a) to protect the radome from rain erosion. What percent of the incident power is reflected at the center frequency?

3-17. Transmission through a multilayered dielectric. (a) Find the three lowest frequencies at which all of the incident power would be transmitted through the three-layer structure shown in Figure 3.51. The permeability of all three media is μ_0. (b) If complete transmission is required for *any* thickness of the center medium, what is the lowest usable frequency? (c) Find the bandwidth of the transmission, defined as the range between the two lowest percentage values adjacent to and on either side of the frequency found in (b). Also find the lowest percentage values of transmission. (d) Why does the reflection from multiply coated optical lenses tend to be purple in color?

3-18. Glass slab. Consider a 1-cm thick slab of crown glass, with index of refraction $n = 1.52$. (a) If a beam of visible light at 500 nm is normally incident from one side of the slab, what percentage of the incident power transmits to the other side? (b) Repeat for 400 and 600 nm.

3-19. Refractive index of a liquid. To measure the refractive index of a liquid, a container is designed as shown in Figure 3.52 to hold the liquid sample.[67] Consider a container made of Teflon ($\epsilon_r \simeq 2.08$ at 10 GHz) with wall thickness of $L_1 \simeq 1.04$ cm on each side. A liquid with refractive index n is poured inside the container's compartment with thickness $L_2 \simeq 1.49$ cm. When a 10-GHz plane wave is normally incident from one side of the container, the effective reflection coefficient at that side is measured to be $\Gamma_{\text{eff}} \simeq -0.39$. (a) Find the refractive index of the liquid (assume lossless case). (b) Recalculate Γ_{eff} at 20 GHz (assume the same material properties apply). (c) Repeat part (b) at 5 GHz.

3-20. Antireflection (AR) coating on a glass slab. A beam of light is normally incident on one side of a 1-cm thick slab of flint glass (assume $n = 1.86$) at 550 nm. (a) What percentage of the incident power reflects back? (b) To minimize reflections, the glass is coated with a thin layer of antireflection coating material on both sides. The material

---

[67]D. Kralj and L. Carin, Wideband dispersion measurements of water in reflection and transmission, *IEEE Trans. Microwave Theory and Techniques,* 42(4), pp. 553–557, April 1994.

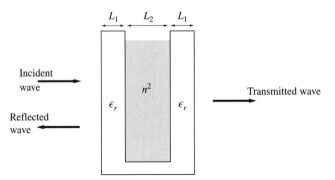

FIGURE 3.52. **Refractive index of a liquid.** Liquid with unknown index of refraction n in a container of known dielectric constant ϵ_r. Problem 3-19.

chosen is magnesium fluoride (MgF$_2$), which has a refractive index around 1.38 at 550 nm. Find the approximate thickness of each coating layer of MgF$_2$ needed. (c) If a beam of light at 400 nm is normally incident on the coated glass with the thickness of the coating layers found in part (b), what percentage of the incident power reflects back? (d) Repeat part (c) for a light beam at 700 nm.

3-21. **AR coating on a glass slab.** A 1-cm thick slab of flint glass ($n = 1.86$) is to be coated only on one side so that when a beam of light is incident on the uncoated side, a sample of the light beam that reflects back from that side can be used to monitor the power of the incident beam. Assuming the light beam to be normally incident at 550 nm and the coating material used on the other side to be MgF$_2$, calculate the percentage of the incident power that reflects back.

3-22. **Infrared antireflection coating.** To minimize reflections at the air–germanium interface in the infrared frequency spectrum, a coating material with an index of refraction of 2.04 is introduced as shown in Figure 3.53. (a) If the thickness of the coating material is adjusted to be a quarter-wavelength in the coating material for a free-space wavelength of 4 μm, find the effective reflection coefficient at free-space wavelengths of 4 and 8 μm. (b) Repeat part (a) if the thickness of the coating material is adjusted to be a quarter-wavelength in the coating material at 8 μm.

3-23. **Superwide infrared antireflection coating.** A wideband antireflection coating system as shown in Figure 3.54 is designed to be used between air and germanium at infrared frequencies. If the thicknesses of the coating layers are each quarter wavelength for operation at $\lambda_1 = 3.5$ μm, find and sketch the magnitude of the effective reflection coefficient over the range from 3 to 12 μm.

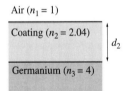

FIGURE 3.53. **Infrared antireflection coating.** Problem 3-22.

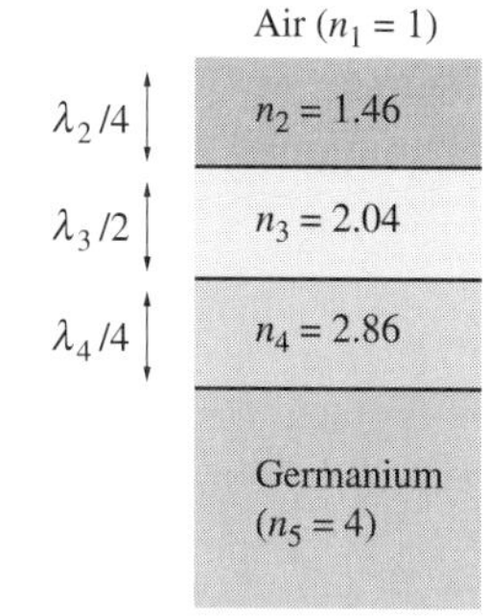

Air ($n_1 = 1$)

$\lambda_2/4$ $n_2 = 1.46$

$\lambda_3/2$ $n_3 = 2.04$

$\lambda_4/4$ $n_4 = 2.86$

Germanium
($n_5 = 4$)

FIGURE 3.54.   **Infrared antireflection coating.** Problem 3-23.

3-24. A snow-covered glacier. Glaciers are huge masses of ice formed in the cold polar regions and in high mountains. Most glaciers range in thickness from about 100 m to 3000 m. In Antarctica, the deepest ice on the polar plateau is 4.7 km. Consider a large glacier in Alaska covered with a layer of snow of 1 m thickness during late winter. A radar signal operating at 56 MHz is normally incident from air onto the air–snow interface. Assume both the snow and the ice to be lossless; for ice, $\epsilon_r = 3.2$, and for snow, ϵ_r can vary between 1.2 and 1.8. Assuming both the snow and the ice to be homogeneous and the ice to be semi-infinite in extent, calculate the reflection coefficient at the air–snow interface for three different permittivity values of snow: $\epsilon_r = 1.2$, 1.5, and 1.8. For which case is the snow layer most transparent (invisible) to the radar signal? Why?

3-25. Minimum ice thickness. Consider a 500-MHz uniform plane wave radiated by an aircraft radar normally incident on a freshwater ($\epsilon_r = 88$) lake covered with a layer of ice ($\epsilon_r = 3.2$), as shown in Figure 3.55. (a) Find the minimum thickness of the ice such that the reflected wave has maximum strength. Assume the lake water to be very deep. (b) What is the ratio of the amplitudes of the reflected and incident electric fields?

3-26. A snow–ice-covered lake. An interior lake in Alaska can be 30 to 100 m deep and is covered with ice and snow on the top, each of which can be about a meter deep in late winter.[68] Consider a 5-GHz C-band radar signal normally incident from air onto the surface of a lake that is 50 m deep, covered with a layer of snow (assume $\epsilon_r = 1.5$) of

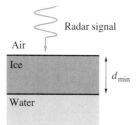

FIGURE 3.55.   **Aircraft radar signal incident on an icy lake surface.** Problem 3-25.

[68] S. A. Arcone, N. E. Yankielun, and E. F. Chacho, Jr., Reflection profiling of Arctic lake ice using microwave FM-CW radar, *IEEE Trans. Geosci. Remote Sensing,* 35(2), pp. 436–443, March 1997.

60 cm thickness over a layer of ice ($\epsilon_r = 3.2$) of 1.35 m. Assume both the snow and the ice to be lossless. Also assume the lake water, with $\epsilon_{cr} = 68 - j35$ at 5 GHz at 0°C, to be slightly brackish (salty), with an approximate conductivity of $\sigma = 0.01$ S-m^{-1}, and the bottom of the lake to consist of thick silt with $\epsilon_r \approx 50$. (a) Calculate the reflection coefficient at the air interface with and without the snow layer. (b) Repeat part (a) at an X-band radar frequency of 10 GHz. Assume all the other parameters to be the same except for the lake water, $\epsilon_{cr} = 42 - j41$ at 10 GHz and 0°C. Use any approximations possible, with the condition that sufficient justifications are provided.

3-27. Air–water–air interface. A uniform plane wave operating at 2.45 GHz in air is normally incident onto a planar water boundary at 20°C ($\epsilon_r = \epsilon_r' - j\epsilon_r'' = 79 - j11$).[69] (a) Calculate the percentage of the incident power that is transmitted into the water (assume the water region to be semi-infinite in extent). (b) What percentage of the incident power is absorbed in the first 1-cm thick layer of water? (c) If the water layer has a finite thickness of 1 cm with air on the other side (i.e., air–water–air interface), calculate the percentage of the incident power that is absorbed in the water layer and compare it with the result of part (b).

3-28. Air–concrete wall–air. A 900-MHz wireless communication signal is normally incident from one side onto a reinforced concrete wall of thickness d having air on both sides. (a) Find the percentages of the incident power that is reflected back and that is transmitted to the other side of the wall for three different wall thicknesses: 10, 20, and 30 cm. (See Problem 2-17 for data on the properties of reinforced concrete wall.) (b) Repeat part (a) at 1.8 GHz.

3-29. Oblique incidence on a perfect conductor. A 53 W-m^{-2} uniform plane wave in air is obliquely incident on a perfect conductor boundary located at the $z = 0$ plane. The electric field of the incident wave is given by

$$\mathbf{E}_i(x, z) = \hat{\mathbf{y}} E_0 e^{-j4.8\pi x} e^{-j6.4\pi z}$$

(a) Find E_0, f, and θ_i. (b) Find $\mathbf{E}_r$. (c) Find the total electric field $\mathbf{E}_1$ and the nearest positions (with respect to the conductor surface) of its minima and maxima.

3-30. Oblique incidence on a perfect conductor. A parallel-polarized (with respect to the plane of incidence) 30 μW-(cm)$^{-2}$, 1.91-GHz wireless communication signal is incident on a perfect conductor surface located at $y = 0$ at an incidence angle of $\theta_i = 30°$ as shown in Figure 3.56. The signal can be approximated as a uniform plane wave. (a) Write the instantaneous expressions for $\overline{\mathscr{E}}_i(y, z, t)$ and $\overline{\mathscr{H}}_i(y, z, t)$. (b) Find $\mathbf{E}_r(y, z)$

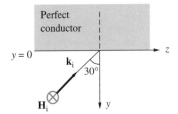

FIGURE 3.56. Oblique incidence on a perfect conductor. Problem 3-30.

[69]W. Fu and A. Metaxas, A mathematical derivation of power penetration depth for thin lossy materials, *J. Microwave Power and Electromagnetic Energy,* Vol. 27, No. 4, pp. 217–222, 1992.

and $\mathbf{H}_r(y, z)$ of the reflected wave. (c) Find the magnitude of the total magnetic field phasor $\mathbf{H}_1(y, z)$ and sketch it as a function of y.

3-31. Plane wave at 45° angle. A plane wave is incident at 45° upon a perfectly conducting surface located at $x = 0$. The plane wave consists of two components as follows:

$$\mathbf{E}_{i\perp} = \hat{\mathbf{y}} E_0 e^{jk(x-z)/\sqrt{2}}$$

$$\mathbf{E}_{i\|} = (\hat{\mathbf{x}} + \hat{\mathbf{z}}) \frac{jE_0}{\sqrt{2}} e^{jk(x-z)/\sqrt{2}}$$

(a) Write the perpendicular and parallel polarization components of the electric fields of the reflected wave, and show that the tangential electric fields satisfy the boundary conditions. (b) What are the polarizations of the incident and the reflected waves?

3-32. Oblique incidence. A uniform plane wave is obliquely incident at an angle θ_i at the interface between two nonmagnetic ($\mu_1 = \mu_2 = \mu_0$) dielectric media as shown in Figure 3.57. The relative permittivity of the second medium is known to be $\epsilon_{2r} = 3$, and the electric field of the incident wave is given by

$$\overline{\mathscr{E}}_i(x, z, t) = \hat{\mathbf{y}} E_0 \cos[12 \times 10^9 t - 40\sqrt{3}(x + z)]$$

(a) Calculate the relative dielectric constant ϵ_{1r} and the angle of incidence θ_i. (b) Write the corresponding expression for the magnetic field of the incident wave (i.e., $\overline{\mathscr{H}}_i(x, z, t)$). (c) Determine the percentage of the incident power that will be transmitted across the interface.

3-33. Reflection from ground. A 6-GHz and 500-μW-m^{-2} microwave communication signal is obliquely incident at $\theta_i = 45°$ onto the ground (assume lossless and $\epsilon_r = 15$) located at $z = 0$. (a) If the incident wave is perpendicularly polarized, write the complete expressions for $\mathbf{E}_i$, $\mathbf{E}_r$, and $\mathbf{E}_t$. (b) Repeat part (a) for a parallel-polarized wave.

3-34. Air–dielectric interface. A uniform plane wave propagating in air has an electric field given by

$$\mathbf{E}_i(x, y) = E_0(0.5\hat{\mathbf{x}} + 0.5\sqrt{3}\hat{\mathbf{y}} - e^{j\pi/2}\hat{\mathbf{z}})e^{-j2\sqrt{3}\pi x + j2\pi y}$$

where E_0 is a real constant. The wave is incident on the planar interface (located at $y = 0$) of a dielectric with $\mu_r = 1$, $\epsilon_r = 3$, as shown in Figure 3.58. (a) What are the values of the wave frequency and the angle of incidence? (b) What is the polarization of the incident wave (i.e., linear, circular, elliptical, right-handed or left-handed)? (c) Write the complete expression for the electric field of the reflected wave in a simplified form. (d) What is the polarization of the reflected wave?

3-35. Brewster angle at the air–water interface. A perpendicularly polarized uniform plane wave is obliquely incident from air onto the surface of a smooth freshwater lake

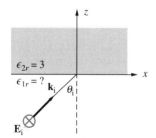

FIGURE 3.57.   Oblique incidence. Problem 3-32.

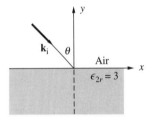

FIGURE 3.58. **Air–dielectric interface.** Figure for 3-34.

(assume lossless with $\epsilon_r \simeq 81$) at the Brewster angle (i.e., $\theta_i = \theta_{iB}$). Calculate the reflection and transmission coefficients.

3-36. Air–ice interface. A 1-W-m^{-2}, 1-GHz radar signal is obliquely incident at an angle $\theta_i = 30°$ from air onto an air–ice interface. (a) Assuming the ice to be lossless and semi-infinite in extent, with $\epsilon_r \simeq 3.17$, calculate the reflection and the transmission coefficients and the average power densities of the reflected and the transmitted waves if the incident wave is perpendicularly polarized. (b) Repeat part (a) for an incident wave that is parallel-polarized. (c) Find the Brewster angle and repeat parts (a) and (b) for an incident wave incident at the Brewster angle.

3-37. Communication over a lake. Consider a ground-to-air communication system as shown in Figure 3.59. The receiver antenna is on an aircraft over a huge lake circling at a horizontal distance of ~8 km from the transmitter antenna as it waits for a landing time. The transmitter antenna is located right at the shore mounted on top of a 50-m tower above the lake surface overlooking the lake and transmits a parallel polarized (with respect to the plane of incidence) signal. The transmitter operates in the VHF band. The pilot of the aircraft experiences noise (sometimes called *ghosting* effect) in his receiver due to the destructive interference between the direct wave and the ground-reflected wave and needs to adjust his altitude to minimize this interference. Assuming the lake to be flat and lossless with $\epsilon_r \simeq 79$, calculate the critical height of the aircraft in order to achieve clear transmission between the transmitter and the receiver.

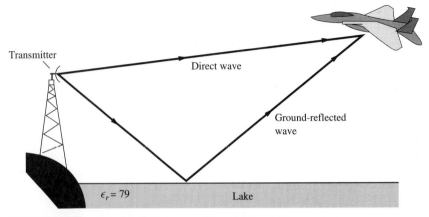

FIGURE 3.59. **Communication over a lake.** Problem 3-37.

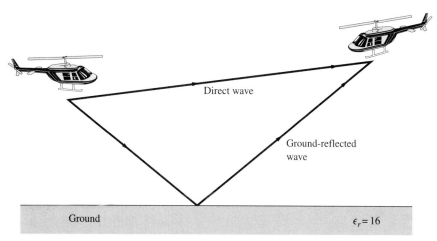

FIGURE 3.60.   Air-to-air communication. Problem 3-38.

3-38.   Air-to-air communication. Consider two helicopters flying in air separated by a horizontal distance of 2 km over a flat terrain as shown in Figure 3.60. The pilot of one of the helicopters, located at an altitude of 100 m, transmits a parallel-polarized (with respect to the plane of incidence) VHF-band signal (assume 200 MHz) to communicate with the other helicopter. The pilot of the other helicopter needs to adjust her altitude to eliminate the noise on her receiver due to the interference of the ground-reflected wave with the direct wave. (a) Assuming the ground to be homogeneous and lossless with $\epsilon_r \simeq 16$, find the critical altitude of the receiver helicopter in order to minimize this interference. (b) Consider another scenario when both helicopters are at 250-m altitude. In this case, what should be the horizontal separation distance between the aircraft in order to achieve clear signal transmission? (c) Repeat both (a) and (b) for the case in which the helicopters try to land at a remote site in Alaska where the ground is covered with permafrost (assume $\epsilon_r \simeq 4$).

3-39.   Dry asphalt roads acting as mayfly traps? Adult mayflies have only a few hours in which to find a mate and reproduce. Direct sunlight reflected off the surface of water is strongly polarized in the horizontal plane (i.e., parallel polarized), and many water-dwelling insects, including mayflies, use this reflected polarized light to identify open stretches of water where they can lay their eggs during their brief mating period. However, researchers discovered that light reflected from dry asphalt roads is also horizontally polarized and visually deceives mayflies into laying their eggs on roads instead of in rivers.[70] The higher the degree of polarization of the reflected light, the more attractive it is for mayflies. Hence, mayflies swarming, mating, and egg-laying on asphalt roads are predominantly deceived by and attracted to the asphalt surface because the largely horizontally polarized reflected light imitates a water surface. Note that

[70]G. Kriska, G. Horvath, and S. Andrikovics, Why do mayflies lay their eggs *en masse* on dry asphalt roads? Water-imitating polarized light reflected from asphalt attracts Ephemeroptera, *The Journal of Experimental Biology,* 201, pp. 2273–2286, 1998; J. Copley, Tar babies: Swarms of mayflies are laying their eggs on roads rather than rivers, *New Scientist,* p. 16, July 25, 1998.

although sunlight has mixed polarization and is incident on the asphalt over a range of angles, the polarization of the reflected light is predominantly horizontal because of the deep minimum for $\rho_\parallel$ in the vicinity of the Brewster angle (see Figure 3.30). (a) The reflected light from asphalt is almost 100% horizontally polarized when the light is incident on the asphalt surface at an incidence angle of about 57.5°. Calculate the effective refractive index of asphalt. (b) Assuming the refractive index of water to be $n_w \simeq 1.33$, find the angle at which the reflected light from the water surface is 100% horizontally polarized.

3-40. Total internal reflection. A uniform plane wave with a magnetic field given by

$$\overline{\mathcal{H}}(y, z, t) = \hat{\mathbf{x}} H_0 \cos[9\pi \times 10^9 t - 45\pi(y + \sqrt{3}z)]$$

is obliquely incident at an interface at $z = 0$ separating two nonmagnetic lossless media as shown in Figure 3.61. (a) Calculate the relative dielectric constant ϵ_{1r} of medium 1 and the angle of incidence θ_i. (b) Find the maximum value of the relative dielectric constant ϵ_{2r} of medium 2 for total internal reflection to occur. (c) Is it possible to achieve total transmission by adjusting the incidence angle? If yes, use the maximum value of ϵ_{2r} found in part (b) to determine the incidence angle at which total transmission would occur.

3-41. Reflection from prisms. Consider the various right-angled prisms shown in Figure 3.62. (a) What is the minimum index of refraction n_1 necessary in each case for there to be no time-average power transmitted across the hypotenuse when the prisms

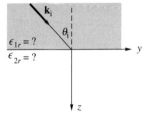

FIGURE 3.61. **Total internal reflection.** Problem 3-40.

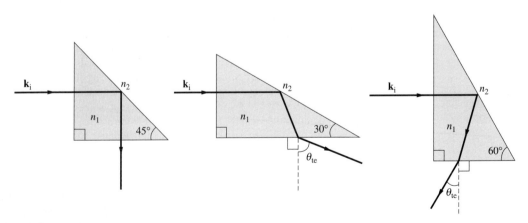

FIGURE 3.62. **Reflection from prisms.** Problem 3-41.

are (i) in free space, (ii) in water (assume $n \simeq 1.33$). (b) At these refractive index values (found in (a)), what are the exit angles θ_{te}?

3-42. MgF$_2$ prism. A 45°–90°–45° prism is constructed from MgF$_2$ ($n = 1.38$) to be used to turn a light beam around by 90° by internal reflection at its hypotenuse. Does the light beam exit at the hypotenuse, and if so, what is its exit angle?

3-43. Refractive index of a prism. An experiment is designed to measure the refractive index of a prism using the principle of total internal reflection.[71] In this experiment, a plane-polarized, collimated, monochromatic beam of light is obliquely incident from one side of the prism at an incidence angle of θ_i as shown in Figure 3.63. The incidence angle θ_i of the beam is adjusted to a critical value ψ_c such that the incidence angle on the other side of the prism is equal to the critical angle of incidence θ_{ic}. Thus, by measuring the refracting angle A of the prism and the critical incidence angle ψ_c, the refractive index n_p of the prism can be calculated. (a) Show that

$$n_p^2 = 1 + \left(\frac{\sin \psi_c + \cos A}{\sin A} \right)^2$$

(b) For a prism under test, the refracting angle of the prism and the critical incidence angle adjusted are measured to be $A = 60°$ and $\psi_c = 42°$ respectively. Calculate the refractive index n_p of this prism.

3-44. Right-angle prism. A new technique is proposed to measure small angles in optical systems using right-angle prisms.[72] Consider a 45°–90°–45° prism as shown in Figure 3.64 made of glass having a refractive index of $n \simeq 1.515$ to be used in an experiment. A light beam that is obliquely incident at an incidence angle θ_1 on the entrance face undergoes reflection and refraction at the entrance face, the hypotenuse face, and the exit face of the prism. (a) For an incidence angle of $\theta_1 = 30°$, find the exit angles θ_4 and θ_6. (b) Find the critical incidence angle θ_1 of the incident beam on the entrance face that results in no transmission at the hypotenuse face. (c) What happens to the critical angle found in part (b) when the hypotenuse face of the prism is coated with an antireflection coating (single or multiple layers)? (d) Repeat parts (a) and (b) if the prism is made of a different type of glass with $n \simeq 1.845$.

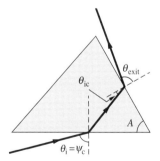

FIGURE 3.63. **Refractive index of a prism.** Problem 3-43.

[71] S. P. Talim, Measurement of the refractive index of a prism by a critical angle method, *Opt. Acta*, 25(2), pp. 157–165, 1978.

[72] P. S. Huang and J. Ni, Angle measurement based on the internal-reflection effect and the use of right-angle prisms, *Appl. Optics*, 34(22), pp. 4976–4981, August 1995.

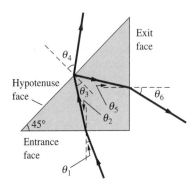

FIGURE 3.64. **Right-angle prism.** Problem 3-44.

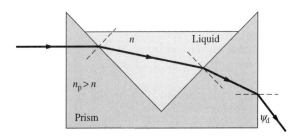

FIGURE 3.65. **A V-shaped prism.** Problem 3-45.

3-45. A V-shaped prism. A V-shaped right-angled prism is designed to measure accurately the refractive index of liquids[73] as shown in Figure 3.65. A laser beam normally incident from one side is refracted at the inclined faces, between which the liquid sample under test is placed, and leaves the prism on the other side at a deflection angle ψ. (a) A liquid of known refractive index $n \simeq 1.512$ is placed in the top compartment of the prism with $n_p \simeq 1.628$. Find the deflection angle ψ_d. (b) Using the same prism, the deflection angle for a different liquid with unknown refractive index is approximately measured to be $\psi \simeq 55.5°$. Calculate the refractive index n of this liquid.

3-46. Turning a perfect corner. An interesting optical phenomenon is the invariance of the angle between the incoming and outgoing light rays passing through a right-angle prism with a silver-coated hypotenuse.[74] For the isosceles right-angle prism shown in Figure 3.66, the incident ray enters at side AB and exits at side AC after being reflected twice and refracted twice. Show that the total deviation angle ψ_d between the incident and the exiting rays is exactly 90°.

3-47. Air–oil–water. Consider a layer of oil (assume $n \simeq 1.6$) about 5 mm thick floating over a body of water ($n \simeq 1.33$). (a) If a light ray is obliquely incident from air onto the oil surface, find the range of incidence angles (if any) that results in total internal reflection at the oil–water interface. (b) If a light ray is obliquely incident from water

[73]E. Moreels, C. de Greef, and R. Finsy, Laser light refractometer, *Appl. Optics,* 23(17), pp. 3010–3013, September 1984.

[74]S. Taylor and E. Hafner, Turning a perfect corner, *Am. J. Phys.,* 47(1), pp. 113–114, January 1979.

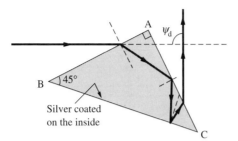

FIGURE 3.66.   Turning a perfect corner. Problem 3-46.

onto the oil surface, find the range of incidence angles (if any) that results in total internal reflection at the oil–air interface.

3-48. An in-line Brewster angle prism. Brewster angle prisms are optical elements that use light at the polarizing angle to obtain perfect transmission of parallel polarized light. An in-line Brewster angle polarizing prism is designed[75] as shown in Figure 3.67, to polarize a light beam without changing its direction. Consider an unpolarized light beam incident on this prism at point A at an incidence angle of $\theta_i = 31.639°$. Given the prism angles to be $\alpha = 63.278°$ and $\beta = 103.939°$ and the prism refractive index to be $n = 1.623$, (a) determine whether light exits the prism at points B, C, and/or D, (b) find the exit angles θ_B, θ_C, and/or θ_D at these exit points, (c) specify the polarization of each exiting beam, and (d) find the angle between each exiting beam and the incident beam. Discuss your results in terms of the stated purpose of this particular prism design.

3-49. V-shaped prism polarizer. A symmetrical three-reflection silicon (Si) polarizer prism is designed as shown in Figure 3.68a, based on the Brewster-angle internal reflection that occurs at the base of the prism.[76] (a) In Figure 3.68a, the prism angle A is adjusted for use at 1.3 μm light wave communication wavelength (the refractive index of silicon at 1.30 μm is $n_{Si} \approx 3.5053$), to $A \approx 52.9613°$. Assuming the incident light

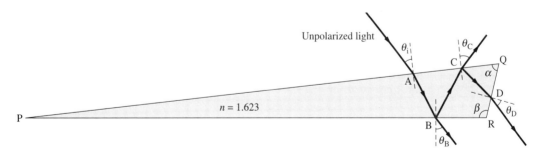

FIGURE 3.67.   **In-line polarizing prism.** Problem 3-48.

[75]R. D. Tewari, A. M. Ghodgaonkar, and V. Bhattacharyya, Modified polarizing prism, *Optics and Laser Technology*, 30, pp. 63–70, 1998.

[76]R. M. A. Azzam and M. M. K. Howlader, Silicon-based polarization optics for the 1.30 and 1.55 μm communication wavelengths, *Journal of Lightwave Technology*, 14(5), pp. 873–878, May 1996.

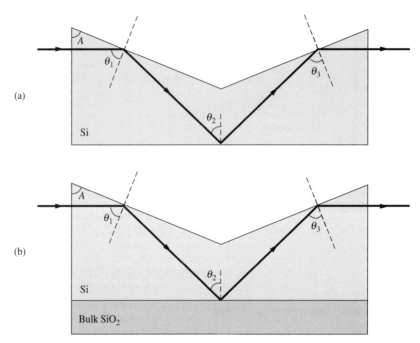

FIGURE 3.68. **V-shaped prism polarizer.** Problem 3-49.

beam entering the prism on the left side to be unpolarized, find the polarization of the beam exiting on the right side. (b) Find the new value of A for the prism to be used as a polarizer at 1.55-μm wavelength (the refractive index of silicon at 1.55 μm is $n_{Si} \simeq 3.4777$). (c) Another interesting design is to coat the base of the silicon prism with silicon dioxide (SiO_2) as shown in Figure 3.68b. At 1.3 μm, the prism angle is adjusted to $A \simeq 56.217°$. Find the refractive index of SiO_2 at 1.3 μm. (d) Repeat part (c) at 1.55 μm when the prism angle is adjusted to $A \simeq 56.277°$.

3-50. Limestone wall versus brick wall. Consider two buildings, one with limestone ($\epsilon_r = 7.51, \sigma = 0.03$ S-m^{-1}) exterior walls and the other with brick ($\epsilon_r = 4.44, \sigma = 0.01$ S-m^{-1}) exterior walls, with the material properties cited measured[77] at 4 GHz. These walls represent some of the typical building surfaces that affect the propagation of mobile radio signals. (a) Assuming the walls to be lossless, semi-infinite in extent, and neglecting the roughness of their surfaces, calculate the reflection coefficients at each surface for both perpendicular and parallel polarizations at three different angles of incidence of $\theta_i = 30°, 45°,$ and $60°$ and compare the results. (b) What is the pseudo Brewster angle in each case?

[77]O. Landron, M. Feuerstein, T. S. Rappaport, A comparison of theoretical and empirical reflection coefficients for typical exterior wall surfaces in a mobile radio environment, *IEEE Transactions on Antennas and Propagation*, 44(3), pp. 341–351, March 1996.

3-51. Refractive index of concrete. Knowledge of the dielectric properties of construction material is important because the reflection and transmission characteristics of buildings and rooms are governed by these properties. The complex refractive index of a plain concrete plate mixed from Portland cement, gravel, sand, and water was measured at 57.5 GHz for use in designing and testing millimeter-wave communication systems.[78] The measured refractive index of the concrete 14 months after concreting was $n = 2.55 - j0.084$. (a) Using the measured values for concrete, calculate and sketch the magnitude of the reflection coefficient at the air–concrete interface at 57.5 GHz as a function of the incidence angle varying between $0°$ to $90°$ for both perpendicular and parallel polarization cases. Assume the concrete to be semi-infinite in extent. (b) For a 5-cm thick concrete wall having air on both sides, calculate the magnitude of the normal incidence reflection coefficient at 57.5 GHz and compare it with the result of part (a). (c) Repeat part (b) for a thickness of 10 cm.

3-52. Oblique incidence on a multiple dielectric interface. Suppose that a parallel polarized uniform plane wave is incident, in air, on a glass slab (assume $\epsilon_r = 2$) of thickness d, as shown in Figure 3.69. The angle of incidence θ_i is chosen to be the air–glass Brewster angle so that there is no reflection at the first interface ($z = 0$). Find the complete expression for the electric field phasor of the wave transmitted into air ($z > d$), i.e., $\mathbf{E}_t(x, z)$. *Hint:* First find the reflection coefficient at the $z = d$ interface.

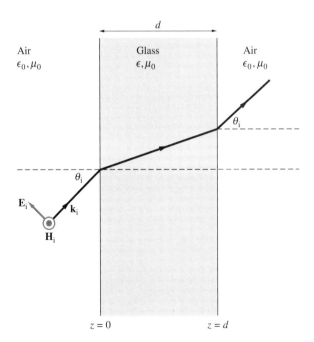

FIGURE 3.69. **Oblique incidence on a multiple interface.** Problem 3-52.

[78]K. Sato, T. Manabe, J. Polivka, T. Ihara, Y. Kasashima, and K. Yamaki, Measurement of complex refractive index of concrete at 57.5 GHz, *IEEE Transactions on Antennas and Propagation*, 44(1), pp. 35–40, January 1996.

3-53. Oblique incidence on a multiple dielectric interface. A perpendicularly polarized uniform plane wave propagating in air is incident obliquely (at an angle θ_i) on a structure consisting of two lossless and nonmagnetic dielectrics: a coating layer with permittivity ϵ_{1r} and thickness d coated on another dielectric with permittivity ϵ_{2r} and of infinite thickness, as shown in Figure 3.70. Derive expressions for the effective reflection and transmission coefficients (Γ_{eff} and $\mathcal{T}_{\text{eff}}$) in terms of parameters of the media ($\epsilon_{1r}, \epsilon_{2r}$), the angle of incidence θ_i, and the slab thickness d.

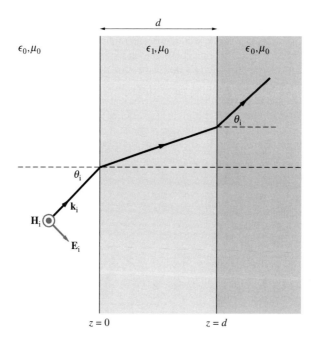

FIGURE 3.70. **Oblique incidence on a multiple interface.** Problem 3-53.

Parallel-Plate and Dielectric Slab Waveguides

4.1 Waves between Parallel Metal Plates

4.2 Dielectric Waveguides

4.3 Wave Velocities and Waveguide Dispersion

4.4 Summary

4.5 Problems

When an electromagnetic wave propagates out from a point source through empty space, it does so as a spherical wave, traveling with equal speed in all directions. At large distances ($r \gg \lambda$) from its source, a spherical electromagnetic wave is well approximated as a uniform plane wave; in Chapter 2, we considered the propagation of uniform plane waves in unbounded media, and in Chapter 3 their reflections from planar interfaces between different simple media. We now consider the guiding of electromagnetic waves from one point to another by means of metallic or dielectric boundaries.

In a wide range of electromagnetic applications it is necessary to convey electromagnetic waves efficiently from one point to another so as to transmit energy or information (signals). The energy or information can be conveyed by means of *unguided* electromagnetic waves propagating in free space. In unguided transmission of electromagnetic energy, the characteristics of the transmitting antennas determine the intensity of the waves radiated in different directions. Some applications, such as line-of-sight microwave (radio, television, and satellite) links, optical links (bar-code readers, infrared smoke detectors, infrared remote controllers), high-power lasers (in materials processing, laser printers, and surgery), and radar, require directional transmission of waves, whereas others, such as navigation, radio communication, and radio and TV broadcast, use omnidirectional transmissions. In addition to the spherical spreading of the available energy, broadcast applications do not represent efficient use of the electromagnetic spectrum, since only one transmitter can operate at a given frequency in a given region.

In many other applications it is necessary to transmit electromagnetic energy or signals by using waves that are *guided* by metallic or dielectric structures to minimize radiation losses and unnecessary spreading of the energy. Guided transmission of waves also facilitates the use of one frequency band to convey multiple signals simultaneously, because most waveguides can be laid close to one another without significant coupling between them. Electromagnetic waves are generally guided by confining them in two dimensions and allowing them to propagate freely in the third dimension. The simplest example is a coaxial line, which completely confines the energy to the region between the inner and outer conductors. However, the principle of operation of the two-wire line is identical; although the electric and magnetic fields are not completely confined in an enclosure, their amplitudes decrease rapidly with transverse distance away from the conductors, so most of the electromagnetic energy is confined to the region in the immediate vicinity of the conductors. The surfaces of the conducting wires provide boundaries on which the electric field lines can terminate, so that the wave can propagate as a plane wave following the conductor from end to end. A power transmission line and a telephone wire are examples of such two-wire lines.

We discuss guiding of electromagnetic waves and different types of waveguides in this and the following chapter, starting with two of the simplest structures. Wave-guiding structures of all types can be called waveguides, although the term *wave-guide* is sometimes used specifically for the particular guiding structure consisting of a metallic tube of the kind considered in Chapter 5. The two guiding systems explicitly considered in this chapter are the parallel-plate waveguide and the dielectric slab waveguide. Both of these are planar structures, guiding electromagnetic waves in one direction (typically taken to be the z direction) by confining the wave energy in one of the transverse directions (e.g., the x direction) while being of infinite extent[1] in the other (e.g., the y direction). We confine our attention in this chapter to these two types of waveguides, using them as examples by means of which we introduce the fundamental principles of guiding of waves by metallic (parallel-plate wave-guide) and dielectric (slab waveguide) boundaries. Primarily because of their infinite extent in one of the two transverse dimensions, these waveguides can be analyzed in a relatively straightforward manner, at a level appropriate for this book. These two waveguides are examples of a class of planar waveguide structures that can be manufactured relatively easily (hence at low cost) using planar integrated circuit technologies. Other planar waveguides, such as the stripline and the microstrip line, may in fact be more commonly utilized in practice than the parallel-plate and dielectric slab waveguides studied here. However, the analysis of these structures is considerably more involved and typically requires approximate or numerical treatments

---

[1]Practical parallel-plate and dielectric slab waveguides necessarily also have a finite extent in the y direction; however, typically the extent of the waveguide in the y direction is much larger than the thickness of the waveguide in the x direction, so the fringing fields can be neglected, and the field structure, propagation characteristics, and attenuation rates can be evaluated by assuming the waveguide to be of infinite extent.

whose description is not appropriate for our coverage of fundamental principles and is thus beyond the scope of this book.[2]

The parallel-plate waveguide is arguably the simplest example of transmission systems that utilize two separate conductors for guiding electromagnetic waves. All of the methodologies used in its analysis are directly applicable to other transmission systems that use two separate conductors symmetrically arranged across the transverse plane (e.g., the coaxial line, two-wire line, twisted-pair, striplines), which can support the propagation of transverse electromagnetic (TEM) waves. At the same time, the well-developed techniques for analysis of voltage and current waves on transmission lines[3] are directly applicable to the parallel-plate waveguide or any other two-conductor structures, as long as the waveguide voltage and current are properly defined (see Section 4.1.5) in terms of the propagating electric and magnetic fields. In most cases, TEM waves guided by two-conductor systems are not *uniform* plane waves, because the amplitudes of the wave field components vary over the planar constant-phase fronts. For example, the amplitudes decrease with distance away from the conductors for the two-wire line; hence the confinement of the energy to the vicinity of the conductors, which "guide" the wave.

In addition to supporting TEM waves, the parallel-plate waveguide also allows the propagation of more complicated electromagnetic waves, for which the wave magnetic or the wave electric field has a component in the direction of propagation. The presence of magnetic or electric field components along the propagation direction requires[4] that the wave field components vary with position along the planar phase fronts. Such waves are respectively called transverse electric (TE) or transverse magnetic (TM) waves. The different types of possible wave solutions are referred to as *modes*, analogous to the different ways (modes) in which a spring with both ends held fixed may vibrate. TE or TM modes are the only types of electromagnetic waves that can propagate in metallic tube waveguides, which confine the wave energy in both of the transverse directions. Our discussion in this chapter of TE and TM modes for the parallel-plate waveguides thus provides the basis for the analysis of these cylindrical metallic tube waveguides covered in Chapter 5.

Dielectric slab waveguides of one form or another are the workhorses of integrated optics,[5] primarily concerned with constructing optical devices and networks on substrates. The field of integrated optics (sometimes also called integrated photonics) combines optics and electronics on a single substrate, often with integrated components having dimensions of the order of the wavelength of light. The dielectric slab waveguide is important in view of its common use as the means by which light

---

[2]For an excellent and brief treatment, see Section 8.6 of S. Ramo, J. R. Whinnery, and T. Van Duzer, *Fields and Waves in Communications Electronics*, 3rd ed., Wiley, 1994.

[3]See Chapters 2 and 3 of U. S. Inan and A. S. Inan, *Engineering Electromagnetics*, Addison Wesley Longman, 1999.

[4]Otherwise we cannot have $\nabla \cdot \mathbf{B} = 0$ or $\nabla \cdot \mathbf{D} = 0$.

[5]R. G. Hunsberger, *Integrated Optics: Theory and Technology*, 4th ed., Springer-Verlag, New York, 1995.

is transferred between components of an integrated optics network. In addition, propagation of light waves in a dielectric slab waveguide is fundamentally similar to the guiding of light in cylindrical dielectric waveguide structures (optical fibers), where the electromagnetic energy is confined in both of the transverse dimensions—a relatively simple form of which is discussed in Section 5.2.3. Thus, our analyses of the dielectric slab waveguide also serve to illustrate and introduce the basic principles of guiding of light waves within dielectric boundaries, with much broader application to the field of fiber optics, beyond the scope of this book but well covered elsewhere.[6]

The mathematical foundation of guided electromagnetic waves and modern transmission line theory was laid down by O. Heaviside, who wrote a series of 47 papers between 1885 and 1887, introducing the vector notation currently in use today and many other concepts that have since proven to be so useful in electrical engineering. Most of the early experimental work on guided waves was carried out by H. Hertz; both Hertz (see biography in section 6.5) and Heaviside exclusively considered two-conductor transmission lines, with Heaviside even declaring that two conductors were definitely necessary for transmission of electromagnetic energy.[7] (Heaviside was wrong in this prophecy, since electromagnetic waves propagate very effectively in hollow single metallic cylindrical guides, as discussed in detail in Chapter 5.) Nevertheless, the series of papers written by Heaviside and his three-volume work published in 1893 stand out as the mathematical foundations of guided electromagnetic waves.

Heaviside, Oliver   *(b. May 18, 1850, London; d. February 3, 1925, Paignton, Devonshire) Heaviside was the son of an artist. Like Edison [1847–1931], he had no formal education past the elementary level, became a telegrapher, and was hampered by deafness. With the encouragement of his uncle (by marriage), Charles Wheatstone [1802–1875], he inaugurated a program of self-education that succeeded admirably. He could concentrate on it more, since he never married and lived with his parents till they died. He did important work in applying mathematics to the study of electrical circuits and extended Maxwell's [1831–1879] work on electromagnetic theory. Perhaps because of his unorthodox education, he made use of mathematical notations and methods of his own that were greeted by the other (and lesser) physicists with disdain. For instance, he used vector notation where many physicists, notably Kelvin [1824–1907], did*

---

[6]For an up-to-date coverage of technology and applications see J. C. Palais, *Fiber Optic Communications*, 4th ed., Prentice-Hall, 1998; for coverage of fundamentals of guiding of light waves in different modes, see A. W. Snyder and J. D. Love, *Optical Waveguide Theory*, Chapman and Hall, London, 1983, and D. B. Keck, in *Fundamentals of Optical Fiber Communications* (M. L. Barnoski, Ed.), Academic Press, San Diego, California, 1976.

[7]On p. 399 of his three-volume work *Electromagnetic Theory,* Benn, London, 1893, or on p. 100 of the reprinted edition by Dover, 1950, Heaviside stated that guided waves needed "two leads as a pair of parallel wires; or if but one be used, there is the earth, or something equivalent, to make another." For an interesting discussion of the early history, see J. H. Bryant, Coaxial transmission lines, related two-conductor transmission lines, connectors, and components: A U.S. historical perspective, *IEEE Trans. Microwave Theory and Techniques,* 32(9), pp. 970–983, September 1984.

*not. It was Kelvin, however, who first brought Heaviside's work to the notice
of the scientific community. Heaviside was forced to publish his papers at his
own expense because of their unorthodoxy. After the discovery of radio waves
by Hertz [1857–1894], Heaviside applied his mathematics to wave motion and
published a large three-volume work,* Electromagnetic Theory. *In it he predicted
the existence of an electrically charged layer in the upper atmosphere just months
after Kennelly [1861–1939] had done so; this layer is now sometimes referred
to as the Kennelly–Heaviside layer. Heaviside spent his last years poor and
alone and died in a nursing home. [Adapted with permission from I. Asimov,*
Biographical Encyclopedia of Science and Technology, *Doubleday, 1982].*

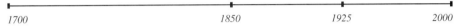

| 1700 | 1850 | 1925 | 2000 |

Our purpose in this and the next chapter is not necessarily to provide comprehensive coverage of electromagnetic wave-guiding systems that are in practical use
at the present time but rather to provide adequate coverage of the fundamental principles of the guiding of waves, using parallel-plate, dielectric slab, rectangular, and
circular waveguides, coaxial lines, and optical fibers as examples. The practical use
of a wave-guiding system in any given application is determined by the frequency
range of operation, available technology, the particular needs of the application in
hand, and engineering feasibility and cost. For example, two-wire (power lines) and
twisted-pair (telephone lines) transmission lines work well at relatively low frequencies, but radiation effects (power losses or interference with neighboring circuits)
become intolerable at higher frequencies. At VHF/UHF frequencies, coaxial lines
or the encapsulated two-wire lines (i.e., the household TV antenna lead wire) are
equally usable, although the latter is generally less desirable because it is an un-
shielded guiding structure. Nevertheless, the encapsulated two-wire line is relatively
inexpensive to manufacture and is widely used in the lower range of the microwave
spectrum. Coaxial lines are more expensive to manufacture but provide excellent
shielding and are thus used in applications where it is important to minimize interference.

Coaxial lines can in principle operate at any frequency, but they become increasingly lossy and difficult to construct for use at higher frequencies. At frequencies
above about 3 GHz, hollow metallic tubes have more practical dimensions, provide
much lower loss, and have substantially larger power capacity. Although the possibility of electromagnetic wave propagation in hollow tubes was first noted[8] in 1897,
practical uses of these types of waveguides were not developed until the 1930s, when
their use with microwave radar and communications systems emerged as a crucially
important application. For some time, coaxial lines and these hollow rectangular and
circular metal tubes were the only guiding systems used at frequencies in the 10- to
100-GHz range. However, in spite of their many desirable properties, waveguides
are often bulky and expensive to manufacture due to the required precision, and their

---

[8]Lord Rayleigh, On the passage of electric waves through tubes, or the vibrations of dielectric cylinders, *Phil. Mag.*, vol. XLIII, pp. 125–132, 1897.

applicability at the higher end of the microwave spectrum has thus been limited. In the 1970s, the stripline and the microstrip line[9] emerged as a practical alternative to waveguides in some applications, primarily because these structures could be manufactured relatively easily (hence at low cost) using planar integrated circuit technologies. On the other hand, recent work on silicon-based metallic waveguides operating at submillimeter frequencies[10] indicates that newly emerging micromachining technologies may significantly extend the upper end of the usable frequency range of hollow metallic waveguides, while also providing for manufacturing ease. Our relatively detailed coverage of propagation of electromagnetic waves in hollow cylindrical metallic structures (Chapter 5) is relevant in view of these current and emerging applications—in addition to serving as a vehicle by which we illustrate the fundamental principles of guiding of waves by metallic boundaries that confine the wave energy in two dimensions.

At optical frequencies ($>10^{12}$ Hz), guiding structures made of conductors become impractical, but dielectric waveguides and optical fibers provide efficient guiding of electromagnetic energy.[11] The practical utilization of optical waveguides is similarly determined by and has evolved on the basis of[12] the available technology and engineering cost tradeoffs. In 1870, British physicist John Tyndall demonstrated that light could be contained within and guided along a thin stream of water. Demonstrations of the channeling of light along other dielectrics, notably glass light pipes and threads of quartz, were soon to follow, taking advantage of the same principle: total internal reflection. The possibilities of long-distance transmission in optical fibers first emerged in the 1960s, shortly after the advent of the laser. The laser provided a coherent source of optical radiation ($\sim 10^{15}$ Hz) suitable for use as a carrier of information, much like radio waves or microwaves (10^9–10^{11} Hz) but allowing much larger bandwidth and information-carrying capacity. However, the losses in glasses available at the time were initially quite high (tens of dB per km), so practical use of fibers for long-distance communication had to wait until low-loss materials were developed. Purified glass materials produced using current technology provide attenuation rates[13] of 0.2 dB-(km)$^{-1}$, substantially below that available with metal waveguides. The complete theory of electromagnetic wave propagation in practical optical fibers is beyond the scope of this book; however, the basic principles of guiding of light waves within dielectric boundaries is well illustrated in the context of our coverage of dielectric slab waveguides in this chapter and the dielectric waveguide with circular cross section in Section 5.2.3.

---

[9]T. Itoh (Ed.), *Planar Transmission Line Structures*, IEEE Press, Piscataway, New Jersey, 1987.

[10]W. R. McGrath, C. Walker, M. Yap, Y. -C. Tai, Silicon micromachined waveguides for millimeter-wave and submillimeter-wave frequencies, *IEEE Microwave and Guided Wave Letters*, 3(3), p. 61–63, March 1993.

[11]For a comparative discussion of optical fibers versus metallic waveguides, see Chapter 9 of W. Tomasi, *Advanced Electronic Communication Systems,* Prentice Hall, 1998.

[12]For an interesting discussion of the history, see Chapter 1: From Copper to Glass, in D. J. H. MacLean, *Optical Line Systems*, Wiley, New York, 1996.

[13]P. E. Green, *Fiber Optic Networks*, Prentice Hall, 1993.

This and the next chapter also constitute an important step in our coverage of electromagnetic wave propagation that we started in Chapter 2, where we constrained our coverage to waves in unbounded, simple, and source-free media. Chapter 3 was our first step in removing the first constraint, where we considered the reflection and refraction of waves from planar boundaries. In this and the next chapter, we investigate the propagation of waves in regions bounded by conducting or dielectric structures, while still assuming that the medium within which the waves propagate is simple and source-free. An important aspect of the guided propagation of electromagnetic waves is that the phase velocity of the propagating modes becomes frequency-dependent, as opposed to the frequency-independent character of uniform plane waves in a simple, unbounded, and lossless medium. This property is important for the guided propagation of an information-carrying signal consisting of a band of frequencies (which occurs at the so-called group velocity, introduced and discussed in Section 4.3). The specific conditions on the propagation constant for guided waves considered in this and the next chapter arise from the presence of the metallic or dielectric boundaries. In Chapter 6, on the other hand, we revert back to considering uniform plane waves in an unbounded medium once again as in Chapter 2, but we now allow the medium not to be simple, so that the medium characteristics (ϵ, μ, and σ) can be functions of frequency and/or direction of propagation. In such cases, special conditions on the propagation constant, wavelength, and phase velocity arise as a result of the material properties, rather than the presence of boundaries.

We now proceed to analyze electromagnetic wave propagation in planar waveguides, starting in Section 4.1 with the simplest case of propagation between two parallel plates. This particular type of waveguide supports the TEM mode as well as TE and TM modes, which are in general more complicated. Electromagnetic wave propagation along purely dielectric structures, in particular the dielectric slab waveguide, is covered in Section 4.2, followed by a general discussion in Section 4.3 of wave propagation velocities and dispersion, applicable to all of the different types of waveguides studied in this and the following chapter.

4.1   WAVES BETWEEN PARALLEL METAL PLATES

The simplest type of waveguide consists of two perfectly conducting plates, between which electromagnetic waves of various types can be guided. Accordingly, we start our considerations of guided waves by analyzing the so-called *parallel-plate waveguide,* as shown in Figure 4.1. The plates have an infinite extent in the z direction, in which the waves are to be guided. In practice the conductor plates would have a finite extent b in the y direction, but it is typically much larger than the plate separation a, so the effects of fringing fields at the edges are negligible. For our purposes here, we assume that the conductor plates have infinite extent ($b \gg a$) in the y direction, so the electric and magnetic field do not vary with y.

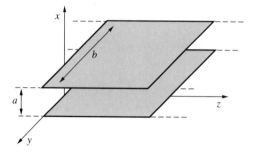

FIGURE 4.1. Parallel-plate waveguide. In an ideal parallel-plate waveguide, the plates are assumed to be infinite in extent in the y and z directions. In practice, the width b of the guide in the y direction is finite but is typically much larger than a. The material between the plates is assumed to be a lossless dielectric.

Just as in the case of waves propagating in unbounded media, the configuration and propagation of the electromagnetic fields in any guiding structure are governed by Maxwell's equations and the wave equations derived from them. However, in the case of metallic waveguides, the solutions of the wave equations are now subject to the following boundary conditions at the conductor surfaces:

$$E_{\text{tangential}} = 0 \qquad H_{\text{normal}} = 0$$

Assuming that the medium between the perfectly conducting plates is source-free, simple (i.e., ϵ and μ are simple constants), and lossless[14] (i.e., $\sigma = 0$ and ϵ is real), the governing equations are the time-harmonic form of Maxwell's equations, namely

$$\nabla \times \mathbf{H} = j\omega\epsilon\mathbf{E} \qquad\qquad [4.1a]$$

$$\nabla \cdot \mathbf{E} = 0 \qquad\qquad [4.1b]$$

$$\nabla \times \mathbf{E} = -j\omega\mu\mathbf{H} \qquad\qquad [4.1c]$$

$$\nabla \cdot \mathbf{H} = 0 \qquad\qquad [4.1d]$$

from which the wave equations for $\mathbf{E}$ and $\mathbf{H}$ can be derived in a straightforward manner following the same procedure as in Section 2.2. We have

$$\nabla^2\mathbf{E} = -\beta^2\mathbf{E} \qquad\qquad [4.2a]$$

$$\nabla^2\mathbf{H} = -\beta^2\mathbf{H} \qquad\qquad [4.2b]$$

where $\beta = \omega\sqrt{\mu\epsilon}$.

Equations [4.1a] and [4.1c] can be written in component form for rectangular coordinates (note that the conductor configuration is rectangular; otherwise, for example for circularly shaped conductors, it is more appropriate to write the equations in terms of cylindrical coordinates r, ϕ, and z) for the region between plates (assumed

---

[14]We shall later relax this requirement and calculate the losses in the dielectric medium using a perturbation solution.

to be a lossless dielectric, with both σ and ϵ'' equal to zero) as

$$\frac{\partial H_z}{\partial y} - \frac{\partial H_y}{\partial z} = j\omega\epsilon E_x \qquad \frac{\partial E_z}{\partial y} - \frac{\partial E_y}{\partial z} = -j\omega\mu H_x$$

$$\underbrace{\frac{\partial H_x}{\partial z} - \frac{\partial H_z}{\partial x} = j\omega\epsilon E_y}_{} \qquad \frac{\partial E_x}{\partial z} - \frac{\partial E_z}{\partial x} = -j\omega\mu H_y \qquad \text{[4.3]}$$

$$\underbrace{\frac{\partial H_y}{\partial x} - \frac{\partial H_x}{\partial y} = j\omega\epsilon E_z}_{\nabla\times\mathbf{H} = j\omega\epsilon\mathbf{E}} \qquad \underbrace{\frac{\partial E_y}{\partial x} - \frac{\partial E_x}{\partial y} = -j\omega\mu H_z}_{\nabla\times\mathbf{E} = -j\omega\mu\mathbf{H}}$$

Note that since the extent of the conductor plates in the y direction is infinite, none of the field quantities vary in the y direction, and as a result, the $\partial(\cdots)/\partial y$ terms in [4.3] are all zero.

We assume propagation in the $+z$ direction, so all field components vary as $e^{-\bar{\gamma}z}$, where $\bar{\gamma} = \bar{\alpha} + j\bar{\beta}$, with $\bar{\alpha}$ and $\bar{\beta}$ being real.[15] In the rest of this chapter, we consider the two special cases for which $\bar{\beta} = 0$ or for which $\bar{\alpha} = 0$. When the phasor fields vary in this manner, the real fields vary as

$$\mathcal{R}e\,\{e^{-\bar{\gamma}z}e^{j\omega t}\} = \begin{cases} e^{-\bar{\alpha}z}\cos(\omega t) & \bar{\gamma} = \bar{\alpha} & \text{Evanescent wave} \\ \cos(\omega t - \bar{\beta}z) & \bar{\gamma} = j\bar{\beta} & \text{Propagating wave} \end{cases}$$

An *evanescent*[16] wave is one that does not propagate but instead exponentially decays with distance.

The wave equation(s) [4.2] can be rewritten as

$$\nabla^2\mathbf{E} = -\omega^2\mu\epsilon\mathbf{E} \qquad \rightarrow \qquad \frac{\partial^2\mathbf{E}}{\partial x^2} + \frac{\partial^2\mathbf{E}}{\partial y^2} + \frac{\partial^2\mathbf{E}}{\partial z^2} = -\omega^2\mu\epsilon\mathbf{E}$$

$$\nabla^2\mathbf{H} = -\omega^2\mu\epsilon\mathbf{H} \qquad \rightarrow \qquad \frac{\partial^2\mathbf{H}}{\partial x^2} + \frac{\partial^2\mathbf{H}}{\partial y^2} + \frac{\partial^2\mathbf{H}}{\partial z^2} = -\omega^2\mu\epsilon\mathbf{H}$$

Once again, we assume that none of the field quantities vary in the y direction. In other words, we have

$$\frac{\partial^2}{\partial y^2}(\cdots) = 0$$

---

[15]Note here that $\bar{\gamma}$ is different from the complex propagation constant of the medium $\gamma = \alpha + j\beta$ used earlier for uniform plane waves. Although γ, α, and β are dependent only on the wave frequency and the medium parameters (ϵ, μ, σ), the waveguide propagation constant $\bar{\gamma}$ depends in general also on the dimensions of the guiding structure. Even though the propagation medium may be nondissipative, $\bar{\gamma}$ may be either imaginary ($\bar{\gamma} = j\bar{\beta}$) or real ($\bar{\gamma} = \bar{\alpha}$), depending on the frequency of operation, as we shall see later. Note also that the quantities $\bar{\gamma}$, $\bar{\alpha}$, and $\bar{\beta}$ are not vectors; the bar in this case is simply used to distinguish them from corresponding quantities for waves propagating in simple, unbounded media. The $\bar{\gamma}$, $\bar{\alpha}$, and $\bar{\beta}$ notation used here is adapted from E. C. Jordan and K. G. Balmain, *Electromagnetic Waves and Radiating Systems,* Prentice Hall, 1968.

[16]The dictionary meaning of the word *evanescent* is "tending to vanish like vapor."

The field variations in the x direction cannot be similarly limited, because the conductor imposes boundary conditions in the x direction at $x = 0$ and $x = a$. Note that for any of the field components, for example for H_y, we can explicitly show the variations in the z direction by expressing $H_y(x, z)$ as a product of two functions having x and z, respectively, as their independent variables:

$$H_y(x, z) = H_y^0(x)e^{-\bar{\gamma}z}$$

where $H_y^0(x)$ is a function only of x. We then have

$$\frac{\partial}{\partial z}H_y(x, z) = -\bar{\gamma}H_y^0(x)e^{-\bar{\gamma}z} = -\bar{\gamma}H_y(x, z) \quad \longrightarrow \quad \frac{\partial}{\partial z} \quad \longrightarrow \quad -\bar{\gamma}$$

The wave equations [4.2] can thus be rewritten as

$$\frac{\partial^2 \mathbf{E}}{\partial x^2} + \bar{\gamma}^2 \mathbf{E} = -\omega^2 \mu \epsilon \mathbf{E} \qquad [4.4]$$

$$\frac{\partial^2 \mathbf{H}}{\partial x^2} + \bar{\gamma}^2 \mathbf{H} = -\omega^2 \mu \epsilon \mathbf{H} \qquad [4.5]$$

Substituting $\partial/\partial z \rightarrow -\bar{\gamma}$ and $\partial/\partial y = 0$ in [4.3], we find

$$\bar{\gamma}E_y = -j\omega\mu H_x \qquad \bar{\gamma}H_y = j\omega\epsilon E_x$$

$$-\bar{\gamma}E_x - \frac{\partial E_z}{\partial x} = -j\omega\mu H_y \qquad -\bar{\gamma}H_x - \frac{\partial H_z}{\partial x} = j\omega\epsilon E_y$$

$$\frac{\partial E_y}{\partial x} = -j\omega\mu H_z \qquad \frac{\partial H_y}{\partial x} = j\omega\epsilon E_z$$

[4.6]

The foregoing relationships between the field components are valid in general for any time-harmonic electromagnetic wave solution for which (i) there are no variations in the y direction (i.e., $\partial(\cdot)/\partial y = 0$) and (ii) the field components vary in the z direction as $e^{-\bar{\gamma}z}$. In general, if a solution of the wave equation subject to the boundary conditions is obtained for any one of the field components, the other components can be found from [4.6].

In some cases, it is convenient to express the various field components explicitly in terms of the axial components (i.e., E_z and H_z). Introducing $h^2 = \bar{\gamma}^2 + \omega^2 \mu \epsilon$, we can rewrite equations [4.6] in the form

$$H_x = -\frac{\bar{\gamma}}{h^2}\frac{\partial H_z}{\partial x} \qquad [4.7a]$$

$$H_y = -\frac{j\omega\epsilon}{h^2}\frac{\partial E_z}{\partial x} \qquad [4.7b]$$

$$E_x = -\frac{\bar{\gamma}}{h^2}\frac{\partial E_z}{\partial x} \qquad [4.7c]$$

$$E_y = +\frac{j\omega\mu}{h^2}\frac{\partial H_z}{\partial x} \qquad [4.7d]$$

These expressions are useful because the different possible solutions of the wave equation are categorized in terms of the axial components (i.e., z components), as discussed in the next section.

4.1.1   Field Solutions for TE and TM Waves

In general, all of the field components, including both E_z and H_z, may need to be nonzero to satisfy the boundary conditions imposed by an arbitrary conductor or dielectric structure. However, it is convenient and customary to divide the solutions of guided-wave equations into three categories:

$$\left.\begin{array}{l} E_z = 0 \\ H_z \neq 0 \end{array}\right\} = \text{Transverse electric (TE) waves}$$

$$\left.\begin{array}{l} E_z \neq 0 \\ H_z = 0 \end{array}\right\} = \text{Transverse magnetic (TM) waves}$$

$$E_z, H_z = 0 \quad \longrightarrow \quad \text{Transverse electromagnetic (TEM) waves}$$

where z is the axial direction of the guide along which the wave propagation takes place. In the following, we separately examine the field solutions for TE, TM, and TEM waves.

Transverse Electric (TE) Waves   We first consider transverse electric (TE) waves, for which we have $E_z = 0, H_z \neq 0$. With $E_z = 0$, the electric field vector of a TE wave is always and everywhere transverse to the direction of propagation. We can solve the wave equation for any of the field components; for the parallel-plate case, it is most convenient[17] to solve for E_y, from which all the other field components can be found using equations [4.6]. Rewriting the wave equation [4.4] for the y component, we have

$$\frac{\partial^2 E_y}{\partial x^2} + \overline{\gamma}^2 E_y = -\omega^2 \mu \epsilon E_y \qquad [4.8]$$

or

$$\frac{\partial^2 E_y}{\partial x^2} = -h^2 E_y \qquad [4.9]$$

where $h^2 = \overline{\gamma}^2 + \omega^2 \mu \epsilon$. We now express $E_y(x, z)$ as a product of two functions, one varying with x and another varying with z as

$$E_y(x, z) = E_y^0(x) e^{-\overline{\gamma} z}$$

Note that z variation of all the field components has the form $e^{-\overline{\gamma} z}$.

---

[17] Since $E_z \equiv 0$ for a TE wave, and since there is no variation with y, it is clear that there can be a y component of the electric field, as long as E_y, which is tangential to the conductors, varies with x such that it is zero at the perfectly conducting walls.

We can now rewrite [4.9] as an ordinary differential equation in terms of $E_y^0(x)$:

$$\frac{d^2 E_y^0(x)}{dx^2} = -h^2 E_y^0(x) \qquad [4.10]$$

Note that [4.10] is a second-order differential equation whose general solution is

$$E_y^0(x) = C_1 \sin(hx) + C_2 \cos(hx) \qquad [4.11]$$

where C_1, C_2, and h are constants to be determined by the boundary conditions. Note that the complete phasor for the y component of the electric field is

$$E_y(x, z) = [C_1 \sin(hx) + C_2 \cos(hx)]e^{-\bar{\gamma}z}$$

We now consider the boundary condition that requires the tangential electric field to be zero at the perfectly conducting plates. In other words,

$$\left.\begin{array}{l} E_y = 0 \text{ at } x = 0 \\ E_y = 0 \text{ at } x = a \end{array}\right\} \text{ for all } y \text{ and } z$$

Thus $C_2 = 0$, and

$$E_y(x, z) = C_1 \sin(hx)e^{-\bar{\gamma}z}$$

results from the first boundary condition. The second condition (i.e., $E_y = 0$ at $x = a$) imposes a restriction on h; that is,

$$h = \frac{m\pi}{a} \qquad m = 1, 2, 3, \ldots$$

This result illustrates the particular importance of the constant h. Waveguide field solutions such as [4.11] can satisfy the necessary boundary conditions only for a discrete set of values of h. These values of h for which the field solutions of the wave equation can satisfy the boundary conditions are known as the *characteristic values* or *eigenvalues*.[18] We shall see in Chapter 5 that the solutions for the field components for the different metallic waveguides with cylindrical cross sections differ primarily in terms of their eigenvalues, which are mathematical manifestations of the physical boundary conditions for the particular guiding structures.

The other wave field components can all be obtained from E_y using [4.6]. We find $E_x = H_y = 0$, and

Parallel-plate TE$_m$, $m = 0, \pm 1, \pm 2, \ldots$

$$
\begin{array}{l}
E_y = C_1 \sin\left(\dfrac{m\pi}{a}x\right)e^{-\bar{\gamma}z} \\[2ex]
H_z = -\dfrac{1}{j\omega\mu}\dfrac{\partial E_y}{\partial x} = -\dfrac{m\pi}{j\omega\mu a}C_1 \cos\left(\dfrac{m\pi}{a}x\right)e^{-\bar{\gamma}z} \\[2ex]
H_x = -\dfrac{\bar{\gamma}}{j\omega\mu}E_y = -\dfrac{\bar{\gamma}}{j\omega\mu}C_1 \sin\left(\dfrac{m\pi}{a}x\right)e^{-\bar{\gamma}z}
\end{array}
\qquad [4.12]
$$

---

[18] *Eigen* is from the German word for "characteristic" or "one's own."

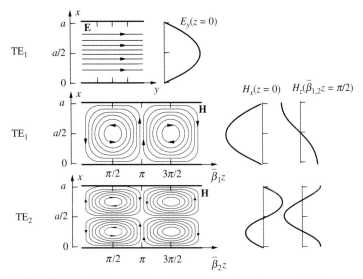

FIGURE 4.2.   TE$_1$ and TE$_2$ modes. The electric and magnetic field distributions for the TE$_1$ and the magnetic field distribution for the TE$_2$ modes in a parallel-plate waveguide. Careful examination of the field structure for the TE$_1$ mode indicates that the magnetic field lines encircle the electric field lines (i.e., displacement current) in accordance with [4.1a] and the right-hand rule. The same is true for the TE$_2$ mode, although the electric field distribution for this mode is not shown.

Note that $m = 0$ makes all fields vanish. Each integer value of m specifies a given field configuration, or *mode,* referred to in this case as the TE$_m$ mode. The configurations of fields for the cases of $m = 1$ and 2 are shown in Figure 4.2.

Transverse Magnetic (TM) Waves   For TM waves the magnetic field vector is always and everywhere transverse to the direction of propagation (i.e., $H_z = 0$), whereas the axial component of the electric field is nonzero (i.e., $E_z \neq 0$). We proceed with the solution in a similar manner as was done for TE waves, except that it is more convenient to use the wave equation for **H**, namely [4.5]. The y component of this equation for H_y is in the form of [4.8] for E_y. Following the same procedure as for the TE case for E_y, the general solution for H_y is

$$H_y = [C_3 \sin(hx) + C_4 \cos(hx)]e^{-\bar{\gamma}z}$$

Since the boundary condition for the magnetic field is in terms of its component normal to the conductor boundary, it cannot be applied to H_y directly. However, using [4.6], we can write the tangential component of the electric field E_z in terms of H_y as

$$E_z = \frac{1}{j\omega\epsilon}\frac{\partial H_y}{\partial x} = \frac{h}{j\omega\epsilon}[C_3 \cos(hx) - C_4 \sin(hx)]e^{-\bar{\gamma}z}$$

We now note that

$$E_z = 0 \quad \text{at } x = 0 \quad \rightarrow \quad C_3 = 0$$

and

$$E_z = 0 \quad \text{at } x = a \quad \rightarrow \quad h = \frac{m\pi}{a} \qquad m = 0, 1, 2, 3, \ldots$$

The only other nonzero field component is E_x, which can be simply found from H_y or E_z. The nonzero field components for this mode, denoted as the TM_m mode, are then

Parallel-plate TM_m, $m = 0, \pm 1, \pm 2, \ldots$

$$H_y = C_4 \cos\left(\frac{m\pi}{a}x\right)e^{-\bar{\gamma}z}$$

$$E_x = \frac{\bar{\gamma}}{j\omega\epsilon}H_y = \frac{\bar{\gamma}}{j\omega\epsilon}C_4 \cos\left(\frac{m\pi}{a}x\right)e^{-\bar{\gamma}z} \qquad [4.13]$$

$$E_z = \frac{jm\pi}{\omega\epsilon a}C_4 \sin\left(\frac{m\pi}{a}x\right)e^{-\bar{\gamma}z}$$

The configurations of the fields for $m = 1$ and 2 (i.e., for the TM_1 and TM_2 modes) are shown in Figure 4.3.

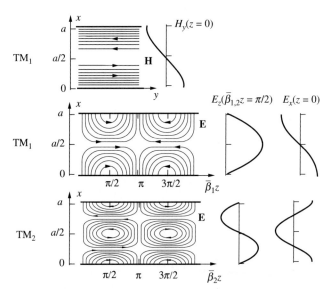

FIGURE 4.3. TM_1 and TM_2 modes. The electric and magnetic field distributions for the TM_1 and TM_2 modes in a parallel-plate waveguide. Note that for TM_1 the electric field lines encircle the magnetic field lines (Faraday's law) in accordance with [4.1c]. The same is true for TM_2, although the magnetic field distribution for TM_2 is not shown.

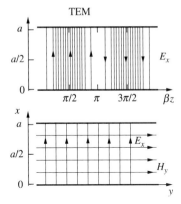

FIGURE 4.4.   TEM mode. Electric and magnetic fields between parallel planes for the TEM (TM$_0$) mode. Only electric field lines are shown in the top panel; the magnetic field lines are out of (or into) the page.

Transverse Electromagnetic (TEM) Waves   Note that contrary to the TE case, the TM solutions do not all vanish for $m = 0$. Since E_z is zero for $m = 0$, the TM$_0$ mode is actually a transverse electromagnetic (or TEM) wave. In this case, we have

$$\text{Parallel-plate TEM}\qquad \boxed{\begin{aligned} H_y &= C_4 e^{-\bar{\gamma}z} \\[2pt] E_x &= \frac{\bar{\gamma}}{j\omega\epsilon}C_4 e^{-\bar{\gamma}z} \\[2pt] E_z &= 0 \end{aligned}} \qquad [4.14]$$

For this case, we have $h = 0$, so $\bar{\gamma} = j\bar{\beta} = j\beta$, where $\beta = \omega\sqrt{\mu\epsilon}$ is the phase constant for a uniform plane wave in the unbounded lossless medium. The field configuration for the TM$_0$ or TEM wave is shown in Figure 4.4.

4.1.2   Cutoff Frequency, Phase Velocity, and Wavelength

Close examination of the solutions for TE and TM waves reveal a number of common characteristics: (i) **E** and **H** have sinusoidal standing-wave distributions in the x direction; (ii) any xy plane is an equiphase plane, or, in other words, the TE and TM waves are plane waves with surfaces of constant phase given by $z = \text{const.}$; and (iii) these equiphase surfaces progress (propagate) along the waveguide with a velocity $\bar{v}_p = \omega/\bar{\beta}$. To see this, consider any of the field components—say, E_y, for TE waves. For a propagating wave, with $\bar{\gamma} = j\bar{\beta}$, and assuming C_1 to be real, we have

$$\mathcal{E}_y(x, z, t) = C_1 \sin\left(\frac{m\pi}{a}x\right)\cos(\omega t - \bar{\beta}z)$$

To determine the conditions under which we can have a propagating wave (i.e., $\bar{\gamma}$ purely imaginary, or $\bar{\gamma} = j\bar{\beta}$), consider the definition of the eigenvalue parameter h:

$$h^2 = \bar{\gamma}^2 + \omega^2\mu\epsilon \quad \rightarrow \quad \bar{\gamma} = \sqrt{h^2 - \omega^2\mu\epsilon} = \sqrt{\left(\frac{m\pi}{a}\right)^2 - \omega^2\mu\epsilon}$$

Note that this expression for h is valid for both TE and TM modes. Thus, the expressions given below for propagation constant, phase velocity, and wavelength are also equally valid for both TE and TM modes.

From the preceding definition of h it is apparent that for each mode (TE or TM) identified by mode index m, there exists a critical or *cutoff* frequency f_{c_m} (or wavelength λ_{c_m}) such that

$$\bar{\gamma} = 0 \quad \rightarrow \quad f_{c_m} = \frac{mv_p}{2a} = \frac{m}{2a\sqrt{\mu\epsilon}} \quad \text{or} \quad \lambda_{c_m} = \frac{v_p}{f_{c_m}} = \frac{2a}{m} \quad [4.15]$$

where $v_p = 1/\sqrt{\mu\epsilon}$ is the *intrinsic* phase velocity for uniform plane waves in the unbounded lossless medium. Note that $v_p = (\mu\epsilon)^{-1/2}$ depends only on ϵ and μ and is thus indeed an intrinsic property of the medium. For $f > f_{c_m}$, the propagation constant $\bar{\gamma}$ is purely imaginary and is given by

$$\bar{\gamma} = j\bar{\beta}_m = j\sqrt{\omega^2\mu\epsilon - \left(\frac{m\pi}{a}\right)^2} = j\beta\sqrt{1 - \left(\frac{f_{c_m}}{f}\right)^2} \qquad f > f_{c_m} \quad [4.16]$$

where we now start to use $\bar{\beta}_m$ instead of $\bar{\beta}$ to underscore the dependence of the phase constant on the mode index m. Note that $\bar{\beta}_m$ is the phase constant (sometimes referred to as the longitudinal phase constant or the propagation constant) for mode m, and $\beta = \omega\sqrt{\mu\epsilon}$ is the intrinsic phase constant, being that for uniform plane waves in the unbounded lossless medium. For $f < f_{c_m}$, $\bar{\gamma}$ is a real number, given by

$$\bar{\gamma} = \bar{\alpha}_m = \sqrt{\left(\frac{m\pi}{a}\right)^2 - \omega^2\mu\epsilon} = \beta\sqrt{\left(\frac{f_{c_m}}{f}\right)^2 - 1} \qquad f < f_{c_m} \quad [4.17]$$

where $\bar{\alpha}_m$ is the attenuation constant, in which case the fields attenuate exponentially in the z direction, without any wave motion. For example, the expression for the $\mathscr{E}_y$ component for TE waves with C_1 real for the case when $f < f_{c_m}$ is

$$\mathscr{E}_y(x, z, t) = C_1 e^{-\bar{\alpha}z} \sin\left(\frac{m\pi}{a}x\right)\cos(\omega t)$$

Such a nonpropagating mode is called an evanescent wave.[19] For $f > f_{c_m}$, the propagating wave has a wavelength $\bar{\lambda}_m$ (sometimes referred to as the guide wavelength) and phase velocity $\bar{v}_{p_m}$ given as

---

[19] It is important to note that the attenuation of the field in the z direction is not due to any energy losses; this condition results simply from the fact that the boundary conditions cannot be satisfied by a TE or TM wave at frequency ω. If a TE or TM field configuration is somehow excited (e.g., by an appropriate configuration of source currents or charges) at a given point z, the fields attenuate exponentially with distance away from the excitation point, with no energy being carried away.

$$\boxed{\begin{aligned} \overline{\lambda}_m &= \frac{2\pi}{\overline{\beta}_m} = \frac{\lambda}{\sqrt{1 - (f_{c_m}/f)^2}} \\[1em] \overline{v}_{p_m} &= \frac{\omega}{\overline{\beta}_m} = \frac{v_p}{\sqrt{1 - (f_{c_m}/f)^2}} \end{aligned}}$$

[4.18]

where $\lambda = 2\pi/\beta = v_p/f$ is the intrinsic wavelength, being the wavelength for uniform plane waves in the unbounded lossless medium, and thus depending only on μ, ϵ, and frequency f. Note that for the TM_0 mode (which is in fact a TEM wave, as discussed above), $\overline{v}_p$ is equal to the intrinsic phase velocity $v_p = (\sqrt{\mu\epsilon})^{-1}$ in the unbounded lossless medium and is thus independent of frequency. Furthermore, the phase constant $\overline{\beta}_m$ and wavelength $\overline{\lambda}_m$ are also equal to their values for a uniform plane wave in an unbounded lossless medium. There is no cutoff frequency for this mode, since the propagating field solutions are valid at any frequency.

For TE and TM waves in general, it is clear from this discussion that the wavelength $\overline{\lambda}_m$, as observed along the guide, is longer than the corresponding wavelength in an unbounded lossless medium by the factor $[1 - (f_{c_m}/f)^2]^{-1/2}$. Also, the velocity of phase progression inside the guide is likewise greater than the corresponding intrinsic phase velocity. Note, however, that $\overline{v}_p$ is not the velocity with which energy or information propagates,[20] so $\overline{v}_p > v_p$ does not pose any particular dilemma.

The difference between $\overline{v}_p$ and v_p can be further understood using an analogy with water waves. Consider water waves approaching a shoreline or a breakwater at an angle θ_i from the line perpendicular to it as shown in Figure 4.5. The velocity

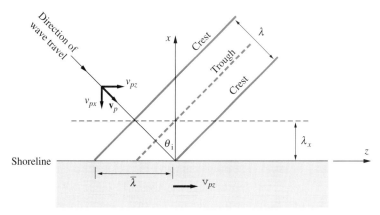

FIGURE 4.5. Water wave. Water wave approaching a shoreline or a breakwater.

---

[20]The velocity at which information (e.g., the envelope of a modulated signal) travels is the so-called *group velocity* v_g, given by $v_g = d\omega/d\beta$, which for metallic waveguides is $v_g = c^2/\overline{v}_p$. The concept of group velocity is discussed extensively in Section 4.3.

of the wave in its direction of propagation can be found from the distance between successive crests (λ) and the frequency (f) with which crests pass a given point, in other words, $v_p = \lambda f$. However, we all remember from our own experiences watching waves hit a shoreline that the velocity of water flow along the shoreline can appear to be much faster. To understand this effect, we can observe from Figure 4.5 that the velocity of flow along the shoreline is given by $v_{pz} = \bar{\lambda} f$ where $\bar{\lambda}$ is the distance between successive crests along the shoreline. We thus have $\bar{\lambda} = \lambda / \sin \theta_i$ or $v_{pz} = v_p / \sin \theta_i$, so that in general v_{pz} is indeed larger then v_p.

Examples 4-1 through 4-3 illustrate the application of the various parallel-plate waveguide relationships derived above.

Example 4-1: Parallel-plate waveguide modes. An air-filled parallel-plate waveguide has a plate separation of 1.25 cm. Find (a) the cutoff frequencies of the TE_0, TM_0, TE_1, TM_1, and TM_2 modes, (b) the phase velocities of those modes at 15 GHz, (c) the lowest-order TE and TM mode that cannot propagate in this waveguide at 25 GHz.

Solution:

(a) As we have seen in earlier sections, the TE_0 mode does not exist in a parallel-plate waveguide. TM_0 is equivalent to the TEM mode, and this mode can propagate at all frequencies (i.e., $f_{c_0} = 0$). The cutoff frequencies of the other modes can be calculated using [4.15] as

$$f_{c_m} = \frac{m}{2a\sqrt{\mu_0\epsilon_0}} = \frac{m(3 \times 10^{10} \text{ cm-s}^{-1})}{2(1.25 \text{ cm})} = 1.2 \times 10^{10} m \text{ Hz} = 12m \text{ GHz}$$

The results are summarized in the following table:

Mode	TM_0	TE_1	TM_1	TE_2	TM_2
f_{c_m}(GHz)	0	12	12	24	24

(b) For the TM_0 mode, $\bar{v}_p = v_p = c$. For the TE_1 and TM_1 modes, we have

$$\bar{v}_p = \frac{c}{\sqrt{1 - (f_{c_1}/f)^2}} = \frac{c}{\sqrt{1 - (12/15)^2}} = \frac{5c}{4}$$

Note that the TE_2 and TM_2 modes do not propagate at 15 GHz; that is, their cutoff frequencies are above 15 GHz.

(c) The lowest-order modes that cannot propagate in this waveguide are TE_3 and TM_3, since

$$f_{c_3} = 12 \times 3 = 36 \text{ GHz} > 25 \text{ GHz}$$

Example 4-2: ELF propagation in the earth–ionosphere waveguide.
Extremely low frequencies (ELF) are ideal for communicating with deeply submerged submarines, because below 1 kHz, electromagnetic waves penetrate into

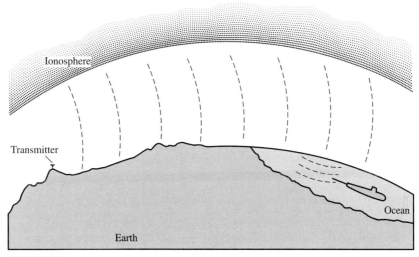

FIGURE 4.6. **ELF propagation and submarine reception.**

seawater.[21] Propagation at these frequencies takes place in the "waveguide" formed between the earth and the ionosphere (Figure 4.6); low propagation losses allow nearly worldwide communication from a single ELF transmitter.

In J. R. Wait's simple model,[22] the surface of the earth and the bottom of the ionosphere form the boundaries of a terrestrial "parallel"-plate waveguide with lossy walls. The ionosphere is approximated by an isotropic layer beginning at a given altitude and extending to infinity with no horizontal variations allowed. Energy is lost through the "walls" either into the finitely conducting ionosphere or into the ground, with the former loss being dominant. The important feature of propagation below 1 kHz is that there is a single propagating mode, a so-called quasi-TEM mode. All the other modes are evanescent and are almost undetectable at distances in excess of 1000 km. In the far field, the wave consists of a vertical electric field and a horizontal magnetic field transverse to the direction of propagation (Figure 4.6). The leakage of energy from this wave into the ocean (see Section 3.8) gives rise to a plane wave propagating downward, and it is this signal that the submarine receiver detects.

Consider an idealized earth–ionosphere waveguide where both the ionosphere and the earth are assumed to be perfect conductors. In addition, neglect the curvature of the waveguide and assume it to be flat. The height of the terrestrial waveguide can vary anywhere from 70 km to 90 km depending on conditions; for our purposes,

---

[21] S. L. Bernstein, et al., Long-range communications at extremely low frequencies, *Proc. IEEE,* 62(3), pp. 292–312, March 1974.

[22] J. R. Wait, Earth-ionosphere cavity resonances and the propagation of ELF radio waves, *Radio Sci.,* 69D, pp. 1057–1070, August 1965.

assume it to be 80 km. Find all the propagating modes at an operating frequency of
(a) 100 Hz (ELF); (b) 1 kHz (ELF); and (c) 10 kHz (VLF).

Solution: As we already know, the TEM mode (or TM_0 mode) exists in all
cases. The cutoff frequencies of other modes can be found using [4.15] as

$$f_{c_m} = \frac{m(3 \times 10^8 \text{ m-s}^{-1})}{2 \times 80 \times 10^3 \text{ m}} = 1875m \text{ Hz}$$

So the cutoff frequencies of some of the lower-order modes are

Mode	TM_0	TE_1	TM_1	TE_2	TM_2	TE_3	TM_3
f_{c_m}(Hz)	0	1875	1875	3750	3750	5625	5625

Therefore, (a) at 100 Hz and (b) at 1 kHz, only one mode (the TEM mode) can
propagate, whereas (c) at 10 kHz, eleven different modes (i.e., TEM, TE_1, TM_1,
TE_2, TM_2, TE_3, TM_3, TE_4, TM_4, TE_5, and TM_5) can propagate in the earth–
ionosphere waveguide.

Example 4-3: Waveguide propagation constant versus frequency.
Sketch the magnitude of the waveguide propagation constant $\overline{\gamma}$ as a function of the
operating frequency f for TM_0, TE_1, TM_1, TE_2, and TM_2 modes in a parallel-plate
waveguide of plate separation a and filled with a lossless dielectric with properties
ϵ and μ.

Solution: For TM_0 or TEM mode, $|\overline{\gamma}| = \overline{\beta} = \omega\sqrt{\mu\epsilon} = 2\pi f\sqrt{\mu\epsilon}$ for all
operating frequencies. For higher-order TE_m and TM_m modes, we have

$$\left.\begin{array}{l} \overline{\gamma} = \overline{\alpha} = 2\pi f\sqrt{\mu\epsilon}\sqrt{(f_{c_m}/f)^2 - 1} \\ \overline{\beta} = 0 \text{ (no propagation)} \end{array}\right\} \quad \text{for } f < f_{c_m}$$

and

$$\left.\begin{array}{l} \overline{\alpha} = 0 \text{ (no attenuation)} \\ |\overline{\gamma}| = \overline{\beta} = 2\pi f\sqrt{\mu\epsilon}\sqrt{1 - (f_{c_m}/f)^2} \end{array}\right\} \quad \text{for } f > f_{c_m}$$

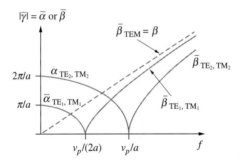

**FIGURE 4.7. Waveguide propagation
constant versus frequency.** Magnitude of the
propagation constant $|\overline{\gamma}|$ versus frequency f for
the five lowest-order modes in a parallel-plate
waveguide. Note that $v_p = 1/\sqrt{\mu\epsilon}$.

where f_{c_m} is the cutoff frequency of the mth-order mode, given by

$$f_{c_m} = \frac{m}{2a\sqrt{\mu\epsilon}}$$

Using these expressions, we can sketch the magnitudes of $\overline{\gamma}$, $\overline{\alpha}$, and $\overline{\beta}$ as a function of the operating frequency f for TM_0, TE_1, TM_1, TE_2, and TM_2 modes, as shown in Figure 4.7.

4.1.3 TE and TM Modes as Superpositions of TEM Waves

It is often instructive to think of the different TE or TM waveguide modes as a superposition of two or more TEM waves. This picture is helpful, as it provides a better physical understanding of the mode field structures as well as concepts such as guide wavelength, cutoff frequency, and losses in the waveguide walls.

To demonstrate the decomposition of the mode fields analytically, consider, for example, the electric field component of the propagating TE_m mode, which can be written from [4.12] as

$$E_y(x, z) = C_1 \sin\left(\frac{m\pi}{a}x\right)e^{-j\overline{\beta}z} = \frac{C_1}{2j}\left[e^{j(m\pi x/a)} - e^{-j(m\pi x/a)}\right]e^{-j\overline{\beta}z}$$

$$= \frac{C_1}{2}\left[e^{j(m\pi x/a - \overline{\beta}z - \pi/2)} - e^{-j(m\pi x/a + \overline{\beta}z + \pi/2)}\right]$$

Using the notation introduced in Section 2.6, and Figure 2.31, the first term in the above expression represents a perpendicularly polarized (with respect to the xz plane) TEM wave (denoted as TEM_1) propagating in the direction $\hat{\mathbf{k}}_1 = \beta^{-1}(-\beta_x\hat{\mathbf{x}} + \beta_z\hat{\mathbf{z}}) = -\hat{\mathbf{x}}\cos\theta_{i_m} + \hat{\mathbf{z}}\sin\theta_i$, where θ_{i_m} is the angle between $\hat{\mathbf{k}}_1$ and the x axis.[23] The phase constants β_x and β_z are given as

$$\beta_x = \beta\cos\theta_{i_m} = \frac{m\pi}{a} = \omega_{c_m}\sqrt{\mu\epsilon}$$

$$\beta_z = \beta\sin\theta_{i_m} = \sqrt{\beta^2 - \left(\frac{m\pi}{a}\right)^2} = \overline{\beta} = \omega\sqrt{\mu\epsilon}\sqrt{1 - \left(\frac{\omega_{c_m}}{\omega}\right)^2}$$

where $\beta = \omega\sqrt{\mu\epsilon}$. Similarly, the second term in the above expression for $E_y(x, z)$ represents another perpendicularly polarized TEM plane wave (denoted as TEM_2) propagating in the direction $\hat{\mathbf{k}}_2 = \beta^{-1}(\beta_x\hat{\mathbf{x}} + \beta_z\hat{\mathbf{z}}) = \hat{\mathbf{x}}\cos\theta_{i_m} + \hat{\mathbf{z}}\sin\theta_{i_m}$ with the same phase constants β_x and β_z. These component TEM waves are shown in Figure 4.8. Note that this picture of the superposition of two TEM waves is identical to

---

[23]In comparison with Figure 3.17 in Chapter 3, it should be noted that the angle θ_i defined in Figure 3.17 is the angle between $\hat{\mathbf{k}}$ and the z axis, or the complement (i.e., $90° - \theta_i$) of the angle θ_{i_m} defined here (Figure 4.8).

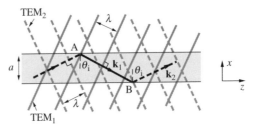

FIGURE 4.8. **Two TEM waves forming a TE$_m$ wave in a parallel-plate waveguide.** Waveguide modes represented as a superposition of two TEM waves. TEM$_1$ propagates in the direction $\hat{\mathbf{k}}_1 = \mathbf{k}_1/|\mathbf{k}_1|$ with surfaces of constant phase (i.e., phase fronts) indicated by solid lines. TEM$_2$ propagates in the direction $\hat{\mathbf{k}}_2 = \mathbf{k}_2/|\mathbf{k}_2|$, with phase fronts shown in dashed lines. The same type of decomposition, and thus the same picture shown here, is also valid for the TM$_m$ modes.

that for TEM waves obliquely incident on a conducting sheet, which was described in Chapter 3 in connection with Figure 3.17. The field solutions in front of the conducting sheet also consisted of a superposition of two TEM waves, propagating in the $\hat{\mathbf{k}}_i$ and $\hat{\mathbf{k}}_r$ directions, at an incidence angle θ_{i_m} with respect to the line perpendicular to the conductor boundary.

The angle θ_{i_m} can be found from the above relationship for β_x as

$$\theta_{i_m} = \cos^{-1}\left[\frac{\omega_{c_m}}{\omega}\right] = \cos^{-1}\left[\frac{f_{c_m}}{f}\right] = \cos^{-1}\left[\frac{m}{2af\sqrt{\mu\epsilon}}\right]$$

$$= \cos^{-1}\left[\frac{m\lambda}{2a}\right] = \cos^{-1}\left[\frac{\lambda}{\lambda_{c_m}}\right] \qquad m = 0, 1, 2, 3, \ldots \qquad [4.19]$$

where we have used the facts that $f_{c_m} = m/(2a\sqrt{\mu\epsilon}) = (\lambda_{c_m}\sqrt{\mu\epsilon})^{-1}$ and $\lambda f = (\sqrt{\mu\epsilon})^{-1}$. Note that each mode (i.e., each integer value of m) corresponds to a discrete value of θ_{i_m}. Note also that a real value for θ_{i_m} exists only when $f > f_{c_m}$ (or $\lambda < \lambda_{c_m}$), which corresponds to a propagating mode. In the special case when $f = f_{c_m}$, we have $\cos\theta_{i_m} = 1$, and $\theta_{i_m} = 0°$, in which case the two component TEM waves that constitute the TE$_m$ wave propagate respectively in the x and $-x$ directions, with no propagation of wave energy along the waveguide in the z direction.

Although we have illustrated the decomposition for the case for one field component of the TE$_m$ mode, a similar decomposition can be done for each of the other field components of the TE$_m$ mode or for the field components of TM$_m$ modes. The only difference for the case of TM$_m$ modes is that the component TEM waves are parallel polarized with respect to the xz plane, and the electric field vectors lie in the xz plane.

The decomposition of TE or TM mode fields into component TEM waves, and the fact that this is possible for only a discrete set of values of θ_{i_m}, can also be understood graphically by considering Figure 4.9. The combination of the two component waves (TEM$_1$ and TEM$_2$) can be thought in terms of a "ray" reflecting back and

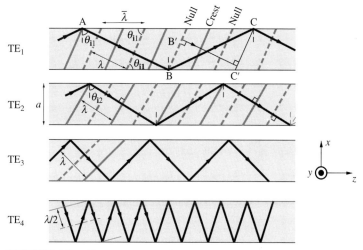

FIGURE 4.9.   **Representation of propagating waveguide modes as "rays" reflecting between the parallel plates.** The dashed lines represent the phase fronts, and the solid lines represent the ray paths.

forth between the walls, with the wave energy regarded as being transported along the ray path.[24] Consider the ray path ABC shown in the top panel of Figure 4.9. The solid (dashed) lines perpendicular to the ray paths represent the constant phase fronts corresponding to the crests (nulls) of the TEM wave during its travel from A to B (the constant phase fronts corresponding to the ray path from B to C are not shown). Before the ray undergoes reflection at point B, the line B–B′ constitutes a constant phase front. In order for the ray to continue down the guide as a uniform plane wave, the line C–C′ must also be a constant phase front. Thus the amount of wave phase change along paths B–C and B′–C′ must be an integer multiple of 2π, that is,

$$\beta(\overline{BC} - \overline{B'C'}) = m2\pi \qquad m = 0, 1, 2, 3, \ldots$$

where $\beta = \omega\sqrt{\mu\epsilon}$ is the propagation constant in the unbounded lossless medium. From geometrical considerations we have

$$\overline{BC} = \frac{a}{\cos\theta_{i_m}} \qquad \overline{B'C'} = \overline{BC'}\sin\theta_{i_m}$$

---

[24]It is sometimes convenient to represent an electromagnetic wave by *rays* rather than by wave fronts (or the surfaces of constant phase). This picture is most appropriate for light rays, and rays were used to describe light long before its wave nature was firmly established. In the context of the wave nature of light, and in an isotropic medium, a ray is an imaginary line along the direction of propagation of the wave and perpendicular to the wave fronts. In the context of the particle theory of light, rays can be thought of as paths of the photons.

where

$$\overline{BC'} = a \tan \theta_{i_m} - \frac{a}{\tan \theta_{i_m}}$$

Substituting in the above and performing algebraic manipulation, we find

$$2\beta a \cos \theta_{i_m} = m2\pi \qquad m = 0, 1, 2, 3, \ldots$$

Noting that $\beta = 2\pi/\lambda$, we can write the angle θ_{i_m} as

$$\cos \theta_{i_m} = \frac{m\lambda}{2a} \quad \rightarrow \quad \theta_{i_m} = \cos^{-1}\left[\frac{m\lambda}{2a}\right] = \cos^{-1}\left[\frac{\lambda}{\lambda_{c_m}}\right] \qquad m = 0, 1, 2, 3, \ldots \quad [4.20]$$

which is an expression identical to that determined above on the basis of an analytical decomposition of the mode fields into component TEM waves.[25] We find once again that a uniform plane wave (represented by a ray) can propagate down the waveguide by undergoing multiple reflections, as long as the angle θ_{i_m} between its ray path and the normal to the waveguide walls is equal to one of the set of discrete values of θ_{i_m} given by [4.20]. Note that each one of these discrete values θ_{i_m} corresponds to a particular waveguide mode, TE_m or TM_m. The ray path representation is therefore equally valid for both TM and TE waves; the difference between the two mode types is the orientation of the wave electric field with respect to the xz plane. For TE modes the ray represents a perpendicularly polarized wave, with the electric field perpendicular to the page (i.e., the y direction) in Figure 4.9. For TM modes, the uniform plane wave represented by the ray is parallel polarized, with the electric field lying in the plane of Figure 4.9 (i.e., on the xz plane).

As defined earlier, the wavelength $\lambda_{c_m} = 2a/m$ is the cutoff wavelength. At this wavelength, $\theta_{i_m} = 0$, and, as a result, there is no energy propagation along the z axis, because the ray path is perpendicular to the z direction and the ray simply reflects up and down between the waveguide walls. From the condition on θ_{i_m}, namely $\cos \theta_{i_m} = m\lambda/(2a)$, it is clear that for a given value of a the ray paths for waves with very short wavelengths $\lambda \ll a$ (very high frequencies) are almost parallel to the z axis. The wavelength parallel to the walls (or along the z direction), which is the guide wavelength $\overline{\lambda}_m$ introduced in the previous section, and the velocity $\overline{v}_{p_m}$ are given by

$$\overline{\lambda}_m = \frac{\lambda}{\sin \theta_{i_m}} = \frac{\lambda}{\sqrt{1 - (m\lambda/2a)^2}} = \frac{\lambda}{\sqrt{1 - (\lambda/\lambda_{c_m})^2}}$$

$$\overline{v}_{p_m} = \frac{v_p}{\sin \theta_{i_m}} = \frac{v_p}{\sqrt{1 - (m\lambda/2a)^2}} = \frac{v_p}{\sqrt{1 - (\lambda/\lambda_{c_m})^2}}$$

---

[25]Note that we could have arrived at the same condition for θ_{i_m} by simply noting that for the picture of the ray paths and phase fronts of Figure 4.9 to be valid, the separation between the waveguide walls (i.e., a) must be an integral multiple of half-wavelengths in the x direction, or

$$\frac{m\lambda_x}{2} = a \qquad \rightarrow \qquad \cos \theta_{i_m} = \frac{m\lambda}{2a} = \frac{\lambda}{\lambda_{c_m}}$$

since $\lambda_x = \lambda/\cos \theta_{i_m}$.

Because of the zigzag path that is followed by the ray, the velocity with which the energy carried by the ray propagates along the guide has a smaller magnitude than that of the free space velocity in the unbounded lossless medium (i.e., v_p). We shall see in Section 4.3 that this velocity is known as the *group velocity* v_g, and is the projection of the ray velocity v_p along the guide given by

$$v_g = v_p \sin\theta_{i_m} = v_p\sqrt{1 - \left(\frac{m\lambda}{2a}\right)^2} = v_p\sqrt{1 - (\lambda/\lambda_{c_m})^2}$$

The identification of the different propagation modes in terms of component TEM waves incident at different values of a discrete set of angles is quantitatively illustrated in Example 4-4.

Example 4-4:   Incidence angle of TEM waves.   Consider an air-filled parallel-plate waveguide having a plate separation of 3 cm. At an operating frequency of 20 GHz, find the oblique incidence angle θ_{i_m} and sketch the ray paths corresponding to the TE_1, TE_2, TE_3, and TE_4 modes and determine the mode which propagates with the highest phase velocity.

Solution: Using $\theta_{i_m} = \cos^{-1}(f_{c_m}/f)$, where $f_{c_m} = m/(2a\sqrt{\mu_0\epsilon_0}) \simeq 5 \times 10^9 m$ Hz $= 5m$ GHz, we find

$$\theta_{i_1} = \cos^{-1}\frac{5}{20} \simeq 75.5° \quad \text{for } TE_1$$

$$\theta_{i_2} = \cos^{-1}\frac{10}{20} = 60° \quad \text{for } TE_2$$

$$\theta_{i_3} = \cos^{-1}\frac{15}{20} \simeq 41.4° \quad \text{for } TE_3$$

$$\theta_{i_4} = \cos^{-1}\frac{20}{20} = 0° \quad \text{for } TE_4$$

Figure 4.10 shows a graphical representation of each TE_m mode in terms of the superposition of two TEM waves. Note that the component TEM waves corresponding to the TE_4 mode bounce back and forth between the waveguide walls without any propagation along the guide length, because its cutoff frequency is equal to the operating frequency.

4.1.4   Attenuation in Parallel-Plate Waveguides

Up to now, we have considered the field structures and other characteristics of propagating modes, assuming the waveguide walls to be perfectly conducting and the dielectric between the plates to be lossless, and thus implicitly neglecting losses.

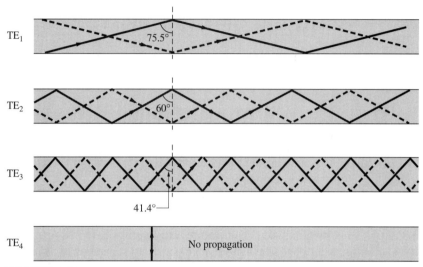

FIGURE 4.10. **Incidence angle of component TEM waves.** Different TE_m modes represented by ray paths propagating inside the parallel-plate waveguide at different angles of incidence θ_{i_m}.

In this section we consider attenuation of the fields due to conduction or dielectric losses.[26]

Attenuation Due to Conduction Losses In all of the TE and TM mode solutions we have discussed, the wave magnetic field has a nonzero component parallel to the waveguide walls. As a result of the boundary condition $\hat{n} \times \mathbf{H} = \mathbf{J}_s$, this parallel magnetic field component causes surface currents to flow in the metallic waveguide walls. If the walls are made of perfectly conducting materials, this current flow does not have any effect on the wave fields. However, in actual waveguides, the fact that the metallic walls have finite conductivity results in losses, and the fields attenuate with distance.[27]

Determining attenuation in a waveguide in the general case, when the waveguide walls consist of an arbitrary lossy material, is a rather difficult problem. When the walls are not perfect conductors, the basic boundary conditions, namely that $E_{tan} = 0$ and $H_{nor} = 0$, no longer hold, so the field configurations (e.g., those given by [4.12] for TE_m waves) derived using these conditions are not valid either. In other words, although the losses can in principle be calculated (e.g., using [3.43] with the surface currents determined from $\mathbf{J}_s = \hat{n} \times \mathbf{H}$) if the field configurations were

---

[26]Attenuation in waveguides can occur because the walls are not perfectly conducting and also if the dielectric material between the plates is lossy (i.e., $\epsilon'' \neq 0$ or $\sigma \neq 0$).

[27]Note that this attenuation is in general quite small, unlike the very rapid attenuation of an evanescent wave. In other words, typically the value of $\bar{\alpha}$ for a nonpropagating mode is much larger than the attenuation constant for conducting losses; that is, α_c.

known, the latter in fact depends on the losses and cannot be calculated without accounting for the losses. We resolve this dilemma by resorting to a so-called "perturbation" solution, which is often used in many engineering problems, and which gives highly accurate results in the case of metallic waveguides. The basic premise of our solution is that the losses are small enough that they have a negligible effect on the field distribution within the waveguide. This assumption makes sense, since the waveguides we work with are presumably designed for and used in applications for which the losses are actually small—otherwise they would not be used for that application. In the context of this basic assumption, we can use the field distributions derived for the lossless case to determine the tangential magnetic field on the conductor surfaces, from which we can in turn find the surface currents and use [3.43] to compute the wall losses. If deemed necessary, a second and improved estimate of losses can then be made, using a field distribution corrected (to first order) to account for the calculated losses.[28] However, since practical waveguides are constructed of high-conductivity metals such as copper or brass or are coated with silver, the original assumption of determining the field configuration as if the walls were perfect conductors yields very accurate results, and a second assumption is rarely necessary.

We now proceed by adopting the common practice of assuming that the attenuation is exponential in form (i.e., the fields vary as $e^{-\alpha_c z}$) and determine the attenuation constant α_c using a perturbation solution. Since the fields vary as $e^{-\alpha_c z}$, the average wave power P_{av} transmitted along the waveguide (i.e., in the z direction) varies as $e^{-2\alpha_c z}$. The rate of decrease of transmitted power along the waveguide is then

$$-\frac{\partial P_{av}}{\partial z} = +2\alpha_c P_{av}$$

The reduction in power per unit length must be equal to the power lost or dissipated per unit length. Therefore we have

$$\frac{\text{Power lost per unit length}}{\text{Power transmitted}} = \frac{2\alpha_c P_{av}}{P_{av}} = 2\alpha_c$$

---

[28] As an example of the manner in which such a correction may be introduced, consider the TEM mode, for which the magnetic field at the conductor surface is in the y direction, so the surface current is in the $+z$ direction on the lower wall and in the $-z$ direction on the upper wall. Since the conductor is not perfect, the presence of a current $\mathbf{J}_s = \hat{\mathbf{z}} J_z$ implies that there is a nonzero z component of the electric field (i.e., E_z) within the conductor. The continuity of the tangential component of an electric field across any boundary between two materials (i.e., equation [1.12]) in turn requires that there must be a nonzero electric field component E_z immediately outside the conductor and inside the parallel-plate waveguide. Our original assumption of a TEM mode, with both electric and magnetic fields completely transverse, clearly cannot be precisely valid. Assuming that the magnitude of E_z is small, we may approximate its variation within the waveguide as $E_z = K(1 - 2x/a)$, where K is a constant, dependent on the surface impedance $Z_s = (1 + j)R_s$ of the conductor. Note that we then have a linear variation of E_z from $E_z = +K$ at the lower wall to $E_z = -K$ at the upper wall. We can now take the total electric field to be $\mathbf{E} = \hat{\mathbf{x}} E_x + \hat{\mathbf{z}} E_z$ and use Maxwell's equations to determine the associated magnetic field, to arrive at the first-order corrected field distribution.

or

$$\alpha_c = \frac{\text{Power lost per unit length}}{2 \times \text{Power transmitted}} = \frac{P_{\text{loss}}}{2 \, P_{\text{av}}}$$

[4.21]

We now use this definition of α_c for the parallel-plate waveguide.

We first consider attenuation due to conductor losses for a TEM wave. Using the solutions [4.14] obtained in the previous subsection, we have

$$H_y = C_4 e^{-j\bar{\beta}z} \qquad E_x = \frac{\bar{\beta}}{\omega\epsilon} C_4 e^{-j\bar{\beta}z}$$

The surface current density on each of the conducting plates is

$$\mathbf{J}_s = \hat{\mathbf{n}} \times \mathbf{H}$$

The surface current in each plate is $J_s = C_4 e^{-j\bar{\beta}z}$, so $|J_s| = C_4$, assuming that C_4 is positive and real.

The total loss (sum of that in upper and lower plates) for a length of 1 meter and a width of b meters of the guide is

$$P_{\text{loss}} = 2\int_0^1 \int_0^b \tfrac{1}{2}|J_s|^2 R_s \, dy \, dz = C_4^2 R_s b$$

where R_s is the resistive part of the surface impedance discussed in Section 3.7. From [3.40], the surface impedance of a conductor is

$$\eta_c = Z_s = R_s + jX_s = (\sigma\delta)^{-1}(1 + j)$$

where $\delta = \sqrt{2/(\omega\mu_0\sigma)}$ is the skin depth, assuming nonmagnetic, conducting walls with permeability $\mu = \mu_0$. Another expression for R_s is $R_s = \sqrt{\omega\mu_0/(2\sigma)}$.

The time-average power density transmitted in the z direction along the guide per unit cross-sectional area is

$$\mathbf{S}_{\text{av}} = \tfrac{1}{2}\mathcal{R}e\{\mathbf{E} \times \mathbf{H}^*\}$$

Since $|E_x| = \eta|H_y|$, with C_4 assumed to be real, we have $|\mathbf{S}_{\text{av}}| = \tfrac{1}{2}\eta C_4^2$. For a spacing of a meters and a width of b meters, the cross-sectional area is ba; thus the total time-average power transmitted through such a cross-sectional area is

$$P_{\text{av}} = \tfrac{1}{2}\eta C_4^2 ba$$

Using the definition of α_c in [4.21], we then have

Parallel-plate

TEM (TM$_0$)

$$\alpha_{c_{\text{TEM}}} = \frac{C_4^2 R_s b}{2(\tfrac{1}{2}\eta C_4^2 ba)} = \frac{R_s}{\eta a} = \frac{1}{\eta a}\sqrt{\frac{\omega\mu_0}{2\sigma}}$$

[4.22]

Next we consider conductor losses for TE waves. Note that the nonzero field components for TE waves in a parallel-plate waveguide are as given in [4.12]. The amplitude of the surface current density is determined by the tangential $\mathbf{H}$ (i.e., H_z) at $x = 0$ and $x = a$ to be (for C_1 real and positive)

$$|J_{sy}| = |H_z| = \frac{m\pi C_1}{\omega \mu a}$$

OPP STROOM·O·LHTH ∈IO

It is interesting to note that J_{sy} is the only nonzero current component and does not flow in the propagation direction (i.e., z direction). The total power loss (sum of the losses in the two plates) for a length of 1 meter and a width of b meters of the guide is

$$P_{\text{loss}} = 2\int_0^1 \int_0^b \left[\frac{1}{2}|J_{sy}|^2 R_s \right] dy\, dz = \frac{2bm^2\pi^2 C_1^2 \sqrt{\omega\mu_0/(2\sigma)}}{2\omega^2\mu^2 a^2}$$

The time-average power density transmitted in the z direction per unit cross-sectional area is

$$|\mathbf{S}_{\text{av}}| = \frac{1}{2}\mathcal{R}e\{\mathbf{E} \times \mathbf{H}^*\} \cdot \hat{\mathbf{z}} = -\frac{1}{2}E_y H_x^* = \frac{\overline{\beta} C_1^2}{2\omega\mu} \sin^2\left(\frac{m\pi}{a}x\right)$$

The total power through a guide with cross-sectional area of width b and a height of a meters can be determined by integrating $|\mathbf{S}_{\text{av}}|$ in x and y:

$$P_{\text{av}} = \int_0^b \int_0^a \frac{\overline{\beta} C_1^2}{2\omega\mu} \sin^2\left(\frac{m\pi}{a}x\right) dx\, dy = \frac{\overline{\beta} C_1^2 ab}{4\omega\mu}$$

The attenuation constant α_c is then given by

Parallel-plate TE$_m$ $$\alpha_{c_{\text{TE}m}} = \frac{2m^2\pi^2 \sqrt{\omega\mu_0/(2\sigma)}}{\overline{\beta}\omega\mu a^3} = \frac{2R_s(f_{c_m}/f)^2}{\eta a\sqrt{1 - (f_{c_m}/f)^2}}$$ [4.23]

where we have used $\overline{\beta} = \beta\sqrt{1 - (f_{c_m}/f)^2}$.

The derivation of a corresponding expression for α_c for TM waves proceeds in the same manner as for TE waves.

Parallel-plate TM$_m$ $$\alpha_{c_{\text{TM}m}} = \frac{2R_s}{\eta a\sqrt{1 - (f_{c_m}/f)^2}}$$ [4.24]

Attenuation constants due to conduction losses for TEM, TE, and TM waves are plotted as a function of frequency in Figure 4.11. It is interesting to note that the losses are generally higher for TM than for TE waves. This is because the surface current $\mathbf{J}_s$ flows in support of the tangential component of $\mathbf{H}$, and for TM waves the transverse component of the magnetic field (H_y) is tangential. For TE waves, the currents are related to H_z, whose magnitude approaches zero as frequency increases,

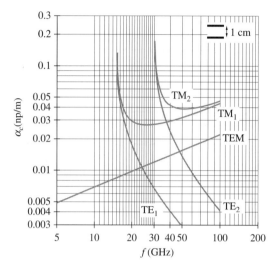

FIGURE 4.11. **Attenuation versus frequency for parallel-plate waveguide.** Attenuation-versus-frequency characteristics of waves guided by parallel plates.

as can be seen by considering the fact that, for a given value of a, the component TEM waves (see Section 4.1.3) of any given mode have increasingly higher values of incident angle θ_i for higher frequencies. The fact that the losses are very large near the cutoff frequency can also be understood in view of the uniform plane wave decomposition of the waveguide modes (Figure 4.9). Near cutoff, the component plane waves have to reflect many more times (per distance z traveled) from the walls, losing power in each reflection.

Example 4-5 compares the numerical values of attenuation constant α_c for different waveguide modes.

Example 4-5: Parallel-copper-plate waveguide. Consider an air-filled parallel-copper-plate waveguide with an inner plate separation distance of 1 cm. Calculate the attenuation constant due to conductor losses, α_c, in decibels per meter (dB-m^{-1}), for TEM, TE$_1$, and TM$_1$ modes at operating frequencies of (a) 20 GHz and (b) 30 GHz.

Solution: For copper, the surface resistance R_s is given by

$$R_s = \sqrt{\frac{\omega \mu_0}{2\sigma}} = \sqrt{\frac{2\pi f \times 4\pi \times 10^{-7}}{2 \times 5.8 \times 10^7}} \simeq 2.61 \times 10^{-7}\sqrt{f} \ \Omega$$

Thus, at 20 GHz we have $R_s \simeq 3.69 \times 10^{-2}\Omega$, whereas at 30 GHz $R_s \simeq 4.52 \times 10^{-2}\Omega$. For the TEM mode, the attenuation constant is given by

$$\alpha_{c_{\text{TEM}}} = \frac{R_s}{\eta a}$$

So at 20 GHz, we have $\alpha_{c_{\text{TEM}}} \simeq 9.79 \times 10^{-3}$ np-m^{-1} (or[29] $\sim 8.50 \times 10^{-2}$ dB-m^{-1}). At 30 GHz, we have $\alpha_{c_{\text{TEM}}} \simeq 0.104$ dB-m^{-1}.

The cutoff frequencies for the TE$_1$ and TM$_1$ modes are $f_{c_1} = (3 \times 10^8)/0.02 = 15$ GHz < 20 GHz. Thus, both TE$_1$ and TM$_1$ modes propagate at 20 GHz and 30 GHz. At 20 GHz for the TE$_1$ mode, we have

$$\alpha_{c_{\text{TE}_1}} = \frac{2R_s(f_{c_1}/f)^2}{\eta a \sqrt{1 - (f_{c_1}/f)^2}} \simeq \frac{2 \times 3.69 \times 10^{-2}(15/20)^2}{377 \times 10^{-2} \times \sqrt{1 - (15/20)^2}} \simeq 1.66 \times 10^{-2} \text{ np-m}^{-1}$$

corresponding to an attenuation rate of ~ 0.145 dB-m^{-1}. Similarly, at 30 GHz the attenuation rate is $\sim 6.01 \times 10^{-2}$ dB-m^{-1}.

For the TM$_1$ mode at 20 GHz, we have

$$\alpha_{c_{\text{TM}_1}} = \frac{2R_s}{\eta a \sqrt{1 - (f_{c_1}/f)^2}} \simeq \frac{2 \times 3.69 \times 10^{-2}}{377 \times 10^{-2} \times \sqrt{1 - (15/20)^2}} \simeq 2.96 \times 10^{-2} \text{ np-m}^{-1}$$

or an attenuation rate of ~ 0.257 dB-m^{-1}. At 30 GHz, the attenuation rate is ~ 0.240 dB-m^{-1}.

Attenuation Due to Dielectric Losses   We have so far considered losses due to imperfectly conducting parallel plate walls. In practice, losses also occur because of imperfections of the dielectric material between the conductors. Such losses can typically be accounted for by taking the permittivity of the dielectric to be complex, or

$$\epsilon_c = \epsilon' - j\epsilon''$$

For the TEM mode of propagation, simple substitution of $\epsilon_c = \epsilon' - j\epsilon''$ for ϵ in all field expressions gives an attenuation constant in the same manner as we obtained the fields for a uniform TEM plane wave in a lossy dielectric (Section 2.3) by making the substitution $\sigma \rightarrow \omega\epsilon''$.

For TE or TM modes, the analysis of the field structures (i.e., mode configurations) was developed by assuming $\bar{\gamma}^2$ to be real ($\bar{\gamma}^2 = \bar{\alpha}^2$ or $-\bar{\beta}^2$). If the dielectric losses are substantial, the field configurations may well have to be different; however, in most cases, the losses are small enough that we can assume that the character of the fields remains the same, and that the modes and their cutoff frequencies can be calculated assuming no losses. Consider the expression [4.16] for the propagation constant

$$\bar{\gamma} = j[\mu\epsilon(\omega^2 - \omega_{c_m}^2)]^{1/2} = j\left[\omega^2\mu\epsilon - \left(\frac{m\pi}{a}\right)^2\right]^{1/2}$$

---

[29] As discussed in Section 2.3, conversion from np-m^{-1} to dB-m^{-1} simply requires multiplication by 20 $\log_{10} e \simeq 8.686$.

where we have used that $\omega_{c_m} = \pi m/(a\sqrt{\mu\epsilon})$ to show the dependence of $\bar{\gamma}$ on ϵ explicitly. Now substituting for $\epsilon \to \epsilon' - j\epsilon''$, we have

$$\bar{\gamma} = j\left[\omega^2\mu\,(\epsilon' - j\epsilon'') - \left(\frac{m\pi}{a}\right)^2\right]^{1/2}$$

$$= j\left[\omega^2\mu\epsilon' - \left(\frac{m\pi}{a}\right)^2\right]^{1/2}\left[1 - \frac{j\,\omega^2\mu\epsilon''}{\omega^2\mu\epsilon' - \left(\frac{m\pi}{a}\right)^2}\right]^{1/2}$$

$$\simeq j\left[\omega^2\mu\epsilon' - \left(\frac{m\pi}{a}\right)^2\right]^{1/2}\left\{1 - \frac{j\,\omega^2\mu\epsilon''}{2\left[\omega^2\mu\epsilon' - \left(\frac{m\pi}{a}\right)^2\right]} + \cdots\right\}$$

$$\boxed{\bar{\gamma} \simeq \frac{\omega^2\mu\epsilon''}{2\left[\omega^2\mu\epsilon' - \left(\frac{m\pi}{a}\right)^2\right]^{1/2}} + j\left[\omega^2\mu\epsilon' - \left(\frac{m\pi}{a}\right)^2\right]^{1/2}}$$

where we have used the binomial expansion, given in footnote 34 of Chapter 2. For a dielectric with complex permittivity, the cutoff frequency ω_{c_m} can naturally be defined in terms of the real part of ϵ_c, namely $\omega_{c_m} = \pi m/(a\sqrt{\mu\epsilon'})$, in which case we can rewrite $\bar{\gamma}$ as

$$\bar{\gamma} \simeq \alpha_d + j\bar{\beta}_m \simeq \frac{\omega\sqrt{\mu\epsilon'}\,\epsilon''/\epsilon'}{2\sqrt{1 - (\omega_{c_m}/\omega)^2}} + j\sqrt{(\omega^2 - \omega_{c_m}^2)\mu\epsilon'} \qquad [4.25]$$

Thus, we see that the phase constant (i.e., the imaginary part of $\bar{\gamma}$) is

$$\bar{\beta}_m = \sqrt{(\omega^2 - \omega_{c_m}^2)\mu\epsilon'} \qquad [4.26]$$

which is simply the waveguide propagation constant for a lossless dielectric with permittivity ϵ'. The attenuation constant due to dielectric losses is the real part of $\bar{\gamma}$ in [4.25]:

$$\alpha_d \simeq \frac{\omega\sqrt{\mu\epsilon'}\,\epsilon''/\epsilon'}{2\sqrt{1 - (\omega_{c_m}/\omega)^2}} \qquad [4.27]$$

Note that in most cases the dielectric materials used are nonmagnetic, so $\mu = \mu_0$. Expressions [4.26] and [4.27] for α_d and β_m are valid for dielectrics with small losses such that $\epsilon'' \ll \epsilon'$. Note that equation [4.27] for α_d is valid for all TE or TM modes, and in fact also for all cross-sectional shapes (including the cylindrical waveguides discussed in Chapter 5) of the waveguide, as long as f_{c_m} is the cutoff frequency for the particular mode under consideration. The attenuation rate due to dielectric losses for the TEM mode can be found from [4.27] by substituting $\omega_{c_m} = 0$, which gives $\alpha_d \simeq \omega\epsilon''\sqrt{\mu_0/\epsilon'}/2$.

4.1.5   Voltage, Current, and Impedance in a Parallel-Plate Waveguide

The TEM wave on a parallel-plate line is the propagation mode that corresponds to the voltage and current waves on two-conductor transmission lines that are the subject of most transmission line analyses and applications.[30] The voltage between the two conductors and the total current flowing through a 1-m wide segment of each of the plates are related to the electric and magnetic fields of the TEM wave as

$$V_{\text{TEM}}(z) = -\int_{x=0}^{a} \mathbf{E}\cdot d\mathbf{l} = \int_{a}^{0} E_x(z)\,dx = \frac{\beta}{\omega\epsilon}C_4 e^{-j\beta z}\int_{a}^{0} dx = V^+ e^{-j\beta z}$$

$$I_{\text{TEM}}(z) = |\mathbf{J}_s|(1\text{ m}) = |\hat{\mathbf{n}}\times\mathbf{H}| = |\pm\hat{\mathbf{x}}\times\hat{\mathbf{y}}[H_x]_{x=0}| = |\hat{\mathbf{x}}\times\hat{\mathbf{y}}C_4 e^{-j\beta z}|$$

$$= C_4 e^{-j\beta z} = \frac{V^+}{Z_{\text{TEM}}}e^{-j\beta z}$$

where we have noted that for the TEM mode $\bar{\beta} = \beta = \omega\sqrt{\mu\epsilon}$, we have defined $V^+ = -\beta C_4 a/(\omega\epsilon) = -\eta a C_4$ as the peak voltage, and we have introduced the quantity Z_{TEM}, called the wave impedance, also discussed in Section 3.3.2. The wave impedance is defined as the ratio of the transverse components of the electric and magnetic fields:

$$Z_{\text{TEM}} \equiv \frac{E_x}{H_y} = \frac{[\beta/(\omega\epsilon)]C_4 e^{-j\beta z}}{C_4 e^{-j\beta z}} = \sqrt{\frac{\mu}{\epsilon}} = \eta$$

If we identify $V(z) = V_{\text{TEM}}$, $I(z) = I_{\text{TEM}}$, and $Z_0 = Z_{\text{TEM}}$, respectively, as the line voltage, line current, and characteristic line impedance, all of the standard transmission line analysis techniques are applicable to parallel-plate waveguides operating in the TEM mode. As an example, the time-average power flowing down the transmission line, given by $\frac{1}{2}|V^+|^2/Z_0$, is equal to the time-average transmitted power P_{av} evaluated in the previous section using the phasor expressions for the electric and magnetic fields. We note that the phase velocity $\bar{v}_p = \omega/\beta = 1/\sqrt{\mu\epsilon}$ and the wave impedance Z_{TEM} for TEM waves are both independent of frequency.

---

[30]See Chapters 2 and 3 of U. S. Inan and A. S. Inan, *Engineering Electromagnetics*, Addison Wesley Longman, 1999.

In general, for TE or TM modes, the definitions of equivalent voltage and current become ambiguous. To see this, consider the propagating TE_m mode, the fields for which are

$$E_y = C_1 \sin\left(\frac{m\pi}{a}x\right)e^{-j\bar{\beta}z}$$

$$H_z = -\frac{m\pi}{j\omega\mu a}C_1 \cos\left(\frac{m\pi}{a}x\right)e^{-j\bar{\beta}z}$$

$$H_x = -\frac{\bar{\beta}}{\omega\mu}C_1 \sin\left(\frac{m\pi}{a}x\right)e^{-j\bar{\beta}z}$$

where $\bar{\beta} = \sqrt{\omega^2\mu\epsilon - (m\pi/a)^2}$. We note that the wave electric field is in the y direction, so the quantity that we would normally define as voltage—that is, the line integral of the electric field from one plate to another—is zero. In other words,

$$-\int_0^a \mathbf{E} \cdot d\mathbf{l} = -\int_0^a \hat{\mathbf{y}}E_y \cdot \hat{\mathbf{x}}\, dx = 0$$

Similarly, the line current, which we may normally think of as a z-directed current in the plates, actually flows in the y direction. To see this, note that the only nonzero magnetic field at the surfaces of the conducting plates is H_z, which requires a surface current of

$$\mathbf{J}_s = \hat{\mathbf{n}} \times \hat{\mathbf{z}}[H_z]_{x=0} = \hat{\mathbf{x}} \times \hat{\mathbf{z}}\left(\frac{-m\pi}{j\omega\mu a}\right)e^{-j\bar{\beta}z} = \hat{\mathbf{y}}\left(\frac{m\pi}{j\omega\mu a}\right)e^{-j\bar{\beta}z}$$

along the $x = 0$ plate. In practice, the lack of unique definitions of voltage and current for TE and TM modes does not preclude the use of transmission line techniques in analyzing the behavior of waveguides or other microwave circuit components. In most cases, useful definitions suitable for the particular problems at hand can be put forth.[31]

It is common practice to define voltage and current in such a way that (i) the line voltage is proportional to the transverse component of the wave electric field, (ii) the line current is taken to be proportional to the transverse component of the wave magnetic field, and (iii) the product of the line voltage and current, $\frac{1}{2}\mathcal{R}e\{VI^*\}$, is equal to the time-average power transmitted P_{av}. The ratio of the voltage to current for a forward propagating wave is defined as the characteristic impedance of the transmission system. This impedance is usually taken to be the wave impedance, defined as the ratio of the transverse components of the electric and magnetic fields:

$$Z_{\text{TM or TE}} \equiv \frac{E_x}{H_y} = -\frac{E_y}{H_x} \qquad\qquad [4.28]$$

[31] See R. E. Collin, *Foundations of Microwave Engineering*, 2nd ed., McGraw-Hill, New York, 1992.

For propagating TE_m modes (i.e., $f > f_{c_m}$ so that $\overline{\gamma} = j\overline{\beta}$), using [4.12] and [4.16] we have

$$Z_{TE_m} = -\frac{E_y}{H_x} = -\frac{C_1 \sin\left(\frac{m\pi}{a}x\right)e^{-j\overline{\beta}z}}{-\frac{\overline{\beta}}{\omega\mu}C_1 \sin\left(\frac{m\pi}{a}x\right)e^{-j\overline{\beta}z}} = \frac{\omega\mu}{\overline{\beta}} = \frac{\eta}{\sqrt{1 - (f_{c_m}/f)^2}} \qquad [4.29]$$

where $\eta = \sqrt{\mu/\epsilon}$ is the intrinsic impedance of the dielectric inside the waveguide and we have used the fact that $\overline{\beta} = \beta\sqrt{1 - (f_{c_m}/f)^2}$, with f_{c_m} being the cutoff frequency of the TE_m mode. We thus see that the wave impedance of propagating TE_m modes in a parallel-plate waveguide with a lossless dielectric is purely resistive and is always *larger* than the intrinsic impedance of the dielectric medium. The higher the operating frequency, the closer Z_{TE_m} is to η, meaning that the wave is more like a TEM wave. This behavior can be understood in terms of the component TEM wave representation of the TE and TM modes discussed in Section 4.1.3. The higher the operating frequency, the smaller is the wavelength, which in turn means that more half-wavelengths "fit" into the waveguide dimension a and that the angle θ_{i_m} that the component TEM wave ray path makes with the normal to the plates is larger. In other words, the higher the frequency, the closer are the ray paths of the component TEM waves to being parallel to the waveguide walls along the axial direction. The frequency dependence of Z_{TE_m} is illustrated in Figure 4.12. At frequencies below cutoff, $f < f_{c_m}$, when we have $\overline{\gamma} = \overline{\alpha}$ and thus an attenuating (evanescent) wave, it is clear from [4.29] that the wave impedance becomes purely imaginary. A purely reactive wave impedance is consistent with the fact that there is no electromagnetic power flow associated with evanescent waves.

The following general relation between the electric and magnetic fields holds for TE_m waves:

$$\mathbf{E} = -Z_{TE_m}(\hat{\mathbf{z}} \times \mathbf{H}) \qquad [4.30]$$

Equation [4.30] is valid for TE waves in any waveguide, including rectangular and circular waveguides, to be studied in Chapter 5.

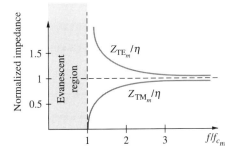

FIGURE 4.12. **TE_m and TM_m wave impedances.** The frequency dependence of normalized wave impedances (normalized to η) for TE_m and TM_m modes. Note that $\eta = \sqrt{\mu/\epsilon}$ and $f_{c_m} = m/(2a\sqrt{\mu\epsilon})$.

For TM_m waves, the wave impedance is defined in the same manner, namely, as the ratio of the transverse components of the wave electric and magnetic fields as given in [4.28]. Using the field expressions derived earlier for propagating TM_m modes, we have

$$Z_{TM_m} = \frac{E_x}{H_y} = \frac{[\bar{\beta}/(\omega\epsilon)]C_4 \cos\left(\frac{m\pi}{a}x\right)e^{-j\bar{\beta}z}}{C_4 \cos\left(\frac{m\pi}{a}x\right)e^{-j\bar{\beta}z}} = \frac{\bar{\beta}}{\omega\epsilon} = \eta\sqrt{1 - \left(\frac{f_{c_m}}{f}\right)^2} \quad [4.31]$$

Equation [4.31] indicates that the wave impedance of propagating TM_m modes in a parallel-plate waveguide with a lossless dielectric is purely resistive and is always *less* than the intrinsic impedance of the dielectric medium. The frequency dependence of Z_{TM_m} is illustrated in Figure 4.12. At frequencies below cutoff ($f \leq f_{c_m}$) the wave impedance Z_{TM_m} is purely imaginary, as expected for an evanescent wave. At frequencies much larger than the cutoff frequency ($f \gg f_{c_m}$), the wave impedance approaches the intrinsic impedance of the medium, as the TM_m wave becomes more and more like a TEM wave.

The following general relation between the electric and magnetic fields holds for TM_m waves:

$$\mathbf{H} = \frac{1}{Z_{TM_m}}(\hat{\mathbf{z}} \times \mathbf{E}) \quad [4.32]$$

As for the case of TE_m waves given by [4.30], equation [4.32] is valid for TM_m waves in any waveguide, including rectangular and circular waveguides, to be studied in Chapter 5.

4.1.6 Electric and Magnetic Field Distributions

The mode field distributions shown in Figures 4.2 and 4.3 are useful tools for visualizing the behavior of the particular modes shown. Using the fact that the electric and magnetic fields are locally tangent to the corresponding lines shown in such a plot, it is possible to derive mathematical expressions for the electric and magnetic field lines. Considering first the TE modes, we note that the time-domain expressions for the propagating TE_m mode can be obtained from the phasor fields found earlier.

$$\mathcal{E}_y(x, z, t) = \mathcal{R}e\{E_y^0(x)e^{-j\bar{\beta}z}\} = C_1 \sin\left(\frac{m\pi}{a}x\right)\cos(\omega t - \bar{\beta}z)$$

$$\mathcal{H}_x(x, z, t) = -\frac{\bar{\beta}}{\omega\mu}C_1 \sin\left(\frac{m\pi}{a}x\right)\cos(\omega t - \bar{\beta}z)$$

$$\mathcal{H}_z(x, z, t) = -\frac{m\pi}{\omega\mu a}C_1 \cos\left(\frac{m\pi}{a}x\right)\sin(\omega t - \bar{\beta}z)$$

Similarly, for the propagating TM_m mode, we have

$$\mathcal{E}_x(x, z, t) = \frac{\overline{\beta}}{\omega\epsilon} C_4 \cos\left(\frac{m\pi}{a}x\right)\cos(\omega t - \overline{\beta}z)$$

$$\mathcal{E}_z(x, z, t) = -\frac{m\pi}{\omega\epsilon a} C_4 \sin\left(\frac{m\pi}{a}x\right)\sin(\omega t - \overline{\beta}z)$$

$$\mathcal{H}_y(x, z, t) = C_4 \cos\left(\frac{m\pi}{a}x\right)\cos(\omega t - \overline{\beta}z)$$

For the TE_m modes, the magnetic field consists of two components, $\mathcal{H}_x(x, z, t)$ and $\mathcal{H}_z(x, z, t)$. We can find equations for the magnetic field lines by noting that

$$\frac{dx}{dz} = \frac{\mathcal{H}_x(x, z, t)}{\mathcal{H}_z(x, z, t)} = \left(\frac{\overline{\beta}a}{m\pi}\right)\frac{\sin\left(\frac{m\pi}{a}x\right)\cos(\omega t - \overline{\beta}z)}{\cos\left(\frac{m\pi}{a}x\right)\sin(\omega t - \overline{\beta}z)}$$

Rearranging terms, we can write

$$\frac{\left(\frac{m\pi}{a}\right)\cos\left(\frac{m\pi}{a}x\right)dx}{\sin\left(\frac{m\pi}{a}x\right)} = \frac{\overline{\beta}\cos(\omega t - \overline{\beta}z)\,dz}{\sin(\omega t - \overline{\beta}z)}$$

Substituting

$$\left(\frac{m\pi}{a}\right)\cos\left(\frac{m\pi}{a}x\right)dx = d\left[\sin\left(\frac{m\pi}{a}x\right)\right]$$

and

$$\overline{\beta}\cos(\omega t - \overline{\beta}z)\,dz = -d[\sin(\omega t - \overline{\beta}z)]$$

we can rewrite the preceding expression as

$$\frac{d\left[\sin\left(\frac{m\pi}{a}x\right)\right]}{\sin\left(\frac{m\pi}{a}x\right)} = -\frac{d[\sin(\omega t - \overline{\beta}z)]}{\sin(\omega t - \overline{\beta}z)}$$

Integrating both sides yields

$$\ln\left[\sin\left(\frac{m\pi}{a}x\right)\right] = -\ln[\sin(\omega t - \overline{\beta}z)] + \text{const.}$$

or

$$\sin\left(\frac{m\pi}{a}x\right)\sin(\omega t - \overline{\beta}z) = \text{const.}$$

which represents the equation describing the locus of the magnetic field lines in the xz plane. Plots of the field lines for TE_1 and TE_2 were shown in Figure 4.2.

Using a similar procedure, we can find the locus in the xz plane of the electric field lines for TM_m modes. We find

$$\cos\left(\frac{m\pi}{a}x\right)\sin(\omega t - \overline{\beta}z) = \text{const.}$$

The resultant field distributions were shown in Figure 4.3.

4.2 DIELECTRIC WAVEGUIDES

We have seen that the parallel-plate waveguide with metallic boundaries can effectively guide electromagnetic energy in the axial (i.e., z) direction. The metallic waveguides are very effective for microwave frequencies, at which the conductivity of the walls is high and the reflectivity is correspondingly good. However, at optical frequencies, the dimensions of metallic waveguides become too small and wall losses become unacceptably large. In addition, it is difficult to fabricate waveguides with the required precision. Because light has a characteristic wavelength on the order of 1 μm, waveguides must have a cross-sectional dimension of around the same order.

Although research work to extend the wavelength range of metallic waveguides to the μm range continues,[32] a low-cost and potentially much more powerful alternative guiding structure that works extremely well at optical frequencies is the optical fiber. Although detailed coverage of optical fiber transmission systems is well beyond the scope of this book,[33] we consider in this section the so-called dielectric waveguide, which consists of a slab of dielectric surrounded by another dielectric. These structures lend themselves to complete analytical solution and also encompass the basic principles of wave propagation in optical fibers. In addition, waveguiding structures consisting of rectangular dielectric slabs or dielectric slabs on a conducting surface are used in microwave integrated circuits, in integrated optics, and in optoelectronics applications. There are no fundamental reasons why dielectric waveguides should not be used at lower microwave frequencies, except that metal waveguides are more practical, provide better shielding from interference, and allow operation at higher power levels.

In a dielectric waveguide, there are no conducting surfaces, and the electromagnetic energy is guided through the structure not by reflections from conducting metal surfaces but by means of total internal reflections from dielectric boundaries. The potential guiding of electromagnetic waves by dielectric structures was recognized in the late nineteenth century,[34] and a theoretical solution was obtained around

---

[32] Two recent papers were cited in the beginning of this chapter; another example of recent work is Y. Matsuura, A. Hongo, and M. Miyagi, Dielectric-coated metallic hollow waveguide for 3-μm Er:YAG, 5-μm CO, and 10.6-μm CO_2 laser light transmission, *Appl. Optics*, 29(15), pp. 2213–2217, May 1990.

[33] For a recent book, see D. J. H. Maclean, *Optical Line Systems*, Wiley, New York, 1996.

[34] The first demonstration of the guiding of light by a thin stream of water was achieved by John Tyndall in 1870.

the turn of the century, in 1910.[35] Early experimental efforts considered water to be a good dielectric candidate and involved attempts to transmit waves along water-filled metallic tubes. Initial results were interpreted as being due to guiding by the dielectric, but it was soon realized that the water was unnecessary and that the wave was actually supported by the metal tube.[36] Although technological demonstrations of transmission of light through optical fibers were first realized in the 1930s, the practical use of fiber technology became feasible only after the invention of the laser in 1960 and the development of low-loss fibers in the late 1970s. At present, most optical fibers operate in the 1.3 μm and the 1.5 μm optical windows. The former has the advantage of being the wavelength of minimum distortion, and the latter provides lower loss (0.2 dB-(km)$^{-1}$ as opposed to 0.4 dB-(km)$^{-1}$ for 1.3 μm).

In this section, we illustrate how dielectric regions with boundaries in the transverse direction can support electromagnetic fields propagating in the axial direction. The guiding action of dielectric waveguides is more complicated than for metal waveguides, and it is often not possible to obtain analytical solutions. For simplicity, we confine our attention in this section to structures with planar geometry, namely, the dielectric slab waveguide. The dielectric slab waveguide is extensively used in integrated optics systems to transfer light between components. The basic principles and techniques of guiding light waves by dielectric boundaries are well illustrated by our coverage of this planar structure. These techniques can be readily extended to cylindrical and other geometries, as we do in Section 5.2.3 to analyze the dielectric waveguide of circular cross section, constituting the simplest example of an optical fiber.

4.2.1   Dielectric Slab Waveguide: Mode Theory

In general, analysis of dielectric waveguides can be based on "mode theory" or "ray theory." Here, we first discuss mode theory by considering wave propagation along a dielectric slab waveguide as shown in Figure 4.13, consisting of a rectangular dielectric slab sandwiched by free space. The procedure we follow to arrive at the field solutions is similar to that used for parallel-plate or rectangular metallic waveguides. We take advantage of the fact that the structure is infinite in the y direction and look for solutions of the type $e^{-\gamma z}$ in terms of the z variation. We also separately investigate the case of TM or TE modes. We first consider TM waves for which equations are written in terms of $E_z(x, z)$.

Unlike the case of metallic waveguides, dielectric waveguides in general have nonzero fields not only inside but also outside the dielectric slab. In the free space region, waves must decay exponentially in the transverse direction, in order for the energy to be in any sense "guided" by the dielectric slab.

---

[35]D. Hondros and P. Debye, Elektromagnetische Wellen an dielektrischen Drähten, *Ann. d. Phys.*, 49(4), p. 465, 1910.

[36]See K. S. Packard, The origin of waveguides: A case of multiple rediscovery, *IEEE Trans. Microwave Theory and Techniques*, 32(9), pp. 961–969, September 1984, and references therein.

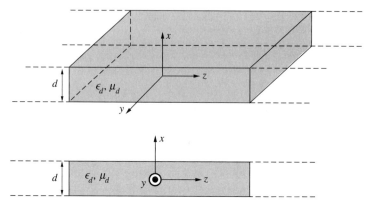

FIGURE 4.13. **Geometry of the dielectric slab waveguide.** (a) Perspective view. (b) Side view.

Transverse Magnetic (TM) Waves Noting that there is no y dependence (i.e., $\partial(\cdot)/\partial y = 0$) for any of the field components, we have

$$E_z(x, z) = E_z^0(x)e^{-\bar{\gamma}z}$$

where $E_z^0(x)$ is that portion of $E_z(x, z)$ that depends on x and is governed by a wave equation similar to [4.10], which was used in Section 4.1 to solve for the E_y component of TE modes in a parallel-plate waveguide. We have

$$\frac{d^2 E_z^0(x)}{dx^2} + h^2 E_z^0(x) = 0 \qquad [4.33]$$

where $h^2 = \bar{\gamma}^2 + \omega^2 \mu \epsilon$. Note that, unlike the case of the parallel-plate waveguide, the wave equation [4.33] must be satisfied both within and outside the dielectric, since the fields are in general nonzero both inside and outside. Also, as in the case of [4.10], we have in general $h^2 = \bar{\gamma}^2 + \omega^2 \mu \epsilon$, so that $h^2 = h_d^2 = \bar{\gamma}^2 + \omega^2 \mu_d \epsilon_d$ in the dielectric region and $h^2 = h_0^2 = \bar{\gamma}^2 + \omega^2 \mu_0 \epsilon_0$ in the surrounding free space.[37] As before, h are the eigenvalues to be determined by the boundary conditions, and $\bar{\gamma}$ can be either purely real ($\bar{\gamma} = \bar{\alpha}$, for evanescent, nonpropagating waves) or purely imaginary ($\bar{\gamma} = j\bar{\beta}$, for propagating waves). We look for forms of solutions appropriate for inside and outside the slab and then match them via boundary conditions at the interfaces. Inside the slab we have

$$E_z^0(x) = C_o \sin(\beta_x x) + C_e \cos(\beta_x x) \qquad |x| \le \frac{d}{2} \qquad [4.34]$$

where β_x is the transverse phase constant and is given by

$$\beta_x^2 = \omega^2 \mu_d \epsilon_d - \bar{\beta}^2 = h_d^2 \qquad [4.35]$$

---

[37] In practice, the medium outside the dielectric waveguide is actually another dielectric with $\epsilon < \epsilon_d$. In such a case, all of the following solutions apply if we replace ϵ_0 with ϵ.

assuming propagating solutions with $\overline{\gamma} = j\overline{\beta}$. The quantities μ_d and ϵ_d are the permeability and permittivity for the dielectric, which is assumed to be lossless. The subscripts on the constants C_o and C_e are the respective coefficients of the "odd" [i.e., $\sin(\beta_x x)$ is an odd function of x] and "even" [i.e., $\cos(\beta_x x)$ is an even function of x] parts of $E_z^0(x)$, which is the basis for classification of the different modes.

In order for the dielectric slab to be "guiding" the wave energy, the wave fields must be confined to the vicinity of the slab. Thus, outside the slab the fields of the propagating wave must decay exponentially with distance from the slab. In other words, we must have

$$E_z^0(x) = \begin{cases} C_a e^{-\alpha_x(x-d/2)} & x \geq \dfrac{d}{2} \\[4mm] C_b e^{\alpha_x(x+d/2)} & x \leq -\dfrac{d}{2} \end{cases} \qquad [4.36]$$

where the subscripts on the constants C_a and C_b are respectively associated with the solutions "above" and "below" the slab, and where α_x is the transverse attenuation constant, given by

$$\alpha_x^2 = \overline{\beta}^2 - \omega^2\mu_0\epsilon_0 = -h_0^2 \quad \text{or} \quad \alpha_x = \sqrt{\overline{\beta}^2 - \omega^2\mu_0\epsilon_0} > 0 \qquad [4.37]$$

with μ_0 and ϵ_0 being the permeability and permittivity of free space. Note that we must have $\alpha_x > 0$ for physically realizable solutions.

To obtain the full solutions, the relationships between C_o, C_e, C_a, C_b need to be determined and the values of β_x and α_x need to be found. In general, the odd and even mode fields have x dependencies as shown in Figure 4.14.

The other field components can be obtained from E_z using [4.7b,c] and noting that $E_z(x, z)$ varies as $E_z^0(x)e^{-j\overline{\beta}z}$. Because E_z depends only on x and z and because we have assumed $H_z = 0$ (i.e., TM mode), from [4.7b], $H_y(x, z)$ is the only nonzero magnetic field component. However, because E_z varies with z, and because there is

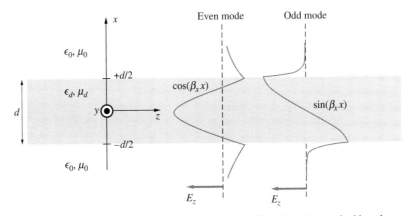

FIGURE 4.14.   Field distributions of even and odd modes. Even and odd mode field distributions in a dielectric slab. The case shown corresponds to 30 GHz operation with $d = 0.75$ cm and $\epsilon_d = 2\epsilon_0$.

no variation of any field quantities with y, the electric field must have an x component in order that $\nabla \cdot \mathbf{E} = 0$.

We now examine the *odd* and *even* TM modes separately.[38]

Odd TM modes: Odd TM modes are the solutions for which we have $C_e = 0$, so that

$$E_z^0(x) = C_o \sin(\beta_x x) \qquad |x| \leq \frac{d}{2} \qquad [4.38]$$

We can use the fact that E_z is continuous across the boundary at $x = \pm d/2$ to determine C_a and C_b in terms of C_o. The field components for odd TM modes are then found to be

Free space above the slab ($x \geq d/2$)

$$E_z^0(x) = \left[C_o \sin\left(\frac{\beta_x d}{2}\right) \right] e^{-\alpha_x(x-d/2)}$$

$$E_x^0(x) = -\frac{j\overline{\beta}}{\alpha_x} \left[C_o \sin\left(\frac{\beta_x d}{2}\right) \right] e^{-\alpha_x(x-d/2)}$$

$$H_y^0(x) = \frac{j\omega\epsilon_0}{\alpha_x} \left[C_o \sin\left(\frac{\beta_x d}{2}\right) \right] e^{-\alpha_x(x-d/2)}$$

Dielectric region ($|x| \leq d/2$)

$$E_z^0(x) = C_o \sin(\beta_x x)$$

$$E_x^0(x) = -\frac{j\overline{\beta}}{\beta_x} C_o \cos(\beta_x x) \qquad [4.39]$$

$$H_y^0(x) = \frac{j\omega\epsilon_d}{\beta_x} C_o \cos(\beta_x x)$$

Free space below the slab ($x \leq -d/2$)

$$E_z^0(x) = \left[C_o \sin\left(\frac{\beta_x d}{2}\right) \right] e^{\alpha_x(x+d/2)}$$

$$E_x^0(x) = -\frac{j\overline{\beta}}{\alpha_x} \left[C_o \sin\left(\frac{\beta_x d}{2}\right) \right] e^{\alpha_x(x+d/2)}$$

$$H_y^0(x) = \frac{j\omega\epsilon_0}{\alpha_x} \left[C_o \sin\left(\frac{\beta_x d}{2}\right) \right] e^{\alpha_x(x+d/2)}$$

---

[38] We choose to define the odd and even designations with respect to the $\sin(\beta_x x)$ and $\cos(\beta_x x)$ functional form of the transverse variation of the axial (E_z for TM and H_z for TE) field component. In general, the odd/even classification can also be made with respect to the functional forms of the other components, such as H_y for TM modes or E_y for TE modes. The modes so identified would in general be different from those identified on the basis of the axial components; in other words, for example, the odd TM$_3$ mode defined on the basis of the oddness/evenness of the H_y component would correspond to a different distribution of fields than the odd TM$_3$ defined on the basis of the functional form of the E_z component. The relative arbitrariness of this nomenclature is important to note here since different definitions are used in different texts.

We observe from [4.39] that the time-average electromagnetic power flow is in the z direction; that is, H_y and E_x are in time phase. Also, E_z and H_y are 90° out of phase, so that no real average power flows in the x direction.

We can now apply the boundary conditions to determine β_x and α_x. Note that the continuity of E_z was used to find C_a and C_b in terms of C_0. The continuity of H_y at the dielectric surface, that is, at $x = \pm d/2$, requires that

$$\text{Odd TM modes:} \qquad \frac{\alpha_x}{\beta_x} = \frac{\epsilon_0}{\epsilon_d} \tan\left(\frac{\beta_x d}{2}\right) \qquad\qquad \text{[4.40]}$$

From expressions [4.35] and [4.37] for β_x and α_x, we have

$$\alpha_x^2 + \beta_x^2 = \omega^2(\mu_d\epsilon_d - \mu_0\epsilon_0) \qquad\qquad \text{[4.41]}$$

or

$$\alpha_x = [\omega^2(\mu_d\epsilon_d - \mu_0\epsilon_0) - \beta_x^2]^{1/2} \qquad\qquad \text{[4.42]}$$

which are valid for both odd and even modes.

For given values of ω, μ_d, ϵ_d, and d, [4.40] and [4.42] can be plotted versus α_x and β_x and a graphical solution for α_x and β_x can be obtained. An example for the case of $\epsilon_d = 2\epsilon_0$ and $d = 1.25\lambda$, where λ is the free space wavelength ($\lambda = c/f$), is illustrated in Figure 4.15. For the case shown, there are three solutions possible, denoted as S1, S2, and S3, which are respectively the odd TM_1, even TM_2, and odd TM_3 modes.

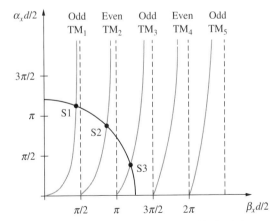

FIGURE 4.15.   Graphical solution for TM modes. Graphical solution representing transverse attenuation and phase constants for a dielectric slab waveguide. Both odd and even TM modes are shown. The quarter-circle is the contour corresponding to [4.42]; the curves marked "odd" correspond to [4.40] and the curves marked "even" correspond to [4.44]. The parameters chosen for this example are $\epsilon_d = 2\epsilon_0$ and $d = 1.25\lambda$, where λ is the free space wavelength.

Even TM modes: Even TM modes (see Figure 4.14) are the solutions obtained from

$$E_z^0(x) = C_e \cos(\beta_x x) \qquad |x| \leq \frac{d}{2} \qquad [4.43]$$

Much of the analysis is the same as above; instead of [4.40], we have

$$\boxed{\text{Even TM modes:} \qquad \frac{\alpha_x}{\beta_x} = -\frac{\epsilon_0}{\epsilon_d} \cot\left(\frac{\beta_x d}{2}\right)} \qquad [4.44]$$

which correspond to the curves marked "even TM" in Figure 4.15.

Propagation and Cutoff Characteristics   Note that the propagation constant $\overline{\beta}$ has values between the intrinsic propagation constant of free space and that of the dielectric, that is,

$$\omega\sqrt{\mu_0\epsilon_0} < \overline{\beta} < \omega\sqrt{\mu_d\epsilon_d}$$

As $\overline{\beta}$ approaches $\omega\sqrt{\mu_0\epsilon_0}$, we note that α_x approaches zero, indicating that there is no attenuation with distance from the dielectric, so that the wave energy is not "guided" by the slab and the wave fields are not confined to the vicinity of the slab. The condition $\alpha_x = 0$ thus defines the cutoff condition, and the cutoff frequencies at which this occurs can be determined from [4.42] by setting $\alpha_x = 0$:

$$\beta_x = \omega_c\sqrt{\mu_d\epsilon_d - \mu_0\epsilon_0}$$

In other words, guided propagation occurs only for frequencies $\omega > \omega_c$, for which α_x is real and positive.

The cutoff frequencies for a given mode can be determined by substituting into the respective equations for that mode. Using [4.40] and [4.44], we find

$$\tan\left(\frac{\beta_x d}{2}\right) = 0 \qquad \text{for odd TM modes}$$

$$\cot\left(\frac{\beta_x d}{2}\right) = 0 \qquad \text{for even TM modes}$$

evaluating at $\beta_x = \omega_c\sqrt{\mu_d\epsilon_d - \mu_0\epsilon_0}$, we find

$$\boxed{f_{c_{TM_m}} = \frac{(m-1)}{2d\sqrt{\mu_d\epsilon_d - \mu_0\epsilon_0}} \qquad \begin{array}{ll} m = 1, 3, 5, \ldots & \text{Odd TM}_m \\ m = 2, 4, 6, \ldots & \text{Even TM}_m \end{array}} \qquad [4.45]$$

as the cutoff frequencies.[39]

---

[39]Note once again that the "odd" and "even" designations simply refer respectively to the $\sin(\beta_x x)$ and $\cos(\beta_x x)$ functional forms of the transverse variation of the axial field (E_z for TM and H_z for TE) component. In addition, however, the dependence of f_{c_m} on the index m is written such that the odd/even values of m correspond respectively to odd/even modes.

For the odd TM_1 mode, $f_{c_{TM_1}} = 0$, so that this lowest-order odd TM mode can propagate for any slab thickness (analogous to the TEM mode on a parallel-plate waveguide).

The calculation of the propagating modes in a dielectric slab waveguide is illustrated in Example 4-6.

Example 4-6:   Propagating modes in a dielectric slab waveguide.   Consider a dielectric slab waveguide with $d = 0.75$ cm, $\mu_r = 1$, $\epsilon_r = 2$ surrounded by air, operating at 30 GHz. (a) Find all TM modes and cutoff frequencies for propagating modes. (b) Calculate β_x, α_x, $\overline{\beta}$, and $(\overline{\beta}/\beta_0)$ for propagating waves.

Solution: The only modes with cutoff frequencies smaller than 30 GHz are TM_1 (odd) and TM_2 (even). We have

$$f_{c_{TM_1}} = 0 \qquad f_{c_{TM_2}} = \frac{1}{(2 \times 0.725)\sqrt{\mu_r\mu_0\epsilon_r\epsilon_0 - \epsilon_0\mu_0}} \simeq 20 \text{ GHz}$$

To find the corresponding phase constants we can either carry out a graphical solution or solve a transcendental equation. Combining [4.40] and [4.42], we have for odd TM modes

$$\alpha_x = \frac{\beta_x\epsilon_0}{\epsilon_d}\tan\left(\frac{\beta_x d}{2}\right) = [\omega^2(\mu_d\epsilon_d - \mu_0\epsilon_0) - \beta_x^2]^{1/2}$$

and for even TM modes

$$\alpha_x = -\frac{\beta_x\epsilon_0}{\epsilon_d}\cot\left(\frac{\beta_x d}{2}\right) = [\omega^2(\mu_d\epsilon_d - \mu_0\epsilon_0) - \beta_x^2]^{1/2}$$

Solution of these equations gives the following roots:

TM_1 odd: $\beta_x \simeq 3.37$ rad-(cm)$^{-1}$ and TM_2 even: $\beta_x \simeq 5.89$ rad-(cm)$^{-1}$

The axial electric field distributions for the odd TM_1 and even TM_2 modes are shown in Figure 4.16.

From β_x, we can directly calculate α_x using [4.42]. We find

Odd TM_1	Even TM_2
$\beta_x \simeq 3.37$ rad-(cm)$^{-1}$	$\beta_x \simeq 5.89$ rad-(cm)$^{-1}$
$\alpha_x \simeq 5.31$ np-(cm)$^{-1}$	$\alpha_x \simeq 2.18$ np-(cm)$^{-1}$
$\simeq 46.16$ dB-(cm)$^{-1}$	$\simeq 18.92$ dB-(cm)$^{-1}$

The propagation constants in the axial direction can be obtained from [4.35], which is

$$\beta_x^2 = \omega^2\mu_d\epsilon_d - \overline{\beta}^2$$

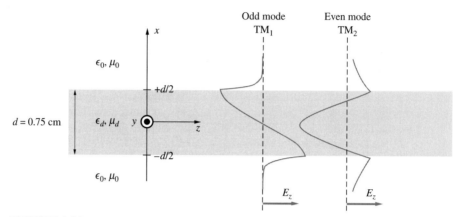

FIGURE 4.16. **The propagating modes for Example 4-6.** The plots show the variation of the axial component E_z over the vertical cross section of the dielectric slab. Parameters are $f = 30$ GHz, $d = 0.75$ cm, and $\epsilon_d = 2\epsilon_0$.

We find

$$\overline{\beta}_m \approx 7.95 \ \text{rad-(cm)}^{-1} \quad \rightarrow \quad \frac{\overline{\beta}}{\beta_0} \approx 1.26 \qquad \text{for TM}_1 \text{ (odd)}$$

$$\overline{\beta}_m \approx 6.48 \ \text{rad-(cm)}^{-1} \quad \rightarrow \quad \frac{\overline{\beta}}{\beta_0} \approx 1.03 \qquad \text{for TM}_2 \text{ (even)}$$

The dependencies of the mode structures on the operating frequency and slab thickness are illustrated in the lower two panels of Figure 4.17, for the same parameter values as in Example 4-6. As seen from the sketches, increasing the operating frequency increases the number of propagating modes in the guide. Also, for a fixed operating frequency, using a thicker slab increases the number of propagating modes.

Figures 4.16 and 4.17 show the transverse variation of the axial electric field for different TM modes. Figure 4.18 shows the complete electric and magnetic field distributions for the even TM$_2$ mode, in a manner similar to that given in Figure 4.3 for TM modes in a parallel-plate waveguide. In this plot, the intensity of the electric field at any point is indicated by the density of the field lines at that location. We see that the electric field intensity falls off (i.e., the density of lines decreases) with distance away from the slab, consistent with the $e^{-\alpha_x(x \pm d/2)}$ variation of the fields as given in [4.39].

Transverse Electric (TE) Modes The solutions for TE modes can be found in a completely analogous manner, for which we start with $H_z^0(x)$ and write similar solutions for inside and outside the slab. Note that we have

$$H_z^0(x) = C_o \sin(\beta_x x) + C_e \cos(\beta_x x) \qquad [4.46]$$

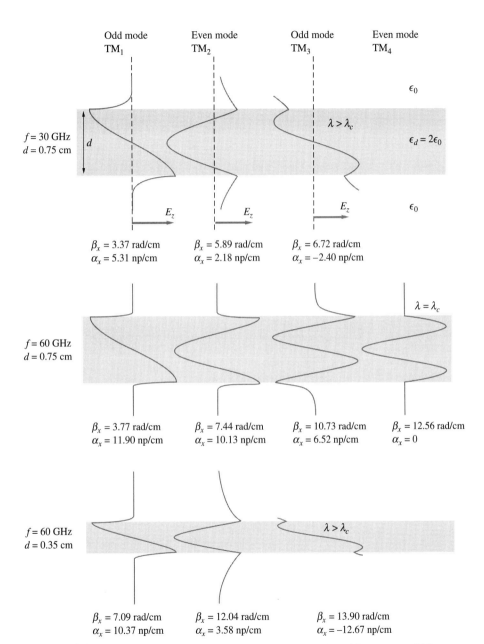

FIGURE 4.17.   Variation of field distributions for various modes with respect to thickness and frequency. Dependence of the dielectric waveguide mode structure on frequency and slab thickness. The plots show the variation of the axial field component E_z over the vertical cross section of the dielectric slab. For all cases, $\epsilon_d = 2\epsilon_0$. Note that the numerical values are the same as those for Example 4-6.

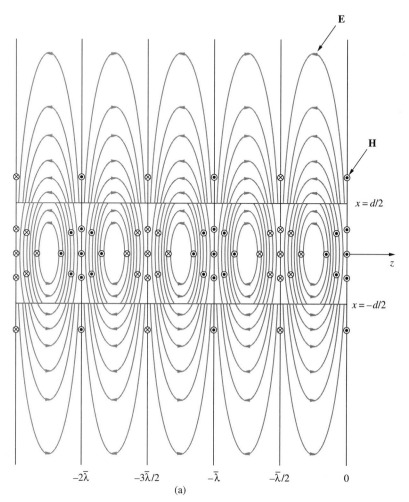

FIGURE 4.18. **Even TM$_2$ mode field distribution.** The electric field lines are shown as solid lines, while the magnetic field lines are orthogonal to the page and are indicated alternately with circles or crosses. Figure taken (with permission) from H. A. Haus, *Waves and Fields in Optoelectronics*, Prentice Hall, Englewood Cliffs, New Jersey, 1984.

and that the odd or even modes are now defined with respect to the functional form of $H_z^0(x)$ being respectively $\sin(\beta_x x)$ or $\cos(\beta_x x)$. The nonzero field components are now H_z, H_x, and E_y. We find the following relations

$$\frac{\alpha_x}{\beta_x} = \frac{\mu_0}{\mu_d} \tan\left(\frac{\beta_x d}{2}\right) \qquad \text{Odd TE modes} \qquad [4.47]$$

and

$$\boxed{\frac{\alpha_x}{\beta_x} = -\frac{\mu_0}{\mu_d}\cot\left(\frac{\beta_x d}{2}\right) \qquad \text{Even TE modes}}$$ [4.48]

Equations [4.47] and [4.48], as well as [4.42], which is valid for both odd and even modes, must be solved simultaneously to determine the values of β_x and α_x for the different modes.

Solutions can once again be obtained either graphically or by numerical solution of the transcendental equations. The cutoff frequencies are determined by the $\alpha_x = 0$ condition and are the same as for TM modes. An illustration of a graphical solution for TE modes is given in Figure 4.19. The three possible solutions correspond to the intersection points S1, S2, and S3 as marked. The case treated here is the same as that for which the TM mode solutions are shown in Figure 4.15, for which $\epsilon_d = 2\epsilon_0$ and $d = 1.25\lambda$, where λ is the free space wavelength.

Note that the cutoff frequencies for the TE_m modes are given by the same expression as that for TM_m modes, namely,

$$\boxed{f_{c_{TE_m}} = \frac{(m-1)}{2d\sqrt{\mu_d\epsilon_d - \mu_0\epsilon_0}} \qquad \begin{array}{ll} m = 1, 3, 5, \ldots & \text{Odd } TE_m \\ m = 2, 4, 6, \ldots & \text{Even } TE_m \end{array}}$$ [4.49]

For example, the lowest-order odd TE_1 mode has no cutoff frequency (similar to the odd TM_1 mode). The next higher mode (even TE_2 mode) has a cutoff frequency given by

$$f_{c_{TE_2}} = \frac{1}{2d\sqrt{\mu_d\epsilon_d - \mu_0\epsilon_0}}$$

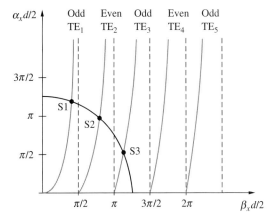

FIGURE 4.19.   **Graphical solution for TE modes.** Graphical solution representing attenuation and phase constants for TE modes in a dielectric slab waveguide. Both odd and even TE lower-order modes are shown. The parameters chosen for this example are $\epsilon_d = 2\epsilon_0$ and $d = 1.25\lambda$, where $\lambda = c/f$ is the free space wavelength.

4.2.2 Dielectric Covered Ground Plane

In the previous section we considered wave propagation in planar dielectric slabs without any other boundaries. In a wide range of applications, especially those involving miniature microwave circuits, guiding structures with dielectrics on a conducting ground plane or sandwiched between two conductors (microstrips) are used. In this subsection, we apply our previous analyses to a simple planar geometry as shown in Figure 4.20.

Note here that the boundary conditions on the metal end require both $E_y = 0$ and $E_z = 0$ at $x = 0$. We can obtain the possible solutions for the fields by examining the solutions obtained earlier for the dielectric slab. The electric field components of the propagating TM modes within the slab were

$$\left. \begin{aligned} E_y &= 0 \\[2mm] E_x &= \frac{j\overline{\beta}}{\beta_x} C_e \sin(\beta_x x)e^{-j\overline{\beta}z} \\[2mm] E_z &= C_e \cos(\beta_x x)e^{-j\overline{\beta}z} \end{aligned} \right\} \text{Even TM} \qquad 0 \le x \le \frac{d}{2}$$

$$\left. \begin{aligned} E_y &= 0 \\[2mm] E_x &= -\frac{j\overline{\beta}}{\beta_x} C_o \cos(\beta_x x)e^{-j\overline{\beta}z} \\[2mm] E_z &= C_o \sin(\beta_x x)e^{-j\overline{\beta}z} \end{aligned} \right\} \text{Odd TM} \qquad 0 \le x \le \frac{d}{2}$$

Note that even TM modes *do not* satisfy boundary conditions at $x = 0$ (i.e., $E_z \ne 0$). For TE modes, we have

$$\left. \begin{aligned} E_y &= \frac{-j\omega\mu}{\beta_x} C_e \sin(\beta_x x)e^{-j\overline{\beta}z} \\[2mm] E_x &= 0 \\[2mm] E_z &= 0 \end{aligned} \right\} \text{Even TE} \qquad 0 \le x \le \frac{d}{2}$$

$$\left. \begin{aligned} E_y &= \frac{j\omega\mu}{\beta_x} C_o \cos(\beta_x x)e^{-j\overline{\beta}z} \\[2mm] E_x &= 0 \\[2mm] E_z &= 0 \end{aligned} \right\} \text{Odd TE} \qquad 0 \le x \le \frac{d}{2}$$

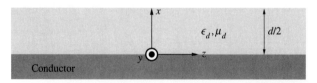

FIGURE 4.20. **Dielectric slab with a ground plane.** Geometry for a dielectric covered ground plane.

Note that odd TE modes also *do not* satisfy the boundary conditions at $x = 0$, since $E_y(x = 0, z) \neq 0$.

It thus appears that the geometry of the dielectric covered ground plane supports only TM odd and TE even modes. The governing equations for these modes are

$$\frac{\epsilon_0}{\epsilon_d} \beta_x \tan\left(\frac{\beta_x d}{2}\right) = \alpha_x \qquad \text{Odd TM}$$

$$-\frac{\mu_0}{\mu_d} \beta_x \cot\left(\frac{\beta_x d}{2}\right) = \alpha_x \qquad \text{Even TE}$$

and for all modes, we have

$$\left.\begin{array}{c} \beta_x^2 + \overline{\beta}^2 = \omega^2 \mu_d \epsilon_d \\ -\alpha_x^2 + \overline{\beta}^2 = \omega^2 \mu_0 \epsilon_0 \end{array}\right\} \quad \rightarrow \quad \alpha_x^2 + \beta_x^2 = \omega^2(\mu_d \epsilon_d - \mu_0 \epsilon_0)$$

The cutoff frequencies are

$$f_{c\text{TM or TE}} = \frac{(m-1)}{2d\sqrt{\mu_d \epsilon_d - \mu_0 \epsilon_0}} \qquad \begin{array}{c} m = 1, 3, 5, \ldots \\ m = 2, 4, 6, \ldots \end{array} \qquad \begin{array}{c} \text{Odd TM}_m \\ \text{Even TE}_m \end{array} \qquad \text{[4.50]}$$

The dominant mode is the odd TM_1 with cutoff frequency of zero. All of the modes are called surface wave modes, and solutions are obtained exactly in the same way as for the dielectric slab discussed before. The only difference between the dielectric slab and the dielectric above a ground plane is that the latter structure cannot support the propagation of the TM even and TE odd modes.

Note that the transverse attenuation rate α_x for $f > f_c$ is given as

$$\alpha_x = \sqrt{\overline{\beta}^2 - \omega^2 \mu_0 \epsilon_0}$$

For very thick dielectrics (d large compared with the transverse wavelength $\lambda_x = 2\pi/\beta_x$), $\overline{\beta}$ approaches $\beta_d = \omega\sqrt{\mu_d \epsilon_d}$ and, as a result, we have

$$\alpha_x \simeq \sqrt{\omega^2(\mu_d \epsilon_d - \mu_0 \epsilon_0)} = \beta\sqrt{\frac{\mu_d \epsilon_d}{\mu_0 \epsilon_0} - 1} \qquad \text{[4.51]}$$

which at microwave and higher frequencies is usually very large, indicating that the fields decrease rapidly with distance from the dielectric, or that the guiding provided by the slab is effective. For very thin dielectrics ($d \ll \lambda_x$), the propagation constant $\overline{\beta}$ approaches that of uniform plane waves in free space, that is, $\overline{\beta} = \omega\sqrt{\mu_0 \epsilon_0}$, and we have

$$\beta_x^2 = \omega^2 \mu_d \epsilon_d - \overline{\beta}^2 \simeq \omega^2(\mu_d \epsilon_d - \mu_0 \epsilon_0)$$

In the limit $(d/\lambda_x) \rightarrow 0$ we can also write

$$\alpha_x = \frac{\epsilon_0}{\epsilon_d} \beta_x \tan\left(\frac{\beta_x d}{2}\right) \simeq \frac{\epsilon_0}{\epsilon_d} \beta_x \left(\beta_x \frac{d}{2}\right) + \cdots \simeq \frac{d}{2} \frac{\epsilon_0}{\epsilon_d} \beta_x^2$$

Substituting in the preceeding expression for β_x, we find

$$\alpha_x \simeq \frac{d}{2} \frac{\epsilon_0}{\epsilon_d} \left[\omega^2 \mu_0 \epsilon_0 \left(\frac{\epsilon_d \mu_d}{\epsilon_0 \mu_0} - 1 \right) \right] = 2\pi\beta \left[\frac{\mu_d}{\mu_0} - \frac{\epsilon_d}{\epsilon_0} \right] \frac{d}{2\lambda} \qquad [4.52]$$

where β and λ are respectively the phase constant and wavelength in free space. The attenuation constant as given above for thin dielectrics (d small compared with λ_x) is usually quite small, indicating that the wave is not well guided (i.e., the fields are loosely confined to the dielectric).

Example 4-7 considers a polystyrene slab waveguide on a ground plane.

Example 4-7: Polystyrene waveguide with ground plane. A ground plane is covered with polystyrene ($\epsilon_r = 2.56$) of thickness ($d/2$). Determine the distance above the dielectric air interface over which the fields decrease to e^{-1} of their value at the interface, for the cases in which (i) d is very large ($d \gg \lambda$) and (ii) d is very small (assume $d \simeq 2 \times 10^{-3}\lambda$, where λ is the free space wavelength).

Solution: Note that the distance at which the fields decrease to e^{-1} of their value at the surface is simply $(\alpha_x)^{-1}$. Thus, we have

$$d \gg \lambda \qquad (\alpha_x)^{-1} = \frac{1}{\beta \sqrt{\dfrac{\mu_d \epsilon_d}{\mu_0 \epsilon_0} - 1}} \simeq 0.127\lambda \qquad \text{Tightly bound wave}$$

$$d \simeq 2 \times 10^{-3}\lambda \qquad (\alpha_x)^{-1} = \frac{\lambda}{\pi\beta d \left(\dfrac{\mu_d}{\mu_0} - \dfrac{\epsilon_0}{\epsilon_d} \right)} \simeq 41.6\lambda \qquad \text{Loosely bound wave}$$

The Lowest-Order Modes In both the case of the dielectric slab waveguide and the dielectric above a conducting ground plane, the existence of a TM_1 mode that persists down to zero frequency (i.e., $f_c = 0$) is quite remarkable. For the dielectric slab in free space, a TE_1 mode also exists with $f_c = 0$. It thus appears that very low-frequency waves can in principle be guided by a dielectric sheet. However, further consideration indicates that as the frequency is reduced, the transverse decay rate α_x decreases and the resultant extension of the fields far beyond the slab requires large amounts of power to be supplied in order to sustain such a mode.

First note that for any TM or TE mode at very high frequencies ($\omega \to \infty$), we have

$$\bar{\beta} = \sqrt{\omega^2 \mu_d \epsilon_d - \beta_x^2} \simeq \omega\sqrt{\mu_d \epsilon_d} \qquad [4.53]$$

so that $\bar{\beta} \to \infty$ as $\omega \to \infty$. Thus, at high frequencies, the phase constant (and thus the phase velocity $\bar{v}_p = \omega/\bar{\beta}$) is the same as that for a uniform plane wave in the unbounded dielectric, so that most of the real axial power flow is confined within the slab. Consistent with this observation is the fact that for $\omega \to \infty$, the transverse attenuation constant α_x is given by

$$\alpha_x = \sqrt{\omega^2(\mu_d\epsilon_d - \mu_0\epsilon_0) - \beta_x^2} \simeq \omega\sqrt{\mu_d\epsilon_d - \mu_0\epsilon_0} \qquad [4.54]$$

It thus appears that the transverse attenuation constant α_x increases without limit proportionally with frequency, indicating that the depth of penetration (proportional to α_x^{-1} in view of the $e^{-\alpha_x x}$ type of variation of the fields) into the free space surrounding the slab approaches zero.

On the other hand, we see that at the cutoff frequency $\omega = \omega_c$, where

$$\omega_c = \frac{(m-1)\pi}{(d\sqrt{\mu_d\epsilon_d - \mu_0\epsilon_0})}$$

we have

$$\overline{\beta} = \omega\sqrt{\mu_0\epsilon_0} = \beta \qquad \text{and} \qquad \alpha_x = 0$$

so that the fields extend to infinity outside the slab. In other words, an infinite amount of real power is carried in the axial direction *outside* the slab! The propagation constant is dominated by free space. The trend, therefore, is for the fields to spread to larger distances from the slab as frequency decreases toward cutoff, even in the case where the cutoff frequency is zero. Thus, whereas the lowest order mode can in principle be excited at very low frequencies, the extension of the fields far beyond the order slab requires unreasonably large amounts of power to be supplied by the excitation.

For the case of the dielectric above a ground plane, the TM_1 solution is maintained, so that it appears that this geometry can guide waves of arbitrarily low frequency. However, we note once again that the decay rate α_x decreases as the frequency is reduced, so that increasingly large amounts of power must be supplied in order to sustain the TM_1 mode, subject to the same power limitation discussed previously. However, note that for the TE case, there is no mode that propagates at very low frequencies for the dielectric above a ground plane.

Typical numerical values for various quantities are given in Example 4-8 for an optical waveguide.

Example 4-8:   An InP–Ga$_x$In$_{1-x}$As–InP optical waveguide.   Consider a Ga$_x$In$_{1-x}$As planar waveguide with a thickness d sandwiched between two InP regions, as shown in Figure 4.21. (a) Find the maximum thickness d_{max} (in μm) so that only the TE_1 mode is guided at $\lambda = 1.65$ μm. (b) If the thickness of the waveguide is $d = 0.3$ μm, find the attenuation constant α_x in the InP region, and the phase constants β_x and $\overline{\beta}$ in the waveguide, all in units of μm.

Solution:
(a) As we know, the TE_1 mode can propagate regardless of the thickness of the guide. The next higher mode (even TE_2) has a cutoff frequency given by

$$f_{c_2} = \left[\frac{(m-1)c}{2d\sqrt{n_d^2 - n_c^2}}\right]_{m=2} \simeq \frac{3\times10^8}{2d\sqrt{(3.56)^2 - (3.16)^2}} \geq \frac{3\times10^8 \text{ m-s}^{-1}}{1.65\mu m}$$

which yields $d_{max} \simeq 0.503$ μm.

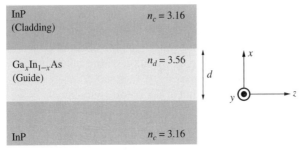

FIGURE 4.21. An InP–Ga$_x$In$_{1-x}$As–InP planar waveguide.

(b) Noting that the TE$_1$ mode is an odd mode, we need to solve the following equations in order to find α_x and β_x:

$$\alpha_x = \beta_x \tan\left(\frac{\beta_x d}{2}\right)$$

$$\alpha_x^2 + \beta_x^2 = \beta(n_d^2 - n_c^2)$$

Using $d = 0.3$ μm, $n_d = 3.56$, and $n_c = 3.16$, we find (by either graphical or numerical solution) $\beta_x \simeq 4.8$ rad-(μm)$^{-1}$ and $\alpha_x \simeq 4.26$ np-(μm)$^{-1}$. To find $\overline{\beta}$, we can use

$$\beta_x^2 = \omega^2 \mu_d \epsilon_d - \overline{\beta}^2 \quad \text{or} \quad \alpha_x^2 = \overline{\beta}^2 - \omega^2 \mu_c \epsilon_c$$

both of which give $\overline{\beta} \simeq 13.12$ rad-(μm)$^{-1}$, which indicates the wavelength in the guide is $\overline{\lambda} = 2\pi/\overline{\beta} \simeq 0.478$ μm.

4.2.3 Dielectric Slab Waveguide: Ray Theory

Our previous analysis of the dielectric slab waveguide was based on waveguide mode theory. It is also possible to analyze the wave propagation in a dielectric slab using ray theory, in a manner similar to the decomposition of waveguide modes into two uniform plane waves for the case of parallel-plate or hollow metallic wave-guides. In some ways, the ray approach provides better physical insight into the guiding properties of dielectric waveguides, since dielectric guiding is a manifesta-tion of the total internal reflection phenomena.

Consider wave propagation in the dielectric slab represented by a ray with a zigzag path, as shown in Figure 4.22, due to a TEM wave launched from the left.[40] Two possible cases are illustrated in Figure 4.22 a. When $\theta_i < \theta_{ic}$, where θ_{ic} is the critical angle of incidence given by $\theta_{ic} = \sin^{-1}\sqrt{\epsilon_0/\epsilon_d}$, the wave refracts out of the dielectric slab and into the surrounding medium, loses a fraction of its power at each reflection, and thus eventually vanishes. In other words, the dielectric does not

---

[40]Actual launching of the wave may be realized using a prism coupler as shown in Figure 3.38.

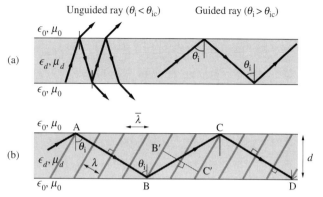

FIGURE 4.22.   Illustration of dielectric waveguide operation using a ray approach. (a) The wave is guided if the incidence angle θ_i is smaller than the critical angle θ_{ic}. (b) Constructive interference condition and geometry for rays in a dielectric waveguide for permitted angles of reflection. The dotted lines represent the wave phase fronts for the ray segments AB and CD.

'guide' the wave or confine the wave energy in any useful sense; to sustain such a wave at steady state, an infinite amount of input wave power would be required, since the waves outside the slab have infinite extent. When $\theta_i > \theta_{ic}$, on the other hand, the waves undergo total reflection at the boundaries and are guided or confined inside the dielectric slab, traveling in the z direction by bouncing back and forth between the slab boundaries without loss of power. This circumstance is very similar to propaga-tion of uniform plane waves between two perfectly conducting parallel plates, with the important difference that the tangential component of the wave electric field does not need to be zero at the dielectric-dielectric interface. Note that, as discussed in Chapter 3 in connection with total internal reflection, the fields outside the dielectric slab decrease rapidly with transverse distance from the slab.

At first it might appear that any wave for which $\theta_i > \theta_{ic}$ propagates in the dielec-tric slab. However, due to the nature (constructive or destructive) of the interference between different rays, only waves at certain discrete angles of incidence can ac-tually propagate, for a given slab thickness d and operating frequency f. Consider the typical ray path ABCD shown in Figure 4.22 b. The dotted lines represent the constant phase fronts of the uniform plane wave during its travel from A to B and from C to D (the constant phase fronts corresponding to path B–C are not shown). Before the ray goes through total internal reflection at B, the line B–B' constitutes a constant phase front. For constructive interference, we impose a self-consistency condition that after two successive reflections, at B and C, the wave front must re-main in phase. For example, waves starting at two different points on the constant phase front B–B' must go through the same phase difference as one moves directly from B' to C' while the other reflects at B, moves from B to C, and reflects at C. This means that any phase difference between these two paths may amount only to an integer multiple of 2π, that is,

$$\beta_d(\overline{BC} - \overline{B'C'}) - 2\phi_r = (m-1)2\pi \qquad m = 1, 2, 3, \ldots \qquad [4.55]$$

where $\beta_d = 2\pi/\lambda_d = (2\pi/\lambda)\sqrt{\epsilon_d/\epsilon_0}$ and ϕ_r is the phase shift that the wave acquires during total internal reflection at point B or C from either of the boundaries of the slab. From geometrical considerations, we have

$$\overline{BC} = \frac{d}{\cos\theta_i} \qquad \text{and} \qquad \overline{B'C'} = \overline{BC'}\sin\theta_i$$

where

$$\overline{BC'} = d\tan\theta_i - \frac{d}{\tan\theta_i}$$

Substituting in [4.55] and using algebraic manipulation, we find

$$2\beta_d d\cos\theta_i - 2\phi_r = (m-1)2\pi \qquad m = 1, 2, 3, \ldots$$

Now we restrict our attention to waves with **E** perpendicular to the plane of incidence (i.e., **E** out of the page). In that case, we know from Chapter 3 that the reflection coefficient is given as

$$\Gamma_\perp = \frac{\cos\theta_i + j\sqrt{\sin^2\theta_i - (\epsilon_0/\epsilon_d)}}{\cos\theta_i - j\sqrt{\sin^2\theta_i - (\epsilon_0/\epsilon_d)}} = 1e^{j\phi_\perp} \qquad [4.56]$$

where $\phi_r = \phi_\perp = +2\tan^{-1}(\sqrt{\sin^2\theta_i - (\epsilon_0/\epsilon_d)}/\cos\theta_i)$. In other words, we must have

$$2\beta_d d\cos\theta_i - 4\tan^{-1}\left[\frac{\sqrt{\sin^2\theta_i - (\epsilon_0/\epsilon_d)}}{\cos\theta_i}\right] = (m-1)2\pi \qquad m = 1, 2, 3, \ldots$$

or

$$\tan\left(\frac{\beta_d d\cos\theta_i}{2} - \frac{(m-1)\pi}{2}\right) = \frac{\sqrt{\sin^2\theta_i - \epsilon_0/\epsilon_d}}{\cos\theta_i} \qquad m = 1, 2, 3, \ldots \quad [4.57]$$

Example 4-9 illustrates the application of ray theory to a glass slab waveguide to identify the propagating modes.

Example 4-9:   Glass slab.   Consider the dielectric slab waveguide shown in Figure 4.21, which consists of glass ($n_d = 1.5$ or $\epsilon_d = 2.25\epsilon_0$) of thickness $d = 1$ cm surrounded by air. For an operating frequency of 30 GHz, find the propagating TE$_m$ modes and their corresponding angles of incidence θ_i.

Solution: The critical angle of incidence for the glass-air interface is

$$\theta_{ic} = \sin^{-1}\sqrt{\frac{\epsilon_0}{\epsilon_d}} = \sin^{-1}\left(\frac{1}{1.5}\right) \simeq 41.8°$$

Thus, for total internal reflection we must have $\theta_i \geq \sim 41.8°$. Using the transcendental equation [4.57], we have

$$\tan\left(\frac{2\pi(1.5)\cos\theta_i}{2} - \frac{(m-1)\pi}{2}\right) = \frac{\sqrt{\sin^2\theta_i - (2.25)^{-1}}}{\cos\theta_i} \qquad m = 1, 2, 3, \ldots$$

or

$$\tan(1.5\pi\cos\theta_i) = \frac{\sqrt{\sin^2\theta_i - (2.25)^{-1}}}{\cos\theta_i} \qquad m = 1, 3, 5, \ldots$$

$$\tan(1.5\pi\cos\theta_i) = -\frac{\cos\theta_i}{\sqrt{\sin^2\theta_i - (2.25)^{-1}}} \qquad m = 2, 4, 6, \ldots$$

The right- and left-hand sides of these equations are plotted in Figure 4.23. The values of the θ_i values corresponding to the different modes can be either read from the graph or numerically evaluated from the equations. We find

$$\text{TE}_1: \quad \theta_i \simeq 75.03°$$

$$\text{TE}_2: \quad \theta_i \simeq 59.47°$$

$$\text{TE}_3: \quad \theta_i \simeq 43.86°$$

The three θ_i values at which the curves intersect are the three solutions, or eigenvalues, for which [4.57] is satisfied. These permitted angles correspond to the three transverse electric modes TE_1, TE_2, and TE_3. Note that all of these modes

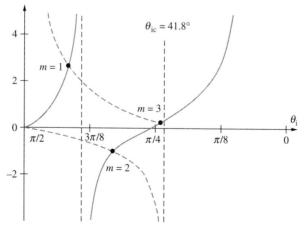

FIGURE 4.23.   The solid line is a plot of $\tan(1.5\pi \cos\theta_i)$, and the dashed lines above and below the horizontal axis, respectively, correspond to $\sqrt{\sin^2\theta_i - (2.25)^{-1}}/\cos\theta_i$ and $-\cos\theta_i/\sqrt{\sin^2\theta_i - (2.25)^{-1}}$. The θ_i values corresponding to the three solutions $m = 1, 2, 3$ are, respectively, $\theta_i \simeq 75.03°$, $59.47°$, and $43.86°$.

may be present simultaneously, but if the thickness d is decreased or the wavelength λ increased (or both), fewer solutions (eigenvalues) or modes are possible. However, we note that one solution, the TE_1 mode, will always exist, so that, at least in theory, waves at zero frequency can be guided by a dielectric slab. Note, however, the caveat in the previous section, namely, that an increasingly large amount of power is required to support such a mode as the frequency is decreased.

Relation between Ray and Mode Theory Parameters   The mode theory treatment of the dielectric slab waveguide discussed in Section 4.2.1 led to transcendental equations that needed to be solved in order to determine the transverse attenuation rate α_x and the transverse wave number β_x, from which we can find the axial phase constant $\overline{\beta}$ using [4.35]. The ray theory treatment discussed in the present section leads to a different type of transcendental equation ([4.57]), which needs to be solved in order to determine the θ_i values corresponding to the different modes. The relationship between these ray and mode theory parameters can be seen by considering the decomposition of the mode solutions into component TEM waves, in the same manner as was discussed in Section 4.1.3. Consider the E_y component of the odd TE mode:

$$
\begin{aligned}
E_y &= \frac{j\omega\mu}{\beta_x}\, C_e\, \cos(\beta_x x)\, e^{-j\overline{\beta}z} \\[2mm]
&= \frac{j\omega\mu}{\beta_x}\, \frac{C_e}{2} \left[e^{j\beta_x x} - e^{-j\beta_x x} \right] e^{-j\overline{\beta}z} \\[2mm]
&= \frac{j\omega\mu}{\beta_x}\, \frac{C_e}{2} \left[e^{j(\beta_x x - \overline{\beta}z)} - e^{-j(\beta_x x + \overline{\beta}z)} \right] \\[2mm]
&= \frac{j\omega\mu}{\beta_x}\, \frac{C_e}{2} \left[e^{j\beta_d(x\cos\theta_i - z\sin\theta_i)} - e^{-j\beta_d(x\cos\theta_i + z\sin\theta_i)} \right]
\end{aligned}
$$

where $\beta_d^2 = \beta_x^2 + \overline{\beta}^2 = \omega^2\mu_d\epsilon_d$. It is thus clear that the component uniform plane (or TEM) waves propagate in the x and z directions, respectively, with phase constants β_x and $\overline{\beta}$, so the tangent of the angle they make with the line that is vertical to the boundary of the dielectric is given by

$$
\tan\theta_i = \frac{\overline{\beta}}{\beta_x}
$$

Thus, once we know the values of β_x and $\overline{\beta}$ for a particular dielectric waveguide mode, the corresponding angle θ_i of the ray model representation of that mode can be readily determined without having to solve an additional transcendental equation such as [4.57]. In practice, one uses either the ray or mode theory description to analyze or design a dielectric waveguide, so the values of β_x and $\overline{\beta}$ are generally not available when we attempt to use a ray analysis, in which case [4.57] can be used to find the θ_i values.

4.3 WAVE VELOCITIES AND WAVEGUIDE DISPERSION

In our study of guided waves, we have exclusively considered a special class of electromagnetic fields that vary harmonically in time t and where the phasors of the field quantities vary with axial distance z as $e^{-\bar{\gamma}z}$, so that for propagating waves ($\bar{\gamma} = j\bar{\beta}$), the field components (e.g., $\mathscr{E}_z$) in the time domain vary with axial distance and time as $\mathscr{E}_z(z, t) = \cos(\omega t \pm \bar{\beta}z)$. The axial direction is the direction in which we want to convey (or guide) the wave energy. Accordingly, all of the wave-guiding structures considered in this and the next chapter have cross-sectional shapes that are identical at different points along the z axis. For such applications, it is natural to define the phase velocity of the propagating wave as that speed with which one has to move along the waveguide in order to keep up with points at which the fields have a specific instantaneous phase. For example, we may want to consider a particular electric field component (e.g., $\mathscr{E}_z$) and want to stay with the crest (or maximum) of the field as it travels along the guide. A specific phase of the wave corresponds to a constant value of $\mathscr{E}_z^+(z, t)$, so that we have

$$\mathscr{E}_z^+(z, t) = \text{const.} \quad \rightarrow \quad \omega t - \bar{\beta}z = \text{const.}$$

$$\xrightarrow{d/dt} \quad \omega - \bar{\beta}\frac{dz}{dt} = 0$$

$$\rightarrow \quad \bar{v}_p = \frac{dz}{dt} = \frac{\omega}{\bar{\beta}}$$

where we exclusively consider propagation in the $+z$ direction. The different wave-guiding systems considered in this and the next chapter have different functional forms for the phase constant $\bar{\beta}$; for example, for parallel-plate metal waveguides, $\bar{\beta} = \omega\sqrt{\mu\epsilon}\sqrt{1 - (f_c/f)^2}$, so that the phase velocity is given as

$$\bar{v}_p = \frac{1}{\sqrt{\mu\epsilon}\sqrt{1 - (f_c/f)^2}} \qquad [4.58]$$

where μ and ϵ are, respectively, the permeability and permittivity of the lossless dielectric material within the waveguide walls. We noted earlier that the phase velocity $\bar{v}_p$ in a waveguide is always greater than the phase velocity $v_p = \omega/\beta = 1/\sqrt{\mu\epsilon}$ of uniform plane waves in the unbounded lossless dielectric filling, and it is clear from [4.58] and other expressions for $\bar{v}_p$ in earlier sections that $\bar{v}_p$ increases toward infinity as the operating frequency f approaches the cutoff frequency (i.e., $f \rightarrow f_c$). The fact that $\bar{v}_p$ is larger than v_p does not pose any dilemma or violate any fundamental principle, since no energy or information can be transmitted at $\bar{v}_p$.

The most important aspect of equation [4.58] is that the phase velocity $\bar{v}_p$ depends on frequency, since the phase constant $\bar{\beta}$ is a nonlinear function of frequency. This result is to be contrasted with the phase velocity of uniform plane waves propagating (with phase constant $\beta = \omega\sqrt{\mu\epsilon}$) in a lossless unbounded dielectric medium, which is a constant independent of frequency,[41] assuming that the dielectric medium between the guide walls is simple (i.e., ϵ and μ are simple constants) and lossless ($\sigma = 0$, $\epsilon'' = 0$, and $\mu'' = 0$). The frequency dependence of $\bar{v}_p$ in [4.58] is not due to the nature of the material filling but is rather a consequence of the configuration of the waveguide as it influences the behavior of the fields within it. Note that for lossy material media, ($\sigma \neq 0$, $\epsilon'' \neq 0$, or $\mu'' \neq 0$), the propagation constant is complex and the phase velocity and attenuation rate are both functions of frequency, as discussed in Section 2.3.

The phase velocity $\bar{v}_p$ as given in [4.58] is for continuous, steady-state waves of a single frequency. In order to convey any information, it is necessary to "modulate" the signal by changing its amplitude, phase, or frequency, or by interrupting it in accordance with the information to be transmitted. Any information-carrying signal consists of a band of Fourier components with different amplitudes and phases such that the signal at the input to the waveguide is constituted by a superposition of these components. If the phase velocity $\bar{v}_p$ is the same for different frequency components and there is no attenuation, the spectral components add in proper phase at all locations in the waveguide, to precisely reproduce the original signal shape. However, when the phase velocity $\bar{v}_p$ is a function of frequency, the different frequency components acquire different phase delays as the signal travels through the waveguide, so that the superposition of the individual Fourier components at some point down the waveguide is no longer a precise replica of the transmitted signal. The signal intended for transmission (e.g., a rectangular pulse riding on a carrier) smears out as a result, losing its sharp edges and broadening as it propagates along the guide. In other words, the fact that $\bar{v}_p$ is a function of frequency leads to the *distortion* of the signal, by *dispersing* its frequency components. This phenomenon is referred to as *dispersion*. This type of dispersion, which is a result of the propagation characteristics imposed by the waveguide structure, is sometimes referred to as *waveguide dispersion*, to distinguish it from *material dispersion*, which occurs because the material media is either lossy or exhibits properties such that ϵ and μ depend on frequency (see Section 6.3).

When the propagation of an electromagnetic signal is subject to either waveguide or material dispersion, an important question arises concerning the velocity of a wave packet containing a band of frequency components. The "information" carried by a modulated signal is represented by the "shape" or the envelope of the wave. Although this shape itself changes somewhat with distance, it is nevertheless possible to obtain a useful measure of its "speed" of propagation. This measure of the speed of propagation of a *group* of frequencies that constitute a wave packet is the *group velocity*, defined in the next section. In many (but not all) cases, the group

---

[41] In many material media, either or both ϵ and μ may be functions of frequency (see Section 6.3). The phase velocity of uniform plane waves in such media is a function of frequency even if the medium is unbounded and lossless ($\sigma = 0$, $\epsilon'' = 0$, and $\mu'' = 0$), and is given by

$$v_p = \frac{\omega}{\beta} = \frac{\omega}{\omega\sqrt{\mu(f)\epsilon(f)}} = \frac{1}{\sqrt{\mu(f)\epsilon(f)}}$$

velocity also represents the velocity of travel of electromagnetic energy. Further discussion of different types of wave velocities and related references can be found elsewhere.[42]

4.3.1   Group Velocity

In introducing the concept of group velocity, we recognize that, in general, its utility is not confined to the subject of waveguides and that the concept is equally applicable in cases of material dispersion, for example, in the case of propagation of uniform plane waves in a lossy dielectric or in cases of more complicated media where the permittivity ϵ and/or permeability μ may be functions of frequency (Section 6.3). To understand dispersion and group velocity, we consider propagating wave solutions for any one of the field components (say, $\mathscr{E}_x$) with the general form

$$\mathscr{E}_x(z, t) = \mathcal{R}e\{Ce^{j(\omega t - \bar{\beta}z)}\} \qquad [4.59]$$

Note that because any electromagnetic field solutions must obey the wave equation derived from Maxwell's equations, the frequency ω and the phase constant $\bar{\beta}$ are related. The phase constant $\bar{\beta}$ can in general be a complicated function of frequency. Note that for uniform plane waves in a lossless unbounded dielectric medium, we have $\bar{\beta} = \beta = \omega\sqrt{\mu\epsilon}$, so that $\bar{\beta}$ is a linear function of frequency, as long as μ and ϵ are both independent of frequency. The phase velocity $\bar{v}_p = \omega/\bar{\beta}$ is then simply $\bar{v}_p = v_p = (\sqrt{\mu\epsilon})^{-1}$, which is independent of frequency. Thus, there is no signal dispersion in cases where the phase constant $\bar{\beta}$ is linearly proportional to frequency. Fundamentally, dispersion occurs when the phase constant $\bar{\beta}$ is a nonlinear function of frequency, that is, $\bar{\beta}(\omega) \neq K\omega$, where K is a constant.

In general, an information-carrying signal is constituted by a superposition of many Fourier components, each of which is given by [4.59], with both $\bar{\beta}$ and C being functions of frequency. Before we consider a general wave packet, we consider the simplest case of the superposition of two harmonic waves of equal amplitudes but slightly different frequencies and phase constants, represented by

$$\mathscr{E}_{1x}(z, t) = C\cos(\omega t - \bar{\beta}z)$$

$$\mathscr{E}_{2x}(z, t) = C\cos[(\omega + \Delta\omega)t - (\bar{\beta} + \Delta\bar{\beta})z]$$

where $\Delta\omega \ll \omega$ and $\Delta\bar{\beta} \ll \bar{\beta}$, and C is a constant. Using the trigonometric identity $\cos\zeta + \cos\xi = 2\cos[(\zeta + \xi)/2]\cos[(\zeta - \xi)/2]$, the sum of these two waves can be written as

$$\mathscr{E}_x(z, t) = \mathscr{E}_{1x} + \mathscr{E}_{2x} = 2C\cos\left[\frac{1}{2}\left(\Delta\omega t - \Delta\bar{\beta}z\right)\right]\cos\left[\left(\omega + \frac{\Delta\omega}{2}\right)t - \left(\bar{\beta} + \frac{\Delta\bar{\beta}}{2}\right)z\right]$$

$$[4.60]$$

---

[42]See Section 5.17 of J. A. Stratton, *Electromagnetic Theory*, McGraw-Hill, New York, 1941.

Examination of [4.60] indicates that the total field structure is similar to a uniform plane wave oscillating at a frequency of $\omega + \Delta\omega/2$, which is negligibly different from ω for $\Delta\omega \ll \omega$. The amplitude of the wave, however, is not a constant, but varies slowly with time and position between $-2C$ and $2C$, giving rise to the well-known phenomenon of *beats*, as illustrated in Figure 4.24. The amplitude envelope is designated $A_{\text{env}}(z, t)$ and is given by

$$A_{\text{env}}(z, t) = 2C \cos\left[\frac{1}{2}\left(\Delta\omega t - \Delta\overline{\beta} z\right)\right]$$

The successive maxima of the slowly varying amplitude function $A_{\text{env}}(z, t)$ at a fixed position occurs every $4\pi/\Delta\omega$ seconds and at a fixed time occurs at distance intervals of $4\pi/\Delta\overline{\beta}$. The velocity at which the amplitude envelope propagates can be found by equating the argument of the cosine term to a constant as

$$\Delta\omega t - \Delta\overline{\beta} z = \text{const.}$$

from which it follows that the envelope propagates with the special velocity

$$v_g = \frac{dz}{dt} = \frac{\Delta\omega}{\Delta\overline{\beta}}$$

or, in the limit of $\Delta\omega \to 0$,

$$v_g \equiv \frac{d\omega}{d\overline{\beta}} \qquad [4.61]$$

which constitutes the definition of the *group velocity*. If the medium is nondispersive, with, for example, $\overline{\beta} = \beta = \omega\sqrt{\mu\epsilon}$, we have

$$v_g = \frac{d\omega}{d\overline{\beta}} = \frac{1}{\dfrac{d\overline{\beta}}{d\omega}} = (\sqrt{\mu\epsilon})^{-1} = v_p$$

so that the group velocity is equal to the phase velocity (i.e., $v_g = v_p$). In a dispersive medium, however, where $\overline{\beta}$ is a nonlinear function of ω, the phase and group velocities are different.

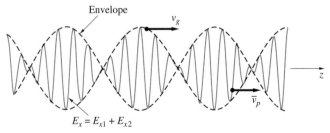

FIGURE 4.24. **Two-wave beat pattern.** The superposition of two harmonic waves of equal amplitude at slightly different frequencies. The envelope shows a beat pattern that propagates at the group velocity v_g.

In general, when the phase constant $\overline{\beta}$ is a nonlinear function of frequency, the phase velocity $\overline{v}_p = \omega/\overline{\beta}(\omega)$ is also a function of frequency. In this case, we can relate v_g to $\overline{v}_p$ as

$$v_g = \frac{d\omega}{d\overline{\beta}} = \frac{1}{\dfrac{d\overline{\beta}}{d\omega}} = \frac{1}{\dfrac{d}{d\omega}\left(\dfrac{\omega}{\overline{v}_p}\right)} = \frac{\overline{v}_p}{1 - \dfrac{\omega}{\overline{v}_p}\dfrac{d\overline{v}_p}{d\omega}} \qquad [4.62]$$

It is clear from equation [4.62] that for $d\overline{v}_p/d\omega = 0$ the medium is not dispersive and $v_g = \overline{v}_p$. It is also clear from [4.62] that whether the group velocity v_g is greater or smaller than the phase velocity $\overline{v}_p$ depends on the sign of $d\overline{v}_p/d\omega$. In cases where the phase velocity $\overline{v}_p$ decreases with increasing frequency (i.e., $d\overline{v}_p/d\omega < 0$), as in a metallic parallel-plate waveguide, we have *normal dispersion*, with $v_g < \overline{v}_p$. The case in which the dependence of $\overline{v}_p$ on frequency is such that $d\overline{v}_p/d\omega > 0$ is called *anomalous dispersion*, with $v_g > \overline{v}_p$.

In the preceding discussion we have considered the superposition of two harmonic waves of equal amplitude as the simplest "group" of waves. A single group or pulse of any desired shape may be constructed using Fourier superposition by noting that the amplitudes (and, in general, also the phases) of the component harmonic waves are functions of frequency. For the general case of nonperiodic signals, the Fourier summation must extend continuously over the entire range of frequencies. Thus, we can represent a wave packet as

$$\mathcal{E}_x(z, t) = \mathcal{R}e\left\{\int_{-\infty}^{\infty} A(\omega)e^{j(\omega t - \overline{\beta}z)}\,d\omega\right\} \qquad [4.63]$$

Since ω and $\overline{\beta}$ are one-to-one related as discussed above, we can equally take $\overline{\beta}$ as the independent variable in the Fourier decomposition of the wave packet:

$$\mathcal{E}_x(z, t) = \mathcal{R}e\left\{\int_{-\infty}^{\infty} A(\overline{\beta})e^{j(\omega(\overline{\beta})t - \overline{\beta}z)}\,d\overline{\beta}\right\} \qquad [4.64]$$

where $A(\overline{\beta})$ is obviously a different function from $A(\omega)$. In general, the concept of group velocity applies only to groups of waves with frequency components that are confined to a narrowband spectrum centered around a frequency ω_0 corresponding to a wave number $\overline{\beta}_0$. In such cases, the amplitude function $A(\overline{\beta})$ has negligible magnitude outside the frequency and phase constant ranges

$$\omega_0 - \Delta\omega \leq \omega \leq \omega_0 + \Delta\omega \quad \text{or} \quad \overline{\beta}_0 - \Delta\overline{\beta} \leq \overline{\beta} \leq \overline{\beta}_0 + \Delta\overline{\beta}$$

where $\Delta\omega \ll \omega$ (or $\Delta\overline{\beta} \ll \overline{\beta}$). We can thus rewrite the superposition integral [4.64] as

$$\mathcal{E}_x(z, t) \simeq \mathcal{R}e\left\{\int_{\overline{\beta}_0 - \Delta\overline{\beta}}^{\overline{\beta}_0 + \Delta\overline{\beta}} A(\overline{\beta})e^{j[\omega(\overline{\beta})t - \overline{\beta}(\omega)z]}\,d\overline{\beta}\right\} \qquad [4.65]$$

which represents a *wave packet*. Since the range $\Delta\overline{\beta}$ is small compared with $\overline{\beta}$ (i.e., $\Delta\overline{\beta} \ll \overline{\beta}$), $\omega(\overline{\beta})$ is only slightly different from its value at $\overline{\beta} = \overline{\beta}_0$. Consider a

Taylor series representation of $\omega(\overline{\beta})$:

$$\omega(\overline{\beta}) = \omega(\overline{\beta}_0) + \left(\frac{d\omega}{d\overline{\beta}}\right)_{\overline{\beta}=\overline{\beta}_0} (\overline{\beta} - \overline{\beta}_0) + \cdots \qquad [4.66]$$

The higher-order terms in [4.66] can be neglected since $\Delta\omega \ll \omega$. Using [4.66] for $\omega(\overline{\beta})$, we can rewrite $[\omega t - \overline{\beta} z]$ as

$$[\omega t - \overline{\beta} z] = \omega_0 t - \overline{\beta}_0 z + (\overline{\beta} - \overline{\beta}_0)\left[t\left(\frac{d\omega}{d\overline{\beta}}\right)_{\overline{\beta}=\overline{\beta}_0} - z\right] + \cdots \qquad [4.67]$$

where $\omega_0 = \omega(\overline{\beta}_0)$. The wave packet of [4.65] can thus be represented as

$$\mathcal{E}_x(z, t) = \mathcal{R}e\left\{A_{\text{env}}(z, t)e^{j(\omega_0 t - \overline{\beta}_0 z)}\right\}$$

where $A_{\text{env}}(z, t)$ is the slowly varying amplitude given approximately by

$$A_{\text{env}}(z, t) \simeq \int_{\overline{\beta}_0 - \Delta\overline{\beta}}^{\overline{\beta}_0 + \Delta\overline{\beta}} A(\overline{\beta})e^{j(\overline{\beta} - \overline{\beta}_0)[t(d\omega/d\overline{\beta})_{\overline{\beta}=\overline{\beta}_0} - z]} \, d\overline{\beta}$$

We note that the amplitude envelope $A_{\text{env}}(z, t)$ is constant over surfaces defined by

$$t\left(\frac{d\omega}{d\overline{\beta}}\right)_{\overline{\beta}=\overline{\beta}_0} - z = \text{const.}$$

from which it is clear that the wave packet (or the "shape" of the envelope of the wave packet, as discussed earlier) propagates at the special velocity called the group velocity, given by

$$v_g = \frac{dz}{dt} = \left(\frac{d\omega}{d\overline{\beta}}\right)_{\overline{\beta}=\overline{\beta}_0}$$

The concepts of phase and group velocities can be further illustrated by considering the superposition of two harmonic waves (e.g., two sidebands) at a certain instant of time as shown in Figure 4.25. If the component waves a and b have the

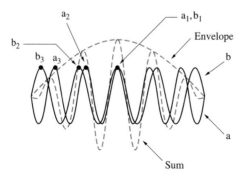

FIGURE 4.25. Physical interpretation of phase versus group velocity. Two waves a and b of slightly different frequencies form an amplitude-modulated "beat" wave pattern. Wave a has a slightly higher frequency than wave b.

same velocity, the two corresponding crests a_1 and b_1 move along together, and the maximum of the modulation envelope (i.e., beat pattern) moves along at the same velocity. Under these circumstances, phase and group velocities are the same. If, however, the lower-frequency (i.e., longer-wavelength) wave b has a velocity slightly greater (normal dispersion) than that of wave a, the crests a_1 and b_1 move apart in time while the crests a_2 and b_2 move closer together. Therefore, at some later instant the maximum of the beat pattern occurs at the point where a_2 and b_2 coincide in time, and at a still later time where a_3 and b_3 coincide. It is evident from Figure 4.25 that the envelope (i.e., amplitude of the sum of the two waves) slips backward with respect to the component waves; in other words, it moves forward with a different velocity, that is, the group velocity v_g, which in this case is less than the phase velocity of either of the two component waves.

If the shorter-wavelength (i.e., higher-frequency) wave (in this case, wave a) has a phase velocity greater than that of wave b (anomalous dispersion), the relative motion of the crests of the two waves is reversed, and the beat pattern slips forward with respect to the component waves. The group velocity is then greater than the phase velocity of the component waves.[43]

4.3.2 Dispersion ($\overline{\beta}$–ω) Diagrams

A useful way of visualizing and analyzing dispersion properties for different types of waves is to plot the phase constant $\overline{\beta}$ as a function of frequency ω, as shown for the case of TE or TM modes in a parallel-plate metal waveguide in Figure 4.26. For these modes, the *dispersion relation* (i.e., the relation between the phase constant $\overline{\beta}$

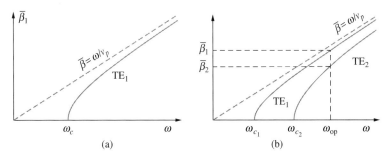

FIGURE 4.26. **Dispersion relation for TE or TM modes in a parallel-plate waveguide.** $\overline{\beta}$ – ω diagram for a parallel-plate waveguide. (a) TE_1 (or TM_1) mode. (b) TE_1 and TE_2 modes.

---

[43]Note that the fact that v_g can be greater than the phase velocity does not violate special relativity by implying that energy or information can travel faster than the speed of light. In the case of anomalous dispersion, group velocity as defined in [4.61] is simply not a useful concept. The fact that $d\overline{\beta}/d\omega$ is large means that the approximations made in connection with [4.66], where we dropped the higher-order terms in characterizing the envelope, do not hold. In other words, the envelope of the signal behaves in a more complicated manner than that described by the first-order term in [4.66].

and frequency ω) has the form

$$\overline{\beta} = \omega\sqrt{\mu\epsilon}\sqrt{1 - \left(\frac{\omega_c}{\omega}\right)^2} \quad \rightarrow \quad \omega^2 = \omega_c^2 + \overline{\beta}^2 v_p^2 \qquad [4.68]$$

where ω_c is the cutoff frequency and $v_p = (\mu\epsilon)^{-1/2}$ is the phase velocity in the unbounded dielectric medium. Equation [4.68] is a rather common form of dispersion relation that holds also for rectangular and circular waveguides (Chapter 5) as well as for other propagation media.[44] The dispersion characteristics of such systems can be understood by examining the dispersion curve, or a $\overline{\beta}$–ω diagram, as shown in Figure 4.26.

The phase velocity in a waveguide at any given frequency is given by [4.18],

$$\overline{v}_p = \overline{v}_p(\omega) = \frac{\omega}{\overline{\beta}} = \frac{v_p}{\sqrt{1 - (\omega_c/\omega)^2}}$$

whereas the group velocity can be found using [4.68] as

$$v_g = v_g(\omega) = \frac{d\omega}{d\overline{\beta}} = \frac{1}{(d\overline{\beta}/d\omega)} = v_p\sqrt{1 - \left(\frac{\omega_c}{\omega}\right)^2} \qquad [4.69]$$

or can be evaluated from [4.62]. Note that for this case, the phase and group velocities are related by

$$\overline{v}_p v_g = v_p^2$$

where $v_p = 1/\sqrt{\mu\epsilon}$. This interesting relationship between group and phase velocity is true only for propagation media for which the $\overline{\beta}$–ω relationship is described by [4.68]. Note from Figure 4.26b that each waveguide mode has its own dispersion curve. If care is not taken in the choice of the operational frequency ω_{op}, different modes with different phase constants $\overline{\beta}$ can coexist. For example, if the operational frequency ω_{op} is chosen to be above the cutoff frequencies of both the TE$_1$ and TE$_2$ modes, as indicated in Figure 4.26b, the dispersion relations of these two modes both allow real values for $\overline{\beta}$ (i.e., propagating solutions). On the other hand, if ω_{op} is chosen such that $\omega_{c_1} < \omega_{op} < \omega_{c_2}$, a real solution for the phase constant $\overline{\beta}$ exists only for the TE$_1$ mode.

---

[44]For example, the dispersion relation for an isotropic plasma is identical. A plasma is an ionized medium in which free electrons exist and are able to move under the influence of the electric and magnetic fields of the wave. This motion of charged particles constitutes a current $\mathbf{J}$ that in turn modifies the fields via $\nabla \times \mathbf{H} = \mathbf{J} + j\omega\mathbf{D}$. It turns out (see Section 6.2) that the physical effects of these charged particle motions can be accounted for by using an effective permittivity $\epsilon_{eff} = \epsilon_0(1 - \omega_p^2/\omega^2)$, where ω_p is a characteristic frequency of the plasma medium that is a function of the density of free electrons. The phase constant for uniform plane waves in an unbounded isotropic plasma is thus given as

$$\overline{\beta} = \omega\sqrt{\mu_0\epsilon_0}\sqrt{1 - \left(\frac{\omega_p^2}{\omega^2}\right)} \quad \rightarrow \quad \omega^2 = \omega_p^2 + \overline{\beta}^2 c^2$$

Thus, the plasma medium allows the propagation of waves above a cutoff frequency $\omega_c = \omega_p$.

The variation with frequency of the phase and group velocities for a waveguide is shown in Figure 4.27. We note that the group velocity is zero at the cutoff frequency, corresponding to the situation where the component uniform plane waves constituting the waveguide mode reflect back and forth between the upper and lower plates with no propagation along the guide. For waveguides, and for all media for which [4.68] holds, we have normal dispersion since $d\bar{\beta}/d\omega$ is always larger than $\bar{\beta}/\omega$, so that $\bar{v}_p$ is always greater than v_g. This fact can be seen graphically from Figure 4.28a, since the slope of the line from the origin to an arbitrary point P (which is equal to the inverse of $\bar{v}_p$) is smaller than the slope of the tangent line at the same point (which is the inverse of v_g). It is also interesting to note that $\bar{v}_p$ and v_g both approach v_p as $\omega \to \infty$.

Under other circumstances the phase velocity can be less than the group velocity. This case of anomalous dispersion ($v_g > \bar{v}_p$) is illustrated in Figure 4.28b, which shows the $\bar{\beta}-\omega$ diagram for uniform plane waves propagating in an unbounded lossy dielectric. For this case, we recall from Section 2.3 that

$$\beta = \omega \sqrt{\frac{\mu\epsilon}{2}\left[\sqrt{1 + \left(\frac{\sigma}{\omega\epsilon}\right)^2} + 1\right]}^{1/2}$$

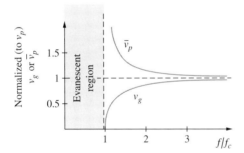

FIGURE 4.27.   Phase and group velocities. Phase and group velocities in a waveguide plotted as a function of frequency. Note that the quantities plotted are normalized by v_p, the phase velocity of uniform plane waves in the unbounded dielectric medium.

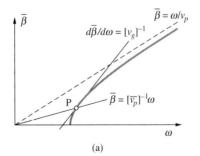
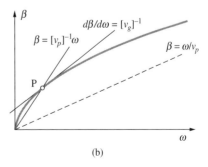

(a)   (b)

FIGURE 4.28. $\bar{\beta}-\omega$ diagrams. (a) Dispersion diagram for metallic waveguides ($v_g < \bar{v}_p$). (b) Dispersion diagram for uniform plane waves in an unbounded lossy dielectric ($v_g > v_p$).

so that the phase constant is clearly a function of frequency, and the medium is thus dispersive. We can see from Figure 4.28a that $d\beta/d\omega$ for this case is always greater than β/ω, so that the phase velocity $\bar{v}_p$ is always less than the group velocity v_g. As noted in footnote 43, group velocity as defined in [4.61] is not a useful concept in such cases. The fact that $d\bar{\beta}/d\omega$ is large means that the approximations made in connection with [4.66], namely, dropping the higher-order terms in characterizing the envelope, do not hold.

4.3.3 Group Velocity as Energy Velocity

We have seen that the envelope of a group of frequency components representing any given field quantity travels at the group velocity. If the magnitude of the quantity involved (e.g., wave electric field) is associated with the energy density of the electromagnetic wave, we can expect that the transport of energy carried by an electromagnetic wave also occurs at the group velocity. In this section, we demonstrate this interpretation of group velocity. Consider the propagating TM_m mode in a parallel-plate waveguide, for which the field components are given by [4.13] as

$$H_y = C \cos\left(\frac{m\pi}{a}x\right)e^{-j\bar{\beta}_m z}$$

$$E_x = \frac{\bar{\beta}}{\omega\epsilon}C \cos\left(\frac{m\pi}{a}x\right)e^{-j\bar{\beta}_m z}$$

$$E_z = \frac{jm\pi}{\omega\epsilon a}C \sin\left(\frac{m\pi}{a}x\right)e^{-j\bar{\beta}_m z}$$

The velocity of energy flow represented by this wave can be expressed in terms of the time-average power flowing through a cross-sectional area of width b (see Figure 4.29) divided by the energy stored per unit length of the guide of the same width. In other words, we can write the velocity of energy flow as

$$v_E = \frac{P_{av}}{\overline{W}_{str}} \qquad \frac{\text{J-s}^{-1}}{\text{J-m}^{-1}} = \text{m-s}^{-1}$$

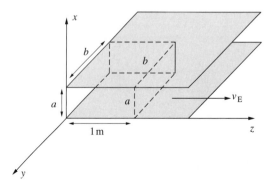

FIGURE 4.29. A parallel-plate waveguide. A segment of a parallel-plate waveguide of height a, width b, and unit axial length.

where $\overline{W}_{str}$ is the energy stored per unit length of waveguide of width b. Note that since the power flow is in the z direction, the energy transport velocity v_E is also in the z direction.

The total time-average power flow through the cross-sectional area of height a and width b is then

$$P_{av} = b \int_0^a \frac{1}{2} [\mathfrak{Re}\{\mathbf{E} \times \mathbf{H}^*\}] \cdot \hat{\mathbf{z}} \, dx = b \int_0^a \frac{1}{2} \mathfrak{Re}(E_x H_y^*) \, dx$$

$$= \frac{b}{2} C^2 \frac{\overline{\beta}_m}{\omega\epsilon} \int_0^a \cos^2\left(\frac{m\pi}{a}x\right) dx = \frac{bC^2 \overline{\beta}_m}{2\omega\epsilon} \frac{a}{2} = \frac{baC^2 \overline{\beta}_m}{4\omega\epsilon}$$

The time-average stored energy per unit length of the same waveguide of height a and width b, including both electric and magnetic fields, is[45]

$$\overline{W}_{str} = \frac{1}{4} \int_0^a \int_0^b \int_0^1 [\epsilon \mathbf{E} \cdot \mathbf{E}^* + \mu \mathbf{H} \cdot \mathbf{H}^*] \, dz \, dy \, dx$$

$$= b \int_0^a \left\{ \frac{\epsilon}{4} [|E_x|^2 + |E_z^2|] + \frac{\mu}{4} |H_y|^2 \right\} dx$$

$$= \frac{bC^2}{4} \int_0^a \left\{ \epsilon \left[\frac{\overline{\beta}_m^2}{\omega^2\epsilon^2} \cos^2\left(\frac{m\pi}{a}x\right) + \frac{m^2\pi^2}{\omega^2\epsilon^2 a^2} \sin^2\left(\frac{m\pi}{a}x\right) \right] + \mu \cos^2\left(\frac{m\pi}{a}x\right) \right\} dx$$

$$= \frac{bC^2}{4} \frac{a}{2} \left[\frac{\overline{\beta}_m^2}{\omega^2\epsilon} + \frac{m^2\pi^2}{\omega^2\epsilon a^2} + \frac{\omega^2\epsilon\mu}{\omega^2\epsilon} \right] = \frac{baC^2}{8\omega^2\epsilon} \left[\overline{\beta}_m^2 + \frac{m^2\pi^2}{a^2} + \omega^2\mu\epsilon \right]$$

$$\overline{W}_{str} = \frac{baC^2\mu}{4}$$

where we have recognized that for TM$_m$ modes,

$$\overline{\beta}_m^2 + (m^2\pi^2/a^2) = \omega^2\mu\epsilon$$

Using $\overline{W}_{str}$ and P_{av} in the expression for v_E, we have

$$v_E = \frac{P_{av}}{\overline{W}_{str}} = \frac{baC^2\overline{\beta}_m/4\omega\epsilon}{baC^2\mu/4} = \frac{\overline{\beta}_m}{\omega\mu\epsilon}$$

---

[45]Note that the instantaneous stored energy in an electric (or magnetic) field is given by $\frac{1}{2}\epsilon|\overline{\mathscr{E}}|^2$ (or $\frac{1}{2}\mu|\overline{\mathscr{H}}|^2$), where $\overline{\mathscr{E}}(z, t) = \hat{\mathbf{x}} E_0 \cos(\omega t - \overline{\beta}z + \phi)$ is the "real" electric field, related to the phasor $\mathbf{E}(z) = \hat{\mathbf{x}} E_0 e^{-j\overline{\beta}z} e^{j\phi}$ through

$$\overline{\mathscr{E}}(z, t) = \mathfrak{Re}\{\mathbf{E}(z)e^{j\omega t}\}$$

The time-average stored energy can be expressed as

$$\frac{1}{T_p} \int_0^{T_p} \frac{1}{2} \epsilon E_0^2 \cos^2(\omega t - \beta z + \phi) \, dt = \frac{1}{4} \epsilon E_0^2$$

where $T_p = 2\pi/\omega$ is the period of the assumed sinusoidal variation.

Noting that the waveguide phase velocity $\bar{v}_p = \omega/\bar{\beta}_m$, we can write the energy velocity as

$$v_{\text{E}} = \frac{\dfrac{1}{(\mu\epsilon)}}{\bar{v}_p} = \frac{v_p^2}{\bar{v}_p} \quad \rightarrow \quad v_{\text{E}}\bar{v}_p = v_p^2 \quad \rightarrow \quad v_{\text{E}} = v_g$$

where we have used the fact that for waveguides $v_g\bar{v}_p = v_p^2$. Also note that since from [4.16] we have $\bar{\beta}_m = \sqrt{\omega^2\mu\epsilon - (m^2\pi^2/a^2)} = \beta\sqrt{1 - (f_c/f)^2}$, the energy velocity can be expressed as

$$v_{\text{E}} = v_g = \frac{\beta\sqrt{1 - (f_c/f)^2}}{\omega\mu\epsilon}$$

$$= v_p\sqrt{1 - \left(\frac{m\lambda}{2a}\right)^2}$$

Note that a similar result can be obtained for any of the other TE or TM modes for the parallel-plate waveguide. More generally, it can be shown (see Problem 5-65) that the group velocity $v_g = d\bar{\beta}/d\omega$ is the velocity of energy transport for any of the TE or TM modes discussed in Chapter 5 for cylindrical waveguides having rectangular or circular cross section. The fact that the group velocity is the velocity of energy transport for TE or TM modes in waveguides is consistent with the decomposition of such modes into component TEM waves, as discussed in Section 4.1.3. In this context, we can think of the energy of the electromagnetic wave as being associated with the component plane waves that move along the waveguide by bouncing back and forth between the waveguide walls, being incident on the walls at an angle θ_{i_m} from the vertical. Although these component TEM waves propagate along at velocity v_p, the component of their velocity in the z direction (i.e., the velocity at which energy travels along the waveguide) is $v_{p_z} = v_p\sin\theta_{i_m} = v_p\sqrt{1 - (\omega_{c_m}/\omega)^2}$, which is identical to the group velocity v_g.

It should be noted that, in general, both the velocity of energy transport and the velocity at which information travels (sometimes called the signal velocity) are difficult to define.[46] The concept of group velocity is particularly useful when a signal consists of component frequencies concentrated in a narrow band around a carrier frequency. In such cases, the approximation made in connection with [4.66] is valid, and the group velocity is a very good approximation of both the signal velocity and the velocity of energy transport. However, in other cases, for example in the case of a step excitation, the frequency content of the signal is not restricted to a narrow range of frequencies, and the concept of a group velocity is not useful.

---

[46] A complete discussion of the underlying issues and of phase, group, and signal velocities is given by L. Brillouin, *Wave Propagation and Group Velocity*, Academic Press, New York, 1960. Although the level of mathematical analysis is quite high, clear discussions of the underlying physics are provided. Also see Section 5.18 of J. A. Stratton, *Electromagnetic Theory*, McGraw-Hill, New York, 1941. For an excellent discussion at a simpler level, see the questions and answers sections of *Am. J. Phys.*, 64(11), p. 1353, November 1996 and 66(8), pp. 656–661, August 1998.

4.4   SUMMARY

This chapter discussed the following topics:

- **General relationships for waves guided by metallic conductors.** The configuration and propagation of electromagnetic waves guided by metallic conductors are governed by the solutions of the wave equations subject to the boundary conditions

$$E_{\text{tangential}} = 0 \qquad H_{\text{normal}} = 0$$

The basic method of solution assumes propagation in the z direction, so that all field components vary as $e^{-\overline{\gamma} z}$, where $\overline{\gamma}$ is equal to either $\overline{\alpha}$ or $j\overline{\beta}$. The z variations of real fields, then, have the form

$$\overline{\gamma} = \overline{\alpha} \qquad e^{-\overline{\alpha} z}\cos(\omega t) \qquad \text{Evanescent wave}$$

$$\overline{\gamma} = j\overline{\beta} \qquad \cos(\omega t - \overline{\beta} z) \qquad \text{Propagating wave}$$

The different type of possible solutions are classified as TE, TM, or TEM waves:

$$E_z = 0, H_z \neq 0 \qquad \text{Transverse electric (TE) waves}$$

$$E_z \neq 0, H_z = 0 \qquad \text{Transverse magnetic (TM) waves}$$

$$E_z, H_z = 0 \qquad \text{Transverse electromagnetic (TEM) waves}$$

Assuming that the wave-guiding structures extend in the z direction, the transverse field components can be derived from the axial (z direction) components, so that the z components act as a generating function for the transverse field components.

- **Parallel-plate waveguide.** The forms of the z components of the propagating TE_m and TM_m modes between parallel plates are

$$TE_m: \quad H_z = H_0\cos\left(\frac{m\pi}{a}x\right)e^{-j\overline{\beta}_m z}$$

$$TM_m: \quad E_z = E_0\sin\left(\frac{m\pi}{a}x\right)e^{-j\overline{\beta}_m z}$$

The cutoff frequency f_c and cutoff wavelength λ_c for the TE_m or TM_m modes are

$$f_{c_m} = \frac{m}{2a\sqrt{\mu\epsilon}} \qquad \lambda_{c_m} = \frac{2a}{m}$$

Note that f_{c_m} determines all the other quantities, such as $\overline{\beta}_m$, $\overline{\lambda}_m$, and $\overline{v}_{p_m}$, via [4.16] and [4.18]. The TEM mode electric and magnetic field phasors for the parallel-plate waveguide are given by

$$\mathbf{E} = \hat{\mathbf{x}}E_0 e^{-j\beta z} \qquad \text{and} \qquad \mathbf{H} = \hat{\mathbf{y}}\frac{E_0}{\eta}e^{-j\beta z}$$

where $\beta = \omega\sqrt{\mu\epsilon}$. TEM modes do not exhibit any cutoff frequency and can in principle propagate at any frequency.

The attenuation rate due to conductor losses typically increases with frequency for TEM and TM_m modes, exhibiting a minimum as a function of frequency for TM_m modes. For TE_m modes, the attenuation rate due to conductor losses decreases with increasing frequency. For both TE_m and TM_m modes, the attenuation increases rapidly as the frequency approaches the cutoff frequency for the particular modes. The attenuation constants due to conduction losses for parallel-plate waveguide modes can be summarized as follows:

$$\text{TEM}: \quad \alpha_c = \frac{1}{\eta a}\sqrt{\frac{\omega\mu}{2\sigma}}$$

$$\text{TM}_m: \quad \alpha_{c_{TM_m}} = \frac{2R_s}{\eta a\sqrt{1-(f_{c_m}/f)^2}}$$

$$\text{TE}_m: \quad \alpha_{c_{TE_m}} = \frac{2R_s(f_{c_m}/f)^2}{\eta a\sqrt{1-(f_{c_m}/f)^2}}$$

where f_{c_m} is the cutoff frequency for the particular mode. The attenuation coefficients for dielectric losses in a parallel-plate waveguide are

$$\text{TEM}: \quad \alpha_d = \frac{\omega\epsilon''\sqrt{\mu/\epsilon'}}{2} \qquad \text{TM}_m \text{ or } \text{TE}_m: \quad \alpha_d \simeq \frac{\pi f\sqrt{\mu_0\epsilon'}\epsilon''/\epsilon'}{\sqrt{1-(f_{c_m}/f)^2}}$$

■ **Dielectric waveguides.** Electromagnetic waves guided by a dielectric slab of thickness d must have electric and magnetic fields that decay exponentially (at a rate of α_x) with transverse distance outside the slab, while having variations of the form $\sin(\beta_x d/2)$ or $\cos(\beta_x d/2)$ inside the slab. The modes are classified as odd or even depending on whether the axial field component (E_z for TM and H_z for TE modes) varies in the form of $\sin(\beta_x d/2)$ or $\cos(\beta_x d/2)$, respectively. The determination of the values of α_x and β_x requires the simultaneous solution of the equation

$$\alpha_x = [\omega^2(\mu_d\epsilon_d - \mu_0\epsilon_0) - \beta_x^2]^{1/2}$$

together with one of the following equations, depending on the mode under consideration:

Odd TM: $\dfrac{\alpha_x}{\beta_x} = \dfrac{\epsilon_0}{\epsilon_d}\tan\left(\dfrac{\beta_x d}{2}\right)$ Odd TE: $\dfrac{\alpha_x}{\beta_x} = \dfrac{\mu_0}{\mu_d}\tan\left(\dfrac{\beta_x d}{2}\right)$

Even TM: $\dfrac{\alpha_x}{\beta_x} = -\dfrac{\epsilon_0}{\epsilon_d}\cot\left(\dfrac{\beta_x d}{2}\right)$ Even TE: $\dfrac{\alpha_x}{\beta_x} = -\dfrac{\mu_0}{\mu_d}\cot\left(\dfrac{\beta_x d}{2}\right)$

The propagation constant $\overline{\beta}$ is related to β_x as

$$\overline{\beta} = \sqrt{\omega^2\mu_d\epsilon_d - \beta_x^2}$$

and, as usual, determines the phase velocity through $\bar{v}_p = \omega/\bar{\beta}$ and wavelength via $\bar{\lambda} = 2\pi/\bar{\beta}$. The cutoff frequencies of the modes are given by

$$f_c = \frac{(m-1)}{2d\sqrt{\mu_d\epsilon_d - \mu_0\epsilon_0}} \quad \begin{array}{l} m = 1,3,5,\ldots \quad \text{Odd TM}_m \text{ or TE}_m \\ m = 2,4,6,\ldots \quad \text{Even TM}_m \text{ or TE}_m \end{array}$$

■ **Group velocity.** When the propagation constant $\bar{\beta}$ is not linearly proportional to frequency, the phase velocity $\bar{v}_p$ is a function of frequency, and the different frequency components of a wave packet travel along a waveguide at different speeds. In such cases, and for electromagnetic wave packets consisting of frequency components closely clustered around a high-frequency carrier such that $\Delta\omega \ll \omega_0$, the signal envelope travels at the group velocity, which for the parallel-plate waveguide is always less than the phase velocity and is given by

$$v_g \equiv \frac{d\omega}{d\bar{\beta}} = v_p\sqrt{1 - \left(\frac{f_c}{f}\right)^2}$$

In the parallel-plate waveguide (and in cylindrical waveguides discussed in Chapter 5), the group velocity is also the velocity at which the electromagnetic energy travels along the waveguide.

4.5 PROBLEMS

4-1. Parallel-plate waveguide modes. A parallel-plate waveguide consists of two perfectly conducting infinite plates spaced 2 cm apart in air. Find the propagation constant $\bar{\gamma}$ for the TM$_0$, TE$_1$, TM$_1$, TE$_2$, and TM$_2$ modes at an operating frequency of (a) 5 GHz, (b) 10 GHz, and (c) 20 GHz.

4-2. Parallel-plate waveguide modes. A parallel-plate air waveguide has a plate separation of 6 mm and width of 10 cm. (a) List the cutoff frequencies of the seven lowest-order modes (TE$_m$ and TM$_m$) that can propagate in this guide. (b) Find all the propagating modes (TE$_m$ and TM$_m$) at 40 GHz. (c) Find all the propagating modes at 60 GHz. (d) Repeat part (c) if the waveguide is filled with polyethylene (assume it is lossless, with $\epsilon_r \simeq 2.25$).

4-3. Parallel-plate waveguide modes. An air-filled parallel-plate waveguide with a plate separation of 1 cm is to be used to connect a 25-GHz microwave transmitter to an antenna. (a) Find all the propagating modes. (b) Repeat part (a) if the waveguide is filled with polyethylene (assume it is lossless, with $\epsilon_r \simeq 2.25$).

4-4. VLF propagation in the earth–ionosphere waveguide. The height of the terrestrial earth–ionosphere waveguide considered in Example 4-2 varies for VLF (3–30 kHz) from 70 km during the day to about 90 km during the night.[47] (a) Find the total number of propagating modes during the day at 12 kHz. (Assume both the ionosphere and the earth to be flat and perfect conductors.) (b) Repeat part (a) for during the night. (c) Find

[47] J. Galejs, *Terrestrial Propagation of Long Electromagnetic Waves*, Pergamon Press, New York, 1972.

the total number of propagation modes during the day at 24 kHz. (d) Repeat part (c) for during the night. (e) Does the TM_{17} mode propagate during the day at 30 kHz? (f) Does the TM_{17} mode propagate during the night at 30 kHz?

4-5. Waveguide in the earth's crust. It has been proposed[48] that radio waves may propagate in a waveguide deep in the earth's crust, where the basement rock has very low conductivity and is sandwiched between the conductive layers near the surface and the high-temperature conductive layer far below the surface. The upper boundary of this waveguide is on the order of 1 km to several kilometers below the earth's surface. The depth of the dielectric layer of the waveguide (the basement rock) can vary anywhere from 2 to 20 km, with a conductivity of 10^{-6} to 10^{-11} S-m^{-1} and a relative dielectric constant of ~6. This waveguide may be used for communication from a shore sending station to an underwater receiving station. Consider such a waveguide and assume it to be an ideal parallel-plate waveguide. (a) If the depth of the dielectric layer of the guide is 2 km, find all the propagating modes of this guide below an operating frequency of 2 kHz. (b) Repeat part (a) (same depth) below 5 kHz. (c) Repeat parts (a) and (b) for a dielectric layer depth of 20 km.

4-6. Single-mode propagation. Consider a parallel-plate air waveguide with plate separation a. (a) Find the maximum plate separation a_{max} that results in single-mode propagation along the guide at 10 GHz. (b) Repeat part (a) for the waveguide filled with a dielectric with $\epsilon_r \simeq 2.54$.

4-7. Evanescent wave attenuator design. A section of a parallel-plate air waveguide with a plate separation of 7.11 mm is constructed to be used at 15 GHz as an evanescent wave attenuator to attenuate all the modes except the TEM mode along the guide. Find the minimum length of this attenuator needed to attenuate each mode by at least 100 dB. Assume perfect conductor plates.

4-8. Evanescent wave filter design. Consider a parallel-plate air waveguide to be designed such that no other mode but the TEM mode will propagate along the guide at 6 GHz. If it is required by the design constraints that the lowest-order nonpropagating mode should face a minimum attenuation of 20 dB-(cm)$^{-1}$ along the guide, find the maximum plate separation a_{max} of the guide needed to satisfy this criterion. (b) Using the maximum value of a found in part (a), find the dB-(cm)$^{-1}$ attenuation experienced by the TE_2 and the TM_2 modes.

4-9. Cutoff and waveguide wavelengths. Consider a parallel-plate waveguide in air with a plate separation of 3 cm to be used at 8 GHz. (a) Determine the cutoff wavelengths of the three lowest-order nonpropagating TM_m modes. (b) Determine the guide wavelengths of all the propagating TM_m modes.

4-10. Guide wavelength of an unknown mode. The waveguide wavelength of a propagating mode along an air-filled parallel-plate waveguide at 15 GHz is found to be $\bar{\lambda} = 2.5$ cm. (a) Find the cutoff frequency of this mode. (b) Recalculate the guide wavelength of this same mode for a waveguide filled with polyethylene (lossless, with $\epsilon_r \simeq 2.25$).

4-11. Guide wavelength, phase velocity, and wave impedance. Consider a parallel-plate air waveguide having a plate separation of 5 cm. Find the following: (a) The cutoff frequencies of the TM_0, TM_1, and TM_2 modes. (b) The phase velocity $\bar{v}_{pm}$, wave-

---

guide wavelength $\bar{\lambda}_m$ and wave impedance Z_{TM_m} of the above modes at 8 GHz. (c) The highest-order TM_m mode that can propagate in this guide at 20 GHz.

4-12. Parallel-plate waveguide design. Design a parallel-plate air waveguide to operate at 5 GHz such that the cutoff frequency of the TE_1 mode is at least 25% less than 5 GHz, the cutoff frequency of the TE_2 mode is at least 25% greater than 5 GHz, and the power-carrying capability of the guide is maximized.

4-13. TEM decomposition of TE_m modes. For an air-filled parallel-plate waveguide with 4 cm plate separation, find the oblique incidence angle θ_i and sketch TE_1, TE_2, TE_3, and TE_4 modes in terms of two TEM waves at an operating frequency of 20 GHz.

4-14. Unknown waveguide mode. The electric field of a particular mode in a parallel-plate air waveguide with a plate separation of 2 cm is given by

$$E_x(y, z) = 10e^{-60\pi y} \sin(100\pi z) \text{ kV-m}^{-1}$$

(a) What is this mode? Is it a propagating or a nonpropagating mode? (b) What is the operating frequency? (c) What is the similar highest-order mode (TE_m or TM_m) that can propagate in this waveguide?

4-15. Unknown waveguide mode. The magnetic field of a particular mode in a parallel-plate air waveguide with a plate separation of 2.5 cm is given by

$$H_z(x, y) = C_1 e^{-j640\pi x/3} \cos(160\pi y)$$

where x and y are both in meters. (a) What is this mode? Is it a propagating or a nonpropagating mode? (b) What is the operating frequency? (c) Find the corresponding electric field $\mathbf{E}(x, y)$. (d) Find the lowest-order similar mode (TE_m or TM_m) that does not propagate in this waveguide at the same operating frequency.

4-16. Maximum power capacity. (a) In Problem 4-15, determine the value of constant C_1 (assumed to be real) which maximizes the power carried by all the propagating modes without causing any dielectric breakdown (use 15 kV-m^{-1} for maximum allowable electric field in air, which is half of the breakdown electric field in air at sea level). (b) Using the value of C_1 found in part (a), find the maximum time-average power per unit width carried by the mode found in part (a) of Problem 4-15.

4-17. Power-handling capacity of a parallel-plate waveguide. A parallel-plate air waveguide with a plate separation of 1.5 cm is operated at a frequency of 15 GHz. Determine the maximum time-average power per unit guide width in units of kW-(cm)$^{-1}$ that can be carried by the TE_1 mode in this guide, using a breakdown strength of air of 15 kV-(cm)$^{-1}$ (safety factor of approximately 2 to 1) at sea level.

4-18. Power capacity of a parallel-plate waveguide. Show that the maximum power-handling capability of a TM_m mode propagating in a parallel-plate waveguide without dielectric breakdown is determined only by the longitudinal component of the electric field for $f_{c_m} < f < \sqrt{2} f_{c_m}$ and by the transverse component of the electric field for $f > \sqrt{2} f_{c_m}$.

4-19. Power capacity of a parallel-plate waveguide. For a parallel-plate waveguide formed of two perfectly conducting plates separated by air at an operating frequency of $f = 1.5 f_{c_m}$, find the maximum time-average power per unit area of the waveguide that can be carried without dielectric breakdown [use 15 kV-(cm)$^{-1}$ for maximum allowable electric field in air, which is half of the breakdown electric field in air at sea level] for the following modes: (a) TEM, (b) TE_1, and (c) TM_1.

4-20. Attenuation in a parallel-plate waveguide. Consider a parallel-plate waveguide with plate separation a having a lossless dielectric medium with properties ϵ and μ. (a) Find the frequency in terms of the cutoff frequency (i.e., find f/f_{c_m}) such that the attenuation constants α_c due to conductor losses of the TEM and the TE$_m$ modes are equal. (b) Find the attenuation α_c for the TEM, TE$_m$, and TM$_m$ modes at that frequency.

4-21. Attenuation in a parallel-plate waveguide. For a TM$_m$ mode propagating in a parallel-plate waveguide, do the following: (a) Show that the attenuation constant α_c due to conductor losses for the propagating TM$_m$ mode is given by

$$\alpha_c = \frac{2\omega \epsilon R_s}{\bar{\beta} a}$$

(b) Find the frequency in terms of f_{c_m} (i.e., find f/f_{c_m}) such that the attenuation constant α_c found in part (a) is minimum. (c) Find the minimum for α_c. (d) For an air-filled waveguide made of copper plates 2.5 cm apart, find α_c for TEM, TE$_1$, and TM$_1$ modes at the frequency found in part (b).

4-22. Parallel-plate waveguide: phase velocity, wavelength, and attenuation. A parallel-plate waveguide is formed by two parallel brass plates ($\sigma = 2.56 \times 10^7$ S-m^{-1}) separated by a 1.6-cm thick polyethylene slab ($\epsilon_r' \approx 2.25$, $\tan \delta \approx 4 \times 10^{-4}$) to operate at a frequency of 10 GHz. For TEM, TE$_1$, and TM$_1$ modes, find (a) the phase velocity $\bar{v}_p$ and the waveguide wavelength $\bar{\lambda}$, and (b) the attenuation constants α_c due to conductor losses and α_d due to dielectric losses.

4-23. Semi-infinite parallel-plate waveguide. Two perfectly conducting and infinitesimally thin sheets in air form a semi-infinite parallel-plate waveguide, with mouth in the plane $z = 0$ and sides parallel to the yz plane as shown in Figure 4.30. Two perpendicularly polarized (i.e., electric field in the y direction) uniform plane waves 1 and 2 of equal strength are incident upon the mouth of the guide at angles θ_i as shown. The two waves are in phase, so their separate surfaces of constant phase intersect in lines lying in the yz plane. The peak electric field strength for each wave is known to be 1 V-m^{-1}. (a) Find an expression (in terms of θ_i) for the time-average power flowing down the inside of the guide, per unit meter of the guide in the y direction. (b) If the wavelength λ of the incident waves is such that $\sin \theta_i = \lambda/(2a) = \sqrt{3}/2$ and $a = 1$ cm, find the numerical value of the time-average power transmitted, per unit guide width in the y direction.

4-24. Unknown dielectric waveguide mode. The electric field inside a certain dielectric-slab waveguide (where $\epsilon_d = 9\epsilon_0$, $\mu_d = \mu_0$) is given by

$$\mathbf{E} = \hat{\mathbf{y}}\, 10 \cos(5000x)\, e^{-j5000z} \quad \text{mV-m}^{-1}$$

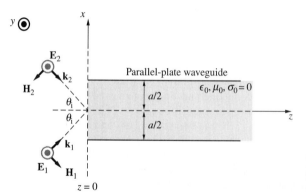

FIGURE 4.30. Semi-infinite parallel-plate waveguide. Problem 4-23.

The thickness of the slab is $d = 2$ mm. (a) What is the waveguide wavelength $\bar{\lambda}$ in meters? What is the wave frequency f? Identify the propagating mode (i.e., specify whether it is TE or TM, odd or even, and find the value of the mode index number m). Specify all other possible propagating modes at frequency f. (b) Considering a ray theory analysis of this propagating mode, find the value of the corresponding incidence angle θ_i.

4-25. Propagating modes in a dielectric slab waveguide. Consider a dielectric slab waveguide with thickness d and refractive indices of 1.5 (for the guide) and 1.48 (for the cladding). (a) Find all the propagating modes at $\lambda = 2\,\mu$m if $d = 5\,\mu$m. (b) Repeat part (a) if $d = 15\,\mu$m. (c) Repeat part (a) if the refractive index of the cladding is 1.49.

4-26. Single TE$_1$ mode dielectric slab design. Consider a dielectric slab waveguide with guide thickness d and refractive index 1.55 sandwiched between two cladding regions each with refractive index 1.5. (a) Design the guide such that only the TE$_1$ mode is guided at $\lambda = 1\,\mu$m. (b) Repeat part (a) for a cladding region with a refractive index of 1.53.

4-27. Millimeter-wave dielectric slab waveguide. Consider a dielectric slab waveguide with $\epsilon_d = 4\epsilon_0$ surrounded by air to be used to guide millimeter waves. Find the guide thickness d such that only the TE$_1$ mode can propagate at frequencies up to 300 GHz.

4-28. TE$_2$ mode in a dielectric slab waveguide. Find the cutoff frequency for the TE$_2$ mode in a dielectric slab waveguide with $\epsilon = 7\epsilon_0$ and thickness 2 cm, which is embedded in another dielectric with $\epsilon = 3\epsilon_0$.

4-29. Dielectric slab waveguide modes. A dielectric slab waveguide is used to guide electromagnetic energy along its axis. Assume that the slab is 2 cm in thickness, with $\epsilon = 5\epsilon_0$ and $\mu = \mu_0$. (a) Find all of the propagating modes for an operating frequency of 4 GHz and specify their cutoff frequencies. (b) Find α_x (in np-m^{-1}) and β_x (in rad-m^{-1}) at 8 GHz for each of the propagating modes. (c) Considering a ray theory analysis of the propagating modes, find the incidence angles θ_i of the component TEM waves within the slab for all of the propagating modes at 8 GHz.

4-30. Dielectric slab waveguide modes. Consider a dielectric slab waveguide made of a dielectric core material of $d = 10\,\mu$m thickness with a refractive index of $n_d = 1.5$ covered with a cladding material of refractive index $n_c = 1.45$, which is assumed to be of infinite extent. (a) Find all the propagating modes at $1\,\mu$m air wavelength. (b) Determine the shortest wavelength allowed in the single-mode transmission.

4-31. Dielectric slab waveguide thickness. A dielectric slab waveguide is made of a dielectric core and cladding materials with refractive indices of $n_d = 1.5$ and $n_c = 1$, respectively. If the number of propagating modes is 100 at a free-space wavelength of 500 nm, calculate the thickness of the core material.

4-32. TEM decomposition of TM$_m$ modes. A dielectric slab waveguide is designed using a core dielectric material of refractive index $n_d = 3$ and thickness $5\,\mu$m covered by a cladding dielectric material of refractive index $n_c = 2.5$. Find all the propagating TM$_m$ modes and corresponding angles of incidence with which they are bouncing back and forth between the two boundaries at $3\,\mu$m air wavelength.

4-33. Dielectric above a ground plane. A planar perfect conductor of infinite dimensions is coated with a dielectric material ($\epsilon_r = 5$, $\mu_r = 1$) of thickness 5.625 cm. (a) Find the cutoff frequencies of the first four TE and/or TM modes, and specify whether they are odd or even. (b) For an operating frequency of 1 GHz, find all of the propagating TE modes. (c) For each of the TE modes found in part (b), find the corresponding propagation constant $\bar{\beta}$. Assume the medium above the coating to be free space.

4-34. Group velocity. Derive the expression (equation [4.69]) for the group velocity for TE_m or TM_m modes in a parallel-plate waveguide.

4-35. Dispersion in seawater. Reconsider Problem 2-25 in view of your knowledge of group velocity. Assuming that the frequency dependence of the phase velocity is as given in Problem 2-25, derive an expression for the group velocity in seawater, and plot both v_p and v_g as a function of frequency between 0.5 and 2.5 kHz.

4-36. Group velocity in a plasma. A cold ionized gas consisting of equal numbers of electrons and protons behaves (see Section 6.2) as a medium with an effective permittivity $\epsilon_{\text{eff}} = \epsilon_0(1 - \omega_p^2/\omega^2)$, where $\omega_p = \sqrt{Nq_e^2/(m_e\epsilon_0)}$ is known as the *plasma frequency*, with N being the volume density of free electrons, $q_e \simeq -1.6 \times 10^{-19}$ C the electronic charge, and $m_e \simeq 9.11 \times 10^{-31}$ kg the electronic mass. (a) Derive the expression for the group velocity v_g in this medium. (b) Evaluate the group velocity (and express it as a fraction of the speed of light in free space) for a 1 MHz radio signal propagating through the earth's ionosphere, where $N \simeq 10^{11}$ m^{-3}. (c) Repeat part (b) for 100 kHz.

4-37. Group velocity in a dielectric slab waveguide. Derive an expression for the group velocity v_g for the odd TM_m modes in a dielectric slab waveguide of slab thickness d and permittivity ϵ_d.

CHAPTER 5

Cylindrical Waveguides and Cavity Resonators

We started our coverage of guided electromagnetic waves and different types of waveguides in Chapter 4, introducing the fundamental principles and mathematical foundations of the guiding of waves and analyzing in detail two of the simplest structures: the parallel-plate waveguide and the dielectric slab waveguide. Both of these are planar structures that guide electromagnetic waves in the z direction by confining the wave energy in the x direction, while being of infinite extent in the y direction. In this chapter, we continue our discussion of guided waves by focusing our attention on cylindrical waveguides, which confine the wave energy in both of the transverse directions (i.e., x and y). In particular, we consider hollow or dielectric-filled metallic tube waveguides having a rectangular or circular cross section. We also briefly discuss the dielectric rod waveguide with a circular cross section.[1] We further consider in this chapter enclosed metallic structures that are designed to confine electromagnetic energy in all three dimensions (i.e., x, y, and z). These so-called cavity resonators can very efficiently trap and store electromagnetic energy within their metallic walls, as long as the excitation frequency is within narrow frequency bands centered around discrete frequencies called *resonances*.

---

[1]This structure is in fact the so-called step-index optical fiber and encompasses a wide range of optical communications and photonics applications. Unfortunately, only a brief treatment can be provided within the scope of this book, since the field solutions for such dielectric waveguides cannot be expressed in analytical form. For up-to-date coverage of optical transmission technology and applications see J. C. Palais, *Fiber Optic Communications,* 4th ed., Prentice Hall, 1998. For coverage of fundamentals of the guiding of light waves in different modes, see A. W. Snyder and J. D. Love, *Optical Waveguide Theory,* Chapman and Hall, London, 1983; and M. L. Barnoski (ed.), *Fundamentals of Optical Fiber Communications,* Academic Press, San Diego, 1976.

The development of electromagnetic wave propagation in hollow-tube waveguides and the practical utilization of such waveguides have an interesting history, involving theoretical denial of the possibility by Heaviside in 1893,[2] discovery and detailed documentation by Rayleigh in 1897 (see the short biography provided in this section), and multiple rediscovery in the mid 1930s after it had been forgotten for nearly four decades. This history constitutes an excellent example of the varying relationships between science and technology.[3]

Experimental demonstration of the propagation of electromagnetic waves in hollow tubes and the practical possibilities for such waveguides as transmission systems was realized independently and almost simultaneously in the 1930s by G. C. Southworth of Bell Laboratories and W. L. Barrow of the Massachusetts Institute of Technology, as a result of a rather interesting sequence of events[4] and technological and scientific developments.[5] The lack of experimental work before the 1930s is probably due partly to the unavailability of signal sources of sufficiently high frequencies for waveguide dimensions to be reasonable, and also to the fact that early long-distance radio communications emphasized frequencies no higher than 15 to 20 MHz. With the advent of World War II and military radar, hollow-tube waveguides operating at frequencies of 1 to 30 GHz were extensively utilized. Especially at frequencies above about 10 GHz, hollow metallic tubes have practical dimensions, provide much lower loss, and have substantially larger power capacity than coaxial

---

[2]Oliver Heaviside (see Chapter 4) undertook a vigorous development of the modern theory of transmission of electromagnetic energy in two-conductor systems, introducing much of the notation and many of the concepts used in modern transmission line theory. In 1893, Heaviside considered various possibilities of guiding of electromagnetic waves by metallic conductors and concluded that single-conductor lines were not possible. Heaviside considered the possible propagation of waves in hollow tubes, raising the question of "whether we cannot transmit an electromagnetic wave along the interior of a tube, in a manner resembling a beam of light." He went on to say, "It does not seem possible to do without the inner conductor, for when it is taken away, we have nothing left upon which the tubes of displacement can terminate internally, and along which they can run" (O. Heaviside, *Electromagnetic Theory*, Benn, London, 1893, p. 400).

[3]An excellent account of the history and origin of waveguides, as well as the underlying relationship between science and technology, is given in K. S. Packard, The origin of waveguides: A case of multiple rediscovery, *IEEE Trans. Microwave Theory and Techniques*, 32(9), pp. 961–969, September 1984.

[4]George C. Southworth worked for nearly fifteen years (between 1920 and 1936) on problems involving the guiding of waves initially in water-filled and then in hollow pipes. When he was finally ready to present his paper, on April 30, 1936, demonstrating propagation of electromagnetic waves in hollow tubes both experimentally and theoretically in a companion paper by J. R. Carson, S. P. Mead, and S. A. Schelkunoff, he realized to his surprise that Wilmer L. Barrow was scheduled to present a similarly titled paper the day after his presentation. He then wrote to W. L. Barrow and suggested that they coordinate their presentations, which they did. For further details of this very interesting path to discovery, see the citation in footnote 3.

[5]Fundamental papers on the subject include W. L. Barrow, Transmission of electromagnetic waves in hollow tubes of metal, *Proc. IRE,* 24(10), pp. 1298–1328, October 1936; G. C. Southworth, Hyperfrequency wave guides—General considerations and experimental results, *Bell System Tech. J.,* 15, pp. 284–309, April 1936; L. J. Chu and W. L. Barrow, Electromagnetic waves of hollow metal tubes of rectangular cross section, *Proc. IRE,* 26(12), pp. 1520–1555, December 1938.

lines. For some time hollow rectangular and circular metal tubes were the only guiding systems used at frequencies in the 10- to 100-GHz range. However, in spite of their many desirable properties, these waveguides are often bulky and expensive to manufacture due to the required precision, and their applicability at the higher end of the microwave spectrum is thus limited. In the 1970s, the stripline and the microstrip line emerged as practical alternatives to waveguides in some applications, primarily because these structures could be manufactured using planar integrated circuit technologies. Microstrips are widely used for interconnecting components within microwave transmitters and receivers and in microwave integrated circuits. However, analyses of microstrips and striplines typically require approximate and/or numerical treatments, the description of which is beyond the scope of this book.[6]

At one time, hollow metallic waveguides were considered for use in long-distance communication, especially with the use of a particularly attractive low-loss mode that exists for waveguides with a circular cross section (see Section 5.2.1). However, the advent of optical fiber technology provided a low-cost alternative for long-distance transmission of much higher bandwidths. At present, waveguides are used for relatively short-distance transmission in applications involving high power and minimal interference requirements, a particularly good example being high-power radar. However, waveguides may see more extensive use in the coming years. Recently emerging silicon-based micromachining technologies might facilitate low-cost fabrication of waveguides with the required precision,[7] and hollow metallic waveguides have been fabricated and tested for operation at optical frequencies.[8]

Rayleigh, John William Strutt   *(English physicist, b. November 12, 1842, Malden, Essex; d. June 30, 1919, Witham, Essex) Strutt acquired his father's title at the age of 31 and thus is almost invariably referred to as Lord Rayleigh. At school, he attended the lectures of Stokes [1819–1903] and was a talented student, especially in mathematics, and in 1865 he finished at the head of his class in Cambridge. He got sick in 1871, visited Egypt, and was elected to the Royal Society in 1873. In 1879, he succeeded Maxwell [1831–1879] as director of the Cavendish Laboratory at Cambridge. Throughout most of his professional life he was interested in wave motion of all kinds. In studying electromagnetic waves, he worked out an equation to account for the variation of the scattering of light with wavelength, confirming earlier suggestions that the sky is blue due to light*

[6]For an excellent brief treatment, see Section 8.6 of S. Ramo, J. R. Whinnery, and T. Van Duzer, *Fields and Waves in Communications Electronics,* 3rd ed., Wiley, New York, 1994.

[7]B. Shenouda, L. W. Pearson, J. E. Harriss, W. Wang, and Y. Guo, Etched-silicon micromachined waveguides and horn antennas at 94 GHz, *IEEE Antennas and Propagation Society International Symposium: 1996 Digest,* 1996, vol. 2, pp. 988–991.

[8]J. N. McMullin, R. Narendra, and C. R. James, Hollow metallic waveguides in silicon-V grooves, *IEEE Photonics Technology Letters,* 5(9), pp. 1080–1082, September 1993.

scattering by atmospheric molecules and dust. He studied sound waves, water waves, and earthquake waves.

Although his work on guided electromagnetic waves was not his most important contribution to science, Rayleigh's 1897 paper on the subject was quite comprehensive indeed. He studied the propagation of electromagnetic waves in hollow cylindrical structures. Solving the boundary value problem for Maxwell's equations, he found that propagating waves existed in a set of well-defined normal modes, much like the vibrations of a membrane. He classified the waves in cylindrical structures into two kinds, one with a longitudinal component of electric field only, the other with a longitudinal component of magnetic field only. He found that such waves could exist only if the frequency exceeded a lower limit, and he gave specific solutions for cylinders with circular and rectangular cross sections.

Rayleigh made many other discoveries in chemistry and physics, and received the Nobel Prize in physics in 1904. [Adapted with permission from I. Asimov, Biographical Encyclopedia of Science and Technology, *Doubleday, 1982].*

| 1700 | 1842 | 1919 | 2000 |

One of the basic elements in a wide variety of dynamical systems is a resonator. Resonant circuits are used in the design of systems that store energy or that selectively detect, filter, or transmit a single frequency or a narrow band of frequencies. In the microwave frequency range of 10–100 GHz, enclosed metallic cavities with rectangular, circular, or coaxial cross sections can be used for highly efficient storage of electromagnetic energy at a discrete set of resonant frequencies. The fundamental theory of electromagnetic energy storage in resonant metallic cavities was developed during Maxwell's time. However, practical engineering interest in resonant cavities did not emerge until the late 1930s, when signal sources and associated test equipment became available for frequencies high enough to allow the use of cavities of reasonable size.[9] Although a comprehensive coverage of different types of practical resonators is beyond the scope of this book,[10] we introduce the fundamental principles of electromagnetic resonators with a discussion in Section 5.3 of metallic cavities with rectangular, circular, and coaxial cross sections. This coverage of metallic cavity resonators naturally follows our discussion of rectangular and circular waveguides, since we can simply consider the cavity as a cylindrical waveguide terminated at both of its open ends.

We now proceed to analyze electromagnetic wave propagation in cylindrical waveguides, starting with rectangular waveguides. Wave propagation in circular

---

[9]The first basic papers on cavity resonators are W. W. Hansen, A type of electrical resonator, *J. Appl. Phys.,* 9(10), pp. 654–663, October 1938; W. W. Hansen, On the resonant frequency of closed concentric lines, *J. Appl. Phys.,* 10(1), pp. 38–45, January 1939.

[10]For treatments at an appropriate level, see Chapter 10 of S. Ramo, J. R. Whinnery, and T. Van Duzer, *Fields and Waves in Communication Electronics,* 3rd ed., John Wiley & Sons, New York, 1994; and Chapter 7 of D. M. Pozar, *Microwave Engineering,* Addison-Wesley, Reading, Massachusetts, 1993.

waveguides, coaxial lines, and dielectric waveguides of circular cross section are covered in Section 5.2, followed in Section 5.3 with a discussion of cavity resonators.

5.1   RECTANGULAR WAVEGUIDES

As for the case of the parallel-plate waveguide, solutions for the electromagnetic fields that can exist in uniform (or dielectric-filled) hollow axial pipes are based on Maxwell's equations (or the wave equation derived from them) subject to boundary conditions at the pipe walls. The following development is very much a direct extension of that followed in the previous chapter on the parallel-plate waveguide. Many of the methods and techniques of analysis apply to waveguides with any type of cross section (Figure 5.1a). For ease of analysis and presentation, we first consider the rectangular waveguide geometry shown in Figure 5.1b, which requires the use of a rectangular coordinate system for the field solutions.

Using the coordinate system shown in Figure 5.1b and assuming perfectly conducting waveguide walls, the boundary conditions at the conductor surfaces are $E_{\text{tan}} = 0$ and $H_{\text{norm}} = 0$, or

$$E_x, E_z = 0 \quad \text{and} \quad H_y = 0 \qquad \text{at } y = 0, b$$

$$E_y, E_z = 0 \quad \text{and} \quad H_x = 0 \qquad \text{at } x = 0, a$$

We assume once again that the waveguide walls are perfect conductors and that all fields vary in the z direction as $e^{-\bar{\gamma}z}$, where in general $\bar{\gamma} = \alpha$ or $\bar{\gamma} = j\beta$ within the waveguide.[11]

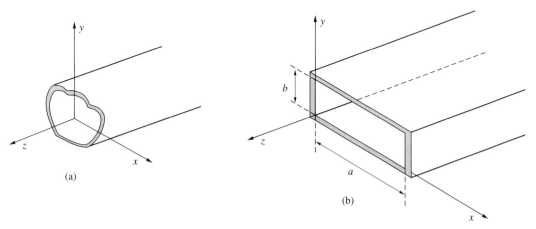

FIGURE 5.1.   Cylindrical waveguides. (a) Uniform axial cylindrical waveguide with arbitrary cross section. (b) Rectangular waveguide.

---

[11] As in the case of parallel-plate waveguide, in the absence of losses we have $\bar{\gamma} = j\beta$ for propagating waves and $\bar{\gamma} = \alpha$ only when the frequency is below cutoff so that there is no propagating wave solution.

With the substitution $\partial/\partial z \to -\overline{\gamma}$, the component form of equations [4.1a] and [4.1c] can be written as follows

$$\frac{\partial H_z}{\partial y} + \overline{\gamma}H_y = j\omega\epsilon E_x \qquad \frac{\partial E_z}{\partial y} + \overline{\gamma}E_y = -j\omega\mu H_x$$

$$\frac{\partial H_z}{\partial x} + \overline{\gamma}H_x = -j\omega\epsilon E_y \qquad \frac{\partial E_z}{\partial x} + \overline{\gamma}E_x = j\omega\mu H_y \qquad [5.1]$$

$$\underbrace{\frac{\partial H_y}{\partial x} - \frac{\partial H_x}{\partial y} = j\omega\epsilon E_z}_{\nabla\times\mathbf{H}=j\omega\epsilon\mathbf{E}} \qquad \underbrace{\frac{\partial E_y}{\partial x} - \frac{\partial E_x}{\partial y} = -j\omega\mu H_z}_{\nabla\times\mathbf{E}=-j\omega\mu\mathbf{H}}$$

and the wave equations for E_z and H_z can be obtained as

$$[\nabla^2\mathbf{E} = -\omega^2\mu\epsilon\mathbf{E}]_z \quad \to \quad \frac{\partial^2 E_z}{\partial x^2} + \frac{\partial^2 E_z}{\partial y^2} + \overline{\gamma}^2 E_z = -\omega^2\mu\epsilon E_z \quad [5.2a]$$

$$[\nabla^2\mathbf{H} = -\omega^2\mu\epsilon\mathbf{H}]_z \quad \to \quad \frac{\partial^2 H_z}{\partial x^2} + \frac{\partial^2 H_z}{\partial y^2} + \overline{\gamma}^2 H_z = -\omega^2\mu\epsilon H_z \quad [5.2b]$$

The various field components can be expressed in terms of E_z and H_z as follows:

$$H_x = -\frac{\overline{\gamma}}{h^2}\frac{\partial H_z}{\partial x} + j\frac{\omega\epsilon}{h^2}\frac{\partial E_z}{\partial y} \qquad [5.3a]$$

$$H_y = -\frac{\overline{\gamma}}{h^2}\frac{\partial H_z}{\partial y} - j\frac{\omega\epsilon}{h^2}\frac{\partial E_z}{\partial x} \qquad [5.3b]$$

$$E_x = -\frac{\overline{\gamma}}{h^2}\frac{\partial E_z}{\partial x} - j\frac{\omega\mu}{h^2}\frac{\partial H_z}{\partial y} \qquad [5.3c]$$

$$E_y = -\frac{\overline{\gamma}}{h^2}\frac{\partial E_z}{\partial y} + j\frac{\omega\mu}{h^2}\frac{\partial H_z}{\partial x} \qquad [5.3d]$$

where $h^2 = \overline{\gamma}^2 + \omega^2\mu\epsilon$.

The preceding relationships between the field components are valid in general for any time-harmonic electromagnetic wave solution for which the field components vary in the z direction as $e^{-\overline{\gamma}z}$. In general, if a solution of the wave equation subject to the boundary conditions is obtained for any one of the field components, the other components can be found from these relationships. Because the possible field solutions are typically classified in terms of TE or TM modes, it is particularly convenient to express the transverse field components explicitly in terms of the axial components (i.e., E_z or H_z), as is done in [5.3]. With either $H_z = 0$ (TM modes) or $E_z = 0$ (TE modes), the wave equation [5.2a] or [5.2b] is typically solved for E_z or H_z, and equations [5.3] are subsequently used to determine the other field components.

5.1.1   Field Solutions for TM and TE Modes

As in the case of the parallel-plate waveguide, we classify the possible field configurations in two categories

$$\text{TM} \longrightarrow H_z = 0 \qquad \text{and} \qquad \text{TE} \longrightarrow E_z = 0$$

Using the transverse nabla operator ∇_{tr},[12] the wave equations [5.2a] and [5.2b] can be written in more general terms and more compactly as

$$\nabla_{tr}^2 E_z + (\bar{\gamma}^2 + \omega^2 \mu \epsilon) E_z = 0 \qquad [5.4a]$$

$$\nabla_{tr}^2 H_z + (\bar{\gamma}^2 + \omega^2 \mu \epsilon) H_z = 0 \qquad [5.4b]$$

where for rectangular coordinates ∇_{tr} is defined as

$$\nabla_{tr} \equiv \hat{\mathbf{x}} \frac{\partial}{\partial x} + \hat{\mathbf{y}} \frac{\partial}{\partial y} \quad \text{and} \quad \nabla_{tr}^2 = \nabla_{tr} \cdot \nabla_{tr}$$

Equations [5.4] also apply for cylindrical coordinates as long as we define the transverse nabla operator as

$$\nabla_{tr} \equiv \hat{\mathbf{r}} \frac{\partial}{\partial r} + \hat{\boldsymbol{\phi}} \frac{1}{r} \frac{\partial}{\partial \phi}$$

The component equations [5.3] can also be rewritten more compactly using the ∇_{tr} notation as follows:

$$\mathbf{E}_{tr} = \hat{\mathbf{x}} E_x + \hat{\mathbf{y}} E_y = -\frac{\bar{\gamma}}{\bar{\gamma}^2 + \omega^2 \mu \epsilon} \nabla_{tr} E_z \qquad [5.5a]$$

$$\mathbf{H}_{tr} = \hat{\mathbf{x}} H_x + \hat{\mathbf{y}} H_y = \frac{j\omega\epsilon}{\bar{\gamma}} \hat{\mathbf{z}} \times \mathbf{E}_{tr} \qquad [5.5b]$$

for TM waves and

$$\mathbf{H}_{tr} = \hat{\mathbf{x}} H_x + \hat{\mathbf{y}} H_y = -\frac{\bar{\gamma}}{\bar{\gamma}^2 + \omega^2 \mu \epsilon} \nabla_{tr} H_z \qquad [5.6a]$$

$$\mathbf{E}_{tr} = \hat{\mathbf{x}} E_x + \hat{\mathbf{y}} E_y = \frac{j\omega\mu}{\bar{\gamma}} \mathbf{H}_{tr} \times \hat{\mathbf{z}} \qquad [5.6b]$$

for TE waves. Note that in the preceding expressions, $\mathbf{E}_{tr}$ and $\mathbf{H}_{tr}$ are the transverse components of the electric and magnetic fields, respectively.

In solving equation [5.4b] for H_z in the case of TE waves, we need to express the boundary conditions in terms of H_z. Since the component of **H** normal to the waveguide walls must vanish at the surface of the perfect conductors, and because

---

[12]Note that with the wave propagation assumed to be in the z direction, ∇_{tr} simply represents differentiation with respect to the *transverse* coordinates only, that is, x, y or r, ϕ, respectively, in rectangular or cylindrical coordinates.

$\mathbf{H}_{tr}$ is proportional to $\nabla_{tr}H_z$ from [5.6a], we must have

$$\hat{\mathbf{n}} \cdot \mathbf{H}_{tr} = 0 \quad \longrightarrow \quad \hat{\mathbf{n}} \cdot \nabla_{tr}H_z = \hat{\mathbf{n}} \cdot \left(\hat{\mathbf{x}}\frac{\partial H_z}{\partial x} + \hat{\mathbf{y}}\frac{\partial H_z}{\partial y} \right) = 0 \qquad [5.7]$$

where $\hat{\mathbf{n}}$ is the outward normal to the waveguide walls. In component form,[13] we must have

$$\frac{\partial H_z}{\partial x} = 0 \quad \text{at} \quad x = 0, a \quad \text{since} \quad \hat{\mathbf{n}} = \pm\hat{\mathbf{x}} \qquad [5.8a]$$

$$\frac{\partial H_z}{\partial y} = 0 \quad \text{at} \quad y = 0, b \quad \text{since} \quad \hat{\mathbf{n}} = \pm\hat{\mathbf{y}} \qquad [5.8b]$$

This requirement can be physically interpreted as follows, considering as an example $\partial H_z/\partial x = 0$ at $x = 0, a$. If H_z varies in the x direction near the boundary, it must have nonzero curl in the y direction, implying a y direction electric field. However, the latter cannot be nonzero at the boundary ($x = 0$ or a) because E_y is tangential to the perfectly conducting waveguide wall and must therefore be equal to zero.

Transverse Magnetic (TM) Modes The wave equations are second-order partial differential equations in two variables and can be solved by using separation of variables. Note that for the TM case, it is convenient to solve the wave equation given by [5.2a] in terms of E_z. We have

$$E_z(x, y, z) = E_z^0(x, y)e^{-\gamma z}$$

where $E_z^0(x, y)$ is that portion of $E_z(x, y, z)$ that depends on x, y only. We now use a method known as separation of variables and look for solutions for $E_z^0(x, y)$ of the form $E_z^0(x, y) = f(x)g(y)$. Note that since the solution of Maxwell's equations for any given set of boundary conditions is unique, any solution we can find with this method has to be the only solution. With $E_z^0(x, y) = f(x)g(y)$, we can rewrite the wave equation as

$$g\frac{d^2f}{dx^2} + f\frac{d^2g}{dy^2} + \gamma^2 fg = -\omega^2\mu\epsilon fg \quad \longrightarrow \quad g\frac{d^2f}{dx^2} + f\frac{d^2g}{dy^2} + h^2 fg = 0$$

Dividing by fg, we find

$$\frac{1}{f}\frac{d^2f}{dx^2} + h^2 = -\frac{1}{g}\frac{d^2g}{dy^2} \qquad [5.9]$$

which is in the form of a function of x equal to a function of y. Since the equality must hold for all values of x, y, both sides of [5.9] must be equal to the same constant.

---

[13]For example, for the waveguide wall at $x = 0, a$, we have $\hat{\mathbf{n}} = \pm\hat{\mathbf{x}}$ and

$$\hat{\mathbf{n}} \cdot \nabla_{tr}H_z = \pm\hat{\mathbf{x}} \cdot \left(\hat{\mathbf{x}}\frac{\partial H_z}{\partial x} + \hat{\mathbf{y}}\frac{\partial H_z}{\partial y} \right) = \pm\frac{\partial H_z}{\partial x}$$

both leading to $\partial H_z/\partial x = 0$.

Equation [5.9] can thus be thought of as two separate differential equations in terms of only x and only y:

$$\frac{1}{f}\frac{d^2 f}{dx^2} + h^2 = A^2 \qquad\qquad\qquad [5.10a]$$

$$\frac{1}{g}\frac{d^2 g}{dy^2} = -A^2 \qquad\qquad\qquad [5.10b]$$

where A is a constant yet to be determined by the boundary conditions. The general solutions of the second-order differential equations [5.10] are

$$f(x) = C_1 \cos(Bx) + C_2 \sin(Bx) \qquad\qquad\qquad [5.11a]$$

$$g(y) = C_3 \cos(Ay) + C_4 \sin(Ay) \qquad\qquad\qquad [5.11b]$$

where $B = \sqrt{h^2 - A^2}$. We thus have

$$\begin{aligned} E_z^0(x, y) &= f(x)g(y) \\ &= C_1 C_3 \cos(Bx)\cos(Ay) + C_1 C_4 \cos(Bx)\sin(Ay) \\ &\quad + C_2 C_3 \sin(Bx)\cos(Ay) + C_2 C_4 \sin(Bx)\sin(Ay) \end{aligned}$$

where the constants C_i, A, and B can be selected to satisfy the boundary conditions $E_z^0 = 0$ at $x = 0$, $x = a$, $y = 0$, and $y = b$. At $x = 0$, we have

$$E_z^0(0, y) = C_1 C_3 \cos(Ay) + C_1 C_4 \sin(Ay) + 0 + 0$$

so that we must have $C_1 = 0$ in order for $E_z^0 = 0$ at $x = 0$. (The other solution, $C_3 = C_4 = 0$, results in zero field everywhere, and is not considered further.) At $y = 0$, noting that $C_1 = 0$, we have

$$E_z^0(x, 0) = 0 + 0 + C_2 C_3 \sin(Bx) + 0$$

In order that $E_z^0(x, 0) = 0$, either C_2 or C_3 must be zero. However, $C_2 = 0$ would give $E_z^0(x, y) = 0$ everywhere, so that for a nontrivial solution we must have $C_3 = 0$. We can thus write $E_z^0(x, y)$ as

$$E_z^0(x, y) = C \sin(Bx)\sin(Ay)$$

where the constants A and B are yet to be determined. Note that $C = C_2 C_4$ is simply a constant multiplier that is determined by the source that establishes the field. Note that we still have two more boundary conditions. At $x = a$ we must have

$$E_z^0(a, y) = C \sin(Ba)\sin(Ay) = 0 \qquad \text{for all } y \longrightarrow B = \frac{m\pi}{a} \qquad m = 1, 2, 3, \ldots$$

Similarly, at $y = b$,

$$E_z^0(x, b) = C \sin(Bx)\sin(Ab) = 0 \qquad \text{for all } x \longrightarrow A = \frac{n\pi}{b} \qquad n = 1, 2, 3, \ldots$$

The final expression for $E_z^0(x, y)$ is thus

$$E_z^0(x, y) = C \sin\left(\frac{m\pi}{a}x\right)\sin\left(\frac{n\pi}{b}y\right) \qquad\qquad\qquad [5.12]$$

and the total phasor $E_z(x, y, z)$ and the corresponding real electric field $\mathcal{E}_z(x, y, z, t)$ are given by

$$E_z(x, y, z) = C \sin\left(\frac{m\pi}{a}x\right)\sin\left(\frac{n\pi}{b}y\right)e^{-j\overline{\beta}_{mn}z}$$

$$\mathcal{E}_z(x, y, z, t) = C \sin\left(\frac{m\pi}{a}x\right)\sin\left(\frac{n\pi}{b}y\right)\cos(\omega t - \overline{\beta}_{mn}z)$$

$\left.\rule{0pt}{40pt}\right\}$ for $\overline{\gamma} = j\overline{\beta}_{mn}$

or

$$E_z(x, y, z) = C \sin\left(\frac{m\pi}{a}x\right)\sin\left(\frac{n\pi}{b}y\right)e^{-\overline{\alpha}_{mn}z}$$

$$\mathcal{E}_z(x, y, z, t) = Ce^{-\overline{\alpha}_{mn}z} \sin\left(\frac{m\pi}{a}x\right)\sin\left(\frac{n\pi}{b}y\right)\cos(\omega t)$$

$\left.\rule{0pt}{40pt}\right\}$ for $\overline{\gamma} = \overline{\alpha}_{mn}$

As before, the solutions for $\overline{\gamma} = j\overline{\beta}_{mn}$ are referred to as propagating waves, whereas those for $\overline{\gamma} = \overline{\alpha}_{mn}$ are referred to as evanescent waves.

Since $H_z = 0$ (TM mode), and assuming propagating modes so that $\overline{\gamma} = j\overline{\beta}_{mn}$, the other field components can be determined from [5.3]. Therefore, the complete set of field components for the propagating TM_{mn} modes is

Rectangular
TM_{mn}
$m = 1, 2, 3, \ldots$
$n = 1, 2, 3, \ldots$

$$E_z = C \sin\left(\frac{m\pi}{a}x\right)\sin\left(\frac{n\pi}{b}y\right)e^{-j\overline{\beta}_{mn}z}$$

$$E_x = -\frac{j\overline{\beta}_{mn}C}{h^2}\frac{m\pi}{a} \cos\left(\frac{m\pi}{a}x\right)\sin\left(\frac{n\pi}{b}y\right)e^{-j\overline{\beta}_{mn}z}$$

$$E_y = -\frac{j\overline{\beta}_{mn}C}{h^2}\frac{n\pi}{b} \sin\left(\frac{m\pi}{a}x\right)\cos\left(\frac{n\pi}{b}y\right)e^{-j\overline{\beta}_{mn}z}$$

$$H_x = \frac{j\omega\epsilon C}{h^2}\frac{n\pi}{b} \sin\left(\frac{m\pi}{a}x\right)\cos\left(\frac{n\pi}{b}y\right)e^{-j\overline{\beta}_{mn}z}$$

$$H_y = -\frac{j\omega\epsilon C}{h^2}\frac{m\pi}{a} \cos\left(\frac{m\pi}{a}x\right)\sin\left(\frac{n\pi}{b}y\right)e^{-j\overline{\beta}_{mn}z}$$

[5.13]

Note that for these TM_{mn} modes, the field components are nonzero only when both m and n are nonzero. Because $h^2 = A^2 + B^2$, and $A = m\pi/a$ and $B = n\pi/b$, the propagation constant is given by

$$\overline{\gamma} = \sqrt{h^2 - \omega^2\mu\epsilon} = \sqrt{\left(\frac{m\pi}{a}\right)^2 + \left(\frac{n\pi}{b}\right)^2 - \omega^2\mu\epsilon} \qquad [5.14]$$

We see from [5.14] that wave propagation is possible (i.e., $\overline{\gamma}$ is purely imaginary) only when $\omega > \omega_{c_{mn}}$, where $\omega_{c_{mn}}$ is the cutoff frequency of the TM_{mn} mode given by

$$\overline{\gamma} = 0 \quad \rightarrow \quad \boxed{\omega_{c_{mn}} = \frac{1}{\sqrt{\mu\epsilon}}\sqrt{\left(\frac{m\pi}{a}\right)^2 + \left(\frac{n\pi}{b}\right)^2}} \qquad [5.15]$$

For $\omega > \omega_{c_{mn}}$, we use $\bar{\gamma} = j\bar{\beta}_{mn}$ to write

$$\bar{\beta}_{mn} = \sqrt{\omega^2 \mu\epsilon - \left(\frac{m\pi}{a}\right)^2 - \left(\frac{n\pi}{b}\right)^2} = \beta\sqrt{1 - \left(\frac{f_{c_{mn}}}{f}\right)^2} \qquad [5.16]$$

where $\beta = \omega\sqrt{\mu\epsilon}$ and $f_{c_{mn}} = \omega_{c_{mn}}/(2\pi)$. The cutoff wavelength $\lambda_{c_{mn}}$ (i.e., propagation is possible for $\lambda < \lambda_{c_{mn}}$) is

$$\lambda_{c_{mn}} = \frac{v_p}{f_{c_{mn}}} = \frac{2}{\sqrt{\left(\frac{m}{a}\right)^2 + \left(\frac{n}{b}\right)^2}} = \frac{2ab}{\sqrt{m^2 b^2 + n^2 a^2}} \qquad [5.17]$$

For propagating waves, the wave phase velocity and waveguide wavelength are given as

$$\bar{v}_{p_{mn}} = \frac{\omega}{\bar{\beta}_{mn}} = \frac{\omega}{\sqrt{\omega^2 \mu\epsilon - \left(\frac{m\pi}{a}\right)^2 - \left(\frac{n\pi}{b}\right)^2}} = \frac{1}{\sqrt{\mu\epsilon}\sqrt{1 - \left(\frac{f_{c_{mn}}}{f}\right)^2}} \qquad [5.18]$$

$$\bar{\lambda}_{mn} = \frac{2\pi}{\bar{\beta}_{mn}} = \frac{2\pi}{\sqrt{\omega^2 \mu\epsilon - \left(\frac{m\pi}{a}\right)^2 - \left(\frac{n\pi}{b}\right)^2}} = \frac{\lambda}{\sqrt{1 - \left(\frac{f_{c_{mn}}}{f}\right)^2}} \qquad [5.19]$$

We note that $\bar{v}_{p_{mn}} > v_p$ and $\bar{\lambda}_{mn} > \lambda$, where $v_p = \omega/\beta = 1/\sqrt{\mu\epsilon}$ and $\lambda = 2\pi/\beta$, are the phase velocity and wavelength, respectively, for a uniform plane wave in the unbounded lossless medium. Another useful expression for $\bar{\lambda}_{mn}$ is

$$\bar{\lambda}_{mn} = \frac{\lambda}{\sqrt{1 - \left(\frac{\lambda}{\lambda_{c_{mn}}}\right)^2}}$$

Although the field expressions for TM_{mn} modes in rectangular waveguides are considerably more complicated that those for parallel-plate metal waveguides studied in Chapter 4, it is interesting to note that the wave impedance Z_{TM} as defined in Section 4.1.5 is similar in form for the two waveguides. Recalling the definition of Z_{TM} as given in [4.28], we have

$$Z_{TM_{mn}} \equiv \frac{E_x}{H_y} = \frac{E_x^0 e^{-j\bar{\beta}z}}{H_y^0 e^{-j\bar{\beta}z}} = \frac{E_x^0}{H_y^0}$$

or, alternatively,

$$Z_{TM_{mn}} \equiv -\frac{E_y}{H_x} = -\frac{E_y^0}{H_x^0}$$

Substituting the expressions for the various field components from [5.13], we have

$$Z_{\text{TM}_{mn}} = \frac{E_x^0}{H_y^0} = \frac{-\dfrac{j\bar{\beta}_{mn}C}{h^2}\dfrac{m\pi}{a}\cos\left(\dfrac{m\pi}{a}x\right)\sin\left(\dfrac{n\pi}{b}y\right)}{-\dfrac{j\omega\epsilon C}{h^2}\dfrac{m\pi}{a}\cos\left(\dfrac{m\pi}{a}x\right)\sin\left(\dfrac{n\pi}{b}y\right)} = \frac{\bar{\beta}_{mn}}{\omega\epsilon}$$

$$= \frac{\beta\sqrt{1-(f_{c_{mn}}/f)^2}}{\omega\epsilon} = \frac{\omega\sqrt{\mu\epsilon}\sqrt{1-(f_{c_{mn}}/f)^2}}{\omega\epsilon}$$

$$Z_{\text{TM}_{mn}} = \eta\sqrt{1-\left(\frac{f_{c_{mn}}}{f}\right)^2} \qquad [5.20]$$

which is identical to the expression [4.31] for the wave impedance for TM_m modes in a parallel-plate metal waveguide, except that the cutoff frequency f_{c_m} is replaced by that for TM_{mn} modes. Note that we would have arrived at the same result if we had worked with E_y^0 and H_x^0.

Transverse Electric (TE) Modes The formulation of the problem and the solution for TE waves can be carried out in a very similar manner as discussed for TM modes. We now solve the wave equation [5.2b] in terms of H_z^0 to obtain

$$H_z(x, y, z) = H_z^0(x, y)e^{-\bar{\gamma}z}$$

where $H_z^0(x, y)$ can be found from the boundary conditions [5.8] as

$$H_z^0(x, y) = C\cos\left(\frac{m\pi}{a}x\right)\cos\left(\frac{n\pi}{b}y\right)$$

Noting that $E_z = 0$ (i.e., TE_{mn} mode), the other field components can be found from [5.3]. The complete set of field components for the TE_{mn} modes is:

Rectangular
TE_{mn}
$m = 0, 1, 2, 3, \ldots$
$n = 0, 1, 2, 3, \ldots$

$$H_z = C\cos\left(\frac{m\pi}{a}x\right)\cos\left(\frac{n\pi}{b}y\right)e^{-j\bar{\beta}_{mn}z}$$

$$H_x = \frac{j\bar{\beta}_{mn}}{h^2}C\frac{m\pi}{a}\sin\left(\frac{m\pi}{a}x\right)\cos\left(\frac{n\pi}{b}y\right)e^{-j\bar{\beta}_{mn}z}$$

$$H_y = \frac{j\bar{\beta}_{mn}}{h^2}C\frac{n\pi}{b}\cos\left(\frac{m\pi}{a}x\right)\sin\left(\frac{n\pi}{b}y\right)e^{-j\bar{\beta}_{mn}z}$$

$$E_x = \frac{j\omega\mu}{h^2}C\frac{n\pi}{b}\cos\left(\frac{m\pi}{a}x\right)\sin\left(\frac{n\pi}{b}y\right)e^{-j\bar{\beta}_{mn}z}$$

$$E_y = -\frac{j\omega\mu}{h^2}C\frac{m\pi}{a}\sin\left(\frac{m\pi}{a}x\right)\cos\left(\frac{n\pi}{b}y\right)e^{-j\bar{\beta}_{mn}z}$$

[5.21]

where we have assumed $\bar{\gamma} = j\bar{\beta}_{mn}$.

The wave number $\overline{\beta}_{mn}$, the cutoff frequency $f_{c_{mn}}$ and cutoff wavelength $\lambda_{c_{mn}}$, the guide wavelength $\overline{\lambda}_{mn}$, and the phase velocity $\overline{v}_{p_{mn}}$ expressions are the same for TE_{mn} modes as for TM_{mn} modes. The wave impedance for TE waves, namely, Z_{TE}, can be evaluated in the same manner as was done for TM waves in the previous section. We find

$$Z_{TE_{mn}} = \frac{\eta}{\sqrt{1 - \left(f_{c_{mn}}/f\right)^2}} \qquad [5.22]$$

which is identical to equation [4.29] for the wave impedance of TE waves in a parallel-plate waveguide, except for the fact that $f_{c_{mn}}$ in [5.22] is the cutoff frequency for TE_{mn} modes in a rectangular waveguide as given in [5.15].

The Dominant TE_{10} Mode   Unlike the TM case, for TE_{mn}, wave expressions for $m = 0$ (or $n = 0$) do not result in a trivial solution. In fact, for a rectangular waveguide with $a > b$, the TE_{10} mode is known as the *dominant* mode, for reasons discussed in the next section.

For this mode, we have

Rectangular
TE_{10}

$$
\begin{aligned}
H_z &= C \cos\left(\frac{\pi x}{a}\right) e^{-j\overline{\beta}_{10}z} \\[2mm]
H_x &= \frac{j\overline{\beta}_{10}aC}{\pi} \sin\left(\frac{\pi x}{a}\right) e^{-j\overline{\beta}_{10}z} \\[2mm]
E_y &= -\frac{j\omega\mu aC}{\pi} \sin\left(\frac{\pi x}{a}\right) e^{-j\overline{\beta}_{10}z} \\[2mm]
E_x &= H_y = 0 \\[2mm]
\overline{\beta}_{10} &= \sqrt{\omega^2\mu\epsilon - \left(\frac{\pi}{a}\right)^2} = \left[\left(\frac{2\pi}{\lambda}\right)^2 - \left(\frac{\pi}{a}\right)^2\right]^{1/2} \\[2mm]
\overline{\lambda}_{10} &= \frac{2\pi}{\overline{\beta}_{10}} = \frac{\lambda}{[1 - (\lambda/2a)^2]^{1/2}}
\end{aligned}
\qquad [5.23]
$$

where λ is the wavelength for the unbounded lossless medium at this frequency [i.e., $\lambda = (f\sqrt{\mu\epsilon})^{-1}$] and $\overline{\lambda}_{10}$ is the wavelength along the waveguide (Figure 5.2). A sketch of the field configuration for TE_{10} is also shown in Figure 5.2.

Note that for the TE_{10} mode, we have $h = \pi/a$, so that the cutoff frequency and wavelength are given by

$$f_{c_{10}} = \frac{1}{2a\sqrt{\mu\epsilon}} \qquad [5.24]$$

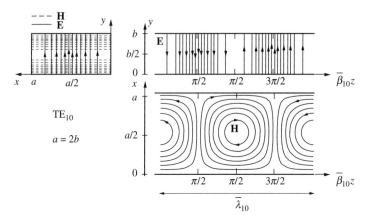

FIGURE 5.2. **TE$_{10}$ mode.** The electric and magnetic field configurations for a TE$_{10}$ mode in a rectangular waveguide.

and

$$\lambda_{c_{10}} = \frac{v_p}{f_{c_{10}}} = 2a$$ [5.25]

Since wave propagation is possible for $f > f_{c_{10}}$ (or $\lambda < \lambda_{c_{10}}$), it is clear that $f_{c_{10}}$ decreases (or $\lambda_{c_{10}}$ increases) if the guide dimension a is enlarged. In other words, for a given wave frequency f, the guide dimension a must be such that $a > v_p/(2f) = \lambda/2$ for propagation to occur in the dominant TE$_{10}$ mode. In simple terms, one-half of the unbounded medium wavelength ($\lambda/2$) must "fit" into the dimension a for propagation of the TE$_{10}$ mode to be possible.[14]

The wave impedance of the dominant TE$_{10}$ mode is given as

$$Z_{TE_{10}} = \frac{\eta}{\sqrt{1 - (f_{c_{10}}/f)^2}} = \frac{2a\mu f}{\sqrt{4a^2 \mu\epsilon f^2 - 1}}$$ [5.26]

Electric and Magnetic Field Configurations The xy plane electric and magnetic field configurations for some of the lower-order modes are illustrated in Figure 5.3. Although the z components of the fields are not shown in the cross-sectional views in Figure 5.3, it is understood that the magnetic fields for TE waves and the electric fields for TM waves also have z components, which extend into and out of the page, in a manner similar to those shown for the TE$_{10}$ mode in Figure 5.2. Although the field structures clearly vary from one mode to another, we note that the

---

[14]For given waveguide dimensions $a > b$, the cutoff frequency $f_{c_{10}}$ is the lowest of all possible TE$_{mn}$ or TM$_{mn}$ modes; thus, if propagation at a given frequency is not possible in the TE$_{10}$ mode, it is not possible in *any* mode.

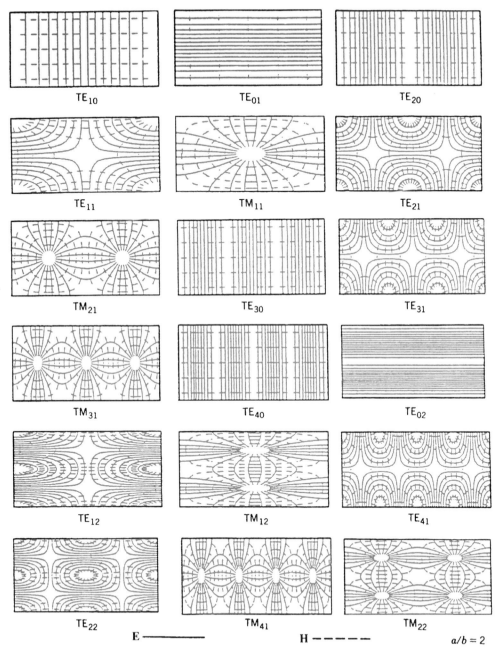

FIGURE 5.3.   Field patterns in a rectangular waveguide. Field configurations in the *xy* plane for selected rectangular waveguide modes. [Taken from C. S. Lee, S. W. Lee, and S. L. Chuang, Plot of modal field distribution in rectangular and circular waveguides, *IEEE Trans. Microwave Theory and Techniques,* 33(3), pp. 271–274, March 1985.]

requirements inherent in Maxwell's equations are met in all cases: the electric field lines must always be perpendicular to the perfectly conducting walls, and the magnetic field lines can never be normal to the walls (i.e., $\nabla \cdot \mathbf{B} = 0$). The electric field lines must "encircle" the magnetic field lines (i.e., $\nabla \times \mathbf{E} = -j\omega\mathbf{B}$), and the magnetic field lines must encircle the electric field lines (i.e., $\nabla \times \mathbf{H} = j\omega\epsilon\mathbf{E}$).

5.1.2   Some Practical Considerations

Excitation of Modes   In practice, whether or not a given propagating mode exists is determined by the excitation and the uniformity of the waveguide. Possible methods for feeding rectangular waveguides with coaxial cables so that a particular mode may be excited are illustrated in Figure 5.4. In order to launch a particular mode, a type of probe is chosen that will produce lines of E and H that are roughly parallel to the lines of E and H for that mode and that also produce maximum electric field at the location in the guide where the field would be maximum for that mode. In general, it is possible for several modes to exist simultaneously in a guide if the frequency is above the cutoff frequency for those particular modes. Note that the solution for each mode separately satisfies the wave equation and the boundary conditions, so that by linear superposition, the total field does also. However, in practice, the physical dimensions of the guide are usually chosen such that only one mode propagates.

The excitation configurations shown in Figure 5.4 can also be used to extract the signal from the waveguide, for example, to couple to a "load" such as an antenna.

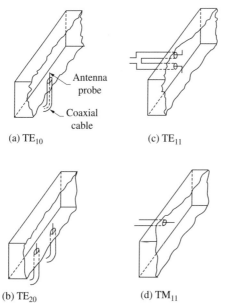

(a) TE$_{10}$

(c) TE$_{11}$

(b) TE$_{20}$

(d) TM$_{11}$

FIGURE 5.4.   Excitation of rectangular waveguide modes. Excitation methods for various rectangular waveguide modes. [Taken from E. C. Jordan and K. G. Balmain, *Electromagnetic Waves and Radiating Systems,* Prentice Hall, Englewood Cliffs, New Jersey, 1968, p. 252.]

Choice of Waveguide Dimensions   In practice, the waveguide dimensions a and b are chosen so that for the frequency band of interest, only a single mode can propagate. As can be seen from Figure 5.5, the cutoff frequencies for a rectangular waveguide are such that for $a > b$, the mode with the lowest cutoff frequency is the dominant mode TE_{10}. For $a = b$, the cutoff frequencies of TE_{10} and TE_{01} are equal, and we have a degenerative situation where modes TE_{10} and TE_{01} differ only by rotation (over the waveguide cross section) through a 90° angle. A square guide of this type is thus prone to undesirable mode conversion and mode coupling.

The ratio of the cutoff frequency of each of the modes to that of the dominant TE_{10} mode is plotted as a function of a/b in Figure 5.5. We note that the separation of cutoff frequencies for different modes is larger for larger values of the dimension ratio (i.e., a/b).

Usually, a is chosen to be approximately equal to $2b$.[15] For $a = 2b$ we have

$$\lambda_{c_{mn}} = \frac{2a}{(m^2 + 4n^2)^{1/2}} \qquad [5.27]$$

Multimode propagation is not desirable in practice because each mode that could propagate has a different phase velocity and a different field configuration. Differences in propagation velocities between modes means that the phase relation between the different portions of the signal power carried by each mode continually varies along the guide and makes it difficult to extract all of the energy from the guide at the receiving end. Differences in field configuration require that a different arrangement of coupling probes (see Figure 5.4) or loops be used to excite each mode in the guide as well as to couple the energy out of the guide. Any irregularities present along the waveguide or in its cross section would tend to excite higher-order modes, in order to satisfy the boundary conditions in the vicinity of the disturbance.

Additional considerations determine the choice of the absolute value of a in terms of a fraction of the unbounded-medium wavelength λ. Clearly we must have $\lambda/2 < a < \lambda$ to ensure the transmission of *only* the TE_{10} mode. Choosing a too close

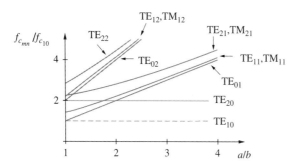

FIGURE 5.5.   Cutoff frequencies of rectangular waveguide modes. Cutoff frequencies of selected modes in rectangular waveguides as a function of a/b.

---

[15]Note that too small a value of b increases attenuation (per unit transmitted power) and also reduces power-handling capabilities of the guide.

TABLE 5.1.   Standard rectangular waveguides

Type	Inside dimensions $a \times b$ (cm)	TE_{10} cutoff (GHz)	Frequency range for TE_{10} mode (GHz)
WR-975	24.765 × 12.383	0.605	0.75–1.12
WR-430	10.922 × 5.461	1.375	1.70–2.60
WR-340	8.636 × 4.318	1.735	2.20–3.30
WR-284	7.214 × 3.404	2.080	2.60–3.95
WR-187	4.755 × 2.215	3.155	3.95–5.85
WR-137	3.485 × 1.580	4.30	5.85–8.20
WR-90	2.286 × 1.016	6.56	8.20–12.4
WR-62	1.580 × 0.790	9.49	12.4–18.0
WR-42	1.067 × 0.432	14.06	17.6–26.7
WR-28	0.711 × 0.356	21.10	26.5–40.0
WR-10	0.254 × 0.127	59.06	75.0–110.0
WR-4	0.102 × 0.0508	137.5	170.0–260.0

to λ does not allow enough separation from the propagating modes with the next lowest cutoff frequency f_c, namely, TE_{20} and TE_{01}, yet a value too close to $\lambda/2$ is not desirable since the phase velocity and wave impedance vary too rapidly with frequency (see Figures 4.12 and 4.27). A value of $a = 0.7\lambda$ is often adopted in practice, leading to the evolution of a range of standard waveguide sizes, examples of which are given in Table 5.1.[16]

Numerical values for the cutoff frequency, wavelength, and phase velocity for K_u-band and C-band rectangular waveguides are illustrated in Examples 5-1 and 5-2, respectively.

Example 5-1: K_u-band rectangular waveguide.   The inner dimensions of an air-filled K_u-band (12–18 GHz) WR-62 rectangular waveguide are $a \simeq 1.58$ cm and $b \simeq 0.79$ cm. (a) Determine the cutoff frequencies of the five lowest-order modes that can propagate in the guide. (b) Which modes propagate in the K_u-band? (c) Determine the phase velocity and the guide wavelength for the TE_{10} mode at 15 GHz.

Solution:

(a) For an air-filled rectangular waveguide, the cutoff frequencies of the TE_{mn} and TM_{mn} modes are given by

$$f_{c_{mn}} = \frac{c}{2\pi}\sqrt{\left(\frac{m\pi}{a}\right)^2 + \left(\frac{n\pi}{b}\right)^2}$$

---

[16]Note that WR-*xxx* refers to a waveguide for which the the larger dimension (*a*) is equal to *xxx*/100 inches.

The results for the five lowest-order modes are as follows:

Mode	TE_{10}	TE_{20}	TE_{01}	TE_{11}	TM_{11}
$f_{c_{mn}}$ (GHz)	~9.49	~19	~19	~21.2	~21.2

(b) Only the TE_{10} mode will propagate in the K_u-band (12-18 GHz).
(c) For the TE_{10} mode in an air-filled rectangular waveguide at 15 GHz, we have

$$\bar{v}_{p_{10}} = \frac{c}{\sqrt{1-(f_{c_{10}}/f)^2}} \simeq \frac{c}{\sqrt{1-(9.49/15)^2}} \simeq 1.29c$$
$$\simeq 3.87 \times 10^8 \text{ m-s}^{-1}$$

and

$$\bar{\lambda}_{10} = \frac{\lambda}{\sqrt{1-(f_{c_{10}}/f)^2}} \simeq 1.29\lambda \simeq 2.58 \text{ cm}$$

Example 5-2:   C-band rectangular waveguide design.   Design an air-filled C-band (4–8 GHz) rectangular waveguide such that the center frequency of this band ($f = 6$ GHz) is at least 25% higher than the cutoff frequency of the TE_{10} mode and at least 25% lower than the cutoff frequency of the next higher mode, so that the dominant mode of propagation is TE_{10}.

Solution: For the TE_{10} mode, $f_{c_{10}} = c/(2a)$. Using the first design criterion, we can write

$$f = 6 \text{ GHz} \geq 1.25 f_{c_{10}} = 1.25\frac{c}{2a}$$

from which we can find the lower bound for the dimension a as

$$a \geq 3.125 \text{ cm}$$

For the TE_{20} mode, $f_{c_{20}} = c/a$. Using the second design criterion, we can write

$$f = 6 \text{ GHz} \leq 0.75 f_{c_{20}} = 0.75\frac{c}{a}$$

from which we can find the upper bound for the dimension a as

$$a \leq 3.75 \text{ cm}$$

Therefore, dimension a is bound by

$$3.125 \leq a \leq 3.75 \text{ cm}$$

For the TE_{01} mode, $f_{c_{01}} = c/(2b)$. Again, using the second design criterion, we can write

$$f = 6 \text{ GHz} \leq 0.75 f_{c_{01}} = 0.75\frac{c}{2b}$$

from which we can find the upper bound for the dimension b as

$$b \leq 2.5 \text{ cm}$$

Although the design criteria do not seem to set a lower bound for the dimension b, in general, a larger b is desirable to increase the power-carrying capability, which is directly proportional to the cross-sectional area of the waveguide. We can find the time-average power density for the TE_{10} mode from the complex Poynting vector as

$$\mathbf{S}_{av} = \frac{1}{2}\mathscr{R}e\{\mathbf{E} \times \mathbf{H}^*\} = -\hat{\mathbf{z}}\frac{1}{2}E_y H_x^* = \hat{\mathbf{z}}\frac{\overline{\beta}_{10}\omega\mu C^2}{2}\left(\frac{a}{\pi}\right)^2 \sin^2\left(\frac{\pi x}{a}\right) \text{ W-m}^{-2}$$

where C is the arbitrary constant in the TE_{10} field solutions, which is the peak amplitude of the axial magnetic field component (H_z). The total time-average power for the TE_{10} mode is

$$P_{av} = \int_0^b \int_0^a \mathbf{S}_{av} \cdot (\hat{\mathbf{z}} \, dx \, dy) = \frac{\overline{\beta}_{10}\omega\mu C^2}{2}\left(\frac{a}{\pi}\right)^2 \frac{ab}{2} \text{ W}$$

Power-Handling Capacity Many radar and communication systems require the transmission of very high microwave power densities through waveguides. The breakdown of the dielectric that fills the waveguide and the dimension of the waveguide are usually the limiting factors in maximizing power-handling capability.

For the dominant TE_{10} mode, the largest electric field occurs along the center line of the broad wall, as shown in Figure 5.6. From [5.23], we have

$$E_y^0 = -\frac{j\omega\mu aC}{\pi}\sin\left(\frac{\pi x}{a}\right)$$

from which the maximum peak value E_0 follows as

$$|E_y|_{max} \equiv E_0 = \frac{\omega\mu aC}{\pi}$$

To avoid dielectric breakdown inside the waveguide, we must have

$$E_0 \leq E_{BR}$$

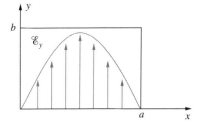

FIGURE 5.6. TE_{10} **mode.** Peak electric field distribution.

where E_{BR} is the dielectric strength of the dielectric material inside the waveguide. Using the result found in Example 5-2, the total peak power for the TE_{10} mode inside the waveguide is

$$P_{peak} = \int_0^b \int_0^a (\mathbf{E} \times \mathbf{H}^*) \cdot (\hat{\mathbf{z}} \, dx \, dy) = \bar{\beta}_{10} \omega \mu C^2 \left(\frac{a}{\pi}\right)^2 \frac{ab}{2}$$

Substituting for $C = \pi E_0/(\omega \mu a)$, we can rewrite this expression in terms of the peak electric field value as

$$P_{peak} = 2P_{av} = \frac{E_0^2}{\eta} \frac{ab}{2} \sqrt{1 - \left(\frac{\lambda}{2a}\right)^2} \qquad [5.28]$$

It is evident from the preceding expression that choosing the dimensions a and b to be as large as possible maximizes the power capacity of the waveguide. In particular, for a given value of a, P_{peak} is proportional to b, so the power capacity can be increased by increasing b. However, as seen in Example 5-2, the fact that it is desirable to have the dominant mode TE_{10} be the only propagating mode provides an upper bound for b, for a given value of a.

Example 5-3 quantitatively illustrates the power-handling capacity of a waveguide operating in the 2–3 GHz range.

Example 5-3:   Power capacity of a rectangular waveguide.   Consider a WR-340 air waveguide with inner dimensions $a = 8.636$ cm and $b = 4.318$ cm. The recommended frequency range for the TE_{10} mode is 2.2 to 3.3 GHz. Calculate the maximum peak power at the low and high ends of the recommended frequency spectrum, taking the breakdown strength of air as 15,000 V-(cm)$^{-1}$ (a safety factor of 2 at sea level is assumed).

Solution: Taking $E_0 = 15$ kV-(cm)$^{-1}$, at $f = 2.2$ GHz, or $\lambda \simeq (3 \times 10^{10}$ cm-s$^{-1})/(2.2 \times 10^9$ Hz$) \simeq 13.64$ cm, the maximum peak power is

$$[P_{peak}]_{max} \simeq \frac{[15 \text{ kV-(cm)}^{-1}]^2}{377 \ \Omega} \frac{(8.636 \text{ cm})(4.318 \text{ cm})}{2} \sqrt{1 - \left(\frac{13.64 \text{ cm}}{2 \times 8.636 \text{ cm}}\right)^2}$$

$$\simeq 6.83 \text{ MW}$$

corresponding to a maximum time-average power of $[P_{av}]_{max} \simeq 3.41$ MW.
    Similarly, at $f = 3.3$ GHz ($\lambda \simeq 9.091$ cm), we have

$$[P_{peak}]_{max} \simeq \frac{(15 \times 10^3)^2}{377} \frac{(8.636)(4.318)}{2} \sqrt{1 - \left(\frac{9.091}{2 \times 8.636}\right)^2} \simeq 9.46 \text{ MW}$$

corresponding to a maximum time-average power of $[P_{av}]_{max} \simeq 4.73$ MW.

5.1.3 Attenuation in Rectangular Waveguides

In general, attenuation in rectangular waveguides may occur due to three different types of losses:

1. Losses in the conducting walls of the waveguide resulting from the flow of surface currents (as discussed for the parallel-plate metallic waveguide in Section 4.1.4).
2. Losses in the dielectric medium between waveguide walls, due to the fact that the dielectric either has nonzero conductivity σ or complex permittivity $\epsilon_c = \epsilon' - j\epsilon''$.
3. Attenuation of the evanescent wave with distance due to the fact that the frequency is below the cutoff frequency.

Next we discuss each of these different losses and derive quantitative expressions for the attenuation rates.

Surface Currents and Attenuation Due to Conduction Losses (α_c) The attenuation due to conducting losses in the waveguide walls is often quite important and is the dominant source of loss, especially for hollow waveguides. To quantitatively evaluate this attenuation, we must first understand the nature of the surface currents that flow in the metallic walls, as a necessary consequence of the nonzero wave magnetic fields tangential to the walls. Expressions for the surface current densities can be obtained by evaluating the tangential magnetic field component solutions at the conductor surfaces. We have

$$\mathbf{J}_s = \hat{\mathbf{n}} \times \mathbf{H}$$

where $\hat{\mathbf{n}}$ is the outward unit normal to the inside waveguide wall.

The configuration of wall currents differs for each mode; however, the procedure for determining them and the evaluation of attenuation due to them is the same. We illustrate these techniques using the important case of the dominant TE_{10} mode.

We examine the surface current density at each waveguide wall using the magnetic field components for TE_{10} derived in the previous section.[17]

$$\mathbf{J}_s^0(x = 0, y) = \hat{\mathbf{x}} \times \hat{\mathbf{z}}H_z^0(x = 0, y) = -\hat{\mathbf{y}}H_z^0(0, y) = -\hat{\mathbf{y}}C$$

$$\mathbf{J}_s^0(x = a, y) = -\hat{\mathbf{x}} \times \hat{\mathbf{z}}H_z^0(x = a, y) = \hat{\mathbf{y}}H_z^0(a, y) = -\hat{\mathbf{y}}C = \mathbf{J}_s^0(x = 0, y)$$

$$\mathbf{J}_s^0(x, y = 0) = \hat{\mathbf{x}}H_z^0(x, 0) - \hat{\mathbf{z}}H_x^0(x, 0)$$

$$= \hat{\mathbf{x}}C \cos\left(\frac{\pi x}{a}\right) - \hat{\mathbf{z}}\frac{j\bar{\beta}_{10}aC}{\pi} \sin\left(\frac{\pi x}{a}\right) = -\mathbf{J}_s^0(x, y = b)$$

The current flow for the TE_{10} mode is shown in Figure 5.7. Note that the pattern shown is for a particular instant of time; as time advances, current flow changes in

---

[17]Note that $\mathbf{J}_s(x, y, z) = \mathbf{J}_s^0(x, y)e^{-j\bar{\beta}z}$, where $\mathbf{J}_s^0(x, y)$ is the portion of $\mathbf{J}_s(x, y, z)$ that depends only on x, y.

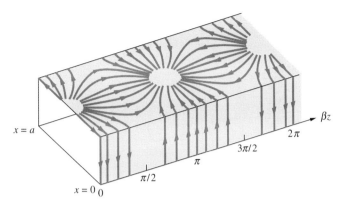

FIGURE 5.7. Surface currents for TE$_{10}$ mode. Wall currents in a rectangular waveguide for the TE$_{10}$ mode. [Taken from D. K. Cheng, *Field and Wave Electromagnetics,* Addison-Wesley, Reading, Massachusetts, 1989.]

accordance with the magnetic field so that the entire pattern propagates down the waveguide. Note also that the points at which the current appears to converge are the places where the electric field normal to the walls has a maximum. Thus, the surface current can be thought to be supplying the surface charge needed to support the normal component of the electric field (i.e., via $\nabla \cdot \mathbf{E} = \rho/\epsilon$ or $\hat{\mathbf{n}} \cdot \mathbf{D} = \rho_s$) in accordance with the continuity equation, $\nabla \cdot \mathbf{J} = -\partial \rho/\partial t$.

We can now proceed to determine the attenuation due to the surface current flow. Using the definition of the attenuation constant as given in [4.21], we have

$$\alpha_c = \frac{\text{Power lost per unit guide length}}{2 \times \text{Power transmitted}} = \frac{P_{\text{loss}}}{2P_{\text{av}}}$$

For the dominant TE$_{10}$ mode, the time-average electromagnetic power flowing through a cross section of the guide can be evaluated as

$$P_{\text{av}}(z) = \int_0^b \int_0^a \underbrace{-\frac{1}{2} E_y^0 (H_x^0)^*}_{\frac{1}{2}\mathfrak{Re}\{\mathbf{E} \times \mathbf{H}^*\} \cdot \hat{\mathbf{z}}} dx\,dy$$

$$= \int_0^b \int_0^a \frac{1}{2} \left(\frac{\omega \mu a C}{\pi} \right) \left(\frac{\overline{\beta}_{10} a C}{\pi} \right) \sin^2 \left(\frac{\pi x}{a} \right) dx\,dy$$

$$= \frac{\omega \mu a^2 \overline{\beta}_{10} C^2}{2\pi^2} b \int_0^a \sin^2 \left(\frac{\pi x}{a} \right) dx$$

$$= \frac{\omega \mu \overline{\beta}_{10} C^2 ba}{\pi^2} \left(\frac{a}{2} \right)^2$$

To evaluate the power lost per unit length, or P_{loss}, we must account for the losses in all four waveguide walls. Because of symmetry, we have

$$P_{loss} = 2[P_{loss_1}]_{x=0} + 2[P_{loss_2}]_{y=0}$$

where P_{loss_2} is the power lost per unit length on the surfaces $y = 0$ or $y = b$ given by

$$[P_{loss_2}]_{y=0} = \int_0^a \frac{1}{2}[\underbrace{|J_{sx}^0(y=0)|^2}_{[C\cos(\frac{\pi x}{a})]^2} + \underbrace{|J_{sz}^0(y=0)|^2}_{\left[\frac{\bar{\beta}_{10}aC}{\pi}\sin(\frac{\pi x}{a})\right]^2}]R_s\,dx = C^2 R_s\left(b + \frac{a}{2} + \frac{\beta_{10}^2 a^2}{2\pi^2}\right)$$

and P_{loss_1} is the power lost per unit length on the surfaces $x = 0$ or $x = a$ given by

$$[P_{loss_1}]_{x=0} = \int_0^b \frac{1}{2}|J_{sy}^0(x=0)|^2 R_s\,dy = \frac{b}{2}C^2 R_s$$

where R_s is the resistive part of the surface impedance of the conducting walls given by $R_s = \sqrt{\omega\mu/(2\sigma)}$. Therefore, the power lost per unit length is

$$P_{loss} = 2[P_{loss_2}]_{x=0} + [P_{loss_1}]_{y=0} = \left[b + \frac{a}{2}\left(\frac{f}{f_{c10}}\right)^2\right]C^2 R_s$$

where we have used

$$\bar{\beta}_{mn} = \sqrt{\omega^2\mu\epsilon - \left(\frac{\pi}{a}\right)^2} = \omega\sqrt{\mu\epsilon}\sqrt{1 - \left(\frac{f_{c10}}{f}\right)^2}$$

After substitution and some manipulation, we find

$$\alpha_{c_{TE_{10}}} = \frac{R_s[1 + (2b/a)(f_{c10}/f)^2]}{\eta b\sqrt{1 - (f_{c10}/f)^2}} \qquad [5.29]$$

For the other TE$_{mn}$ modes, it can be shown by straightforward manipulation that

$$\alpha_{c_{TE_{mn}}} = \frac{2R_s/(b\eta)}{\sqrt{1 - (f_{c_{mn}}/f)^2}}$$
$$\times \left\{\left(1 + \frac{b}{a}\right)\left(\frac{f_{c_{mn}}}{f}\right)^2 + \left[1 - \left(\frac{f_{c_{mn}}}{f}\right)^2\right]\left[\frac{(b/a)[(b/a)m^2 + n^2]}{(b^2m^2/a^2) + n^2}\right]\right\} \qquad [5.30]$$

It should be noted that [5.30] does not reduce to the correct expression for either m or n equal to zero. It is off by a factor of 2 in these cases because the field is uniform in either one or the other of the two directions. Note, however, that [5.29] is valid and may be used for the TE$_{01}$ mode by simply interchanging a and b.

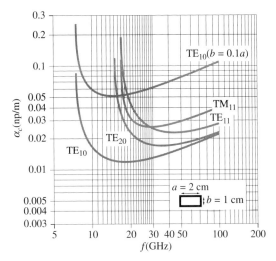

FIGURE 5.8.   Conduction losses for various modes. Attenuation versus frequency curves for various modes in a typical rectangular waveguide. The walls are assumed to be copper ($\sigma = 5.8 \times 10^7$ S-m^{-1}), and the dielectric inside the guide is assumed to be free space. Results are given for $a = 2$ and $b = 1$ cm. Attenuation in dB-m^{-1} can be obtained by multiplying α_c by 8.69.

Following a very similar analysis, it can be shown that for TM$_{mn}$ modes

$$\alpha_{c_{\text{TM}mn}} = \frac{2R_s}{b\eta\sqrt{1 - (f_{c_{mn}}/f)^2}} \frac{m^2(b/a)^3 + n^2}{m^2(b/a)^2 + n^2}$$ [5.31]

Computed values of α_c for a few TE$_{mn}$ and TM$_{mn}$ modes are given in Figure 5.8.

Attenuation Due to Dielectric Losses (α_d)   As noted in Section 4.1.4, losses in a waveguide transmission system also occur because of imperfections in the dielectric material between the conductors. Such losses can be accounted for by taking the permittivity of the dielectric filling to be complex (i.e., $\epsilon_c = \epsilon' - j\epsilon''$) and by deriving an approximate expression for the complex propagation constant $\bar{\gamma}$, as given by equation [4.25]. The same analysis used in Section 4.1.4 for the parallel-plate waveguide is entirely valid for rectangular waveguides. The complex propagation constant $\bar{\gamma}$ for any of the TE$_{mn}$ or TM$_{mn}$ modes is thus given by

$$\bar{\gamma} = \alpha_d + j\bar{\beta}_{mn} \simeq \frac{\omega\sqrt{\mu_0\epsilon'}\,\epsilon''/\epsilon'}{2\sqrt{1 - (\omega_{c_{mn}}/\omega)^2}} + j\sqrt{(\omega^2 - \omega_{c_{mn}}^2)\mu\epsilon'}$$ [5.32]

from which we identify the attenuation constant α_d as

$$\alpha_d \simeq \frac{\omega\sqrt{\mu_0\epsilon'}\,\epsilon''/\epsilon'}{2\sqrt{1 - (\omega_{c_{mn}}/\omega)^2}}$$ [5.33]

and the propagation constant $\bar{\beta}_{mn}$ as

$$\bar{\beta}_{mn} = \sqrt{(\omega^2 - \omega_{c_{mn}}^2)\mu\epsilon'}$$ [5.34]

which is simply the waveguide propagation constant for a lossless dielectric with permittivity ϵ'.

Expressions [5.33] and [5.34] for α_d and $\bar{\beta}_{mn}$ are valid for dielectrics with small losses so that $\epsilon'' \ll \epsilon'$. Note that equation [5.33] for α_d is valid for all TE or TM modes, and in fact also for all cross-sectional shapes of the waveguide, as long as $f_{c_{mn}}$ is the cutoff frequency for the particular mode under consideration.

It is often useful to express the attenuation coefficient in terms of wavelength in the guide $\bar{\lambda}$, and the wavelength λ for uniform plane waves in the unbounded dielectric medium. We find

$$\alpha_d \simeq 3.13 \frac{\epsilon''}{\lambda \epsilon_0} \left(\frac{\bar{\lambda}}{\lambda}\right) \text{np-(cm)}^{-1} = 27 \frac{\epsilon'}{\lambda \epsilon_0} \frac{\bar{\lambda}}{\lambda} \text{ dB-(cm)}^{-1}$$

Attenuation of Evanescent Modes   At frequencies below the cutoff frequency of a particular mode, the electric and magnetic fields of the mode are severely attenuated with distance down the waveguide. In such a case, we have

$$\bar{\gamma} = \bar{\alpha}_{mn} = \frac{2\pi}{\lambda} \sqrt{\left(\frac{f_{c_{mn}}}{f}\right)^2 - 1} \quad \text{or} \quad \bar{\alpha}_{mn} = \frac{2\pi}{\lambda_{c_{mn}}} \sqrt{1 - \left(\frac{f}{f_{c_{mn}}}\right)^2} \quad \text{np-m}^{-1}$$

Note that for $f \ll f_{c_{mn}}$ we have $\bar{\alpha}_{mn} \simeq (6.28/\lambda_{c_{mn}})$ np-m^{-1}, which for the TE$_{10}$ mode ($\lambda_{c_{10}} = 2a$) corresponds to an attenuation rate of ~27 dB per waveguide dimension a. The attenuation rate of nonpropagating modes is typically much higher than the attenuation rates α_c and α_d due to conduction and dielectric losses.

Example 5-4 provides a quantitative comparison of attenuation rates due to wall losses and dielectric losses for a polyethylene-filled waveguide.

Example 5-4:   A copper waveguide filled with polyethylene.   Consider a TE$_{10}$ wave at 10 GHz in a copper ($\sigma_{Cu} = 5.8 \times 10^7$ S-m^{-1}) WR-75 rectangular waveguide with $a = 1.905$ cm and $b = 0.9525$ cm filled with polyethylene ($\epsilon'_r = 2.25$, $\mu_r = 1$, tan $\delta = 4 \times 10^{-4}$ at 10 GHz). Find $\bar{\beta}_{10}$, $\bar{\lambda}_{10}$, $\bar{v}_{p_{10}}$, $\alpha_{c_{TE_{10}}}$, and $\alpha_{d_{TE_{10}}}$.

Solution: At 10 GHz, the phase velocity and the wavelength of a uniform plane wave traveling in an unbounded medium of polyethylene are given as

$$v_p \simeq \frac{c}{\sqrt{\epsilon'_r}} \simeq \frac{3 \times 10^8 \text{ m-s}^{-1}}{\sqrt{2.25}} = 2 \times 10^8 \text{ m-s}^{-1}$$

$$\lambda = \frac{v_p}{f} \simeq \frac{2 \times 10^8 \text{ m/s}}{10^{10} \text{ Hz}} = 2 \text{ cm}$$

The cutoff frequency for the TE$_{10}$ mode is

$$f_{c_{10}} = \frac{v_p}{2a} \simeq \frac{2 \times 10^8 \text{ m-s}^{-1}}{2 \times (1.905 \times 10^{-2} \text{ m})} \simeq 5.25 \text{ GHz}$$

Thus we have

$$\bar{\beta}_{10} = \frac{\omega}{v_p}\sqrt{1 - \left(\frac{f_{c_{10}}}{f}\right)^2} \simeq \frac{2\pi \times 10^{10}\ \text{rad-s}^{-1}}{2 \times 10^8\ \text{m-s}^{-1}}\sqrt{1 - (5.25/10)^2} \simeq 267\ \text{rad-m}^{-1}$$

$$\bar{\lambda}_{10} = \frac{\lambda}{\sqrt{1 - (f_{c_{10}}/f)^2}} \simeq \frac{2\ \text{cm}}{0.851} \simeq 2.35\ \text{cm}$$

$$\bar{v}_{p_{10}} = \frac{v_p}{\sqrt{1 - \left(f_{c_{10}}/f\right)^2}} \simeq \frac{2 \times 10^8\ \text{m-s}^{-1}}{0.851} \simeq 2.35 \times 10^8\ \text{m-s}^{-1}$$

The surface resistance of copper at 10 GHz is given by

$$R_s = \sqrt{\frac{\pi f \mu_0}{\sigma_{\text{Cu}}}} = \sqrt{\frac{\pi \times 10^{10}\ \text{rad-s}^{-1} \times 4\pi \times 10^{-7}\ \text{H-m}^{-1}}{5.8 \times 10^7\ \text{S-m}^{-1}}} \simeq 2.61 \times 10^{-2}\,\Omega$$

Hence we have

$$\begin{aligned}
\alpha_{c_{\text{TE}_{10}}} &= \frac{R_s[1 + (2b/a)(f_{c_{10}}/f)^2]}{\eta b\sqrt{1 - (f_{c_{10}}/f)^2}} \\
&\simeq \frac{2.61 \times 10^{-2}\,\Omega[1 + (2 \times 0.9525/1.905)(0.525)^2]}{251\Omega \times 0.9525 \times 10^{-2}\ \text{m} \times 0.851} \\
&\simeq 0.0163\ \text{np-m}^{-1} \simeq 0.142\ \text{dB-m}^{-1}
\end{aligned}$$

and

$$\begin{aligned}
\alpha_{d_{\text{TE}_{10}}} &= \frac{\omega\sqrt{\mu_0\epsilon'}\tan\delta}{2\sqrt{1 - (f_{c_{10}}/f)^2}} \simeq \frac{\pi \times 10^{10}\ \text{rad-s}^{-1} \times \sqrt{2.25} \times 4 \times 10^{-4}}{3 \times 10^8\ \text{m-s}^{-1}\sqrt{1 - (5.25/10)^2}} \\
&\simeq 0.0738\ \text{np-m}^{-1} \\
&\simeq 0.641\ \text{dB-m}^{-1}
\end{aligned}$$

5.2 CYLINDRICAL WAVEGUIDES WITH CIRCULAR CROSS SECTION

In this section, we consider cylindrical waveguides with circular cross section. The fundamental principles of analysis of wave guiding in such structures are similar to those for rectangular waveguides, except for the use of the wave equation written out in cylindrical coordinates. We specifically consider three different cylindrical waveguides with circular cross section: the circular hollow (or dielectric-filled) tube waveguide, the coaxial line, and the dielectric waveguide with circular cross section.

5.2.1 Circular Waveguide

Hollow cylindrical waveguides of circular cross section (Figure 5.9) are used in a number of applications, including those in which circularly polarized waves are to be transmitted to certain classes of antennas. Some circular waveguide modes, such as the TE_{0l} modes, are interesting because of their low attenuation at high frequencies. Short circular waveguide sections are used as parts of rotary joints in tracking radar systems and as cavity resonators.[18]

Transverse Magnetic (TM) Modes The analysis of circular waveguides is straightforward and follows much the same procedure as for rectangular waveguides, although mathematical expressions for the field components of the various modes need to be expressed in terms of Bessel functions. For TM waves ($H_z = 0$), all field components can be derived from E_z using expressions from the previous section, namely, [5.3] or [5.5], and the wave equation for E_z is given by [5.4a]. However, the transverse nabla operator ∇_{tr}^2 must now be written in terms of cylindrical coordinates:

$$\underbrace{\frac{\partial^2 E_z}{\partial r^2} + \frac{1}{r}\frac{\partial E_z}{\partial r} + \frac{1}{r^2}\frac{\partial^2 E_z}{\partial \phi^2}}_{\nabla_{tr}^2 E_z} + (\bar{\gamma}^2 + \omega^2 \mu\epsilon)E_z = 0 \qquad [5.35]$$

Noting that $E_z(r, \phi, z) = E_z^0(r, \phi)e^{-\bar{\gamma}z}$, we look for a solution using the method of separation of variables,

$$E_z^0(r, \phi) = f(r)g(\phi)$$

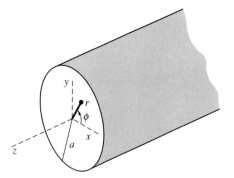

FIGURE 5.9. **Cylindrical waveguide with circular cross section.** Coordinate system for a circular metal pipe waveguide.

---

[18]Both rectangular and circular waveguides are also used as integral parts of many other microwave circuit components, including phase shifters, isolators, and couplers. For a discussion at an appropriate level, see R. S. Elliott, *An Introduction to Guided Waves and Microwave Circuits,* Prentice Hall, Englewood Cliffs, New Jersey, 1993.

Substituting in [5.35] and manipulating, we have

$$\frac{r}{f}\frac{d}{dr}\left(r\frac{df}{dr}\right) + h^2 r^2 = -\frac{1}{g}\frac{d^2 g}{d\phi^2} \qquad [5.36]$$

where $h^2 = \bar{\gamma}^2 + \omega^2 \mu\epsilon$.

For the equality in [5.36] to hold for all values of r and ϕ, both sides must be equal to the same constant, which we take to be equal to n^2, where n is an integer.[19] We thus have two separate differential equations, one in terms of $g(\phi)$ given as

$$\frac{d^2 g}{d\phi^2} + n^2 g = 0 \qquad [5.37]$$

and the other in terms of $f(r)$, which after some algebraic manipulations can be written as

$$\frac{d^2 f}{dr^2} + \frac{1}{r}\frac{df}{dr} + \left(h^2 - \frac{n^2}{r^2}\right)f = 0 \qquad [5.38]$$

The general solution of [5.37] is of the form

$$g(\phi) = C_1 \cos(n\phi) + C_2 \sin(n\phi) \qquad [5.39]$$

Equation [5.38] is known as *Bessel's equation* and has the general solution

$$f(r) = C_3 J_n(hr) + C_4 Y_n(hr) \qquad [5.40]$$

where $J_n(\cdot)$ and $Y_n(\cdot)$ are, respectively, nth-order Bessel functions of the first kind and the second kind.[20] The function $Y_n(\cdot)$ has a singularity at the origin ($r = 0$),

---

[19]This is required since, in view of the azimuthal symmetry, the fields must be periodic (i.e., must involve sine and cosine solutions) in ϕ.

[20]The reader should not be disturbed or intimidated by the appearance of new functions. All of these functions are widely tabulated and plotted and can often be expressed as infinite series. For example, $J_0(\cdot)$ is given by

$$J_0(x) = 1 - \frac{(x)^2}{2^2} + \frac{(x)^4}{2^4(2!)^2} - \frac{(x)^6}{2^6(3!)^2} + \cdots$$

and appears often in problems involving cylindrical symmetry, a classic example being the vibration of a round diaphragm. In this context, we offer the following advice from H. H. Skilling (*Electric Transmission Lines,* McGraw-Hill, New York, 1951, p. 151): "From time to time, a physical problem finds mathematical expression in a differential equation that has no solution in terms of known functions; the solution can be expressed as an infinite series, but not otherwise. Numerical results can be computed from the infinite series, but this is tedious; therefore, if the problem is of enough importance to justify the work, tables of values are computed. Thus a new function is born, and is given a name to identify it and a symbol for convenient use. Friedrich Wilhelm Bessel was a German astronomer [1784–1846]; in 1817 he introduced the functions now known as Bessel functions in the solution of a problem relating to the motion of the planets. In the next few years he studied the properties of these functions, and computed tables of values. It is notable that earlier scientists had solved the differential equation that leads to Bessel functions [Bernoulli in 1732 for the problem of a swinging chain and Euler in 1764 for a stretched membrane], but Bessel was the first to study the functions systematically. There is nothing to prevent the reader from developing a new function. You may take a differential equation, find a solution as a series, compute numerical values, and tabulate them. This new function you can name after yourself. If it is sufficiently useful to enough people it will, perhaps, survive."

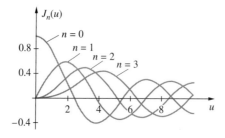

FIGURE 5.10. **Bessel functions of the first kind, $J_n(u)$.** Sketches of Bessel functions of the first kind, $J_n(u)$, of order $n = 0$ to 3.

TABLE 5.2. lth roots (t_{nl}) of $J_n(\cdot) = 0$.

				n				
l	0	1	2	3	4	5	6	7
1	2.405	3.832	5.136	6.380	7.588	8.771	9.936	11.086
2	5.520	7.016	8.417	9.761	11.065	12.339	13.589	14.821
3	8.654	10.173	11.620	13.015	14.372	15.700	17.004	18.288
4	11.792	13.323	14.796	16.223	17.616	18.980	20.321	21.642

inconsistent with physical fields expected in a hollow or dielectric-filled wave-guide,[21] so that $C_4 = 0$. Thus, the general solution for $E_z^0(r, \phi)$ is

$$E_z^0(r, \phi) = C_3 J_n(hr)[C_1 \cos(n\phi) + C_2 \sin(n\phi)] \qquad [5.41]$$

Since we can arbitrarily choose the origin for ϕ in view of the symmetrical geometry, the term in square brackets in [5.41] can be replaced by $\cos(n\phi)$ without any loss of generality. Thus, we have

$$E_z^0(r, \phi) = C_n J_n(hr) \cos(n\phi) \qquad [5.42]$$

where C_n is a constant.

The boundary condition $E_{\tan} = 0$ requires that $E_z^0(r, \phi) = 0$ for $r = a$. As for the case of rectangular waveguides, a nontrivial solution is obtained only for discrete eigenvalues h of the boundary value problem in hand. These values must be such that $J_n(ha) = 0$, so that ha must be a root of the nth-order Bessel function. The form of $J_n(u)$ for a few values of n is plotted in Figure 5.10, and the values t_{nl} such that $J_n(t_{nl}) = 0$ (i.e., lth root of J_n) are given in Table 5.2.[22]

Bessel, Friedrich Wilhelm *(German astronomer, b. July 22, 1784, Minden, Prussia; d. March 17, 1846, Königsberg, Prussia, now Kaliningrad, Russia) Bessel, the son of a civil servant, began life as an accountant in Bremen but taught himself astronomy and mathematics. In 1804, at the age of 20, Bessel recalculated the orbit of Halley's comet and sent the results to Olbers [1758–1840], who*

---

[21] If the region of interest for the problem at hand does not include $r = 0$, as in the case of a coaxial line, the solution $f(r)$ can also include terms with $Y_n(hr)$.

[22] For a more extensive table of roots of J_n and J_n', see C. L. Beattie, Table of First 700 Zeros of Bessel Functions-$J_l(x)$ and $J_l'(x)$, *The Bell System Tech. J.*, pp. 689–697, May 1958.

was sufficiently impressed to obtain a post at an observatory for the young man in 1806. In later years, Olbers would say that his discovery of Bessel was his greatest astronomical achievement. By 1810 Bessel had grown prominent enough to attract the attention of King Frederick William III of Prussia, who appointed him to superintend the construction of an observatory at Königsberg. He remained as director of that observatory until his death. In order to qualify for the post and for the professorial dignity that went with it, Bessel needed a doctor's degree, and he was awarded it for the astronomical work he had already done.

At the new observatory, Bessel worked industriously and in 1818 produced a new and excellent star catalog containing fifty thousand stars. He also introduced numerous refinements into astronomical calculations and worked out a method of analysis involving what are still called Bessel functions, applicable to the solution of many problems both in and outside of astronomy.

His most renowned feat was being the first to achieve a three-century-long dream of astronomers, the determination of the parallax of a star. This result extended the known size of the universe from two to six light-years and provided visible evidence of earth's motion through space. Bessel's discovery marked the beginning of the shift of astronomers' attention from the solar system to the outer universe of the stars. Nevertheless, Bessel's last few years were concerned with the solar system, as he took up the question of the anomalous motion of Uranus and the possibility that an undiscovered planet might exist beyond it. He calculated the masses of Jupiter and Saturn with greater precision than anyone had before and showed that the irregularities in the motion of Uranus could not be explained by the gravitational attractions of those two giant planets. He died of cancer, however, before he could carry his research further, and too soon to witness the success of Leverrier [1811–1877] and John C. Adams [1819–1892] in solving this problem by discovering Neptune. [Adapted with permission from I. Asimov, Biographical Encyclopedia of Science and Technology, *Doubleday, 1982.]*

1700	*1784*	*1846*	*2000*

We may now specify the doubly infinite set of eigenvalues h_{nl} for the TM modes of a circular waveguide as

$$h_{TM_{nl}} = \frac{t_{nl}}{a}$$ [5.43]

where each choice of n and l specifies a particular solution or mode, which is designated as TM_{nl}. Note that n is related to the number of circumferential variations, and l describes the number of radial variations of the field.

The propagation constant for the nlth propagating TM mode is

$$\overline{\beta}_{TM_{nl}} = \left[\omega^2 \mu\epsilon - \left(\frac{t_{nl}}{a} \right)^2 \right]^{1/2}$$ [5.44]

exhibiting a similar cutoff property to that for rectangular waveguides. Propagation occurs for $\lambda < \lambda_{c_{TM_{nl}}}$ or $f > f_{c_{TM_{nl}}}$, where the cutoff wavelength and the frequency can be found from $\overline{\gamma} = 0$ as

$$\lambda_{c_{TM_{nl}}} = \frac{2\pi a}{t_{nl}}$$

$$f_{c_{TM_{nl}}} = \frac{t_{nl}}{2\pi a\sqrt{\mu\epsilon}}$$

[5.45]

Inspection of Table 5.2 shows that the TM_{nl} mode with the lowest cutoff frequency is TM_{01}.

The other field components can be obtained from E_z using [5.5]. The complete set of field components for the TM_{nl} mode is

Circular
TM_{nl}
$n = 0, 1, 2, 3, \ldots$
$l = 1, 2, 3, \ldots$

$$E_z = C_n J_n\left(\frac{t_{nl}}{a}r\right)\cos(n\phi)e^{-j\overline{\beta}_{nl}z}$$

$$E_r = -\frac{ja\overline{\beta}_{TM_{nl}}}{t_{nl}}C_n J_n'\left(\frac{t_{nl}}{a}r\right)\cos(n\phi)e^{-j\overline{\beta}_{nl}z}$$

$$E_\phi = \frac{ja^2 n\overline{\beta}_{TM_{nl}}}{t_{nl}^2 r}C_n J_n\left(\frac{t_{nl}}{a}r\right)\sin(n\phi)e^{-j\overline{\beta}_{nl}z}$$

$$H_r = -\frac{j\omega\epsilon na^2}{t_{nl}^2 r}C_n J_n\left(\frac{t_{nl}}{a}r\right)\sin(n\phi)e^{-j\overline{\beta}_{nl}z}$$

$$H_\phi = -\frac{j\omega\epsilon a}{t_{nl}}C_n J_n'\left(\frac{t_{nl}}{a}r\right)\cos(n\phi)e^{-j\overline{\beta}_{nl}z}$$

$$H_z = 0$$

[5.46]

where

$$J_n'(\zeta) = \frac{dJ_n(\zeta)}{d\zeta}$$

Transverse Electric (TE) Modes The solution for TE modes can be found in a similar manner, except for the fact that we now solve for $H_z^0(r, \phi)$. We find

$$H_z^0(r, \phi) = C_n J_n(hr)\cos(n\phi)$$

[5.47]

TABLE 5.3. *l*th roots (s_{nl}) of $J_n'(\cdot) = 0$.

				n				
l	0	1	2	3	4	5	6	7
1	3.832	1.841	3.054	4.201	5.317	6.416	7.501	8.578
2	7.016	5.331	6.706	8.015	9.282	10.520	11.735	12.932
3	10.173	8.536	9.969	11.346	12.682	13.987	15.268	16.529
4	13.324	11.706	13.170	14.586	15.964	17.313	18.637	19.942

and apply the boundary condition $E_{\text{tan}} = 0$, which requires that $E_\phi^0(r, \phi) = 0$ for $r = a$; and since from [5.6] E_ϕ^0 is proportional to $\partial H_z^0/\partial r$, then $\partial H_z^0/\partial r = 0$ at $r = a$. Equivalently, we can note from [5.7] that the outward normal coordinate from the waveguide walls for the circular waveguide is r, so that we must have

$$\hat{\mathbf{n}} \cdot \nabla_{\text{tr}} H_z = \partial H_z^0/\partial r = 0 \quad \text{at} \quad r = a$$

The eigenvalues are then specified by the zeros of $J_n'(u)$, given as s_{nl} in Table 5.3. The propagation constant and the cutoff frequency and wavelength have the same expressions as for TM, except for the substitution $t_{nl} \longrightarrow s_{nl}$. The propagation constant for the nlth TE mode is

$$\boxed{\overline{\beta}_{\text{TE}_{nl}} = \left[\omega^2 \mu \epsilon - \left(\frac{s_{nl}}{a}\right)^2\right]^{1/2}}$$

with propagation possible for $\lambda < \lambda_{c_{\text{TE}_{nl}}}$ or $f > f_{c_{\text{TE}_{nl}}}$, where

$$\boxed{\begin{aligned} \lambda_{c_{\text{TE}_{nl}}} &= \frac{2\pi a}{s_{nl}} \\[2mm] f_{c_{\text{TE}_{nl}}} &= \frac{s_{nl}}{2\pi a \sqrt{\mu \epsilon}} \end{aligned}}$$

Inspection of Table 5.3 shows that the TE mode with the lowest cutoff frequency is TE_{11}, which actually is the *dominant* mode, since it has the lowest cutoff frequency of all possible TE or TM modes. For $\sim 2.61a < \lambda < \sim 3.41a$, only the dominant TE_{11} mode can propagate in a circular waveguide. The relative values of the cutoff frequencies of waves in a circular waveguide are shown in Figure 5.11.

The other field components can be obtained from H_z using [5.5]. The complete set of field components for the TE_{nl} mode is:

Circular
TE_{nl}
$n = 0, 1, 2, 3, \ldots$
$l = 1, 2, 3, \ldots$

$$\boxed{\begin{aligned} H_z &= C_n J_n\left(\frac{s_{nl}}{a}r\right)\cos(n\phi)e^{-j\overline{\beta}_{nl}z} \\[2mm] H_r &= -\frac{ja\overline{\beta}_{\text{TE}_{nl}}}{s_{nl}}C_n J_n'\left(\frac{s_{nl}}{a}r\right)\cos(n\phi)e^{-j\overline{\beta}_{nl}z} \\[2mm] H_\phi &= \frac{jna^2\overline{\beta}_{\text{TE}_{nl}}}{s_{nl}^2 r}C_n J_n\left(\frac{s_{nl}}{a}r\right)\sin(n\phi)e^{-j\overline{\beta}_{nl}z} \\[2mm] E_r &= \frac{ja^2\omega\mu n}{s_{nl}^2 r}C_n J_n\left(\frac{s_{nl}}{a}r\right)\sin(n\phi)e^{-j\overline{\beta}_{nl}z} \\[2mm] E_\phi &= -\frac{ja\omega\mu}{s_{nl}}C_n J_n'\left(\frac{s_{nl}}{a}r\right)\cos(n\phi)e^{-j\overline{\beta}_{nl}z} \\[2mm] E_z &= 0 \end{aligned}}$$

[5.48]

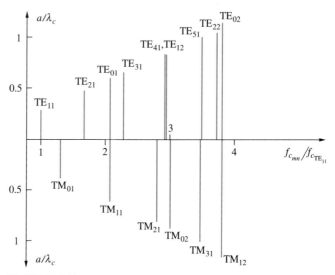

FIGURE 5.11. **Cutoff frequencies.** Cutoff frequencies for low-order modes in circular waveguides, in terms of the cutoff frequency of the TE_{11} mode.

The electric and magnetic field configurations for some of the lower-order TM and TE modes are illustrated in Figure 5.12.

Examples 5-5 and 5-6 illustrate the design and properties of single-mode circular waveguides.

Example 5-5:   Single-mode circular waveguide design.   Design an air-filled circular waveguide such that only the dominant mode will propagate over a bandwidth of 10 GHz.

Solution: The cutoff frequency of the dominant TE_{11} mode is the lower bound of the bandwidth given by

$$f_{c_{TE_{11}}} = \frac{1.8412c}{2\pi a}$$

The cutoff frequency of the next higher mode, TM_{01}, is the upper bound of the bandwidth given by

$$f_{c_{TM_{01}}} = \frac{2.4049c}{2\pi a}$$

The bandwidth is the difference between these two frequencies, given by

$$\text{Bandwidth} = f_{c_{TM_{01}}} - f_{c_{TE_{11}}} = \frac{c}{2\pi a}(2.4049 - 1.8412) = 10 \text{ GHz}$$

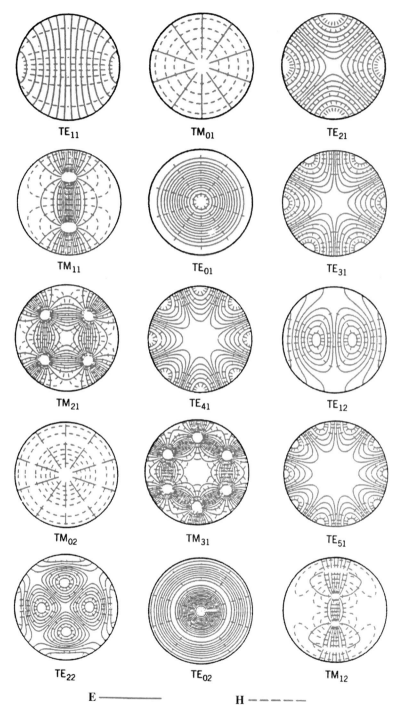

TE$_{11}$	TM$_{01}$	TE$_{21}$
TM$_{11}$	TE$_{01}$	TE$_{31}$
TM$_{21}$	TE$_{41}$	TE$_{12}$
TM$_{02}$	TM$_{31}$	TE$_{51}$
TE$_{22}$	TE$_{02}$	TM$_{12}$

E ——————   H – – – – –

FIGURE 5.12.   Different modes of a circular waveguide. Field configurations for some circular waveguide modes. [Taken from C. S. Lee, S. W. Lee, and S. L. Chuang, Plot of modal field distribution in rectangular and circular waveguides, *IEEE Trans. Microwave Theory and Techniques,* 33(3), pp. 271–274, March 1985.]

from which we find $a \simeq 0.269$ cm. Substituting this back into the expressions for the cutoff frequencies of the TE_{11} and TM_{01} modes, we find $f_{c_{TE_{11}}} \simeq 32.7$ GHz and $f_{c_{TM_{01}}} \simeq 42.76$ GHz. (Note that the recommended frequency range for TE_{11} mode propagation in a WC-25 circular waveguide with an inside diameter of 0.635 cm is 31.8 GHz to 43.6 GHz.)

Example 5-6: An X-band circular waveguide. An air-filled, X-band (8–12 GHz), WC-94 circular waveguide has an inner diameter of 2.383 cm. (a) Determine the cutoff frequencies of the TE_{11}, TM_{01}, and TE_{21} modes. (b) Find the modes that will propagate through this guide at 10 GHz. (c) Find the guide wavelengths for all the propagating modes at 10 GHz. (d) Find the frequency range within which only the TE_{11} mode propagates.

Solution:
(a) For the TE_{11} mode, we have

$$f_{c_{TE_{11}}} = \frac{1.8412c}{2\pi a} = \frac{1.8412 \times 3 \times 10^{10}}{2.383\pi} \simeq 7.38 \text{ GHz}$$

For the TM_{01} and TE_{21} modes, we have

$$f_{c_{TM_{01}}} = \frac{2.4049}{1.8412} f_{c_{TE_{11}}} \simeq 9.64 \text{ GHz}$$

$$f_{c_{TE_{21}}} = \frac{3.0542}{1.8412} f_{c_{TE_{11}}} \simeq 12.2 \text{ GHz}$$

(b) At 10 GHz, the only modes that will propagate through this guide are TE_{11} and TM_{01}.
(c) Using [5.19],

$$\bar{\lambda} = \frac{\lambda}{\sqrt{1 - (f_c/f)^2}}$$

we have

$$\bar{\lambda}_{TE_{11}} \simeq \frac{3 \times 10^{10}/(10^{10})}{\sqrt{1 - (7.38/10)^2}} \simeq 4.44 \text{ cm}$$

and

$$\bar{\lambda}_{TM_{01}} = \frac{3}{\sqrt{1 - (9.64/10)^2}} \simeq 11.2 \text{ cm}$$

(d) The frequency range over which only the dominant TE_{11} mode propagates along the guide can be found as

Frequency range $= f_{c_{TM_{01}}} - f_{c_{TE_{11}}} \simeq 9.64 - 7.38 = 2.26$ GHz

Attenuation in Circular Waveguides   The attenuation produced by the imperfectly conducting walls of a circular waveguide can be calculated in the same manner as for rectangular waveguides. We find that at frequencies above the cutoff frequency, the attenuation constant is

$$\alpha_{c_{TM_{nl}}} = \frac{R_s}{a\eta}\left[1 - \left(\frac{f_{c_{TM_{nl}}}}{f}\right)^2\right]^{-1/2}$$

[5.49]

for TM modes and

$$\alpha_{c_{TE_{nl}}} = \frac{R_s}{a\eta}\left[1 - \left(\frac{f_{c_{TE_{nl}}}}{f}\right)^2\right]^{-1/2}\left[\left(\frac{f_{c_{TE_{nl}}}}{f}\right)^2 + \frac{n^2}{s_{nl}^2 - n^2}\right]$$

[5.50]

for TE modes. Computed values of α_c for some of the lower-order TE and TM modes are plotted as a function of frequency in Figure 5.13. The attenuation constant for TM waves decreases with frequency until $f = \sqrt{3}f_c$ and then increases indefinitely. The same behavior occurs for TE modes except for TE_{0l} modes. For the latter, the attenuation constant decreases indefinitely with frequency, making these modes suitable for long-haul communication links.

We have seen that TE_{11} is the "dominant" mode in a circular waveguide, exhibiting the lowest cutoff frequency for a given guide radius. We see from Figure 5.12 that the field pattern has a general similarity to the dominant TE_{10} mode in a rectangular waveguide. Note also that the TE_{11} mode in a circular guide looks very different from the TE_{11} mode in a rectangular guide.

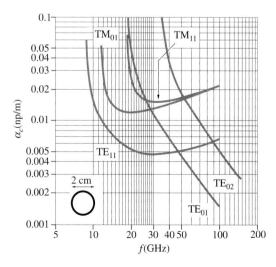

FIGURE 5.13.   Conduction losses for various modes. Attenuation versus frequency curves for various modes in a typical circular waveguide. The walls are assumed to be copper, and the dielectric filling is taken to be air. Results are given for a guide radius of $a = 1$ cm. Attenuation in dB can be obtained by multiplying α_c by 8.69.

In circular waveguides, the second subscript in the mode identification scheme (i.e., l) identifies the order of the root of the Bessel function at the guide wall ($r = a$) for a TM mode. For a TE mode, l is the number of the root of the derivative of the Bessel function; it represents the number of zero values of E_ϕ along a radius, counting the zero on the wall but not a zero at the center. For a TM mode, l represents the number of zero values of E_z. The first subscript, n, determines the order of the Bessel function and is also the coefficient of the variations in ϕ. The value of n can be determined by counting the number of zero values of E_r around the circumference of the guide, and dividing by 2. The attenuation characteristics of most modes are similar. Attenuation is high near cutoff, exhibits a minimum near two or three times the cutoff frequency, and increases as $\sim\sqrt{f}$ for frequencies well above the cutoff frequency. The latter is due to the increasing R_s (proportional to $\sqrt{f}$) as frequency increases, that is, extreme skin effect.

The TE_{01} mode exhibits the very interesting property of decreasing loss with increasing frequency. For this particular mode, all flux lines for the electric field (as well as the magnetic field) are closed loops. Thus, no electric field lines terminate at the walls, so no induced charge or current is required to support it. The only current in the wall is due to the axial magnetic field, which decreases in magnitude as frequency increases. Although at one time this property of the TE_{01} mode made it attractive for long-haul communication,[23] excitation of other modes (with lower cutoff frequencies) caused difficulties, and the optical fiber transmission technology proved to provide a much greater potential at much lower cost.

Note that to obtain practically useful (for long-haul transmission) attenuation rates of <2 dB-(km)$^{-1}$, we must have frequencies many times higher than the cutoff frequency. For this mode, we have

$$\alpha_{c_{TE_{01}}} = \frac{R_s}{a}\sqrt{\frac{\epsilon}{\mu}}\left(\frac{f_{c_{TE_{01}}}}{f}\right)^2\left[1 - \left(\frac{f_{c_{TE_{01}}}}{f}\right)^2\right]^{1/2} \qquad [5.51]$$

---

[23]It was recognized as early as 1933 by both Schelkunoff and Mead that the axially symmetric TE_{01} mode in a circular waveguide would have an attenuation that decreased with frequency, thus offering great potential for wideband multichannel systems. Although development of practical systems was the focus of much work at Bell Laboratories, it was soon realized that microwave radio links using highly directive antennas would be preferable to long circular waveguides. [See G. C. Southworth, *Forty Years of Radio Research,* New York, Gordon and Breach, 1962. Also see K. S. Packard, The origin of waveguides: A case of multiple rediscovery, *IEEE Trans. Microwave Theory and Techniques,* 32(9), pp. 961–969, September 1984.] Nevertheless, the development of a low-loss long-haul communication system operating in the frequency range of 30–110 GHz was considered in the 1950s and 1960s, and a 14.2-km test system built by the British Post Office was measured to have an attenuation of ~2 dB-(km)$^{-1}$, which was substantially better than the optical fiber attenuation rates at the time, although the waveguide system was quite expensive. However, low-loss fiber materials used at present provide rates of 0.2 dB-(km)$^{-1}$, facilitating transmission over distances of hundreds of miles without repeaters.

5.2.2   Coaxial Lines

The coaxial line, illustrated in Figure 5.14, is one of the most common two-conductor transmission systems and is widely used at frequencies mostly below about 5 GHz.[24] However, specially constructed coaxial lines are used when possible at frequencies up to 20 GHz. The physical structure of a coaxial line consists of a braided outer conductor with many strands of thin copper wire wound in a helix shape. The inner conductor is also made out of many strands of wire in order to provide flexibility. The dielectric filling used in most microwave applications is polyethylene or Teflon. A commonly used coaxial line, the RG58, has a nominal attenuation rate at 1 GHz of ~0.76 dB-m^{-1}. Coaxial lines for use in the 5 to 20 GHz range typically have smaller overall diameters and silver-coated conductors, and may have attenuation rates of a few dB-m^{-1}.

From the point of view of the distribution of the electric and magnetic fields, the coaxial line can essentially be viewed as a cylindrical-coordinate version of the parallel-plate waveguide studied in Section 4.1. In this section, we discuss the dominant TEM mode in a coaxial transmission line and briefly comment on the higher-order TE or TM modes.

TEM Mode in a Coaxial Line   As in the case of other two-conductor systems with symmetry in the plane transverse to the direction of propagation, the dominant mode of propagation for the coaxial line is the TEM mode. The fields for this mode can be derived by solving the wave equation [4.4] written in cylindrical coordinates. However, we can arrive at the TEM solution more simply by considering the boundary conditions on the inner and outer conductor surfaces. The fact that the tangential electric field and the normal magnetic field components must be zero on the conductor surfaces requires that $E_\phi = 0$ and $H_r = 0$ at $r = a, b$. Thus, nonzero E_ϕ and H_r components of a TEM solution could exist in the space between the two conductors only if these components varied with r. However, an r variation of a ϕ-directed

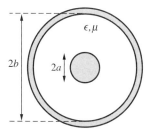
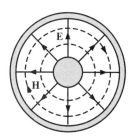

FIGURE 5.14.   Coaxial transmission line. Coordinate system and dimensions for a coaxial line.

---

[24]For an excellent discussion of coaxial lines, including the history of their development, see J. H. Bryant, Coaxial transmission lines, related two-conductor transmission lines, connectors, and components: A U.S. historical perspective, *IEEE Trans. Microwave Theory and Techniques,* 32(9), pp. 970–983, September 1984.

electric field would necessitate the presence of axial components,[25] contradicting the premise of a TEM wave. Thus, a TEM solution can exist only with $\mathbf{E} = \hat{\mathbf{r}}E_r(r, z)$ and $\mathbf{H} = \hat{\boldsymbol{\phi}}H_\phi(r, z)$. Note that there is no ϕ dependence, because of azimuthal symmetry. Under these conditions, we can write [4.1a] in component form as

$$-\frac{\partial H_\phi}{\partial z} = j\omega\epsilon E_r \quad \rightarrow \quad j\beta H_\phi^0(r) = j\omega\epsilon E_r^0(r) \qquad [5.52a]$$

$$\frac{1}{r}H_\phi + \frac{\partial H_\phi}{\partial r} = 0 \quad \rightarrow \quad \frac{1}{r}H_\phi^0(r) + \frac{\partial H_\phi^0(r)}{\partial r} = 0 \qquad [5.52b]$$

where we have assumed solutions propagating in the z direction, with E_r and H_ϕ expressed as products of two functions, respectively, of r and z—namely, $\mathbf{H} = \hat{\boldsymbol{\phi}}H_\phi(r, z) = \hat{\boldsymbol{\phi}}H_\phi^0(r)e^{-j\beta z}$ and $\mathbf{E} = \hat{\mathbf{r}}E_r^0(r)e^{-j\beta z}$, with $\beta = \omega\sqrt{\mu\epsilon}$. Equation [5.52b] can be directly solved for $H_\phi^0(r)$, from which the associated wave electric field can be found from [5.52a], yielding

$$\mathbf{H} = \hat{\boldsymbol{\phi}}H_\phi(r, z) = \hat{\boldsymbol{\phi}}\frac{H_0}{r}e^{-j\beta z} \qquad [5.53a]$$

$$\mathbf{E} = \hat{\mathbf{r}}E_r(r, z) = \hat{\mathbf{r}}\frac{H_0\eta}{r}e^{-j\beta z} \qquad [5.53b]$$

where H_0 is a constant. Note that there are no cutoff conditions inherent in any of these discussions, since the phase constant β is that of a uniform plane wave in the unbounded dielectric medium. Thus the TEM mode can propagate at all frequencies.

Attenuation of the TEM Mode in a Coaxial Line The nonzero tangential magnetic fields on the surfaces of the inner and outer conductors necessitate the flow of surface currents on the conductors. We have

$$\mathbf{J}_s(z) = \hat{\mathbf{n}} \times \mathbf{H} = \begin{cases} \hat{\mathbf{z}}H_\phi(a, z) & \text{on the inner conductor } (\hat{\mathbf{n}} = \hat{\mathbf{r}}) \\ -\hat{\mathbf{z}}H_\phi(b, z) & \text{on the outer conductor } (\hat{\mathbf{n}} = -\hat{\mathbf{r}}) \end{cases}$$

These surface currents lead to power loss in the inner and outer conductors and hence attenuation of the TEM wave as it propagates along the line. The attenuation constant α_c due to conduction losses in a coaxial line can be found as usual from [4.21]. Following a procedure similar to that in previous sections, and assuming H_0 is real, we first find the power transmitted as

$$P_{av} = \frac{1}{2}\Re e\left[\int_a^b\int_0^{2\pi} E_r H_\phi^* r \, d\phi \, dr\right] = \pi\eta H_0^2[\ln r]_a^b = \pi\eta H_0^2 \ln\left(\frac{b}{a}\right)$$

---

[25]To see this, consider that

$$[\nabla \times \mathbf{E}]_z = \frac{1}{r}\left[\frac{\partial}{\partial r}(rE_\phi) - \frac{\partial E_r}{\partial \phi}\right]$$

and the power lost per unit length is given as

$$P_{\text{loss}}(z) = \frac{R_s}{2} \int_S |J_s|^2 \, ds = \frac{R_s}{2} \int_0^1 \left[\int_0^{2\pi} |H_\phi(a, z)|^2 a \, d\phi + \int_0^1 \int_0^{2\pi} |H_\phi(b, z)|^2 b \, d\phi \right] dz$$

$$= \pi R_s H_0^2 \left[\frac{1}{a} + \frac{1}{b} \right]$$

The attenuation constant α_c is then given by

$$\boxed{\alpha_{\text{CTEM}} = \frac{P_{\text{loss}}(z)}{2P_{\text{av}}} = \frac{\pi R_s H_0^2 \left[\dfrac{1}{a} + \dfrac{1}{b} \right]}{2\pi \eta H_0^2 \ln (b/a)} = \frac{R_s}{2\eta \ln (b/a)} \left[\frac{1}{a} + \frac{1}{b} \right]} \qquad [5.54]$$

Noting that $R_s = \sqrt{\omega \mu_0/(2\sigma)}$, we see that α_{CTEM} for a coaxial line increases with frequency in the same manner as for the parallel-plate line as given by [4.22].

Attenuation of TEM waves propagating in a coaxial line can also occur because of dielectric losses. To determine the attenuation constant α_d due to dielectric losses, we can use the general expression for α_d derived in Section 4.1.4. Noting that the cutoff frequency for the TEM is $f_c = 0$ (i.e., there is no cutoff), we have

$$\boxed{\alpha_d \simeq \left[\frac{\omega \sqrt{\mu_0 \epsilon' \epsilon''/\epsilon'}}{2\sqrt{1 - (\omega_c/\omega)^2}} \right]_{\omega_c = 0} = \frac{\omega \epsilon'' \sqrt{\mu_0}}{2\sqrt{\epsilon'}} \simeq \frac{\omega \epsilon'' \eta}{2}} \qquad [5.55]$$

The intrinsic impedance of a low-loss nonmagnetic dielectric is $\eta \simeq \sqrt{\mu_0/\epsilon'}$.

Coaxial Transmission Line Voltage, Current, and Characteristic Impedance   As was discussed previously in connection with the parallel-plate waveguide, the TEM wave on a two-conductor line is the propagation mode that corresponds to the voltage and current waves on two-conductor transmission lines. The voltage between the two conductors and the total current on the inner conductor of the coaxial line are related to the electric and magnetic fields of the TEM wave as

$$V(z) = -\int_a^b \mathbf{E} \cdot d\mathbf{l} = -\int_a^b E_r \hat{\mathbf{r}} \cdot \hat{\mathbf{r}} \, dr = -\int_a^b E_r(r, z) \, dr$$

$$= -\eta H_0 \ln (b/a) e^{-j\beta z} = V^+ e^{-j\beta z}$$

and

$$I(z) = \oint_C \mathbf{H} \cdot d\mathbf{l}$$

$$= \int_0^{2\pi} H_\phi(a, z) \hat{\boldsymbol{\phi}} \cdot \hat{\boldsymbol{\phi}} a \, d\phi = 2\pi H_0 e^{-j\beta z} = \frac{2\pi V^+ e^{-j\beta z}}{\eta \ln b/a} = \frac{V^+}{Z_0} e^{-j\beta z}$$

where we have defined the peak voltage to be $V^+ = \eta H_0 \ln (b/a)$, and Z_0 is the characteristic impedance of the coaxial line given by

$$Z_0 = \frac{\eta \ln b/a}{2\pi} \simeq 60 \ln \frac{b}{a} \qquad [5.56]$$

since $\eta \simeq 120\pi$. Note that [5.56] is identical to the expression for the characteristic impedance of a coaxial line.[26] Upon substitution of $V^+ = \eta H_0 \ln(b/a)$ in the expressions for P_{av} and P_{loss}, the transmitted power flowing down the coaxial line and the power lost per unit line can be shown to be

$$P_{av}(z) = \frac{1}{2} \frac{|V^+|^2}{Z_0} \qquad\qquad P_{loss}(z) = \frac{R_s |V^+|^2}{4\pi Z_0^2} \left(\frac{1}{a} + \frac{1}{b} \right)$$

Optimum Coaxial Lines   The (b/a) for a coaxial line can be varied to achieve different objectives, such as maximizing power- or voltage-handling capacity or minimizing α_c. Because most of these characteristics can be improved by using a larger coaxial line, the optimum dimensions are generally determined for a fixed value of the outer radius b. For this purpose, we assume the line to be terminated in its characteristic impedance, treat b as a constant, and find the ratio (b/a) for the optimum property desired. Note from [5.56] that the characteristic impedance of an air-filled coaxial line is completely determined by the value of (b/a).

As an example, we can determine the best Z_0 for handling maximum power in the TEM mode. The maximum power capacity is determined by the maximum value of the electric field beyond which dielectric breakdown and sparking occur in an air-filled line. Since the electric field varies as r^{-1}, the maximum field occurs at minimum radius, or at $r = a$. The maximum voltage V_{max} and the maximum electric field E_{max} are related by

$$E_{max} = \frac{V_{max}}{a \ln (b/a)} = \frac{V_{max}\zeta}{b \ln \zeta} \quad \rightarrow \quad V_{max} = \frac{E_{max} b \ln \zeta}{\zeta}$$

where $\zeta = b/a$. The time-average power transmitted by the coaxial transmission line at this maximum voltage is given by

$$P_{av} = \frac{V_{max}^2}{2Z_0} = \frac{(E_{max}b)^2 (\ln \zeta)^2}{120\zeta^2 \ln \zeta} = K \frac{\ln \zeta}{\zeta^2}$$

where K is a constant. To determine the value of ζ for maximum P_{av}, we set $dP_{av}/d\zeta = 0$:

$$\frac{dP_{av}}{d\zeta} = K \left(\frac{1}{\zeta^2} - \frac{2 \ln \zeta}{\zeta^2} \right) = 0 \quad \rightarrow \quad \ln \zeta = \frac{1}{2} \quad \rightarrow \quad \frac{b}{a} \simeq 1.65$$

which gives a characteristic impedance of $Z_0 \simeq 60 \ln (1.65) \simeq 30\Omega$.

---

[26]See Table 2.2 of U. S. Inan and A. S. Inan, *Engineering Electromagnetics*, Addison Wesley Longman, 1999.

Similar analyses indicate that the maximum voltage-handling capacity (i.e., maximum $V(z) = \int_a^b \mathbf{E} \cdot d\mathbf{l}$) is achieved with a 60Ω line and that minimum conduction losses (minimum $\alpha_{c_{\text{TEM}}}$) occur for a 77Ω line.

TE and TM Modes in a Coaxial Line   Transverse electric and transverse magnetic propagation modes are in principle possible on any transmission system with boundaries such that the field components for single TM or TE modes can satisfy the boundary conditions. As in the case of the parallel-plate waveguide, TE and TM modes may also exist in a coaxial line, in addition to the principal (dominant) TEM mode discussed earlier. Although these modes are generally undesirable and attempts are made in practice to avoid their excitation, it is nevertheless useful to understand their structure and properties.

The mathematical procedure for determination of the TE and TM modes on a coaxial cable is very similar to that used for finding the TE and TM modes in a circular waveguide. The existence of the inner and outer conductors and the boundary conditions require that the tangential component of the electric field must vanish at $r = a$ and $r = b$. In the general solution of the wave equation, the Bessel function of the second kind [i.e., $Y_n(\cdot)$] is also admissible even though it dictates infinite fields at $r = 0$, since the origin is within the inner conductor and is therefore excluded from consideration.

Consider the solution of the r-dependent portion of the wave equation [5.36] in cylindrical coordinates that was obtained for circular waveguides, namely,

$$f(r) = C_3 J_n(hr) + C_4 Y_n(hr)$$

where $h^2 = \bar{\gamma}^2 \omega^2 \mu\epsilon$. Note that for TM modes, this $f(r)$ is the r-dependent portion of the axial component of the electric field, namely, $E_z(r, \phi, z) = E_z^0(r, \phi)e^{-\bar{\gamma}z}$ and $E_z^0(r, \phi) = f(r)g(\phi)$. For circular waveguides, we had to conclude that $C_4 = 0$, since otherwise the function $Y_n(hr)$ dictated infinite fields at $r = 0$. For coaxial lines, no fields exist at $r < a$, so that C_4 may well be nonzero. For TM waves we thus have

$$E_z^0(r, \phi) = [C_3 J_n(hr) + C_4 Y_n(hr)] \cos (n\phi)$$

For the TE case a similar expression can be written for $H_z^0(r, \phi)$:

$$H_z^0(r, \phi) = [C_3' J_n(hr) + C_4' Y_n(hr)] \cos (n\phi)$$

Using the boundary conditions at $r = a, b$ of

$$E_z(r, \phi) = 0 \qquad \text{for TM modes}$$

$$\frac{\partial H_z}{\partial r} = 0 \qquad \text{for TE modes}$$

we find

$$J_n(ha)Y_n(hb) = J_n(hb)Y_n(ha) \qquad \text{TM modes}$$

and

$$J_n'(ha)Y_n'(hb) = J_n'(hb)Y_n'(ha) \qquad \text{TE modes}$$

The solution of these transcendental equations determine the eigenvalues h for given a, b. Once h is determined, the complete expressions for H_z or E_z are established, from which all the other field components can be found. In general, these transcendental equations are difficult to solve analytically; however, tabulated solutions are readily available.[27] Also, the eigenvalues of such equations can be readily obtained numerically using any one of the commonly available numerical analysis software packages. The electric and magnetic field distributions for a few of the lowest-order TE and TM modes are shown in Figure 5.15.

As in the case of circular waveguides, the modes for coaxial waveguides are denoted as TE_{nl} or TM_{nl}, where the index l refers respectively to the lth zero of the Bessel function $J_n'(\cdot)$ or $J_n(\cdot)$ and the number of its zero crossings in the radial direction, and the index n refers to the number of half-cycles in the circumferential direction. The mode with the lowest nonzero cutoff frequency is the TE_{11} mode, which has the smallest eigenvalue h determined by the first zero of $J_n'(\cdot)$. For the TE_{11} mode, the eigenvalue h is approximately[28] given by

$$h \simeq \frac{2}{a + b}$$

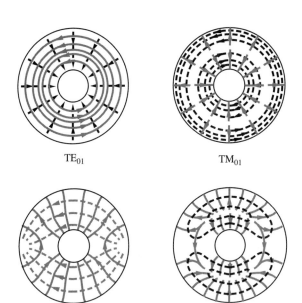

TE$_{01}$ TM$_{01}$

TE$_{11}$ TM$_{11}$

FIGURE 5.15.   TE and TM mode fields in a coaxial line. The electric (magnetic) field lines are shown as solid (dashed) lines. [Taken from H. A. Atwater, *Introduction to Microwave Theory,* p. 76, McGraw-Hill, New York, 1962.]

[27] A table of solutions and detailed descriptions of modes are given in N. Marcuvitz, *Waveguide Handbook,* IEEE Press, Piscataway, New Jersey, 1986.

[28] See Section 8.10 of S. Ramo, J. R. Whinnery, and T. Van Duzer, *Fields and Waves in Communication Electronics,* John Wiley & Sons, New York, 1994.

provided that $b \leq 5a$. This eigenvalue leads to a cutoff wavelength and cutoff frequency of

$$\lambda_{c_{11}} = \frac{2\pi}{h} \simeq \pi(a+b) \quad \text{and} \quad f_{c_{11}} \simeq \frac{1}{\pi(a+b)\sqrt{\mu\epsilon}} \qquad [5.57]$$

respectively. The TE_{11} mode in a coaxial cable is analogous to the TE_{10} mode in a rectangular waveguide. Thus, it is not surprising that the cutoff condition occurs when the average circumference is approximately equal to a wavelength.

In order for the higher-order modes to propagate in coaxial lines, it is necessary that the frequency of operation exceed the cutoff frequencies corresponding to these modes. However, in practice, it is usually desirable to have only the dominant TEM mode propagate and to have the TE or TM modes be attenuated. For a given frequency of operation, the dimensions of the coaxial line are usually chosen to be sufficiently small that $\lambda_{c_{11}}$ is less than $\lambda = (f\sqrt{\mu\epsilon})^{-1}$. This criterion restricts the value of b for a fixed inner radius a, and also determines the practical upper limit of frequency for which TEM is the only propagating mode on the coaxial line.[29]

5.2.3   Dielectric Waveguide with Circular Cross Section

The principles of guiding electromagnetic waves by dielectric structures were introduced in Chapter 4 with a discussion of planar dielectric slab waveguides. Many practical dielectric waveguides used in optical communications (i.e., optical fibers) have circular geometry, consisting of a dielectric rod of permittivity ϵ_d embedded in another dielectric (referred to as cladding) of permittivity ϵ_{cl}. The fundamental principle of operation of such a *cladded circular waveguide* (or a step-index optical fiber, as it is alternatively referred to) is similar to that for the planar dielectric waveguide considered in Chapter 4, except that the circular geometry imposes Bessel-function solutions. For effective guiding of the wave along the dielectric rod, the field in the cladding must decrease[30] radially faster than r^{-1}. To simplify the analysis, the thickness of the cladding is assumed to be infinite as implied in Figure 5.16; in practice, this assumption introduces negligible error, because the fields decay rapidly in the cladding and are reduced to very small values at its edge. Although in principle a dielectric rod in air (i.e., with no cladding) could also be used, the cladding serves two important purposes: (i) Since the fields of a dielectric waveguide are not completely confined within the core ($r \leq a$), any external object in contact with the bare guide

---

[29]For very similar reasons, it can be shown that a two-wire transmission line would propagate only the TEM modes if the wavelength were smaller than twice the spacing between the wires. However, the practical upper limit of frequency for the two-wire line is usually much lower due to radiation losses.

[30]Note that the surface area of a cylinder of unit axial length and with radius r is $2\pi r$. Thus, if the electric and magnetic fields vary radially precisely as r^{-1}, the electric flux density remains constant with radial distance, corresponding to the case of $\alpha_x = 0$ for planar dielectric waveguides. In other words, to confine the fields to the vicinity of the guide in the same manner as for $\alpha_x > 0$, the electric and magnetic fields outside the dielectric rod must decay with radial distance faster than r^{-1}.

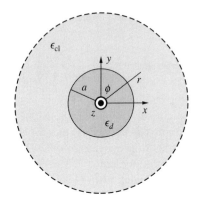

FIGURE 5.16.   Cross section of a step-index optical fiber. The cladding can usually be assumed to be infinite in extent, since for propagating solutions the fields must decay with radial distance in the cladding, and they are thus usually at negligible levels at its outer edge.

can alter the boundary conditions and distort the field distribution in the waveguide. The cladding minimizes such effects, because the fields at its outer edge are at negligible levels. (ii) The number of propagating modes within a dielectric waveguide depends on its radius a and the ratio of the permittivities of the core and cladding (i.e., ϵ_d/ϵ_{cl}). With a proper choice of ϵ_d and ϵ_{cl}, the waveguide can be designed so that only one mode propagates.

At the boundary between the two dielectrics ($r = a$), the tangential components of **E** and **H** must be continuous. In general, neither E_z nor H_z can be zero, so that modes cannot be classified as TE and TM. However, as before, both E_z and H_z satisfy the cylindrical wave equation [5.35]:

$$\nabla_{tr}^2 E_z + h^2 E_z = 0 \quad \rightarrow \quad \frac{1}{r}\frac{\partial}{\partial r}\left(r\frac{\partial E_z}{\partial r}\right) + \frac{1}{r^2}\frac{\partial^2 E_z}{\partial \phi^2} + h^2 E_z = 0 \quad [5.58a]$$

$$\nabla_{tr}^2 H_z + h^2 H_z = 0 \quad \rightarrow \quad \frac{1}{r}\frac{\partial}{\partial r}\left(r\frac{\partial H_z}{\partial r}\right) + \frac{1}{r^2}\frac{\partial^2 H_z}{\partial \phi^2} + h^2 H_z = 0 \quad [5.58b]$$

where, as before,

$$h^2 = h_d^2 = \overline{\gamma}^2 + \omega^2 \mu_d \epsilon_d \quad \text{or} \quad h^2 = -h_{cl}^2 = \overline{\gamma}^2 + \omega^2 \mu_{cl} \epsilon_{cl} \quad [5.59]$$

depending on whether [5.58a] and [5.58b] refer to the fields in the dielectric guide (h_d) or the cladding (h_{cl}).

As was done in Section 5.2.1 for circular waveguides, we assume separable solutions such as

$$\left.\begin{array}{c} E_z(r, \phi, z) \\ \text{or} \\ H_z(r, \phi, z) \end{array}\right\} = \left.\begin{array}{c} E_z^0(r, \phi)e^{-\overline{\gamma}z} \\ \text{or} \\ H_z^0(r, \phi)e^{-\overline{\gamma}z} \end{array}\right\} = f(r)g(\phi)e^{-\overline{\gamma}z}$$

In both regions, the solutions of [5.58a] and [5.58b] can be written in terms of various Bessel functions. Inside the dielectric rod ($r \leq a$), ordinary Bessel functions $J_n(\cdot)$ and $Y_n(\cdot)$ are suitable since we want $\overline{\gamma} = j\overline{\beta}$ or $\omega^2 \mu_d \epsilon_d - h_d^2 > 0$ for propagating field solutions. In the cladding region ($r > a$), we look for field solutions that decay with radial distance, as provided by the modified Bessel functions $K_n(\cdot)$ and $I_n(\cdot)$. Note that the most general form of the azimuthal variation of the solutions is $g(\phi) = A\cos(n\phi) + B\sin(n\phi)$, assuming the proper choice of origin for the ϕ coordinate. Thus, in general we have

$$
\left\{
\begin{array}{c}
E_z^0(r, \phi) \\
\text{or} \\
H_z^0(r, \phi)
\end{array}
\right\}
=
\left\{
\begin{array}{ll}
[C_d J_n(h_d r) + C_d' Y_n(h_d r)][A_d \cos(n\phi) + B_d \sin(n\phi)] & r \le a \\
[C_{cl} K_n(h_{cl} r) + C_{cl}' I_n(h_{cl} r)][A_{cl} \cos(n\phi) + B_{cl} \sin(n\phi)] & r > a
\end{array}
\right.
$$

[5.60]

where C_d, C_d', C_{cl}, C_{cl}' are constants (in general, different for E_z and H_z) to be determined by matching the boundary conditions. Note that because $Y_n(h_d r)$ has a singularity at $r = 0$, we must have $C_d' = 0$ in order for the fields to remain finite. Similarly, for fields to decay (rather than increase) with radial distance, we must have $C_{cl}' = 0$. Thus, the forms of the solutions for propagating waves ($\overline{\gamma} = j\overline{\beta}$) are

$$
\left\{
\begin{array}{c}
E_z(r, \phi, z) \\
\text{or} \\
H_z(r, \phi, z)
\end{array}
\right\}
=
\left\{
\begin{array}{ll}
C_d J_n(h_d r)[A_d \cos(n\phi) + B_d \sin(n\phi)]e^{-j\overline{\beta}z} & r \le a \\
C_{cl} K_n(h_{cl} r)[A_{cl} \cos(n\phi) + B_{cl} \sin(n\phi)]e^{-j\overline{\beta}z} & r > a
\end{array}
\right.
\quad [5.61]
$$

A graphical representation of the modified Bessel function $K_n(u)$ is given in Figure 5.17, clearly showing the decaying nature of the solutions with radial distance. The reader is encouraged to verify (by simple scaling at a few values of u) that this decay is more rapid than r^{-1}, as required for effective guiding. In contrast, asymptotic behavior of the ordinary Bessel function $J_n(hr)$ is proportional to $r^{-1/2}$ as r becomes very large, which is why $J_n(\cdot)$ solutions are not suitable for the cladding region.

For a propagating wave ($\overline{\gamma} = j\overline{\beta}$), we have, from [5.59] and [5.60],

$$
\omega^2 \mu_{cl} \epsilon_{cl} \le \overline{\beta}^2 \le \omega^2 \mu_d \epsilon_d
$$

[5.62]

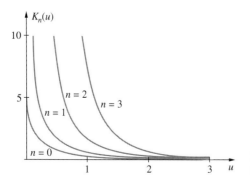

FIGURE 5.17.   The modified Bessel functions of the second kind, $K_n(u)$.

For optical media, we have $\epsilon = n^2\epsilon_0$, where n is the refractive index, so that, for nonmagnetic media ($\mu_{\mathrm{cl}} = \mu_d = \mu_0$), [5.62] can be rewritten

$$\frac{\omega n_{\mathrm{cl}}}{c} \leq \overline{\beta} \leq \frac{\omega n_d}{c} \qquad [5.63]$$

where c is the velocity of light in free space.

The expressions for H_z and E_z can be normalized so that they are continuous at $r = a$. In other words, for propagating fields ($\overline{\gamma} = j\overline{\beta}$), we can write

$$H_z = C\frac{J_n(h_d r)}{J_n(h_d a)}\cos{(n\phi)}e^{-j\overline{\beta}z} \qquad 0 < r \leq a \qquad [5.64a]$$

$$= C\frac{K_n(h_{\mathrm{cl}} r)}{K_n(h_{\mathrm{cl}} a)}\cos{(n\phi)}e^{-j\overline{\beta}z} \qquad r > a \qquad [5.64b]$$

$$E_z = C'\frac{J_n(h_d r)}{J_n(h_d a)}\sin{(n\phi)}e^{-j\overline{\beta}z} \qquad 0 < r \leq a \qquad [5.64c]$$

$$= C'\frac{K_n(h_{\mathrm{cl}} r)}{K_n(h_{\mathrm{cl}} a)}\sin{(n\phi)}e^{-j\overline{\beta}z} \qquad r > a \qquad [5.64d]$$

where C and C' are constants. Note that an arbitrary sum of $\cos(n\phi)$ and $\sin(n\phi)$ also satisfies the wave equation, but by appropriate choice of origin for ϕ we can have H_z proportional to $\cos(n\phi)$, as given. Once this choice is made, E_z must have the $\sin(n\phi)$ form in order for both E_ϕ and H_ϕ to also be continuous at $r = a$ (see expressions for E_ϕ and H_ϕ below).

Using the expressions for the axial components, the transverse field components can be found using the general relationships (i.e., equation [5.3] with $\overline{\gamma} = j\overline{\beta}$):

$$\mathbf{E}_{\mathrm{tr}} = \frac{-j\overline{\beta}}{h^2}\nabla_{\mathrm{tr}}E_z + \frac{j\omega\mu}{h^2}\hat{\mathbf{z}} \times \nabla_{\mathrm{tr}}H_z \qquad [5.65a]$$

$$\mathbf{H}_{\mathrm{tr}} = -\frac{j\omega\epsilon}{h^2}\hat{\mathbf{z}} \times \nabla_{\mathrm{tr}}E_z - \frac{j\overline{\beta}}{h^2}\nabla_{\mathrm{tr}}H_z \qquad [5.65b]$$

where we should have $h^2 = \overline{\gamma}^2 + \omega^2\mu_d\epsilon_d$ in the dielectric rod and $h^2 = \overline{\gamma}^2 + \omega^2\mu_{\mathrm{cl}}\epsilon_{\mathrm{cl}}$ in the cladding region. Therefore, using the cylindrical coordinates, the other field components within the dielectric rod guide can be found from

$$E_\phi = \frac{j\omega\mu}{h_d^2}\frac{\partial H_z}{\partial r} - \frac{j\overline{\beta}}{h_d^2 r}\frac{\partial E_z}{\partial \phi}$$

$$E_r = \frac{-j\overline{\beta}}{h_d^2}\frac{\partial E_z}{\partial r} - \frac{j\omega\mu_0}{h_d^2 r}\frac{\partial H_z}{\partial \phi} \qquad \qquad r \leq a \qquad [5.66]$$

$$H_\phi = -\frac{j\overline{\beta}}{h_d^2 r}\frac{\partial H_z}{\partial \phi} - \frac{j\omega\epsilon_d}{h_d^2}\frac{\partial E_z}{\partial r}$$

$$H_r = \frac{-j\overline{\beta}}{h_d^2}\frac{\partial H_z}{\partial r} + \frac{j\omega\epsilon_d}{h_d^2 r}\frac{\partial E_z}{\partial \phi}$$

in which E_z and H_z are given by equations [5.64c] and [5.64a]. For $r > a$, we use [5.64b] and [5.64d] with $h^2 = -h_{cl}^2$; the azimuthal components (which are the only ones needed for imposing the boundary conditions) are given by

$$E_\phi = \frac{-j\omega\mu}{h_{cl}^2}\frac{\partial H_z}{\partial r} + \frac{j\overline{\beta}}{h_{cl}^2}\frac{\partial E_z}{r\partial\phi}$$

$$\qquad\qquad\qquad\qquad\qquad r > a \qquad\qquad [5.67]$$

$$H_\phi = \frac{j\overline{\beta}}{h_{cl}^2}\frac{\partial H_z}{r\partial\phi} + \frac{j\omega\epsilon_{cl}}{h_{cl}^2}\frac{\partial E_z}{\partial r}$$

Note that both E_ϕ and H_ϕ must be continuous at $r = a$. Because of the choice of $\sin(n\phi)$ in equations [5.64c] and [5.64d] for E_z, each term in equations [5.66] and [5.67] has the same azimuthal behavior. Explicit expressions for the ϕ and r components are given elsewhere.[31]

Once the field expressions are determined, the characteristic equation for determining $\overline{\beta}$ can be derived. The basic requirement that we need to impose to determine C and C' is the continuity of the tangential components at the core-cladding interface (i.e., at $r = a$). Using the continuity of E_ϕ at $r = a$, we have

$$C\frac{\omega\mu_0}{s}f_n(s) - C'\frac{n\overline{\beta}}{s^2} = -C\frac{\omega\mu_0}{t}g_n(t) + C'\frac{n\overline{\beta}}{t^2} \qquad [5.68]$$

The continuity of H_ϕ at $r = a$ gives

$$C\frac{n\overline{\beta}}{s^2} - C'\frac{\omega\epsilon_2}{s}f_n(s) = -C\frac{n\overline{\beta}}{t^2} + C'\frac{\omega\epsilon_2}{t}g_n(t) \qquad [5.69]$$

where $s = h_d a$ and $t = h_{cl} a$ and the functions $f_n(\cdot)$ and $g_n(\cdot)$ are defined as

$$f_n(h_d a) \equiv \frac{1}{h_d}\left[\frac{d}{dr}\left\{\frac{J_n(h_d r)}{J_n(h_d a)}\right\}\right]_{r=a} = h_d\frac{J_n'(h_d a)}{J_n(h_d a)} \qquad [5.70a]$$

$$g_n(h_{cl} a) \equiv \frac{1}{h_{cl}}\left[\frac{d}{dr}\left\{\frac{K_n(h_{cl} r)}{K_n(h_{cl} a)}\right\}\right]_{r=a} = h_{cl}\frac{K_n'(h_{cl} a)}{K_n(h_{cl} a)} \qquad [5.70b]$$

Each of these equations yields a solution for C/C' in terms of $\overline{\beta}$, but only for certain values of $\overline{\beta}$ are the two solutions identical. Thus, by equating C/C' solutions for the two equations, we obtain an equation for $\overline{\beta}$ as

$$\boxed{\omega^2\mu_0\left(\frac{f_n(s)}{s} + \frac{g_n(t)}{t}\right)\left(\epsilon_d\frac{f_n(s)}{s} + \epsilon_c\frac{g_n(t)}{t}\right) = (n\overline{\beta})^2\left(\frac{1}{s^2} + \frac{1}{t^2}\right)^2} \qquad [5.71]$$

---

[31] See, for example, Chapter 14 of S. Ramo, J. R. Whinnery, and T. Van Duzer, *Fields and Waves in Communications Electronics,* 3rd ed., John Wiley & Sons, New York, 1994; see also Chapter 8 of G. H. Owyang, *Foundations of Optical Waveguides,* Elsevier North Holland, New York, 1981.

In general, the solution of equation [5.71] for $\overline{\beta}$ requires numerical techniques, and the roots correspond to the so-called hybrid modes for which neither E_z nor H_z is zero.[32]

However, solutions can be separated into TE or TM modes in the special case of axial symmetry (i.e., $n = 0$). For TE modes, equation [5.69] is satisfied by $C' = 0$, leaving equation [5.68] as

$$\frac{1}{s} f_n(s) = -\frac{1}{t} g_n(t) \qquad [5.72]$$

By definition of s and t, we have

$$s^2 + t^2 = \omega^2 \mu_0 (\epsilon_d - \epsilon_{cl}) a^2 = u^2 \qquad [5.73]$$

so that for a given ω, s and t can be determined. Similarly, TM modes are given by $C = 0$, satisfying equation [5.68] together with

$$\frac{\epsilon_d}{s} f_n(s) = -\frac{\epsilon_c}{t} g_n(t) \qquad [5.74]$$

The recurrence relations for Bessel functions[33] can be used to rewrite equations [5.72] and [5.73] in the form

$$\frac{1}{s} \frac{J_1(s)}{J_0(s)} = -\frac{1}{t} \frac{K_1(t)}{K_0(t)} \qquad [5.75a]$$

$$\frac{\epsilon_d}{s} \frac{J_1(s)}{J_0(s)} = -\frac{\epsilon_{cl}}{t} \frac{K_1(t)}{K_0(t)} \qquad [5.75b]$$

Equations [5.75a], [5.75b], and [5.73] can be solved either numerically or graphically for s and t. Once these eigenvalues are found, the cutoff frequencies and other properties can be determined.

In practice, the characteristic equations for s and t and the classification of modes can be greatly simplified by considering the weak-guidance case, when ϵ_d and ϵ_{cl} differ by only a small amount, typically less than 1%. This is, in fact, the situation of practical interest in some optical fiber applications.[34] In other practical applications, graded-index fibers in which the refractive index gradually varies with radial distance are used for more effective guiding of light.[35]

Dielectric-Coated Metal Rod An interesting extension of the dielectric and dielectric-above-metal types of guiding structures is the "one-wire transmission" application, which was first demonstrated by G. Goubau.[36] The problem of waves

---

[32]D. B. Keck, in M. L. Barnoski (ed.), *Fundamentals of Optical Fiber Communications,* Academic Press, San Diego, 1976.

[33]See Chapter 9 in M. Abramowitz and I. A. Stegun, eds., *Handbook of Mathematical Functions,* Dover, New York, 1970.

[34]A. W. Snyder and J. D. Love, *Optical Waveguide Theory,* Chapman and Hall, London, 1983.

[35]See Section 10.4 of S. Ramo, J. R. Whinnery, and T. Van Duzer, *Fields and Waves in Communications Electronics,* John Wiley & Sons, New York, 1994.

[36]G. Goubau, Surface waves and their application to transmission lines, *J. Appl. Phys.,* 21(11), pp. 1119–1128, 1950, and Single conductor surface wave transmission lines, *Proc. IRE,* 39(6), pp. 619–624, June 1951.

on a wire is as old as electromagnetic radiation itself; experiments of Heinrich Hertz dealt with "surface waves" along a wire as well as "space waves" traveling through air.[37] The phenomenon of "waves on a single wire" was investigated in great detail before it was realized that extremely efficient transmission can be achieved by using a simple two-wire line. Sommerfeld was the first to investigate in detail the propagation of waves on a single round metal wire of finite conductivity and discovered that an axially symmetric TM mode exists with very low attenuation.[38] However, the fields associated with such a wave would have to extend very far from the surface of the wire, and thus a single round wire did not appear to be an efficient structure for guiding electromagnetic waves.

Goubau, on the other hand, considered the structure shown in Figure 5.18 and showed that a thin dielectric coating on the wire confines the fields much closer to the surface. The presence of the dielectric coating also slows the wave considerably, so that the structure in Figure 5.18 constitutes a "slow-wave" transmission line. The propagating mode considered by Goubau is the cylindrical analog of the lowest-order TM mode considered for the planar dielectric-above-metal configuration in Section 4.2. It turns out that just rust or corrosion on the metal wire suffices as a thin dielectric layer, although it is more efficient to actually coat the wire with a uniform dielectric layer. The nature of the propagating field components are shown in Figure 5.18b. Because the coating is thin, the field usually spreads out to many wire diameters in the transverse direction. Most of the electromagnetic power is carried *outside* the wire and coating.[39]

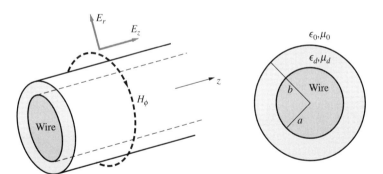

FIGURE 5.18.   **Dielectric-coated metal rod waveguide.** The fields shown are those for the lowest-order TM mode.

[37] H. Hertz, On the finite velocity of propagation of electromagnetic actions, *Sitzb. d. Berl. Akad. d. Wiss.,* February 2, 1888; for a collected English translation of this and other papers by H. Hertz, see H. Hertz, *Electric Waves,* Macmillan, London, 1893.

[38] A. Sommerfeld, *Ann. d. Physik,* 67, pp. 233–290, 1899; also see Sections 22 and 23 of A. Sommerfeld, *Electrodynamics,* Academic Press, New York, 1952.

[39] For further description of the use of single-wire transmission see Section 7.5.3 of R. B. Adler, L. J. Chu, and R. M. Fano, *Electromagnetic Energy Transmission and Radiation,* John Wiley & Sons, New York, 1960.

The formal solution of the fields for the dielectric-covered metal rod waveguide can be based on much the same type of formulation used in the previous section. The presence of the metal at $r = 0$ means that the $Y_n(h_d r)$ terms in [5.60] can also be retained, and only a subset of the solutions are possible due to the requirement that the tangential electric fields be zero at the conductor surface ($r = a$). After matching the boundary conditions, the characteristic equation for the azimuthally symmetric TM mode can be shown to be

$$h_0 b \epsilon_d \frac{K_0(h_0 b)}{K_1(h_0 b)} = h_d b \frac{J_0(h_d b)Y_0(h_d a) - J_0(h_d a)Y_0(h_d b)}{J_0(h_d a)Y_1(h_d b) - J_1(h_d b a)Y_0(h_d a)}$$

where

$$h_0^2 = \overline{\beta}^2 - \omega^2 \mu_0 \epsilon_0 \quad \text{and} \quad h_d^2 = \omega^2 \mu_d \epsilon_d - \overline{\beta}^2$$

For any given value of b/a, values of $h_0 b$ and $h_d b$ can be found graphically or numerically. For $(b - a)/a \ll 1$, we have

$$h_0 b \frac{\epsilon_d}{\epsilon_0} \frac{K_0(h_0 b)}{K_1(h_0 b)} = h_d b \tan[h_d(b - a)]$$

If we further have $h_0 b \gg 1$, the solution reduces to that of a dielectric coating on a semi-infinite planar conductor:

$$h_0 b \frac{\epsilon_d}{\epsilon_0} = h_d b \tan[h_d(b - a)]$$

5.3 CAVITY RESONATORS

We have considered uniform plane waves propagating in unbounded dielectric media, as well as guided waves that are bounded in one or both of the transverse dimensions but are free to propagate in the the z direction. In this section, we consider storage of high-frequency electromagnetic energy in enclosed metallic structures called "cavities." Any dielectric medium enclosed by metallic boundaries constitutes a "cavity resonator." Electromagnetic energy can exist in such structures by satisfying Maxwell's equations and the boundary conditions, but typically only at a discrete set of *resonant frequencies*. The electromagnetic energy trapped in a resonator is only weakly coupled to the outside world, and it decays gradually due to both conductor and dielectric losses.[40]

---

[40]In this section, we consider only resonant cavities that are assumed to be completely enclosed. It is clear that such a cavity cannot be of any practical interest, because there would be no way of setting up waves in it or making use of any wave energy enclosed inside the cavity. In practice, there must be one or more small opening(s) to get the energy into (or out of) the cavity. This is usually achieved by using small probes, loops, or other devices similar to those shown in Figure 5.4. In our discussions, we neglect the effects of these coupling devices on the mode structures within the cavity, a well-justified assumption as long as the dimensions of the coupling loops or probes are small in comparison with the other dimensions of the cavity and the wavelength.

One of the basic elements in a wide variety of electronic systems is the resonant circuit. At high frequencies, the utility of lumped elements for achieving resonant behavior decreases due to energy leakage and radiation losses. At UHF frequencies, cavity resonators are used for energy storage, precise frequency measurements, filtering, impedance matching, and the production of high field strengths with limited input signal resources. Most common cavity resonators can be constructed by closing the ends of a rectangular or circular waveguide with conducting walls. The conducting end walls reflect any waves propagating along the waveguide at each end with a phase reversal (i.e., reflection coefficient $\Gamma = -1$), and if the dimensions of the cavity are such that a standing wave is set up within the cavity, it continues to exist indefinitely if there are no dissipative losses (conversion of electromagnetic energy into heat, through conduction losses in the imperfectly conducting walls and dielectric losses in the material inside the cavity). An important property of any resonator is the so-called quality factor Q, which is a measure of the bandwidth of the system around the resonant frequency. Whereas resonant lumped electrical circuits may have Q values on the order of up to 300, extremely high Q values (of up to ~30,000) can be achieved with cavity resonators. An important application of such high-Q devices is accurate measurement of frequency, by making use of the measurable absorption of energy when the excitation signal supplied to the cavity is at the resonant frequency.

The three different types of resonators that we discuss in this section are shown in Figure 5.19. Other important types of resonators not discussed here include the dielectric resonator, typically consisting of a small disc- or cube-shaped dielectric such as can be obtained by cutting a short piece of a dielectric rod waveguide (Figure 5.16), and the strip resonator, consisting of microstrips short-circuited at both ends. Strip and dielectric resonators can be easily incorporated into microwave and

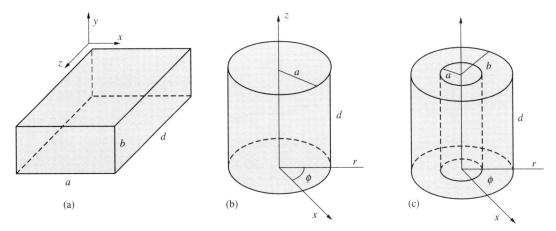

FIGURE 5.19.   Cavity resonators. (a) A rectangular cavity resonator consists of a rectangular waveguide constrained with two new walls at $z = 0, d$. (b) A circular cavity resonator of radius a and height d is shown for future reference. (c) A coaxial cavity resonator consists of a coaxial line of length d capped at both ends.

millimeter-wave integrated circuits and coupled to various forms of planar transmission lines. However, although the principles of operation of such resonators are very similar to those discussed here, their analysis does not lend itself to compact analytical treatment and is thus beyond the scope of this book.[41] We now proceed to consider the rectangular cavity resonator, followed by the circular and coaxial cavity resonators.

5.3.1 Rectangular Resonator

Consider a cavity with dimensions a, b, d such that $a > b$, as shown in Figure 5.19a. The cavity walls are metallic, and the material inside the cavity is typically a dielectric with constants μ and ϵ. We can find the solutions for the fields by considering the cavity as a section of a standard waveguide with short-circuited terminations and using the known wave mode solutions to produce a standing wave field by applying the boundary conditions at $z = 0$ and $z = d$. Solutions can be found from the expressions for any of the rectangular waveguide modes. Also, since d can be multiple wavelengths long, an infinite number of possible cavity solutions can be found for each waveguide mode.

Let us first consider the TE_{10} mode and the case with $d = \overline{\lambda}_{10}/2$. Consider the axial magnetic field, which in general can be written as a superposition of two oppositely traveling waves as

$$H_z = C^+ \cos\left(\frac{\pi x}{a}\right) e^{-j\overline{\beta}_{10} z} + C^- \cos\left(\frac{\pi x}{a}\right) e^{+j\overline{\beta}_{10} z}$$

where C^+ and C^- are constants to be determined by the boundary conditions at the new walls at $z = 0, d$. Because the axial magnetic field must vanish at $z = 0, d$ (where it is normal to the conducting surface), applying $H_z = 0$ at $z = 0$ yields $C^- = -C^+$ and applying $H_z = 0$ at $z = d$ leads to

$$H_z(z = d) = C^+ \cos\left(\frac{\pi x}{a}\right)\left[e^{-j\overline{\beta} z} - e^{+j\overline{\beta} z}\right] = 0$$

Thus, a nontrivial (i.e., $C^+ \neq 0$) solution of the preceding equation is possible only for $\overline{\beta} d = p\pi$, $p = 1, 2, 3, \ldots$. For $p = 1$, the axial magnetic field can be written as

$$H_z = H_0 \cos\left(\frac{\pi x}{a}\right)\sin\left(\frac{\pi z}{d}\right) \qquad [5.76]$$

where $H_0 = -2jC^+$ is a new constant. The other field components can be found from H_z using [5.3]. The complete set of field components for the TE_{101} mode is

---

[41] For an excellent discussion and relevant references, see Chapter 10 of S. Ramo, J. R. Whinnery, and T. Van Duzer, *Fields and Waves for Communication Electronics,* 3rd ed., John Wiley & Sons, New York, 1994; see also Chapter 9 of R. S. Elliott, *An Introduction to Guided Waves and Microwave Circuits,* Prentice Hall, 1993.

$$H_z = H_0 \cos\left(\frac{\pi x}{a}\right)\sin\left(\frac{\pi z}{d}\right)$$

Rectangular
TE$_{101}$

$$H_x = -\frac{a}{d}H_0 \sin\left(\frac{\pi x}{a}\right)\cos\left(\frac{\pi z}{d}\right)$$

[5.77]

$$E_y = -j\frac{\omega\mu a}{\pi}H_0 \sin\left(\frac{\pi x}{a}\right)\sin\left(\frac{\pi z}{d}\right)$$

The field structure for this cavity resonator mode (designated as the TE$_{101}$ mode) is shown in Figure 5.20. We see that the electric field lines pass between the top and the bottom inner surfaces of the cavity, while the magnetic field lines lie in horizontal (xz) planes and encircle the electric field lines. The magnetic field lines are tangential to the walls. These tangential magnetic fields are supported by surface currents, which lead to energy dissipation and ultimately limit the ability of the cavity to store energy. Note that the magnetic field is in the z direction on the side walls ($x = 0$, $x = a$), so that the wall currents are in the vertical (i.e., y) direction. On the front and back ($z = 0$, $z = d$) walls the magnetic field is in the x direction, with the associated surface currents also being in the vertical (y) direction. On the top and bottom walls ($y = 0$, $y = b$), both H_x and H_z are in general nonzero, so that the wall

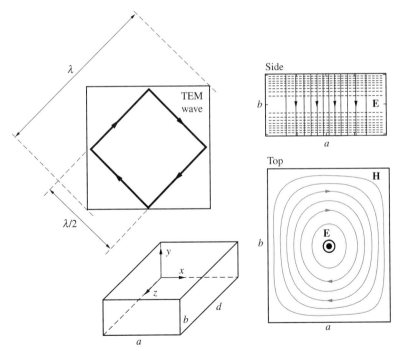

FIGURE 5.20. **TE$_{101}$ mode fields.** Electric and magnetic field distributions for the cavity resonator mode TE$_{101}$.

currents have corresponding z and x components. The structure of the current on the top and bottom walls is similar to the surface currents for the TE_{10} mode, illustrated in Figure 5.7.

As illustrated in Figure 5.20, the TE_{101} mode fields exhibit a single half-period variation (or a single half-wavelength) in the x direction, no variation in the y direction, and one half-period variation in the z direction. In general, for a resonant TE_{mnp} mode, the indices m, n, and p indicate the number of half-period variations along the respective directions. For any given set of box dimensions a, b, and d, each mode can exist only at a single frequency, defined as the *resonant frequency* for that mode. The resonant frequency can be determined by substituting any of the field components into the wave equation. Consider, for example, substituting E_y into the y component of the wave equation, given by

$$\nabla^2 E_y = -\omega^2 \mu \epsilon E_y$$

We find

$$\left[\frac{\pi^2}{a^2} + \frac{\pi^2}{d^2} \right] = \omega^2 \mu \epsilon$$

The resonant frequency of the TE_{101} mode is thus

$$\omega_{101} = \frac{1}{\sqrt{\mu\epsilon}} \sqrt{\frac{\pi^2}{a^2} + \frac{\pi^2}{d^2}} \quad \text{or} \quad \boxed{f_{101} = \frac{\sqrt{a^2 + d^2}}{2ad\sqrt{\mu\epsilon}}} \qquad [5.78]$$

Note that the resonant frequency for the TE_{101} mode depends only on the properties of the dielectric medium and the dimensions a and d. It is independent of dimension b of the cavity since the fields do not vary in the y direction, just as the cutoff frequency of the TE_{10} mode in a rectangular waveguide is independent of the dimension b. For $b < a < d$, the TE_{101} mode has the lowest resonant frequency and is referred to as the dominant resonant mode.

For the TE_{10p} mode, with p half-wavelengths in the axial direction, the resonant frequency is

$$\boxed{\omega_{10p} = \frac{1}{\sqrt{\mu\epsilon}} \sqrt{\frac{\pi^2}{a^2} + \frac{p^2\pi^2}{d^2}}} \qquad [5.79]$$

Similarly, if we were to choose an arbitrary TE_{mn} or TM_{mn} mode to construct the fields in the cavity, we would find the fields for the TE_{mnp} or TM_{mnp} modes, the resonant frequency for which is

$$\boxed{\omega_{mnp} = \frac{1}{\sqrt{\mu\epsilon}} \sqrt{\frac{m^2\pi^2}{a^2} + \frac{n^2\pi^2}{b^2} + \frac{p^2\pi^2}{d^2}}} \qquad [5.80]$$

where m, n, and p are integers. A distinct resonance exists for each combination of m, n, and p, with the qualification that either m or n may be zero for TE_{mnp} modes and none can be zero for TM_{mnp} modes. Hence, an infinite number of resonance modes is theoretically possible. Usually only the lowest-order modes are of practical interest in view of the extremely high frequencies that are otherwise involved. In most applications, a cavity resonator is designed and excited in such a manner that only one mode of resonance is obtained over a limited frequency range.

The Quality Factor Q and Conduction Losses   The frequency selectivity or the quality factor Q of the cavity resonator is defined in an analogous manner as for resonant circuits or transmission line resonators.[42] The quality factor is in general a measure of the ability of a resonator to store energy in relation to time-average power dissipation. Specifically, the Q of a resonator is defined as

$$Q \equiv 2\pi \frac{\text{Maximum energy stored}}{\text{Energy dissipated per cycle}} = \omega_0 \frac{\overline{W}_{str}}{P_{wall}} \qquad [5.81]$$

where $\overline{W}_{str}$ is the total time-average stored energy ($\overline{W}_{str} = \overline{W}_e + \overline{W}_m$, with $\overline{W}_e$ and $\overline{W}_m$ being, respectively, the time-average stored electric and magnetic energies) and P_{wall} is the time-average power dissipated due to conduction losses in the walls, so that $P_{wall} f_0^{-1}$ is the energy dissipated in one cycle. We can find the total time-average stored energy either by separately evaluating the time-average energies $\overline{W}_e$ and $\overline{W}_m$ or by just finding the energy stored in the electric fields at the instant when these are a maximum, since at that instant, energy stored in the magnetic field is zero and the electric energy is the total energy in the system. For the TE_{101} mode, we choose to evaluate the time-average energies and start with the time-average stored electric energy:

$$\overline{W}_e = \frac{\epsilon}{4} \int_V |E_y|^2 \, dv = \frac{\epsilon}{4} \left(\frac{\omega \mu a}{\pi} \right)^2 H_0^2 \int_0^d \int_0^b \int_0^a \sin^2 \left(\frac{\pi x}{a} \right) \sin^2 \left(\frac{\pi z}{d} \right) dx\,dy\,dz$$

$$\overline{W}_e = \frac{abd\,\mu H_0^2}{16} \left[\frac{a^2}{d^2} + 1 \right]$$

where we have used

$$\omega^2 = \omega_{101}^2 = \frac{\pi^2}{\mu\epsilon} \left[\frac{1}{a^2} + \frac{1}{d^2} \right]$$

and

$$\int_0^a \sin^2 \left(\frac{\pi x}{a} \right) dx = \frac{a}{2}$$

---

[42] For a brief discussion of resonance and transmission line resonators at an appropriate level, see Section 3.9 of U. S. Inan and A. S. Inan, *Engineering Electromagnetics,* Addison Wesley Longman, 1999.

The time-average stored magnetic energy $\overline{W}_m$ can be found as

$$\overline{W}_m = \frac{\mu}{4} \int \left[|H_x|^2 + |H_z|^2 \right] dv$$

$$= \frac{\mu H_0^2}{4} \int_0^d \int_0^b \int_0^a \left[\frac{a^2}{d^2} \sin^2\left(\frac{\pi x}{a}\right) \cos^2\left(\frac{\pi z}{d}\right) + \cos^2\left(\frac{\pi x}{a}\right) \sin^2\left(\frac{\pi z}{d}\right) \right] dx\, dy\, dz$$

$$\overline{W}_m = \frac{\mu H_0^2 abd}{16} \left[\frac{a^2}{d^2} + 1 \right]$$

We note that the time-average electric and magnetic energies are precisely equal (i.e., $\overline{W}_e = \overline{W}_m$). That this equality should be true in general simply follows from the complex Poynting's theorem discussed in Section 2.4. Physically, the fact that $\overline{W}_e = \overline{W}_m$ is an indication of the fact that over the course of one period, the stored energy cycles between being purely electric, partly electric and partly magnetic, and purely magnetic storage, such that on the average over one period (i.e., time-average) it is shared equally between the electric and magnetic forms. The total time-average stored energy in the system is thus

$$\overline{W}_{str} = \overline{W}_e + \overline{W}_m = \frac{\mu H_0^2 abd}{8} \left[\frac{a^2}{d^2} + 1 \right]$$

We now need to evaluate the power dissipated in the cavity walls. This dissipation will be due to the surface currents on each of the six walls as induced by the tangential magnetic fields, that is, $\mathbf{J}_s = \hat{\mathbf{n}} \times \mathbf{H}$. Note that the power dissipated is given by $\frac{1}{2}|J_s|^2 R_s$ and that $|J_s| = |H_{tan}|$, where H_{tan} is the field component tangential to the particular wall, and $R_s = \sqrt{\pi f \mu_m / \sigma}$ is the surface resistance, with σ and μ_m being respectively the conductivity and magnetic permeability of the metallic material constituting the cavity walls. We have

$$P_{wall} = \frac{R_s}{2} \int_{walls} |H_{tan}|^2\, ds$$

$$P_{wall} = \frac{R_s}{2} \left[2\underbrace{\int_0^b \int_0^a |H_x|^2_{z=0}\, dx\, dy}_{\text{front, back}} + 2\underbrace{\int_0^d \int_0^b |H_z|^2_{x=0}\, dy\, dz}_{\text{right, left}} + 2\underbrace{\int_0^d \int_0^a [|H_z|^2 + |H_x|^2]\, dx\, dz}_{\text{top, bottom}} \right]$$

where we have relied on the symmetry of the cavity in doubling the contributions from the walls $x = 0$, $y = 0$, and $z = 0$ to account for the contributions from the walls at $x = a$, $y = b$, and $z = d$, respectively. Upon manipulation, we find

$$P_{wall} = \frac{R_s H_0^2 d^2}{4} \left[\left(\frac{a}{d}\right)\left(\frac{a^2}{d^2} + 1\right) + \left(\frac{2b}{d}\right)\left(\frac{a^3}{d^3} + 1\right) \right]$$

Therefore, the quality factor Q_c (subscript c indicates that this quality factor is that due to losses in the conducting walls) is

$$Q_c = \frac{\omega_0 \overline{W}_{str}}{P_{wall}} = \frac{\pi \mu f_{101} ab}{R_s d} \frac{\left(\dfrac{a^2}{d^2} + 1\right)}{\left[\left(\dfrac{a}{d}\right)\left(\dfrac{a^2}{d^2} + 1\right) + \left(\dfrac{2b}{d}\right)\left(\dfrac{a^3}{d^3} + 1\right)\right]}$$

which, upon substitution of f_{101} from [5.78], gives

Rectangular
TE$_{101}$

$$Q_c = \frac{\pi^2 \eta b}{2 R_s d} \frac{\left(\dfrac{a^2}{d^2} + 1\right)^{3/2}}{\left[\left(\dfrac{a}{d}\right)\left(\dfrac{a^2}{d^2} + 1\right) + \left(\dfrac{2b}{d}\right)\left(\dfrac{a^3}{d^3} + 1\right)\right]}$$

[5.82]

Note that the preceding two forms for Q_c are equivalent since the resonant frequency is

$$f_{101} = \frac{1}{2\sqrt{\mu\epsilon}} \sqrt{\frac{1}{a^2} + \frac{1}{d^2}}$$

For a cubical resonator with $a = b = d$, we have $f_{101} = a^{-1}\sqrt{1/(2\mu\epsilon)}$ and

$$Q_{c_{cube}} = \frac{\pi \mu f_{101} a}{3 R_s} = \frac{a\mu}{3\mu_m \delta}$$

[5.83]

where $\delta = (\pi f \mu_m \sigma)^{-1/2}$ is the skin depth of the surrounding metallic walls, where μ_m is the permeability of the metallic walls. For nonmagnetic metallic materials, $\mu_m = \mu_0$.

The quality factor of an air-filled cubical cavity operating at 10 GHz is calculated in Example 5-7.

Example 5-7:   Air-filled cubical cavity.   Consider an air-filled cubical cavity ($a = b = d$) designed to be resonant in the TE$_{101}$ mode at 10 GHz (free space wavelength $\lambda = 3$ cm) with silver-plated surfaces ($\sigma = 6.14 \times 10^7$ S-m^{-1}, $\mu_m = \mu_0$). Find the quality factor.

Solution: Using the preceding expression for the resonant frequency and the fact that the cavity is air-filled, we have

$$f_{101} = a^{-1}\sqrt{\frac{1}{2\mu\epsilon}} \quad \rightarrow \quad a = \frac{1}{f_{101}}\sqrt{\frac{1}{2\mu_0\epsilon_0}} = \frac{c}{f_{101}\sqrt{2}} = \frac{\lambda}{\sqrt{2}} \approx 2.12 \text{ cm}$$

At 10 GHz, the skin depth for silver is given by

$$\delta = \left(\pi \times 10 \times 10^9 \times 4\pi \times 10^{-7} \times 6.14 \times 10^7\right)^{-1/2} \simeq 0.642 \ \mu m$$

and the quality factor is

$$Q_{c_{\text{cube}}} = \frac{a}{3\delta} \simeq \frac{2.12 \text{ cm}}{3 \times 0.642 \ \mu m} \simeq 11{,}000$$

which is a very large value compared with those achievable with lumped resonant circuit elements.

Example 5-7 shows that very large quality factors can be achieved with resonant cavities. The Q_c evaluated for a cubical cavity is in fact representative of cavities of other simple shapes. Slightly higher Q_c values may be possible in resonators with other simple shapes, such as an elongated cylinder or a sphere, but the Q_c values are generally on the order of magnitude of the volume-to-surface ratio divided by the skin depth. To see this, note that

$$Q_c = \omega_0 \frac{\overline{W}_{\text{str}}}{P_{\text{wall}}} = \frac{\omega_0 2\overline{W}_m}{P_{\text{wall}}} = \frac{(2\pi f_0)\dfrac{\mu}{2}\displaystyle\int_V H^2 \, dv}{\dfrac{R_s}{2}\displaystyle\oint_S H_t^2 \, ds} \simeq \frac{2}{\delta}\frac{V_{\text{cavity}}}{S_{\text{cavity}}} \qquad [5.84]$$

where S_{cavity} is the cavity surface enclosing the cavity volume V_{cavity}.

Although very large Q_c values are possible in cavity resonators, disturbances caused by the coupling system (loop or aperture coupling), surface irregularities, and other perturbations (e.g., dents on the walls) in practice act to increase losses and reduce Q_c. Dielectric losses and radiation losses from small holes may be especially important in reducing Q_c. The resonant frequency of a cavity may also vary due to the presence of a coupling connection. It may also vary with changing temperature due to dimensional variations (as determined by the thermal expansion coefficient). In addition, for an air-filled cavity, if the cavity is not sealed, there are changes in resonant frequency because of the varying dielectric constant of air with changing temperature and humidity.[43] These types of considerations become important when the frequency selectivity (i.e., Q) of the device is so high.

Dielectric Losses   Additional losses in a cavity occur due to the fact that at microwave frequencies for which resonant cavities are used most dielectrics have a complex dielectric constant $\epsilon_c = \epsilon' - j\epsilon''$. As discussed in Section 2.3, a dielectric material with complex permittivity draws an effective current of $\mathbf{J}_{\text{eff}} = \omega_0\epsilon''\mathbf{E}$, leading to losses that occur effectively due to $\mathbf{E} \cdot \mathbf{J}_{\text{eff}}^*$. The power dissipated in the

---

[43]T. Moreno, *Microwave Transmission Design Data,* Artech House, Norwood, Massachusetts, 1989.

dielectric filling is

$$P_{\text{dielectric}} = \frac{1}{2} \int_V \mathbf{E} \cdot \mathbf{J}^*_{\text{eff}} \, dv = \frac{1}{2} \int_V \mathbf{E} \cdot \omega \epsilon'' \mathbf{E}^* \, dv$$

$$= \frac{\omega_0 \epsilon''}{2} \int_0^a \int_0^b \int_0^d |E_y|^2 \, dy \, dx \, dz$$

Using the expression for E_y for the TE_{101} mode, we have

$$P_{\text{dielectric}} = \frac{\epsilon''}{\epsilon'} \omega_0 \frac{\mu H_0^2 abd}{8} \left[\frac{a^2}{d^2} + 1 \right]$$

so that the quality factor due to dielectric losses (i.e., Q_d) is

$$\boxed{Q_d = \omega_0 \frac{\overline{W}_{\text{str}}}{P_d} = \frac{\epsilon'}{\epsilon''}} \qquad [5.85]$$

Note that this a very simple result, but it is not surprising since we have

$$\overline{W}_{\text{str}} = 2\overline{W}_m = \frac{\epsilon'}{2} \int_V |E_y|^2 \, dv \quad \text{and} \quad P_{\text{dielectric}} = \frac{\omega_0 \epsilon''}{2} \int_V |E_y|^2 \, dv$$

The total Q of the system is given[44] by

$$\boxed{\frac{1}{Q} = \frac{1}{Q_d} + \frac{1}{Q_c}} \qquad [5.86]$$

The Q of a Teflon-filled 10-GHz cavity is calculated in Example 5-8.

Example 5-8: Q of a cubic Teflon-filled cavity.   Consider a cubic-shaped air-filled cavity resonant at 10 GHz ($a \simeq 2.12$ cm). We found earlier that $Q_c \simeq 11{,}000$, for silver-plated walls. Now consider a Teflon-filled cavity, with $\epsilon = \epsilon_0(2.05 - j0.0006)$. Find the total quality factor Q of this cavity.

Solution: At the same frequency, the dimensions of the cavity will be different since for a cubic resonator, the resonant frequency is

$$f_0 = \left[f_{101} \right]_{a=d} \simeq \frac{1}{2\sqrt{\mu \epsilon'}} \sqrt{\frac{2}{a^2}} = \frac{c}{a\sqrt{2\mu_r \epsilon_r}} \quad \rightarrow \quad a = \frac{c}{\sqrt{2} f_0 \sqrt{\epsilon_r'}}$$

---

[44]This follows because the quality factor is defined as $Q \equiv \omega_0 \overline{W}_{\text{str}} / P_{\text{loss}}$ and the total time-average power dissipated is $P_{\text{loss}} = P_{\text{wall}} + P_{\text{dielectric}}$.

since $\mu_r = 1$ for Teflon. In other words, the cavity is $\sqrt{\epsilon_r'}$ times smaller, or $a = b = d \simeq 1.48$ cm. Thus, we have

$$Q_c = \frac{a}{3\delta} \simeq 7684$$

or $\sqrt{\epsilon_r'}$ times lower than that of the air-filled cavity. The quality factor Q_d due to the dielectric losses is given by

$$Q_d = \frac{\epsilon'}{\epsilon''} = \frac{2.05}{0.0006} \simeq 3417$$

The total Q is then

$$Q = \frac{Q_d Q_c}{Q_d + Q_c} \simeq 2365$$

Thus, the presence of the Teflon dielectric substantially reduces the quality factor of the resonator. Note, however, that the Teflon-filled resonator is smaller in size than an air-filled one, and the dielectric strengths of plastics such as Teflon are usually substantially higher than that of air, thus allowing for greater energy storage capacity.

Fields for Other Cavity Modes The electric and magnetic fields for the other TE_{mnp} and TM_{mnp} modes can be obtained in a straightforward manner, by simply considering the TE_{mn} or TM_{mn} mode field expressions for a rectangular waveguide and imposing the boundary conditions at $z = 0$ and $z = d$. These are given below for the sake of completion.

For TE_{mnp} modes, we have

Rectangular TE_{mnp}

$$H_z = C \cos\left(\frac{m\pi}{a}x\right)\cos\left(\frac{n\pi}{b}y\right)\sin\left(\frac{p\pi}{d}z\right)$$

$$H_x = -\frac{1}{h^2}\left(\frac{m\pi}{a}\right)\left(\frac{p\pi}{d}\right)C \sin\left(\frac{m\pi}{a}x\right)\cos\left(\frac{n\pi}{b}y\right)\cos\left(\frac{p\pi}{d}z\right)$$

$$H_y = -\frac{1}{h^2}\left(\frac{n\pi}{b}\right)\left(\frac{p\pi}{d}\right)C \cos\left(\frac{m\pi}{a}x\right)\sin\left(\frac{n\pi}{b}y\right)\cos\left(\frac{p\pi}{d}z\right)$$

$$E_x = \frac{j\omega\mu}{h^2}\left(\frac{n\pi}{b}\right)C \cos\left(\frac{m\pi}{a}x\right)\sin\left(\frac{n\pi}{b}y\right)\sin\left(\frac{p\pi}{d}z\right)$$

$$E_y = -\frac{j\omega\mu}{h^2}\left(\frac{m\pi}{a}\right)C \sin\left(\frac{m\pi}{a}x\right)\cos\left(\frac{n\pi}{b}y\right)\sin\left(\frac{p\pi}{d}z\right)$$

[5.87]

where

$$h^2 = \left(\frac{m\pi}{a}\right)^2 + \left(\frac{n\pi}{b}\right)^2$$

Similarly, for the TM_{mnp} modes,

Rectangular
TM_{mnp}

$$E_z = C \sin\left(\frac{m\pi}{a}x\right)\sin\left(\frac{n\pi}{b}y\right)\cos\left(\frac{p\pi}{d}z\right)$$

$$E_x = -\frac{1}{h^2}\left(\frac{m\pi}{a}\right)\left(\frac{p\pi}{d}\right)C\cos\left(\frac{m\pi}{a}x\right)\sin\left(\frac{n\pi}{b}y\right)\sin\left(\frac{p\pi}{d}z\right)$$

$$E_y = -\frac{1}{h^2}\left(\frac{n\pi}{b}\right)\left(\frac{p\pi}{d}\right)C\sin\left(\frac{m\pi}{a}x\right)\cos\left(\frac{n\pi}{b}y\right)\sin\left(\frac{p\pi}{d}z\right)$$

$$H_x = \frac{j\omega\epsilon}{h^2}\left(\frac{n\pi}{b}\right)C\sin\left(\frac{m\pi}{a}x\right)\cos\left(\frac{n\pi}{b}y\right)\cos\left(\frac{p\pi}{d}z\right)$$

$$H_y = -\frac{j\omega\epsilon}{h^2}\left(\frac{m\pi}{a}\right)C\cos\left(\frac{m\pi}{a}x\right)\sin\left(\frac{n\pi}{b}y\right)\cos\left(\frac{p\pi}{d}z\right)$$

[5.88]

Different waveguide modes having the same resonant frequency f_{mnp} are called *degenerate modes,* and the mode with the lowest resonant frequency is referred to as the *dominant mode.* Close examination of the electric and magnetic field expressions indicates that the longitudinal (E_z) and transverse (E_x, E_y) electric field components are in time quadrature with their magnetic field counterparts. Hence, the time-average power transmitted in any direction is identically zero, as required in a lossless cavity.

We have noted that, in general, the resonant frequencies for the TE_{mnp} or TM_{mnp} modes are given by [5.80],

$$\omega_{mnp} = \frac{1}{\sqrt{\mu\epsilon}}\sqrt{\frac{m^2\pi^2}{a^2} + \frac{n^2\pi^2}{b^2} + \frac{p^2\pi^2}{d^2}}$$

It is interesting to note that for a given cavity (fixed a, b, d), the resonant frequencies are not harmonically related but consist of a set of irregularly spaced eigenvalues of the wave equation. This circumstance is quite unlike the case of a lumped resonant circuit (e.g., a parallel *RLC* circuit), which is resonant at $\omega_0 = (\sqrt{LC})^{-1}$ as well as at $2\omega_0$, $3\omega_0$, and so on. To be resonant at a fixed frequency, the cavity must be larger for higher-order modes, since more half–sine waves need to be fit in each dimension. As the box becomes larger, it has higher Q (greater volume-to-surface ratio); thus, higher-order cavity modes generally have higher Q values.

In most applications, the objective in the design of resonant cavities is to achieve high values of Q and precise control of resonant frequency. Since the presence of the dielectric in general reduces Q, the highest Q values are obtained with hollow metal cavities. For a hollow cavity, physically larger cavities provide larger Q factors (higher volume-to-surface ratios). However, large cavities can support a number of modes at different frequencies, and it is necessary to determine all the possible resonant frequencies of the cavity. Rectangular cavities are easier to analyze but are difficult to manufacture to precise dimensions. Circular cavities are easier to make to precise dimensions and have higher Q factors than rectangular cavities of the same size. It is usually necessary to design a cavity so that it supports only one mode over the proposed operating frequency band. Circular cavity modes are discussed in the next section.

Example 5-9 illustrates the resonant frequencies of different modes in a microwave oven.

Example 5-9: Microwave oven design. A certain microwave oven is designed from WR-1150 aluminum rectangular waveguide with dimensions $a = 29.21$ cm and $b = a/2 = 14.605$ cm. The length of the guide is d with both open ends closed with aluminum sheets. (a) For $d = a = 29.21$ cm, find the resonant frequencies of the two nontrivial TE_{mnp} and TM_{mnp} modes that have the lowest resonant frequencies. (b) Redesign the oven by readjusting the length d such that the resonant frequency of the TE_{101} mode is 1 GHz. (c) In part (b), what are the resonant frequencies of the TE_{102} and TE_{011} modes?

Solution:
(a) Using

$$f_{mnp} = \frac{c}{2}\sqrt{\left(\frac{m}{a}\right)^2 + \left(\frac{n}{b}\right)^2 + \left(\frac{p}{d}\right)^2}$$

the two TE_{mnp} modes with the lowest resonant frequencies are TE_{101} and TE_{102} or TE_{011}, with

$$f_{101} = \frac{c}{2}\sqrt{\frac{1}{a^2} + \frac{1}{a^2}} = \frac{\sqrt{2}c}{2a} \simeq \frac{\sqrt{2} \times 3 \times 10^8}{2 \times 0.2921} \simeq 726 \text{ MHz}$$

and

$$f_{102 \text{ or } 110} = \frac{c}{2}\sqrt{\frac{1}{a^2} + \frac{4}{a^2}} = \frac{\sqrt{5}c}{2a} \simeq \frac{\sqrt{5} \times 3 \times 10^8}{2 \times 0.2921} \simeq 1.15 \text{ GHz}$$

Similarly, the two TM_{mnp} modes with the lowest resonant frequencies are TM_{110} and TM_{210}, with

$$f_{110} = f_{102} \simeq 1.15 \text{ GHz}$$

and

$$f_{210} = \frac{c}{2}\sqrt{\frac{4}{a^2} + \frac{4}{a^2}} = \frac{2\sqrt{2}c}{2a} \simeq \frac{\sqrt{2} \times 3 \times 10^8}{0.2921} \simeq 1.45 \text{ GHz}$$

(b) For TE_{101} mode, we have

$$f_{101} = \frac{c}{2}\sqrt{\frac{1}{a^2} + \frac{1}{d^2}} \simeq \frac{3 \times 10^8}{2}\sqrt{\frac{1}{(0.2921)^2} + \frac{1}{d^2}} = 10^9$$

Solving, we find $d \simeq 17.5$ cm.

(c) The resonant frequencies of the TE_{102} and TE_{011} modes are

$$f_{102} \simeq \frac{c}{2}\sqrt{\frac{1}{(0.2921)^2} + \frac{4}{(0.1748)^2}} \simeq 1.79 \text{ GHz}$$

and

$$f_{011} \simeq \frac{c}{2}\sqrt{\frac{4}{(0.2921)^2} + \frac{1}{(0.1748)^2}} \simeq 1.34 \text{ GHz}$$

5.3.2   Circular Cavity Resonators

As in the case of rectangular cavities, a circular cavity resonator can be constructed by closing a section of a circular waveguide at both ends with conducting walls, as shown in Figure 5.19b. The resonator mode in an actual case depends on the way the cavity is excited and the application for which it is used. In the following discussion we provide a detailed analysis of the TE_{011} mode, which has particularly high Q and interesting properties, and also briefly consider other modes.

The Low-Loss TE_{011} Circular Cavity Mode   A common use of a circular resonator (see Figure 5.19b) is as a wavemeter or frequency meter,[45] for which the TE_{011} mode is particularly useful. For this mode, we can construct the fields by considering the low-loss TE_{01} mode in a cylindrical waveguide and allow for one half-period variation ($p = 1$) in the z direction.

For the general TE_{nl} mode in a circular waveguide, the axial magnetic field has the form

$$H_z \sim J_n\left(s_{nl}\frac{r}{a}\right)\cos\left(n\phi\right)$$

where $h^2 = -\overline{\beta}_{TE_{nl}}^2 + \omega^2 \mu\epsilon$, assuming propagating modes. For the TE_{01} mode, the basic radial dependence of H_z has the form $H_z \sim J_0(s_{10}r/a)$. Allowing for waves

---

[45]For use as a frequency meter, the cavity is constructed with a movable top wall to allow mechanical tuning of the resonant frequency, and weak coupling to a waveguide is provided via a small aperture. In operation, power is absorbed by the cavity as it is tuned to the operating frequency of the system; this absorption of power is monitored with a detector along the waveguide past the point of coupling. The mechanical tuning dial is usually calibrated in frequency. The precision by which the frequency is measured is determined by the Q of the cavity.

traveling in both the $+z$ and $-z$ directions, we can write the z component of the magnetic field as

$$H_z = C^+ J_0\left(s_{nl}\frac{r}{a}\right)e^{-j\bar{\beta}_{\text{TE}_{01}}z} + C^- J_0\left(s_{nl}\frac{r}{a}\right)e^{+j\bar{\beta}_{\text{TE}_{01}}z}$$

Using the boundary conditions at $z = 0$ and $z = d$ yields

$$H_z = 0\Big|_{z=0,d} \quad\rightarrow\quad C^+ = -C^- \quad\text{and}\quad \bar{\beta}d = p\pi \qquad p = 1, 2, 3, \ldots$$

For $p = 1$ (i.e., the TE$_{011}$ mode), we have $\bar{\beta} = \pi/d$. In addition, using [5.7] for the boundary condition at $r = a$,

$$\left[\frac{\partial H_z}{\partial r}\right]_{r=a} = 0$$

This condition results in $J_0'(s_{nl}(a/a)) = 0$, which requires $s_{01} = 3.832$, where s_{01} is the first nonzero root of $J_0'(\cdot)$. We have

$$H_z = H_0 J_0\left(\frac{3.832r}{a}\right)\sin\left(\frac{\pi z}{d}\right) \tag{5.89}$$

where, as before, $H_0 = -2jC^+$ is a new constant.

The other nonzero field components can now be found in the usual manner by substituting the expression for H_z into [5.6]. The complete set of field components for the circular TE$_{011}$ cavity mode is

Circular
TE$_{011}$

$$\boxed{\begin{aligned} H_z &= H_0 J_0\left(\frac{3.832r}{a}\right)\sin\left(\frac{\pi z}{d}\right) \\[2mm] H_r &= \frac{\pi a H_0}{3.832d}J_1\left(\frac{3.832r}{a}\right)\cos\left(\frac{\pi z}{d}\right) \\[2mm] E_\phi &= \frac{-j\omega\mu a H_0}{3.832d}J_1\left(\frac{3.832r}{a}\right)\sin\left(\frac{\pi z}{d}\right) \end{aligned}} \tag{5.90}$$

where we have used the identity $J_1(hr) \equiv -J_0'(hr)$. The configuration of the electric and magnetic fields for the TE$_{011}$ mode is shown in Figure 5.21.

Note that the electric field lines form closed circular loops centered around the cylinder axis. The electric field lines are threaded with closed loops of magnetic field lines in the radial planes. No surface charges appear on any of the cavity walls, since normal electric field is zero everywhere on the walls. However, surface currents $\mathbf{J}_s = \hat{n} \times \mathbf{H}$ do flow in the walls due to the tangential magnetic fields. On the curved surface of the cylinder we have J_{s_ϕ} due to H_z given by

$$J_{s_\phi} = H_0 J_0(3.832)\sin\left(\frac{\pi z}{d}\right) \simeq -0.403 H_0\sin\left(\frac{\pi z}{d}\right) \qquad \text{at } r = a$$

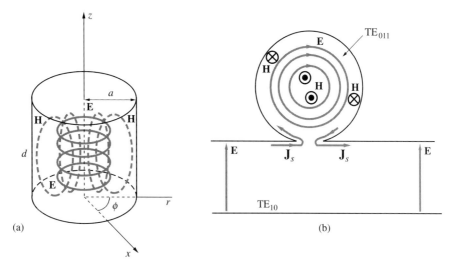

FIGURE 5.21.   The TE$_{011}$ mode. Electric and magnetic fields for TE$_{011}$ in a circular cavity resonator. (a) Configuration of electric (solid lines) and magnetic (dashed lines) fields. (b) Excitation of the TE$_{011}$ mode in a circular cavity via coupling from a TE$_{10}$ mode in a rectangular waveguide.

On the flat end surfaces, we have J_{s_ϕ} due to H_r:

$$J_{s_\phi} = \pm \frac{\pi a H_0}{3.832 d} J_1 \left(\frac{3.832 r}{a} \right) \qquad \text{at } z = 0 \quad \text{and} \quad z = a$$

It is interesting to note that the surface currents are entirely circumferential (i.e., in the ϕ direction). No surface current flows between the flat walls and the curved walls. Hence, if one end of the cavity is mounted on micrometers and moved to change the length of the cavity, the TE$_{011}$ mode can still be fully supported, since the current flow is not interrupted. Movable construction of the end faces also suppresses other modes, particularly TM$_{111}$, which has the same resonant frequency[46] but lower Q. The currents that are required to support TE$_{111}$ are interrupted by the space between the movable end and the side walls.

As in the case of the rectangular cavity resonator, the resonant frequency of the TE$_{011}$ mode can be found by substituting the expressions for any one of the field components into the wave equation.

$$\omega_{011} = \frac{1}{\sqrt{\mu \epsilon}} \sqrt{\left(\frac{\pi}{d} \right)^2 + \left(\frac{s_{01}}{a} \right)^2} \qquad\qquad [5.91]$$

---

[46]This is because eigenvalues for TM$_{nlp}$ modes are determined by the roots of $J_n'(t_{nl}) = 0$, and the root $t_{11} = 3.832$ of $J_0(\cdot) = 0$ (see Table 5.2) is equal to the root s_{01} of $J_0'(\cdot) = 0$ (see Table 5.3).

The resonant free-space wavelength of the cavity corresponding to the resonant frequency is given by

$$\lambda_{011} = \frac{c}{f_{011}} = \frac{2\sqrt{\mu_r\epsilon_r}}{\sqrt{\left(\frac{1}{d}\right)^2 + \left(\frac{s_{01}}{\pi a}\right)^2}}$$ [5.92]

For the more general case of the TE_{nlp} mode, the resonant free-space wavelength λ_{nlp} is given by

$$\lambda_{nlp} = \frac{c}{f_{nlp}} = \frac{2\sqrt{\mu_r\epsilon_r}}{\sqrt{\left(\frac{p}{d}\right)^2 + \left(\frac{s_{nl}}{\pi a}\right)^2}}$$ [5.93]

Following the same procedure as that for rectangular cavities, we can determine the Q of the cavity for the TE_{nlp} mode. For a half-wavelength-long resonator (i.e., $d = \bar{\lambda}_{TE_{nl}}/2$ or $\bar{\beta}_{TE_{nl}}d = \pi$), it can be shown[47] that

Circular
TE_{nlp}

$$Q = \frac{1}{\delta}\frac{\frac{d}{2}\left[s_{nl}^2 + \left(\frac{\pi a}{d}\right)^2\right]\left[1 - \left(\frac{l}{s_{nl}}\right)^2\right]}{\frac{d}{2a}s_{nl}^2 + \left(\frac{a\pi}{d}\right)^2 + \left[\frac{a(d-2a)}{2d^2}\right]\left[\frac{\pi l}{s_{nl}}\right]^2}$$ [5.94]

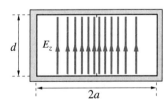

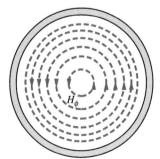

FIGURE 5.22. The electric and magnetic fields for the cylindrical TM_{010} mode.

[47] See T. Moreno, *Microwave Transmission Design Data,* Artech House, Norwood, Massachusetts, 1989.

where δ is the skin depth of the metallic walls. For TE_{011}, and using $s_{01} \simeq 3.832$, we have

Circular
TE_{011}

$$Q \simeq \frac{0.61\lambda}{\delta}\sqrt{1 + 0.168\,(2a/d)^2}\left[\frac{1 + 0.168\,(2a/d)^2}{1 + 0.168\,(2a/d)^3}\right] \qquad [5.95]$$

The design of an X-band circular cavity operating at 10 GHz is considered in Example 5-10.

Example 5-10:   X-band circular cavity design.   Design a circular cavity with $d = 2a$ such that its TE_{011} resonates at 10 GHz. Determine the Q of the cavity.

Solution: For the TE_{011} mode to be resonant at 10 GHz, we must have

$$\lambda_{011} = \frac{c}{f_{011}} \simeq \frac{3 \times 10^8 \text{ m-s}^{-1}}{10 \times 10^9 \text{ Hz}} = \frac{2}{\sqrt{\left(\dfrac{1}{2a}\right)^2 + \left(\dfrac{3.832}{\pi a}\right)^2}} \qquad \rightarrow \qquad a \simeq 1.98 \text{ cm}$$

which can be compared with $a = 2.12$ cm for the cubical resonator and the TE_{101} rectangular mode in Example 5-8. From the above expression for Q, and once again assuming silver-plated inner walls ($\sigma = 6.14 \times 10^7$ S-m^{-1}), so that $\delta = (\sqrt{\pi f \mu_0 \sigma})^{-1} \simeq 6.42 \times 10^{-7}$ m, we have

$$Q = \frac{0.61\lambda_0}{\delta}\sqrt{1 + 0.168} \simeq 30{,}792 \quad !$$

The very large value of Q found in Example 5-10 indicates that the TE_{011} mode in a circular cavity is especially suited for use in precision frequency meters. It is reminiscent of the low-loss TE_{01} mode in circular waveguides. In practice, the resultant Q is lowered by the transfer of energy to modes with lower Q but identical resonant frequency, such as the TM_{111} mode.

Other Circular Cavity Modes   A mode that is commonly discussed and that has the lowest resonant frequency is the TM_{010} mode, the field structure of which is independent of the height of the cylinder d. The nonzero fields for this TM_{010} cylindrical mode (see Figure 5.22) are

Circular
TM_{010}

$$E_z = E_0 J_0\left(\frac{2.405r}{a}\right)$$

$$H_\phi = \frac{jE_0}{\eta} J_1\left(\frac{2.405r}{a}\right) \qquad [5.96]$$

where $\eta = \sqrt{\mu/\epsilon}$ and $t_{01} = 2.405$ is the first nonzero root of $J_0(\cdot) = 0$.

In general, for the TM_{nlp} mode the resonant free-space wavelength is

$$\lambda_{nlp} = \frac{c}{f_{nlp}} = \frac{2}{\sqrt{\left(\frac{p}{d}\right)^2 + \left(\frac{t_{nl}}{\pi a}\right)^2}}$$ [5.97]

where t_{nl} is the *l*th root of $J_n(\cdot) = 0$. The Q of the TM_{nlp} mode for $d = \overline{\lambda}_{TE_{nl}}$ is

$$Q = \frac{1}{\delta}\frac{a}{1 + (2a/d)} \qquad \text{for } n \neq 0$$

$$Q = \frac{1}{\delta}\frac{a}{1 + (a/d)} \qquad \text{for } n = 0$$

Example 5-11 considers the design of a 10-GHz circular cavity operating in the TM_{010} mode.

Example 5-11:   X-band circular cavity with TM_{010} mode.   Calculate the Q of a circular cavity operating at 10 GHz in the TM_{010} mode.

Solution: For the TM_{010} to be resonant at 10 GHz, we must have

$$\lambda_{010} = \frac{c}{f_{010}} \simeq \frac{3 \times 10^8}{10 \times 10^9} = \frac{2}{\sqrt{0 + (2.405/\pi a)^2}} \qquad \rightarrow \qquad a \simeq 1.15 \text{ cm}$$

For silver-plated inside walls, we have $\delta \simeq 0.642 \text{ μm}$, and Q is

$$Q = \left(\frac{1}{\delta}\right)\frac{a}{1 + [a/(2a)]} \simeq 11{,}919$$

which compares well with the cubic cavity of $a = 2.12$ cm. Note that the volume of the cubic cavity is ~9.56 cm^3, whereas that of the cylindrical cavity is ~9.53 cm^3.

Note that for the TM_{010} mode, with fields E_z and H_ϕ, the currents on the walls flow in the z direction on curved walls and connect with radial currents on the end plates. Thus a movable end plate (as is desirable for operation as a frequency meter) would break the currents and suppress this mode.

5.3.3   Coaxial Resonators

Coaxial lines of appropriate length and terminated with short circuits (i.e., metal plates) at both ends constitute efficient cavity resonators at frequencies up to about 5 GHz. At frequencies above 5 GHz, coaxial lines become impractical; their physical dimensions become too small if unwanted higher-order modes are to be avoided,

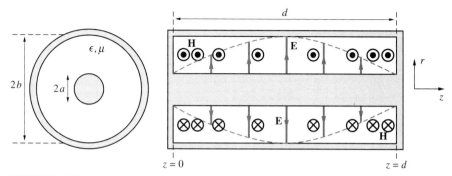

FIGURE 5.23. **Coaxial resonator.** The TEM$_1$ mode in a coaxial resonator. Note that what is shown here is only one type of coaxial resonator with $d = \lambda/2$, where λ is the wavelength for a uniform plane wave in the unbounded dielectric medium with parameters ϵ and μ.

and losses become greater (because of the small physical dimensions), lowering the attainable Q values. In this section we consider a coaxial resonator such as that shown in Figure 5.19c and Figure 5.23, operating in the TEM$_1$ mode. Note that the coaxial line is but one example of two-conductor transmission lines that support a TEM mode and that can also be used as resonant circuit elements,[48] even though such resonators may not necessarily be in the form of enclosed cavities (e.g., a two-wire line short-circuited at both ends) and thus typically have lower quality factors.

We start by recalling from [5.54] the expressions for the electric and magnetic fields for the TEM mode in a coaxial line:

$$\mathbf{E}_{\text{TEM}} = \hat{\mathbf{r}} \frac{C\eta}{r} e^{-j\beta z}$$

$$\mathbf{H}_{\text{TEM}} = \hat{\boldsymbol{\phi}} \frac{C}{r} e^{-j\beta z}$$

where $\beta = \omega \sqrt{\mu\epsilon} = 2\pi/\lambda$ and C is a constant. In the presence of the end plates, the electric field $\mathbf{E}$ within the coaxial cavity can in general be written as the superposition of two oppositely traveling waves:

$$E_r(r, z) = \frac{C^+\eta}{r} e^{-j\beta z} + \frac{C^-\eta}{r} e^{+j\beta z}$$

Noting that the electric field is tangential to the end plates at $z = 0$ and $z = d$, we must have $E_r(r, 0) = 0$ so that $C^+ = -C^-$. Furthermore, we must have $E_r(r, d) = 0$, so that

$$E_r(r, d) = \frac{C^+\eta}{r} e^{-j\beta d} - \frac{C^+\eta}{r} e^{+j\beta d} = 0 \quad \rightarrow \quad \beta d = p\pi \quad p = 1, 2, 3, \ldots$$

---

[48]For a discussion of transmission lines as resonant circuit elements, see Section 3.9 of U. S. Inan and A. S. Inan, *Engineering Electromagnetics,* Addison Wesley Longman, 1999.

For $p = 1$, we find

$$d = \frac{\pi}{\beta} = \frac{\pi}{\omega\sqrt{\mu\epsilon}} = \frac{\lambda}{2} \quad \rightarrow \quad \boxed{\omega_0 = \frac{\pi}{d\sqrt{\mu\epsilon}}} \qquad [5.98]$$

as the resonant frequency.

We can rewrite the electric field expression as

$$E_r(r, z) = \frac{H_0\,\eta}{r}\,\sin\beta z$$

where $H_0 = -j2C^+$ is another constant. The corresponding magnetic field can then be found from [5.52a]:

$$H_\phi(r, z) = \frac{H_0}{r}\,\cos\beta z$$

To determine the Q_c of the coaxial cavity due to conduction losses, we need to find the time-average stored energy $\overline{W}_{str}$ and the total power lost in the cavity walls P_{wall}. The time-average stored electric energy is

$$\overline{W}_e = \frac{\epsilon}{4}\int_V |E_r|^2\, dv = \int_0^{2\pi}\int_0^d\int_a^b \frac{H_0^2\eta^2}{r^2}\sin^2(\beta z)\,r\,dr\,dz\,d\phi$$

$$= \frac{\pi\epsilon H_0^2\eta^2}{2}\int_0^d \sin^2(\beta z)\,dz\int_a^b \frac{1}{r}\,dr = \frac{\pi d\epsilon H_0^2\eta^2\ln(b/a)}{4}$$

Similarly, the time-average stored magnetic energy can be found by integrating $\frac{1}{4}\mu|H_\phi|^2$. As with the case of the rectangular cavity resonator, we find $\overline{W}_m = \overline{W}_e$. The total time-average stored energy is then given as

$$\overline{W}_{str} = \overline{W}_e + \overline{W}_m = \frac{\pi d\epsilon H_0^2\eta^2\ln(b/a)}{2}$$

The total wall loss P_{wall} is determined by the tangential magnetic field at the walls of the cavity, which in this case consist of the two end plates at $z = 0$ and $z = d$, the inner conductor at $r = a$, and the outer conductor at $r = b$. We have

$$H_\phi(r, 0) = H_\phi(r, d) = \frac{H_0}{r} \quad H_\phi(a, z) = \frac{H_0}{a}\cos(\beta z) \quad H_\phi(b, z) = \frac{H_0}{b}\cos(\beta z)$$

The total wall loss P_{wall} is given by

$$P_{wall} = \frac{R_s}{2}\left[\underbrace{2\int_0^{2\pi}\int_a^b |H_\phi|^2_{z=0,d}\,r\,dr\,d\phi}_{\text{ends}} + \underbrace{\int_0^{2\pi}\int_0^d |H_\phi|^2_{r=a}\,r\,dz\,d\phi}_{\text{inner conductor}} + \underbrace{\int_0^{2\pi}\int_0^d |H_\phi|^2_{r=b}\,r\,dz\,d\phi}_{\text{outer conductor}}\right]$$

Upon manipulation we find

$$P_{\text{wall}} = \frac{d R_s \pi H_0^2}{2}\left[\frac{1}{a} + \frac{1}{b} + \frac{2}{d}\ln\left(\frac{b}{a}\right)\right]$$

The quality factor Q_c due to conduction losses is then found to be

$$Q_c = \frac{\omega_0 \overline{W}_{\text{str}}}{P_{\text{wall}}} = \frac{\omega_0 \, \mu \, \ln(b/a)}{R_s\left[\dfrac{1}{a} + \dfrac{1}{b} + \dfrac{2}{d}\ln\left(\dfrac{b}{a}\right)\right]}$$

Noting that $R_s = \sqrt{\pi f_0 \mu_m/\sigma}$, where $f_0 = \omega_0/(2\pi)$, the quality factor Q_c can be rewritten in terms of the skin depth $\delta = (\pi f_0 \mu_m \sigma)^{-1/2}$, as

Coaxial

TEM₁

$$Q_c = \frac{2\,\mu\,\ln(b/a)}{\delta\,\mu_m\left[\dfrac{1}{a} + \dfrac{1}{b} + \dfrac{2}{d}\ln\left(\dfrac{b}{a}\right)\right]}$$

[5.99]

where μ_m is the permeability of the metallic walls.

Note that the formula for the quality factor due to dielectric losses for a coaxial cavity resonator is identical to that for any other (i.e., rectangular or circular) cavity and is given by

$$Q_d = \frac{\omega_0 \overline{W}_{\text{str}}}{P_{\text{dielectric}}} = \frac{\epsilon'}{\epsilon''}$$

In the presence of both dielectric and conducting losses, the resultant Q of the cavity is determined by [5.86].

The quality factor of an air-filled, short-circuited, half-wavelength coaxial resonator is calculated in Example 5-12.

Example 5-12: Q of an air-filled coaxial line.   Determine the length d and the quality factor Q_c due to conduction losses of an air-filled half-wavelength coaxial line shorted at both ends and designed to be resonant in the TEM₁ mode at 10 GHz. The inner and outer conductors are made of copper ($\sigma = 5.8 \times 10^7$ S-m^{-1}, $\mu_m = \mu_0$) and have respective dimensions of $a = 0.25$ cm and $b = 1$ cm.

Solution: At $f = 10$ GHz, the wavelength in air is $\lambda = c/f \simeq 3$ cm, so that the length of the resonant coaxial line must be $d = \lambda/2 \simeq 1.5$ cm. The skin depth for copper at 10 GHz is

$$\delta = \left(\pi \times 10 \times 10^9 \times 4\pi \times 10^{-7} \times 5.8 \times 10^7\right)^{-1/2}$$
$$\simeq 0.661 \;\mu\text{m}$$

Noting that $\mu = \mu_0$ and $\mu_m = \mu_0$, the quality factor is

$$Q_c = \frac{2 \ln\left(\dfrac{1.0}{0.25}\right)}{(0.661 \; \mu\text{m}) \underbrace{\left[\dfrac{1}{0.25} + \dfrac{1}{1.0} + \dfrac{2}{1.5} \ln\left(\dfrac{1.0}{0.25}\right)\right]}_{\text{cm}^{-1}}} \approx 6{,}126$$

which is once again a rather large value, compared with the quality factors attainable with lumped resonant circuit elements.

5.4 SUMMARY

This chapter discussed the following topics:

- **General relationships for parallel-plate, rectangular, and circular waveguides.** The configuration and propagation of electromagnetic waves guided by metallic conductors are governed by the solutions of the wave equations subject to the boundary conditions

$$E_{\text{tangential}} = 0 \qquad H_{\text{normal}} = 0$$

The basic method of solution assumes propagation in the z direction, so all field components vary as $e^{-\bar{\gamma}z}$, where $\bar{\gamma}$ is equal to either $\bar{\alpha}$ or $j\bar{\beta}$. The z variations of real fields then have the form

$$\bar{\gamma} = \bar{\alpha} \qquad e^{-\bar{\alpha}z} \cos(\omega t) \qquad \text{Evanescent wave}$$
$$\bar{\gamma} = j\bar{\beta} \qquad \cos(\omega t - \bar{\beta}z) \qquad \text{Propagating wave}$$

The different types of possible solutions are classified as TE, TM, or TEM waves:

$$E_z = 0, H_z \neq 0 \qquad \text{Transverse electric (TE) waves}$$
$$E_z \neq 0, H_z = 0 \qquad \text{Transverse magnetic (TM) waves}$$
$$E_z, H_z = 0 \qquad \text{Transverse electromagnetic (TEM) waves}$$

Assuming that the wave-guiding structures extend in the z direction, the transverse field components can be derived from the axial (z direction) components, so that the z components act as a generating function for the transverse field components. In other words, for any one of the TE_{mn} or TM_{mn} modes, we have

$$\mathbf{E} = \mathbf{E}_{\text{tr}} + \hat{\mathbf{z}}E_z$$
$$\mathbf{H} = \mathbf{H}_{\text{tr}} + \hat{\mathbf{z}}H_z$$

where the subscript "tr" indicates the transverse field quantities. Note, for example, that in rectangular coordinates, $\mathbf{E}_{\mathrm{tr}} = \hat{\mathbf{x}}E_x + \hat{\mathbf{y}}E_y$ and $\mathbf{H}_{\mathrm{tr}} = \hat{\mathbf{x}}H_x + \hat{\mathbf{y}}H_y$. The transverse field components can be directly obtained from the axial components in the following manner:

TE: $$\mathbf{H}_{\mathrm{tr}} = \frac{j\bar{\beta}}{h^2}\nabla_{\mathrm{tr}}H_z$$

$$\mathbf{E}_{\mathrm{tr}} = \frac{j\omega\mu}{h^2}\hat{\mathbf{z}} \times \nabla_{\mathrm{tr}}H_z$$

TM: $$\mathbf{E}_{\mathrm{tr}} = \frac{j\bar{\beta}}{h^2}\nabla_{\mathrm{tr}}E_z$$

$$\mathbf{H}_{\mathrm{tr}} = -\frac{j\omega\epsilon}{h^2}\hat{\mathbf{z}} \times \nabla_{\mathrm{tr}}E_z$$

where the quantities $h = \omega^2\mu\epsilon - \bar{\beta}^2$ and $\bar{\beta}$ are those that correspond to the particular modes under study. The propagation constant, wavelength, and phase velocity of all propagating modes are given by

$$\bar{\beta} = \beta\sqrt{1 - (f_c/f)^2}$$

$$\bar{\lambda} = \frac{2\pi}{\bar{\beta}} = \frac{\lambda}{\sqrt{1 - (f_c/f)^2}}$$

$$\bar{v}_p = \frac{\omega}{\bar{\beta}} = \frac{v_p}{\sqrt{1 - (f_c/f)^2}}$$

where f_c is the cutoff frequency of the particular mode under consideration. The attenuation rate α_c due to conduction losses is defined as

$$\alpha_c = \frac{P_L}{2P_{\mathrm{av}}}$$

where P_L is the power lost per unit length and P_{av} is the average power transmitted. The attenuation rate α_d due to dielectric losses is given by

$$\alpha_d \approx \frac{\omega\sqrt{\mu_0\epsilon'}\,\epsilon''/\epsilon'}{2\sqrt{1 - (f_c/f)^2}}$$

where f_c is once again the cutoff frequency for the mode in question.

■ **Rectangular and circular waveguides.** The general forms of the axial field components are

TE: $$H_z = \begin{cases} \cos\left(\dfrac{m\pi}{a}x\right)\cos\left(\dfrac{n\pi}{b}y\right)e^{-j\bar{\beta}_{mn}z} & \text{Rectangular} \\[2ex] J_n\left(s_{nl}\dfrac{r}{a}\right)\cos(n\phi)e^{-j\bar{\beta}_{\mathrm{TE}_{nl}}z} & \text{Circular} \end{cases}$$

$$
\text{TM:} \quad E_z = \begin{cases} \sin\left(\dfrac{m\pi}{a}x\right)\sin\left(\dfrac{n\pi}{b}y\right)e^{-j\bar{\beta}_{mn}z} & \text{Rectangular} \\[2em] J_n\left(t_{nl}\dfrac{r}{a}\right)\cos(n\phi)e^{-j\bar{\beta}_{\text{TM}_{nl}}z} & \text{Circular} \end{cases}
$$

The cutoff frequency f_c for rectangular waveguides is given by

$$
f_{c\text{TM}_{mn} \text{ or TE}_{mn}} = \frac{1}{2\sqrt{\mu\epsilon}}\sqrt{\left(\frac{m}{a}\right)^2 + \left(\frac{n}{b}\right)^2}
$$

whereas that for circular waveguides is

$$
f_{c\text{TM}_{nl}} = \frac{t_{nl}}{2\pi a\sqrt{\mu\epsilon}} \quad \text{and} \quad f_{c\text{TE}_{nl}} = \frac{s_{nl}}{2\pi a\sqrt{\mu\epsilon}}
$$

where t_{nl} and s_{nl} are the roots of $J_n(\cdot)$ and $J_n'(\cdot)$, respectively. Note once again that the cutoff frequencies f_c determine all the other quantities, such as $\bar{\beta}$, $\bar{\lambda}$, and $\bar{v}_p$, for the different modes.

■ **Cavity resonators.** Complete enclosed metallic boxes of circular or rectangular shape allow extremely efficient storage of electromagnetic energy, with quality factors of up to few tens of thousands. The Q of a cavity due to conduction losses in the walls is given by

$$
Q_c \simeq \frac{2V_{\text{cavity}}}{\delta S_{\text{cavity}}}
$$

where V_{cavity} and S_{cavity} are respectively the volume and the surface area of the cavity, and $\delta = (\pi f \mu_m \sigma_m)^{-1/2}$ is the skin depth of the metallic walls. The Q due to losses in the dielectric material within the cavity is given by $Q_d = \epsilon'/\epsilon''$. The total Q of the system is then given by

$$
\frac{1}{Q} = \frac{1}{Q_d} + \frac{1}{Q_c}
$$

5.5 PROBLEMS

5-1. Frequency range of the dominant mode. An air-filled hollow WR-10 rectangular waveguide has dimensions $a = 0.254$ cm and $b = 0.127$ cm. (a) What is the cutoff frequency of the dominant (TE$_{10}$) mode? (b) What is the cutoff frequency of the next higher-order mode(s)? (c) What is the frequency band over which only the dominant mode can propagate?

5-2. AM waves through a tunnel? Consider the propagation of electromagnetic waves at 1 MHz through a tunnel of rectangular cross section. (Note that AM radio frequencies range from 535 to 1605 kHz.) Find the minimum tunnel dimensions needed for the

1 MHz AM signal to propagate and, based on your results, comment on the feasibility of AM transmission through tunnels that exist in practice.

5-3. Propagation experiment in a road tunnel. Tunnels are often found in metropolitan cities and mountainous areas or under the sea. With the increasing demands of wireless communication applications, research and investigation of radio-wave propagation in tunnels has intensified in recent years. In one study,[49] experiments were performed in various tunnels to investigate electromagnetic wave propagation at frequencies between 1 MHz and 1 GHz. One of the tunnels investigated was a road tunnel located in Brussels, Belgium, with rectangular cross section 17 m wide, 4.9 m high, and about 600 m long, with walls made from concrete. A vertical antenna was used to excite the TE_{10} mode, and a horizontal antenna was used to excite the TE_{01} mode. Assuming the waveguide to be ideal with perfectly conducting walls, (a) find the cutoff frequencies of both the TE_{10} and TE_{01} modes, and (b) find the lowest-order modes that do not propagate at the FM radio frequency of 100 MHz.

5-4. Radio-wave propagation in railway tunnels. A recent study has been carried out concerning radio-wave propagation in railway tunnels.[50] In this study, two railway tunnels of rectangular cross section are considered. Tunnel 1 has transverse dimensions of width $a = 8$ m and height $b = 4$ m, and tunnel 2 has the same height but a narrower width, $a = 6$ m. Assuming the tunnels to be empty with perfectly conducting walls, find the total number of propagation modes possible in each tunnel at 100 MHz. Which tunnel allows more modes to propagate?

5-5. Rectangular waveguide modes. Consider a WR-137 rectangular air waveguide with inside dimensions 3.484 cm $\times$ 1.58 cm, recommended for use in the dominant mode for a frequency range of 5.85–8.20 GHz. Find (a) the cutoff frequencies of the four lowest-order modes, and (b) the phase velocity $\bar{v}_p$ and the guide wavelength $\bar{\lambda}$ for the dominant mode at the two ends of the specified frequency range.

5-6. Dominant mode in a rectangular waveguide. For the waveguide described in Problem 5-5, determine the propagation constant $\bar{\gamma}$ and the wave impedance (Z_{TE} or Z_{TM}) for the dominant mode at (a) 3.5 GHz, (b) 7 GHz.

5-7. Waveguide wavelength. (a) Plot the relationship between waveguide wavelength and frequency for a TE_{10} mode propagating in a rectangular air waveguide with an aspect ratio of 2:1 and a wide dimension of 7.214 cm. (b) Repeat part (a) if the waveguide is filled with water (assume $\epsilon_r \simeq 81$). (c) Repeat parts (a) and (b) for TE_{20} mode.

5-8. Rectangular waveguide modes. Consider a WR-42 rectangular air waveguide with inside dimensions 1.067 cm $\times$ 0.432 cm operating at 40 GHz. (a) Calculate the propagation constant $\bar{\gamma}$ and wave impedance Z_{TE} for the TE_{10}, TE_{01}, TE_{11}, TE_{20}, and TE_{02} modes. (b) Calculate the phase velocity $\bar{v}_p$ and the guide wavelength $\bar{\lambda}$ for all the propagating modes.

5-9. Dielectric-filled waveguide. A standard WR-22 rectangular waveguide (0.569 cm $\times$ 0.284 cm) of 1 cm length is filled with a lossless dielectric material. If the dominant propagating mode signal experiences a phase shift of approximately 354.6° at 13.2 GHz,

[49]L. Deryck, Natural propagation of electromagnetic waves in tunnels, *IEEE Trans. Vehicular Technology,* 27(3), pp. 145–150, August 1978.

[50]Y. P. Zhang and Y. Hwang, Theory of the radio-wave propagation in railway tunnels, *IEEE Trans. Vehicular Technology,* 47(3), pp. 1027–1036, August 1998.

(a) find the relative dielectric constant of the dielectric filling, (b) find all the propagating modes at 13.2 GHz, if any, (c) find all the propagating modes (if any) at 13.2 GHz if the dielectric filling is replaced with air.

5-10. Rectangular waveguide modes. A WR-1500 rectangular air waveguide has inner dimensions 38.1 cm × 19.05 cm. Find the following: (a) the cutoff wavelength for the dominant mode; (b) the phase velocity, guide wavelength, and wave impedance for the dominant mode at a wavelength of 0.8 times the cutoff wavelength; (c) the modes that will propagate in the waveguide at a wavelength of 40 cm; (d) the modes that will propagate in the waveguide at a wavelength of 30 cm.

5-11. Polyethylene-filled waveguide. Repeat Problem 5-10 for the case in which the waveguide is filled with polyethylene (assume it is lossless, with $\epsilon_r \simeq 2.25$).

5-12. Unknown rectangular waveguide mode. The y component of the phasor electric field of a particular mode in a 4 cm × 2 cm (the longer transverse dimension being in the z direction) air-filled rectangular waveguide oriented along the y direction is given by

$$E_y = C_1 \sin(25\pi z) \sin(50\pi x) e^{-j41\pi y}$$

(a) Find this mode. Is it a propagating or an evanescent mode? (b) Find the operating frequency. (c) Find which modes can propagate at this frequency.

5-13. A large waveguide for deep hyperthermia. To apply deep regional hyperthermia, a part of a patient's body is to be exposed to TE_{10} mode waves in a rectangular waveguide through holes in the broad walls made to fit the body cross section, as shown in Figure 5.24.[51] A frequency of operation between 25 and 100 MHz is used for heating, depending on the size of the patient's body. The waveguide dimensions are $a = 6$ m and $b \simeq 30$ cm. (a) Assuming the waveguide to be filled with air, calculate the cutoff frequencies of the five lowest-order modes. (b) Repeat part (a) if the waveguide is filled with water (assume it is lossless with $\epsilon_r \simeq 81$). (c) Repeat part (b) if the waveguide dimensions are changed to $a = 70$ cm and $b = 30$ cm. Comment on the differences.

5-14. Square waveguide modes. A 3 cm × 3 cm square waveguide is operated at frequencies below 8 GHz. Find all the modes (both TE and TM) that can propagate through this waveguide if (a) it is filled with air, and (b) it is filled with polystyrene (assume it is lossless, with $\epsilon_r \cong 2.56$).

5-15. SAR calibration using a rectangular waveguide. The radiation safety tests for handheld mobile phones require a precise calibration of small electric probes used to measure the specific absorption rate (SAR) in phantom human tissues. In a recent study,[52] a calibration system based on a rectangular waveguide was developed for SAR calibrations at 900 MHz. The central component of the system is a nonstandard rectangular air-filled waveguide with dimensions of $a = 19$ cm and $b = 14$ cm respectively. (a) Determine the modes excited at 900 MHz. (b) Determine the frequency range over which only the dominant mode is excited. (c) Repeat part (b) if the TE_{01} mode is a very weak excitation and can be neglected.

5-16. Millimeter-wave rectangular waveguide. Commercial applications in the millimeter-wave frequencies (such as automotive radar, passive imaging, and remote sensing)

[51]S. Mizushina, Y. Xiang, and T. Sugiura, A large waveguide applicator for deep regional hyperthermia, *IEEE Trans. Microwave Theory and Techniques,* 34(5), pp. 644–648, May 1986.

[52]K. Jokela, P. Hyysalo and L. Puranen, Calibration of specific absorption rate (SAR) probes in waveguide at 900 MHz, *IEEE Trans. on Instrumentation and Measurement,* 47(2), pp. 432–438, April 1998.

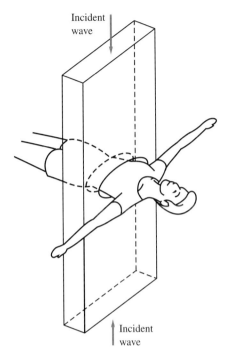

Incident
wave

Incident
wave

**FIGURE 5.24.   A large waveguide for
deep hyperthermia.** Problem 5-13. (Taken
from S. Mizushina, Y. Xiang, and T. Sugiura,
A large waveguide applicator for deep regional
hyperthermia, *IEEE Trans. Microwave Theory and
Techniques,* 34(5), pp. 644–648, May 1986.)

are rapidly expanding, and exploration of the submillimeter-wave range is currently
underway because of the larger bandwidths available. The rectangular waveguide is
commonly used in the millimeter and submillimeter-wave ranges because of its low
loss, although it is more difficult and expensive to manufacture waveguides of such
small sizes. Recently, a W-band rectangular waveguide of 17 mm length, with cross-
sectional dimensions of 2.54 mm by 700 μm (~half height), has been machined using
a new technique.[53] Calculate the cutoff frequencies of the three lowest-order modes of
this waveguide, (b) What is the frequency range of the dominant mode?

5-17. **TE$_{10}$ mode in a rectangular waveguide.** One of the components of the magnetic field
of a TE$_{10}$ mode propagating in an air-filled rectangular waveguide with $a = 2b$ used
for a microwave plasma experiment is given by

$$H_x \simeq 26\cos(29y)e^{-j42.3x} \text{ A-m}^{-1}$$

(a) Calculate the dimensions of the waveguide and the operation frequency. (b) Calculate the time-average power carried by this mode.

5-18. **Dominant mode in rectangular waveguide design.** (a) Design a rectangular air
waveguide to transmit only the dominant mode in the frequency range 15–22 GHz
such that the lower end of the frequency band is at least 25% higher than the cutoff
frequency of the dominant mode, the cutoff frequency of the next higher mode is at
least 25% higher than the center frequency of this frequency band, and the power

[53]C. E. Collins, et. al., Technique for micro-machining millimeter-wave rectangular waveguide, *Electronics Letters,* 34(10), pp. 996–997, May 1998.

transmission in the dominant mode is maximized. (b) What is the cutoff frequency of the nearest mode(s) to the dominant mode? (c) Calculate the maximum time-average power transmitted in the dominant mode at the two extreme ends of the frequency band, taking the breakdown strength of air as 15 kV-(cm)$^{-1}$ (a safety factor of approximately 2 at sea level is assumed).

5-19. Rectangular waveguide design. Design a rectangular air waveguide such that the dominant mode has the largest possible bandwidth (over which no other mode but the dominant mode can propagate) with a cutoff frequency of 2.58 GHz. (b) What is the cutoff frequency of the next higher mode(s)? (c) What is the maximum time-average power capacity of the dominant mode at a frequency of 4 GHz? [Take the breakdown electric field of air to be 15 kV-(cm)$^{-1}$.]

5-20. Maximum field in a rectangular waveguide. A commercially available high-power microwave (HPM) source is capable of providing a peak power of 2 GW at a frequency of 1.1 GHz. Two rectangular waveguides are proposed to transport this power in the TE$_{10}$ mode. These are WR-650 (16.51 cm $\times$ 8.255 cm) and WR-975 (24.765 cm $\times$ 12.383 cm) waveguides. (a) For each waveguide, find how many modes propagate at 1.1 GHz. (b) For each waveguide, find the peak electric field in MV-m^{-1} corresponding to a peak power of 2 GW. (c) Determine which waveguide should be chosen in terms of higher power-handling capability.

5-21. Power capacity of a rectangular waveguide. (a) Consider a WR-284 rectangular air waveguide with inner dimensions 7.214 cm $\times$ 3.404 cm to operate in the dominant mode at 2.45 GHz. Assuming the breakdown electric field of air to be 15 kV-(cm)$^{-1}$ (with a safety factor of 2 at sea level), calculate the maximum time-average power that can be carried by this waveguide. (b) Another waveguide (WR-430 air-filled, with inner dimensions 10.922 cm $\times$ 5.461 cm) is proposed to increase the power-handling capability at 2.45 GHz. Calculate the maximum time-average power that can be carried by this waveguide and compare it with the previous one.

5-22. TE$_{11}$ mode in a rectangular waveguide. The z component of the magnetic field of the TE$_{11}$ mode traveling in an air-filled rectangular waveguide along the z direction is

$$H_z = C \cos(50\pi x) \cos(100\pi y) e^{-j100\pi z}$$

(a) Find the waveguide dimensions and the operating frequency. (b) Find the constant C if the time-average power carried by TE$_{11}$ mode in the z direction is 10 kW. (c) Find all the other nonzero field components of the TE$_{11}$ mode. (d) Find the maximum time-average power-carrying capacity of this waveguide, taking the breakdown electric field of air to be 15 kV-(cm)$^{-1}$ (with a safety factor of 2 at sea level).

5-23. No TM$_{01}$ or TM$_{10}$ modes in a rectangular waveguide. Show that the TM$_{10}$ and TM$_{01}$ modes cannot exist in a rectangular waveguide.

5-24. No TEM wave in single-conductor waveguide. Show that the TEM wave cannot exist in a single-conductor waveguide.

5-25. Power capacity of a rectangular waveguide. A WR-102 rectangular air waveguide with inner dimensions 2.591 cm $\times$ 1.295 cm is used to transport a time-average power of 100 kW in the dominant TE$_{10}$ mode at an operating frequency of 9 GHz. (a) Find the peak value of the electric field occurring in the waveguide. (b) Calculate the maximum time-average power that can be carried by the TE$_{10}$ mode, taking the breakdown strength of air to be 15 kV-(cm)$^{-1}$ (a safety factor of 2 at sea level is assumed).

5-26. Cutoff frequency for a circular waveguide. An air-filled circular waveguide has a diameter of 7 cm. Find the frequency below which waves cannot be transmitted through this waveguide.

5-27. TE_{11} mode in a circular waveguide. Find the smallest diameter that an air-filled circular waveguide must have to allow the propagation of TE_{11} mode at (a) 40 MHz, (b) 40 GHz.

5-28. Square-to-circular waveguide. Consider a 2.4 cm × 2.4 cm square air waveguide designed to be used at 7 GHz so that only the dominant modes propagate. (a) Find all the propagation modes possible at 7 GHz. (b) If the square waveguide is gradually reshaped to become a circular waveguide, find all the new propagation modes possible at 7 GHz. (c) Draw a rough sketch of the electric field lines of each propagating mode found in parts (a) and (b) over the transverse cross section of the waveguide, and comment on their relationships.

5-29. Southworth's experiments. In 1919 George C. Southworth began experimenting with circular waveguides using sources operating at air wavelengths between 1 and 2 m, but these wavelengths were too long for meaningful transmission experiments on air-filled waveguides. As a solution, he filled his waveguides with water (assume $\epsilon_r \simeq 81$) so that the waveguide diameter required for any mode to propagate was reduced by a factor of 9. About 10 years later, microwave sources operating at 15-cm air wavelengths became available when Southworth set up a similar experiment using a 12.7-cm diameter air-filled waveguide. (In 1933, a 6.1-m long, 12.7-cm diameter circular air waveguide was built and used to demonstrate transmission of telegraph signals. The first message was "Send money," symbolic of the economic depression at that time.)[54] (a) Calculate the minimum waveguide diameter required for both air- and water-filled waveguides to allow any propagation at 1 m wavelength (in air). Based on your results, comment on which waveguide is more suitable for an experiment at 1 m wavelength. (b) Find the minimum diameter of a circular air-filled waveguide required for any propagation mode to exist at a wavelength of 15 cm. (c) Find all the possible propagation modes on the 12.7-cm diameter air-filled waveguide used by Southworth at 15 cm wavelength.

5-30. Barrow's experiments. (a) In 1935 Wilmer L. Barrow, a professor of electrical engineering at the Massachusetts Institute of Technology (M.I.T.), conducted an experiment to demonstrate the propagation of electromagnetic waves in hollow tubes. The experiment consisted of a 50-cm signal source used to generate electromagnetic waves in a long, hollow, cylindrical brass pipe with an inner diameter of 4.5 cm. The experiment was not successful. Why? (b) Soon Professor Barrow developed a theory and understood what went wrong in his 1935 experiment. In 1936 he repeated the experiment using a 40-cm signal source to excite a section of an air duct, with an inner diameter of approximately 45.7 cm and a length of approximately 5 m. Was this second experiment successful? Why?

5-31. The Channel Tunnel. The Channel Tunnel, linking Great Britain and France, was completed around the middle of 1994 at a total cost of around $17 billion and is considered by many to be one of the major engineering achievements and symbols of man's

---

[54]H. Sobol, Microwave communications—a historical perspective, *IEEE Trans. Microwave Theory and Techniques,* 32(9), pp. 1170–1181, September 1984.

imagination since the pyramids.[55] The tunnel consists of three 38-km long parallel tunnels, two of which are rail tunnels with circular cross sections of inner diameter 7.6 m, and the third is a service tunnel with an inner diameter of 4.8 m. (a) Assuming the tunnels to be perfect circular waveguides, find the maximum number of modes that can propagate along each of the 7.6-m diameter tunnels at an AM radio frequency (535–1605 kHz). (b) Repeat part (a) for an FM radio frequency (88–108 MHz). (c) Repeat parts (a) and (b) for the 4.8 m-diameter service tunnel.

5-32. Unknown circular waveguide mode. The ϕ component of the phasor magnetic field of a particular mode in an 8-cm diameter air-filled circular waveguide is given by

$$H_\phi \simeq C_1 J_0'(138r)e^{j48\pi z}$$

(a) Identify this mode. Is it a propagating or nonpropagating mode? (b) Find the operating frequency. (c) Write all the other phasor field components of this mode. (d) Find the total number of propagating modes possible at the same operating frequency.

5-33. Unknown circular waveguide mode. The phasor z component of the magnetic field of a particular mode propagating in the z direction in a 7-cm diameter circular air waveguide is given by

$$H_z \simeq C_1 J_3(120r)\cos(3\phi)e^{-j116.9z}$$

where both r and z are in meters. (a) Find this mode. Is it a propagating or an evanescent mode? Why? (b) What is the operating frequency? (c) What is the highest-order mode that can propagate in this waveguide at the same operating frequency? (d) Write the phasor z components of the fields (i.e., both E_z and H_z) for the mode found in part (c) at the same frequency.

5-34. Smallest circular waveguide dimension. Find the smallest diameter that an air-filled circular waveguide must have to allow the propagation of (a) TM_{21} mode at 25 GHz; (b) TM_{12} mode at 25 GHz.

5-35. Circular waveguide modes. A WC-80 circular air waveguide with an inner diameter of 2.024 cm is operated at 16 GHz. (a) Find all the possible propagating modes in the waveguide. (b) Find the phase velocity, guide wavelength, and wave impedance corresponding to each propagating mode.

5-36. Low-loss TE$_{01}$ mode in a circular waveguide. A cylindrical waveguide has a diameter of 10 cm. Find the cutoff frequency of the TE_{01} (low-loss) mode if the waveguide is filled with (a) air, and (b) water (assume $\epsilon_r \cong 81$).

5-37. Low-loss TE$_{01}$ mode in a highly overmoded waveguide. The phasor electric field of the low-loss TE_{01} mode propagating in a highly overmoded air-filled cylindrical waveguide used for a plasma-heating application and oriented along the z direction is given by

$$\mathbf{E} \simeq \hat{\phi} C J_0'(238r)e^{-j716\pi z}$$

(a) Find the waveguide diameter and the frequency at which this mode is excited. (b) How many total distinct modes can propagate in this waveguide at the same frequency?

5-38. A highly overmoded circular waveguide. Gyrotrons are an efficient source of high-power millimeter-wavelength energy. With output powers that reach above 500 kW and frequencies that range from 28 GHz to 150 GHz and beyond, gyrotrons have been used

---

[55]C. J. Kirkland, *Engineering the Channel Tunnel,* Routledge, New York, 1995.

very successfully for heating magnetically confined plasmas at the electron cyclotron resonance frequency.[56] Commercial gyrotrons launch at their output an axially symmetric TE_{0n} mode into a circular waveguide. Because of the large amounts of power produced by the gyrotron, the circular waveguide must have a large diameter and thus is highly overmoded. Consider a gyrotron with a 6.35-cm diameter waveguide that is then tapered to a 2.78-cm diameter in order to achieve mode conversion between TE_{0n} modes. (a) Assuming all possible modes to be excited, find the total number of modes that can propagate in the 6.35-cm diameter waveguide at 60 GHz. (b) Repeat part (a) for the 2.78-cm diameter waveguide. (c) Repeat parts (a) and (b) for 140 GHz. [*Hint:* For this problem, you need to use a reference source such as C. L. Beattie, Table of first 700 zeros of Bessel functions $J_1(x)$ and $J_1'(x)$, *Bell System Tech. J.,* pp. 689–697, May 1958.]

5-39. TE_{11} to TM_{11} mode converter. The use of high-power gyrotron tubes in plasma fusion research has opened up some interesting challenges in the area of microwaves, particularly in the development of mode converters for conversion of higher-order gyrotron modes to lower-order modes for efficient plasma heating. These high-power microwave applications of mode converters typically involve circular waveguide structures. A compact, low-cost, circular air waveguide TE_{11}-to-TM_{11} mode converter is designed to operate at 9.94 GHz.[57] The input waveguide has a radius of $\sim$1.19 cm, and the output waveguide has a radius of $\sim$2.06 cm. (a) Find all the possible modes of propagation in the input waveguide at the operating frequency. (b) Repeat part (a) for the output waveguide.

5-40. Circular waveguide modes. An air-filled circular waveguide has an inner diameter of 1 cm. Find the frequency range over which (a) only the dominant mode can propagate; (b) only TE_{11} and TE_{01} modes can propagate (excluding the possibility of the TM_{01} and TE_{21} modes).

5-41. Circular waveguide design. (a) Design a circular air waveguide to operate at 6.4 GHz such that the cutoff frequency of the dominant mode is at least 20% lower than the operating frequency, the cutoff frequency of the TE_{01} mode is at least 10% higher than the operating frequency, and the waveguide has maximum power-carrying capability. (b) Find all the possible propagating modes in this waveguide at 6.4 GHz. (c) Find all the possible propagating modes if this waveguide is operated at 9.6 GHz.

5-42. Circular waveguide modes. A WC-109 circular waveguide (diameter 2.779 cm) filled with polystyrene (assume $\epsilon_r \cong 2.56$) is operated at a frequency of 10 GHz. (a) Find all the possible propagating modes. (b) Find the phase velocity, guide wavelength, and wave impedance of the dominant mode.

5-43. A circular waveguide used for microwave telecommunications. Due to its being a very cost-effective means of combining multiple-frequency bands into one antenna feeder system with very low insertion loss, WC-281 air circular waveguide ($\sim$7.14 cm inside diameter) has been in commercial use carrying 4, 6, 8, and 11 GHz

[56]M. Thumm et al., Very high power mm-wave components in oversized waveguides, *Microwave J.,* pp. 103–121, November 1986; M. J. Buckley, D. A. Stein, and R. J. Vernon, A single-period TE_{02}–TE_{01} mode converter in a highly overmoded circular waveguide, *IEEE Trans. Microwave Theory and Techniques,* 39(8), pp. 1301–1306, August 1991.

[57]T. Haq, K. J. Webb, and N. C. Gallagher, TE_{11} to TM_{11} compact mode converter for circular waveguide, *IEEE Microwave Theory and Techniques Symp. Dig.,* pp. 1613–1616, 1995.

telecommunications traffic since 1954.[58] Find all the possible propagating modes for this guide at each of the specified microwave frequencies.

5-44. Orthogonally polarized TE_{11} modes. TE_{11} operations in circular waveguides are widely used for commercial microwave terrestial telecommunication systems to allow smaller antennas (and thus less expensive towers) and to provide multiband and dual polarized operation, where two identical but orthogonal modes use the same waveguide for transmission and reception with a single antenna.[59] Consider an air-filled circular waveguide system operating in the orthogonally polarized TE_{11} modes. The r component of the total electric field phasor in the waveguide is given by

$$E_r \simeq \frac{C_1}{r} J_1(80r)(\sin\phi + \cos\phi)e^{-j47z}$$

where both r and z are in meters. (a) Find the inner diameter of the waveguide and the operating frequency. (b) Verify that the waveguide is indeed operating in the dual polarized dominant modes. (c) Find all the other nonzero field components corresponding to the orthogonally polarized TE_{11} modes. (d) Explain why the operation with dual polarized modes works for circular waveguides but not for rectangular waveguides.

5-45. Circular waveguide-to-microstrip transition. Waveguide-to-microstrip transitions are fundamental components in microwave circuits used to couple power from a waveguide to a microstrip integrated circuit. A new type of circular waveguide-to-microstrip transition consisting of a microstrip probe inserted in an air circular waveguide with an inner diameter of 2.1 cm has been developed.[60] If this transition system is designed to be used in the frequency range 8.72–10.66 GHz, find all the propagation modes possible over this frequency band.

5-46. Power capacity of circular versus rectangular waveguides. A circular waveguide has the same internal cross-sectional area as a WR-284 rectangular guide. The two guides are carrying equal powers at 3 GHz. Compare the maximum values of the electric fields. Which guide could carry the highest power without internal sparking?

5-47. Circular versus rectangular waveguides. A 10-m long waveguide needs to be built for a 25 GHz radar system. In hand is a 3-mm thick copper sheet of dimensions 6.3 cm ×10 m from which the waveguide is to be built. Neglecting dielectric losses and fringing fields (for parallel-plate), what type of waveguide would you build to minimize losses: parallel-plate, rectangular, or circular?

5-48. Attenuation in a coaxial line. An air-filled coaxial line has thick walls made of copper. Its attenuation constant is $\alpha_c = 0.003$ np-m^{-1} at 300 MHz. (a) What is its attenuation constant at 600 MHz? (b) If all of its cross-sectional linear dimensions (i.e., inner and outer radii a, b, and thickness d of the outer conductor) are doubled, what is the new value of α_c at 300 MHz? (c) Suppose the original line is filled with a dielectric for which $\epsilon' = 2\epsilon_0$ and the loss tangent $\tan\delta_c = (15/\pi) \times 10^{-4}$ at 300 MHz. Find the total attenuation constant $(\alpha_c + \alpha_d)$ at 300 MHz. (d) Repeat part (c) for 600 MHz, assuming that ϵ' and $\tan\delta_c$ remain the same.

[58]R. Walker, Analysis and correction of echo due to mode conversion in WC-281 waveguide, *IEEE Trans. Microwave Theory and Techniques,* 43(3), pp. 592–600, March 1995.

[59]T. N. Anderson, State of the waveguide art, *Microwave J.,* 25(12), pp. 22–48, December 1982.

[60]L. Fan, M. Y. Li, and K. Chang, Circular waveguide-to-microstrip transitions, *Electronics Letters,* 31(4), pp. 294–295, February 16, 1995.

5-49. Optimum coaxial lines. Consider a coaxial line with an outer conductor of fixed inside radius b. (a) Show that α_c is a minimum when $b = 3.6a$, where a is the outside radius of the inner conductor. (b) What is the characteristic impedance of such a coaxial line designed for minimum attenuation? (c) If you start instead with a coaxial line of fixed inner conductor radius a, can you find a value of outer conductor radius b for which α_c is minimized? If so, what is the minimizing value of b, and what is the characteristic impedance of such a coaxial line?

5-50. Power capacity of a coaxial line. A coaxial transmission line has an inner radius of 0.5 mm and an outer radius of 2.5 mm. What is the maximum power that may be carried in the TEM mode? Assume the breakdown voltage to be 10 kV-(mm)$^{-1}$ and the maximum allowed surface current density to be 10 A-(mm)$^{-1}$. Discuss design changes that could be made to increase the power-handling capacity.

5-51. Higher-order modes on a coaxial line. A coaxial line test fixture with $a \simeq 1.49$ cm and $b \simeq 3.429$ cm is used to measure the properties of low-loss materials based on the assumption that only TEM mode propagates along the line.[61] (a) When the coaxial line is filled with air, calculate the approximate value of the lowest frequency above which modes other than TEM can propagate. (b) Repeat part (a) if the coaxial line is filled with Noryl EN265 (a low-loss nonmagnetic material manufactured by General Electric Corporation). For this material, assume $\epsilon_r' \simeq 2.65$ and neglect ϵ_r''. (c) The coaxial line test fixture is filled with Noryl EN265, and measurements are carried out to determine the properties of this material. However, the plots obtained for the material properties using the TEM approximation over the frequency range of 50 MHz to 10 GHz have an oscillatory behavior and are incorrect above ~ 2 GHz. Can you speculate on the cause of this error?

5-52. Rectangular cavity resonator design. (a) Calculate the length of an air-filled cavity resonator made of a standard WR-340 rectangular waveguide (8.636 cm $\times$ 4.318 cm) to resonate at 2.45 GHz in the TE_{101} mode. (b) Repeat part (a) for the TE_{102} mode. (c) Repeat part (a) for the TE_{101} mode if the cavity is filled with water (assume $\epsilon_r \simeq 10$ at 2.45 GHz).

5-53. Rectangular cavity resonator design. Design a rectangular cavity resonator that will resonate in the TE_{101} mode at 10 GHz and resonate in the TM_{110} mode at 20 GHz.

5-54. Rectangular cavity resonator. A rectangular hollow-pipe waveguide has dimensions such that TE_{21} is the highest-order mode that can propagate in it. A section of this waveguide of length 1.5λ is closed at both ends by metallic caps so that it forms a cavity. What is the highest-frequency TE cavity resonance mode that can exist in this cavity?

5-55. Rectangular cavity resonator modes. Determine the resonant frequencies of the four lowest-order modes of an air-filled rectangular resonator that measures 1 cm $\times$ 2 cm $\times$ 3 cm.

5-56. Rectangular cavity for dielectric measurements. A rectangular cavity is designed from a standard WR-90 brass waveguide nearly 13.5 cm in length to carry out measurements of the complex permittivity of dielectric materials.[62] (a) Calculate the

---

[61] C. C. Courtney, Time-domain measurement of the electromagnetic properties of materials, *IEEE Trans. Microwave Theory and Techniques,* 46(5), pp. 517–522, May 1998.

[62] R. Thomas and D. C. Dube, Extended technique for complex permittivity measurement of dielectric films in the microwave region, *Electronics Letters,* 33(3), pp. 218–220, January 1997.

resonant frequency and the Q of the TE_{101} mode when the cavity is empty. For brass, use $\sigma \simeq 2.564 \times 10^7$ S-m^{-1}. (b) The cavity is excited in the TE_{10p} modes to measure the permittivity of the material under test over the X-band (8–12 GHz) frequencies. For an empty cavity, determine all the TE_{10p} modes that can be used for this purpose.

5-57. Rectangular cavity for water pollution study. A rectangular cavity is constructed using a section of a brass waveguide ($a = 7.20$ cm, $b = 3.35$ cm, and $d = 35.3$ cm) for detecting pollutants in water at microwave frequencies.[63] The polluted water sample is introduced in the cavity via a thin capillary tube, resulting in a shift in the resonant frequencies and the Q factors of the cavity. Using these shifts, the complex permittivity of the water sample can be calculated, from which the effects of the pollutants can be determined. The rectangular waveguide cavity is excited in the TE_{10p} mode. All measurements are carried out at ~2.69 GHz. (a) Determine the mode which resonates at this frequency when the cavity is empty. (b) Calculate the quality factor Q_c corresponding to the mode found in part (a).

5-58. Rectangular cavity resonator. An ideal (lossless) rectangular cavity resonator encloses the volume $0 \le x \le a$, $0 \le y \le b$, and $0 \le z \le d$ and is excited with the TE_{123} mode. The peak instantaneous value of H_z is 1 A-m^{-1}. (a) Calculate the frequency of the signal being supported if the cavity is air-filled and measures $a = b = d = 10$ cm. (b) Find the total energy stored in the cavity. (c) Repeat parts (a) and (b) for a cavity filled with a lossless dielectric with $\epsilon_r = 3$.

5-59. Modes of an X-band rectangular cavity. Consider an X-band (8–12 GHz) rectangular cavity resonator constructed with dimensions $a = 2b = 0.2d$ operating in a specific TE_{mnp} mode having an electric field represented by

$$E_y = C \sin\left(\frac{\pi x}{a}\right) \sin\left(\frac{1.2\pi z}{a}\right)$$

to be used for measuring the dielectric constant of materials. (a) Determine this TE_{mnp} mode. (b) If the resonant frequency of the mode found in part (a) is ~12.3 GHz, determine the dimensions a, b, and d. (Assume the cavity to be empty.) (c) Calculate the resonant frequency and the Q of the TE_{101} mode in this cavity. Assume the cavity to be made of copper with $\sigma = 5.8 \times 10^{-7}$ S-m^{-1}.

5-60. An overmoded rectangular cavity. A resonant overmoded rectangular cavity is designed to operate in a specific TE_{mnp} mode at 11.8 GHz to be used to couple power to a 16-element helical antenna array.[64] Different design techniques are implemented to prevent the formation of or to suppress the other unwanted modes. If the cavity is $12.6 \times 12.6 \times 1$ cm^3 in size, find the TE_{mnp} mode chosen to feed the antenna array. *Hint:* For this mode, $m = n$.

5-61. Circular cavity resonator modes. Determine the resonant frequencies of the four lowest-order modes of an air-filled circular resonator that measures 3 cm in diameter by 6 cm in length.

---

[63]U. Raveendranath and K. T. Mathew, Microwave technique for water pollution study, *J. Microwave Power and Electromagnetic Energy*, 30(3), pp. 188–194, 1995.

[64]S. S. Chana and C. W. Turner, Computer-aided design of an overmoded cavity-backed helical antenna array, *Electronics Letters*, 29(15), pp. 1378–1380, July 1993.

5-62. Circular cavity for cellular radio components. A high-Q circular waveguide cavity resonant in the TE_{111} mode, for testing cellular radio base station components over the frequency range of 1.5–1.8 GHz, is made from glass (Pyrex) with an internal conductive coating so that its resonant frequency is less sensitive to temperature variations than a metal cavity.[65] The cavity has a length of 13.25 cm and a diameter of 14 cm. (a) Calculate the theoretical resonant frequency of this cavity. (b) Assuming a copper cavity, determine the Q at the resonant frequency. For copper, use $\sigma = 5.8 \times 10^{-7}$ S-m^{-1}.

5-63. Tunable circular cavity resonator. A circular cavity resonator is designed to resonate in the TE_{011} mode over a frequency range of 8 GHz to 12 GHz. The tuning is accomplished by moving one end plate, which effectively changes the length of the resonator. If the radius of the circular resonator is 3 cm, calculate the range displacement of the movable end that is required to meet the design specifications.

5-64. A circular cavity to measure ϵ. TE_{01p} circular cavity resonators provide reasonably accurate measurements of the properties of low-loss materials.[66] A range of frequency measurements can be made by use of a number of TE_{01p} modes or by changing the cavity length. A 60-mm diameter mode-filtered cavity is chosen to carry out measurements at 10 GHz. (a) Find the length d of the cavity such that it resonates in the TE_{011} mode and calculate the Q. Assume the cavity to be empty and made of copper. (b) Repeat part (a) for the TE_{014} mode.

5-65. Coaxial line resonator. A half-wavelength resonator is constructed using a piece of copper coaxial line, with an inner conductor of 2 mm diameter and an outer conductor of 8 mm diameter. If the resonant frequency is 3 GHz, find the physical length and quality factor Q of (a) an air-filled coaxial line and (b) a coaxial line filled with Teflon ($\epsilon_r' \simeq 2.1$, $\epsilon_r'' \simeq 3.15 \times 10^{-4}$), and compare the results.

5-66. Coaxial resonator. A half-wavelength coaxial cavity is constructed of perfectly conducting walls and has dimensions $a = 0.4$ cm, $b = 1.6$ cm, and $d = 10$ cm. The space between the two conductors of the coaxial cavity is filled with two different dielectrics: the first 6 cm of its length is filled with dielectric 1 ($\epsilon_{1r}' \simeq 2.2$, $\epsilon_{1r}'' \simeq 0.001$) while the remaining 4 cm is filled with dielectric 2 ($\epsilon_{2r}' \simeq 2.2$, $\epsilon_{2r}'' \simeq 0.005$). Find the resonant frequency and the quality factor of this cavity.

5-67. Group velocity for the TE_{10} mode. Following a procedure similar to that given in Section 4.3.3, show that the group velocity is the velocity of energy transport for the TE_{10} mode in a rectangular waveguide.

---

[65]D. P. Howson and G. Hamer, Glass cavity resonators for low microwave frequencies, *Electronics Letters,* 31(10), pp. 811–812, May 1995.

[66]J. Baker-Jarvis, R. G. Geyer, J. H. Grosvenor, Jr., M. D. Janezic, C. A. Jones, B. Riddle, C. M. Weil, J. Krupka, Dielectric characterization of low-loss materials: A comparison of techniques, *IEEE Trans. Dielectrics and Electrical Insulation,* 5(4), pp. 571–577, August 1998.

CHAPTER 6

Field–Matter Interactions and Elementary Antennas

In Chapter 2 we considered electromagnetic wave propagation in an *unbounded, simple,* and *source-free* medium. In Chapters 3, 4, and 5 we essentially removed the first of these three conditions, considering how waves reflect and refract in the presence of boundaries, and how they can be guided by conducting or dielectric structures. In particular, we saw in Chapters 4 and 5 that the presence of boundaries necessitates that the wave phase velocity be a function of frequency, leading to waveguide dispersion. In this chapter, we revert to considering the case of an *unbounded* medium (i.e., no boundaries) but one that is not *simple,* and illustrate that material properties alone can also lead to dispersion effects. The chapter ends with an elementary discussion of *sources* of electromagnetic waves: namely, elementary antennas.

Electromagnetic fields interact with matter via their influence on *charged particles* (electrons and ions), whether these particles are free to move about (e.g., electrons in a conductor or electrons and ions in a low-density gas) or are tightly bound (e.g., atoms or molecules of a dielectric). In Sections 6.1 and 6.2 we consider the

414

interaction between *freely mobile charged particles* and electromagnetic fields. To illustrate the basic physical principles in the context of an interesting and important example, we discuss the case of wave propagation in a relatively dilute ionized gas in which particles can move freely. It should be noted, however, that the dynamics of charged particles studied in Sections 6.1 and 6.2 are also equally applicable to current carriers in solids, such as electrons and holes in semiconductors. We shall see that the behavior of the freely mobile particles under the influence of the wave fields can be represented in terms of an effective permittivity ϵ_{eff}, which is in general a function of frequency, so an ionized gas is a good example of a medium that exhibits *material dispersion.* When the effects of a static magnetic field are included, an ionized gas behaves further as an *anisotropic* medium, for which the effective permittivity needs to be expressed as a *tensor.*[1]

Interactions of electromagnetic waves with solid, liquid, or densely gaseous media are in general more difficult to treat on the microscopic level (i.e., by describing the motions of individual particles), because material properties play a significant role. At one level, we have already accounted for the interaction of such media with electromagnetic fields by describing their behavior in terms of *macroscopic* parameters ϵ, σ, and μ. These macroscopic parameters represent the response of the free or bound electrons within the material to the fields on a *microscopic* scale. To the degree that these parameters can be considered to be simple (real or complex) constants, the formulations used in previous chapters are adequate. However, in many important applications, the microscopic response of the particles to the fields becomes complicated, so the macroscopic parameters ϵ, μ, and σ need to be formulated as functions of frequency, intensity, and orientation of the electromagnetic fields. Although a general study of such effects are well beyond the scope of this book, we devote Section 6.3 to the discussion of material dispersion in dielectrics and conductors (for which ϵ and σ depend on frequency) and Section 6.4 to wave propagation in magnetically anisotropic media or ferrites (for which μ depends on the orientation of the applied electromagnetic fields and is a tensor).

Maxwell's equations indicate that electromagnetic waves originate in sources, which are either arrangements of electric charge ($\tilde{\rho}$), charges in motion ($\overline{\mathcal{J}}$), or both. Neither static distributions of charge ($\partial\tilde{\rho}/\partial t = 0$) nor steady currents ($\nabla \cdot \overline{\mathcal{J}} = 0$) can generate electromagnetic waves. Only electric charges that accelerate or electric currents that change in time can *radiate* electromagnetic energy. The simplest form of electromagnetic sources, or *antennas,* consists of arrangements of wires carrying alternating currents. Accordingly, Section 6.5 provides a brief discussion of radiation of electromagnetic waves by simple elementary wire antennas.

---

[1]In other words, applied electric field $\overline{\mathcal{E}}$ in a given direction (e.g., x direction) may produce electric displacement $\overline{\mathcal{D}}$ in another direction (e.g., y direction), so that ϵ must be written in a matrix, or tensor, form.

6.1 CHARGED-PARTICLE MOTION DRIVEN BY ELECTRIC AND MAGNETIC FIELDS

Our starting point is the Lorentz force equation[2] describing the electrical and magnetic forces acting on a charged particle of rest mass m_0 and charge q:

$$\overline{\mathcal{F}}_{elec} + \overline{\mathcal{F}}_{mag} = q\overline{\mathcal{E}} + q\tilde{\mathbf{v}} \times \overline{\mathcal{B}} = q[\overline{\mathcal{E}} + \tilde{\mathbf{v}} \times \overline{\mathcal{B}}] \qquad [6.1]$$

where $\overline{\mathcal{F}}_{elec}$ and $\overline{\mathcal{F}}_{mag}$ are, respectively, the electric and magnetic Lorentz forces. For freely mobile elementary particles such as an electron, proton, or even heavier ions (e.g., O^+), the electromagnetic forces on these particles are generally much stronger than other forces (e.g., gravity) for reasonable field intensities (e.g., an electric field of > 0.1 nV-m^{-1} for electrons). Thus, the acceleration of the free charge is dominated by the electromagnetic force, so its motion is described by Newton's second law as

$$\overline{\mathcal{F}} = \frac{d(m\tilde{\mathbf{v}})}{dt} = m\frac{d^2\mathbf{r}}{dt^2}$$

where $\tilde{\mathbf{v}}$ and $\mathbf{r}$ are the velocity and position vectors of the charged particle respectively, and $m = m_0(1 - \tilde{v}^2/c^2)^{-1/2}$ is the mass of the charged particle to allow for a relativistic correction, with m_0 being the rest mass and $\tilde{v} = |\tilde{\mathbf{v}}|$. The notation $\tilde{\mathbf{v}}$ is used for the velocity to distinguish it from the time-harmonic velocity $\mathbf{v}$, although we do not always make such a distinction for position variables $\mathbf{r}$, x, y, or z.

In this chapter we exclusively consider[3] nonrelativistic motion ($\tilde{v} \ll c$), so $m \simeq m_0$ is a constant, and we thus have

$$q[\overline{\mathcal{E}} + \tilde{\mathbf{v}} \times \overline{\mathcal{B}}] = m_0\frac{d\tilde{\mathbf{v}}}{dt} = m_0\frac{d^2\mathbf{r}}{dt^2} \qquad [6.2]$$

as the equation of motion for a charged particle under the influence of external electric and magnetic fields. Accordingly, we drop the subscript distinguishing mass from rest mass and simply denote the particle mass as m.

---

[2]H. A. Lorentz was the first to study the motion of electrons in the presence of electric and magnetic fields. In particular, he applied equation [6.1] to put forth an explanation for the Zeeman effect, involving the splitting of spectral lines in the presence of a magnetic field [P. Zeeman, Doublets and triplets in the spectrum produced by external magnetic forces, *Phil. Mag.,* (5)43, p. 226, 1897.]. Zeeman was a student of Lorentz, and the two later shared the Nobel Prize in Physics in 1902. Lorentz's famous theory of electrons is described in the extensive compilation of lectures that he gave at Columbia University in the spring of 1906: *The Theory of Electrons and Its Applications to the Phenomena of Light and Radiant Heat,* Dover Publications, Inc., 1952.

[3]By the same token, we neglect any radiation produced by the acceleration of charged particles. At nonrelativistic velocities, such radiation is quite negligible; the radiated electric field at a distance R from the particle is proportional to $q^2a^2/(c^2R^2)$, where q is the charge of the particle, c is the speed of light in free space, and a is the acceleration. For a discussion at an appropriate level, see Chapter 7 of J. B. Marion, *Classical Electromagnetic Radiation,* Academic Press, New York, 1965.

Lorentz, Hendrik Antoon   *(Dutch physicist, b. July 18, 1853, Arnhem, Netherlands; d. February 4, 1928, Haarlem, Netherlands) Lorentz attended Leiden University, obtaining his doctor's degree, summa cum laude, in 1875 and returned three years later as professor of theoretical physics, a post he held until his death. Lorentz's doctoral thesis dealt with the theory of electromagnetic radiation, which Maxwell [1831–1879] had advanced a little over a decade before. Lorentz refined the theory to take account of the manner of the reflection and refraction of light, points concerning which Maxwell's own work had been somewhat unsatisfactory. He went further in his search into the implications of Maxwell's work.*

According to Maxwell, electromagnetic radiation was produced by the oscillation of electric charges. Hertz [1857–1894] showed this to be true for radio waves, which in 1887 he formed by causing electric charges to oscillate. But if light was an electromagnetic radiation after the fashion of radio waves, where were the electric charges that did the oscillating? By 1890 it seemed quite likely that the electric current was made up of charged particles, and Lorentz thought it quite possible that the atoms of matter might also consist of charged particles. Lorentz suggested then that it was the charged particles within the atom that oscillated, producing visible light. If this was so, then placing a light in a strong magnetic field ought to affect the nature of the oscillations and therefore of the wavelength of the light emitted. This was demonstrated experimentally in 1896 by Zeeman [1865–1943], a pupil of Lorentz who shared with him the 1902 Nobel Prize in physics.

Lorentz also tackled the negative results of the experiment conducted by Michelson [1852–1931] and Morley [1838–1923] and came to the same conclusion as FitzGerald [1851–1901]. He too postulated that there are contractions of length with motion. It seemed to him, further, that the mass of a charged particle such as the electron depends on its volume; the smaller the volume, the greater the mass. Since the Lorentz–FitzGerald contraction reduced the volume of an electron as it sped along and reduced it the more as it moved more rapidly, it must also increase its mass with velocity. At 161,000 miles a second, the mass of an electron is twice its "rest mass," according to the Lorentz formulation, and at 186,282 miles a second (the velocity of light) the mass must be infinite, since the volume becomes zero. This was another indication that the velocity of light in a vacuum is the greatest velocity at which any material object can travel.

By 1900, mass measurements on speeding subatomic particles did indeed show that Lorentz's equation describing how mass varied with velocity was followed exactly. And in 1905 Einstein [1879–1955] advanced his special theory of relativity, from which the Lorentz–FitzGerald contraction could be deduced and from which it could be shown that the Lorentz mass increase with velocity held not only for charged particles but for all objects, charged and uncharged.

In later life Lorentz ably supervised the enclosure of the Zuider Zee, an ambitious Dutch project to make more agricultural land out of a shallow basin of the sea. [Adapted with permission from I. Asimov, Biographical Encyclopedia of Science and Technology, *Doubleday, 1982.]*

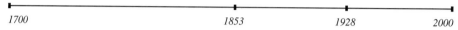

| 1700 | 1853 | 1928 | 2000 |

Equation [6.2] is applicable in a wide variety of situations. In the simplest case, $\overline{\mathscr{E}}$ and $\overline{\mathscr{B}}$ are externally applied, and the number of charged particles is small enough that their collective motion does not alter the fields (by, for example, creating currents $\overline{\mathscr{J}}$, which in turn generate magnetic fields through $\nabla \times \overline{\mathscr{H}} = \overline{\mathscr{J}}$, which in turn generate electric fields and further modify the particle motion). In such a case, $\overline{\mathscr{F}}$ is a specified function and $\tilde{\mathbf{v}}(t)$ and $\mathbf{r}(t)$ can be obtained by successive integration of [6.2]. The trajectory of the particle (i.e., the path it traces in physical space) can then be described by eliminating time from the position relations. In general, when the charged particle density is sufficiently high, the moving charges constitute electric current, which in turn influences the electromagnetic field, so the equations of motion must be solved simultaneously with Maxwell's equations. Proper treatment of the electromagnetic wave–charged particle interaction is the subject of *plasma physics* and is well beyond the scope of this book. Nevertheless, we provide an introduction to the fundamental principles involved, in the context of the propagation of electromagnetic waves in an ionized medium discussed in Section 6.2.

It is often useful to decompose [6.2] in component form in different coordinate systems. For rectangular coordinates we have

$$m\left[\hat{\mathbf{x}}\frac{d\tilde{v}_x}{dt} + \hat{\mathbf{y}}\frac{d\tilde{v}_y}{dt} + \hat{\mathbf{z}}\frac{d\tilde{v}_z}{dt}\right] = q[\hat{\mathbf{x}}(\mathscr{E}_x + \tilde{v}_y\mathscr{B}_z - \tilde{v}_z\mathscr{B}_y) + \hat{\mathbf{y}}(\mathscr{E}_y + \tilde{v}_z\mathscr{B}_x - \tilde{v}_x\mathscr{B}_z)$$
$$+ \hat{\mathbf{z}}(\mathscr{E}_z + \tilde{v}_x\mathscr{B}_y - \tilde{v}_y\mathscr{B}_x)] \qquad [6.3]$$

whereas in cylindrical coordinates (i.e., $\hat{\mathbf{r}}$, $\hat{\boldsymbol{\phi}}$, and $\hat{\mathbf{z}}$) we have[4]

$$m\left[\overbrace{\hat{\mathbf{r}}\frac{d\tilde{v}_r}{dt} + \hat{\boldsymbol{\phi}}\frac{d\tilde{v}_\phi}{dt} + \hat{\mathbf{z}}\frac{d\tilde{v}_z}{dt}}^{\text{Rotational}} - \overbrace{\hat{\mathbf{r}}r\omega^2}^{\text{Centrifugal}} + \overbrace{\hat{\boldsymbol{\phi}}\tilde{v}_r\omega}^{\text{Coriolis}}\right] = \qquad [6.4]$$

$$q[\hat{\mathbf{r}}(\mathscr{E}_r + \tilde{v}_\phi\mathscr{B}_z - \tilde{v}_z\mathscr{B}_\phi) + \hat{\boldsymbol{\phi}}(\mathscr{E}_\phi + \tilde{v}_z\mathscr{B}_r - \tilde{v}_r\mathscr{B}_z) + \hat{\mathbf{z}}(\mathscr{E}_z + \tilde{v}_r\mathscr{B}_\phi - \tilde{v}_\phi\mathscr{B}_r)]$$

where the first three terms represent the *rotational* acceleration in the frame of reference rotating with the particle at an angular velocity ω with $\tilde{v}_\phi = r\omega$, the fourth term $mr\omega^2 = m(\omega r)^2/r = m\tilde{v}_\phi^2/r$ is the *centrifugal* force due to the rotation of the particle,

---

[4]For a derivation of [6.4], see Chapter 9 of E. C. Jordan and K. G. Balmain, *Electromagnetic Waves and Radiating Systems,* Prentice Hall, Englewood Cliffs, New Jersey, 1968.

and the fifth term is the *Coriolis*[5] force due to $\tilde{v}_r$, representing acceleration in the ϕ direction due to motion in the r direction.

As a charged particle moves from point 1 to point 2 in the presence of external fields, the amount of energy expended (or work done) by the fields acting on the particle can be expressed as

$$W_{12} = \int_1^2 \overline{\mathscr{F}} \cdot d\mathbf{l} = m \int_1^2 \frac{d\tilde{\mathbf{v}}}{dt} \cdot d\mathbf{l} = m \int_1^2 \frac{d\tilde{v}}{dt} \, dl = m \int_1^2 \tilde{v} \frac{d\tilde{v}}{dl} \, dl$$

$$= \tfrac{1}{2} m \int_{\tilde{v}_1}^{\tilde{v}_2} d(\tilde{v}^2) = \tfrac{1}{2} m(\tilde{v}_2^2 - \tilde{v}_1^2)$$

where the force $\overline{\mathscr{F}}$ exerted on the particle by the external fields is in general given by [6.1]; $\tilde{v}_1$ and $\tilde{v}_2$ are, respectively, the velocities of the particle at points 1 and 2; and we have used the facts that $d\mathbf{l}$ and $\tilde{\mathbf{v}}$ are in the same direction (since we are integrating along the trajectory that the particle follows in space) and that by the chain rule we have $(d\tilde{v}/dt) = [(d\tilde{v}/dl)(dl/dt)] = \tilde{v}[d\tilde{v}/dl]$. We see that the work done by the field in moving the particle from point 1 to point 2 is simply equal to the difference in kinetic energy of the particle at points 1 and 2. If the force $\overline{\mathscr{F}}$ has both electric and magnetic components, we then have

$$W_{12} = q \int_1^2 [\overline{\mathscr{E}} + \tilde{\mathbf{v}} \times \overline{\mathscr{B}}] \cdot d\mathbf{l}$$

However, since $d\mathbf{l}$ and $\tilde{\mathbf{v}}$ are in the same direction, we have

$$(\tilde{\mathbf{v}} \times \overline{\mathscr{B}}) \cdot d\mathbf{l} = d\mathbf{l} \cdot (\tilde{\mathbf{v}} \times \overline{\mathscr{B}}) = \overline{\mathscr{B}} \cdot (d\mathbf{l} \times \tilde{\mathbf{v}}) = 0$$

where we have used the vector identity $\mathbf{A} \cdot (\mathbf{B} \times \mathbf{C}) \equiv \mathbf{B} \cdot (\mathbf{C} \times \mathbf{A}) \equiv \mathbf{C} \cdot (\mathbf{A} \times \mathbf{B})$. Thus,

$$W_{12} = q \int_1^2 \overline{\mathscr{E}} \cdot d\mathbf{l}$$

In other words, the work done on the particle is entirely due to the electric field. The magnetic force is always perpendicular to the motion of the particle (i.e., to $\tilde{\mathbf{v}}$)

---

[5]Coriolis force derives its name from the French physicist Gaspard-Gustave de Coriolis [1792–1843], who in 1835 undertook experimental and theoretical studies of motion on a spinning surface. Coriolis force appears to act on a body when it is moving relative to a rotating reference frame. It acts to deflect the body sideways, because in a rotating system, points farther away from the rotation axis have higher linear speeds. A common manifestation of the Coriolis force is the fact that ice skaters accelerate their spin by pulling their arms in. The Coriolis effect affects the movement of air masses and influences weather. Since the earth spins as a whole, a point at the equator must travel a distance of $\sim$40,000 km in 24 hours, whereas points farther north or south travel in smaller circles and so move more slowly. A body such as a mass of air moving away from the equator will therefore be deflected eastward. Due to this effect, converging flows of air or water (e.g., cyclones) take up circular motions that are counterclockwise in the Northern Hemisphere and clockwise in the Southern Hemisphere.

and thus cannot do work on a charged particle (i.e., the kinetic energy of a charged particle cannot increase from a magnetic force alone).

We now consider the motion of charged particles under the influence of externally applied electric and magnetic fields. We consider different cases in the order of increasing complexity, starting with particle motion in an electric field.

6.1.1   Motion of a Charged Particle in an Electric Field

The motion of charged particles under the influence of an electric field is relatively simple, since the particle velocity is in the same direction as the applied electric field. An example of such motion, involving deflection of electrons in a cathode-ray oscilloscope, is provided below in Example 6-1, where the deflection is achieved by means of a *stationary* (constant in time) and *uniform* (constant in space) electric field between the plates. When the electric field is a constant, the motion equation [6.2] can be directly integrated twice to obtain the trajectory of the particle, as is done in Example 6-1. When the electric field varies with time or space, the motion of the charged particles would be more complicated, and in general it might not be possible to obtain an analytical solution for the particle trajectories. A relatively simple example is provided in Example 6-2.

Example 6-1:   Deflection of charged particles.   Cathode-ray oscilloscopes use an electrostatic deflection system as shown in Figure 6.1. Electrons emerge from the cathode, each carrying a charge of q_e, where $q_e \simeq -1.6 \times 10^{-19}$ C, with an initial velocity of $\tilde{\mathbf{v}} = \hat{\mathbf{z}} v_0$. Two metal plates of length l are separated by a gap h. A uniform electric field $\overline{\mathscr{E}}$ pointing in the $-x$ direction ($\overline{\mathscr{E}} = -\hat{\mathbf{x}} E_0$, with $E_0 > 0$) is maintained between the plates, as shown. The electric field is assumed to exist only in the region $0 < z < l$. Find the deflection d of the electron on the screen at $z = l + L$ in terms of the electric field intensity E_0.

> **Solution:**   Note that we proceed by ignoring gravitational effects, since even a minuscule electric field exerts a force much larger than gravity does on an electron, for which the electric force $q_e |\overline{\mathscr{E}}|$ is comparable to the gravity force $m_e g$ only if the electric field is $< 10^{-10}$ V-m^{-1}, where $m_e \simeq 9.11 \times 10^{-31}$ kg is the electron mass. Since the electric force is only in the x direction and is given

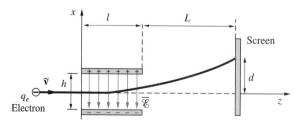

FIGURE 6.1.   Electrostatic deflection system. Electrostatic deflection system for an electron (q_e) initially moving with a speed $\tilde{\mathbf{v}} = \hat{\mathbf{z}} v_0$.

by $\overline{\mathscr{F}} = q\overline{\mathscr{E}} = q_e(-\hat{\mathbf{x}}E_0) = \hat{\mathbf{x}}|q_e|E_0$, we have

$$\overline{\mathscr{F}} = q_e\overline{\mathscr{E}} \quad \rightarrow \quad m_e\frac{d\tilde{v}_x}{dt} = |q_e|E_0 \quad \rightarrow \quad \tilde{v}_x = \frac{|q_e|}{m_e}E_0t + C_1$$

where $C_1 = 0$, since $\tilde{v}_x = 0$ at $t = 0$, which is the time when the electron is at position $z = 0$ (i.e., just entering the region between the deflection plates). Note that the motion of the electron in the z direction is not affected, since it experiences no force in that direction. Thus, $\tilde{v}_z$ remains equal to v_0. We can find the displacement x by integrating the above,

$$x = \frac{1}{2}\frac{|q_e|}{m_e}E_0t^2 + C_2$$

where $C_2 = 0$, since $x = 0$ at $t = 0$; note that we lose no generality by assuming that the electron starts at $z = 0$ at $t = 0$. Since $t = z/v_0$, the displacement and velocity of the electron at the point when it leaves the region between the plates (i.e., $z = l$) are

$$x(z = l) = \frac{1}{2}\frac{|q_e|}{m_e}E_0\left(\frac{l}{v_0}\right)^2 \quad \text{and} \quad \tilde{v}_x(z = l) = \frac{|q_e|}{m_e}E_0\frac{l}{v_0}$$

Since there is no acceleration outside the plates, the velocity $\tilde{v}_x$ remains constant [$\tilde{v}_x = \tilde{v}_x(l)$ between $l < z < l + L$], and the additional deflection is

$$x(l + L) - x(l) = \tilde{v}_x(l)\frac{L}{v_0} = \frac{|q_e|}{m_e}E_0\frac{l}{v_0}\frac{L}{v_0}$$

The total deflection on the screen is then given as

$$d = x(l + L) = \frac{|q_e|}{m_e}E_0\left(\frac{l}{v_0}\right)^2\left(\frac{1}{2} + \frac{L}{l}\right)$$

Considering practical values of $v_0 = 10^7$ m-s^{-1}, $l = 5$ cm, $L = 30$ cm, and $E_0 = 1$ kV-m^{-1}, and noting that $|q_e|/m_e \simeq 1.6 \times 10^{-19}$ C/$(9.11 \times 10^{-31}$ kg$) \simeq 1.757 \times 10^{11}$ C-kg^{-1} for an electron, we find $d \simeq 2.85$ cm.

Example 6-2:   Particle near two stationary charges.   Consider the charge configuration shown in Figure 6.2a, consisting of two stationary positive point charges Q_1 and Q_2 respectively at $x = -a$ and $x = a$, and a positive point test charge q located at $x = x_0$ at $t = 0$, which is initially at rest and free to move. Find the trajectory $x(t)$ (i.e., its position as a function of time) followed by the test particle.

Solution: The force $\overline{\mathscr{F}}_q$ acting on the test particle is the superposition of the forces due to the two charges. At any point x, the total electric field due to the two charges Q_1 and Q_2 is

$$\overline{\mathscr{E}} = \frac{Q_1\hat{\mathbf{x}}}{4\pi\epsilon_0(x + a)^2} - \frac{Q_2\hat{\mathbf{x}}}{4\pi\epsilon_0(x - a)^2}$$

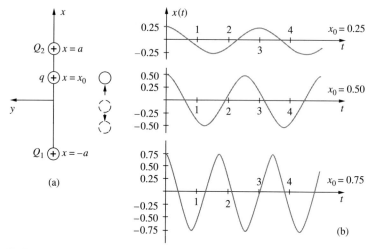

FIGURE 6.2. **A system of three point charges.** (a) Charge configuration and definition of variables. Charges Q_1 and Q_2 are stationary, whereas charge q, initially at $x = x_0$, is free to move. (b) Solutions for the particle trajectory $x(t)$ for $a = 1$, $Q_1 = Q_2 = Q$, and $qQ/(4\pi\epsilon_0 m_q) = 1$ for three different values of x_0.

Since the electric field is in the x direction at every point along the x axis, the particle q can move only along the x axis. Its equation of motion can be written as the x component of [6.3]

$$\hat{\mathbf{x}} m_q \frac{d^2 x}{dt^2} = q\overline{\mathscr{E}} = \frac{qQ_1\hat{\mathbf{x}}}{4\pi\epsilon_0(x + a)^2} - \frac{qQ_2\hat{\mathbf{x}}}{4\pi\epsilon_0(x - a)^2}$$

which is a nonlinear second-order differential equation for which an analytical solution does not exist. To see this better, we can assume $Q_1 = Q_2 = Q$ and rearrange terms as

$$\frac{d^2 x}{dt^2} = \frac{qQ}{4\pi\epsilon_0 m_q} \left[\frac{1}{(x + a)^2} - \frac{1}{(x - a)^2} \right]$$

The numerical solution of the above equation is quite straightforward using any of the modern numerical analysis packages[6]; results for $qQ/(4\pi\epsilon_0 m_q) = 1$ and $a = 1$ are shown in Figure 6.2b. We note that the solution $x(t)$ for the trajectory has an oscillatory character, with the particle moving back and forth between $x = \pm x_0$, where $x = x_0$ is its initial position at $t = 0$. The period of the oscilla- tion depends nonlinearly on the initial position x_0, being shorter for higher val-

---

[6]In producing the solutions plotted in Figure 6.2b, the following simple *Mathematica* statement was used:

```
NDSolve[{ y''[x] + (1/(1-y[x])²) - (1/(1+y[x])²)==0,
         y[0] == 0.75, y'[0] == 0}, y[x], {x,0,5}]
```

ues of x_0. The shape of the solution function $x(t)$ also varies with x_0, being more like a sinusoidal function for smaller x_0 and having a shape closer to a triangular waveform for higher x_0. Note that the peak-to-peak amplitude of the oscillatory motion decreases as x_0 is reduced—as expected, since with $Q_1 = Q_2 = Q$ and $x_0 = 0$, the test charge q does not move, and therefore there is no oscillatory motion.

The nature of the solutions in Figure 6.2b can be physically interpreted as follows. When the particle is at an initial point $x = x_0$, it experiences an acceleration that pushes it towards the origin (the equilibrium point). However, because the particle has inertia, it moves past the origin to the other side and begins to experience a net force in the opposite direction, again toward the origin. Since there are no losses in the system to damp its motion, the particle oscillates back and forth. For higher values of x_0, the particle is closer to one of the charges and thus experiences a larger initial net force (and thus acceleration) and completes its traverse at higher speed (i.e., shorter period). Although it does cover a larger distance in doing so, the nonlinear dependence of the force on distance results in a smaller oscillation period.

This simple example demonstrates how complicated the motion of charged particles can be when the driving electric field (in this case imposed by the two stationary charges) is nonuniform (i.e., a function of spatial coordinates). If the electric field were, in addition, time-dependent (a circumstance that could be modeled, for example, by simply making the charge values Q each functions of time), the particle motion would be even more complicated. Nevertheless, the formulation used above—namely, the numerical solution of the equations of motion—would be entirely adequate to determine the particle trajectories even in the case of a nonuniform and time-varying electric field as long as any time variations of the electric field are relatively slow, so that they do not lead to the production of significant magnetic fields via [1.3], which may then influence the particle motion through [6.1].

6.1.2   Motion of a Charged Particle in a Magnetic Field

We now consider the motion of a charged particle in a steady magnetic field. Consider a constant magnetic field in the z direction (i.e., $\overline{\mathscr{B}} = \hat{\mathbf{z}}B_0$) and a charged particle q with a constant initial velocity $\tilde{v}_\phi$ in the ϕ direction. Since all the other components of $\tilde{\mathbf{v}}$ and $\overline{\mathscr{B}}$ are zero, the motion equation [6.4] becomes

$$\hat{\mathbf{r}}q\tilde{v}_\phi\mathscr{B}_z = -\hat{\mathbf{r}}m_q r\omega^2 \quad \rightarrow \quad q\tilde{v}_\phi B_0 + m_q r\omega^2 = 0$$

Noting that $\tilde{v}_\phi = v_0 = r\omega$, we have

$$qB_0 + m_q\omega = 0 \quad \rightarrow \quad \omega \equiv \omega_c = -\frac{qB_0}{m_q}$$

where the particular angular frequency ω_c with which the charged particle rotates in its circular trajectory is defined as the *cyclotron frequency* or *gyrofrequency*.

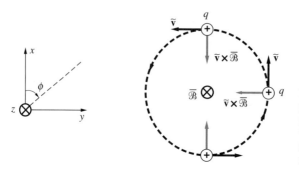

FIGURE 6.3. **Motion of a particle in a magnetic field.** A particle with positive charge q and with a velocity $\tilde{\mathbf{v}}$ experiences a force of $q\tilde{\mathbf{v}} \times \overline{\mathscr{B}}$ in the presence of a magnetic field $\overline{\mathscr{B}} = \hat{\mathbf{z}}B_0$.

According to the above, a charged particle with a constant linear velocity $\tilde{v}_\phi = v_0 = r\omega_c$ executes a circular motion around the magnetic field direction. Note that in a right-handed coordinate system, with ϕ measured from the x axis as shown in Figure 6.3, positive angular velocity ω corresponds to clockwise rotation when viewed in the positive z direction as in Figure 6.3. Thus, a negative value of ω simply indicates that the rotation is in the counterclockwise direction. To appreciate the above result physically, consider the coordinate system and the forces on the particle (assumed to be a positive charge q) as shown in Figure 6.3 at different points along its orbit. It is clear that the particle experiences a $q\tilde{\mathbf{v}} \times \overline{\mathscr{B}}$ force directed inward (towards the center of the circular path) at all times, which balances the centrifugal force, resulting in a circular motion. For a z-directed magnetic field, electrons rotate in the right-hand sense (i.e., have a positive value of ω_c); in other words, if the right thumb points in the direction of the magnetic field, the fingers curl in the direction of the electron motion.

The radius of the circular trajectory can be determined by considering the fact that the $q\tilde{\mathbf{v}} \times \overline{\mathscr{B}}$ force is balanced by the centrifugal force, so we have

$$\frac{m_q v_0^2}{r} = q\tilde{\mathbf{v}} \times \overline{\mathscr{B}} = qv_0 B_0 \quad \rightarrow \quad r_c = \frac{m_q v_0}{qB_0}$$

where r_c is the *gyroradius*. Note that the magnitude of the particle velocity (i.e., $|\tilde{\mathbf{v}}| = v_0$) remains constant, since the magnetic field force is at all times perpendicular to the motion. The magnetic field cannot change the kinetic energy of the particle; however, it does change the direction of its momentum (i.e., its velocity). It is important to note that for a given charged particle (i.e., for fixed q/m_q) the gyrofrequency ω_c of a charged particle does not depend on its velocity (or kinetic energy) and is only a function of the intensity of the magnetic field. Particles with higher velocities (and thus higher energies) orbit in circles with larger radii but complete one revolution in the same time as particles with lower velocities which orbit over smaller circles. Particles with larger masses also orbit in circles with larger radii; however, they complete one revolution in a longer time compared with smaller masses. The gyromotion of a charged particle in a magnetic field is the basis for the cyclotron, used as a particle accelerator or a charged particle separator. Similarly, a magnetron (see Example 6-6), a common source of microwave radiation for microwave ovens and radar systems, emits radiation at a frequency equal to the gyrofrequency of electrons in a vacuum chamber.

Example 6-3 illustrates the use of the circular motion of a charged particle in a magnetic deflection system.

Example 6-3:   Magnetic deflection system.   A static magnetic field can be used to produce electron beam deflection in a cathode-ray tube. As shown in Figure 6.4, the magnetic field $\overline{\mathcal{B}} = \hat{y}B_0$ is approximately uniform over the length of the deflection coil and is directed normal to the plane of the paper. As was the case in Example 6-1, electrons carrying charge q_e each emerge from the cathode with an initial velocity of $\tilde{v} = \hat{z}v_0$. Determine the deflection d on the screen in terms of the given parameters.

Solution: Assuming that the electrons enter at right angles to the magnetic field region ($\overline{\mathcal{B}} = \hat{y}B_0$ for $0 < x < l$) with a velocity $\tilde{v} = \hat{z}v_0$, they follow a circular path with radius

$$r_c = \frac{m_e v_0}{|q_e|B_0}$$

where q_e is the electronic charge. Referring to the geometry in Figure 6.4, we have

$$d_0 = r_c - \sqrt{r_c^2 - l^2}$$

$$d_1 = L\tan\theta = \frac{lL}{\sqrt{r_c^2 - l^2}}$$

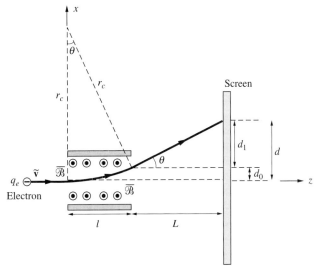

FIGURE 6.4.   **Magnetic deflection system.**

and

$$d = d_0 + d_1$$

where we have used $\tan \theta = l/(r_c - d_0) = l/\sqrt{r_c^2 - l^2}$. For small angles of deflection θ, we have $r_c \gg l$, so that

$$\sqrt{r_c^2 - l^2} = r_c\sqrt{1 - \left(\frac{l}{r_c}\right)^2} \simeq r_c\left[1 - \frac{1}{2}\left(\frac{l}{r_c}\right)^2\right]$$

and thus

$$d \simeq \frac{l(L + l/2)}{r_c} = \frac{l(L + l/2)|q_e|B_0}{m_e v_0}$$

It is interesting to note that the deflection is directly proportional to the applied magnetic field B_0, just as the deflection was linearly proportional to the applied electric field in the electric field deflection system in Example 6-1. Although the above solution is for a static magnetic field B_0, the result is also valid for quasistatic cases in which the magnetic field varies slowly enough that it does not change significantly over times comparable to the gyroperiod $T_c = 2\pi/\omega_c = 2\pi m_e/(|q_e|B_0)$.

In cases when the magnetic field $\overline{\mathscr{B}}$ is nonuniform (i.e., variable in space), time-varying, or both, the particle motion becomes more complicated, and in general it might not be possible to obtain an analytical solution for the trajectories. In Example 6-4 we consider a relatively simple example of a charged particle motion in a nonuniform magnetic field.

Example 6-4: Charged particle near a current-carrying wire. Consider a particle of charge q and mass m moving in the neighborhood of an infinitely long, straight, filamentary wire carrying a constant current I. At time $t = 0$, the particle is at a radial distance of r_0 from the wire and has a velocity v_0 parallel to the wire. Determine the trajectory of the particle.

Solution: With no loss of generality, we can take the wire to be oriented along the z direction, in which case the magnetic field around the wire is $\overline{\mathscr{B}} = \hat{\phi}\mu_0 I/(2\pi r)$, as shown in Figure 6.5a. Note that the charged particle experiences no magnetic force perpendicular to the plane of the paper, so its motion is confined to the plane of the paper. The initial particle velocity is parallel to the wire, i.e., is in the z direction ($\tilde{\mathbf{v}} = \hat{\mathbf{z}}v_0$) producing a $q\tilde{\mathbf{v}} \times \overline{\mathscr{B}}$ force in the $-\hat{\mathbf{r}}$ direction. From [6.4], the equation of motion is

$$q(\tilde{\mathbf{v}} \times \overline{\mathscr{B}}) = -q\tilde{v}_z \mathscr{B}_\phi \hat{\mathbf{r}} = \hat{\mathbf{r}}m\frac{d\tilde{v}_r}{dt}$$

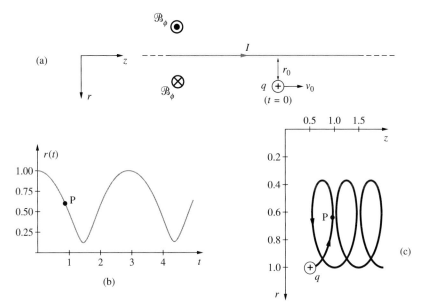

FIGURE 6.5.   Particle near a current-carrying wire. (a) Coordinate system. (b) Variation of the radial distance of the particle from the wire as a function of time. Note that we have used the values $r_0 = 1$ m, $z_0 = 0.5$ m, and $v_0 = 0.5$ m-s^{-1}. (c) The particle trajectory for the same r_0, z_0, and v_0 values.

In addition, the induced radial motion ($\tilde{v}_r$) will produce a force of $q\tilde{\mathbf{v}} \times \overline{\mathcal{B}} = q\tilde{v}_r \hat{\mathbf{r}} \times \hat{\boldsymbol{\phi}} \mathcal{B}_\phi$ that is in the $-z$ direction because $\tilde{v}_r < 0$. In other words, we have

$$m\frac{d\tilde{v}_z}{dt} = q\tilde{v}_r \mathcal{B}_\phi = \frac{q\mu_0 I}{2\pi r}\tilde{v}_r \qquad \rightarrow \qquad \frac{d\tilde{v}_z}{dt} = \frac{q}{m}\frac{\mu_0 I}{2\pi r}\frac{dr}{dt}$$

which can be integrated to find

$$\tilde{v}_z = \frac{q}{m}\frac{\mu_0 I}{2\pi}\ln r + C$$

To determine the constant C, we can use the initial condition that $\tilde{v}_z(r = r_0) = v_0$, which results in

$$v_0 = \frac{q}{m}\frac{\mu_0 I}{2\pi}\ln r_0 + C \qquad \rightarrow \qquad C = v_0 - \frac{q}{m}\frac{\mu_0 I}{2\pi}\ln r_0$$

so that its velocity in the z direction at an arbitrary radial distance r is

$$\tilde{v}_z(r) = \frac{q}{m}\frac{\mu_0 I}{2\pi}\ln r - \frac{q}{m}\frac{\mu_0 I}{2\pi}\ln r_0 + v_0$$

$$= \frac{q}{m}\frac{\mu_0 I}{2\pi}\ln\frac{r}{r_0} + v_0$$

To find the radial acceleration at a radial distance r, we can use the first equation

$$-q\tilde{v}_z\mathcal{B}_\phi = m\frac{d\tilde{v}_r}{dt} \quad \rightarrow \quad -q\tilde{v}_z\mathcal{B}_\phi = ma_r \quad \rightarrow \quad a_r = -\frac{q\tilde{v}_z\mathcal{B}_\phi}{m} = -\frac{q\tilde{v}_z}{m}\frac{\mu_0 I}{2\pi r}$$

which gives

$$a_r(r) = -\frac{q}{m}\frac{\mu_0 I}{2\pi r}\left(\frac{q\mu_0 I}{m2\pi}\ln\frac{r}{r_0} + v_0\right)$$

Note that the actual motion of the particle in this nonuniform magnetic field is in fact quite complicated. The radial motion is described by the nonlinear differential equation

$$\frac{d\tilde{v}_r}{dt} = \frac{d^2 r}{dt^2} = -\frac{q}{m}\frac{\mu_0 I}{2\pi r}\left(\frac{q\mu_0 I}{m2\pi}\ln\frac{r}{r_0} + v_0\right)$$

which can be solved numerically, as was done in the case of Example 6-1. Normalizing the parameters by choosing

$$\frac{q}{m}\frac{\mu_0 I}{2\pi} = 1, \qquad r_0 = 1 \text{ m}, \qquad v_0 = 0.5 \text{ m-s}^{-1}$$

we have

$$\frac{d^2 r}{dt^2} = -\frac{\ln r + 0.5}{r}$$

the solution of which is shown in Figure 6.5b. We note that the radial motion is oscillatory in character. This can be understood by considering the coupled set of equations that describe the complete motion of the charged particle:

$$\frac{d\tilde{v}_r}{dt} = -\left(\frac{q\mu_0 I}{2\pi m}\right)\frac{\tilde{v}_z}{r}$$

$$\frac{d\tilde{v}_z}{dt} = \left(\frac{q\mu_0 I}{2\pi m}\right)\frac{\tilde{v}_r}{r}$$

Noting that the particle starts with $\tilde{v}_z = v_0$ and $\tilde{v}_r = 0$, it initially experiences negative acceleration in the r direction, with $\tilde{v}_r$ being negative. As it develops its radial velocity, its motion in the z direction slows down (since $d\tilde{v}_z/dt$ is negative) until it eventually turns around (at point P in Figure 6.5b). By numerically solving[7] the coupled differential equations given above, the trajectory of the particle

---

[7]Such a solution can be readily achieved using a variety of modern numerical solution tools. The trajectory shown in Figure 6.5c was obtained using *Mathematica* with the following set of simple statements:

```
eq1=r''[t]+(1/r[t]) z'[t] ==0
eq2=z''[t] - (1/r[t]) r'[t] ==0
NDSolve[{eq1,eq2, r[0] == 1, z[0] == 0.5, r'[0]==0, z'[0]==0.5}, {r,z}, {t,0,15}]
```

providing a solution corresponding to $\tilde{v}_z(0) = v_0 = 0.5$ and a starting point for the particles of $z(0) = 0.5$ and $r(0) = r_0 = 1$, all in normalized units.

in the rz plane can be determined as shown in Figure 6.5c. The particle trajectories are in the form of nonuniform "circles," with the radius of curvature being smaller near the wire and larger away from the wire, resulting in the "drift" of the particle in the z direction. This drift occurs because the magnetic field intensity has a gradient in the r direction; it is known as the *gradient drift*.

6.1.3   Motion of a Charged Particle in Crossed $\overline{\mathscr{E}}$ and $\overline{\mathscr{B}}$ Fields

We now consider the motion of a charged particle under the influence of steady and uniform electric and magnetic fields oriented perpendicular (i.e., crossed) to one another. We can choose our coordinate system so that both the electric and magnetic fields are aligned with a principal axis of a rectangular coordinate system. We choose

$$\overline{\mathscr{B}} = \hat{\mathbf{z}}B_0 \qquad \text{and} \qquad \overline{\mathscr{E}} = \hat{\mathbf{y}}E_0$$

as shown in Figure 6.6a. We can use the equation of motion [6.3] to determine the trajectory (i.e., the velocity and position as a function of time) of a particle of charge q and mass m, initially moving with velocity $\tilde{\mathbf{v}} = \hat{\mathbf{x}}v_0$. From [6.3] we have

$$m\frac{d\tilde{\mathbf{v}}}{dt} = q[\overline{\mathscr{E}} + \tilde{\mathbf{v}} \times \overline{\mathscr{B}}] = q[\hat{\mathbf{x}}\tilde{v}_y B_0 + \hat{\mathbf{y}}(E_0 - \tilde{v}_x B_0)]$$

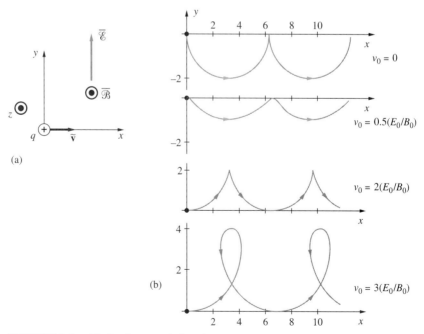

FIGURE 6.6.   **Motion in crossed electric and magnetic fields.** (a) The three vectors are $\overline{\mathscr{E}} = \hat{\mathbf{y}}E_0$, $\overline{\mathscr{B}} = \hat{\mathbf{z}}B_0$, and $\tilde{\mathbf{v}} = \hat{\mathbf{x}}v_0$. (b) The trajectories of the charged particle shown are for different values of v_0, specified in terms of multiples of (E_0/B_0).

Using the cyclotron frequency $\omega_c = -qB_0/m$, we can write this equation in component form as

$$\frac{d\tilde{v}_x}{dt} = -\omega_c \tilde{v}_y \qquad [6.5a]$$

$$\frac{d\tilde{v}_y}{dt} = \frac{q}{m} E_0 + \omega_c \tilde{v}_x \qquad [6.5b]$$

$$\frac{d\tilde{v}_z}{dt} = 0 \qquad [6.5c]$$

Note that since there is no acceleration in the z direction and no initial $\tilde{v}_z$, we have $\tilde{v}_z = 0$ for all time. Assuming that the particle is initially (at $t = 0$) at the origin, we have $z = 0$ at all points along the trajectory, and we need only consider the particle motion in the xy plane. Taking the time derivative of [6.5b], we find

$$\frac{d^2 \tilde{v}_y}{dt^2} = \omega_c \frac{d\tilde{v}_x}{dt}$$

and, upon substitution from [6.5a], we have

$$\frac{d^2 \tilde{v}_y}{dt^2} = -\omega_c^2 \tilde{v}_y \qquad \text{or} \qquad \frac{d^2 \tilde{v}_y}{dt^2} + \omega_c^2 \tilde{v}_y = 0$$

The general solution of this second-order differential equation is

$$\tilde{v}_y(t) = C_1 \sin(\omega_c t) + C_2 \cos(\omega_c t)$$

where C_1 and C_2 are to be determined using the initial conditions. Since $\tilde{v}_y = 0$ at $t = 0$, we must have $C_2 = 0$, so that

$$\tilde{v}_y(t) = C_1 \sin(\omega_c t)$$

The x component of the particle velocity can be found by substituting into the equation for $d\tilde{v}_y/dt$ in terms of $\tilde{v}_x$ (i.e., [6.5b]). We find

$$\frac{d\tilde{v}_y}{dt} = \omega_c C_1 \cos(\omega_c t) = \frac{q}{m} E_0 + \omega_c \tilde{v}_x$$

Since $\tilde{v}_x = v_0$ at $t = 0$, we find C_1 to be

$$C_1 = \frac{q}{m\omega_c} E_0 + v_0 = -\frac{E_0}{B_0} + v_0$$

so the complete expressions for the two velocity components are

$$\tilde{v}_x(t) = \frac{E_0}{B_0} + \left(v_0 - \frac{E_0}{B_0}\right)\cos(\omega_c t)$$

$$\tilde{v}_y(t) = \left(v_0 - \frac{E_0}{B_0}\right)\sin(\omega_c t)$$

These equations reduce to simple circular motion for $E_0 = 0$, as derived in the previous subsection.

To find the position of the particle, we can integrate $\tilde{v}_x$ and $\tilde{v}_y$ with respect to time:

$$x = \frac{E_0}{B_0}t + \frac{1}{\omega_c}\left(v_0 - \frac{E_0}{B_0}\right)\sin(\omega_c t)$$

$$y = -\frac{1}{\omega_c}\left(v_0 - \frac{E_0}{B_0}\right)\cos(\omega_c t) + C_3$$

We choose the constants of integration C_2 and C_3 such that $x = 0$ and $y = 0$ at $t = 0$. Thus,

$$x = \frac{E_0}{B_0}t + \frac{1}{\omega_c}\left(v_0 - \frac{E_0}{B_0}\right)\sin(\omega_c t)$$

$$y = \frac{1}{\omega_c}\left(v_0 - \frac{E_0}{B_0}\right)[1 - \cos(\omega_c t)]$$

where we have also recognized that the arbitrary integration constant for x is zero since $x = 0$ at $t = 0$.

The preceding equations describe the trajectories of the particle in the xy plane, as shown in Figure 6.6b for different values of v_0 (expressed in terms of the ratio E_0/B_0). We note that the motion of the particle is the sum of a straight-line motion in the x direction (due to the electric field) superposed on circular motion due to the magnetic field. For $v_0 = 0$, the particle trajectories have the form of a *cycloid*, which is the path traced by a point on the rim of a rolling wheel. The trajectories pass through fixed points on the x axis regardless of v_0; the time of flight between these *focal points*, which are the points for which $\cos(\omega_c t) = 1$ or $\omega_c t = n2\pi$, is $t = n2\pi/\omega_c$, where n is an integer given by $n = 0, 1, 2, \ldots$. Thus the focal points are located at $x = n2\pi E_0/(\omega_c B_0)$, independent of v_0. Finally we note that $v_0 = E_0/B_0$ results in a straight-line motion, since the $q\tilde{\mathbf{v}} \times \overline{\mathscr{B}}$ force is completely balanced by the $q\mathscr{E}$ force.

Examples 6-5 and 6-6 illustrate crossed-field motion in a parallel-plate configuration and in a magnetron.

Example 6-5: Crossed-field motion between parallel plates. Two parallel metal plates lying at $y = 0$ and $y = d$ are maintained at electrostatic potentials $\Phi = 0$ and $\Phi = -V_0$, respectively. In addition, a magnetic field $\overline{\mathscr{B}} = \hat{\mathbf{z}}B_0$ is applied in the region between the plates. Find the magnitude B_0 such that a particle of positive charge q starting at the lower plate ($y = 0$) with initial velocity $\tilde{\mathbf{v}} = \hat{\mathbf{x}}v_0$ (with $v_0 > 0$) will just graze the upper plate at $y = d$.

Solution: The electrostatic potential difference between the two plates produces an electric field of $\overline{\mathscr{E}} = \hat{\mathbf{y}}E_0 = \hat{\mathbf{y}}V_0/d$. Using the expressions [6.6] for crossed-field motion, we have

$$\tilde{v}_x - \frac{E_0}{B_0} = \left(\frac{E_0}{B_0} - v_0\right)\cos(\omega_c t)$$

$$\tilde{v}_y = \left(v_0 - \frac{E_0}{B_0}\right)\sin(\omega_c t)$$

Squaring and combining terms, we have

$$\left(\tilde{v}_x - \frac{E_0}{B_0}\right)^2 + \tilde{v}_y^2 = \left(\frac{E_0}{B_0} - v_0\right)^2$$

For the charged particle to graze the upper plate, we must have $\tilde{v}_y = 0$ at $y = d$. Thus, the x component of the particle velocity at $y = d$ must be

$$[\tilde{v}_x^2]_{y=d} - \frac{E_0}{B_0} = \frac{E_0}{B_0} - v_0 \qquad \rightarrow \qquad [\tilde{v}_x^2]_{y=d} = \frac{2E_0}{B_0} - v_0$$

Since only the electric field can do work on the particle, the particle's kinetic energy at $y = d$ must be equal to its initial energy plus what it acquires by going through a potential difference of V_0. In other words, we have

$$\frac{1}{2}mv_0^2 + qV_0 = \frac{1}{2}m[\tilde{v}_x^2]_{y=d}$$

Substituting into the previous equation and solving for B_0, we find

$$B_0 = \frac{2V_0/d}{v_0 + \sqrt{(2qV_0/m) + v_0^2}}$$

Example 6-6: Magnetron. A magnetron[8] consists of two coaxial cylindrical electrodes with radii $r = a$ and $r = b$ ($a < b$) inside a region permeated by a magnetic field $\overline{\mathscr{B}} = \hat{z}B_0$. Suppose that the inner electrode (cathode) is at zero potential and the outer electrode (anode) is at potential V_0. Find the value of B_0 such that an electron leaving the cathode with zero initial velocity will just graze the surface of the anode. What is the angular velocity of the electron for this value of B_0?

Solution: Note that the electric field $\overline{\mathscr{E}}$ is not constant in the region between the coaxial cylinders, but instead varies as

$$\overline{\mathscr{E}}(r) = -\hat{r}\frac{V_0}{r \ln b/a}$$

---

[8]The magnetron tube is one of the most efficient devices for generating microwave radiation and is widely used in a variety of applications, the most common being the microwave oven. A microwave oven consists of a rectangular cavity that is fed (via a waveguide) with microwave energy generated by a magnetron tube as the high-power source. Most ovens operate at 2.45 GHz at power output levels of about 1 kW. The first uses of magnetrons were during World War II, as power sources for military radar. Presently available magnetrons can provide up to 100 MW of pulsed power. For more information, see C. D. Taylor and D. V. Giri, *High-Power Microwave Systems and Effects,* Taylor & Francis, 1994.

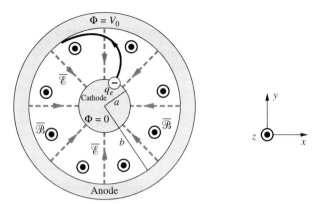

FIGURE 6.7.   The magnetron. The potential difference
between the cathode and the anode creates an electric field directed
radially inward.

With the electric field and the magnetic field respectively in the r and z directions, the equation of motion [6.4] is

$$m_e \frac{d\tilde{\mathbf{v}}}{dt} = \hat{\mathbf{r}} \frac{d\tilde{v}_r}{dt} + \hat{\boldsymbol{\phi}} \frac{d\tilde{v}_\phi}{dt} = q_e[\hat{\mathbf{r}}(\overline{\mathscr{E}}_r + \tilde{v}_\phi \mathscr{B}_z) - \hat{\boldsymbol{\phi}} \tilde{v}_r \mathscr{B}_z]$$

or

$$\hat{\mathbf{r}} \frac{d\tilde{v}_r}{dt} + \hat{\boldsymbol{\phi}} \frac{d\tilde{v}_\phi}{dt} = \hat{\mathbf{r}} \left(-\omega_c \tilde{v}_\phi + \frac{A}{r} \right) + \hat{\boldsymbol{\phi}} \omega_c \tilde{v}_r$$

where

$$\omega_c = -\frac{q_e B_0}{m_e} \qquad A = -\frac{q_e}{m_e} \frac{V_0}{\ln(b/a)}$$

We find the components of $\tilde{\mathbf{v}}$ by integrating the components of $d\tilde{\mathbf{v}}/dt$ separately, since the right-hand side is a function of time. Considering the ϕ component of the equation of motion, we have

$$\frac{d\tilde{v}_\phi}{dt} = \frac{d\tilde{v}_\phi}{dr} \frac{dr}{dt} = \frac{d\tilde{v}_\phi}{dr} \tilde{v}_r = \omega_c \tilde{v}_r$$

or

$$\frac{d\tilde{v}_\phi}{dr} = \omega_c \qquad \rightarrow \qquad \tilde{v}_\phi = \omega_c r + C$$

where C is a constant to be determined by the initial or final (at the edge) conditions. At $r = a$ we have $\tilde{v}_\phi = 0$, so

$$\tilde{v}_\phi(r = a) = 0 \qquad \rightarrow \qquad C = -\omega_c a$$

Thus

$$\tilde{v}_\phi = \omega_c(r - a)$$

At the edge $r = b$, we have $\tilde{\mathbf{v}} = \hat{\boldsymbol{\phi}}\tilde{v}_\phi$ and the potential is V_0. Thus

$$\frac{1}{2}m_e\tilde{v}_\phi^2 = q_eV_0 \quad \rightarrow \quad \tilde{v}_\phi^2 = \frac{2q_eV_0}{m_e}$$

and we have

$$\frac{2q_eV_0}{m_e} = \omega_c^2(b - a)^2$$

and since $\omega_c = -q_eB/m$, the maximum magnetic field is

$$\boxed{B_0 = \frac{\sqrt{2m_eV_0/q_e}}{b - a}}$$

For magnetic fields larger than this value, the particle would not reach the anode; thus this value is the "cutoff" value of B_0, beyond which the magnetron is not useful as an accelerator. At this value of B_0 we have $\tilde{v}_\phi = \omega_0b$, and the angular frequency is

$$\omega_0 = \frac{\tilde{v}_\phi}{b} = \sqrt{\frac{2q_eV_0}{m_e}}\frac{1}{b}$$

The maximum energy to which the magnetron can accelerate the electron is

$$(1/2)m_e\tilde{v}_\phi^2 = (1/2)m_e\omega_0^2b^2 = q_eV_0$$

6.1.4   The Hall Effect

In 1879, Edwin H. Hall, at that time conducting an experiment as a student at Baltimore, discovered[9] a new effect of a magnetic field on electric currents. Hall placed a strip of gold leaf mounted on glass between the poles of an electromagnet so that the plane of the strip was perpendicular to the magnetic flux lines and passed an electric current through the strip. The two ends of a sensitive voltmeter were then placed in contact with different parts of the strip until two points at the same potential were found, so that the voltmeter reading was zero. When the magnetic field (i.e., the current of the electromagnet) was turned on or off, a deflection of the galvanometer needle was observed, indicating a potential difference between the voltmeter leads. In this way, Hall showed that the magnetic field produces a new electromotive force in the strip of gold, at right angles to the primary electromotive force (i.e., the electric

---

[9]E. H. Hall, *Phil. Mag.* ix, p. 225, and x, p. 301, 1880. For a detailed discussion of the Hall effect and its applications, see R. S. Popović, *Hall Effect Devices: Magnetic Sensors and Characterization of Semiconductors,* IOP Publishing Ltd., England, 1991.

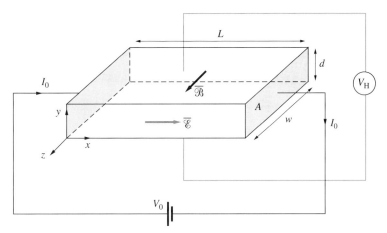

FIGURE 6.8.   **Illustration of the Hall effect.**

field) and to the magnetic field, and proportional to the product of these two forces.[10] Physically, we can regard this effect as an additional electromotive force generated by the action of a magnetic field on the current. This phenomena, called the Hall effect, is extremely useful in practice in the determination of charge densities in materials, especially in semiconductors.

Consider a rectangular semiconductor bar as shown in Figure 6.8. Assume that this is a p-type semiconductor, so that the charge carriers are holes[11] with charge $|q_e|$, where $q_e \simeq -1.6 \times 10^{-19}$ C is the charge of an electron. A potential difference of V_0 applied between the two side faces of the bar sets up an applied electric field E_0 in the x direction, which causes a current $I_0 = J_0 A = wd\sigma E_0$ to flow in the x direction, where σ is the conductivity of the semiconductor. In general, the conductivity for a semiconductor is given by

$$\sigma = |q_e|(\mu_e N_e + \mu_p N_p)$$

where N_e and N_p are respectively the densities of electrons and holes, and μ_e and μ_p are respectively the *mobilities* of the electrons and holes. Mobility is the ratio between the drift velocity of the charge carriers and the applied electric field. For

---

[10]In his excellent historical account *A History of the Theories of Aether and Electricity* (Thomas Nelson and Sons Ltd, 1951), E. T. Whittaker notes that in the early 1870s Oliver Lodge had done experiments on the flow of electricity in a metallic sheet and had come very close to discovering the Hall effect. However, he was deterred from making the crucial test upon reading a passage in Article 501 in Vol. ii of Maxwell's *Electricity and Magnetism*: "It must be carefully remembered, that the mechanical force which urges a conductor carrying a current across the lines of magnetic force acts, not on the electric current, but on the conductor which carries it."

[11]A hole is a positive charge carrier that represents a missing electron in a bond between the atoms of a semiconductor. Electrons and holes are the primary charge carriers that participate in the operation of most semiconductor devices. For further discussion, see R. F. Pierret, *Semiconductor Device Fundamentals,* Addison-Wesley, 1996.

a p-type semiconductor, we have $N_p \gg N_e$, so $\sigma \simeq |q_e|\mu_p N_p$ and the drift velocity is $\tilde{v}_d = \mu_p E_0$, where $\mu_p = \sigma/(|q_e|n_p)$. If a magnetic field B_0 is applied in the z direction, the holes in the p-type semiconductor deflect in the $-y$ direction. Using vector notation, the total force on a single hole due to the electric and magnetic fields is given by [6.1], namely

$$\overline{\mathscr{F}} = |q_e|(\overline{\mathscr{E}} + \tilde{\mathbf{v}}_d \times \overline{\mathscr{B}}) = |q_e|(\hat{\mathbf{x}}E_0 + \hat{\mathbf{x}}\tilde{v}_d \times \hat{\mathbf{z}}B_0)$$

$$= |q_e|(\hat{\mathbf{x}}E_0 - \hat{\mathbf{y}}\tilde{v}_d B_0)$$

According to this result, the holes in the semiconductor experience a magnetic force of $-\hat{\mathbf{y}}(|q_e|\tilde{v}_d B_0)$ and thus an acceleration in the $-y$ direction. However, the displacement of the holes in the $-y$ direction leads to the creation of an electric field $\overline{\mathscr{E}}_y$ between the positively charged region into which the holes move and the negatively charged atoms that are left behind. Thus, this restoring electric field points in the y direction and at steady state must have a magnitude $\tilde{v}_d B_0$ to balance the magnetic force $|q_e|\tilde{v}_d B_0$. The formation of this electric field $\overline{\mathscr{E}}_y$ sets up a potential difference V_H between the top and bottom ends of the bar, which is given by

$$V_H = \mathscr{E}_y d = \tilde{v}_d B_0 d$$

and is called the Hall voltage.

Substituting for $\tilde{v}_d = \mu_p E_0$ and $\mu_p = r/(|q_e|N_p)$, and noting that $\overline{\mathscr{F}} = \sigma\overline{\mathscr{E}} = \hat{\mathbf{x}}\sigma E_0 = \hat{\mathbf{x}}J_0$, the electric field $\overline{\mathscr{E}}_y$ can be written in terms of the current density as

$$\overline{\mathscr{E}}_y = \tilde{v}_d B_0 = \frac{\sigma E_0}{|q_e|N_p}B_0 = R_H J_0 B_0$$

where $R_H = (|q_e|N_p)^{-1}$ is called the Hall coefficient and N_p is the hole concentration in the p-type semiconductor. The hole concentration N_p is given as

$$N_p = \frac{1}{|q_e|R_H} = \frac{J_0 B_0}{|q_e|\mathscr{E}_y} = \frac{[I_0/(wd)]B_0}{|q_e|(V_H/d)} = \frac{I_0 B_0}{|q_e|wV_H}$$

That is, because I_0, B_0, w, $|q_e|$, and V_H are either known or can be measured, the Hall effect can be used to determine (quite accurately) the hole concentration N_p in the p-type semiconductor. Note that the same would be true for an n-type semiconductor, with the only difference being that the charge carriers would be electrons.

The Hall effect can also be used to determine the conductivity of the semiconductor material. If the resistance of the semiconductor bar is measured ($R = V_0/I_0$), the conductivity is given by

$$\sigma = \frac{L}{RA} = \frac{LI_0}{wdV_0}$$

But since the conductivity of the p-type bar is $\sigma \simeq |q_e|\mu_p N_p$, we can write

$$\mu_p \simeq \frac{\sigma}{|q_e|N_p}$$

so that by measuring the hole concentration N_p and the conductivity of the bar (inferred from the resistance of the bar), the mobility of the holes in the p-type material can be determined. Such measurements are very important in the analysis of semiconductor materials.

6.1.5 Electromagnetic Forces on Material Bodies and Radiation Pressure

In general, any material body occupying a volume element dv experiences a force whenever the volume is permeated by electric or magnetic fields. Energy and momentum are transferred from electromagnetic fields as the fields do work on charges. In any region of space where both electric and magnetic fields are present, the total force on a charged particle moving with velocity $\tilde{\mathbf{v}}$ is given by the Lorentz force equation [6.1], repeated here for convenience:

$$\overline{\mathscr{F}} = \underbrace{q\overline{\mathscr{E}}}_{\overline{\mathscr{F}}_{elec}} + \underbrace{q\,\tilde{\mathbf{v}} \times \overline{\mathscr{B}}}_{\overline{\mathscr{F}}_{mag}} \qquad [6.1]$$

Whereas [6.1] is valid for point charges, it is also useful to derive an analogous expression for a continuous distribution of moving charges. Consider an incremental volume Δv in a moving continuous charge distribution represented by charge density $\tilde{\rho}$. Let the total charge in the elemental volume be $\Delta Q = \tilde{\rho}\Delta v$, and let the velocity of the charges in that volume be $\tilde{\mathbf{v}}$. By applying [6.1] to the incremental volume Δv, we can write

$$\Delta\overline{\mathscr{F}} = \tilde{\rho}\Delta v\,(\overline{\mathscr{E}} + \tilde{\mathbf{v}} \times \overline{\mathscr{B}})$$

where $\Delta\overline{\mathscr{F}}$ is the total electromagnetic force exerted on the charge ΔQ in the incremental volume Δv. Since a collection of charges represented by the charge density $\tilde{\rho}$ and moving with velocity $\tilde{\mathbf{v}}$ constitutes a current density $\overline{\mathscr{J}} = \tilde{\rho}\tilde{\mathbf{v}}$, we have $\tilde{\rho}\tilde{\mathbf{v}}\Delta v = \overline{\mathscr{J}}\Delta v$, so that

$$\Delta\overline{\mathscr{F}} = (\tilde{\rho}\overline{\mathscr{E}} + \overline{\mathscr{J}} \times \overline{\mathscr{B}})\,\Delta v$$

It is now convenient to define a volume force density, or force per unit volume, $\overline{\mathscr{F}}_{\Delta v}$, such that

$$\overline{\mathscr{F}}_{\Delta v} = \lim_{\Delta v \to 0} \frac{\Delta\overline{\mathscr{F}}}{\Delta v} \quad \to \quad \overline{\mathscr{F}}_{\Delta v} = \underbrace{\tilde{\rho}\overline{\mathscr{E}}}_{\overline{\mathscr{F}}_{elec}} + \underbrace{\overline{\mathscr{J}} \times \overline{\mathscr{B}}}_{\overline{\mathscr{F}}_{mag}} \text{ N-m}^{-3} \qquad [6.6]$$

The preceding equation gives the total volume force density due to $\overline{\mathscr{E}}$ and $\overline{\mathscr{B}}$ at a field point where the parameters $\tilde{\rho}$, $\overline{\mathscr{J}}$, $\overline{\mathscr{E}}$ and $\overline{\mathscr{B}}$ are defined. The total force $\overline{\mathscr{F}}$ on a macroscopic volume V containing a charge density $\tilde{\rho}$ and current density $\overline{\mathscr{J}}$, in the presence of the fields $\overline{\mathscr{E}}$ and $\overline{\mathscr{B}}$, can be found by integrating this equation over the volume V as

$$\overline{\mathscr{F}} = \int_V (\tilde{\rho}\overline{\mathscr{E}} + \overline{\mathscr{J}} \times \overline{\mathscr{B}})\,dv \quad \text{N}$$

With the field force exerted on a collection of particles thus defined, we can now consider the energy transfer between fields and particles, by considering the work done by the fields on charges. The time rate of doing work, or power, is given in mechanics as $\overline{\mathscr{F}} \cdot \tilde{\mathbf{v}}$ (watts). For a continuous charge distribution, the volume force density exerted on an incremental volume of charge needs to be considered, so that the volume power density delivered to the charges, denoted as $P_{\Delta v}$, is given by

$$P_{\Delta v} = \overline{\mathscr{F}}_{\Delta v} \cdot \tilde{\mathbf{v}} = (\tilde{\rho}\overline{\mathscr{E}} + \overline{\mathscr{J}} \times \overline{\mathscr{B}}) \cdot \tilde{\mathbf{v}}$$
$$= (\tilde{\rho}\overline{\mathscr{E}} + \tilde{\rho}\tilde{\mathbf{v}} \times \overline{\mathscr{B}}) \cdot \tilde{\mathbf{v}}$$
$$= \tilde{\rho}\overline{\mathscr{E}} \cdot \tilde{\mathbf{v}} = \overline{\mathscr{E}} \cdot \overline{\mathscr{J}} \quad \text{W-m}^{-3}$$

where we have used $\overline{\mathscr{J}} = \tilde{\rho}\tilde{\mathbf{v}}$ and the second term drops out because $(\tilde{\mathbf{v}} \times \overline{\mathscr{B}}) \cdot \tilde{\mathbf{v}} = 0$.

Note that because the magnetic force $\overline{\mathscr{F}}_{\text{mag}} = q\tilde{\mathbf{v}} \times \overline{\mathscr{B}}$ is always perpendicular to the velocity, only the electric force $\overline{\mathscr{F}}_{\text{elec}} = q\overline{\mathscr{E}}$ is involved in energy transfer to the charged particles. The preceding equation is a fundamental relation describing the power delivered to a charged medium (characterized by a current density $\overline{\mathscr{J}}$) by the field $\overline{\mathscr{E}}$. The power delivered to the charged particles by the field $\overline{\mathscr{E}}$ may result in their acceleration or, if collisions occur, in heating of the medium by means of particle impact. If the current density $\overline{\mathscr{J}}$ is in time phase with $\overline{\mathscr{E}}$, the power loss is real, a characteristic of Joule heating of a resistive medium. On the other hand, if the current $\overline{\mathscr{J}}$ is 90° out of phase with $\overline{\mathscr{E}}$, the power delivered is imaginary, which is a characteristic of energy storage in the medium.

Radiation Pressure Due to the Electromagnetic Body Force   Equation [6.6] can be applied to determine the radiation pressure exerted by an electromagnetic wave normally incident on a reflective surface, such as a metallic conductor. In causing the electromagnetic energy to be turned around (i.e., reflected), the conductor suffers a reaction. This reaction can also be computed using the magnetic Lorentz force exerted on a volume element dv of the material carrying current density $\overline{\mathscr{J}}$ and given by

$$\overline{\mathscr{F}}_{\Delta v} = \overline{\mathscr{J}} \times \overline{\mathscr{B}} \, dv$$

If the current consists of a surface current $\overline{\mathscr{J}}_s = \overline{\mathscr{J}} \, dz$, we have

$$\overline{\mathscr{F}}_{\Delta v} = \overline{\mathscr{J}}_s \times \overline{\mathscr{B}} \, ds$$

which means that the pressure (force per unit area) on the conductor due to the wave is

$$\mathbf{P} = \frac{\overline{\mathscr{F}}_{\Delta v}}{ds} = \overline{\mathscr{J}}_s \times \overline{\mathscr{B}}$$

Physically, the electromagnetic pressure exerted on the surface of a highly reflective material by an electromagnetic wave that is normally incident on the reflective surface can be understood as follows. The incident wave has an electric field in the x direction (see Figure 3.7), tangential to the surface of the conductor. This means that the electric field immediately inside the conductor surface is also in the x direction, but with a magnitude equal to a small fraction of the wave electric field. For example, consider a normally incident plane wave on the surface ($z = 0$) of a highly conducting material such that $\tan \delta_c = \sigma/(\omega\epsilon) \gg 1$. Assuming the material properties to be σ, ϵ_0, and μ_0, and the intrinsic impedance of the material to be such that $|\eta_c| \ll \eta_{air}$ (e.g., good conductor), the transmission coefficient from air into such a material can be approximately evaluated (see Section 3.7) as

$$\mathcal{T} = \frac{2\eta_c}{\eta_c + \eta_{air}} \simeq \frac{2\eta_c}{\eta_{air}} \simeq \frac{2\sqrt{\omega\mu_0/\sigma}\,e^{j\pi/4}}{\sqrt{\mu_0/\epsilon_0}} = 2\sqrt{\frac{\omega\epsilon_0}{\sigma}}\,e^{j\pi/4}$$

Using the preceding transmission coefficient, the electric field transmitted into the material at $z = 0$ follows as

$$E_{t0} \simeq \left[2\left(\frac{\epsilon_0\omega}{\sigma}\right)^{1/2} e^{-j\pi/4}\right] E_{i0}$$

and the transmitted magnetic field at $z = 0$ is $H_{t0} \simeq 2H_{i0}$. This transmitted electric field exerts an x-directed force on the free conduction electrons in the metal, which, as a result, acquire a drift velocity of $\tilde{\mathbf{v}}_d = \mu_e \mathbf{E}_t$, where μ_e is the mobility of the conduction electrons. As the electrons move, they now experience a force due to the wave magnetic field, i.e.,

$$\overline{\mathcal{F}} = q_e \tilde{\mathbf{v}}_d \times \overline{\mathcal{B}} = q_e\,(\hat{\mathbf{x}}\,\tilde{v}_d) \times (\hat{\mathbf{y}}B_{t0}) = \hat{\mathbf{z}}\mu_e(q_e E_{t0} B_{t0})$$

which is directed in the propagation direction or *into* the conducting material. As individual electrons are pushed, they in turn push the conductor in the process of colliding with atoms on their path, thus giving rise to the radiation pressure. The force F_z represents the time rate of change of momentum delivered to each electron, which is related to the rate of energy transfer (or work done) on each electron, as given by $q_e\overline{\mathcal{E}} \cdot \tilde{\mathbf{v}}_d = (q_e E_{t0})(\mu_e E_{t0}) = \mu_e q_e E_{t0}^2$, noting that no work is done by the wave magnetic field, since $\tilde{\mathbf{v}}_d \times \overline{\mathcal{B}}$ is at all times orthogonal to $\tilde{\mathbf{v}}_d$.

To determine the total force exerted on the highly conducting material considered above, we can also consider the time-average electromagnetic force acting on a macroscopic volume element of the material as given by [6.6]. Considering that a current of density $\mathbf{J} = \sigma\mathbf{E}$ flows in the conductor in response to the wave electric field $\mathbf{E}$, the time-average force per unit volume in the $+z$ direction is given by the phasor form of [6.6]

$$[\mathbf{F}_{\Delta v}]_{av} = \underbrace{\rho\mathbf{E}}_{\mathbf{F}_{elec}} + \underbrace{\mathbf{J}\times\mathbf{B}}_{\mathbf{F}_{mag}} \qquad \text{N-m}^{-3}$$

The total time-average force in the z direction can be found by integrating the time-average magnetic Lorentz force component $\mathbf{F}_{mag}$ over the volume of the conducting slab (having a surface area A). We have

$$\mathbf{F}_{av} = \int_0^\infty \frac{1}{2} \mathscr{R}e\{\mathbf{J} \times \mathbf{B}^*\} A \, dz$$

$$= \frac{A}{2} \mathscr{R}e \left\{ \int_0^\infty \sigma \mathbf{E} \times \mu_0 \mathbf{H}^* \right\}$$

$$\simeq \frac{A\sigma\mu_0}{2} \mathscr{R}e \left\{ \int_0^\infty \left[\hat{\mathbf{x}} 2 \left(\frac{\epsilon_0 \omega}{\sigma} \right)^{1/2} e^{j\pi/4} e^{-\alpha_t z} e^{-j\beta_t z} E_{i0} \right] \times \left[\hat{\mathbf{y}} 2 e^{-\alpha_t z} e^{+j\beta_t z} H_{i0} \right] dz \right\}$$

$$= \hat{\mathbf{z}} \frac{A\sigma\mu_0}{2} (4E_{i0}H_{i0}) \sqrt{\frac{\omega\epsilon_0}{\sigma}} \cos\left(\frac{\pi}{4}\right) \underbrace{\int_0^\infty e^{-2\alpha_t z} dz}_{1/(2\alpha_t)}$$

$$\simeq \hat{\mathbf{z}} \frac{A\sigma\mu_0}{2} (4E_{i0}H_{i0}) \sqrt{\frac{\omega\epsilon_0}{\sigma}} \left(\frac{1}{\sqrt{2}}\right) \frac{\sqrt{2}}{2\sqrt{\omega\mu_0\sigma}}$$

$$= \hat{\mathbf{z}} A \sqrt{\epsilon_0 \mu_0}\, E_{i0} H_{i0}$$

so that the time-average radiation pressure is

$$\mathbf{P}_{av} = \frac{\mathbf{F}_{av}}{A} = \hat{\mathbf{z}}(\mu_0\epsilon_0)^{1/2} E_{i0}H_{i0} = \hat{\mathbf{z}} \frac{(\mu_0\epsilon_0)^{1/2} E_{i0}^2}{\eta_{air}} = \hat{\mathbf{z}} \epsilon_0 E_{i0}^2 \qquad \text{N-m}^{-2}$$

where we have substituted $\alpha_t \simeq \sqrt{\omega\mu_0\sigma/2}$, $H_{i0} = E_{i0}/\eta_{air}$, and $\eta_{air} = \sqrt{\mu_0/\epsilon_0}$.

Note that the radiation pressure found here for a highly reflective material is identical to that calculated in Section 3.1.1 for a perfect conductor from another point of view. At the same time, this radiation pressure exerted on a highly reflective material is exactly twice the time-average radiation pressure exerted on a completely absorbing material as given in equation [2.44]. This result is expected since the reflective material recoils in the process of reflecting the incident wave, as was also discussed in Section 3.1.1.

6.2 WAVE PROPAGATION IN IONIZED GASES (PLASMAS)

In our everyday environment we observe matter in solid, liquid, or gaseous form. However, these three states of matter, which are found on the surfaces of Earth and some other planets, are not typical in the universe at large. More than 99% of the matter in the universe exists as the "fourth state of matter," a *plasma*. Plasmas are a special class of gases that include a large number of electrons and ionized atoms and molecules as well as the neutral atoms and molecules that are present in a normal gas. Although one might think of an ionized collection of positive and negatively charged particles as a gas, the bulk behavior of plasmas more closely resembles that of electrically conducting solids (metals) and can often be effectively described using formulations developed for fluids. The most important distinction between a plasma and a normal gas is the fact that mutual Coulomb interactions between charged particles are important in the dynamics of a plasma and cannot be disregarded. When a neutral gas is raised to a sufficiently high temperature, or when it is subjected to electric fields of sufficient intensity, the atoms and molecules of the gas may become ionized as electrons are stripped off by collisions as a result of the heightened thermal agitation of the particles. Ionization in gases can also be produced by illumination with ultraviolet light or X rays, by bombarding the substance with energetic electrons and ions, and in other ways. When a gas is ionized, even to rather small degree, its dynamical behavior is typically dominated by the electromagnetic forces acting on the free ions and electrons, and it begins to conduct electricity. The charged particles in such an ionized gas interact with electromagnetic fields, and the organized motion of these charge carriers (i.e., electric currents, fluctuations in charge density) can in turn produce electromagnetic fields.

During the 1920's, I. Langmuir and colleagues first showed that characteristic electrical oscillations of very high frequency can exist in an ionized gas that is neutral or quasi-neutral, and they introduced the terms *plasma*[12] and *plasma oscillations* in recognition of the fact that these oscillations resembled those of jellylike substances. When subjected to a static electric field, the charge carriers in an ionized gas rapidly redistribute themselves in such a way that most of the gas is shielded from the field, in a manner quite similar to the redistribution of charge that occurs within a metallic conductor placed in an electric field, resulting in zero electric field everywhere inside.

The plasma medium is often referred to as the fourth state of matter because it has properties profoundly different from those of the gaseous, liquid, and solid states. All states of matter represent different degrees of organization, corresponding to certain values of binding energy. In the solid state, the important quantity is the binding energy of molecules in a crystal. If the average kinetic energy of a molecule exceeds the binding energy (typically a small fraction of an electron volt), the crystal structure breaks up, either into a liquid or directly into a gas (e.g., iodine). Similarly, a certain minimum kinetic energy is required in order to break the bonds of the van der Waals forces to change a liquid into a gas. For matter to pass transition into its fourth state and exist as a plasma, the kinetic energy per plasma particle must exceed the ionizing potential of atoms (typically a few electron volts). Thus, the state of matter is basically determined by the average kinetic energy per particle. Using water as a convenient example, we note that at low temperatures the bond between the H_2O molecules holds them tightly together against the low energy of molecular motion, so the matter is in the solid state (ice). At room temperature, the

---

[12] I. Langmuir, *Phys. Rev.*, 33, p. 954, 1929; L. Tonks and I. Langmuir, *Phys. Rev.*, 33, p. 195, 990, 1929. The word *plasma* first appeared as a scientific term in 1839, when the Czech biologist J. Purkynie coined the term *protoplasma* to describe the jellylike medium, containing a large number of floating particles, that makes up the interior of the cells. The word *plasma* thus means a mold or form and is also used for the liquid part of blood in which corpuscles are suspended.

increased molecular energy permits more widespread movements and currents of molecular motion, and we have the liquid state (water). Since the particle motions are random, not all particles have the same energy, and the more energetic ones escape from the liquid surface to form a vapor above it. As the temperature of the water is further increased, a larger fraction of molecules escapes, until the whole substance is in the gaseous phase (steam). If steam is subjected to further heating, illumination by ultraviolet or X rays, or bombardment by energetic particles, it becomes ionized (plasma).

Although we live in a bubble of essentially un-ionized gas in the midst of an otherwise ionized environment, examples of partially ionized gases or plasmas have long been part of our natural environment, including flaming fire, aurora borealis, and lightning. The immediate environment of our planet Earth, including the radiation belts and the vast upper atmospheric regions referred to as the *ionosphere* and the *magnetosphere,* are also in a plasma state.[13] Early natural philosophers held that the material Universe is built of four "roots": earth, water, air and fire, curiously resembling our modern terminology of solid, liquid, gas, and plasma states of matter. A transient plasma exists in Earth's atmosphere every time a lightning stroke occurs, but it is clearly not very much at home and is short-lived. Early work on electrical discharges included generation of electric sparks by rubbing a large rotating sphere of sulfur against a cloth [O. von Guericke, 1672], production of sparks by harnessing atmospheric electricity in rather hazardous experiments [B. Franklin, 1751], and studies of dust patterns left by a spark discharge passing through a surface of an insulator [G. C. Lichtenberg, 1777]. However, only when electrical and vacuum techniques were developed to the point that long-lived and relatively stable electrical discharges were available did the physics of ionized gases emerge as a field of study. In 1879, W. Crookes published the results of his investigations of discharges at low pressure and remarked: "The phenomena in these exhausted tubes reveal to physical science a new world, a world where matter may exist in a fourth state ..." A rich period of discoveries followed, leading to Langmuir's coining the word *plasma* in 1929, and continuing on as a most fascinating branch of physics until now.

Langmuir, Irving   *(American chemist, b. January 3, 1881, New York; d. August 16, 1957, Massachusetts) Langmuir, the son of an insurance executive, graduated from Pratt Institute in 1898 and obtained degrees in metallurgical engineering at Columbia University in 1903 and a Ph.D. in chemistry in 1906 at the University of Göttingen in Germany. He taught chemistry at the Stevens Institute of Technology and then joined the staff of General Electric Company at Schenectady, New York, in 1909, remaining there until his retirement in 1950.*

His first task at General Electric was to extend the life of the light bulb, which was then very short. The tungsten filaments used were typically enclosed

---

[13]J. K. Hargreaves, *The Solar-Terrestrial Environment,* Cambridge University Press, 1992. Also see M. C. Kelley, *The Earth's Ionosphere,* Academic Press, 1989.

in vacuum, since the presence of air rapidly oxidizes tungsten once it is heated. It was thus thought that to lengthen the life of the bulb, one must improve the vacuum. Langmuir showed, however, that in a vacuum, tungsten atoms slowly evaporated from the wire at the white-hot temperature of the glowing bulb. The wire grew thinner and eventually broke. The rate of evaporation decreased significantly in the presence of a gas with which tungsten does not interact. Thereafter, light bulbs were filled with nitrogen (and later with the still less reactive argon), and lifetimes were multiplied.

Langmuir then studied the effect of hot metal surfaces on all sorts of gases, pursuing research in many directions that had nothing to do with lighting. General Electric was happy to give him complete freedom, since they had already profited immensely from his work with the incandescent bulb. In the mid 1920s, Langmuir developed an atomic hydrogen blowtorch that could produce temperatures as hot as the surface of the sun. His interest in the vacuums within electric bulbs led him to devise methods of producing high-vacuum tubes, which proved essential for radio broadcasting. His work on the gas films that formed on metal wires led him to consider how atoms formed bonds with one another.

For his work on surface chemistry, Langmuir received the 1932 Nobel Prize in chemistry, the first American industrial scientist to do so. [Adapted with permission from I. Asimov, Biographical Encyclopedia of Science and Technology, Doubleday, 1982].

1700   1881   1957   2000

Although plasma is often considered to be the fourth state of matter, it has many properties in common with the gaseous state. However, at the same time, the plasma is an ionized gas in which the long range of Coulomb forces gives rise to collective interaction effects, resembling a fluid with a density higher than that of a gas. In its most general sense, a plasma is any state of matter that contains enough number of free charged particles for its dynamical behavior to be dominated by electromagnetic forces.[14] Most applications of plasma physics is concerned with ionized gases. It turns out that a very low degree of ionization is sufficient for a gas to exhibit electromagnetic properties and behave as a plasma; a gas achieves an electrical conductivity of about half of its possible maximum at about 0.1% ionization and has a conductivity nearly equal to that of a fully ionized gas at 1% ionization. The degree of ionization of a gas can be defined as the ratio $N_e/(N_e + N_0)$, where N_e is the free electron density and N_0 is the density of neutral molecules. As an example, the degree of ionization in a fluorescent tube is $\sim 10^{-5}$, with $N_0 \simeq 10^{16}$ (cm)$^{-3}$ and $N_e \simeq 10^{11}$ (cm)$^{-3}$. Typically, a gas is considered to be a weakly (strongly) ionized gas if the degree of ionization is less than (greater than) 10^{-4}.

---

[14]Plasma physics therefore encompasses the solid state, since electrons in metals and semiconductors fall into this category. However, the redistribution of charge and the screening of the inner regions of a metal occurs extremely quickly (typically $\sim 10^{-19}$ s), in view of the very high density of free charges.

The sun and the stars are hot enough to be almost completely ionized with enormous densities ($N_e \simeq 10^{27}$ (cm)$^{-3}$), and the interstellar gas is sparse enough to be almost completely ionized by stellar radiation. Starting at about 60 km altitude, the sun bathes our atmosphere in a variety of radiations, and the energy in the ultraviolet part of the spectrum is absorbed by the atmospheric gas. In the process, a significant number of air molecules and atoms receive enough energy to become ionized. The resulting free electrons and positive ions constitute the *ionosphere*. Maximum ionization density occurs in the ionosphere at about 350 km altitude, where $N_e = 10^6$ (cm)$^{-3}$. The atmospheric density at 350 km altitude is $N_0 \simeq 3.3 \times 10^8$ (cm)$^{-3}$, so the degree of ionization is $\sim 3 \times 10^{-3}$. At even higher altitudes, air is thin enough (i.e., density of air is low enough) that it is almost completely ionized, and the motions of charged particles are dominated by the earth's magnetic field in a region known as the *magnetosphere*.

In the last few decades an important objective of plasma physics research has been to reproduce thermonuclear fusion under controlled conditions on earth, for possible use as an environmentally "clean" source of energy.[15] In terms of practical engineering applications, plasma etching is now widely used in integrated circuit manufacturing and other materials processing.[16]

6.2.1 Plasma Oscillations and the Electron Plasma Frequency

The most fundamental and elementary collective behavior of plasmas consists of the so-called *plasma oscillations*. If a plasma is disturbed, powerful electrical restoring forces are set up, leading to oscillatory motion of the particles around their equilibrium positions at a characteristic frequency, referred to as the plasma frequency. To describe this behavior quantitatively, we consider a plasma that consists of an equal number of electrons, each with charge $q_e \simeq -1.6 \times 10^{-19}$ C, and ions (atomic nuclei) with charge $q_i = -q_e$. We assume that these particles are initially uniformly distributed, so that the plasma is electrically neutral over any macroscopic spatial scale. We further assume that the plasma is "cold"—that is, that the thermal motion of the electrons and ions is negligible, so they are initially at rest.

We now perturb this system by transferring a group of electrons (assumed for simplicity to be in a one-dimensional slab) from a given region of space to a nearby region, leaving net positive charge behind (i.e., the ions), as shown in Figure 6.9. This local charge separation gives rise to an electric field $\overline{\mathscr{E}}$, which exerts a force on the electrons and ions. Since the electrons are much lighter[17] than the ions, they respond more rapidly to the electric field $\overline{\mathscr{E}}$, so the motion of the ions can be neglected. The electric field $\overline{\mathscr{E}}$ acts to reduce the charge separation by pulling the electrons

[15] See Chapter 9 of F. F. Chen, *Introduction to Plasma Physics*, 1st ed., Plenum Press, 1977.

[16] M. A. Lieberman and A. J. Lichtenberg, *Principles of Plasma Discharges and Materials Processing*, John Wiley & Sons, Inc., New York, 1995.

[17] Note that even if the ions consisted simply of the hydrogen ions (H$^+$), or protons, the mass of the proton is $m_{H^+} \simeq 1831 m_e$.

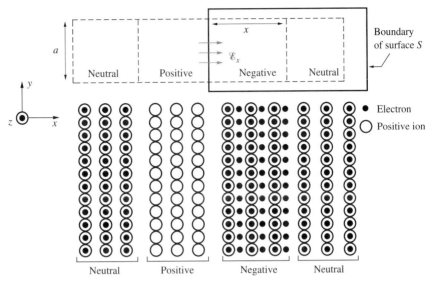

FIGURE 6.9.   **A one-dimensional plasma slab.** The positively charged region consists of atoms which lost one electron while the negatively charged region has excess electrons.

back to their initial locations. The electrons are thus accelerated by the $\overline{\mathscr{E}}$ field toward their initial positions. However, as they acquire kinetic energy in this process, their inertia carries them past their neutral positions. The plasma once again becomes non-neutral, and again an electric field is set up (now pointing in the direction opposite to that shown in Figure 6.9) to retard their motion. Now the electrons accelerate toward the right and go past their equilibrium positions due to their inertia, and once again the charge displacement depicted in Figure 6.9 is set up. In the absence of any damping (for example, due to collisions of the electrons with ions or other electrons), this oscillatory motion continues forever. In relatively tenuous plasmas, collisional damping can be neglected, so any slight disturbance of the system leads to the oscillatory process just described.

We now consider the frequency of this oscillation, or the period of time in which the electrons move from one extreme to the other and back. Intuitively, we expect that the restoring electric field force depends on the amount of the charge displaced, and hence the electronic charge q_e and the density or number of electrons per unit volume. Since the inertia the particles exhibit depends on their mass, the oscillation frequency should also depend on the electron mass m_e. The frequency with which displaced electrons oscillate in a plasma is known as the *plasma frequency* and is denoted ω_p. We can determine ω_p using the equation of motion [6.1] for a single electron in the presence of only an electric field $\overline{\mathscr{E}} = \hat{\mathbf{x}}\mathscr{E}_x$ given by

$$m_e \frac{d^2 x}{dt^2} = q_e \mathscr{E}_x \qquad [6.7]$$

where x is the direction parallel to the electric field as shown in Figure 6.9. Consider Gauss's law, applied to a closed rectangular box-shaped surface as indicated in Figure 6.9, noting that only the boundary of the surface in the xy plane is shown. We have

$$\oint_S \mathscr{E} \cdot d\mathbf{s} = \frac{Q}{\epsilon_0}$$

where Q is the total charge contained within the closed surface S. If the equilibrium density of electrons is N_e, we must have $Q = AxN_eq_e$, where A is the cross-sectional area, and x denotes the displacement of the electrons as shown in Figure 6.9. Assuming that a is the dimension of the rectangular-box surface in the y direction and that the depth of the box is Δz, we have

$$\oint_S \mathscr{E} \cdot d\mathbf{s} = -a\Delta z\mathscr{E}_x$$

$$= \frac{Q}{\epsilon_0} = \frac{ax\Delta zN_eq_e}{\epsilon_0} \quad \rightarrow \quad \mathscr{E}_x = -\frac{xN_eq_e}{\epsilon_0}$$

Substituting back into the equation of motion [6.7], we have

$$m_e\frac{d^2x}{dt^2} + \frac{N_eq_e^2}{\epsilon_0}x = 0 \quad \rightarrow \quad \frac{d^2x}{dt^2} + \omega_p^2x = 0 \qquad [6.8]$$

where $\omega_p \equiv \sqrt{N_eq_e^2/(m_e\epsilon_0)}$ is the oscillation frequency (the plasma frequency).

Equation [6.8] is a second-order differential equation that we have encountered many times before. Its general solution is

$$x = C_1 \sin(\omega_pt) + C_2 \cos(\omega_pt) \qquad [6.9]$$

where C_1 and C_2 are to be determined by initial conditions. Equation [6.9] describes the displacement for free oscillations of the plasma slab. This is the natural frequency of oscillation of the plasma. Oscillations at other frequencies can only occur if the plasma slab is driven by an external field.

As expected, the oscillation frequency depends on the total amount of displaced electrons, having a total charge of N_eq_e, which determines the magnitude of the restoring electric field. The dependence on mass m_e is also expected, since if the electron had no mass, the electrostatic energy could not be transformed into kinetic energy, and the particle would have zero inertia. The fact that the mass of the electrons is small accounts for their rapid response to the electric field. In typical laboratory plasmas, the value of ω_p is in the microwave range. For example, for a plasma with an electron density of $N_e = 10^{12}$ cm^{-3}, $f_p = \omega_p/(2\pi) \simeq 9$ GHz. In the earth's ionosphere, where the typical maximum density of free electrons is $\sim10^{10}$ to $\sim10^{12}$ m^{-3}, the corresponding plasma frequency f_p is ~1 to 10 MHz. Note that since q_e, m_e, and ϵ_0 are fundamental constants, the plasma frequency is essentially a direct measure of the density of electrons N_e. An often useful approximate expression for the plasma frequency is $f_p = \omega_p/(2\pi) \simeq 9\sqrt{N_e}$.

6.2.2 Electromagnetic Wave Propagation in a Plasma

To describe the propagation of electromagnetic waves in a plasma medium, we use Maxwell's equations together with the equations of motion and determine which oscillations are possible, assuming, as usual, sinusoidal steady-state or time-harmonic solutions with $e^{j\omega t}$ type of time dependence. We continue to assume the plasma to be made up solely of electrons and ions, with the ions being relatively immobile (because of their heavier mass). The motion of the electrons under the influence of the wave electric and magnetic fields constitute a current, which must be accounted for in Maxwell's equations via the $\overline{\mathscr{J}}$ term, and their displacement creates localized regions of finite charge, represented by the volume charge density term $\tilde{\rho}$. In general, we have

$$\overline{\mathscr{J}} = \mathscr{N}_e q_e \tilde{\mathbf{v}} \qquad \text{and} \qquad \tilde{\rho} = \mathscr{N}_e q_e$$

where $\mathscr{N}_e$ is the total density of electrons given by

$$\mathscr{N}_e = N_e + \mathscr{R}e\{n_e e^{j\omega t}\}$$

with N_e being the steady or ambient density of electrons, and n_e being the "phasor" electron density representing the time-harmonic variations. The electron velocity $\tilde{\mathbf{v}}$ is assumed to vary harmonically with no steady (or drift) components so that we have

$$\tilde{\mathbf{v}} = \mathscr{R}e\{\mathbf{v}e^{j\omega t}\}$$

where $\mathbf{v}$ is the velocity phasor. Note that the quantities n_e, $\mathbf{J}$, and $\mathbf{v}$ are phasors, while $\tilde{\mathbf{v}}$, $\overline{\mathscr{J}}$, $\tilde{\rho}$, and $\mathscr{N}_e$ are time-varying quantities. Substituting in the expression for the current density, we find

$$\overline{\mathscr{J}} = [N_e + \mathscr{R}e\{n_e e^{j\omega t}\}]q_e[\mathscr{R}e\{\mathbf{v}e^{j\omega t}\}]$$
$$= N_e q_e \mathscr{R}e\{\mathbf{v}e^{j\omega t}\} + q_e[\mathscr{R}e\{n_e e^{j\omega t}\}][\mathscr{R}e\{\mathbf{v}e^{j\omega t}\}]$$

Here we make the assumption that we consider the case of small oscillations, so $|n_e| \ll N_e$. As a result, the second term above can then be neglected compared to the first, and we have

$$\overline{\mathscr{J}} \simeq N_e q_e \mathscr{R}e\{\mathbf{v}e^{j\omega t}\} \qquad [6.10]$$

Since $\overline{\mathscr{J}} = \mathscr{R}e\{\mathbf{J}e^{j\omega t}\}$, in phasor form we have

$$\boxed{\mathbf{J} \simeq N_e q_e \mathbf{v}} \qquad [6.11]$$

This is the current density resulting from the motions of the electrons, which we now have to consider in Maxwell's equations. Note that similar relations can also be written for the current due to the motion of the ions. However, the current density term due to the ions is generally small and negligible because they are heavy and relatively immobile. It can be shown that the charge density term due to ions cancels out the steady part of the charge density (N_e) due to the electrons, so that there are no steady (dc) charge density or currents and electric fields. In other words, $\tilde{\rho} \simeq \mathscr{R}e\{n_e e^{j\omega t}\}$, or $\rho \simeq n_e$.

We can now write the time-harmonic (or phasor) form of Maxwell's equations [1.11] and the equations of motion [6.2] as

$$\nabla \times \mathbf{H} \simeq j\omega\epsilon_0 \mathbf{E} + N_e q_e \mathbf{v} \qquad [6.12a]$$

$$\nabla \times \mathbf{E} = -j\omega\mu_0 \mathbf{H} \qquad [6.12b]$$

$$\nabla \cdot \mathbf{E} \simeq \frac{n_e q_e}{\epsilon_0} \qquad [6.12c]$$

$$\nabla \cdot \mathbf{H} = 0 \qquad [6.12d]$$

$$q_e \mathbf{E} \simeq j\omega m_e \mathbf{v} \qquad [6.12e]$$

whereas the continuity of equation (contained within the above) is

$$\nabla \cdot (N_e \mathbf{v}) \simeq -j\omega n_e \qquad [6.13]$$

Note that equation [6.12e] is the phasor form of the equation of motion, the complete form of which is

$$m_e \frac{d\tilde{\mathbf{v}}}{dt} = m_e \left[\frac{\partial \tilde{\mathbf{v}}}{\partial t} + (\tilde{\mathbf{v}} \cdot \nabla)\tilde{\mathbf{v}} \right] \simeq m_e \frac{\partial \tilde{\mathbf{v}}}{\partial t} = q_e[\tilde{\mathscr{E}} + \tilde{\mathbf{v}} \times \tilde{\mathscr{B}}] \qquad [6.14]$$

where the $(\tilde{\mathbf{v}} \cdot \nabla)\tilde{\mathbf{v}}$ term represents the change in velocity that an observer moving with the particle would observe (since in general the velocity would be varying with space coordinates). Note that the $\tilde{\mathscr{B}}$ in [6.14] is the magnetic field of the electromagnetic wave that must necessarily accompany a time-harmonic electric field $\tilde{\mathscr{E}}$. However, for the small-oscillation case considered here, the $\tilde{\mathbf{v}} \times \tilde{\mathscr{B}}$ term is negligible, since it represents the cross product of two small quantities. For the time-harmonic case, [6.14] can be written in terms of phasor quantities as

$$j\omega m_e \mathbf{v} \simeq q_e[\mathbf{E} + \mathbf{v} \times \mathbf{B}]$$

where the $\mathbf{v} \times \mathbf{B}$ term also represents the cross product of two small terms (both $\mathbf{v}$ and $\mathbf{B}$ are sinusoidal oscillations each with a small amplitude), so that it can also be neglected,[18] resulting in [6.12e].

We now consider the solutions of equations [6.12] for propagating waves. By eliminating $\mathbf{v}$ from equations [6.12e] and [6.12a] we find

$$\nabla \times \mathbf{H} = j\omega\epsilon_0 \left(1 - \frac{N_e q_e^2}{\omega^2 m_e \epsilon_0} \right) \mathbf{E} \qquad [6.15]$$

The similarity of [6.15] with the fundamental Maxwell's equation [1.11c], i.e., $\nabla \times \mathbf{H} = j\omega\epsilon\mathbf{E}$, suggests that the plasma can be represented by an *effective dielectric permittivity* given by

$$\boxed{\epsilon_{\text{eff}} = \epsilon_0 \left(1 - \frac{\omega_p^2}{\omega^2} \right)} \qquad [6.16]$$

---

[18]This term cannot be neglected when a steady magnetic field $\mathbf{B}_0$ (such as the magnetic field of Earth) is present, since then $\mathbf{v} \times \mathbf{B}_0$ is no longer a cross product of two small quantities.

where $\omega_p = \sqrt{N_e q_e^2/(m_e \epsilon_0)}$ is the plasma frequency introduced in the previous subsection. Note that equation [6.12c] can also be rewritten[19] using this effective permittivity as

$$\nabla \cdot (\epsilon_{\text{eff}} \mathbf{E}) = 0$$

With the effects of the plasma accounted for in terms of an effective permittivity, *all* of our previous solutions for wave propagation in a dielectric with permittivity ϵ can be used. For example, from [2.9], the phasor form of the wave equation for uniform plane waves in an unbounded medium for a single component of $\mathbf{E}$ is

$$\frac{d^2 E_x}{dz^2} + \omega^2 \mu_0 \epsilon_{\text{eff}} E_x = 0$$

and has the general solution [2.10]

$$E_x(z) = C_1 e^{-j\beta z} + C_2 e^{+j\beta z}$$

where $\beta = \omega\sqrt{\mu_0 \epsilon_{\text{eff}}} = \omega\sqrt{\mu_0 \epsilon_0 (1 - \omega_p^2/\omega^2)}$. Introducing the notation $X = \omega_p^2/\omega^2$, we have

$$E_x(z) = C_1 e^{-jz\omega\sqrt{\mu_0 \epsilon_0 (1-X)}} + C_2 e^{+jz\omega\sqrt{\mu_0 \epsilon_0 (1-X)}} \qquad [6.17]$$

We see that the physical properties of the plasma medium—more specifically, the fact that electrons move under the influence of the wave fields, constituting current, which in turn influences the wave fields through [6.12a]—require that uniform plane waves can propagate in the plasma only if they have a phase constant of $\beta = \omega\sqrt{\mu_0 \epsilon_0 (1 - X)}$. For $X < 1$, or $\omega > \omega_p$, the propagation constant β is real, and wave propagation in a plasma is like that in an ordinary dielectric. At $X = 1$, $\beta = 0$, and plane waves cease to propagate. By analogy to a waveguide, ω_p is the *cutoff* frequency of the plasma medium. It is interesting to note that in the case of the plasma, the cutoff behavior (that is, that β is a function of frequency) is brought about by the physical properties of the medium, rather than by the presence of the boundaries as in the case of waveguides. For $X > 1$, or $\omega < \omega_p$, β is imaginary, and plane waves do not propagate. In that case, we can write [6.17] as

$$E_x(z) = C_1 e^{-z\omega\sqrt{\mu_0 \epsilon_0 (X-1)}} + C_2 e^{+z\omega\sqrt{\mu_0 \epsilon_0 (X-1)}}$$

so the waves are evanescent and attenuate with distance.

In a similar manner, the intrinsic impedance of the plasma medium is

$$\eta_p = \sqrt{\frac{\mu_0}{\epsilon_{\text{eff}}}} = \frac{\sqrt{\mu_0/\epsilon_0}}{\sqrt{1 - \omega_p^2/\omega^2}} \qquad [6.18]$$

and the wave magnetic field corresponding to the electric field given in [6.17] is

$$H_y(z) = \frac{C_1}{\eta_p} e^{-jz\omega\sqrt{\mu_0 \epsilon_0 (1-X)}} + \frac{C_2}{\eta_p} e^{+jz\omega\sqrt{\mu_0 \epsilon_0 (1-X)}} \qquad [6.19]$$

---

[19]Using [6.12e], [6.12c] and the continuity equation [6.13].

Note that for $\omega > \omega_p$, the impedance η_p is purely real, so the electric and magnetic fields are in phase, and their product represents real time-average power flow. However, for $\omega < \omega_p$, the waves are evanescent, and we note from [6.18] that η_p is imaginary. Thus, no power is carried by evanescent waves, since $\mathbf{E}$ and $\mathbf{H}$ are out of phase by 90°. Note also that the evanescent attenuation of the wave for $\omega < \omega_p$ does not represent any absorption of power and conversion into heat, since any such damping (for example due to collisions) was neglected.

Considering only a wave propagating in the $+z$ direction, and $\omega > \omega_p$, the real electric and magnetic fields are given by

$$\overline{\mathscr{E}}(z, t) = \hat{\mathbf{x}} C_1 \cos[\omega t - z\omega\sqrt{\mu_0\epsilon_0(1 - X)}] \qquad [6.20a]$$

$$\overline{\mathscr{H}}(z, t) = \hat{\mathbf{y}}\frac{C_1\sqrt{1 - X}}{\sqrt{\mu_0/\epsilon_0}} \cos[\omega t - z\omega\sqrt{\mu_0\epsilon_0(1 - X)}] \qquad [6.20b]$$

where we have assumed C_1 to be real. It is interesting to note that for a constant peak electric field amplitude given as $C_1 = E_0$, the amplitude of the magnetic field is frequency-dependent. The impedance of the plasma medium is high for frequencies in the vicinity of the cutoff frequency ω_p. In other words, for $\omega = \omega_p$, the plasma behaves like an open circuit. For frequencies much higher than ω_p, i.e., $\omega \gg \omega_p$, we have $\eta_p \simeq \sqrt{\mu_0/\epsilon_0}$, so that the plasma behaves much like free space. Note that the time-average power flow for uniform plane waves in a unmagnetized plasma (i.e., a plasma without a steady magnetic field $\mathbf{B}_0$) is given by

$$\mathbf{S}_{\text{av}} = \hat{\mathbf{z}}\frac{E_0^2}{2\eta_p} = \frac{E_0^2}{2\sqrt{\mu_0/\epsilon_0}}\sqrt{1 - \omega_p^2/\omega^2} \qquad [6.21]$$

The phase velocity for uniform plane waves in an ionized medium is given by

$$v_p = \frac{\omega}{\beta} = \frac{\omega}{\sqrt{\omega^2\mu_0\epsilon_0(1 - \omega_p^2/\omega^2)}} = \frac{c}{\sqrt{1 - \omega_p^2/\omega^2}} \qquad [6.22]$$

Thus we see that vibration of electrons in the ionized region results in a wave phase velocity greater than that of the speed of light in free space. The group velocity can be found in the usual manner, namely

$$v_g = \frac{d\omega}{d\beta} = c\sqrt{1 - \frac{\omega_p^2}{\omega^2}} \qquad [6.23]$$

We note that $v_p v_g = c^2$, as in the case of waveguides. This result is simply a consequence of the fact that the functional frequency dependence of the propagation constant β for an ionized gas is similar to that for β in a waveguide.

Total Reflection from Free Space–Plasma Interface An interesting and important consequence of the fact that uniform plane waves can propagate in a plasma only at frequencies $\omega > \omega_p$ involves the total reflection of a uniform plane wave at the interface between a dielectric and a plasma medium or an ionized gas when $\omega < \omega_p$. A natural example of such an interface is that between free space and

Earth's ionosphere (although the electron density in the ionosphere increases relatively gradually with height, so the interface is not a single sharp interface between two media like those considered in Chapter 3). It is this type of reflection that makes radio waves "bounce" off the ionosphere, making long-range radio communication possible, with reflection occurring when ω is less than ω_p. For the ionosphere, the peak value of f_p is approximately 10 MHz; thus, AM radio broadcast frequencies are reflected from the ionospheric conducting layer. Microwave, television, and FM radio signals are typically above 40 MHz, and are thus easily transmitted through the conducting ionospheric layer with negligible reflection. To illustrate the basic concept of total reflection at such an interface, we consider a sharp single interface between free space and an ionized region (characterized with plasma frequency ω_p) as illustrated in Figure 6.10.

Assuming that the incident wave in Figure 6.10 is at a frequency ω, that the first medium is free space, and that the ionized region behaves as a dielectric with dielectric constant $\epsilon_{\text{eff}} = \epsilon_0 \sqrt{1 - \omega_p^2/\omega^2}$, the reflection and transmission coefficients for the case of normal incidence are given by the expressions similar to those that were given in Chapter 3 for the case of reflection from the interface between two ordinary dielectrics, namely

$$\Gamma = \frac{\sqrt{\epsilon_1} - \sqrt{\epsilon_2}}{\sqrt{\epsilon_1} + \sqrt{\epsilon_2}} = \frac{\sqrt{\epsilon_0} - \sqrt{\epsilon_0(1 - \omega_p^2/\omega^2)}}{\sqrt{\epsilon_0} + \sqrt{\epsilon_0(1 - \omega_p^2/\omega^2)}} = \frac{\omega - \sqrt{\omega^2 - \omega_p^2}}{\omega + \sqrt{\omega^2 - \omega_p^2}} \quad [6.24a]$$

$$\mathcal{T} = \frac{2\sqrt{\epsilon_1}}{\sqrt{\epsilon_1} + \sqrt{\epsilon_2}} = \frac{2\omega}{\omega + \sqrt{\omega^2 - \omega_p^2}} \quad\quad\quad\quad\quad [6.24b]$$

For $\omega > \omega_p$, we note from the above that both Γ and $\mathcal{T}$ are real, and portions of the incident wave energy are reflected and transmitted accordingly. However, for

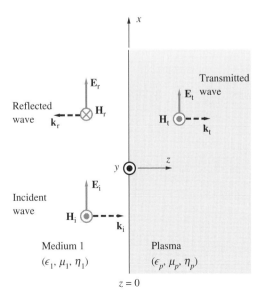

FIGURE 6.10.   **Reflection from free space–plasma interface.** Normal incidence at a sharp interface between free space and an ionized medium (plasma) characterized by a plasma frequency of ω_p.

$\omega < \omega_p$, Γ becomes imaginary, in which case we can write it as

$$\Gamma = \frac{\omega - j\sqrt{(\omega_p^2 - \omega^2)}}{\omega + j\sqrt{(\omega_p^2 - \omega^2)}} = 1e^{j\phi_\Gamma} \qquad [6.25]$$

We note that [6.25] is similar to the form of the reflection coefficients $\Gamma_\parallel$ and $\Gamma_\perp$ derived in Chapter 3 for total internal reflection upon oblique incidence at a dielectric interface (i.e., [3.32] and [3.33]). This similarity indicates that an electromagnetic wave passing from free space into a plasma with $\omega < \omega_p$ is thus totally reflected, even at normal incidence. The amplitude of the reflected wave is equal to that of the incident wave, but the wave acquires phase upon reflection, so the phase of the reflected wave is different from that of the incident one by an amount ϕ_Γ, which depends on frequency as given in [6.25]. Note that—as with the refracted wave in the case of total internal reflection at oblique incidence discussed in Section 3.6.2— the transmission coefficient $\mathcal{T}$ also becomes complex, so the transmitted wave is also out of phase with the incident wave. More importantly, the transmitted wave is evanescent, decays rapidly with distance in the z direction, and carries no real power, since for $\omega < \omega_p$ the intrinsic impedance of the plasma η_p is purely imaginary. As a result the electric and magnetic fields of the wave in the ionized medium are 90° out of phase.

Attenuation Due to Collisions   In practice, some electromagnetic power is always lost in a plasma, because the electrons frequently collide with gas molecules, ions, and even other electrons. These collisions cause electromagnetic power to be transformed into heat. For $\omega > \omega_p$, the collisions cause the wave to be attenuated with distance. Also, for $\omega < \omega_p$, the losses due to collisions lead to partial reflection instead of the total reflection just discussed.

Collisional effects may be taken into account by including in the equations of motion a frictional force term in [6.12e] such as

$$q_e\mathbf{E} = j\omega m_e\mathbf{v} + m_e\nu\mathbf{v} = j\omega m_e\left(1 - j\frac{\nu}{\omega}\right)\mathbf{v} \qquad [6.26]$$

where ν is the effective collision frequency of electrons with other particles, in units of collisions per second or simply s^{-1}.

With the collision term included, we can once again eliminate $\mathbf{v}$ from the equation [6.12a] to obtain

$$\nabla \times \mathbf{H} = j\omega\epsilon_0\mathbf{E} + \frac{N_e q_e^2\mathbf{E}}{j\omega m_e\left(1 - j\dfrac{\nu}{\omega}\right)}$$

or

$$\nabla \times \mathbf{H} = j\omega\epsilon_0\left(1 - \frac{X}{1 - jZ}\right)\mathbf{E}$$

where $X = \omega_p^2/\omega^2$ and $Z = \nu/\omega$. Note that X and Z are dimensionless quantities.

The effective permittivity of the plasma with collisions included is thus

$$\epsilon_{\text{eff}} = \epsilon_0 \left(1 - \frac{X}{1 - jZ} \right) = \epsilon'_{\text{eff}} - j\epsilon''_{\text{eff}}$$

As was shown in Chapter 2 for lossy dielectrics, the imaginary part ϵ''_{eff} represents power loss and results in attenuation of the wave. The expressions for uniform plane waves in a collisional plasma can be obtained by using the general form of uniform plane waves in a lossy medium represented by σ, ϵ, and μ, as was discussed in Section 2.3, by replacing ϵ with ϵ'_{eff} and σ with $\omega\epsilon''_{\text{eff}}$.

6.2.3   Wave Propagation in a Plasma with a Static Magnetic Field

In the previous section we saw that the electromagnetic behavior of a plasma can be expressed in terms of an equivalent complex permittivity. When an applied steady magnetic field permeates the plasma, the medium becomes anisotropic, and the permittivity must now be expressed as a tensor (a matrix), that relates the three components of **E** to three components (not necessarily the same ones) of **D**. The formulation given below[20] is important not only in understanding electromagnetic wave propagation through the ionosphere in the presence of Earth's magnetic field but also in some other applications.

Appleton, Sir Edward Victor   *(English physicist, b. September 6, 1892, Bradford, Yorkshire; d. April 21, 1965, Edinburgh, Scotland) Appleton, the son of a millworker, had an early ambition to become a professional cricket player but won a scholarship, which took him to Cambridge and to science, where he studied under J. J. Thomson [1856–1940] and Ernest Rutherford [1871–1937]. Appleton served as a radio officer during World War I, which interrupted his studies but introduced him to the problem of the fading of radio signals.*

After the war he looked into the problem in earnest and was helped by the fact that by 1922 commercial broadcasting had started in Great Britain, so there were plenty of powerful signals to play with. Appleton found that fading took place at night, and he wondered whether this might not be due to reflection from the upper atmosphere, which took place chiefly at night. If so, such reflection might set up interference, since the same radio beam would reach a particular spot by two different routes: one direct and the other bounced off the layers of charged particles postulated by Kennelly [1861–1939] and Heaviside [1850–1925] twenty years earlier. If so, the two beams might arrive out of phase, with partial cancellation of the wave.

---

[20]The formulation of electromagnetic wave propagation in a magnetized plasma is generally referred to as the magnetoionic theory and was put forth in two papers published in 1927 and 1932 by Sir Edward V. Appleton: The existence of more than one ionized layer in the upper atmosphere, *Nature,* 130, p. 330, 1927; Wireless studies of the ionosphere, *J. I. E. E.,* 71, pp. 642–650, 1932.

Appleton began to experiment by using a transmitter and receiver that were about seventy miles apart, altering the wavelength of the signal, and noting when it was in phase so that the signal was strengthened and when out of phase so that it was weakened. From this he could calculate the minimum height of reflection. Further experiments over the next few years detailed the manner in which the charged layers altered in behavior with the position of the sun and with the changes in the sunspot cycle.

By 1924 Appleton had become a professor of physics at the University of London, and in 1936 he was appointed professor of natural philosophy at Cambridge, succeeding Wilson [1869–1959]. During World War II he was in charge of British atomic bomb research, and in 1941 he was knighted. In 1944 he became vice-chancellor of Edinburgh University, but the climax of his career came in 1947, when he was awarded the Nobel Prize in physics. [Adapted with permission from I. Asimov, Biographical Encyclopedia of Science and Technology, *Doubleday, 1982.]*

```
├──────────────────────────────────────┼────────┼────┤
1700                                   1892    1965  2000
```

We follow a development similar to that in the previous subsection for the case of the plasma without a magnetic field, in which we neglect second-order terms (i.e., products of time-harmonic terms) and express the current as

$$\mathbf{J} \simeq Nq_e\mathbf{v} \qquad [6.27]$$

where $\mathbf{J}$ and $\mathbf{v}$ are both phasor quantities. The electron velocity $\tilde{\mathbf{v}} = \mathcal{R}e\{\mathbf{v}e^{j\omega t}\}$ can be obtained from the Lorentz force equation [6.1], namely

$$\frac{d\tilde{\mathbf{v}}}{dt} = \frac{q_e}{m_e}[\overline{\mathscr{E}} + \tilde{\mathbf{v}} \times (\mathbf{B}_0 + \overline{\mathscr{B}})] \qquad [6.28]$$

where $\mathbf{B}_0 = \hat{\mathbf{z}}B_0$ is the externally applied steady magnetic field and $\overline{\mathscr{E}}$ and $\overline{\mathscr{B}}$ are the electric and magnetic fields of the electromagnetic wave. Since we consider small-signal oscillations, $\tilde{\mathbf{v}} \times \overline{\mathscr{B}}$ is of second order in magnitude, so the only important terms in the right-hand side of [6.28] are the $\overline{\mathscr{E}}$ and $\tilde{\mathbf{v}} \times \mathbf{B}_0$ terms.

To first order,[21] we thus have

$$\frac{d\tilde{\mathbf{v}}}{dt} = \frac{d}{dt}\left[\mathcal{R}e\{\mathbf{v}e^{j\omega t}\}\right] \simeq j\omega\mathcal{R}e\{\mathbf{v}e^{j\omega t}\}$$

so [6.28] can be rewritten in terms of phasor quantities as

$$j\omega\mathbf{v} \simeq \frac{q_e}{m_e}[\mathbf{E} + \mathbf{v} \times (\mathbf{B}_0)] \qquad [6.29]$$

---

[21] Note that $\mathbf{v}$ also varies with x, y, z, so that we have $d\tilde{\mathbf{v}}/dt = \partial\tilde{\mathbf{v}}/\partial t + (\tilde{\mathbf{v}} \cdot \nabla)\tilde{\mathbf{v}}$. However, the second term is of higher order because it involves products of two small quantities. Thus $d\tilde{\mathbf{v}}/dt \simeq j\omega\mathbf{v}e^{j\omega t}$ constitutes a linearization.

In component form we have from [6.3]

$$j\omega v_x = \frac{q_e}{m_e}E_x - \frac{q_e}{m_e}B_{0z}v_y \qquad [6.30a]$$

$$j\omega v_y = \frac{q_e}{m_e}E_y - \frac{q_e}{m_e}B_{0z}v_x \qquad [6.30b]$$

$$j\omega v_z = \frac{q_e}{m_e}E_z \qquad [6.30c]$$

Equations [6.30] may be solved explicitly for the components of the velocity

$$v_x = \frac{-j\omega(q_e/m_e)E_x + (q_e/m_e)\omega_c E_y}{\omega_c^2 - \omega^2} \qquad [6.31a]$$

$$v_y = \frac{(q_e/m_e)\omega_c E_x \quad j\omega(q_e/m_e)E_y}{\omega_c^2 - \omega^2} \qquad [6.31b]$$

$$v_z = \frac{j(q_e/m_e)}{\omega}E_z \qquad [6.31c]$$

where $\omega_c = -(q_e/m_e)B_0$ is the cyclotron frequency. The velocity $\mathbf{v}$ as defined in [6.31] can now be used in [6.27] to express the current density $\mathbf{J}$ in terms of the fields. As before, it is convenient to include the effects of the plasma in terms of an equivalent electric flux density $\mathbf{D}$. Note that for the case of plasma with no magnetic field we had

$$j\omega\mathbf{D} = j\omega\epsilon_0\mathbf{E} + \mathbf{J} = j\omega\epsilon_{\text{eff}}\mathbf{E} \qquad [6.32]$$

with $\epsilon_{\text{eff}} = \epsilon_0(1 - \omega_p^2/\omega^2)$. In the case in hand, where there is a steady magnetic field, it is clear from [6.31] that, for example, the x component of the current density $\mathbf{J}$ is related to not just the x component of $\mathbf{E}$ but to both E_x and E_y. To account for this property, we can use matrix notation to write the equivalent of [6.32] as

$$j\omega\begin{bmatrix} D_x \\ D_y \\ D_z \end{bmatrix} = j\omega\epsilon_0\begin{bmatrix} E_x \\ E_y \\ E_z \end{bmatrix} + \begin{bmatrix} J_x \\ J_y \\ J_z \end{bmatrix}$$

$$= j\omega\begin{bmatrix} \epsilon_{\text{eff}}^{11} & \epsilon_{\text{eff}}^{12} & 0 \\ \epsilon_{\text{eff}}^{21} & \epsilon_{\text{eff}}^{22} & 0 \\ 0 & 0 & \epsilon_{\text{eff}}^{33} \end{bmatrix}\begin{bmatrix} E_x \\ E_y \\ E_z \end{bmatrix} \qquad [6.33]$$

where $[\epsilon_{\text{eff}}]$ is the equivalent permittivity matrix (tensor) with nonzero entries of

$$\epsilon_{\text{eff}}^{11} = \epsilon_{\text{eff}}^{22} = \epsilon_0\left[1 + \frac{\omega_p^2}{\omega_c^2 - \omega^2}\right] \qquad [6.34a]$$

$$\epsilon_{\text{eff}}^{12} = (\epsilon_{\text{eff}}^{21})^* = \frac{j\omega_p^2(\omega_c/\omega)\epsilon_0}{\omega_c^2 - \omega^2} \qquad [6.34b]$$

$$\epsilon_{\text{eff}}^{33} = \epsilon_0\left[1 - \frac{\omega_p^2}{\omega^2}\right] \qquad [6.34c]$$

where ω_p and ω_c are the characteristic frequencies, being respectively the plasma and cyclotron frequencies.

For plane waves in a plasma with no magnetic field, we found in the previous section that ω_p was a characteristic frequency, with no wave propagation being possible for $\omega < \omega_p$. In a plasma with an imposed magnetic field, we see from [6.33] that the cyclotron frequency ω_c is also a characteristic frequency. The singularity at ω_c results from the fact that velocities produced by the wave become unbounded at ω_c; as can be seen from [6.31a] and [6.31b]. A moving electron in a static magnetic field rotates at an angular frequency ω_c even in the absence of the wave. If an applied electric field is at a frequency $\omega = \omega_c$, it is synchronized with the electron motion, so it can continually pump the electron to higher and higher velocities (in practice, this process would be limited by collisional damping).

In general, a variety of propagating wave modes are found in a plasma with a magnetic field, with propagation directions parallel, perpendicular, or at a nonzero angle to $\mathbf{B}_0$. It is found that a TEM wave solution is possible but that no pure TM or TE modes can exist. In the following, we only briefly consider the relatively simple TEM wave. Detailed analysis of other wave modes can be found elsewhere.[22]

Consider uniform plane (i.e., TEM) waves propagating parallel to the magnetic field $\mathbf{B}_0$, so that the spatial dependence of field quantities is of the form

$$e^{\pm j\beta z}$$

and further assume that there are no z components of either $\mathbf{E}$ or $\mathbf{H}$ and the phasor field quantities do not vary with the transverse spatial coordinates of x or y (i.e., a *uniform* plane wave).

Taking the curl of $\nabla \times \mathbf{E} = -j\omega\mu_0\mathbf{H}$ equation and substituting for $\nabla \times \mathbf{H} = j\omega\mathbf{D}$, we find

$$\nabla \times \nabla \times \mathbf{E} = \nabla^2\mathbf{E} - \nabla(\nabla \cdot \mathbf{E}) + \omega^2\mu_0\mathbf{D} = 0$$

For a uniform plane wave we have no z component, and the transverse components do not vary with any coordinate other than z, so $\nabla \cdot \mathbf{E} = 0$, and the above reduces to

$$\nabla^2\mathbf{E} + \omega^2\mu_0\mathbf{D} = 0$$

Using the relationship $[\mathbf{D}] = [\epsilon_{\text{eff}}][\mathbf{E}]$, we can write the above as

$$\frac{d^2}{dz^2}\begin{bmatrix} E_x \\ E_y \\ 0 \end{bmatrix} + \omega^2\mu_0\begin{bmatrix} \epsilon_{\text{eff}}^{11} & \epsilon_{\text{eff}}^{12} & 0 \\ \epsilon_{\text{eff}}^{21} & \epsilon_{\text{eff}}^{22} & 0 \\ 0 & 0 & \epsilon_{\text{eff}}^{33} \end{bmatrix}\begin{bmatrix} E_x \\ E_y \\ 0 \end{bmatrix} = 0 \qquad [6.35]$$

where we have also recognized that $E_z = 0$ for a uniform plane wave.

Examination of the wave equation [6.35] indicates that a linearly polarized wave—with, for example, only one component of $\mathbf{E}$ (say $\mathbf{E} = \hat{\mathbf{x}}E_x$)—is not a possible solution, because if we substitute such an assumed solution in [6.35] (i.e.,

---

[22] See, for example, F. F. Chen, *Introduction to Plasma Physics and Controlled Fusion,* 2nd ed., Plenum Press, 1984.

$E_y = 0$), the vector in the first term has only an x component, but the vector in the second term has both x and y components. In other words, an x component of **E** leads to both x and y components of **D**. Since $\epsilon_{\text{eff}}^{21}$ is in general nonzero, such a solution is not possible. Thus, a plasma with a magnetic field does not support linearly polarized TEM waves propagating along the magnetic field.

Consider the possibility of a solution in the form of circularly polarized waves. For a right-hand (negative sign) or left-hand (positive sign) circularly polarized wave the electric field is of the form

$$\mathbf{E}_{\text{RH}} = C(\hat{\mathbf{x}} \mp j\hat{\mathbf{y}})$$

Substituting in [6.35] and making use of [6.32] and [6.33], we find

$$\beta = \omega\sqrt{\mu_0}\left(\epsilon_{\text{eff}}^{11} - j\epsilon_{\text{eff}}^{12}\right)^{1/2}$$

$$= \omega\sqrt{\mu_0\epsilon_0}\left(1 \pm \frac{\omega_p^2/\omega}{\omega_c \mp \omega}\right)^{1/2}$$

It thus appears that circularly polarized uniform plane waves are possible propagating-mode solutions. It is often useful to express the propagation constant separately for the right-hand (RH) and left-hand (LH) polarized waves. For the RH wave we have

$$\beta_{\text{RH}} = \omega\sqrt{\mu_0\epsilon_0}\left(1 + \frac{\omega_p^2/\omega}{\omega_c - \omega}\right)^{1/2} \qquad [6.36]$$

Thus for RH waves, cutoff occurs at the point where $\beta = 0$, for which we have

$$\left(1 + \frac{\omega_p^2/\omega}{\omega_c - \omega}\right) = 0 \longrightarrow \omega = \omega_{\text{R}} = \left(\omega_p^2 + \frac{\omega_c^2}{4}\right)^{1/2} + \frac{\omega_c}{2}$$

so that for $\omega > \omega_c$, propagation is possible only for $\omega > \omega_{\text{R}}$. However, we note from the above that for $\omega < \omega_c < \omega_p$, the term in brackets in [6.36] is always positive (i.e., β is real) and propagation is possible. This propagation mode, which occurs for $\omega_p > \omega_c$, is a particularly interesting solution that is referred to as the *whistler mode*[23] wave. Note that at $\omega = \omega_c$ we have $\beta \to \infty$, so the wave loses all of its energy to the electrons by continually accelerating them as discussed above.

For LH waves we have

$$\beta_{\text{LH}} = \omega\sqrt{\mu_0\epsilon_0}\left(1 - \frac{\omega_p^2/\omega}{\omega_c + \omega}\right)^{1/2} \qquad [6.37]$$

---

[23]This terminology is adopted in recognition of the naturally occurring radio signal known as the "whistling atmospheric" or *whistler*. Whistlers occur primarily in the audio-frequency range, which is why they were discovered early as descending whistling tones on radio and telephone equipment. Whistlers are impulsive radio waves, launched by lightning discharges, that couple into the earth's magnetosphere and propagate from one hemisphere to another along the earth's magnetic field lines. The dispersion of the impulsive signal as a result of its propagation through the magnetized plasma transforms it into a descending tone. For further discussion, see R. A. Helliwell, *Whistlers and Related Ionospheric Phenomena,* Stanford University Press, Stanford, California, 1965.

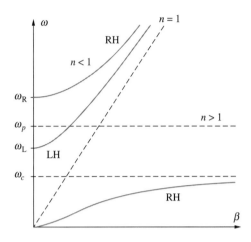

FIGURE 6.11. *ω-β* **diagram for TEM wave propagation along the z direction in a plasma with a steady magnetic field** $\mathbf{B_0} = \hat{\mathbf{z}}B_{0z}$. The branches corresponding to RH and LH solutions are indicated.

The cutoff frequency, again found by equating β to zero, is

$$\omega = \omega_\mathrm{L} = \left(\omega_p^2 + \frac{\omega_c^2}{4}\right)^{1/2} - \frac{\omega_c}{2}$$

so propagation is possible in the left-hand mode only for $\omega > \omega_\mathrm{L}$.

The various propagation and cutoff relationships become clearer when viewed on a plot of ω versus β, as shown in Figure 6.11 for the case of $\omega_p > \omega_c$. We see that for the RH and LH waves for $\omega > \omega_c$ the refractive index ($n = (\beta c/\omega) < 1$) is less than unity, and the cutoff frequencies (i.e., points where $\beta = 0$) are ω_L and ω_R. For the whistler mode (i.e., the RH mode for $\omega < \omega_c$), the refractive index n is greater than unity.

Faraday Rotation The fact that the propagation constant and therefore the phase velocity is different for the RH and LH circularly polarized waves causes the plane of polarization of a linearly polarized wave to rotate[24] as the wave propagates through the medium. This result can be easily seen by decomposing the linearly polarized wave into a sum of LH and RH circularly polarized components and solving for the propagation of each component separately.

---

[24]This phenomenon is called Faraday rotation because it was Faraday who discovered, in 1845, that when a plane-polarized ray of light was passed through a piece of glass in the direction of the lines of force of an imposed magnetic field, the plane of polarization was rotated by an amount proportional to the thickness of the glass traversed and the strength of the magnetic field [M. Faraday, On the magnetization of light and the illumination of magnetic lines of force, *Phil. Trans.*, I. p. 1, 1846]. The interaction between the wave and the bound electrons within the glass material is more complex than the interaction with a free-electron plasma considered here. Nevertheless, the presence of a static magnetic field imposes a preferential direction for the electrons and makes the medium anisotropic.

At $z = 0$, consider the electric field of a linearly polarized wave propagating in the z direction to be

$$\mathbf{E}(z = 0) = \hat{\mathbf{x}} E_0 e^{j\beta z}\Big|_{z=0} = \hat{\mathbf{x}} E_0$$

so the plane of polarization includes the x axis. This wave can be written as a super-position of two circularly polarized waves in the following form

$$\mathbf{E}(z = 0) = \frac{E_0}{2}(\hat{\mathbf{x}} - j\hat{\mathbf{y}}) + \frac{E_0}{2}(\hat{\mathbf{x}} + j\hat{\mathbf{y}})$$

After the component waves propagate in the magnetized plasma for a distance d, the wave electric field at $z = d$ is given by

$$\mathbf{E}(z = d) = \frac{E_0}{2}(\hat{\mathbf{x}} - j\hat{\mathbf{y}})e^{-j\beta_{\mathrm{RH}}d} + \frac{E_0}{2}(\hat{\mathbf{x}} + j\hat{\mathbf{y}})e^{-j\beta_{\mathrm{LH}}d}$$

which can be rewritten as

$$\mathbf{E} = E_0 e^{-j(\beta_{\mathrm{RH}}+\beta_{\mathrm{LH}})d/2}\left\{ \hat{\mathbf{x}}\cos\left[\frac{(\beta_{\mathrm{RH}} - \beta_{\mathrm{LH}})d}{2}\right] - \hat{\mathbf{y}}\sin\left[\frac{(\beta_{\mathrm{RH}} - \beta_{\mathrm{LH}})d}{2}\right]\right\} \qquad [6.38]$$

which is a linearly polarized wave, but with the plane of polarization making an angle of θ_{F} with the x axis, where

$$\theta_{\mathrm{F}} = \tan^{-1}\frac{E_y}{E_x} = -\tan^{-1}\left\{\tan\left[\frac{(\beta_{\mathrm{RH}} - \beta_{\mathrm{LH}})d}{2}\right]\right\} = \frac{(\beta_{\mathrm{LH}} - \beta_{\mathrm{RH}})d}{2} \qquad [6.39]$$

Thus, it appears that the plane of polarization rotates by an amount $\theta_{\mathrm{F}} = (\beta_{\mathrm{LH}} - \beta_{\mathrm{RH}})/2$ per unit distance. The wave propagates with an effective propagation constant of $(\beta_{\mathrm{RH}} + \beta_{\mathrm{LH}})/2$, as is apparent from the $e^{-j(\beta_{\mathrm{RH}}+\beta_{\mathrm{LH}})d/2}$ term in equation [6.39], and undergoes rotation at the same time. Note that this phenomena occurs only for frequencies where both the LH and RH components propagate; from Figure 6.11 and the related discussion, this condition is satisfied for $\omega > \omega_{\mathrm{R}}$.

It is interesting to express the rotation in terms of the physical parameters of the plasma. Using the definitions of ω_p and ω_c we find

$$\theta_{\mathrm{F}} = \frac{|q_e|^3 B_0}{2m_e^2 \epsilon_0 \omega^2 c} N_e d$$

where c is the speed of light in free space. If the electron density varies with distance and if B_0 is approximately constant, the net rotation in angle is given by

$$\theta_{\mathrm{F}} = \frac{|q_e|^3 B_{0z}}{2m_e^2 \epsilon_0 \omega^2 c} \int N_e(z)\, dz$$

where the integration is along the entire propagation path. For signals in the several hundred megahertz range most of the Faraday rotation occurs in the 90–1000-km altitude range. Measurements of the *total electron content* (in a one-square-meter column extending through the most highly ionized part of the ionosphere) are made regularly using satellite-to-earth transmissions.[25]

6.3 FREQUENCY RESPONSE OF DIELECTRICS AND CONDUCTORS

When an external electric field is applied to a dielectric, it polarizes the material by displacing particles. The simplest type of polarization, referred to as electronic polarization, occurs when electrons are displaced with respect to their nuclei. The electric permittivity ϵ of a material is a macroscopic parameter that represents the combined effect of microscopic electric dipoles produced by such displacement.[26] Since electrons or other displaced particles have finite mass, it is clear that the amount of polarization must be a function of the frequency of the applied field. At a qualitative level, we expect that if the electric field alternates rapidly (i.e., high frequencies), inertia of the electrons will tend to prevent them from following the rapid oscillations. At low enough frequencies, the electron can vibrate in phase with the applied field; however, at very high frequencies, the displacements of the electrons are out of phase with the applied field, because of their slow inertial response. In between the two regimes, a resonance phenomenon occurs, leading to significant absorption of electromagnetic energy. These microscopic frequency-dependent behaviors of the materials are generally represented in terms of a complex permittivity, as discussed in Section 2.3. In this section, we provide a first-order description of the frequency dependence of electronic polarization and note that other types of polarization exhibit generally similar behavior.

Electrical conductivities of materials also vary with frequency, primarily because of the inertia of the electrons, whose average drift motion results in electric current flow. In most metals, the frequency dependence of conductivity becomes significant only at optical frequencies. In the following, we provide a very brief discussion of frequency response of conducting materials.

A good example of a material that exhibits frequency-dependent polarization and conductivity is pure water.[27] At low frequencies, distilled water is essentially nonconducting, and its permittivity is $\epsilon \simeq 81\epsilon_0$. At microwave frequencies, however, the permittivity of water ranges from $10\epsilon_0$ to $20\epsilon_0$, and furthermore it becomes quite lossy. Similarly, a salt solution is a good conductor at low frequencies but be-

---

[25]For further information, see K. Davies, *Ionospheric Radio,* Peter Peregrinus Ltd., Exeter, England, 1990.

[26]For a relatively simple discussion of atomic polarization and the permittivity concept, see Section 4.10 of U. S. Inan and A. S. Inan, *Engineering Electromagnetics,* Addison Wesley Longman, 1999. For further discussion on ionic, orientational, and other polarizations, see Chapter 13 of C. Kittel, *Introduction to Solid State Physics,* 5th ed., Wiley, 1976.

[27]An excellent brief discussion of the variation of the refractive index (and hence ϵ) of water as a function of frequency is given in Section 7.5 of J. D. Jackson, *Classical Electrodynamics,* 2nd ed., Wiley, 1975.

haves as an insulator at optical frequencies (i.e., such a solution is transparent to visible light). Thin films of silver are good conductors up through optical frequencies but become transparent in the ultraviolet region. At X ray frequencies, metals and insulators react quite similarly to electromagnetic radiation.

As noted in Section 2.3.1, some magnetic materials also exhibit frequency-dependent properties and relaxation effects, leading to frequency-dependent permeability μ. Such effects need to be considered only in a limited set of applications, and their consideration is beyond the scope of this book.

6.3.1 Frequency Response of Dielectric Materials

To analyze the frequency dependence of polarization quantitatively, we start by considering a nonconducting ($\sigma = 0$) isotropic gas. At the simplest level, gases and solids are made up of molecules, each of which can be considered a classical oscillator. In gases, the separation of the oscillators is large enough so that they do not interact with each other. To approximate the response to an applied electromagnetic field of a real gas, we consider the motion of an electron of a typical gas atom. Since the nucleus is relatively heavy, we consider it to be fixed in space and to be surrounded by an electron cloud of charge q_e and mass m_e that oscillates back and forth in response to the applied electromagnetic field. Consider the Lorentz force acting on an electron in a plane electromagnetic wave in a lossless medium:

$$\overline{\mathscr{F}} = q_e \overline{\mathscr{E}} + q_e \tilde{\mathbf{v}} \times \overline{\mathscr{B}}$$

where $\overline{\mathscr{E}}$ and $\overline{\mathscr{B}}$ are, respectively, the electric and magnetic fields of an electromagnetic wave and $\tilde{\mathbf{v}}$ is the velocity of the electron. We can recall from Chapter 2 that for a uniform plane wave, we have $|\overline{\mathscr{B}}| = \mu|\overline{\mathscr{H}}| = \mu|\overline{\mathscr{E}}|/\eta = (v_p)^{-1}|\overline{\mathscr{E}}|$, so that unless the electrons have a velocity approaching that of the phase velocity v_p of the electromagnetic wave (i.e., the speed of light, in the gaseous medium under consideration), the force due to the magnetic field can be neglected.[28] The electron cloud is thus primarily driven by the electric field of the applied electromagnetic wave. For the rest of this discussion, we assume the wave to be polarized in the x direction, so that $\overline{\mathscr{E}} = \hat{\mathbf{x}}\mathscr{E}_x$, in which case the electron cloud is displaced in the x direction. Once the electron cloud is displaced with respect to the nucleus, it is acted upon by a restoring force due to the Coulomb attraction of the positive nucleus. This restoring force is proportional to the displacement $\tilde{x}$ and is given[29] by $-k\tilde{x}$, where the

---

[28]That is, if $|\tilde{\mathbf{v}}| \ll v_p$, then

$$|\tilde{\mathbf{v}} \times \overline{\mathscr{B}}| = |\tilde{\mathbf{v}}||\overline{\mathscr{B}}| \sin \theta = (v_p)^{-1}|\tilde{\mathbf{v}}||\overline{\mathscr{E}}| \sin \theta \ll |\overline{\mathscr{E}}|$$

and therefore $\overline{\mathscr{F}} = q_e \overline{\mathscr{E}} + q_e \tilde{\mathbf{v}} \times \overline{\mathscr{B}} \simeq q_e \overline{\mathscr{E}}$.

[29]The restoring force can be simply calculated by considering the charge q_e of the electron to be distributed in a negatively charged spherical cloud of atomic radius a, with charge density $\rho = 3q_e/(4\pi a^3)$ and assuming that the center of this cloud is displaced by an amount $\tilde{x}$ with respect to a positive charge (nucleus) of $+|q_e|$. Gauss's law can then be used to determine the electric field E_x at a distance $\tilde{x}$ from the center of the cloud that exerts the restoring force of $+|q_e|E_x$ on the positive charge.

constant k is $k = q_e^2/(4\pi\epsilon_0 a^3)$, with a being the radius of the electron cloud. In effect, k is an elastic constant, and the electron can be considered to be elastically bound to its nucleus. The energy that is transferred by the electron to its surroundings during its rapid motion can be represented by a viscous (or frictional) damping force proportional to the mass and to the velocity of the electron, i.e., $-m_e\kappa(d\tilde{x}/dt)$, where κ is a positive constant. What is described here is naturally a rather crude classical model; the damping term is often due to the reradiation of energy, and this system should actually be described using quantum mechanics. Nevertheless, more detailed considerations lead to a model very similar to the classical model described here, the validity of which is based on its success in describing the propagation of electromagnetic waves through gases.

The equation of motion for the electron, under the influence of the three forces described above, is thus given by

$$m_e\frac{d^2\tilde{x}}{\partial t^2} = -k\tilde{x} - m_e\kappa\frac{d\tilde{x}}{\partial t} + q_e\mathscr{E}_x \qquad [6.40]$$

We can simplify this equation by considering sinusoidal oscillations and rewriting it in terms of a phasor displacement x such that $\tilde{x} = \mathcal{R}e\{xe^{j\omega t}\}$. We have

$$-m_e\omega^2 x = -kx - j\omega m_e\kappa x + q_e E_x \qquad [6.41]$$

where E_x is the electric field phasor; that is, $\mathscr{E} = \mathcal{R}e\{\hat{\mathbf{x}}E_x e^{j\omega t}\}$. Equation [6.41] can be readily solved for the displacement x to find

$$x = \frac{(q_e/m_e)E_x}{\omega_0^2 - \omega^2 + j\omega\kappa} \qquad [6.42]$$

where $\omega_0 = \sqrt{k/m_e}$ is the characteristic angular frequency of the elastically bound electron. The constants ω_0 and κ depend on the atomic and molecular structure of the particular material and must in general be determined from experimental data.

The displacement of a single electron by an amount x creates an electric dipole moment of $\mathbf{p} = \hat{\mathbf{x}}q_e x$. If there are N_e electrons of this type per unit volume, the dipole moment per unit volume, or the polarization, is $\mathbf{P} = \hat{\mathbf{x}}Nq_e x = \hat{\mathbf{x}}P_x$ where P_x is given by

$$P_x = \frac{N_e(q_e^2/m_e)E_x}{\omega_0^2 - \omega^2 + j\omega\kappa} \qquad [6.43]$$

Recalling[30] that the susceptibility χ_e is defined by the relation $P_x = \epsilon_0\chi_e E_x$, and that the permittivity is given by $\epsilon = \epsilon_0(1 + \chi_e)$, we have

$$\chi_e = \frac{N_e q_e^2}{m_e\epsilon_0}\left[\frac{1}{\omega_0^2 - \omega^2 + j\omega\kappa}\right]$$

---

[30]See Section 4.10 of U. S. Inan and A. S. Inan, *Engineering Electromagnetics,* Addison Wesley Longman, 1999.

and thus the relative complex permittivity follows as

$$\epsilon_r = \frac{\epsilon}{\epsilon_0} = 1 + \chi_e = 1 + \frac{N_e q_e^2}{m_e \epsilon_0} \left[\frac{1}{\omega_0^2 - \omega^2 + j\omega\kappa} \right] \qquad [6.44a]$$

$$= 1 + \frac{N_e q_e^2}{m_e \epsilon_0} \left[\frac{\omega_0^2 - \omega^2}{(\omega_0^2 - \omega^2)^2 + \omega^2\kappa^2} - j\frac{\omega\kappa}{(\omega_0^2 - \omega^2)^2 + \omega^2\kappa^2} \right] \qquad [6.44b]$$

$$= \epsilon_r' - j\epsilon_r'' \qquad [6.44c]$$

We thus see that the underlying physical cause of the complex permittivity concept that was introduced in Chapter 2 is the elastically bound nature of the electrons to their nuclei.[31] The variations of $\epsilon' = \epsilon_r'\epsilon_0$ and $\epsilon'' = \epsilon_r''\epsilon_0$ as a function of frequency are illustrated in Figure 6.12. The frequency behavior of the real and imaginary parts of the permittivity are not independent of one another; instead, they are related to one another through the so-called Kramers–Kronig relations,[32] which are similar to the relations between the real and imaginary parts of the complex impedance of an electrical circuit. At low frequencies, ϵ is essentially real, but in the vicinity of $\omega \simeq \omega_0$ a resonance takes place and ϵ becomes complex. The width of the resonance is determined by the damping coefficient κ. From earlier discussions in Section 2.3.1, we know that the imaginary part of the permittivity is associated with power loss or

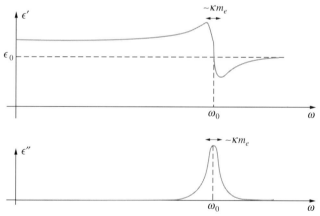

FIGURE 6.12.   Resonance effect in electronic polarization.
At low frequencies, the permittivity is different from ϵ_0 by a constant multiplier. In the vicinity of the resonance, ϵ'' goes through a pronounced peak.

---

[31] Actually, complex permittivity comes about not only from the displacement of electrons but also from the displacement of charged ions or from the change in orientation of existing permanent dipole moments in response to the applied field. Although we have chosen to discuss only electronic polarization, the frequency response of other polarizations is quite similar, as indicated in Figure 6.13.

[32] See p. 311 of J. D. Jackson, *Classical Electrodynamics,* 2nd ed., John Wiley & Sons., 1975.

local field acting to polarize the material is $E_{\text{loc}} = E + P/(3\epsilon_0)$, which in turn leads to the so-called *Clausius–Mossotti* relation,[35] defining the frequency dependence of the complex permittivity ϵ:

$$\frac{\epsilon - 1}{\epsilon + 2} = \frac{1}{3}\left[\frac{Nq_e^2}{m_e\epsilon_0}\right]\left[\frac{1}{\omega_0^2 - \omega^2 + j\omega\kappa}\right]$$

6.3.2 Frequency Response of Metals

We now consider conducting materials and imagine a cloud of freely circulating electrons under statistical equilibrium, which slowly drift in the direction of an applied field.[36] The drift velocity v_d is the solution of an equation of motion with a damping term, namely

$$m_e\frac{dv_d}{dt} + \frac{m_e}{t_c}v_d = q_e\mathscr{E}_x \qquad [6.45]$$

where we have once again assumed that the electric field is in the x direction, so that the drift velocity is $\mathbf{v}_d = \hat{\mathbf{x}}v_d$. Basically, the free electrons are acted upon by an external force $q_e\mathscr{E}_x$, and their motion is opposed by a force $-(m_e/t_c)v_d$ proportional to their velocity, which empirically accounts for the dissipative effect of collisions, where t_c is the mean free time between collisions and is typically of order 10^{-14} s. Note that since $v_d = dx/dt$, equation [6.45] is similar to [6.40], except that the electrons in a metal are not elastically bound so that $k = 0$, and thus $\omega_0 = 0$. For sinusoidal oscillations, the solution for the displacement is similar to [6.41], i.e.,

$$x = \frac{(q_e/m_e)E_x}{-\omega^2 + j(\omega/t_c)} \qquad [6.46]$$

If there are n_c conduction electrons per unit volume, the current density $\mathscr{J}_x$ is then given by

$$\mathscr{J}_x = n_cq_ev_d = n_cq_e\frac{d\tilde{x}}{dt} \quad \rightarrow \quad J_x = n_cq_e(j\omega)x = \frac{(j\omega n_cq_e^2/m_e)E_x}{-\omega^2 + j(\omega/t_c)}$$

where $\mathscr{J}_x = \mathcal{R}e\{J_xe^{j\omega t}\}$, so J_x is the current density phasor. Simplifying further, and by analogy with the static conductivity defined by $\mathbf{J} = \sigma\mathbf{E}$, we have

$$\sigma = \frac{n_cq_e^2}{(m_e/t_c) + jm_e\omega} \qquad [6.47]$$

Note that for $\omega \to 0$ we have $\sigma = n_cq_e^2t_c/m_e$, which is the classic expression for the the static conductivity of a material. In view of the fact that t_c is of order 10^{-14} s, the

---

[35]See Section I-B of *Dielectric Materials and Applications*, A. von Hippel (ed.), Artech House, Boston, 1995.

[36]For a simple and brief discussion of conduction, see Section 5.1 of U. S. Inan and A. S. Inan, *Engineering Electromagnetics*, Addison Wesley Longman, 1999. For further discussion, see Section 8.5 of R. L. Sproull, *Modern Physics: The Quantum Physics of Atoms, Solids, and Nuclei*, 3rd ed., Wiley, 1980.

second term in the denominator of [6.47] is negligible compared to the first term at frequencies well into the infrared region. It has been experimentally verified[37] that conductivity values measured under static conditions can be used for wavelengths greater than ~25 μm. For wavelengths ranging from ~25 μm to the visible region, however, strong frequency dependence is observed, and the current and electric field are no longer in phase. In the far ultraviolet and X ray regions, the conductivity σ approaches zero, as can be seen from [6.47].

To determine the effects of frequency dependence of the conductivity of metals on the propagation constant β and attenuation constant α of a uniform plane wave in a metal, we recall from Section 8.3 that the propagation constant γ in a conducting medium with $\epsilon = \epsilon_0$ and $\mu = \mu_0$ is given by

$$\gamma^2 = j\omega\mu(\sigma + j\omega\epsilon) = \omega^2\mu_0\epsilon_0\left(j\frac{\sigma}{\omega\epsilon_0} - \epsilon_r\right) \quad \rightarrow \quad \gamma = j\omega\sqrt{\mu_0\epsilon_0}\sqrt{\epsilon_r - j\frac{\sigma}{\omega\epsilon_0}}$$

Assuming the frequency dependence of ϵ_r to be of the form [6.44a] and σ of the form [6.47], we have

$$\gamma = j\omega\sqrt{\mu_0\epsilon_0}\sqrt{1 + \frac{Nq_e^2}{m_e\epsilon_0}\left[\frac{1}{\omega_0^2 - \omega^2 + j\omega\kappa}\right] + \frac{n_cq_e^2}{m_e\epsilon_0}\left[\frac{1}{-\omega^2 + j(\omega/t_c)}\right]} \quad [6.48]$$

The second term on the right of [6.48] is due to the bound electrons, whose resonance frequencies for metallic atoms lie well in the violet or ultraviolet frequency range. Thus, in the visible, red, infrared, microwave, and lower frequencies, metals can be treated as if their dielectric constant is ϵ_0. The third term in [6.48] is that due to the free conduction electrons, whose frequency dependence is significant in the low infrared through visible range. As frequency is further increased into the far ultraviolet and X ray regions, the third term in [6.48] approaches zero, so the propagation constant for uniform plane waves in most metals becomes similar to that in dielectrics.

6.4   ELECTROMAGNETIC WAVE PROPAGATION IN FERRITES

Ferrites are a group of chemical compounds that exhibit special and very important electromagnetic properties. The mineral ore lodestone, whose discovery[38] led to much of the early work on magnetism, is a magnetic oxide of iron ($FeO\cdot Fe_2O_3$) and belongs to this group. The general formula for ferrites is $MO\cdot Fe_2O_3$, where M

---

[37]Extensive measurements of reflectivity of a wide range of metals were conducted by Hagen and Rubens in 1903 for frequencies in the range 10^{13} to 10^{14} Hz. See Chap. IX, p. 508 of *Ann. Physik,* 11, p. 873, 1903.

[38]First noted by Roman poet and philosopher Lucretius [99?–55? B.C.], in his philosophical and scientific poem titled *De rerum natura (On the Nature of Things).*

is a divalent metal ion such as manganese, zinc, nickel, or iron. Although ordinary ferromagnetic materials[39] are of little use at high frequencies because of their large conductivities and hence excessive losses, ferrites are usable with low loss at frequencies up to and including the microwave range.

Ferrites are anisotropic in the presence of a magnetic field and thus have characteristics similar to a plasma with a stationary magnetic field. In Section 6.2, we saw that a magnetized plasma is anisotropic because its free electrons respond to a high-frequency electric field in such a way that the applied field gives rise to field components perpendicular to its own direction. In ferrites, the bound electrons produce magnetic dipoles, which respond to a high-frequency magnetic field in such a way that the applied magnetic field gives rise to magnetic fields in directions perpendicular to its own direction. Whereas a tensor permittivity describes the electromagnetic response of a magnetized plasma, a tensor permeability describes the behavior of ferrites.

Ferrites are ceramiclike materials and have relatively low conductivities, in the range of 10^{-4} to 1 S-m^{-1}, compared with $\sim 10^7$ S-m^{-1} for iron and $\sim 10^{-15}$ S-m^{-1} for mica. They have high permeabilities of order a few thousand (which nevertheless is not as high as that for ferromagnetic materials, for which μ can be several tens of thousands). Ferrites have the unique combination of low conductivity and high permeability, in contrast with ferromagnetic materials, which exhibit large losses. The low conductivity makes it possible for ferrites to support electric fields, so electromagnetic wave motion can occur within them.

Ferromagnetism and ferrimagnetism occur because of the interactions of spinning electrons in a material with an alternating applied magnetic field. Such an interaction can in principle occur in all ferromagnetic materials; however, only in ferrites, which are electrical insulators, can useful interaction occur between the magnetic properties of the material and electromagnetic waves. At low radio frequencies, this interaction is not significant, and ferrite materials are used simply as low-loss magnetic core materials for inductors. On the other hand, at microwave frequencies and above, the interaction leads to many unique properties. Ferrites have many useful microwave applications due to their interesting characteristics. In a plasma, free electrons are excited into cyclotron resonance when an electric field at the electron cyclotron frequency is applied perpendicular to a static magnetic field (Section 6.2). In ferrites, bound spinning electrons or magnetic moments are excited into resonance when a high-frequency magnetic field is applied perpendicular to a static magnetic field. The spinning electron in the presence of a static magnetic field is analogous to a gyroscope, with the static magnetic field corresponding to the gravitational field around which the electronic magnetic moment precesses. When a high-frequency

---

[39]The elements of iron, cobalt, and nickel, as well as their alloys and compounds, exhibit ferromagnetic properties resulting in very large values of magnetic permeability μ. At the same time, most of these materials are conducting metals, so electromagnetic fields cannot penetrate them. On the other hand, ferrite materials containing these elements are electrical insulators and allow good penetration of electromagnetic fields while also exhibiting ferromagnetic properties, albeit to a lesser degree.

magnetic field is applied perpendicular to the static magnetic field, a periodic side-wise force is exerted on the electron, which excites the precessing motion.

6.4.1   Magnetic Dipole Moment and Magnetization in Ferrites

The magnetic properties of ferrites are due principally to the magnetic dipole moment **m** and the angular momentum $\overline{\mathscr{L}}$ associated with a spinning electron. In nonmagnetic materials, the arrangement of the electrons in the molecule and of the molecules in the solid structure is such that all electron spins cancel. In ferrites, on the other hand, there exists a resultant spin of electrons. If we imagine the electron as a negatively charged sphere (Figure 6.14a) with a uniform charge distribution, rotating about an axis through its center, then it possesses a fixed angular momentum due to its rotating mass, and constitutes a current loop (i.e., a magnetic dipole) due to its rotating charge. Since the electronic charge is negative, the angular momentum vector $\overline{\mathscr{L}}$ and the magnetic dipole moment **m** are antiparallel,[40] as shown in Figure 6.14a. Based on experimental results and quantum theory, the electron has an angular momentum $|\overline{\mathscr{L}}| = h/4\pi$, or ~$0.527 \times 10^{-34}$ J-m, where h is Planck's constant, and a magnetic moment $|\mathbf{m}| \simeq 9.27 \times 10^{-24}$ A-m^2. The ratio $\gamma_m = |\mathbf{m}|/|\overline{\mathscr{L}}|$ is called the *gyromagnetic ratio.* It has been established from quantum theory that $\gamma_m = q_e/m_e$, where q_e and m_e are respectively the electronic charge and mass. The numerical value of $\gamma_m \simeq 1.76 \times 10^{11}$ rad-s^{-1}-T^{-1} or ~28 GHz-T^{-1}.

We consider a spin angular-momentum vector $\overline{\mathscr{L}}$ placed in a magnetic field $\overline{\mathscr{H}}$, as shown in Figure 6.14b. The magnetic field exerts a torque $\overline{\mathscr{T}}$ on the magnetic

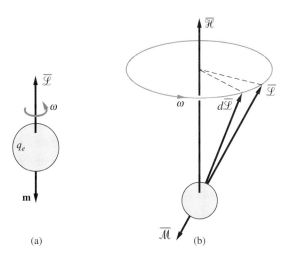

FIGURE 6.14.   Precession of a magnetic dipole in a ferrite medium. (a) The angular momentum $\overline{\mathscr{L}}$ and the magnetic dipole moment **m** associated with a spinning electron. (b) Precession of the magnetization vector $\overline{\mathscr{M}}$ around the field vector $\overline{\mathscr{H}}$.

[40]Note that the direction of the angular momentum $\overline{\mathscr{L}}$ is defined by the right-hand rule and is in the direction of the thumb when the motion is in the direction of the fingers. The direction of the magnetic dipole moment is also defined by the right-hand rule and is in the direction of the thumb when the fingers describe the direction of current flow. Since current carried by electrons is in the opposite direction to their motion, $\overline{\mathscr{L}}$ is oriented in the direction opposite to **m**.

moment **m** of the spin, tending to rotate it into alignment with the field, in keeping with the equation of angular motion:

$$\overline{\mathcal{T}} = \frac{d\overline{\mathcal{L}}}{dt} \qquad [6.49]$$

Since it has angular momentum, the spin does not change its angle with respect to the field but precesses gyroscopically around $\overline{\mathcal{H}}_0$. It is this tendency toward precession in the presence of a polarizing field that accounts for the characteristic behavior of magnetic materials in alternating fields. The torque acting upon **m**, due to the magnetic field $\overline{\mathcal{H}}$, is proportional to $\mu_0\overline{\mathcal{H}}$ and the sine of the angle between **m** and $\overline{\mathcal{H}}$, namely

$$\overline{\mathcal{T}} = \mu_0\mathbf{m} \times \overline{\mathcal{H}} \qquad [6.50]$$

Combining [6.49] and [6.50], and using $\mathbf{m} = \gamma_m\overline{\mathcal{L}}$, we have

$$\frac{d\mathbf{m}}{dt} = \gamma_m\mu_0\mathbf{m} \times \overline{\mathcal{H}}$$

We can picture the precession of the spin in terms of a gyroscope. We know that if the axis of a spinning gyroscope is in a slightly off-vertical position, it sweeps a cone about the vertical. This precession is a result of the torque produced by the mechanical mass of the gyroscope interacting with the vertical gravitational field. In our case, the torque is magnetic in origin (rather than mechanical) because it arises from the interaction of the magnetic dipole with the magnetic field. However, the result is a mechanical torque, which interacts with the angular momentum and produces precession about the magnetic field. From elementary mechanics we know that the velocity of precession of a gyroscope with an angular momentum $\overline{\mathcal{L}}$ under the influence of a torque $\overline{\mathcal{T}}$ is $|\overline{\mathcal{T}}|/(|\overline{\mathcal{L}}|\sin\theta)$, where θ is the angle between $\overline{\mathcal{L}}$ and the gravity vector, which in our case is the magnetic field vector $\overline{\mathcal{H}}$ as shown in Figure 6.14b. Hence the angular velocity of the precession is given by

$$\omega_0 = \frac{|\overline{\mathcal{T}}|}{|\overline{\mathcal{L}}|\sin\theta} = \frac{\mu_0|\mathbf{m}||\overline{\mathcal{H}}|\sin\theta}{(|\mathbf{m}|/\gamma_m)\sin\theta} = \mu_0\gamma_m|\overline{\mathcal{H}}|$$

This motion is known as *Larmor precession,* and the frequency ω_0 is called the gyromagnetic resonance frequency or simply the *Larmor frequency.* Note that ω_0 is independent of the precession angle θ. Since, as mentioned above, $\gamma_m = q_e/m_e$, ω_0 is in fact equal to the electron gyrofrequency ω_c.

If there are N_e electrons per unit volume, the magnetic moment per unit volume, or the *magnetization,* is $\overline{\mathcal{M}} = N_e\mathbf{m}$; thus, we have

$$\frac{d\overline{\mathcal{M}}}{dt} = \gamma_m\mu_0\overline{\mathcal{M}} \times \overline{\mathcal{H}} \qquad [6.51]$$

Equation [6.51] is valid for any general functional form of $\overline{\mathcal{M}}$ and $\overline{\mathcal{H}}$.

6.4.2   Magnetic Permeability Tensor for Time-Harmonic Fields

Most applications of ferrites exploit their interesting properties at frequencies in the microwave range and higher, as mentioned earlier. For all such applications, a time-harmonic or sinusoidal steady-state analysis is appropriate. For time-harmonic fields, for which the magnetization $\overline{\mathcal{M}}$ is given by $\overline{\mathcal{M}} = \mathcal{R}e\{\mathbf{M}e^{j\omega t}\}$, and $\overline{\mathcal{H}} = \mathcal{R}e\{\mathbf{H}e^{j\omega t}\}$, [6.51] can be rewritten as

$$\mathbf{M} = \frac{\gamma_m \mu_0}{j\omega}\mathbf{M} \times \mathbf{H}_0 \qquad [6.52]$$

Note that [6.52] is an equation relating $\mathbf{H}$ and $\mathbf{M}$, which we can solve to obtain the relationships between the components of $\mathbf{H}$ and $\mathbf{B}$. We shall consider the specific case of the magnetic field $\mathbf{H}$ within the ferrite[41] consisting of a small alternating vector field $\mathbf{H}_1$ superimposed on a large polarizing field $\mathbf{H}_0 = \hat{\mathbf{z}}H_0$ in the z direction. The response of the medium to this $\mathbf{H}$ is in turn assumed to be a large constant magnetization $\mathbf{M}_0$ in the z direction plus a small alternating component $\mathbf{M}_1$. Note that since we have $\overline{\mathcal{B}} = \mu_0(\overline{\mathcal{H}} + \overline{\mathcal{M}}) = \mu_0\mu_r\overline{\mathcal{H}}$ the constant magnetization $\mathbf{M}_0$ is related to the large polarizing field $\mathbf{H}_0$ as $\mathbf{M}_0 = \mu_0(\mu_r - 1)\mathbf{H}_0$. Thus we assume

$$\overline{\mathcal{H}} \simeq \mathcal{R}e\{(\hat{\mathbf{x}}H_{1x} + \hat{\mathbf{y}}H_{1y})e^{j\omega t}\} + \hat{\mathbf{z}}H_0 \qquad [6.53a]$$

$$\overline{\mathcal{M}} \simeq \mathcal{R}e\{(\hat{\mathbf{x}}M_{1x} + \hat{\mathbf{y}}M_{1y})e^{j\omega t}\} + \hat{\mathbf{z}}M_0 \qquad [6.53b]$$

where the time-varying components in the $\hat{\mathbf{z}}$ direction have been omitted, since they are negligible compared to the large constant polarizing fields $\overline{\mathcal{H}}_0$ and $\overline{\mathcal{M}}_0$. In phasor form we have

$$\mathbf{H} = \hat{\mathbf{x}}H_{1x} + \hat{\mathbf{y}}H_{1y} + \hat{\mathbf{z}}H_0 \qquad [6.54a]$$

$$\mathbf{M} = \hat{\mathbf{x}}M_{1x} + \hat{\mathbf{y}}M_{1y} + \hat{\mathbf{z}}M_0 \qquad [6.54b]$$

We can now substitute [6.54a] and [6.54b] in [6.52] to find

$$j\omega(\hat{\mathbf{x}}M_{1x} + \hat{\mathbf{y}}M_{1y}) = \gamma_m\mu_0\hat{\mathbf{z}} \times [M_0(\hat{\mathbf{x}}H_{1x} + \hat{\mathbf{y}}H_{1y}) - H_0(\hat{\mathbf{x}}M_{1x} + \hat{\mathbf{y}}M_{1y})]$$

Equating individual components and solving for M_{1x} and M_{1y}, we find

$$M_{1x} = \frac{\omega_0\omega_M}{\omega_0^2 - \omega^2}H_{1x} + j\frac{\omega\omega_M}{\omega_0^2 - \omega^2}H_{1y} \qquad [6.55a]$$

$$M_{1y} = -j\frac{\omega\omega_M}{\omega_0^2 - \omega^2}H_{1x} + \frac{\omega_0\omega_M}{\omega_0^2 - \omega^2}H_{1y} \qquad [11.55b]$$

where[42] $\omega_0 = -\mu_0\gamma_m H_0$ and $\omega_M = -\mu_0\gamma_m M_0$. Equations [6.55a] and [6.55b] can be used to obtain the magnetic permeability of the medium for time-varying fields.

---

[41]In a ferromagnetic material, the magnetic intensity $\mathbf{H}$ is the value averaged over many molecules within the material. The relation between $\mathbf{H}$ and the externally applied field depends upon the material properties and the shape of the ferrite body.

[42]Note that the minus sign simply indicates that the sense of the precession is opposite to that defined by the right-hand rule. In other words, if the thumb of the right hand is pointed in the direction of $\mathbf{H}_0$, the precession is opposite to that indicated by the fingers.

The general relation between the $\mathbf{B}_1$ and $\mathbf{H}_1$ fields is

$$\mathbf{B}_1 = \mu_0(\mathbf{H}_1 + \mathbf{M}_1)$$

We can see from [6.55] that magnetization in the x direction (M_{1x}) can be produced by magnetic fields in both the x and y directions (H_{1x} and H_{1y}). Thus, it is clear that $\mathbf{B}_1$ and $\mathbf{H}_1$ cannot be related by a simple permeability value; instead we can write the permeability as a tensor $[\mu_{\text{eff}}]$ so that $[B] = [\mu_{\text{eff}}][H]$ or

$$\begin{bmatrix} B_x \\ B_y \\ B_z \end{bmatrix} = \begin{bmatrix} \mu_{\text{eff}}^{11} & \mu_{\text{eff}}^{12} & 0 \\ \mu_{\text{eff}}^{21} & \mu_{\text{eff}}^{22} & 0 \\ 0 & 0 & \mu_0 \end{bmatrix} \begin{bmatrix} H_x \\ H_y \\ H_z \end{bmatrix} \qquad [6.56]$$

where

$$\mu_{\text{eff}}^{11} = \mu_{\text{eff}}^{22} = \mu_0 \left[1 + \frac{\omega_0 \omega_{\text{M}}}{\omega_0^2 - \omega^2} \right] \qquad [6.57a]$$

$$\mu_{\text{eff}}^{12} = \left(\mu_{\text{eff}}^{21} \right)^* = j \frac{\mu_0 \omega_{\text{M}}/\omega}{\omega_0^2 - \omega^2} \qquad [6.57b]$$

The similarity of the permeability tensor for ferrites and the permittivity tensor for a magnetized plasma (Section 6.2) is clearly evident.

6.4.3 Electromagnetic Waves in Ferrites

In general, a variety of propagating wave modes are found in ferrites, with propagation directions parallel, perpendicular, or at an oblique angle to $\mathbf{H}_0$. Just as in the case of magnetized plasmas, it is found that a TEM wave solution is possible, but that no pure TM or TE modes can exist. Here we only briefly consider the relatively simple TEM wave.

Consider a uniform plane wave propagating in the z direction through an unbounded ferrite medium that is magnetized by a static magnetic field $\mathbf{H}_0 = \hat{\mathbf{z}} H_0$. The spatial dependence of wave field quantities is of the form

$$e^{\pm j\beta z}$$

Note that because this is a TEM wave, there are no z components of either $\mathbf{E}$ or $\mathbf{H}$, and the field quantities do not vary with the transverse spatial coordinates of x or y. Taking the curl of the $\nabla \times \mathbf{H} = j\omega\epsilon\mathbf{E}$ equation and substituting into the $\nabla \times \mathbf{E} = -j\omega\mathbf{B}$ equation, we find

$$\nabla \times \nabla \times \mathbf{H} = \nabla^2\mathbf{H} - \nabla(\nabla \cdot \mathbf{H}) + \omega^2\epsilon\mathbf{B} = 0$$

Because for a TEM wave we have no variation in x or y and because in a homogeneous ferrite we have $\nabla \cdot \mathbf{H} = 0$, the above reduces to

$$\nabla^2\mathbf{H} + \omega^2\epsilon\mathbf{B} = 0 \qquad [6.58]$$

Since we have $[\mathbf{B}] = [\mu_{\text{eff}}][\mathbf{H}]$ we can write [6.58] as

$$\frac{d^2}{dz^2}\begin{bmatrix} H_x \\ H_y \\ 0 \end{bmatrix} + \omega^2\epsilon \begin{bmatrix} \mu_{\text{eff}}^{11} & \mu_{\text{eff}}^{12} & 0 \\ \mu_{\text{eff}}^{21} & \mu_{\text{eff}}^{22} & 0 \\ 0 & 0 & \mu_{\text{eff}}^{33} \end{bmatrix}\begin{bmatrix} H_x \\ H_y \\ 0 \end{bmatrix} = 0 \qquad [6.59]$$

where we have also recognized the fact that $H_z = 0$ for a TEM wave.

Examination of [6.59] indicates that a linearly polarized wave, such as one having only one component of $\mathbf{H}$ (say $\mathbf{H} = \hat{\mathbf{x}}H_x$), is not a possible solution. This is because, if we substitute such an assumed solution in [6.59] (i.e., with $H_y = 0$), the vector in the first term would only have an x component but the vector in the second term would have both x and y components (in other words, an x component of $\mathbf{H}$ leads to both x and y components of $\mathbf{B}$). The entries μ_{eff}^{21} and μ_{eff}^{12} of the permeability tensor are in general not zero (see [11.57]), so such a solution is not possible. Thus a ferrite medium does not support linearly polarized TEM waves propagating along the magnetic field.

Consider the possibility of a solution in the form of circularly polarized waves. For a right-hand or left-hand circularly polarized wave the magnetic field is of the form

$$\mathbf{H}_{\text{RH}} = C(\hat{\mathbf{x}} - j\hat{\mathbf{y}})e^{-j\beta_{\text{RH}}z} \quad \text{or} \quad \mathbf{H}_{\text{LH}} = C(\hat{\mathbf{x}} + j\hat{\mathbf{y}})e^{-j\beta_{\text{LH}}z}$$

Substituting in [6.59] and making use of [6.56] and [6.57], we find

$$\beta_{\text{RH}} = \omega\sqrt{\mu_0\epsilon}\left(1 + \frac{\omega_{\text{M}}}{\omega_0 - \omega}\right)^{1/2} \qquad [6.60a]$$

$$\beta_{\text{LH}} = \omega\sqrt{\mu_0\epsilon}\left(1 + \frac{\omega_{\text{M}}}{\omega_0 + \omega}\right)^{1/2} \qquad [6.60b]$$

It thus appears that circularly polarized waves are indeed possible TEM wave solutions. Once again, the similarity between expressions [6.60] and the corresponding ones for a magnetized plasma (Section 6.2) is remarkable.

The ω–β diagrams for the two waves are shown in Figure 6.15. At low frequencies the RH and LH waves propagate at the same speed, with $v_p \simeq c\sqrt{\omega_0}/\sqrt{\omega_0 + \omega_{\text{M}}}$. The LH wave propagates almost unaffected by the ferrite material, except at lower frequencies. The RH wave, on the other hand, excites the electrons into resonance, and β_{RH} has a pole at $\omega = \omega_0$. A stopband (no propagation possible) exists between ω_0 and ω_{M}.

Faraday Rotation in Ferrites   The different phase constants for the RH and LH waves leads to Faraday rotation, which forms the basis for several microwave ferrite devices. The Faraday rotation that occurs in ferrites is completely analogous to the rotation that occurs in a magnetized plasma, as discussed in Section 6.2.3. If we start at any point $z = 0$ in the ferrite with a linearly polarized wave propagating in the z

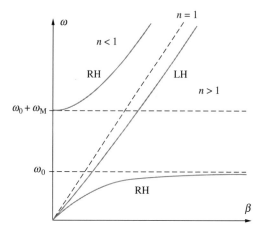

FIGURE 6.15. ω–β **diagram for ferrites.** Note the similarity with the ω–β diagram for a magnetized plasma shown in Figure 6.11.

direction whose magnetic field is given by

$$\mathbf{H}(z = 0) = \hat{\mathbf{x}}H_0 e^{-j\beta z}\bigg|_{z=0} = \hat{\mathbf{x}}\,H_0$$

so that the plane of polarization includes the x axis. This wave can be written as a superposition of two circularly polarized waves in the following form

$$\mathbf{H} = \frac{H_0}{2}(\hat{\mathbf{x}} - j\hat{\mathbf{y}})e^{-j\beta_{RH}z}\bigg|_{z=0} + \frac{H_0}{2}(\hat{\mathbf{x}} + j\hat{\mathbf{y}})e^{-j\beta_{LH}z}\bigg|_{z=0}$$

After the component waves propagate in the ferrite for a distance d, the wave magnetic field at $z = d$ is given by

$$\mathbf{H}(z = d) = \frac{H_0}{2}(\hat{\mathbf{x}} - j\hat{\mathbf{y}})e^{-j\beta_{RH}d} + \frac{H_0}{2}(\hat{\mathbf{x}} + j\hat{\mathbf{y}})e^{-j\beta_{LH}d}$$

which can be rewritten as

$$\mathbf{H}(z = d) = H_0 e^{-j(\beta_{RH} + \beta_{LH})d/2}\left\{\hat{\mathbf{x}}\cos\left[\frac{(\beta_{RH} - \beta_{LH})d}{2}\right] - \hat{\mathbf{y}}\sin\left[\frac{(\beta_{RH} - \beta_{LH})d}{2}\right]\right\}$$

which is a linearly polarized wave, but with the plane of polarization making an angle of θ_F with the x axis, where

$$\theta_F = \tan^{-1}\frac{H_y}{H_x} = -\tan^{-1}\left\{\tan\left[\frac{(\beta_{RH} - \beta_{LH})d}{2}\right]\right\} = \frac{(\beta_{LH} - \beta_{RH})d}{2}$$

$$\theta_F = \omega\sqrt{\mu_0\epsilon}\left(\sqrt{1 + \frac{\omega_M}{\omega_0 + \omega}} - \sqrt{1 + \frac{\omega_M}{\omega_0 - \omega}}\right)\left(\frac{d}{2}\right) \qquad [6.61]$$

Thus, it appears that the plane of polarization rotates by an amount $\theta_F = (\beta_{LH} - \beta_{RH})/2$ per unit distance. The wave propagates with an effective propagation constant of $(\beta_{RH} + \beta_{LH})/2$, as is apparent from the $e^{-j(\beta_{RH} + \beta_{LH})d/2}$ term in equation [6.61], and undergoes rotation at the same time. Note that this phenomena occurs only for frequencies at which both the LH and RH components propagate; from Figure 6.15, we see that real values of β exist (i.e., propagation is possible) for both the RH and LH branches for $\omega < \omega_0$ or for $\omega > (\omega_0 + \omega_M)$.

At frequencies well above the stopband for the RH wave, both waves propagate with low loss. For $\omega \gg \omega_0$ and $\omega \gg \omega_M$, [6.61] reduces to

$$\theta_F \simeq \frac{1}{2}\omega_M d \sqrt{\mu_0 \epsilon_0}$$

It is interesting to note that in this regime the angle of rotation is not a function of frequency, so a device that can operate over a wideband of frequencies is practical. In other words, all frequency components of an information-carrying signal are rotated by the same amount. In addition, the fact that θ_F is independent of frequency indicates that ferrite-based Faraday rotation devices can be used at millimeter wave and optical frequencies as well as at microwave frequencies.

When the foregoing analysis is carried out for a wave traveling in the negative z direction, we find that the same amount of rotation is incurred in the same direction over the same distance d. Thus the polarization of the wave is not restored to its original direction when it travels back through the ferrite. The angle by which the polarization is rotated is doubled, not canceled. Reciprocity[43] requires that the rotations be equal in magnitude but opposite in sign; thus, the ferrite medium is non-reciprocal. In other words, the direction of rotation of the electric field vector of the electromagnetic wave is dependent on the direction of propagation of the wave.

The interesting anisotropic and nonreciprocal properties of ferrites form the basis of many different microwave devices, using both waveguides and microstrips, which are described elsewhere.[44]

Attenuation Due to Damping and Relaxation Effects

or plasmas, we saw that collisions of the electrons with neutral molecules, ions, or other electrons provided damping whose overall effects could be represented by a complex permittivity. In plasmas, collisional losses moderate the unbounded responses that would otherwise occur at those excitation frequencies for which some of the entries of the

---

[43]The reciprocity theorem is of fundamental importance in network and antenna theory. To understand this theorem in simple terms, consider a ferrite-filled circular waveguide section with the two open ends labeled A and B, respectively, at $z = 0$ and $z = d$. If we were to place a generator at A and measure the fields at B, we would find a different result than if the generator were at B and the measurement were made at A. As applied to two-port electrical networks, the reciprocity theorem may be briefly stated as: *The positions of an impedanceless generator and an ammeter may be interchanged without affecting the ammeter reading.* For further discussion, see Section 11.3 of S. Ramo, J. R. Whinnery, and T. Van Duzer, *Fields and Waves in Communication Electronics,* 3rd ed., John Wiley & Sons, New York, 1994.

[44]For a brief discussion at an appropriate level, see A. J. Baden Fuller, An introduction to the use of ferrite and garnet materials at microwave and optical frequencies, *Electronics & Communication Engineering Journal,* pp. 72–80, April 1995. For comprehensive coverage, see A. J. Baden Fuller, *Ferrites at Microwave Frequencies,* Peter Pegrinus, London, 1987. Also see Chapter 14 of R. S. Elliott, *An Introduction to Guided Waves and Microwave Circuits,* Prentice Hall, New Jersey, 1993.

permittivity matrix become infinite. In ferrites, it is clear from [6.56] that some of the entries of the permeability tensor become infinite at $\omega = \omega_0$, implying unbounded response. All real ferrites, however, exhibit various magnetic loss mechanisms that damp out such effects. One of the principal loss mechanisms is the excitation of spin waves.[45] The magnetic dipoles along a ferrite sample are coupled so that they are all parallel and precess in phase. Therefore, if a high-frequency excitation is introduced at one end of the sample, the disturbance propagates to the other end in the form of spin waves. Other loss mechanisms[46] include spin–lattice and spin–spin relaxation times as well as hysteresis losses. Although the underlying physics of these processes is highly complex, it has been found that effects of damping can be accounted for by simply substituting ω_0 by $\omega_0 + j/\tau_r$, where τ_r is the relaxation time (usually determined experimentally) for the particular ferrite material.

6.5 ELECTROMAGNETIC RADIATION AND ELEMENTARY ANTENNAS

Up to now we have studied sinusoidal steady-state (time-harmonic) solutions of Maxwell's equations to determine the characteristics of electromagnetic waves which can exist under various conditions, as propagating or standing waves, or as evanescent waves when propagation at the frequencies of interest is prevented by boundary conditions (as in waveguides) or the physical properties of the medium (as in plasmas). In this section, we consider how the fields are related to their sources, i.e., currents ($\bar{\mathcal{J}}$) and charges ($\tilde{\rho}$) that produce them.

The ultimate sources of all electromagnetic radiation are moving charges. More specifically, radiation can be produced only when a charge undergoes acceleration or when electric currents vary with time. The general subject of electromagnetic radiation is very broad and encompasses a wide variety of physical circumstances, including the production of X rays by decelerating relativistic electron beams, the production of infrared radiation by a heated object, the production of monochromatic light in a laser source, and the production of radio waves by alternating currents flowing in conductors. In this section we confine our attention to a brief discussion of the latter: namely electromagnetic radiation produced by currents flowing in systems of wires or other conductors.

Any system of conductors in open space carrying a time-varying current radiates electromagnetic energy. In many engineering applications, electromagnetic radiation may be intentionally maximized for the purpose of transmitting electrical energy from one point to another without using guiding structures linking the source and the load. In such cases, a system of conductors referred to as the *antenna* is ar-

[45] See Chap. 5 of B. Lax and K. J. Button, *Microwave Ferrites and Ferrimagnetics,* McGraw-Hill, New York, 1962.

[46] See Section 7-16 of R. S. Elliott, *Electromagnetics,* IEEE Press, 1992. Also see Chapters 16 and 17 of C. Kittel, *Introduction to Solid State Physics,* 5th ed., John Wiley & Sons, New York, 1971.

ranged so as to optimize the radiation of energy into open space, either isotropically (i.e., equally in all directions) or only in particular directions. In other engineering applications, electromagnetic radiation may be undesirable, leading to leakage losses and unwanted interference (for example, imperfectly shielded circuit components or interconnects). In both cases, the desired engineering goals can be achieved only by means of a quantitative understanding of electromagnetic radiation: the link between the time-varying currents or charges and the resultant electromagnetic fields.

Hertz, Heinrich Rudolf   *(German physicist, b. February 22, 1857, Hamburg, Germany; d. January 1, 1894, Bonn, Germany) Hertz studied physics at the University of Berlin under Helmholtz [1821–1894] and Kirchhoff [1824–1887], obtained his Ph.D. magna cum laude in 1880, and stayed on for two years more as an assistant to Helmholtz, his lifelong friend and mentor. Working at the University of Kiel in 1883, Hertz grew interested in Maxwell's equations.*

In 1888, Hertz set up an electrical circuit that oscillated, surging into first one, then another, of two metal balls separated by an air gap. Each time the potential reached a peak in one direction or the other, it sent a spark across the gap. With such an oscillating spark, Maxwell's equations predicted that electromagnetic radiation should be generated. Hertz used a simple loop of wire with a small air gap at one point to detect the possible radiation. Just as current gave rise to radiation in the first coil, so the radiation ought to give rise to a current in the second coil. Sure enough, Hertz was able to detect small sparks jumping across the gap in his detector coil. By moving his detector coil to various points in the room, Hertz could tell the shape of the waves by the intensity of spark formation and also managed to show that the waves involved both an electric and a magnetic field and were therefore electromagnetic in nature.

Hertz's experiments were quickly confirmed in England by Lodge [1852–1940], while Righi [1850–1920] in Italy demonstrated the relationship of these "Hertzian waves" to light. In 1889 Hertz became a professor of physics at the University of Bonn. Hertz died tragically early, after a long illness due to chronic blood poisoning, before his thirty-seventh birthday. [Adapted with permission from I. Asimov, Biographical Encyclopedia of Science and Technology, *Doubleday, 1982.]*

1700   *1857*   *1894*   *2000*

The first artificial antenna was designed and used by H. Hertz and operated at about 37 MHz. However, it was G. Marconi who first put antennas into practical use for long-distance wireless communication. Nowadays, antennas are an integral part of our everyday environment. The broad subject of antennas is well beyond the scope of this book, but is covered extensively elsewhere.[47] Our purpose in this section is

---

[47]An excellent textbook, covering many of the fundamentals and applications of antennas as well as a brief discussion of the history of antennas, is J. D. Kraus, *Antennas,* McGraw-Hill, 1988. Also see C. A. Balanis, *Antenna Theory,* 2nd ed., Wiley, New York, 1997.

to provide an elementary understanding of the process by which electromagnetic energy emerges from the antenna and propagates away as an electromagnetic wave, without any discussion of the many different types of antennas.

Marconi, Marchese Guglielmo   *(Italian electrical engineer, b. April 25, 1874, Bologna, Italy; d. July 20, 1937, Rome, Italy) Marconi came of a well-to-do family and was privately tutored. He studied physics under well-known Italian professors but without formally enrolling in any university. In 1894 he came across an article on the electromagnetic waves discovered eight years earlier by Hertz [1857– 1894], and it occurred to him that these might be used in signaling. By the end of the year he was ringing a bell at a distance of thirty feet. As time went on, Marconi sent signals across greater and greater distances. In 1895 he sent one from his house to his garden and later for the distance of a mile and a half. In 1986, when the Italian government was interested in his work, he went to England (his mother was Irish and he could speak English perfectly) and sent a signal over a nine-mile distance. He then applied for and obtained the first patent in the history of radio. In 1897, again in Italy, he sent a signal from land to a warship twelve miles away, and in 1898 (back in England) he covered eighteen miles. By then he was beginning to make his system commercial.*

Marconi obtained a key patent (number 7,777) in 1900 and then, in 1901, reached the denouement of his drama. His experiments had already convinced him that the Hertzian radio waves would follow the curve of the earth instead of radiating straight outward as electromagnetic waves might be expected to do. (The theoretical explanation as to why this happens is the presence of the ionosphere, as was suggested the next year by Kennelly [1861–1939] and Heaviside [1850–1925] and demonstrated to be correct by Appleton [1892– 1965].) For this reason he made elaborate preparations for sending a radio signal from the southwest tip of England to Newfoundland, using balloons to lift his antennae as high as possible. The day he succeeded in doing so, December 12, 1901, is considered as good a date as any for the invention of radio, although it was still useful only for sending signals in Morse code. In 1909 Marconi shared the Nobel Prize in physics with Braun [1850–1918] and in later years experimented extensively with the use of short-wave radio for signaling.

Marconi was in charge of Italy's radio service during World War I, and perfected the "radio beam" along which a pilot could fly blind. Marconi served as one of the Italian delegates to the peace conference that concluded the war. After that, he was an enthusiastic supporter of Mussolini's Fascist government. In 1929 he was made a noble, with the rank of marchese, by the Italian government. When Marconi died, he was given a state funeral by the Italian government. [Adapted with permission from I. Asimov, Biographical Encyclopedia of Science and Technology, *Doubleday, 1982.]*

1700	*1874*	*1937*	*2000*

The formulation of the radiation problem must include the source that establishes the electromagnetic waves in the medium in which it is embedded. The electromagnetic waves thus radiated propagate outward from the source. The uniform plane waves that we have considered in previous chapters are examples of electromagnetic radiation; however, generation and support of such waves would in principle require a current sheet of infinite extent and capable of supplying an infinite amount of power. Nevertheless, the uniform plane waves which we have studied are not only of academic interest. We shall see below that in the region of physical space well removed from the sources, the radiation fields can be quite accurately represented as uniform plane waves.

In principle, radiation problems can be formulated as an exact solution of Maxwell's equations subject to the boundary conditions of the antenna system, in a manner similar to what was done in Chapters 4 and 5 for wave-guiding systems. However, because of the geometrical forms of practical antennas, such solutions do not in most cases lend themselves to simple analytical treatments. As we shall see below, the complete solution of the electric and magnetic fields in the vicinity of a wire antenna are quite complicated even in the case of the simplest wire geometries. A fundamental difficulty is that the distribution of the electrical current on the antenna wires is in general not known and is influenced by the interaction of the wire with the electromagnetic fields it generates. Fortunately, the electromagnetic fields at great distances from an antenna are relatively insensitive to small changes in the current distribution, so good approximations to the source current distribution can often be made. The electromagnetic fields at any point can then be calculated in terms of this source current distribution. Once the fields are calculated, the Poynting theorem can be used to determine the total radiated power that escapes outward from the source into the surrounding medium. The determination of the total power radiated by an antenna, and the distribution of this power in different directions away from the source, are the ultimate objectives in most antenna analysis problems.

Assuming that the source current ($\mathscr{J}$) and/or charge ($\tilde{\rho}$) distributions are known, the electromagnetic fields can in principle be determined from the original forms of Maxwell's equations, namely [1.11]. However, it is more common and usually easier to evaluate the scalar and/or vector potentials ($\tilde{\Phi}, \bar{\mathscr{A}}$) as an intermediate step and derive the electric and magnetic fields from them. Accordingly, we start with a discussion of the relationships between potential functions and source currents and charges.

6.5.1   Retarded Potentials

At a heuristic level, we note that the electrostatic potential $\tilde{\Phi}(\mathbf{r})$ at any point $\mathbf{r}$ is related to charge distributions $\tilde{\rho}(\mathbf{r}')$ in its vicinity as

$$\tilde{\Phi}(\mathbf{r}) = \frac{1}{4\pi\epsilon} \int_{V'} \frac{\tilde{\rho}(\mathbf{r}')}{R} dv'$$

where the primed quantities refer to the sources [i.e., the integration above is over the volume V' containing the source charges represented by $\tilde{\rho}(\mathbf{r}')$] and where $R =$

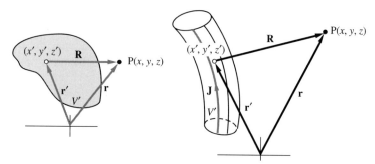

FIGURE 6.16. **Coordinate systems for the evaluation of the scalar and vector potentials.** The relationship between the source coordinates (primed) and observation point (**r**) for (a) volume distribution of charge $\tilde{\rho}$, and (b) volume distribution of current density $\overline{\mathcal{J}}$.

$|\mathbf{r}-\mathbf{r}'|$, as shown in Figure 6.16. Note that the above integral also constitutes a general solution of Poisson's equation[48] $\nabla^2\tilde{\Phi} = -\tilde{\rho}/\epsilon$. Similarly, the magnetostatic vector potential $\overline{\mathcal{A}}(\mathbf{r})$ at an observation point **r** is related to source current distributions $\overline{\mathcal{J}}(\mathbf{r}')$ as

$$\overline{\mathcal{A}}(\mathbf{r}) = \frac{\mu}{4\pi} \int_{V'} \frac{\overline{\mathcal{J}}(\mathbf{r}')}{R} dv'$$

The above integral constitutes a general solution of the vector form of Poisson's equation $\nabla^2\overline{\mathcal{A}} = -\mu\overline{\mathcal{J}}$, when we choose[49] to have $\nabla \cdot \overline{\mathcal{A}} = 0$.

For time-varying electromagnetic fields, the sources $(\overline{\mathcal{J}}, \tilde{\rho})$ vary with time, and we might reasonably expect the solutions for the potentials to be similar

$$\overline{\mathcal{A}}(\mathbf{r}, t) = \frac{\mu}{4\pi} \int_{V'} \frac{\overline{\mathcal{J}}(\mathbf{r}', t)}{R} dv'$$

$$\tilde{\Phi}(\mathbf{r}, t) = \frac{1}{4\pi\epsilon} \int_{V'} \frac{\tilde{\rho}(\mathbf{r}', t)}{R} dv'$$

where $R = |\mathbf{r} - \mathbf{r}'|$ as before. However, since electromagnetic waves have finite propagation times, there is a time delay (retardation) between the sources and the potentials (and therefore the fields) at a distance from the sources. In other words, for an electromagnetic wave propagating with a finite phase velocity, the variation of the source current at position $\mathbf{r}'$ is conveyed to the observation point at a slightly later

---

[48] Poisson's equation as well as Laplace's equation (i.e., $\nabla^2\tilde{\Phi} = 0$) are used in electrostatic problems for the solution of electrostatic potential distributions subject to boundary conditions, typically involving conductors held (by batteries or other sources) at given potentials. For further discussion see Section 4.8 of U. S. Inan and A. S. Inan, *Engineering Electromagnetics,* Addison Wesley Longman, 1999.

[49] Note that the vector potential is defined so that $\overline{\mathcal{B}} = \nabla \times \overline{\mathcal{A}}$. However, for $\overline{\mathcal{A}}$ to be uniquely specified, we must also define its divergence. For static fields, it is common to choose that $\overline{\mathcal{A}}$ (among all those for which $\nabla \times \overline{\mathcal{A}} = \overline{\mathcal{B}}$) for which $\nabla \cdot \overline{\mathcal{A}} = 0$.

time because of the finite velocity. Hence, for the time-varying current source, we must modify the expressions for $\overline{\mathcal{A}}$ and $\tilde{\Phi}$ by making the substitution $t \leftrightarrow (t - R/v_p)$. Namely,

$$\overline{\mathcal{A}}(\mathbf{r}, t) = \frac{\mu}{4\pi} \int_{V'} \frac{\overline{\mathcal{J}}(\mathbf{r}', t - R/v_p)}{R} dv' \qquad [6.62a]$$

$$\tilde{\Phi}(\mathbf{r}, t) = \frac{1}{4\pi\epsilon} \int_{V'} \frac{\tilde{\rho}(\mathbf{r}', t - R/v_p)}{R} dv' \qquad [6.62b]$$

where R/v_p is the propagation time, with v_p being the phase velocity of uniform plane waves in the medium in which the sources are embedded. We note that the time t in this equation is the time at the point of observation, while $t' = t - R/v_p$ is the time at the source point. Thus, the relations [6.62] indicate that the sources that had the configurations of $\overline{\mathcal{J}}$ and $\tilde{\rho}$ at t' produced the potentials $\overline{\mathcal{A}}$ and $\tilde{\Phi}$ at a later time t. Because of this time-delayed aspect of the solutions, the potentials $\overline{\mathcal{A}}(\mathbf{r}, t)$ and $\tilde{\Phi}(\mathbf{r}, t)$ are referred to as the *retarded potentials*.

Although we arrived at the foregoing solutions rather heuristically, the retarded potential expressions are actually quite correct, and they can be rigorously derived.[50] However, in moving from the static case to the time-varying case, the relationship between the potentials and the fields also needs to be modified. In other words, if we now try to use the potentials given in [6.62] to derive the fields using $\mu\overline{\mathcal{H}} = \nabla \times \overline{\mathcal{A}}$ and $\overline{\mathcal{E}} = -\nabla\tilde{\Phi}$, we find that the latter cannot be correct, since we would otherwise have $\nabla \times \overline{\mathcal{E}} = -\nabla \times \nabla\tilde{\Phi} \equiv 0$, which is obviously not the case for time-varying fields. The reasons for this discrepancy are that the electric and magnetic fields are not independent but must be related to one another through Maxwell's equations and that the source distributions $\tilde{\rho}$ and $\overline{\mathcal{J}}$ cannot be specified independently but must be related through the continuity equation.

We can start with Maxwell's equations and formally derive the differential equations for $\tilde{\Phi}$ and $\overline{\mathcal{A}}$ and the relationship between them and the fields. Using $\nabla \times \overline{\mathcal{A}} = \overline{\mathcal{B}}$, we have

$$\nabla \times \overline{\mathcal{E}} = -\frac{\partial\overline{\mathcal{B}}}{\partial t} \qquad \rightarrow \qquad \nabla \times (\overline{\mathcal{E}} + \frac{\partial\overline{\mathcal{A}}}{\partial t}) = 0$$

Noting that the scalar potential for the electrostatic case was defined on the basis of the fact that $\nabla \times \overline{\mathcal{E}} = 0$ as $\overline{\mathcal{E}} = -\nabla\tilde{\Phi}$, we can now define the *scalar potential* $\tilde{\Phi}$ for time-varying fields as

$$\overline{\mathcal{E}} + \frac{\partial\overline{\mathcal{A}}}{\partial t} = -\nabla\tilde{\Phi} \qquad \rightarrow \qquad \overline{\mathcal{E}} = -\nabla\tilde{\Phi} - \frac{\partial\overline{\mathcal{A}}}{\partial t}$$

Note that this expression for $\overline{\mathcal{E}}$ recognizes two possible sources for the electric field, namely charges (represented by the $\tilde{\Phi}$ term) and time-varying magnetic fields (represented by the $\partial\overline{\mathcal{A}}/\partial t$ term).

---

[50]The derivation is given in many advanced electromagnetics texts, including Chapter 8 of J. A. Stratton, *Electromagnetic Theory,* McGraw-Hill, New York, 1941.

In order for the vector potential $\overline{\mathcal{A}}$ to be uniquely specified, we must specify its divergence (in addition to its curl, which is already specified as $\nabla \times \overline{\mathcal{A}} = \overline{\mathcal{B}}$). The choice that leads to the simplest differential equations for $\overline{\mathcal{A}}$ and $\tilde{\Phi}$, and to solutions of $\overline{\mathcal{A}}$ and $\tilde{\Phi}$ in the form of retarded potentials as given in [6.62], is

$$\nabla \cdot \overline{\mathcal{A}} = -\mu\epsilon\frac{\partial\tilde{\Phi}}{\partial t} \qquad [6.63]$$

This condition, known as the *Lorentz condition* or the *Lorentz gauge,* is also consistent with the source charge and current distributions being related through the continuity equation. Using [6.63] and Maxwell's equations, the wave equations in terms of $\overline{\mathcal{A}}$ and $\tilde{\Phi}$ can be shown to be

$$\nabla^2\overline{\mathcal{A}} - \mu\epsilon\frac{\partial^2\overline{\mathcal{A}}}{\partial t^2} = -\mu\overline{\mathcal{J}} \qquad [6.64a]$$

$$\nabla^2\tilde{\Phi} - \mu\epsilon\frac{\partial^2\tilde{\Phi}}{\partial t^2} = -\tilde{\rho}/\epsilon \qquad [6.64b]$$

The fact that the retarded potentials as given in [6.62] are indeed the solutions of [6.64] can be verified by direct substitution[51] of [6.62] into the respective differential equations given by [6.64], although this process is rather involved mathematically.

In typical antenna problems, it is customary to avoid use of the scalar potential $\tilde{\Phi}$ and to cast the problem in hand in terms of only the vector potential $\overline{\mathcal{A}}$. With the vector potential $\overline{\mathcal{A}}$ described as in [6.62a] in terms of the source current $\overline{\mathcal{J}}$, solutions of antenna problems reduce to the following procedure:

1. Find $\overline{\mathcal{A}}$ from $\overline{\mathcal{J}}$ by integrating [6.62a].
2. Find $\overline{\mathcal{H}}$ from $\mu\overline{\mathcal{H}} = \nabla \times \overline{\mathcal{A}}$.
3. Find $\overline{\mathcal{E}}$ from $\overline{\mathcal{H}}$ using [1.11c] as $\nabla \times \overline{\mathcal{H}} = \epsilon\partial\overline{\mathcal{E}}/\partial t$. Note that if we were to find $\overline{\mathcal{E}}$ directly from the potentials, we would have to use $\overline{\mathcal{E}} = -\nabla\Phi - \partial\overline{\mathcal{A}}/\partial t$.

In the case when the sources vary sinusoidally in time, the steady-state wave equation [6.64a] for $\overline{\mathcal{A}}$ becomes

$$\nabla^2\mathbf{A} + \omega^2\mu\epsilon\mathbf{A} = -\mu\mathbf{J} \qquad [6.65]$$

where $\mathbf{A}$ and $\mathbf{J}$ are phasor quantities. Note that for this time-harmonic case, the phase delay over a distance R is given by $e^{-j\beta R}$ (for a uniform plane wave), so that the retarded potential solution [6.62a] becomes

$$\boxed{\mathbf{A}(\mathbf{r}) = \frac{\mu}{4\pi}\int_{V'}\mathbf{J}(\mathbf{r}')\frac{e^{-j\beta R}}{R}dv'} \qquad [6.66]$$

---

[51] Such substitution must be done with considerable care; see footnote on page 164 of H. H. Skilling, *Fundamentals of Electric Waves,* R. E. Krieger Publishing Co., New York, 1974. Expressions for $\mathbf{A}$ and Φ that satisfy [6.64] can also be written with $(t + R/v_p)$ instead of $(t - R/v_p)$ in [6.64] and are called *advanced potentials.* However, their use would clearly violate causality, implying that the potentials at a given time and place depend on *later* events elsewhere.

dissipation of electromagnetic energy within the dielectric. At any given frequency, ϵ'' produces the same macroscopic effect as the conductivity σ. Since the values of ϵ'' are often determined by measurement, losses due to nonzero conductivity σ and the dielectric losses discussed here are often indistinguishable. In practice, losses are represented by σ for metals and by ϵ'' for dielectrics.

For simplicity, we have only considered electronic (sometimes also called atomic) polarization, for which the resonance frequency ω_0 is typically quite high, usually in the ultraviolet range. However, total polarizability of a material under the influence of an electric field can be separated into three parts: electronic, ionic, and dipolar (or orientational) polarization. The electronic polarization arises from the displacement of the electron cloud with respect to the nucleus, as discussed above. The ionic polarization contribution comes from the displacement of a charged ion with respect to other ions. In some materials, two different atoms may join together as a molecule by forming a chemical bond. We can think of such molecules as consisting of positively and negatively charged ions, with the Coulomb forces between them serving as the binding force. Examples of such materials are HCl, CO_2, and H_2O. Depending on whether the electrons are transferred from one atom to the other or shared between atoms, the bond can be *ionic* or *covalent*. In either case the material may possess a permanent dipole moment, and furthermore, the application of an electric field to any such molecule displaces the positive ions with respect to negative ones and thereby induces a dipole moment. The dipolar or orientational polarizability is due to molecules with a permanent electric dipole that change orientation in response to the applied electromagnetic field. Some polyatomic molecules, such as H_2O, may be at least partially ionic and may consist of *polar* molecules, which carry a permanent dipole moment. With no electric field, the individual dipole moments point in random directions, so that the net dipole moment is zero. When an electric field is applied, such materials exhibit the electronic and ionic polarization effects discussed above. In addition, the electric field tends to line up the individual dipoles to produce an additional net dipole moment per unit volume. If all the dipoles in a material were to line up, the polarization would be very large; however, at ordinary temperatures and relatively small electric field levels, the collisions of the molecules in their thermal motion allow only a small fraction of the dipoles to line up with the field.

The frequency response of the dipolar and ionic polarizations is qualitatively similar to that of electronic polarization, but they exhibit much lower resonance frequencies. The reason for this behavior is the fact that the effective masses of the microscopic bodies that are displaced are substantially larger, consistent with the fact that for electronic polarization, the resonant frequency increases with decreasing particle mass (we have $\omega_0 = \sqrt{k/m_e}$). Although the elastic constant k is different for the different types of polarizations, the resonant frequency is nevertheless inversely proportional to the square root of the particle mass. The frequency response of a hypothetical dielectric is shown in Figure 6.13, with the contributions from the different types of polarizations identified in terms of the frequency ranges in which they significantly contribute.

Once again this solution can be verified by substitution into the time-harmonic form of the wave equation [6.64a]. Note that for the time-harmonic case, the Lorentz condition [6.63] is

$$\nabla \cdot \mathbf{A} = -j\omega\mu\epsilon\Phi$$

where $\mathbf{A}$ and Φ are phasor quantities.

6.5.2 Elementary Electric Dipole (Hertzian Dipole)

We now proceed with the quantitative description of electromagnetic radiation from isolated sources in free space. We first consider the simplest imaginable source: the elementary electrical dipole, also commonly known as the *Hertzian dipole*. Some more practical and complicated configurations of antennas can be considered to be made up of a large number of Hertzian dipoles, with magnitudes and phases of their currents appropriately selected. The elementary dipole antenna is taken to be so short (compared to the wavelength in the medium in which the antenna resides) that the current I on it may be considered to be uniform at all points over its length. The length of the infinitesimal dipole is denoted to be dl. The coordinate system is shown in Figure 6.17a. We consider only the case in which the current varies sinusoidally with time, so that [6.66] can be used to determine the phasor vector potential $\mathbf{A}$.

For cases such as the one in hand, where the current is confined to a wire whose cross-sectional area (ds') is small, we have $\mathbf{J}\, dv' = \mathbf{J}\, ds'\, dl' = I\, dl'\hat{\mathbf{z}}$, where $I = |\mathbf{J}|\, ds'$ is the total current in the wire in units of amperes (note that $\mathbf{J}$ is in units of A-m^{-2}). In this case, the integral [6.66] for the retarded vector potential $\mathbf{A}$

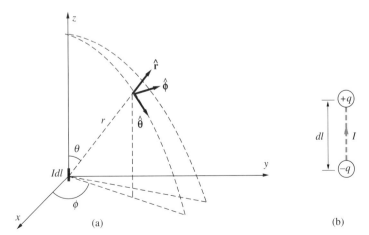

FIGURE 6.17. **The geometry of the Hertzian dipole.** (a) Coordinate system with the Hertzian dipole located at the origin with its current oriented in the z direction. (b) Hertzian dipole viewed in terms of charges accumulated at the ends of the wire.

reduces to

$$\mathbf{A}(\mathbf{r}) = \frac{\mu_0}{4\pi} \frac{\mathbf{J}(\mathbf{r})e^{-j\beta r}}{r} dv' = \hat{\mathbf{z}} \frac{\mu_0 Idl'}{4\pi} \left(\frac{e^{-j\beta r}}{r} \right)$$

where $\beta = \omega\sqrt{\mu_0\epsilon_0} = \omega/v_p = 2\pi/\lambda$. Note that we have chosen to locate the Hertzian dipole at the origin so that $\mathbf{r}' = 0$, and $R = |\mathbf{r}| = r$. Note also that no integration is required, since the source region is confined to the differential element itself. In order to express the phasor $\mathbf{A}$ in spherical coordinates we note that

$$\hat{\mathbf{z}} = \hat{\mathbf{r}}\cos\theta - \hat{\boldsymbol{\theta}}\sin\theta$$

so the spherical coordinate components of $\mathbf{A} = \hat{\mathbf{r}}A_r + \hat{\boldsymbol{\theta}}A_\theta + \hat{\boldsymbol{\phi}}A_\phi$ are

$$A_r = A_z\cos\theta = \frac{\mu_0 Idl}{4\pi} \left(\frac{e^{-j\beta r}}{r} \right) \cos\theta$$

$$A_\theta = -A_z\sin\theta = -\frac{\mu_0 Idl}{4\pi} \left(\frac{e^{-j\beta r}}{r} \right) \sin\theta$$

$$A_\phi = 0$$

The phasor magnetic field $\mathbf{H}$ can be obtained using $\mathbf{H} = \mu_0^{-1}\nabla \times \mathbf{A}$ as

$$\mathbf{H} = \frac{1}{\mu_0}\nabla \times \mathbf{A} = \hat{\boldsymbol{\phi}}\frac{1}{\mu_0 r}\left[\frac{\partial}{\partial r}(rA_\theta) - \frac{\partial A_r}{\partial \theta} \right]$$

$$= -\hat{\boldsymbol{\phi}}\frac{Idl}{4\pi}\beta^2\sin\theta\left[\frac{1}{j\beta r} + \frac{1}{(j\beta r)^2} \right]e^{-j\beta r} \qquad [6.67]$$

The phasor electric field $\mathbf{E}$ can be found from $\mathbf{H}$ as

$$\mathbf{E} = \frac{1}{j\omega\epsilon_0}\nabla \times \mathbf{H} = \frac{1}{j\omega\epsilon_0}\left[\hat{\mathbf{r}}\frac{1}{r\sin\theta}\frac{\partial}{\partial \theta}(H_\phi\sin\theta) - \hat{\boldsymbol{\theta}}\frac{1}{r}\frac{\partial}{\partial r}(rH_\phi) \right]$$

or

$$E_r = -\frac{Idl}{4\pi}\eta\beta^2 2\cos\theta\left[\frac{1}{(j\beta r)^2} + \frac{1}{(j\beta r)^3} \right]e^{-j\beta r} \qquad [6.68a]$$

$$E_\theta = -\frac{Idl}{4\pi}\eta\beta^2\sin\theta\left[\frac{1}{(j\beta r)} + \frac{1}{(j\beta r)^2} + \frac{1}{(j\beta r)^3} \right]e^{-j\beta r} \qquad [6.68b]$$

$$E_\phi = 0 \qquad [6.68c]$$

where $\eta = \sqrt{\mu_0/\epsilon_0}$.

We can find the instantaneous expressions for the electric and magnetic fields from these phasors in the usual manner. We have

$$\mathcal{H}_\phi = \frac{Idl\sin\theta}{4\pi}\left[\frac{-\beta\sin(\omega t - \beta r)}{r} + \frac{\cos(\omega t - \beta r)}{r^2} \right] \qquad [6.69]$$

and

$$\mathscr{E}_\theta = \frac{Idl \sin\theta}{4\pi\epsilon_0} \left[\frac{-\beta \sin(\omega t - \beta r)}{r\omega} + \frac{\beta \cos(\omega t - \beta r)}{r^2\omega} + \frac{\sin(\omega t - \beta r)}{\omega r^3} \right] \qquad [6.70a]$$

$$\mathscr{E}_r = \frac{2Idl \cos\theta}{4\pi\epsilon_0} \left[\frac{\beta \cos(\omega t - \beta r)}{r^2\omega} + \frac{\sin(\omega t - \beta r)}{\omega r^3} \right] \qquad [6.70b]$$

We note that even for this simplest current configuration the complete field expressions for the exact total field are quite complicated. Fortunately, it is seldom necessary to consider the total field. Note that the various terms involve dependencies on distance from the source in the form of r^{-1}, r^{-2}, and r^{-3}. For large distances from the current element, we can neglect the higher-order terms. For example, the r^{-1} and r^{-2} terms in E_θ and H_ϕ are equal in magnitude when $r = \beta^{-1} = \lambda/(2\pi) \simeq \lambda/6$. Thus, the r^{-2} rapidly becomes negligible at distances $r > \lambda/6$. Fundamentally, since the primary function of the antenna is to convey electromagnetic energy to distant points, in most applications we are able to ignore those terms that do not contribute to the field at large distances from the source. Furthermore, only the r^{-1} terms carry real time-average power away from the source.

In view of these considerations, the field represented by the r^{-1} terms in [6.69] and [6.70] is referred to as the *radiation field.* The field associated with the r^{-2} terms is usually referred to as the *induction field,* as this field is predicted by the Biot–Savart law. The r^{-3} term in [6.67] resembles the electric field intensity of an electric dipole and is sometimes referred to as the *electrostatic field* term.

We now examine the various terms in detail to determine their significance.

Near-Zone (Induction and Electrostatic) Fields

In the near field, defined as $\beta r = (2\pi r/\lambda) \ll 1$, or $r \ll \lambda/(2\pi) \simeq 0.16\lambda$, the dominant term (i.e., the terms with the highest power of r in the denominator) for the magnetic field phasor from [6.67] is

$$H_\phi \simeq \frac{Idl}{4\pi r^2} \sin\theta\, e^{-j\beta r} \simeq \frac{Idl}{4\pi r^2} \sin\theta \qquad [6.71]$$

where we have used the fact that

$$e^{-j\beta r} \simeq 1 - j\beta r - \frac{(\beta r)^2}{2} + \cdots \simeq 1 \qquad [6.72]$$

We note that this magnetic field is the same as the magnetostatic field due to an infinitesimal current element, as given by the Biot–Savart law,[52] if that law were

---

[52]The Biot–Savart law specifies the magnetostatic field at $d\mathbf{B}$ produced at point $\mathbf{r}$ by a current carrying element $d\mathbf{l}'$ located at $\mathbf{r}'$ as

$$d\mathbf{B}(\mathbf{r}) = \frac{\mu_0 I d\mathbf{l}' \times \hat{\mathbf{R}}}{4\pi R^2}$$

where $\mathbf{R} = \mathbf{r} - \mathbf{r}'$, $\hat{\mathbf{R}} = \mathbf{R}/|\mathbf{R}|$, and $R = |\mathbf{R}|$. For further discussion see Section 6.2 of U. S. Inan and A. S. Inan, *Engineering Electromagnetics,* Addison Wesley Longman, 1999.

extended to cover the case of an alternating current element $I \cos(\omega t)$. The instantaneous expression for this so-called *induction* term is

$$\mathcal{H}_\phi \simeq \frac{Idl}{4\pi r^2} \sin\theta \cos(\omega t - \beta r)$$

Note that at points very close to the current element, where the induction term dominates, r/v_p is very small, so we have $(t - r/v_p) \simeq t$, which is equivalent to $e^{-j\beta r} \simeq 1$. The curl of [6.71], i.e., $\nabla \times \mathbf{H} = j\omega\epsilon_0\mathbf{E}$, gives the electric field components that accompany the induction field H_ϕ, namely the r^{-2} terms in [6.67].

The dominant terms in the near field for the electric field components are the r^{-3} terms, which resemble the fields for an electrostatic dipole (i.e., the electric field at large distances from a dipole consisting of two charges $+q$ and $-q$ separated by a distance l). Taking the separation distance to be dl, we have

$$
\begin{aligned}
E_r &\simeq \frac{Idl\eta}{j4\pi\beta r^3} 2\cos\theta & \qquad E_r &\simeq \frac{qdl}{4\pi\epsilon r^3} 2\cos\theta \\
E_\theta &\simeq \frac{Idl\eta}{j4\pi\beta r^3} \sin\theta & \qquad E_\theta &\simeq \underbrace{\frac{qdl}{4\pi\epsilon r^3} \sin\theta}_{\text{Electric dipole}}
\end{aligned}
\qquad \text{[6.73]}
$$

The correspondence of the near-zone electric fields to those of an electric dipole can be physically understood as follows. Although the infinitesimal current element of length dl was only specified in terms of its current, the equation of continuity requires that there be accumulation of charge at the ends of the current element, given by

$$\frac{dq}{dt} = I \cos(\omega t) \qquad \longrightarrow \qquad q = \frac{I \sin(\omega t)}{\omega}$$

so the current flow in the z direction as depicted in Figure 6.17b corresponds to an increase of charge at one end and a corresponding decrease at the other end.

As a physical approximation, then, we can consider the current element as if it were terminated in two small spheres on which charges can accumulate, as shown in Figure 6.17b. As the current sinusoidally fluctuates in time, the amount of charge at both ends will also similarly oscillate. In phasor form, the oscillating charge is equivalent to a small electric dipole with dipole moment $qdl = -jIdl/\omega$. The terms on the right-hand side of [6.73] were indeed obtained from those on the left-hand side by substituting qdl for $-jIdl/\omega$.

Note that both of the near field-components (E_θ, E_r) of the electric field are out of phase by 90° (note the j in the electric field phasors) with respect to the induction magnetic field component H_ϕ, so these near fields do not represent any real power flow. The r^{-2} terms in $\mathcal{H}_\phi$ and those in $\mathcal{E}_r$, $\mathcal{E}_\theta$ are in time phase (i.e., they are purely real, as can be seen from [6.69] and [6.70]), which might lead one to think that the cross product of these terms represents real power flow. However, the real part of the Poynting vector resulting from the product of the r^{-2} terms is exactly canceled out by the product of the r^{-1} terms in $\mathcal{H}_\phi$ and the r^{-3} terms in $\mathcal{E}_r$ and $\mathcal{E}_\theta$, so no real power flow results from any of the near-zone fields.

Far-Zone (Fraunhofer) Fields   In the far field, also often called the Fraunhofer[53] field, defined as $\beta r \gg 1$ or $r \gg (\lambda/2\pi) \simeq 0.16\lambda$, the dominant terms are those that vary with radial distance as r^{-1}. From [6.69] and 6.70] we have

$$H_\phi \simeq j\frac{Idl}{4\pi}\left(\frac{e^{-j\beta r}}{r}\right)\beta \sin\theta \qquad\qquad [6.74a]$$

$$E_\theta \simeq j\frac{Idl}{4\pi}\left(\frac{e^{-j\beta r}}{r}\right)\eta\beta \sin\theta \qquad\qquad [6.74b]$$

$$E_r \simeq 0$$

where we see that the far-zone electric and magnetic fields are orthogonal to one another and are in time phase (i.e., both have a j as a multiplier). The ratio of the electric and magnetic field magnitudes is η; in other words, $|E_\theta/H_\phi| = \eta$.

To underscore the similarity of the field expressions [6.74] to those for uniform plane waves, we can rewrite the phasors for the far fields by absorbing all of the amplitude factors and the 90° phase factor represented by j into a maximum electric field amplitude $E_m = |(jI\eta dl\beta)/(4\pi)|$. We then have

$$E_\theta \simeq \frac{E_m \sin\theta}{r}e^{-j\beta r}$$

$$H_\phi \simeq \frac{E_m \sin\theta}{\eta r}e^{-j\beta r}$$

As a result, the instantaneous expressions for the far-field electric and magnetic fields can be written as

$$\overline{\mathcal{E}}(r, \theta, t) = \frac{E_m \sin\theta}{r}\cos(\omega t - \beta r)$$

$$\overline{\mathcal{H}}(r, \theta, t) = \frac{E_m \sin\theta}{\eta r}\cos(\omega t - \beta r)$$

Since the phasor fields vary with radial distance r as $r^{-1}e^{-j\beta r}$, the surfaces of constant phase, which are defined by $r^{-1}e^{-j\beta r} = $ const., are spheres, and E_θ and H_ϕ lie in a surface perpendicular to the direction of propagation (radial direction), as depicted in Figure 6.18. In other words, the far-zone or radiation fields have all the characteristics of transverse electromagnetic (TEM) waves, or uniform plane waves, except that the surfaces of constant phase are spheres instead of planes. Therefore,

---

[53]Joseph von Fraunhofer [1787–1826] was a German optician and physicist. He was orphaned at the age of eleven, but educated himself to become an accomplished lens maker and inventor of optical instruments, among which was a diffraction grating for the accurate measurement of wavelengths. In testing prisms for his glasses Fraunhofer found that the solar spectrum contained numerous dark lines, and led the way for the later discovery by Kirchhoff [1824–1887] of the constitution of the sun by means of spectroscopic measurements. Despite his many interesting findings, Fraunhofer was scorned by most scientists as a mere technician, and although he might attend scientific meetings, he was not allowed to address them. He never married, and died of tuberculosis before he was forty.

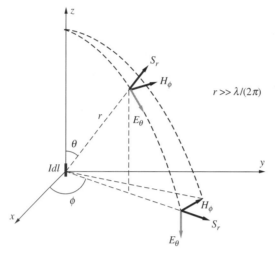

FIGURE 6.18. **Far-field components of the Hertzian dipole.** At large distances from the source, the electric and magnetic field components are both perpendicular to the direction of propagation, and the Poynting vector S_{av} points in the radial direction. In other words, $S_{av} = \hat{\mathbf{r}} S_r$, where S_r is the radial component as shown.

at large distances from the source, the radiation field is indistinguishable from a uniform plane wave. This result is an approximation that recognizes the fact that if the wave is observed over a planar region of small transverse extent tangent to a point on a large sphere centered at the source, neither the $\sin\theta$ nor r^{-1} terms can change appreciably over the region under observation, so the fields are approximately constant over the planar observation region.

For different antennas, the far-zone field expressions have similar forms, except that their dependence on direction (i.e., ϕ and θ) may be more complex, and the electric and magnetic fields may point in directions other than $\hat{\boldsymbol{\theta}}$ and $\hat{\boldsymbol{\phi}}$. In general, assuming an antenna structure centered at the origin, we can write the far-zone electric and magnetic field phasors as

$$\mathbf{E} = \mathbf{E}_0(\theta, \phi)\frac{e^{-j\beta r}}{r} \qquad [6.75a]$$

$$\mathbf{H} = \frac{1}{\eta}\hat{\mathbf{r}} \times \mathbf{E} \qquad [6.75b]$$

where $\mathbf{E}_0(\theta, \phi)$ is a vector that in general is a function of θ and ϕ.

The Electric Field Lines Figure 6.19 shows the configuration of the electric field lines[54] in the vicinity of the Hertzian dipole at different instances of time. A cross section is shown in the yz plane containing the antenna wire. The magnetic field lines

---

[54]To find the electric field lines we can set $(E_r/dr) = E_\theta/(rd\theta)$, because an elemental field line with components dr and $rd\theta$ is parallel to the local electric field intensity, with components E_r and E_θ. The equation of the field lines can be shown to be

$$\frac{1}{\lambda}\sin^2\theta\left(\frac{\lambda^2}{r^2} + 1\right)\cos\left[\omega t - \frac{r}{\lambda} + \tan^{-1}\left(\frac{r}{\lambda}\right)\right] = \text{const.}$$

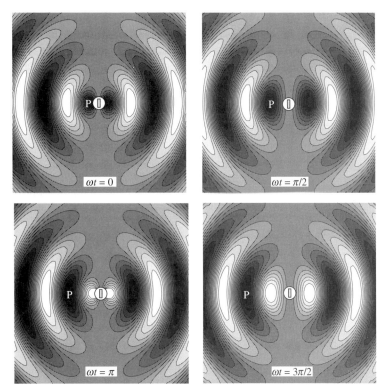

FIGURE 6.19.   **The electric field lines of the Hertzian dipole.** The electric field lines are shown at $\omega t = 0, \pi/2, \pi, 3\pi/2$. The darkest shading corresponds to positive electric field peaks, whereas the lightest areas correspond to negative peaks.

are much simpler in form; they are circles perpendicular to and centered on the axis of the dipole. The electric field lines in the induction field terminate on the equivalent dipole charges (see Figure 6.17b) on the antenna. Note that the electric field lines shown are not valid in the immediate vicinity of the Hertzian dipole, since the field solutions as given by [6.74] grow without limit as $r \rightarrow 0$. To accurately determine the electric field values as $r \rightarrow 0$, the finite length and the diameter of the antenna wire must also be included in our analysis.

Figure 6.19 indicates that the electric field lines of the radiated field do not terminate on charges but instead close on themselves. This is to be expected, since the electric field has zero divergence at points away from the source. It is instructive to follow a particular feature of the field, for example the positive peak marked "P" in Figure 6.19. At $\omega t = 0$, the electric field lines associated with P are part of the induction field, and they originate at charges on the antenna. A quarter of a period later (i.e., at $\omega t = \pi/2$), the loops of the field lines begin to move away from the source. At $\omega t = \pi$, the lines have "snapped off" and now close on themselves. After this time, they are cut adrift from their source and progressively move outward. The electric field lines forming the peak P no longer result from the presence of nearby charges but are instead produced by the time-changing magnetic field of the radiated wave. The radiated magnetic field in turn results not from an actual flow of current on the wire but from the time-changing electric field. The radiated wave could of course not have originated if there had not been charges or currents somewhere; however, once produced, the wave can propagate any distance for an unlimited time, even if its source ceases to exist. As an example, light waves from a supernova explosion can reach earth millions of years after the brightness that created them has been extinguished.

Time-Average Radiated Power The power flow per unit area at any point $P(r, \theta, \phi)$ is given by the Poynting vector at that point. For the electric and magnetic fields as given in [6.69] and [6.70], the complex Poynting vector $\mathbf{S} = \mathbf{E} \times \mathbf{H}^*$ has both θ and r components. These are

$$S_\theta = -E_r H_\phi = \left(\frac{Idl}{4\pi}\right)^2 \eta\beta \sin 2\theta \left[\frac{j}{r^3} - \frac{j}{\beta^2 r^5}\right] \qquad [6.76a]$$

$$S_r = \left(\frac{Idl \sin\theta}{4\pi}\right)^2 \eta\beta^2 \left[\frac{1}{r^2} - \frac{j}{\beta r^5}\right] \qquad [6.76b]$$

Note that the only *real* term in [6.76] is the r^{-2} term in S_r, which is simply the product of the r^{-1} terms in E_θ and H_ϕ. All other terms in [6.76] are imaginary and thus represent reactive power involving energy alternating back and forth between the electric and magnetic field. Note that at large distances from the source the radiation fields are the only terms that have appreciable value, and their product constitutes real time-average outward power flow. However, even close to the current element, where the induction and the electrostatic field terms are dominant, the r^{-1} terms are the only ones that represent outward time-average power flow emanating from the source.

The fact that only the r^{-1} terms contribute to real outward power flow is to be expected on the basis of physical reasoning. Consider two concentric spherical shells of radii r_a and r_b centered at the origin. The outward (i.e., in the r direction) rate of energy flow (i.e., power) through one of the shells must be the same as that through the other shell, since otherwise there would be continuous accumulation (or depletion) of energy stored in the region between them. Since the surface areas of the two concentric shells increase as r^2, the power density must correspondingly decrease as r^{-2}. Since electromagnetic power density is proportional to the product of electric and magnetic fields, the only field terms that can contribute to real outward power flow are the radiation fields that vary with distance as r^{-1}. Other field components can contribute to the instantaneous flow of energy into the region between the shells during parts of the sinusoidal cycle; however, this energy must be returned to the source during other parts of the cycle, since it cannot accumulate in the finite volume between the shells. Note that since we consider sinusoidal steady-state solutions, any such accumulation would have been ongoing for all time, leading to an infinite amount of energy in a finite volume.

The total time-average power P_{rad} radiated into the surrounding space by the current element can be found by integrating the real part of [6.76b] over a sphere. We have the time-average Poynting vector given by

$$\mathbf{S}_{av} = \frac{1}{2}\mathscr{R}e\{\mathbf{E} \times \mathbf{H}^*\} = \frac{1}{2}|E_\theta||H_\phi|$$

$$P_{rad} = \oint_{all\ sphere} \mathbf{S}_{av} \cdot d\mathbf{S} = \frac{1}{2}\int_0^{2\pi}\int_0^\pi E_\theta H_\phi^* \underbrace{R^2 \sin\theta d\theta d\phi}_{volume\ element}$$

$$\simeq \frac{1}{2}\int_0^{2\pi}\int_0^\pi \left[j\frac{Idl}{4\pi}\left(\frac{e^{kJ\beta R}}{R}\right)\eta\beta\sin\theta\right]\left[j\frac{Idl}{4\pi}\left(\frac{e^{kJ\beta R}}{R}\right)\beta\sin\theta\right]R^2\sin\theta d\theta d\phi$$

$$= \frac{I^2(dl)^2}{32\pi^2}\eta\beta^2\int_0^{2\pi}\underbrace{\int_0^\pi \sin^3\theta d\theta}_{\left[-\frac{\cos\theta}{3}(\sin^2\theta+2)\right]_0^\pi} d\phi$$

$$P_{rad} = \frac{(Idl)^2}{12\pi}\eta\beta^2$$

It is customary to express the total time-average power radiated by the antenna in terms of average power absorbed in an equivalent resistance called the *radiation resistance*. For the infinitesimal current element the radiation resistance R_{rad} is defined by

$$\frac{1}{2}R_{rad}I^2 = P_{rad} \qquad \longrightarrow \qquad R_{rad} = \frac{2P_{rad}}{I^2} \simeq \frac{2}{I^2}\left[\frac{(Idl)^2}{12\pi}\eta\beta^2\right]$$

Substituting the values for the permittivity and permeability for free space (i.e., $\eta = 120\pi\ \Omega$), and noting that $\beta = 2\pi/\lambda$, we have

$$[R_{rad}]_{\text{Hertzian dipole}} \simeq 80\pi^2\left(\frac{dl}{\lambda}\right)^2 \simeq 790\left(\frac{dl}{\lambda}\right)^2 \qquad [6.77]$$

as the radiation resistance of a Hertzian dipole in free space. As an example, for an infinitesimal dipole of length one-hundredth of a wavelength (i.e., $dl = 0.01\lambda$), we have $R_{rad} \simeq 0.079\Omega$. Since the radiation resistance in this case is very small, significant power will be radiated only for very large current amplitudes, which lead to large amounts of power dissipation in the conductor (i.e., the $I^2 R_{wire}$ losses in the antenna wire due to the electrical resistance R_{wire} of the wire), and therefore very low efficiency. It thus appears that an infinitesimally short linear current element is not an efficient electromagnetic radiator.

The Hertzian dipole is not an efficient radiator because its dimensions are much smaller than a wavelength. That an efficient radiator or antenna must have dimensions comparable to or greater than a wavelength can be heuristically understood as follows: For wire antennas, the currents and their associated charges are the sources for the distant fields. At any given time, there must be equal numbers of positive and negative sources (for example, consider the positive and negative charges in Figure 6.17b), and when the distance between them is small compared to a wavelength, their effects at large distances tend to cancel. On the other hand, when the positive and negative sources are separated by distances comparable to a wavelength, the phase shift (or retardation) between different points of the source distribution and the distant observation point can be large enough that the positive and negative source contributions do not cancel, and may even add constructively in certain directions. In the next subsection, we will briefly discuss dipole antennas of larger size, including the highly practical and commonly used half-wavelength dipole antenna.

Radiation Pattern We now consider the directional properties of the Hertzian dipole antenna. In this context we also introduce several important antenna concepts such as radiation intensity, directive gain, and directivity of an antenna. In assessing the directional properties of an antenna, we are interested in the relative amount of power radiated in a given direction as specified by the polar angle θ and the azimuthal angle ϕ. Noting that there are r^2 square meters of spherical surface area for each unit solid angle, the *radiation intensity* $U(\theta, \phi)$ in a given direction is defined as the power per unit solid angle in that direction. Namely,

$$U(\theta, \phi) \equiv r^2 |\mathbf{S}_{av}| \qquad \text{watts/unit solid angle} \qquad [6.78]$$

where $\mathbf{S}_{av}$ is the time-average Poynting vector. For the Hertzian dipole, we have

$$U(\theta, \phi) = \frac{1}{2} \left(\frac{Idl\beta}{4\pi} \right)^2 \eta \sin^2 \theta \qquad \text{watts/unit solid angle} \qquad [6.79]$$

As expected from the symmetry involved, the radiation intensity for an infinitesimal wire antenna oriented in the z direction and located at the origin is independent of ϕ. As a function of θ, the radiation intensity $U(\theta)$ is a maximum in the $\theta = \pi/2$ direction and zero in the $\theta = 0, \pi$ directions.

The *directive gain* $g_d(\theta, \phi)$ of an antenna *in a given direction* is defined as the ratio of the radiation intensity $U(\theta, \phi)$ in that direction and the total time-average power radiated per unit solid angle, denoted as U_{av}. Since the total radiated average power is P_{rad} and there are 4π steradians on a sphere, the time-average power radiated per unit solid angle is

$$U_{av} = \frac{P_{rad}}{4\pi}$$

A fictitious isotropic (nondirectional) antenna radiating a total time-average power of P_{rad} uniformly in all directions radiates an amount of power, per unit solid angle, as given above. The directive gain is defined as

$$g_d(\theta, \phi) \equiv \frac{U(\theta, \phi)}{U_{av}} = \frac{4\pi U(\theta, \phi)}{P_{rad}}$$

The time-average Poynting vector $\mathbf{S}_{av}$ is simply related to the radiation intensity and is thus given by

$$\mathbf{S}_{av}(\theta, \phi) = \hat{\mathbf{r}}\frac{U(\theta, \phi)}{r^2} = \hat{\mathbf{r}}\frac{g_d(\theta, \phi)P_{rad}}{4\pi r^2}$$ [6.80]

For the Hertzian dipole, the far-zone fields are azimuthally symmetric (i.e., they do not depend on ϕ), so the directive gain is a function only of θ and is given by

$$U_{av} = \frac{P_{rad}}{4\pi} \simeq \frac{(Idl)^2\beta^2\eta}{48\pi^2} \qquad \text{watts/steradian}$$

The directive gain is then

$$g_d(\theta) = \frac{U(\theta, \phi)}{U_{av}} \simeq \frac{\frac{1}{2}\left(\frac{Idl\beta}{4\pi}\right)^2\eta\sin^2\theta}{\frac{(Idl)^2\beta^2\eta}{48\pi^2}} = \frac{3}{2}\sin^2\theta$$ [6.81]

The maximum value of the directive gain for an antenna is commonly referred to as the *directivity* of an antenna. The directivity is a measure of the ability of an antenna to concentrate power in a given direction. Noting that the directive gain $g_d(\theta)$ for the Hertzian dipole is maximum at $\theta = 90°$, with its value given by $g_d(\pi/2) = 1.5$.

In general, the directive gain $g_d(\theta, \phi)$ defines a three-dimensional surface called the *polar radiation pattern* of the radiator. Figure 6.20a shows the radiation pattern for the Hertzian dipole. In any plane defined by $\phi = $ constant, the beamwidth of the radiation pattern between the half-power points is 90°, as determined by solving $g_d(\theta) = 0.5g_d(\pi/2)$ or $\sin^2\theta = 0.5$, as illustrated in Figure 6.20b.

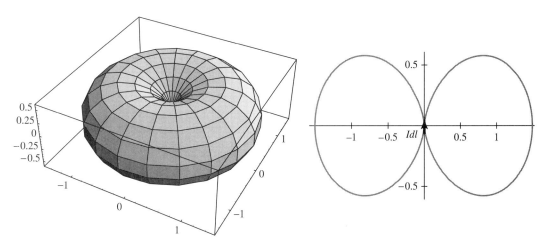

FIGURE 6.20.   **Polar radiation pattern for an infinitesimal dipole.** (a) The three dimensional plot shows the normalized radiation intensity $[U(\theta, \phi)/U_{max}] = \sin^2\theta$ for an infinitesimal current carrying element at the origin. The radial distance from the origin to the surface shown is proportional to the radiation intensity in the corresponding direction. Most of the energy is radiated near the equatorial plane at $\theta = \pi/2$ or in the xy plane; none is radiated along the vertical (z) axis. (b) Two-dimensional cross section in any plane defined by $\phi = $ constant.

Example 6-7 illustrates the calculation of time-average radiated power for a very short electric dipole antenna (i.e., the Hertzian dipole).

Example 6-7: **Radiated power.** A very short electric dipole antenna is to be used by an amateur enthusiast for local communications within his house and yard. The FCC regulations indicate that the radiated field at a distance of $\lambda/(2\pi)$ and at the direction of maximum intensity must be less than $15\sqrt{2}\ \mu\text{V-m}^{-1}$. Determine the maximum time-average radiated power that the user may achieve with his antenna.

Solution: If E_{max} is the maximum intensity of the electric field, then the time-average power density is $\mathbf{S}_{av} = \hat{\mathbf{r}}\frac{1}{2}E_{max}^2/\eta$. Note that the time-average Poynting flux is related to the radiated power P_{rad} through [6.80]. At the direction of maximum radiation, we have $[g_d(\theta)]_{max} = 1.5$ for the Hertzian dipole, so we have

$$\mathbf{S}_{av}(\theta,\phi) = \frac{g_d(\theta,\phi)P_{rad}}{4\pi r^2} \quad \rightarrow \quad \frac{[g_d(\theta)]_{max}P_{rad}}{4\pi r^2} = \frac{E_{max}^2}{2\eta}$$

Since E_{max} is specified at $r = \lambda/(2\pi)$, we then have

$$P_{rad} = 4\pi\left(\frac{\lambda}{2\pi}\right)^2\frac{E_{max}^2}{2\eta}\frac{1}{[g_d(\theta)]_{max}} = \frac{\lambda^2 E_{max}^2}{2\pi\eta[g_d(\theta)]_{max}}$$

$$= \frac{(10)^2(15\sqrt{2}\times10^{-6})^2}{2\pi\times120\times\pi\times1.5} \simeq 1.27\times10^{-11}\ \text{W}$$

Note, however, that at $r = \lambda/(2\pi)$, we have $\beta r = 1$, so we are at the inner edge of the far field. Thus, the far-field expressions for the fields are not very accurate. If we use the full expressions [6.68] for the electric field, we find that the field at $\beta r = 1$ is contributed by E_θ as well as by E_r and that, in fact, even for E_θ the terms other than the r^{-1} term are not insignificant. As a result, in order to meet the FCC requirement at $r = \lambda/(2\pi)$, the magnitude of the E_θ component must be smaller than the E_{max} value used above, thus leading to a radiated power lower than the P_{rad} determined above. Calculations using [6.68] indicate that the actual value of the maximum radiated power that can be achieved while still meeting the FCC requirement is $\sim 1.58\times10^{-12}$ W.

6.5.3 Short Linear Antenna and the Half-Wave Dipole

The Hertzian dipole is a useful theoretical tool but is not a practical antenna. Furthermore, our assumption of constant current throughout the length of the wire is physically impossible unless the conductor is part of a larger circuit. If, for example, such an infinitesimal element were a part of a longer antenna wire, there would be radiation from each part of the longer wire. Total radiation from the antenna can then be found by superposition (or integration) of the components of radiation from the many short sections of the antenna.

The Short Linear Antenna   As a more practical dipole antenna, consider a wire of finite length but still much shorter than a wavelength. Assume this wire to be isolated in space and center-fed by a source, as shown in Figure 6.21, so as to produce current in the wire. The current amplitude on such a short antenna can be assumed to decrease linearly, from a maximum at the center to zero at the ends. Note that we expect the current to be zero at the open-circuited ends purely on physical grounds. The assumption of a linear current distribution is approximately valid for antenna lengths of $L \leq \lambda/10$. With the assumed current distribution, the radiation fields for the short dipole antenna can be obtained by using [6.66] to find **A** and use it to determine the **E** and **B** fields. Alternatively, we can derive the result for the short antenna by viewing it in terms of a superposition of many differential sections, each short enough so that the current is substantially constant through its length. Each differential section of the short antenna contributes to the electric and magnetic fields at a distant point in accordance with [6.74]. Consider the instantaneous far-field electric field

$$\overline{\mathcal{E}}(r, \theta, t) = \hat{\mathbf{\theta}} \frac{(I\,dl)\eta\beta \sin\theta}{4\pi r} \cos(\omega t - \beta r) \qquad [6.82]$$

The total field at the observation point is simply a summation of the amplitudes of the contributions from each differential section of the antenna. The only factor in this expression for $\overline{\mathcal{E}}$ that varies significantly from one point of the antenna to another is the current I. Since $\overline{\mathcal{E}}(r, \theta, \phi)$ is linearly proportional to I, the radiation from a section near the end of the antenna contributes much less to the field at the observation point than that from a section of equal length near the middle. For the same current I_0 at the terminals, the far-zone electric field for the short dipole of length L is equal to what would be received from an antenna half as long[55] but having a current over its

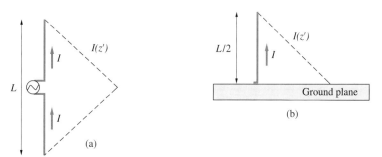

FIGURE 6.21.   **Short antenna.** (a) Current distribution on a short antenna. (b) A short monopole above conducting ground.

---

[55]This result is often expressed by stating that the *effective length* of the short dipole antenna is 1/2. The effective length is defined as

$$L_{\text{eff}} = \frac{1}{I_0} \int_{-L/2}^{L/2} I(z)\, dz$$

where I_0 is the current at the antenna terminals.

entire length equal to the current I_0 at its terminals. By the same token, a short dipole of length L radiates only one-quarter as much power as the Hertzian dipole of the same length but having current I_0 throughout its length. This is because the fields at every observation point are reduced by a factor of two, and the power density is accordingly reduced by a factor of four. Thus, the radiation resistance of a practical short dipole of length L should be four times that of a Hertzian dipole. In other words,

$$[R_{\text{rad}}]_{\text{short dipole}} \simeq 20\pi^2(L/\lambda)^2$$

If the short antenna were a monopole of height $L/2$ above a ground reflecting plane, as shown in Figure 6.21b, it would produce the same field distributions above the plane when fed with the same current at its terminals. However, since it radiates only to the hemisphere above the ground plane, its radiation resistance is half that of the dipole:

$$[R_{\text{rad}}]_{\text{short monopole}} \simeq 10\pi^2(L/\lambda)^2$$

The short dipole and monopole antennas are not particularly good radiators, because their radiation resistance is still quite small; for example, for $L = 0.1\lambda$, the radiation resistance of the the short dipole is $R_{\text{rad}} \simeq 2\Omega$, which is a small value for typical transmission line feeds and thus presents significant problems in matching. We also note that it is not accurate to think of R_{rad} to increase with wire length as L^2, since the expression for R_{rad} was derived by assuming $L \ll \lambda$. Nevertheless, the short dipole antenna is useful in practice, especially in applications involving relatively low frequencies (or large wavelengths), where the size of the radiators must necessarily be small compared to a wavelength.

The Half-Wavelength Dipole Antenna   An antenna configuration widely used in many applications is the half-wavelength dipole or the corresponding quarter-wavelength monopole. In addition to being rather efficient radiators, these antennas have very desirable input impedance properties and so can be efficiently driven by standard transmission lines.[56] The half-wavelength antenna is a special case of center-fed thin-wire electric dipole antennas having lengths L comparable to a wavelength. When the length of the antenna wire is comparable to the wavelength, current on the antenna can no longer be assumed to decrease linearly from a maximum to zero at the ends. This type of an antenna behaves in a manner similar to an open-circuited transmission line[57] with distributed capacitance, since the current on a short open-circuited transmission line section is substantially linear.[58] However, for a longer open-circuited transmission line section, the current is proportional to the sine of the distance from the open-circuited end. The correspondence

[56]See Chapter 14 of E. C. Jordan and K. G. Balmain, *Electromagnetic Waves and Radiating Systems,* Prentice-Hall, Englewood Cliffs, New Jersey, 1968.

[57]See Section 11.14 of E. C. Jordan and K. G. Balmain, *Electromagnetic Waves and Radiating Systems,* Prentice-Hall, Englewood Cliffs, New Jersey, 1968.

[58]See Section 3.2.1 of U. S. Inan and A. S. Inan, *Engineering Electromagnetics,* Addison Wesley Longman, 1999.

between the open-circuited line with a uniform distributed capacitance and the dipole antenna is not precise, since the distributed capacitance along the antenna is not exactly constant. However, assuming the antenna current to be sinusoidal is in fact quite reasonable, especially in cases where the diameter of the wire is much smaller than its length. Accordingly, we proceed by assuming a sinusoidal current distribution on the antenna, as illustrated in Figure 6.22.

Assuming the length of the dipole antenna to be $2L$, the current is thus taken to be

$$I(z') = \begin{cases} I_0 \sin \beta(L - z') & z' > 0 \\ I_0 \sin \beta(L + z') & z' < 0 \end{cases} \qquad [6.83]$$

The electric and magnetic fields around the antenna can be determined by following a similar procedure as that used for the Hertzian dipole, namely starting with [6.66] to find $\mathbf{A}$, from which we can find $\mathbf{E}$ and $\mathbf{B}$. However, since we are only interested in the far-zone fields, we can use the principle of superposition to extend the results we already have for the Hertzian dipole. The contribution dE_θ to the far-zone electric field of each current element $I\,dz$ along the antenna in Figure 6.22, and the corresponding far-zone magnetic field dH_ϕ are given by the Hertzian dipole relationships, namely

$$dE_\theta = \eta\,dH_\phi = j\frac{I(z')dz'}{4\pi}\left(\frac{e^{-j\beta R}}{R}\right)\eta\beta \sin \theta \qquad [6.84]$$

where R is the distance from the source element $I\,dz'$ to the observation point P. For points in the far zone (i.e., $r \gg \lambda$), R is only slightly different from the radial distance r from the origin to point P. Thus, we have

$$R = [r^2 + (z')^2 - 2rz' \cos \theta]^{1/2} \simeq r - z' \cos \theta \qquad [6.85]$$

The difference between R and r is important primarily for the phase term ($e^{-j\beta R}$) in [6.84]; for the amplitude term, we can simply assume $R \simeq r$.

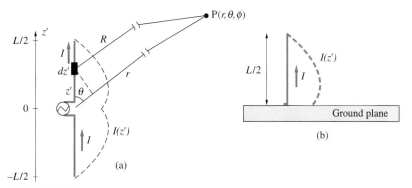

FIGURE 6.22.   Dipole antenna with sinusoidal current distribution. (a) Dipole antenna isolated in space and fed through its terminals. (b) Corresponding monopole above ground.

Integrating [6.84] we obtain

$$
E_\theta = j\frac{I_0\eta\beta\sin\theta}{4\pi r}e^{-j\beta r}\left[\int_{-L}^{0}\sin\beta(L+z')e^{j\beta z'\cos\theta}dz' + \int_{0}^{L}\sin\beta(L-z')e^{j\beta z'\cos\theta}dz'\right]
$$

$$
= j\frac{I_0\eta\beta\sin\theta}{4\pi r}e^{-j\beta r}\left[2\frac{\cos(\beta L\cos\theta)-\cos\beta L}{\sin^2\theta}\right]
$$

$$
E_\theta = \frac{j60I_0}{r}e^{-j\beta r}\left[2\frac{\cos(\beta L\cos\theta)-\cos\beta L}{\sin^2\theta}\right] \tag{6.86}
$$

Note that the far-field magnetic field is simply $H_\phi = E_\theta/\eta$. For the half-wave dipole, with $L = \lambda/4$, [6.86] reduces to

$$
E_\theta = \frac{jI_0\eta}{2\pi r}e^{-j\beta r}\left[\frac{\cos\left(\frac{\pi}{2}\cos\theta\right)}{\sin\theta}\right] = \frac{j60I_0}{r}e^{-j\beta r}\left[\frac{\cos\left(\frac{\pi}{2}\cos\theta\right)}{\sin\theta}\right] \tag{6.87}
$$

Noting that $H_\phi = E_\theta/\eta$, the complex time-average Poynting vector for the half-wave dipole is

$$
\mathbf{S}_{\text{av}} = \frac{1}{2}\mathcal{R}e\{\mathbf{E}\times\mathbf{H}^*\} = \frac{1}{2}E_\theta H_\phi^* = \frac{15I_0^2}{\pi r^2}\left[\frac{\cos(\frac{\pi}{2}\cos\theta)}{\sin\theta}\right]^2
$$

The total time-average radiated power P_{rad} can be obtained by integrating $\mathbf{S}_{\text{av}}$ over a sphere of radius r. We have

$$
P_{\text{rad}} = \int_{0}^{2\pi}\int_{0}^{\pi}[\mathbf{S}_{\text{av}}]R^2\sin\theta\,d\theta\,d\phi
$$

$$
= \frac{I_0^2\eta}{4\pi}\int_{0}^{\pi}\frac{\cos^2(\frac{\pi}{2}\cos\theta)}{\sin\theta}d\theta \tag{6.88}
$$

The above integral can be numerically evaluated or written analytically in terms of the cosine integral function.[59] We find

$$
P_{\text{rad}} \simeq 36.57I_0^2 \qquad \text{or} \qquad \longrightarrow \qquad [R_{\text{rad}}]_{\text{half-wave}} \simeq 73.13\Omega
$$

where we have used $P_{\text{rad}} = (1/2)I_0^2 R_{\text{rad}}$. Note that the radiation resistance of the half-wave dipole is quite large.

The directive gain of the half-wave dipole is given by

$$
g_d(\theta) \simeq \frac{60}{36.57}\left[\frac{\cos\left(\frac{\pi}{2}\cos\theta\right)}{\sin\theta}\right]^2 \simeq 1.64\left[\frac{\cos\left(\frac{\pi}{2}\cos\theta\right)}{\sin\theta}\right]^2
$$

The maximum value of $g_d(\theta)$, or the directivity of the half-wave dipole antenna, is ~ 1.64, which is only slightly larger than that for the Hertzian dipole.

---

[59]Defined as

$$
\text{Ci}(x) \equiv -\int_{x}^{\infty}u^{-1}\cos u\,du
$$

A monopole antenna of length L above a ground reflecting plane as shown in Figure 6.22b, produces the same field distributions above the plane as a dipole of length $2L$ when fed with the same current at its terminals. However, since the quarter-wave monopole radiates only to the hemisphere above the ground plane, its radiation resistance is half that of the half-wave dipole. Thus we have

$$[R_{\text{rad}}]_{\text{quarter-wave monopole}} \simeq 36.5\Omega$$

Although analysis of the input impedance of wire antennas is beyond the scope of this book, the reactive part of the input impedance of the half-wave dipole and the quarter-wave monopole antennas turn out to be approximately zero. In fact, the reactance can be made[60] to be exactly zero by choosing the length L to be slightly less than $\lambda/4$. Since the resistive part of the input impedance is just the radiation resistance, the half-wave dipole antenna can be readily matched using standard transmission lines with characteristic impedances of 50Ω or 75Ω. This property of the half-wave dipole antenna accounts for its popular use.

Example 6-8 compares the effective length and radiation resistance of a dipole antenna in air and water.

Example 6-8:   Electric dipole antenna in air and in water.   A certain electric dipole antenna has an effective length of $L_{\text{eff}} = \lambda/10$ in air. It is to be used at the same frequency under fresh lake water, which for the purposes of this problem may be treated as an unbounded lossless dielectric with $\epsilon = 9\epsilon_0$. (a) Estimate the radiation resistance of this antenna under water, stating any assumptions. (b) Suppose the original dipole had an effective length of $L_{\text{eff}} = \lambda/3$ in air. How does your answer to (a) change?

Solution:

(a) We have

$$\lambda_{\text{H}_2\text{O}} = \frac{\lambda_{\text{air}}}{\sqrt{\epsilon_r}} = \frac{\lambda_{\text{air}}}{3}$$

In air we had $L_{\text{eff}} = \lambda_{\text{air}}/10$. Assuming a linear current distribution as shown in Figure 6.21a, and a physical antenna length of l, we have

$$L_{\text{eff}} = \frac{1}{I_0} \int_{-l/2}^{l/2} I_0 \left(1 - \frac{2|z|}{l}\right) dz = \frac{2}{I_0} I_0 \left[z - \frac{z^2}{l}\right]_0^{l/2} = \frac{l}{2}$$

Thus, the total length of the dipole is

$$l = 2L_{\text{eff}} = \frac{\lambda_{\text{air}}}{5} = \frac{3\lambda_{\text{H}_2\text{O}}}{5}$$

---

[60] See Chapter 14 of E. C. Jordan and K. G. Balmain, *Electromagnetic Waves and Radiating Systems,* Prentice-Hall, Englewood Cliffs, New Jersey, 1968.

If, on the other hand, we assumed a sinusoidal current distribution as in Figure 6.22a, we have

$$L_{\text{eff}} = \frac{1}{I_0} \int_{-l/2}^{l/2} I_0 \cos\left(\frac{\pi z}{l}\right) dz = \frac{2l}{\pi}$$

or

$$l = \frac{\pi L_{\text{eff}}}{2} = \frac{\pi}{20}\lambda_{\text{air}} = \frac{3\pi}{10}\lambda_{H_2O} \simeq \frac{\lambda_{H_2O}}{2}$$

In either case, the radiation resistance of this antenna should be very close to that of a half-wave dipole antenna in water. Thus, we can use [6.88] to determine P_{rad} for a given current magnitude I_0 or the radiation resistance R_{rad}. Note from [6.88] that P_{rad} is proportional to $\eta = \sqrt{\mu/\epsilon}$ or inversely proportional to $\sqrt{\epsilon_r}$, which also means that R_{rad} is also proportional to $\epsilon^{-1/2}$. Thus, we have

$$(R_{\text{rad}})_{\text{in } H_2O} = \frac{1}{\sqrt{\epsilon_r}}(R_{\text{rad}})_{\text{in air}} = \frac{73.13}{3} \simeq 24 \ \Omega$$

(b) We have

$$L_{\text{eff}} = \frac{\lambda_{\text{air}}}{3} \quad \rightarrow \quad l = \frac{\pi\lambda_{\text{air}}}{2 \times 3} = \frac{\pi\lambda_{H_2O}}{2} \simeq 1.6\lambda_{H_2O}$$

Note that we cannot use the linear current approximation in this case, since the antenna length is comparable to wavelength. Determination of the radiation resistance requires integration[61] of [6.86] over θ and ϕ.

6.6 SUMMARY

This chapter discussed the following topics:

■ **Charged particle motion.** Electromagnetic waves interact with matter via their action on the charged particles that constitute matter. The dynamics of a charged particle under the influence of electric and magnetic fields is described by the Lorentz force equation:

$$\overline{\mathscr{F}} = q[\overline{\mathscr{E}} + \tilde{\mathbf{v}} \times \overline{\mathscr{B}}]$$

A steady magnetic field cannot do any work on a charged particle, since the magnetic force $q\tilde{\mathbf{v}} \times \overline{\mathscr{B}}$ is always perpendicular to the particle motion. The natural

---

[61] See Section 10.07 of E. C. Jordan and K. G. Balmain, *Electromagnetic Waves and Radiating Systems,* 2nd ed., Prentice-Hall, Englewood Cliffs, New Jersey, 1968.

motion of a charged particle with charge q, mass m, and velocity $\mathbf{v}$ under the influence of a constant magnetic field $\mathbf{B}_0$ is to gyrate around the magnetic field with the gyrofrequency ω_c and radius of gyration r_c given by

$$\omega_c = -\frac{q|\mathbf{B}_0|}{m} \qquad r_c = \frac{m|\tilde{\mathbf{v}}|}{q|\mathbf{B}_0|}$$

Charged particles under the influence of electric and magnetic fields generally follow highly complicated trajectories, especially when the electric or magnetic fields are nonuniform, and when both crossed electric and magnetic fields are simultaneously present.

■ **Wave propagation in a plasma.** In terms of propagation of uniform plane electromagnetic waves, an ionized medium with electron density N_e behaves as if it has an effective permittivity

$$\epsilon_{\text{eff}} = \epsilon_0 \left(1 - \frac{f_p^2}{f^2}\right)$$

where $f_p = (2\pi)^{-1}\sqrt{N_e q_e^2/(m_e\epsilon_0)} \simeq 9\sqrt{N_e}$ is the plasma frequency. The cutoff frequency of uniform plane waves in a plasma is $f_c = f_p$, with propagation being possible only for frequencies $f > f_p$.

■ **Wave propagation in a plasma with a steady magnetic field.** The wave equation for electromagnetic waves propagating in the z direction in a plasma with a steady magnetic field $\mathbf{B}_0 = \hat{\mathbf{z}}B_0$ is given by

$$\frac{d^2}{dz^2}\begin{bmatrix} E_x \\ E_y \\ 0 \end{bmatrix} + \omega^2\mu_0 \begin{bmatrix} \epsilon_{\text{eff}}^{11} & \epsilon_{\text{eff}}^{12} & 0 \\ \epsilon_{\text{eff}}^{21} & \epsilon_{\text{eff}}^{22} & 0 \\ 0 & 0 & \epsilon_{\text{eff}}^{33} \end{bmatrix}\begin{bmatrix} E_x \\ E_y \\ 0 \end{bmatrix} = 0$$

so that the medium behaves as if its permittivity were a tensor, the entries of which are determined by the wave frequency ω, the electron plasma frequency ω_p, and the electron gyrofrequency $\omega_c = |q_e|B_0/m_e$. Both right- and left-hand circularly polarized waves are natural solutions of the above wave equation, leading to waves with propagation constants

$$\beta_{\text{RH}} = \omega\sqrt{\mu_0\epsilon_0}\left(1 + \frac{\omega_p^2/\omega}{\omega_c - \omega}\right)^{1/2} \quad \text{and} \quad \beta_{\text{LH}} = \omega\sqrt{\mu_0\epsilon_0}\left(1 - \frac{\omega_p^2/\omega}{\omega_c + \omega}\right)^{1/2}$$

A linearly polarized wave propagating for a distance d in a plasma with a magnetic field undergoes Faraday rotation, with its plane of polarization rotating by an amount $\theta_F = (\beta_{\text{LH}} - \beta_{\text{RH}})d/2$.

■ **Frequency response of dielectrics and metals.** Microscopic response to a dielectric to an applied electric field is highly frequency-dependent, with resonances occurring due to electronic, ionic, and dipolar polarizations. In the vicinity of such resonances, the imaginary part of the complex permittivity peaks, leading to losses and absorption of an electromagnetic wave. Similar frequency-dependent microscopic behavior in metals makes the conductivity of metals frequency dependent, typically at infrared and higher frequencies. At far ultraviolet and higher frequencies, metals respond to uniform plane waves in a similar manner as dielectrics.

■ **Wave propagation in ferrites.** The wave equation for electromagnetic waves propagating in the z direction in a ferrite with a steady magnetic field $\mathbf{H}_0 = \hat{\mathbf{z}} H_0$ is given by

$$\frac{d^2}{dz^2}\begin{bmatrix} H_x \\ H_y \\ 0 \end{bmatrix} + \omega^2 \epsilon \begin{bmatrix} \mu_{\text{eff}}^{11} & \mu_{\text{eff}}^{12} & 0 \\ \mu_{\text{eff}}^{21} & \mu_{\text{eff}}^{22} & 0 \\ 0 & 0 & \mu_{\text{eff}}^{33} \end{bmatrix}\begin{bmatrix} H_x \\ H_y \\ 0 \end{bmatrix} = 0$$

so the medium behaves as if its permeability was a tensor, whose entries are determined by the wave frequency ω, the gyromagnetic resonance frequency $\omega_0 = |q_e| H_0/m_e$, and $\omega_{\text{M}} = |q_e| M_0/m_e$, with $\mathbf{M}_0 = \hat{\mathbf{z}} M_0$ being the constant magnetization produced by $\mathbf{H}_0$. Both right- and left-hand circularly polarized waves are natural solutions of the above wave equation, leading to waves with propagation constants

$$\beta_{\text{RH}} = \omega\sqrt{\mu_0\epsilon}\left(1 + \frac{\omega_{\text{M}}}{\omega_0 - \omega}\right)^{1/2} \quad \text{and} \quad \beta_{\text{LH}} = \omega\sqrt{\mu_0\epsilon}\left(1 + \frac{\omega_{\text{M}}}{\omega_0 + \omega}\right)^{1/2}$$

A linearly polarized wave propagating for a distance d in a ferrite with a magnetic field undergoes Faraday rotation, with its plane of polarization rotating by an amount $\theta_{\text{F}} = (\beta_{\text{LH}} - \beta_{\text{RH}})d/2$.

■ **Elementary antennas.** The far-zone ($r \gg \lambda/(2\pi)$) electric and magnetic fields radiated by a Hertzian dipole (an electric dipole of length $l \ll \lambda$) are

$$E_\theta \simeq \frac{E_{\text{m}}\sin\theta}{r}e^{-j\beta r} \quad \text{and} \quad H_\phi \simeq \frac{E_{\text{m}}\sin\theta}{\eta r}e^{-j\beta r}$$

constituting a spherical wave front, which at large distances from the source constitute (for all practical purposes) a uniform plane electromagnetic wave. The time-average Poynting flux produced by an antenna is given by

$$\mathbf{S}_{\text{av}}(\theta, \phi) = \hat{\mathbf{r}}\frac{g_{\text{d}}(\theta, \phi)P_{\text{rad}}}{4\pi r^2}$$

where $g_{\text{d}}(\theta, \phi)$ is the directive gain of the antenna and P_{rad} is the total radiated power. It is customary to express P_{rad} in terms of a radiation resistance R_{rad} as $P_{\text{rad}} = (1/2)I^2 R_{\text{rad}}$. The radiation resistances of the Hertzian dipole of length dl, the short dipole of length L, and the half-wave ($L = \lambda/2$) dipole antennas are given as

$$[R_{\text{rad}}]_{\text{Hertzian}} \simeq 80\pi^2 (dl/\lambda)^2 \ \Omega$$

$$[R_{\text{rad}}]_{\text{short dipole}} \simeq 20\pi^2 (L/\lambda)^2 \ \Omega$$

$$[R_{\text{rad}}]_{\text{half-wave}} \simeq 73.13 \ \Omega$$

6.7   PROBLEMS

6-1. Two charges. Consider two protons 1 μm apart from each other in free space and at rest at $t = 0$. Considering only the electric forces, describe the motion of the protons and calculate the final velocity attained by them.

6-2. 10-keV proton. A proton with 100-keV (note that $1\,\text{eV} \approx 1.6\times19^{-19}$ J) energy suffers a "head-on" collision with the nucleus of a lead (Pb) atom. What is the distance of closest approach? What is the mutual force of repulsion at this separation?

6-3. Field of a point charge. A point charge q is constrained to the vicinity of the origin of a coordinate system by a force $\mathbf{F} = -k\mathbf{r}$, where k is a constant and $\mathbf{r}$ is a vector from the origin to the point where q is located. As an example, the force $\mathbf{F}$ can be the restoring force of a nucleus. (a) If a constant external electric field $\mathbf{E}_0 = E_0\hat{\mathbf{x}}$ is applied, find the resultant total electric field at a given point on the x axis, namely at $(x_0, 0, 0)$. (b) Find the difference between this field and the resultant total that would exist if the charge q were not movable but were fixed at the origin. Explain why the total resultant field is different when q can move versus when it cannot.

6-4. Charged particle in an electrostatic potential. The electrostatic potential near the origin of a coordinate system is given as $\Phi(x, y) = a^2 x^2 + b^2 y^2$, where a and b are constants. (a) Find the components of the electric field. (b) Where is the minimum of the electrostatic potential? Where is the minimum electric field? (c) If a small charged particle of mass m and charge q is released at rest from point $(1, 0)$, how does it move? (d) Repeat (c) if the particle is released from rest at $(0, 1)$. (e) Find the equations of the trajectory followed by the particle if it is released at $(1, 0)$ with an initial velocity of $\tilde{\mathbf{v}} = \hat{\mathbf{y}}v_0$.

6-5. Electron in an electrostatic potential. An electron moves along the x axis, where the electrostatic potential is given by

$$\Phi(x) = \frac{5400}{100x + 12} - \frac{600}{100x + 4} \quad \text{volts}$$

The expression for $\Phi(x)$ is valid for $x > -0.04$ m. The electron starts from $x = +\infty$ with negligible speed. (a) Find the velocity of the electron at $x = 0$. (b) At which point does the electron reverse its direction? What is the value of the electric field at that point?

6-6. Electron between parallel metal plates. Two parallel metal plates lie in the planes $z = 0$ and $z = d$. The plates are at the potentials $\Phi = 0$ and $\Phi = -V_0$ respectively. An electron enters the region between the plates at the origin and at $t = 0$ and with an initial velocity $\tilde{\mathbf{v}} = \hat{\mathbf{y}}v_a + \hat{\mathbf{z}}v_b$. (a) Find the velocity and position of the electron as a function of time. (b) Determine the electron orbit and show that it is in the form of a parabola. (c) Find the point of deepest penetration into the parallel-plate region, both from the orbit and from energy considerations.

6-7. Particle in a magnetic field. A positively charged particle has a velocity $\tilde{\mathbf{v}} = -\hat{\mathbf{x}}v_0$ at the origin in the presence of the magnetic field given by $\mathbf{B} = \hat{\mathbf{z}}B_0$. (a) Find the Cartesian components of the velocity and the position of the particle as functions of time by integrating the equation of motion in Cartesian coordinates. (b) Find the particle's orbit in Cartesian coordinates and check your result with the orbit you expect from the discussion of cyclotron motion in Section 6.1.

6-8. Electron approaching a positive charge. Consider an electron approaching a stationary positive point charge Q located at the origin. At infinity, the electron has an initial velocity of $\tilde{\mathbf{v}} = -\hat{\mathbf{r}} v_0$. In the absence of the positive charge the electron would pass within a distance d of the origin. Show that the presence of the positive charge causes the electron to follow a hyperbolic orbit.

6-9. Gradient drift. Consider static and uniform fields $\mathbf{B} = B_1 \hat{\mathbf{y}}$ for $z > 0$ and $\mathbf{B} = B_2 \hat{\mathbf{y}}$ for $z < 0$. A particle of mass m and charge q approaches the boundary plane $z = 0$ with a velocity $\tilde{\mathbf{v}}$ at an angle θ relative to the x axis; that is, $\tilde{\mathbf{v}} = v_0 \hat{\mathbf{x}} \cos \theta + v_0 \hat{\mathbf{z}} \sin \theta$. Show that the resultant motion of the particle is such as to cause it to creep along the boundary in the x direction with an average drift velocity of

$$v_d = \frac{2v_0(B_1 - B_2)\sin\theta}{(2\pi - 2\theta)B_2 + 2\theta B_1}$$

6-10. Helical path. An electron with 2-keV energy is injected into a magnetic field $B_0 = 1$ mT at an angle of $87°$ with respect to $\mathbf{B}$, resulting in a helical path. Find (a) helix radius, (b) distance between turns of the helix, (c) circular or orbital velocity of the electron, (d) axial velocity of electron, and (e) electron's orbital frequency.

6-11. Two charges at right angles. Consider two charges moving in paths at right angles to each other at the instant they are released, as shown in Figure 6.23. (a) One might at first expect from Newton's third law of motion that the force at each charge due to the influence of the other should be equal. Show that this is not true. (b) How do you explain the apparent violation of this basic physical principle?[62]

6-12. A Hall effect experiment. An n-type silicon (Si) sample doped with phosphorus (P) donor atoms of density $N_d = 10^{17}$ cm^{-3} resulting in an effective electron charge density of $N_e \simeq N_d = 10^{17}$ cm^{-3} and a negligible hole charge density (i.e., $N_p \simeq 0$) and with dimensions of $L = 5$ mm, $w = 0.25$ mm, and $d = 50$ μm is designed to be used in a Hall effect experiment oriented in a $B_0 = 1$ kgauss dc magnetic field applied along the w dimension as shown in Figure 6.8. (a) If a dc voltage of $V_0 = 1$ V is applied across the $L = 5$ mm dimension, find the Hall voltage that will be induced across the $d = 50$ μm dimension. (b) Repeat part (a) if the same dc magnetic field is applied along the $d = 50$ μm dimension and the Hall voltage V_H is measured across the $w = 0.25$ mm dimension.

6-13. Hall experiment to determine n-type or p-type. A silicon (Si) sample with unknown doping is tested in a Hall effect experiment with $B_0 = 1$ T, as shown in Figure 6.8. The sample dimensions are $L = 3$ mm, $w = 0.5$ mm, and $d = 0.1$ mm. A dc voltage of $V_0 = 100$ mV (with its plus polarity on the left as shown in Figure 6.8) is applied, which results in a current flow of $I_0 \simeq 5$ mA. If a Hall voltage of $V_H \simeq 3$ mV is measured with its plus terminal being at the top terminal, find (a) the type of the semiconductor (n-type or p-type); (b) the approximate impurity concentration ($\sim N_e$ or $\sim N_p$); and (c) the mobility of the impurity carriers (μ_e or μ_p).

FIGURE 6.23. Two charges at right angles. Problem 6-11.

---

[62]See L. Page and N. I. Adams, Action and reaction between moving charges, *Am. J. Phys.*, 13, p. 141, 1945.

6-14. **GPS signal transmission through the ionosphere.** The Global Positioning System (GPS) uses frequencies of 1.228 and 1.575 GHz. The highest electron density in the earth's ionosphere occurs near the so-called F-region (at altitudes of ~300 km) of the ionosphere, reaching values of $N_e = 10^7$ cm^{-3} during periods of high solar activity. (a) By modeling the ionosphere as a single layer of 50-km thickness with uniform electron density N_e, assess the effect of the ionosphere on the GPS signal. Neglect the effect of the earth's magnetic field. What fraction of a GPS signal vertically incident from space onto the ionospheric layer is reflected? (b) Noting that the maximum value of the earth's magnetic field is ~0.5 gauss, estimate the maximum amount of Faraday rotation suffered by the GPS signal during its transmission through the ionospheric layer.

6-15. **Space vehicle reentry.** The intense friction around a space vehicle reentering the atmosphere generates a plasma sheath which is 1 meter thick and is characterized by an electron density of $N_e = 10^{13}$ cm^{-3}, and a collision frequency of $\nu = 10^{11}$ s^{-1}. What frequency is required in order to transmit plane waves through the plasma sheath with a minimum power loss of 10 dB?

6-16. **Gyrofrequency of an electron.** (a) Calculate the gyrofrequency of an electron in Earth's magnetic field, having a value of ~35 A-m^{-1}. (b) Determine the electron density values required to produce a zero refractive index for the case of propagation parallel to Earth's magnetic field at a frequency of 3 MHz.

6-17. **Reflection from the ionosphere.** Consider the propagation of waves by reflection from the ionosphere as shown in Figure 6.24. (a) On a particular date and time of day, and above a particular geographic location, the highest electron density occurs at an altitude of ~400 km, and the plasma frequency at this altitude is $f_p \simeq 5$ MHz. Neglecting any ionization below this densest layer (i.e., assume electron density to be zero below 400 km altitude), determine the greatest possible angle of incidence θ_i at which an electromagnetic wave originating at a ground-based transmitter could possibly strike the layer. (b) How far away from the transmitter will the reflected radiation return to

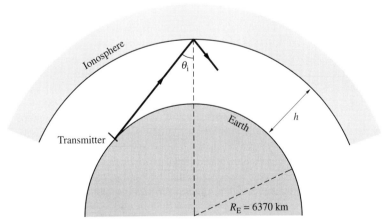

FIGURE 6.24. **Reflection from the ionosphere.** Problem 6-17. The sketch shown is not to scale, since $h = 400$ km while the radius of the earth is ~6370 km as indicated.

earth? (c) What is the highest frequency at which this obliquely incident radiation will be totally reflected?

6-18. Circularly polarized plane wave incident on a plasma with a magnetic field. Consider a stationary plasma filling the half-space $z > 0$ in the presence of a static magnetic field in the $+z$ direction (i.e., $\mathbf{B} = \hat{z}B_0$). The other half-space $z < 0$ is vacuum. Assume the density of electrons for the plasma to be 10^{17} m^{-3} and the magnetic field intensity to be $B_0 = 0.1$ T. A right-hand circularly polarized (RHCP) plane wave of frequency 2 GHz propagating in the $+z$ direction is normally incident on the plasma. (a) What is the magnitude of the electric field of the wave transmitted into the plasma as a fraction of the incident field? (b) What is the polarization of the reflected field? (c) How do your results change if the incident wave is left-hand circularly polarized (LHCP)?

6-19. Plasma-filled waveguide. Consider a rectangular waveguide of dimensions a, b filled with an isotropic plasma with plasma frequency ω_p.(a) Find the cutoff frequency for the dominant TE$_{10}$ mode in terms of a, b, and ω_p. Are there any special conditions on a, b, ω_p for TE$_{10}$ to propagate? (b)Determine the Q of a plasma-filled cubical resonator for the TE$_{101}$ mode.

6-20. Reflection from magnetized plasma. Repeat Problem 6-18 for a linearly polarized incident wave. Determine the magnitudes of the reflected and transmitted fields.

6-21. Attenuation rate in the ionosphere. Show that the effective permittivity of an isotropic (unmagnetized) plasma with collisions can be expressed as

$$\epsilon_{\text{eff}}(\omega) = \epsilon_0 \left[1 - \frac{\omega_p^2}{\omega^2 + \nu^2} + j\frac{\nu\omega_p}{\omega(\omega^2 + \nu^2)} \right]$$

where ν is the collision frequency and ω_p is the plasma frequency. Using this expression to find a simple approximation for the attenuation rate (in dB-m^{-1}) for a 30 MHz wave passing through the lower ionosphere (assume $\nu \simeq 10^7$ s^{-1}, $N_e \simeq 10^8$ m^{-3}).

6-22. Group and phase velocities in a magnetized plasma. Consider the case of a stationary plasma with a static magnetic field $\mathbf{B} = \hat{z}B_0$, such that the gyrofrequency is $f_c = 10$ kHz and the electron plasma frequency is $f_p = 100$ kHz. (a) For the RH mode propagating along z at $f < f_c$ (i.e., the whistler mode discussed in connection with Figure 6.11), derive an expression for the group velocity v_g in terms of f_p, f_c, and the speed of light in free space c. (b) Plot the group and phase velocity as a function of frequency in the range 0 to 10 kHz. Express the velocities in terms of multiples of c. What is the product of v_p and v_g?

6-23. Faraday rotation in interstellar space. Interstellar space can be considered to be a weakly ionized medium with a very low average electron density of $N = 0.01$ cm^{-3} permeated with the interstellar (or intergalactic) magnetic field of $B_0 \simeq 10^{-6}$ gauss. Find the distance that a 10 GHz signal must travel to experience a Faraday rotation of $90°$. You may want to express your result in light years (i.e., $\sim 9.46 \times 10^{15}$ m), since the distance to be traveled is likely to be quite large.

6-24. Reflection from a ferrite. A linearly polarized uniform plane wave in free space is normally incident upon a ferrite interface at $z = 0$. A semi-infinite ferrite medium exists in the region $z > 0$, with a stationary magnetic field of $\mathbf{B} = \hat{z}B_0$. What is the polarization of the reflected field?

6-25. Faraday rotation in ferrite. A ferrite slab is magnetized in the direction of propagation with a magnetic field of $B_0 = 500$ gauss. The material is characterized by a permeability of $\mu = 10^5 \mu_0$ and a gyromagnetic ratio of $\gamma_m = |q_e|/m_e \approx 1.8 \times 10^{11}$ C-$(kg)^{-1}$. (a) How thick must the ferrite slab be in order to achieve 90° Faraday rotation for a 30-GHz signal. (b) Over what frequency band does the rotation angle lie between 88° to 92°?

6-26. Hertzian dipole antenna. A Hertzian dipole antenna is excited by a sinusoidal current of 10-A amplitude. If the length of the dipole is $L = 10$ cm $= 0.01\lambda$, find the maximum radiated time-average power density at the following distances: (a) 100 m. (b) 1 km. (c) 10 km.

6-27. Antenna gain. The time-average power density measured along the main beam of an antenna of 25 W input power at a distance of 10 km from it is about 20 pW-(cm)$^{-2}$. Find the gain of this antenna. Assume isotropic radiation.

6-28. Hertzian dipole antenna. A 2-m long dipole antenna is excited by a sinusoidal current of amplitude 10 A and frequency 1 MHz. Find the time-average power density radiated by the dipole at a distance of 5 km in a direction which is (a) 0°, (b) 45°, and (c) 90° from the axis of the dipole.

6-29. Hertzian dipole antenna. What fraction of the total power radiated by the Hertzian dipole antenna is radiated between ±45° of the equatorial plane?

6-30. Half-wave dipole antenna. A 75-cm long half-wave dipole antenna operating at 200 MHz is used in the VHF TV broadcast band. Determine (a) the radiation resistance of the antenna, and (b) the required current amplitude so that the time-average power density at 10 km away from the center of the antenna perpendicular to its axis is 10 μW-m^{-2}.

6-31. Two Hertzian dipole antennas. Two Hertzian dipoles carry currents of equal magnitude but 90° out of phase. In other words, $I_1 = I_0 \cos(\omega t)$ while $I_2 = I_0 \cos(\omega t + \zeta)$, with $\zeta = \pi/2$. The elements form a cross at the origin, one oriented along the z axis, and the other along the x axis. (a) On the surface of a sphere of large radius ($\beta r \gg 1$), what is the nature of the polarization of the electric field at the points: (i) $\theta = 0$; (ii) $\theta = \pi/2, \phi = 0$; (iii) $\theta = \pi/2, \phi = \pi/2$; (iv) $\theta = \pi/4, \phi = 0$. (b) What is the locus of all points on the sphere at which the polarization is linear? (c) What fraction of the total power radiated is supplied by each antenna? (d) Repeat (a), (b), (c) if the two elements are fed in phase (i.e., $\zeta = 0$).

6-32. Two Hertzian dipole antennas. Repeat parts (a), (b), and (c) of Problem 6-31, if the two dipoles are fed in phase ($\zeta = 0$) but were both oriented in the z direction and located respectively at $(0, \lambda/8, 0)$ and $(0, \lambda/4, 0)$.

APPENDIX A

Vector Analysis

Vector analysis is used extensively in electromagnetics and in other fields of engineering and physics in which the physical quantities involved have both direction and magnitude; that is, they are *vectors*. Examples of physical quantities that are vectors include velocity, momentum, force, displacement, electric field, current, and magnetic field. Vector quantities are inherently different from *scalars*, which are physical quantities that are specified entirely by one value, such as the number of coins in your pocket, time, mass, the temperature in a room, electric charge density, electric potential and energy. In this appendix, we review some basic rules and techniques of vector analysis that are particularly relevant to this book. More general topics of vector algebra or vector calculus are covered elsewhere[1] and are not discussed here.

Many of the essential aspects of vector manipulations and definitions needed to work with the material in this book were covered either in the body of the text or in footnotes in those sections where they first appeared. This appendix simply collects these vector rules and techniques in one handy place. The appendix also covers some of the most basic definitions that were taken for granted, such as the definition of vectors, position vectors, unit vectors, and coordinate systems. Some of the more important concepts of vector calculus, including gradient, divergence, curl, and the associated theorems (the divergence theorem and Stokes's theorem), are covered extensively in the text and are not included here.

Throughout this text, vector quantities are often written either using boldface symbols (e.g., **G**) or with a bar above the symbol (e.g., $\overline{\mathcal{G}}$). The latter notation is used to represent real physical quantities and the former for their corresponding complex

---

[1] Simpler treatments are available in most textbooks; for a complete treatment of the subject, see Chapter 1 of G. B. Arfken and H. J. Weber, *Mathematical Methods for Physicists,* Academic Press, 1995.

phasors, as needed for sinusoidal (or time-harmonic) applications. In this appendix, we represent vectors using boldface symbols.

To accurately describe vectors requires that we specify their direction as well as their magnitude. To do this, we need a framework to orient the vector in three-dimensional space and show its different projections or components. The most common means of doing this is to use the so-called rectangular or Cartesian coordinate system, which we use in Sections A.1 and A.2 to describe the basic aspects of vector addition and multiplication. We then briefly review, in Section A.3, two other commonly used coordinate systems, namely the cylindrical and spherical coordinate systems. Section A.4 presents some commonly encountered vector identities.

A.1 VECTOR COMPONENTS, UNIT VECTORS, AND VECTOR ADDITION

The Cartesian or rectangular coordinate system is illustrated in Figure A.1. To describe a vector **A** in the rectangular coordinate system, we can represent it as extending outward from the origin in a direction determined by the magnitudes of each of its three components, A_x, A_y, and A_z, as shown in Figure A.1a. It is customary to write the vector **A** as

$$\mathbf{A} = \hat{\mathbf{x}}A_x + \hat{\mathbf{y}}A_y + \hat{\mathbf{z}}A_z$$

where $\hat{\mathbf{x}}$, $\hat{\mathbf{y}}$, and $\hat{\mathbf{z}}$ are the rectangular coordinate unit vectors, as shown in Figure A.1b. In this text, we represent unit vectors with the "hat," or circumflex, notation.

The magnitude of vector **A** is denoted $|\mathbf{A}|$ (or sometimes simply A) and is given by

$$A = |\mathbf{A}| = \sqrt{A_x^2 + A_y^2 + A_z^2}$$

Sometimes it is convenient to define a unit vector that is in the direction of a given vector **A**. The unit vector in the direction of vector **A** is denoted $\hat{\mathbf{A}}$ and is defined as

$$\hat{\mathbf{A}} = \frac{\mathbf{A}}{|\mathbf{A}|} = \frac{\mathbf{A}}{\sqrt{A_x^2 + A_y^2 + A_z^2}}$$

Note that $|\hat{\mathbf{A}}| = 1$. An alternative way of expressing vector **A** is $\mathbf{A} = \hat{\mathbf{A}}|\mathbf{A}| = \hat{\mathbf{A}}A$.

Two vectors **A** and **B** can be added together to produce another vector $\mathbf{C} = \mathbf{A} + \mathbf{B}$. In rectangular coordinates, vector addition can be carried out component by component. In other words,

$$\mathbf{C} = \mathbf{A} + \mathbf{B} = \hat{\mathbf{x}}(A_x + B_x) + \hat{\mathbf{y}}(A_y + B_y) + \hat{\mathbf{z}}(A_z + B_z) = \hat{\mathbf{x}}C_x + \hat{\mathbf{y}}C_y + \hat{\mathbf{z}}C_z$$

Geometrically, the orientation and magnitude of the sum vector **C** can be determined using the parallelogram method depicted in Figure A.2a. Subtraction of two vectors

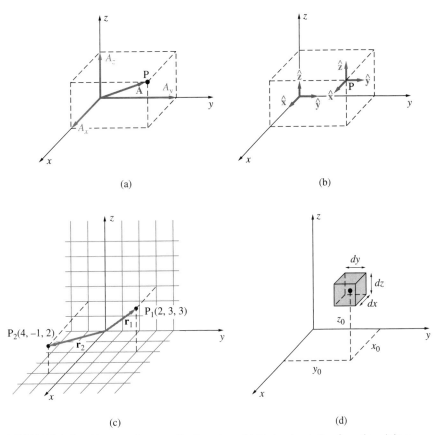

FIGURE A.1. **Rectangular coordinate system.** (a) Vector **A** pointing from the origin to point $P(A_x, A_y, A_z)$. (b) The three rectangular coordinate unit vectors. (c) Position vectors $\mathbf{r}_1 = \hat{\mathbf{x}}2 + \hat{\mathbf{y}}3 + \hat{\mathbf{z}}3$ and $\mathbf{r}_2 = \hat{\mathbf{x}}4 - \hat{\mathbf{y}} + \hat{\mathbf{z}}2$. (d) The differential volume element in a rectangular coordinate system.

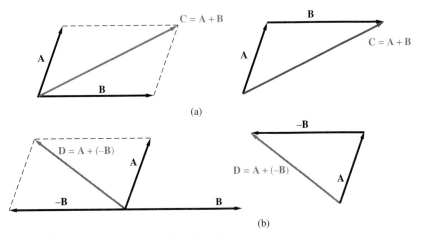

FIGURE A.2. **Addition and subtraction of vectors.**

is done similarly; subtraction of vector **B** from vector **A** gives the same result as the addition of vectors **A** and −**B**, as depicted in Figure A.2b.

A point P in a rectangular coordinate system is represented by its coordinates P(x, y, z). The position vector **r** of a point P is defined as the directed distance from the origin to point P. For example, the position vector of point P_1 in Figure A.1c is $\mathbf{r}_1 = \hat{\mathbf{x}}2 + \hat{\mathbf{y}}3 + \hat{\mathbf{z}}3$. The position vector of point P_2 in Figure A.1c is $\mathbf{r}_2 = \hat{\mathbf{x}}4 - \hat{\mathbf{y}} + \hat{\mathbf{z}}2$. The vector pointing from point P_1 to point P_2 is known as the distance vector (or separation vector) and is given by

$$\mathbf{R} = \mathbf{r}_2 - \mathbf{r}_1 = \hat{\mathbf{x}}4 - \hat{\mathbf{y}} + \hat{\mathbf{z}}2 - (\hat{\mathbf{x}}2 + \hat{\mathbf{y}}3 + \hat{\mathbf{z}}3)$$
$$= \hat{\mathbf{x}}2 - \hat{\mathbf{y}}4 - \hat{\mathbf{z}}$$

Note that the magnitude of **R** is $R = |\mathbf{R}| = \sqrt{2^2 + 4^2 + 1^2} = \sqrt{21}$. The unit vector in the direction of **R** is thus

$$\hat{\mathbf{R}} = \frac{\mathbf{R}}{R} = \hat{\mathbf{x}}\frac{2}{\sqrt{21}} - \hat{\mathbf{y}}\frac{4}{\sqrt{21}} - \hat{\mathbf{z}}\frac{1}{\sqrt{21}}$$

A.2 VECTOR MULTIPLICATION

A.2.1 The Dot Product

The *dot product* of two vectors **A** and **B** is a scalar denoted by **A** · **B**. It is equal to the product of the magnitudes |**A**| and |**B**| of vectors **A** and **B** and the cosine of the angle ψ_{AB} between vectors **A** and **B**. Namely,

$$\mathbf{A} \cdot \mathbf{B} \equiv |\mathbf{A}||\mathbf{B}| \cos \psi_{AB}$$

(The dot product is sometimes referred to as the *scalar product* since the result is a scalar quantity.) Noting that in rectangular coordinates we have $\mathbf{A} = \hat{\mathbf{x}}A_x + \hat{\mathbf{y}}A_y + \hat{\mathbf{z}}A_z$ and $\mathbf{B} = \hat{\mathbf{x}}B_x + \hat{\mathbf{y}}B_y + \hat{\mathbf{z}}B_z$, an alternative expression for the dot product is

$$\mathbf{A} \cdot \mathbf{B} = (\hat{\mathbf{x}}A_x + \hat{\mathbf{y}}A_y + \hat{\mathbf{z}}A_z) \cdot (\hat{\mathbf{x}}B_x + \hat{\mathbf{y}}B_y + \hat{\mathbf{z}}B_z)$$
$$\mathbf{A} \cdot \mathbf{B} = A_x B_x + A_y B_y + A_z B_z$$

where we have used the fact that the dot product of one rectangular unit vector with a different rectangular unit vector is zero, whereas the dot product of any rectangular unit vector with itself is unity. For example, $\hat{\mathbf{x}} \cdot \hat{\mathbf{y}} = 0$ and $\hat{\mathbf{x}} \cdot \hat{\mathbf{z}} = 0$, but $\hat{\mathbf{x}} \cdot \hat{\mathbf{x}} = 1$. This feature of the dot product facilitates determining the component of a vector in a given direction, that is, determining the *projection* of the vector along a given direction. For example, the projection along the y axis of any arbitrary vector $\mathbf{B} = \hat{\mathbf{x}}B_x + \hat{\mathbf{y}}B_y + \hat{\mathbf{z}}B_z$ is simply given by

$$\mathbf{B} \cdot \hat{\mathbf{y}} = (\hat{\mathbf{x}}B_x + \hat{\mathbf{y}}B_y + \hat{\mathbf{z}}B_z) \cdot \hat{\mathbf{y}} = B_y$$

Similarly, the projection of **B** along any other arbitrary unit vector **â** is represented by **B** · **â**, with the resulting scalar being equal to the length of the vector **B** projected

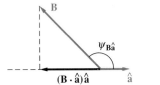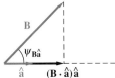

FIGURE A.3.   The dot product. The dot product of **B** with any arbitrary unit vector **â** gives the projection of **B** along **â**, which is negative if $|\psi_{B\hat{a}}| > \pi/2$ or positive if $|\psi_{B\hat{a}}| < \pi/2$.

on vector **â**. Note that, depending on whether the angle $\psi_{B\hat{a}}$ is acute or obtuse, the projection could be positive or negative, as illustrated in Figure A.3.

The dot product is both commutative and distributive; in other words,

$$\mathbf{A} \cdot \mathbf{B} = \mathbf{B} \cdot \mathbf{A}$$

$$\mathbf{A} \cdot (\mathbf{B} + \mathbf{C}) = \mathbf{A} \cdot \mathbf{B} + \mathbf{A} \cdot \mathbf{C}$$

A.2.2   The Cross Product

The *cross product* of two vectors **A** and **B** is a vector **C** denoted by $\mathbf{C} = (\mathbf{A} \times \mathbf{B})$. The magnitude of **C** is equal to the product of the magnitudes of vectors **A** and **B** and the sine of the angle ψ_{AB} between vectors **A** and **B**. The direction of **C**, according to the right-hand rule, follows that of the thumb of the right hand when the fingers rotate from **A** to **B** through the angle ψ_{AB}, as depicted in Figure A.4. In other words,

$$\mathbf{A} \times \mathbf{B} \equiv \hat{\mathbf{n}}|\mathbf{A}||\mathbf{B}| \sin \psi_{AB}$$

where $\hat{\mathbf{n}}$ is the unit vector normal to both **A** and **B** and points in the direction in which a right-handed screw advances as **A** is turned toward **B**. (The cross product is sometimes referred to as the *vector* product since the result is a vector quantity.)

Noting that $\hat{\mathbf{x}} \times \hat{\mathbf{y}} = \hat{\mathbf{z}}, \hat{\mathbf{y}} \times \hat{\mathbf{z}} = \hat{\mathbf{x}}$, and $\hat{\mathbf{z}} \times \hat{\mathbf{x}} = \hat{\mathbf{y}}$, in rectangular coordinates we can use the distributive property of the cross product to write

$$\mathbf{C} = \mathbf{A} \times \mathbf{B} = (\hat{\mathbf{x}}A_x + \hat{\mathbf{y}}A_y + \hat{\mathbf{z}}A_z) \times (\hat{\mathbf{x}}B_x + \hat{\mathbf{y}}B_y + \hat{\mathbf{z}}B_z)$$

$$= \hat{\mathbf{x}}(A_yB_z - A_zB_y) + \hat{\mathbf{y}}(A_zB_x - A_xB_z) + \hat{\mathbf{z}}(A_xB_y - A_yB_x)$$

$$= \hat{\mathbf{x}}C_x + \hat{\mathbf{y}}C_y + \hat{\mathbf{z}}C_z$$

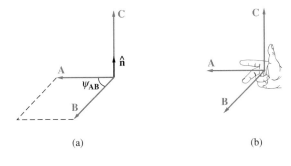

(a)   (b)

FIGURE A.4.   The cross product. The cross product of two vectors **A** and **B** is another vector $\mathbf{C} = \mathbf{A} \times \mathbf{B}$, which is perpendicular to both **A** and **B** and is related to them via the right-hand rule.

It is often convenient to write the cross product in determinant form:

$$\mathbf{A} \times \mathbf{B} = \begin{vmatrix} \hat{\mathbf{x}} & \hat{\mathbf{y}} & \hat{\mathbf{z}} \\ A_x & A_y & A_z \\ B_x & B_y & B_z \end{vmatrix}$$

Using the definition of the cross product and following the right-hand rule, it is clear that

$$\mathbf{B} \times \mathbf{A} = -\mathbf{A} \times \mathbf{B}$$

so that the cross product is not commutative. The cross product is also not associative since we have

$$\mathbf{A} \times (\mathbf{B} \times \mathbf{C}) \neq (\mathbf{A} \times \mathbf{B}) \times \mathbf{C}$$

It is clear, however that the cross product is distributive; in other words,

$$\mathbf{A} \times (\mathbf{B} + \mathbf{C}) = \mathbf{A} \times \mathbf{B} + \mathbf{A} \times \mathbf{C}$$

A.2.3 Triple Products

Given three vectors **A**, **B**, and **C**, there are in general two different ways we can form a triple product. The *scalar triple product* is defined as

$$\mathbf{A} \cdot (\mathbf{B} \times \mathbf{C}) = \mathbf{B} \cdot (\mathbf{C} \times \mathbf{A}) = \mathbf{C} \cdot (\mathbf{A} \times \mathbf{B})$$

Note that the result of the scalar triple product is a scalar.

The *vector triple product* is defined as $\mathbf{A} \times (\mathbf{B} \times \mathbf{C})$ and can be expressed as the difference of two simpler vectors as follows:

$$\mathbf{A} \times (\mathbf{B} \times \mathbf{C}) = \mathbf{B}(\mathbf{A} \cdot \mathbf{C}) - \mathbf{C}(\mathbf{A} \cdot \mathbf{B})$$

A.3 CYLINDRICAL AND SPHERICAL COORDINATE SYSTEMS

The physical properties of electromagnetic fields and waves as studied in this book do not depend on the particular coordinate system used to describe the vector quantities. An electric field exists, and may or may not have nonzero divergence or curl, depending on its physical properties with no regard to any system of coordinates. Scalar products, cross products, and other vector operations are all independent of the mathematical frame of reference used to describe them. In practice, however, description, manipulation, and computation of vector quantities in a given problem may well be easier in coordinate systems other than the rectangular system, depending on the symmetries involved. In this section we describe the two most commonly used additional coordinate systems: namely, the circular cylindrical and the spherical coordinate systems. These two systems are examples of orthogonal curvilinear coordinate systems; other examples not covered here include the elliptic cylindrical, parabolic cylindrical, conical, and prolate spheroidal systems.

A.3.1   Cylindrical Coordinates

The circular cylindrical system is an extension to three dimensions of the polar coordinate system of analytical geometry. In view of its common use, it is often referred to simply as the cylindrical coordinate system, although cylindrical systems with other cross-sectional shapes, such as elliptic or hyperbolic, are also used in special applications. The three cylindrical coordinates are r, ϕ, and z. The radial coordinate r of any point P is the closest distance from the z axis to that point. A given value of r specifies a circular cylindrical surface on which the point P resides, with the particular position of P on the sphere further specified by ϕ and z. The convention is to measure ϕ from the x axis, in the right-handed sense. In other words, with the thumb pointed in the z direction, the fingers trace the direction of increasing ϕ. The angle ϕ is commonly called the azimuthal angle, and it varies from 0 to 2π.

The cylindrical coordinate unit vectors $\hat{\mathbf{r}}$, $\hat{\boldsymbol{\phi}}$, and $\hat{\mathbf{z}}$ are shown in Figure A.5b. Note that the unit vectors are mutually orthogonal and that $\hat{\mathbf{r}} \times \hat{\boldsymbol{\phi}} = \hat{\mathbf{z}}$. A good way

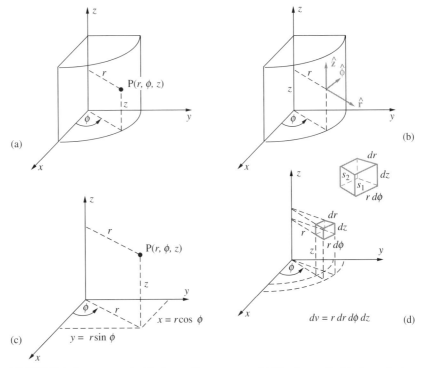

FIGURE A.5.   The cylindrical coordinate system. (a) The three cylindrical coordinates. (b) The three cylindrical coordinate unit vectors. (c) The relationships between rectangular and cylindrical coordinates. (d) The differential volume element in a cylindrical coordinate system.

to remember the orientation of the unit vectors is to note that they obey the right-hand rule, with the thumb, forefinger, and the middle finger representing the directions of $\hat{\mathbf{r}}$, $\hat{\boldsymbol{\phi}}$, and $\hat{\mathbf{z}}$, respectively.

A vector $\mathbf{A}$ in cylindrical coordinates is specified by means of the values of its components A_r, A_ϕ, and A_z and is typically written as

$$\mathbf{A} = \hat{\mathbf{r}}A_r + \hat{\boldsymbol{\phi}}A_\phi + \hat{\mathbf{z}}A_z$$

The magnitude of $\mathbf{A}$ is given by $|\mathbf{A}| = \sqrt{\mathbf{A} \cdot \mathbf{A}} = \sqrt{A_r^2 + A_\phi^2 + A_z^2}$.

The expressions for the scalar and vector products of vectors for cylindrical coordinates are very similar to those for rectangular coordinates. We have

$$\mathbf{A} \cdot \mathbf{B} = A_r B_r + A_\phi B_\phi + A_z B_z$$

$$\mathbf{A} \times \mathbf{B} = \begin{vmatrix} \hat{\mathbf{r}} & \hat{\boldsymbol{\phi}} & \hat{\mathbf{z}} \\ A_r & A_\phi & A_z \\ B_r & B_\phi & B_z \end{vmatrix}$$

Cylindrical-to-Rectangular Transformations A vector specified in cylindrical coordinates can be transformed into rectangular coordinates and vice versa. The relationships between the (r, ϕ, z) and (x, y, z) are illustrated in Figure A.5c. Consider a vector

$$\mathbf{A} = \hat{\mathbf{r}}A_r + \hat{\boldsymbol{\phi}}A_\phi + \hat{\mathbf{z}}A_z$$

where in general each of the component values A_r, A_ϕ, and A_z may themselves be functions of spatial coordinates r, ϕ, and z. The rules for transforming from cylindrical to rectangular coordinates are given as follows, where the set of expressions on the left are for transforming scalar quantities (such as A_r), and those on the right are for transforming the vectors.

$$r = \sqrt{x^2 + y^2} \qquad \hat{\mathbf{r}} = \hat{\mathbf{x}}\cos\phi + \hat{\mathbf{y}}\sin\phi$$

$$\phi = \tan^{-1}\left[\frac{y}{x}\right] \qquad \hat{\boldsymbol{\phi}} = -\hat{\mathbf{x}}\sin\phi + \hat{\mathbf{y}}\cos\phi$$

$$z = z \qquad \hat{\mathbf{z}} = \hat{\mathbf{z}}$$

Note that the preceding relationships can be deduced from careful examination of Figure A.5b (unit vectors) and Figure A.5c (coordinates). A similar set of rules can be given for transformation from rectangular to cylindrical coordinates:

$$x = r\cos\phi \qquad \hat{\mathbf{x}} = \hat{\mathbf{r}}\cos\phi - \hat{\boldsymbol{\phi}}\sin\phi$$

$$y = r\sin\phi \qquad \hat{\mathbf{y}} = \hat{\mathbf{r}}\sin\phi + \hat{\boldsymbol{\phi}}\cos\phi$$

$$z = z \qquad \hat{\mathbf{z}} = \hat{\mathbf{z}}$$

The transformation of a vector given in the cylindrical (rectangular) coordinate system can be accomplished by direct substitution of the preceding unit vector and coordinate variable expressions. However, doing so in fact amounts to taking the dot

TABLE A.1.   Dot products of unit vectors

	Cylindrical coordinates			Spherical coordinates		
	$\hat{\mathbf{r}}$	$\hat{\boldsymbol{\phi}}$	$\hat{\mathbf{z}}$	$\hat{\mathbf{r}}$	$\hat{\boldsymbol{\theta}}$	$\hat{\boldsymbol{\phi}}$
$\hat{\mathbf{x}}\cdot$	$\cos\phi$	$-\sin\phi$	0	$\sin\theta\cos\phi$	$\cos\theta\cos\phi$	$-\sin\phi$
$\hat{\mathbf{y}}\cdot$	$\sin\phi$	$\cos\phi$	0	$\sin\theta\sin\phi$	$\cos\theta\sin\phi$	$\cos\phi$
$\hat{\mathbf{z}}\cdot$	0	0	1	$\cos\theta$	$-\sin\theta$	0

product of the vector as written in cylindrical (rectangular) coordinates with the rectangular (cylindrical) coordinate unit vectors and substituting coordinate variables, as illustrated in Example A-1. The dot products of the cylindrical coordinate unit vectors with their rectangular coordinate counterparts are provided in Table A.1.

Example A-1:   Cylindrical-to-rectangular and rectangular-to-cylindrical transformations.   (a) Transform a vector

$$\mathbf{A} = \hat{\mathbf{r}}\left(\frac{1}{r}\right) + \hat{\boldsymbol{\phi}}\left(\frac{2}{r}\right)$$

given in cylindrical coordinates to rectangular coordinates. (b) Transform a vector $\mathbf{B} = \hat{\mathbf{x}}x^2 + \hat{\mathbf{y}}xy$ given in rectangular coordinates to cylindrical coordinates.

Solution:

(a) We set out to find A_x, A_y, and A_z. We have

$$A_x = \mathbf{A}\cdot\hat{\mathbf{x}} = \left[\hat{\mathbf{r}}\left(\frac{1}{r}\right) + \hat{\boldsymbol{\phi}}\left(\frac{2}{r}\right)\right]\cdot\hat{\mathbf{x}} = \frac{\cos\phi}{r} - \frac{2\sin\phi}{r} = \frac{x - 2y}{x^2 + y^2}$$

where we have used the unit vector dot products as given in Table A.1 and substituted $\cos\phi = x/r$, $\sin\phi = y/r$, and $r = \sqrt{x^2 + y^2}$. Similarly, we have

$$A_y = \mathbf{A}\cdot\hat{\mathbf{y}} = \left[\hat{\mathbf{r}}\left(\frac{1}{r}\right) + \hat{\boldsymbol{\phi}}\left(\frac{2}{r}\right)\right]\cdot\hat{\mathbf{y}} = \frac{\sin\phi}{r} + \frac{2\cos\phi}{r} = \frac{y + 2x}{x^2 + y^2}$$

Note that $A_z = 0$, since $\mathbf{A}\cdot\hat{\mathbf{z}} = 0$. Thus we have

$$\mathbf{A} = \hat{\mathbf{x}}\frac{x - 2y}{x^2 + y^2} + \hat{\mathbf{y}}\frac{y + 2x}{x^2 + y^2}$$

(b) We set out to find B_r, B_ϕ, and B_z. We have

$$B_r = \mathbf{B}\cdot\hat{\mathbf{r}} = (\hat{\mathbf{x}}x^2 + \hat{\mathbf{y}}xy)\cdot\hat{\mathbf{r}} = x^2\cos\phi + xy\sin\phi$$

$$= [r^2\cos^2\phi\cos\phi + (r\cos\phi)(r\sin\phi)\sin\phi] = r^2\cos\phi[\cos^2\phi + \sin^2\phi] = r^2\cos\phi$$

where we have used the unit vector dot products as given in Table A.1 and substituted $x = r\cos\phi$ and $y = r\sin\phi$. Similarly, we have

$$B_\phi = \mathbf{B}\cdot\hat{\boldsymbol{\phi}} = (\hat{\mathbf{x}}x^2 + \hat{\mathbf{y}}xy)\cdot\hat{\boldsymbol{\phi}} = -x^2\sin\phi + xy\cos\phi$$
$$= [-r^2\cos^2\phi\sin\phi + (r\cos\phi)(r\sin\phi)\cos\phi] = 0$$

Note that $B_z = 0$ since $\mathbf{B}\cdot\hat{\mathbf{z}} = 0$. Thus we have

$$\mathbf{B} = \hat{\mathbf{r}}r^2\cos\phi$$

Length, Surface, and Volume Elements The general expression for differential length $d\mathbf{l}$ in cylindrical coordinates is

$$d\mathbf{l} = \hat{\mathbf{r}}\,dr + \hat{\boldsymbol{\phi}}r\,d\phi + \hat{\mathbf{z}}\,dz$$

as is evident from Figure A.5d. By inspection of the volume element sketched in Figure A.5d, we can see that the cylindrical coordinate surface elements in the three coordinate directions are

$$(ds)_r = r\,d\phi\,dz$$
$$(ds)_\phi = dr\,dz$$
$$(ds)_z = r\,dr\,d\phi$$

and the volume element in cylindrical coordinates is:

$$dv = r\,dr\,d\phi\,dz$$

A.3.2 Spherical Coordinates

The three spherical coordinates are r, θ, and ϕ. The radial coordinate r of any point P is simply the distance from the origin to that point. A given value of r specifies a sphere on which the point P resides, with the particular position of P on the sphere further specified by θ and ϕ. The best way to think of the spherical coordinate system and especially the coordinates θ and ϕ is in terms of the latitude/longitude system of identifying a point on the earth's surface. The earth is a sphere with radius $r \simeq$ 6370 km. With respect to Figure A.6a, latitude corresponds to θ while longitude corresponds to ϕ. The convention is that the positive z axis points toward north, and ϕ is measured from the x axis, in the right-handed sense. In other words, with the thumb pointed in the z direction, the fingers trace the direction of increasing ϕ. The Greenwich meridian, from which positive (east) longitude is measured, coincides with the positive x axis. The angle ϕ, commonly called the azimuthal angle, varies from 0 to 2π. Unlike geographic latitude, which is commonly measured with respect to the equatorial plane and which thus varies in the range from $-\pi/2$ to $+\pi/2$, the spherical coordinate θ is measured from the positive z axis (see Figure A.6) and thus varies between $\theta = 0$ (north pole) and $\theta = \pi$ (south pole).

The spherical coordinate unit vectors $\hat{\mathbf{r}}$, $\hat{\boldsymbol{\phi}}$, and $\hat{\boldsymbol{\theta}}$ are shown in Figure A.6b. Note that $\hat{\mathbf{r}}$ points radially outward, $\hat{\boldsymbol{\phi}}$ points "east," and $\hat{\boldsymbol{\theta}}$ points "south." Note that the unit vectors are mutually orthogonal and that $\hat{\mathbf{r}}\times\hat{\boldsymbol{\theta}} = \hat{\boldsymbol{\phi}}$. A good way to remember

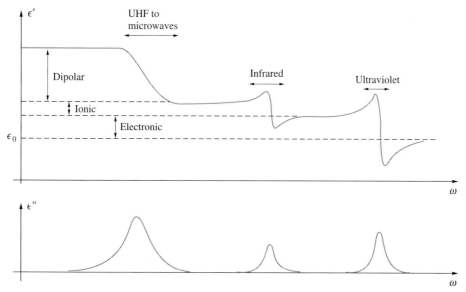

FIGURE 6.13.   **Dielectric constant as a function of frequency.** The behavior of the dipolar and ionic polarization effects as a function of frequency are similar to that of electronic polarization. However, the resonances for these polarizations occur at lower frequencies.

The frequency response of dielectric materials and the various resonances illustrated in Figure 6.13 have many important practical implications. The broad dipolar resonance in the microwave region is responsible for microwave heating applications. A resonance in water vapor molecules at 1.25-cm wavelength places a limit on the frequency of long-range radar. Another strong resonance at 0.5 cm from oxygen, together with other higher-frequency absorptions, prevents propagation of signals above 0.5 cm wavelength. Very strong resonances, due mostly to ozone and nitrogen at high altitudes, keep most of the ultraviolet energy of the sun from penetrating to lower altitudes. Another example of an important resonance occurs in sodium vapor at a wavelength of about 589 nm. Different gases exhibit a large variety of such resonances that have their origins in the basic energy level structure of the atoms and molecules.[33] In most gases, there exist many resonant frequencies for each type of polarization, each exhibiting different damping coefficients κ.

The above discussion is applicable for gases, for which the separation of the different elastically bound oscillators is large enough so that they do not interact with one another. Similar phenomena take place in solids, except that the oscillators are packed much more closely together, so that the effect of neighboring dipoles must be included. Relatively simple analysis of this problem[34] indicates that the effective

---

[33] As an example of the kind of complexity that is possible, see an interesting discussion of the properties of liquid water as a function of frequency in Section 7.5 of J. D. Jackson, *Classical Electrodynamics,* 2nd ed., John Wiley & Sons, New York, 1975.

[34] See Chapter 13 of C. Kittel, *Introduction to Solid State Physics,* 5th ed., John Wiley & Sons, New York, 1971.

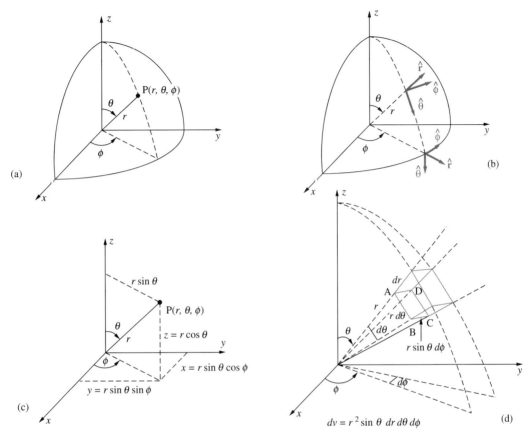

FIGURE A.6.   The spherical coordinate system. (a) The three spherical coordinates. (b) The three spherical coordinate unit vectors. (c) The relationships between rectangular and spherical coordinates. (d) The differential volume element in a spherical coordinate system.

the orientation of the unit vectors is to note that they obey the right-hand rule, with the thumb, forefinger, and middle finger representing the directions of $\hat{\mathbf{r}}$, $\hat{\boldsymbol{\theta}}$, and $\hat{\boldsymbol{\phi}}$, respectively.

A vector $\mathbf{A}$ in spherical coordinates is specified by means of the values of its components A_r, A_θ, and A_ϕ and is typically written as

$$\mathbf{A} = \hat{\mathbf{r}} A_r + \hat{\boldsymbol{\theta}} A_\theta + \hat{\boldsymbol{\phi}} A_\phi$$

The magnitude of $\mathbf{A}$ is given by $|\mathbf{A}| = \sqrt{\mathbf{A} \cdot \mathbf{A}} = \sqrt{A_r^2 + A_\theta^2 + A_\phi^2}$.

The expressions for the scalar and vector products of vectors for spherical coordinates are very similar to those for rectangular coordinates. We have

$$\mathbf{A} \cdot \mathbf{B} = A_r B_r + A_\theta B_\theta + A_\phi B_\phi$$

$$\mathbf{A} \times \mathbf{B} = \begin{vmatrix} \hat{\mathbf{r}} & \hat{\boldsymbol{\theta}} & \hat{\boldsymbol{\phi}} \\ A_r & A_\theta & A_\phi \\ B_r & B_\theta & B_\phi \end{vmatrix}$$

Spherical to Rectangular Transformations   A vector specified in spherical coordinates can be transformed into rectangular coordinates and vice versa. The relationship between (r, θ, ϕ) and (x, y, z) is illustrated in Figure A.6c. Consider a vector

$$\mathbf{A} = \hat{\mathbf{r}}A_r + \hat{\boldsymbol{\theta}}A_\theta + \hat{\boldsymbol{\phi}}A_\phi$$

where in general each of the component values A_r, A_θ, and A_ϕ may themselves be
functions of spatial coordinates r, θ, and ϕ. The rules for transformation from spherical coordinates to rectangular coordinates are given as follows, where the set of
expressions on the left-hand side are for transforming scalar quantities (such as A_r),
and those on the right are for transforming vectors.

Coordinate variables:	Unit vectors:
$r = \sqrt{x^2 + y^2 + z^2}$	$\hat{\mathbf{r}} = \hat{\mathbf{x}}\sin\theta\cos\phi + \hat{\mathbf{y}}\sin\theta\sin\phi + \hat{\mathbf{z}}\cos\theta$
$\theta = \cos^{-1}\left[\dfrac{z}{\sqrt{x^2 + y^2 + z^2}}\right]$	$\hat{\boldsymbol{\theta}} = \hat{\mathbf{x}}\cos\theta\cos\phi + \hat{\mathbf{y}}\cos\theta\sin\phi - \hat{\mathbf{z}}\sin\theta$
$\phi = \tan^{-1}\left[\dfrac{y}{x}\right]$	$\hat{\boldsymbol{\phi}} = -\hat{\mathbf{x}}\sin\phi + \hat{\mathbf{y}}\cos\phi$

Note that the preceding relationships can be deduced from careful examination of
Figures A.6b (unit vectors) and A.6c (quantities). A similar set of rules can be given
for transformation from rectangular to spherical coordinates:

Coordinate variables:	Unit vectors:
$x = r\sin\theta\cos\phi$	$\hat{\mathbf{x}} = \hat{\mathbf{r}}\sin\theta\cos\phi + \hat{\boldsymbol{\theta}}\cos\theta\cos\phi - \hat{\boldsymbol{\phi}}\sin\phi$
$y = r\sin\theta\sin\phi$	$\hat{\mathbf{y}} = \hat{\mathbf{r}}\sin\theta\sin\phi + \hat{\boldsymbol{\theta}}\cos\theta\sin\phi + \hat{\boldsymbol{\phi}}\cos\phi$
$z = r\cos\theta$	$\hat{\mathbf{z}} = \hat{\mathbf{r}}\cos\theta - \hat{\boldsymbol{\theta}}\sin\theta$

The transformation of a vector given in the spherical (rectangular) coordinate
system can be accomplished by direct substitution of the unit vector and the preceding coordinate variable expressions. However, doing so in fact amounts to taking
the dot products of the vector, as written in spherical (rectangular) coordinates,
with the rectangular (spherical) coordinate unit vectors and substituting coordinate
variables, as illustrated in Example A-1. The dot products of the spherical coordinate unit vectors with their rectangular coordinate counterparts are provided in
Table A.1.

Example A-2:   Spherical-to-rectangular and rectangular-to-spherical transformations.   (a) Transform the vector

$$\mathbf{E} = \hat{\mathbf{r}}\frac{2\cos\theta}{r^3} + \hat{\boldsymbol{\theta}}\sin\theta$$

given in spherical coordinates to rectangular coordinates. (b) Transform a vector $\mathbf{B} = \hat{\mathbf{x}}x^2 + \hat{\mathbf{y}}xy$ given in rectangular coordinates to spherical coordinates.

Solution:

(a) Note that this vector $\mathbf{E}$ has the same form as the distant electric field of an electric dipole (Equation [4.27]) or the distant magnetic field of a magnetic dipole (Equation [6.14]). We set out to determine E_x, E_y, and E_z. We have

$$E_x = \mathbf{E}\cdot\hat{\mathbf{x}} = \frac{2\cos\theta}{r^3}\sin\theta\cos\phi + \sin\theta\cos\theta\cos\phi = \left(\frac{2}{r^3}+1\right)\sin\theta\cos\theta\cos\phi$$

$$= \left(\frac{2+r^3}{r^3}\right)\left(\frac{x}{r}\right)\left(\frac{z}{r}\right) = \frac{[2+(x^2+y^2+z^2)^{3/2}]xz}{(x^2+y^2+z^2)^{5/2}}$$

where we have used the unit vector dot products as given in Table A.1 and substituted $x = r\sin\theta\cos\phi$ and $z = r\cos\theta$. Similarly we have

$$E_y = \mathbf{E}\cdot\hat{\mathbf{y}} = \frac{2\cos\theta}{r^3}\sin\theta\sin\phi + \sin\theta\cos\theta\sin\phi = \left(\frac{2}{r^3}+1\right)\cos\theta\sin\theta\cos\phi$$

which is identical to E_x. The z component is given by

$$E_z = \mathbf{E}\cdot\hat{\mathbf{z}} = \frac{2\cos\theta}{r^3}\cos\theta - \sin\theta\sin\theta = \frac{2\cos^2\theta}{r^3} - \sin^2\theta = \frac{2\cos^2\theta}{r^3} - (1-\cos^2\theta)$$

$$= \frac{2z^2}{r^5} - 1 + \frac{z^2}{r^2} = \frac{2z^2}{(x^2+y^2+z^2)^{5/2}} + \frac{z^2}{x^2+y^2+z^2} - 1$$

Thus we have

$$\mathbf{E} = [\hat{\mathbf{x}}+\hat{\mathbf{y}}]\left\{\frac{[2+(x^2+y^2+z^2)^{3/2}]xz}{(x^2+y^2+z^2)^{3/2}}\right\} + \hat{\mathbf{z}}\left[\frac{2z^2}{(x^2+y^2+z^2)^{5/2}} + \frac{z^2}{x^2+y^2+z^2} - 1\right]$$

The rather complicated nature of the rectangular form of vector $\mathbf{E}$ underscores the usefulness of spherical coordinates for problems involving spherical symmetry, such as the electric and magnetic dipoles.

(b) Note that the vector $\mathbf{B}$ is the same as that considered in Example A-1. We set out to find B_r, B_θ, and B_ϕ. We have

$$B_r = \mathbf{B}\cdot\hat{\mathbf{r}} = x^2\sin\theta\cos\phi + xy\sin\theta\sin\phi$$

$$= r^2\sin^2\theta\cos^2\phi\sin\theta\cos\phi + (r\sin\theta\cos\phi)(r\sin\theta\sin\phi)\sin\theta\sin\phi$$

$$= r^2\sin^3\theta\cos\phi(\cos^2\phi + \sin^2\phi) = r^2\sin^3\theta\cos\phi$$

where we have used the unit vector dot products as given in Table A.1 and substituted $x = r \sin\theta \cos\phi$ and $y = r \sin\theta \sin\phi$. Similarly, we have

$$B_\theta = \mathbf{B} \cdot \hat{\boldsymbol{\theta}} = x^2 \cos\theta \cos\phi + xy \cos\theta \sin\phi$$

$$= r^2 \sin^2\theta \cos^2\phi \cos\theta \cos\phi + (r \sin\theta \cos\phi)(r \sin\theta \sin\phi) \cos\theta \sin\phi$$

$$= r^2 \sin^2\theta \cos\theta \cos\phi(\cos^2\phi + \sin^2\phi)$$

$$= r^2 \sin^2\theta \cos\theta \cos\phi$$

And the ϕ component is given by

$$B_\phi = \mathbf{B} \cdot \hat{\boldsymbol{\phi}} = x^2(-\sin\phi) + xy \cos\phi$$

$$= -r^2 \sin^2\theta \cos^2\phi \sin\phi + (r \sin\theta \cos\phi)(r \sin\theta \sin\phi) \cos\phi = 0$$

Thus we have

$$\mathbf{B} = \hat{\mathbf{r}} r^2 \sin^3\theta \cos\phi + \hat{\boldsymbol{\theta}} r^2 \sin^2\theta \cos\theta \cos\phi$$

Length, Surface, and Volume Elements The general expression for differential length $d\mathbf{l}$ in spherical coordinates is

$$d\mathbf{l} = \hat{\mathbf{r}} \, dr + \hat{\boldsymbol{\theta}} r \, d\theta + \hat{\boldsymbol{\phi}} r \sin\theta \, d\phi$$

as is evident from Figure A.6d. By inspection of the volume element sketched in Figure A.6d, we can see that the spherical coordinate surface elements in the three coordinate directions are

$$(ds)_r = r^2 \sin\theta \, d\theta \, d\phi$$

$$(ds)_\theta = r \sin\theta \, dr \, d\phi$$

$$(ds)_\phi = r \, dr \, d\theta$$

and the volume element in spherical coordinates is:

$$dv = r^2 \sin\theta \, dr \, d\theta \, d\phi$$

A.4 VECTOR IDENTITIES

We start with some general vector relations involving dot and cross products of arbitrary vectors **A**, **B**, **C**, and **D**:

$$\mathbf{A} \cdot \mathbf{B} \times \mathbf{C} = \mathbf{A} \times \mathbf{B} \cdot \mathbf{C} = \mathbf{B} \cdot \mathbf{C} \times \mathbf{A}$$

$$\mathbf{A} \times (\mathbf{B} \times \mathbf{C}) = (\mathbf{A} \cdot \mathbf{C})\mathbf{B} - (\mathbf{A} \cdot \mathbf{B})\mathbf{C}$$

$$\mathbf{A} \times (\mathbf{B} \times \mathbf{C}) + \mathbf{B} \times (\mathbf{C} \times \mathbf{A}) + \mathbf{C} \times (\mathbf{A} \times \mathbf{B}) = 0$$

$$(\mathbf{A} \times \mathbf{B}) \cdot (\mathbf{C} \times \mathbf{D}) = \mathbf{A} \cdot [\mathbf{B} \times (\mathbf{C} \times \mathbf{D})] = (\mathbf{A} \cdot \mathbf{C})(\mathbf{B} \cdot \mathbf{D}) - (\mathbf{A} \cdot \mathbf{D})(\mathbf{B} \cdot \mathbf{C})$$

$$(\mathbf{A} \times \mathbf{B}) \times (\mathbf{C} \times \mathbf{D}) = (\mathbf{A} \times \mathbf{B} \cdot \mathbf{D})\mathbf{C} - (\mathbf{A} \times \mathbf{B} \cdot \mathbf{C})\mathbf{D}$$

A.4.1   Identities Involving ∇

The nabla (del or grad) operator is defined in rectangular coordinates as

$$\nabla \equiv \hat{\mathbf{x}}\frac{\partial}{\partial x} + \hat{\mathbf{y}}\frac{\partial}{\partial y} + \hat{\mathbf{z}}\frac{\partial}{\partial z}$$

and has been used extensively in this book. A number of vector identities involving the nabla operator are quite useful and are given in this section. Each of these identities can be verified by direct reduction of both sides of the equation. Arbitrary vectors are represented as $\mathbf{A}$ and $\mathbf{B}$, whereas Φ and Ψ denote scalars.

Note that the preceding definition of the nabla operator is valid and useful only for a rectangular coordinate system. In cylindrical or spherical coordinate systems, the orientation of the unit vectors depends on their position, so that the quantity $\nabla \times \mathbf{A}$ is not simply the cross product of a corresponding nabla operator and the vector $\mathbf{A}$. If any of the following vector identities were used in a cylindrical or spherical coordinate system, the quantities $\nabla \cdot \mathbf{A}$ and $\nabla \times \mathbf{A}$ should be taken to be symbolic representations of the cylindrical- or spherical-coordinate divergence and curl expressions, given in Sections 4.6 and 6.4. For example, in spherical coordinates, the quantity $\nabla \cdot \mathbf{A}$ is the quantity div $\mathbf{A}$ given by equation [4.37].

$$\nabla(\Phi + \Psi) \equiv \nabla\Phi + \nabla\Psi$$

$$\nabla(\Phi\Psi) \equiv \Phi\nabla\Psi + \Psi\nabla\Phi$$

$$\nabla \cdot (\Phi\mathbf{A}) \equiv \mathbf{A} \cdot \nabla\Phi + \Phi\nabla \cdot \mathbf{A}$$

$$\nabla \cdot (\mathbf{A} \times \mathbf{B}) \equiv \mathbf{B} \cdot (\nabla \times \mathbf{A}) - \mathbf{A} \cdot (\nabla \times \mathbf{B})$$

$$\nabla \times (\Phi\mathbf{A}) \equiv \nabla\Phi \times \mathbf{A} + \Phi\nabla \times \mathbf{A}$$

$$\nabla \times (\mathbf{A} \times \mathbf{B}) \equiv \mathbf{A}(\nabla \cdot \mathbf{B}) - \mathbf{B}(\nabla \cdot \mathbf{A}) + (\mathbf{B} \cdot \nabla)\mathbf{A} - (\mathbf{A} \cdot \nabla)\mathbf{B}$$

$$\nabla(\mathbf{A} \cdot \mathbf{B}) \equiv (\mathbf{A} \cdot \nabla)\mathbf{B} + (\mathbf{B} \cdot \nabla)\mathbf{A} + \mathbf{A} \times (\nabla \times \mathbf{B}) + \mathbf{B} \times (\nabla \times \mathbf{A})$$

$$\nabla \cdot \nabla\Phi \equiv \nabla^2\Phi$$

$$\nabla \cdot \nabla \times \mathbf{A} \equiv 0$$

$$\nabla \times \nabla\Phi \equiv 0$$

$$\nabla \times \nabla \times \mathbf{A} \equiv \nabla(\nabla \cdot \mathbf{A}) - \nabla^2\mathbf{A}$$

The operator formed by the dot product of a vector $\mathbf{A}$ and the del operator forms a new operator that is a scalar operation of the form

$$\mathbf{A} \cdot \nabla = A_x\frac{\partial}{\partial x} + A_y\frac{\partial}{\partial y} + A_z\frac{\partial}{\partial z}$$

which, when applied to a vector **B**, gives

$$(\mathbf{A} \cdot \nabla)\mathbf{B} = \hat{\mathbf{x}}\left[A_x \frac{\partial B_x}{\partial x} + A_y \frac{\partial B_x}{\partial y} + A_z \frac{\partial B_x}{\partial z}\right] + \hat{\mathbf{y}}\left[A_x \frac{\partial B_y}{\partial x} + A_y \frac{\partial B_y}{\partial y} + A_z \frac{\partial B_y}{\partial z}\right]$$

$$+ \hat{\mathbf{z}}\left[A_x \frac{\partial B_z}{\partial x} + A_y \frac{\partial B_z}{\partial y} + A_z \frac{\partial B_z}{\partial z}\right]$$

Special caution is needed in interpreting the Laplacian $\nabla^2 \mathbf{A}$, which in rectangular coordinates can be simply expanded as $\nabla^2 \mathbf{A} = \hat{\mathbf{x}}\nabla^2 A_x + \hat{\mathbf{y}}\nabla^2 A_y + \hat{\mathbf{z}}\nabla^2 A_z$. In spherical and cylindrical coordinates, the last identity given here, namely $\nabla^2 \mathbf{A} \equiv \nabla(\nabla \cdot \mathbf{A}) - \nabla \times \nabla \times \mathbf{A} = \text{grad}(\text{div } \mathbf{A}) - \text{curl}(\text{curl } \mathbf{A})$ is in fact the defining relation for the quantity $\nabla^2 \mathbf{A}$.

Special relations involving r. Some special relations involving the position vector $\mathbf{r} = \hat{\mathbf{x}}x + \hat{\mathbf{y}}y + \hat{\mathbf{z}}z$, with $r = |\mathbf{r}|$, and any arbitrary constant vector **K** are

$$\nabla \cdot \mathbf{r} = 3; \qquad \nabla \times \mathbf{r} = 0; \qquad \nabla r = \frac{\mathbf{r}}{r}$$

$$\nabla\left(\frac{1}{r}\right) = \frac{-\mathbf{r}}{r^3}; \qquad \nabla\left(\frac{\mathbf{r}}{r^3}\right) = -\nabla^2\left(\frac{1}{r}\right) = 0 \qquad \text{if } r \neq 0$$

$$\nabla \cdot \left(\frac{\mathbf{K}}{r}\right) = \mathbf{K} \cdot \left[\nabla\left(\frac{1}{r}\right)\right] = \frac{-(\mathbf{K} \cdot \mathbf{r})}{r^3}$$

$$\nabla^2\left(\frac{\mathbf{K}}{r}\right) = \mathbf{K}\nabla^2\left(\frac{1}{r}\right) = 0 \qquad \text{if } r \neq 0$$

$$\nabla \times (\mathbf{K} \times \mathbf{B}) = \mathbf{K}(\nabla \cdot \mathbf{B}) + \mathbf{K} \times (\nabla \times \mathbf{B}) - \nabla(\mathbf{K} \cdot \mathbf{B})$$

Integral relations. For any arbitrary vector field **G** and a scalar field Φ we have

$$\oint_S \mathbf{G} \cdot d\mathbf{s} = \int_V (\nabla \cdot \mathbf{G})\, dv \qquad\qquad \text{Divergence theorem}$$

$$\oint_S \Phi\, d\mathbf{s} = \int_V (\nabla\Phi)\, dv$$

$$\oint_S \mathbf{G} \times d\mathbf{s} = -\int_V (\nabla \times \mathbf{G})\, dv$$

where V is a volume bounded by a closed surface S, with the surface element $d\mathbf{s}$ defined outward from the enclosed volume.

If S' is an open surface bounded by contour C, the line element of which is $d\mathbf{l}$, we have

$$\oint_C \Phi\, d\mathbf{l} = -\int_{S'} \nabla\Phi \times d\mathbf{s}$$

$$\oint_C \mathbf{G} \cdot d\mathbf{s} = \int_{S'} (\nabla \times \mathbf{G}) \cdot d\mathbf{l} \qquad\qquad \text{Stokes' theorem}$$

where the defined direction of $d\mathbf{s}$ and the sense of the contour integration (i.e., direction of $d\mathbf{l}$) are related by the right-hand rule.

A.5 GRADIENT, DIVERGENCE, AND CURL

In this section we briefly define and describe[2] the vector derivative operations gradient, divergence, and curl.

A.5.1 Gradient of a Scalar

The gradient of a scalar function at any given point is the maximum spatial rate of change (i.e., the steepest slope) of that function at that point. Gradient is thus naturally a vector quantity, because the maximum rate of change with distance would occur in a given direction. The direction of the gradient vector is that in which the scalar function (e.g., temperature) changes most rapidly with distance. A good analogy here is the gravitational potential. A marble placed on the slope of a mountain rolls down (i.e., its velocity vector is oriented) in the direction of the gradient of the gravitational potential, its speed (i.e., the magnitude of its velocity) being determined by the spatial derivative in that direction. Mathematically, the gradient of a scalar function Φ is defined in rectangular coordinates as:

$$\text{grad } \Phi \equiv \left(\hat{\mathbf{x}} \frac{\partial \Phi}{\partial x} + \hat{\mathbf{y}} \frac{\partial \Phi}{\partial y} + \hat{\mathbf{z}} \frac{\partial \Phi}{\partial z} \right)$$

Using the nabla operator ∇, the gradient of a scalar Φ is typically written in a rectangular coordinate system as

$$\text{grad } \Phi = \left(\hat{\mathbf{x}} \frac{\partial}{\partial x} + \hat{\mathbf{y}} \frac{\partial}{\partial y} + \hat{\mathbf{z}} \frac{\partial}{\partial z} \right) \Phi = \nabla \Phi$$

In rectangular coordinates, the concept of the gradient of a scalar field can be understood in terms of the motion of a test charge in electrostatic potential field, in which case the negative gradient of the potential $\Phi(x, y, z)$ is the electric field $\mathscr{E}(x, y, z)$. The magnitude of the electric field in any one of the rectangular directions is given by the spatial rate of change of the potential in that direction. If we consider the motion of such a test charge in a cylindrical (r, ϕ, z) coordinate system, we have to realize that in moving by a differential amount $d\phi$ in the ϕ direction the particle spans a distance of $r d\phi$ (see Figure A.5d), so the value of E_ϕ is determined by $r^{-1} \partial \Phi / \partial \phi$ rather than just $\partial \Phi / \partial \phi$. Motion in the r and z directions spans differential distances of dr and dz, so the gradient of the field in those directions is

---

[2]More detailed discussion of these vector derivatives, their physical interpretation, and many examples of scalar and vector fields with gradient, divergence, and curl properties are available elsewhere. Specifically, see Sections 4.4 (gradient), 4.6 (divergence), and 6.4 (curl) of U. S. Inan and A. S. Inan, *Engineering Electromagnetics*, Addison Wesley Longman, 1999.

determined respectively by $\partial\Phi/\partial r$ and $\partial\Phi/\partial z$. Thus, the gradient of a scalar function $\Phi(r, \phi, z)$ in cylindrical coordinates is

$$\text{grad } \Phi = \left[\hat{\mathbf{r}}\frac{\partial\Phi}{\partial r} + \hat{\boldsymbol{\phi}}\frac{1}{r}\frac{\partial\Phi}{\partial\phi} + \hat{\mathbf{z}}\frac{\partial\Phi}{\partial z} \right]$$

Similar considerations for a spherical coordinate system (r, θ, ϕ) indicate that in moving by an amount $d\theta$ in latitude we span a distance of $r d\theta$, whereas the distance we span in moving by an amount $d\phi$ in azimuth (longitude) depends on our latitude and is $r \sin\theta d\phi$, as shown in Figure A.6d. Accordingly, the gradient of a scalar function $\Phi(r, \theta, \phi)$ in spherical coordinates is

$$\text{grad } \Phi = \left[\hat{\mathbf{r}}\frac{\partial\Phi}{\partial r} + \hat{\boldsymbol{\theta}}\frac{1}{r}\frac{\partial\Phi}{\partial\theta} + \hat{\boldsymbol{\phi}}\frac{1}{r\sin\theta}\frac{\partial\Phi}{\partial\phi} \right]$$

A.5.2 Divergence of a Vector Field

The concept of the divergence of a vector field is important with respect to understanding the sources of fields. In the electrostatic context, the divergence of the electric flux density $\mathbf{D}$ is directly related to the source charge density ρ. The definition and meaning of divergence can be understood by applying Gauss's law to a very small (infinitesimal) volume.

If one envisions a closed surface S enclosing a source for any vector field $\mathbf{A}$, then the strength or magnitude of the source is given by the net outward flow or flux of $\mathbf{A}$ through the closed surface S, or $\oint_S \mathbf{A} \cdot d\mathbf{s}$. However, a more suitable measure of the *concentration* or *density* of the source is the outward flux per unit volume, or

$$\frac{\oint_S \mathbf{A} \cdot d\mathbf{s}}{V}$$

where V is the volume enclosed by the surface S. By reducing V to a differential volume, we obtain a point relation that gives the source density per unit volume at that point. The source density in general varies from point to point, so this measure of the source density is a scalar field and is called the *divergence* of the vector field $\mathbf{A}$. The fundamental definition of the divergence of a vector field $\mathbf{A}$ at any point is then

$$\boxed{\text{div } \mathbf{A} \equiv \lim_{V \to 0} \frac{\oint_S \mathbf{A} \cdot d\mathbf{s}}{V}}$$

where S is the closed surface enclosing the volume V.

For the electrostatic field, we know from Gauss's law that the total outward flux through a surface S is equal to the total charge enclosed. For a differential volume element V, we can assume the volume charge density to be constant at all points within the infinitesimal volume, so the total charge enclosed is ρV. Accordingly, we have

$$\text{div } \mathbf{D} \equiv \lim_{V \to 0} \frac{\oint_S \mathbf{D} \cdot d\mathbf{s}}{V} = \rho$$

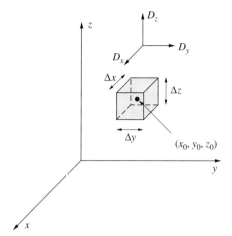

FIGURE A.7. Divergence.   A cubical volume element used to derive the differential expression for the divergence of **D**.

so that the *divergence* of the electric flux density at any point is equal to the volume charge density at that point.

We now derive the proper differential expression for div **D**. Although the divergence of a vector is clearly independent of any coordinate system, for simplicity we use a rectangular coordinate system and consider a small differential cubical volume element as shown in Figure A.7. To determine the net outward flux from this volume, we can consider each pair of parallel faces separately, namely left–right, front–back, and bottom–top. We assume that the flux density **D** at the center of the cube is given by $\mathbf{D}(x_o, y_o, z_o) = \hat{\mathbf{x}}D_{xo} + \hat{\mathbf{y}}D_{yo} + \hat{\mathbf{z}}D_{zo}$. We first consider the net flux through the left and right faces, due to the y component (i.e., D_y) of **D**. Note that on the left face, we have $y = y_o - \Delta y/2$, whereas on the right face $y = y_o + \Delta y/2$. On the left face of the cubical volume element shown in Figure A.7, $d\mathbf{s} = -\hat{\mathbf{y}}\,dx\,dz$, so the flux leaving the left face is, to the first order,[3]

$$\left[\hat{\mathbf{y}}D_y\left(x_o, y_o - \frac{\Delta y}{2}, z_o\right)\right] \cdot (-\hat{\mathbf{y}}|\Delta z\Delta x|) = -(\Delta z\Delta x)D_y\left(x_o, y_o - \frac{\Delta y}{2}, z_o\right)$$

$$= -(\Delta z\Delta x)\left(D_{yo} - \frac{\Delta y}{2}\frac{\partial D_y}{\partial y}\right)$$

That from the right face[4] is similarly

$$+(\Delta z\Delta x)\left(D_{yo} + \frac{\Delta y}{2}\frac{\partial D_y}{\partial y}\right)$$

---

[3] At a more formal level, we can expand D_y in a Taylor series around $y = y_o$, so that

$$D_y(x_o, y, z_o) = D_y(x_o, y_o, z_o) + (y - y_o)\frac{\partial D_y}{\partial y} + (y - y_o)^2\frac{\partial^2 D_y}{\partial y^2} + \text{higher-order terms}$$

Note that for a differential volume element with $\Delta y = (y - y_o) \rightarrow 0$, the higher-order terms involving $(\Delta y)^k$ will be negligible compared to the first-order term multiplied by Δy.

[4] Note that we now have $d\mathbf{s} = +\hat{\mathbf{y}}\,dx\,dz$.

Thus, the net outward flux between these two faces is

$$\Delta z \Delta x \Delta y \frac{\partial D_y}{\partial y}$$

The net outward flux between the other two pairs of faces, namely front–back and bottom–top, can be similarly evaluated to find the net total outward flux from the cubical surface:

$$\Delta z \Delta x \Delta y \left(\frac{\partial D_x}{\partial x} + \frac{\partial D_y}{\partial y} + \frac{\partial D_z}{\partial z} \right)$$

Since divergence was defined as the total outward flux per unit volume, we have

$$\text{div } \mathbf{D} = \frac{\partial D_x}{\partial x} + \frac{\partial D_y}{\partial y} + \frac{\partial D_z}{\partial z}$$

Using nabla operator ∇, the divergence of a vector $\mathbf{D}$ is typically written in a rectangular coordinate system as

$$\text{div } \mathbf{D} = \nabla \cdot \mathbf{D} = \left(\hat{\mathbf{x}} \frac{\partial}{\partial x} + \hat{\mathbf{y}} \frac{\partial}{\partial y} + \hat{\mathbf{z}} \frac{\partial}{\partial z} \right) \cdot (\hat{\mathbf{x}} D_x + \hat{\mathbf{y}} D_y + \hat{\mathbf{z}} D_z)$$

$$= \frac{\partial D_x}{\partial x} + \frac{\partial D_y}{\partial y} + \frac{\partial D_z}{\partial z}$$

Divergence in Other Coordinate Systems Although we use the notation $\nabla \cdot \mathbf{D}$ to indicate the divergence of a vector $\mathbf{D}$, it should be noted that a vector *del* operator as given above is useful only in a rectangular coordinate system. In other coordinate systems, we still denote the divergence of $\mathbf{D}$ by $\nabla \cdot \mathbf{D}$, but note that the specific scalar derivative expressions to be used will need to be derived from the physical definition of divergence using a differential volume element appropriate for that particular coordinate system.

To derive the differential expression for divergence in the cylindrical coordinate system, we consider the cylindrical cuboid volume element shown in Figure A.5d. Note that the volume of this element is $r \Delta r \Delta \phi \Delta z$. The flux of the vector field $\mathbf{F}$ through the face marked as S_1 is

$$\int_{S_1} \mathbf{F} \cdot d\mathbf{s} = \int_{S_1} F_r \, ds \simeq F_r \left(r + \frac{\Delta r}{2}, \phi, z \right) \left(r + \frac{\Delta r}{2} \right) \Delta \phi \Delta z$$

while that through face s_2 is

$$\int_{S_2} \mathbf{F} \cdot (-\hat{\mathbf{r}} \, ds) = -\int_{S_2} F_r \, ds \simeq -F_r \left(r - \frac{\Delta r}{2}, \phi, z \right) \left(r - \frac{\Delta r}{2} \right) \Delta \phi \Delta z$$

Adding these two and dividing by the volume $V = r \Delta r \Delta \phi \Delta z$ gives the net flux per unit volume out of the cube due to the r component of the vector field, namely

$$\frac{1}{\Delta V} \int_{S_1 + S_2} \mathbf{F} \cdot d\mathbf{s} \simeq \frac{1}{r \Delta r} \left[\left(r + \frac{\Delta r}{2} \right) F_r \left(r + \frac{\Delta r}{2}, \phi, z \right) - \left(r - \frac{\Delta r}{2} \right) F_r \left(r - \frac{\Delta r}{2}, \phi, z \right) \right]$$

which in the limit, as $\Delta r \to 0$ (and thus $\Delta V \to 0$), becomes

$$\frac{1}{r}\frac{\partial}{\partial r}(rF_r)$$

Evaluating the contributions from the other two pairs of faces in a similar fashion, we find the expression for divergence in cylindrical coordinates:

$$\text{div } \mathbf{F} = \nabla \cdot \mathbf{F} = \frac{1}{r}\frac{\partial}{\partial r}(rF_r) + \frac{1}{r}\frac{\partial F_\phi}{\partial \phi} + \frac{\partial F_z}{\partial z}$$

Consideration of the flux in and out of the different opposing faces of the spherical cuboid shown in Figure A.6d leads to the divergence expression in spherical coordinates:

$$\text{div } \mathbf{F} = \nabla \cdot \mathbf{F} = \frac{1}{r^2}\frac{\partial}{\partial r}(r^2F_r) + \frac{1}{r\sin\theta}\frac{\partial}{\partial\theta}(\sin\theta F_\theta) + \frac{1}{r\sin\theta}\frac{\partial F_\phi}{\partial\phi}$$

A.5.3   Curl of a Vector Field

Another very important property of a vector field is its circulation. We can observe circulation as we watch water drain out of a bathtub or sink. For a perfectly symmetrical circulating fluid with angular velocity ω, a measure of the circulation may be defined as the product of angular velocity and circumference [i.e., $\omega(2\pi r)$]. We can extend this concept of circulation to any vector field $\mathbf{G}$ by defining total or net circulation about any arbitrary closed path C as

$$\text{Circulation} \equiv \oint_C \mathbf{G} \cdot d\mathbf{l}$$

which is called the circulation of $\mathbf{G}$ around C. Since the closed loop C would enclose a surface S, the net circulation of a vector field $\mathbf{G}$ over the surface S is defined as the circulation of $\mathbf{G}$ around C divided by the area of the surface. Thus, the net circulation per unit surface area is

$$\frac{\oint_C \mathbf{G} \cdot d\mathbf{l}}{S}$$

The *curl* of a vector field $\mathbf{G}$ is defined as an axial vector whose magnitude is the maximum (as the direction of $\hat{\mathbf{s}}$ is varied) net circulation of vector $\mathbf{G}$ per unit area as the area approaches to zero. In other words,

$$\text{curl } \mathbf{G} \equiv \left[\lim_{S\to 0}\frac{\oint_C \mathbf{G} \cdot d\mathbf{l}}{S}\right]_{\max} \hat{\mathbf{s}} \qquad [A.1]$$

Note that the curl is clearly a directional quantity; by convention, the direction of the curl is chosen to be the direction of the surface element $\mathbf{S}$ that gives the maximum value for the magnitude of the net circulation per unit area in the above defining expression. Thus (curl $\mathbf{G}$) is a vector; for example, the x component of the vector (curl **G**) would represent the line integral of **G** along an infinitesimally small (i.e., $S \to 0$) closed path lying in the yz plane.

For the magnetostatic **B** field, we know from Ampère's law that the circulation of **B** along any closed contour C (i.e., the line integral of **B** along C) is equal to the total current passing through the surface area S enclosed by C times μ_0. For a differential surface element, we can assume that the current density is constant at all points over the surface, so the total current is $|\mathbf{J}|S$, if the surface is chosen to be orthogonal to the direction of the current flow so that the net circulation is maximized. Accordingly, we have

$$\text{curl } \mathbf{B} \equiv \left[\lim_{S \to 0} \frac{\oint_C \mathbf{B} \cdot d\mathbf{l}}{S} \right]_{\max} \hat{\mathbf{s}} = \mu_0 \mathbf{J}$$

Curl in the Rectangular Coordinate System We now utilize the definition of curl to obtain a convenient expression in terms of partial derivatives. For this, we simply need to evaluate curl **B** at a general point in a given coordinate system. With reference to Figure A.8, we can evaluate the component of curl **B** in the x direction by conducting a line integral along the path abcda, which, as shown, lies entirely in the yz-plane. Note that this is an infinitesimally small path with dimensions Δy and Δz, so that the magnitude of **B** at different sides can be simply related to the value of the field at the center point x_o, y_o, z_o. We have

$$\oint_{abcda} \mathbf{B} \cdot d\mathbf{l} = B_{1y}\Delta y + B_{2z}\Delta z - B_{3y}\Delta y - B_{4z}\Delta z = \mu_o J_x \Delta y \Delta z$$

$$\underbrace{(B_{1y} - B_{3y})}_{-(\partial B_y/\partial z)\Delta z} \Delta y + \underbrace{(B_{2z} - B_{4z})}_{(\partial B_z/\partial y)\Delta y} \Delta z = \mu_o J_x \Delta y \Delta z$$

$$\left[\frac{\partial B_z}{\partial y} - \frac{\partial B_y}{\partial z} \right] \Delta y \Delta z = \mu_o J_x \Delta y \Delta z$$

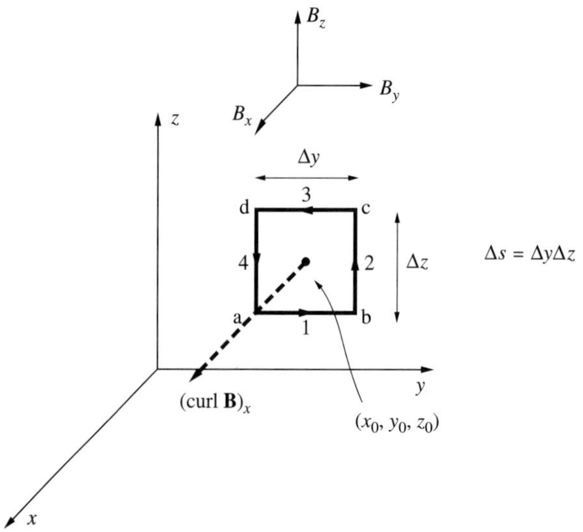

FIGURE A.8. Evaluation of the x component of curl B.

from which

$$(\text{curl } \mathbf{B})_x = \lim_{\Delta s \to 0} \frac{\oint_{abcda} \mathbf{B} \cdot d\mathbf{l}}{\Delta s} = \frac{\partial B_z}{\partial y} - \frac{\partial B_y}{\partial z} = \mu_o J_x$$

By taking line integrals along other contours lying in the xz and xy planes, we can show that

$$\text{curl } \mathbf{B} = \hat{\mathbf{x}}\left(\frac{\partial B_z}{\partial y} - \frac{\partial B_y}{\partial z}\right) + \hat{\mathbf{y}}\left(\frac{\partial B_x}{\partial z} - \frac{\partial B_z}{\partial x}\right) + \hat{\mathbf{z}}\left(\frac{\partial B_y}{\partial x} - \frac{\partial B_x}{\partial y}\right)$$

$$= \mu_o(\hat{\mathbf{x}}J_x + \hat{\mathbf{y}}J_y + \hat{\mathbf{z}}J_z)$$

in rectangular coordinates. Using the *nabla* operator, we can express the curl of $\mathbf{B}$ as

$$\text{curl } \mathbf{B} = \nabla \times \mathbf{B} = \mu_o \mathbf{J}$$

which is the differential form of Ampère's law in magnetostatics. A convenient method for remembering the expression for $\nabla \times \mathbf{B}$ in rectangular coordinate system is to use the determinant form

$$\nabla \times \mathbf{B} = \begin{vmatrix} \hat{\mathbf{x}} & \hat{\mathbf{y}} & \hat{\mathbf{z}} \\ \dfrac{\partial}{\partial x} & \dfrac{\partial}{\partial y} & \dfrac{\partial}{\partial z} \\ B_x & B_y & B_z \end{vmatrix}$$

Curl in Other Coordinate Systems   Although we use the notation $\nabla \times \mathbf{A}$ to indicate the curl of a vector $\mathbf{A}$, the expression of curl as a cross product with a vector *nabla* operator (in the form $\nabla = \hat{\mathbf{x}}(\partial/\partial x) + \hat{\mathbf{y}}(\partial/\partial y) + \hat{\mathbf{z}}(\partial/\partial z)$ for rectangular coordinates) is useful only in the rectangular coordinate system. In other coordinate systems we still denote the curl of $\mathbf{A}$ by $\nabla \times \mathbf{A}$, but note that the specific derivative expressions for each component of curl will need to be derived from the physical definition of curl as given in [A.1], using a differential surface element appropriate for that particular coordinate system.

To derive the differential expression for curl in the spherical coordinate system, we consider the cuboid volume element shown in Figure A.6d. To determine the r component of the curl, we consider the curvilinear contour ABCDA. Note that this contour is perpendicular to $\hat{\mathbf{r}}$, and the sense is related to it by the right-hand rule, if we go around the contour in the $\mathrm{A} \to \mathrm{B} \to \mathrm{C} \to \mathrm{D} \to \mathrm{A}$ direction. The lengths of the sides are $\mathrm{AB} = r\,d\theta$, $\mathrm{CD} = r\,d\theta$, $\mathrm{BC} = r\sin(\theta+d\theta)\,d\phi$, and $\mathrm{DA} = r\sin\theta\,d\phi$. The line integral is given by

$$\oint_{\text{ABCDA}} \mathbf{A} \cdot d\mathbf{l} = A_\theta r\,d\theta + \left(A_\phi + \frac{\partial A_\phi}{\partial \theta}d\theta\right)r\sin(\theta + d\theta)\,d\phi$$

$$- \left(A_\theta + \frac{\partial A_\theta}{\partial \phi}d\phi\right)r\,d\theta - A_\phi r\sin\theta\,d\phi$$

$$= -\frac{\partial A_\theta}{\partial \phi}r\,d\theta\,d\phi + \frac{\partial}{\partial \theta}(A_\phi \sin\theta)r\,d\theta\,d\phi$$

where we have retained small quantities up to second order. The area of the contour ABCDA is $\Delta s = r^2 \sin\theta \, d\theta \, d\phi$, so that

$$[\nabla \times \mathbf{A}]_r = \frac{\oint_{ABCDA} \mathbf{A} \cdot d\mathbf{l}}{\Delta s} = \frac{1}{r\sin\theta}\left[\frac{\partial}{\partial\theta}(A_\theta \sin\theta) - \frac{\partial A_\phi}{\partial\phi}\right]$$

The other two components can be obtained by considering the other sides of the cuboid in Figure A.6d, namely the contour ABFEA for the ϕ component and the contour BCGFB for the θ component. The complete expression for the curl in spherical coordinates is

$$\text{curl } \mathbf{A} = \nabla \times \mathbf{A} = \hat{\mathbf{r}}\frac{1}{r\sin\theta}\left[\frac{\partial}{\partial\theta}(A_\phi \sin\theta) - \frac{\partial A_\theta}{\partial\phi}\right]$$

$$+ \hat{\boldsymbol{\theta}}\frac{1}{r}\left[\frac{1}{\sin\theta}\frac{\partial A_r}{\partial\phi} - \frac{\partial}{\partial r}(rA_\phi)\right]\hat{\boldsymbol{\phi}}\frac{1}{r}\left[\frac{\partial}{\partial r}(rA_\theta) - \frac{\partial A_r}{\partial\theta}\right]$$

Similar reasoning in a cylindrical coordinate system using a volume element such as that shown in Figure A.5d leads to the curl expression in cylindrical coordinates:

$$\text{curl } \mathbf{A} = \nabla \times \mathbf{A} = \hat{\mathbf{r}}\left[\frac{1}{r}\frac{\partial A_z}{\partial\phi} - \frac{\partial A_\phi}{\partial z}\right] + \hat{\boldsymbol{\phi}}\left[\frac{\partial A_r}{\partial z} - \frac{\partial A_z}{\partial r}\right] + \hat{\mathbf{z}}\frac{1}{r}\left[\frac{\partial}{\partial r}(rA_\phi) - \frac{\partial A_r}{\partial\phi}\right]$$

APPENDIX B

Useful Tables

TABLE B.1. Transmission line parameters for some uniform two-conductor transmission lines surrounded by air

	Coaxial	Two-wire	Parallel-plate*
L (μH-m^{-1})	$0.2\ln(b/a)$	$0.4\ln\left[\dfrac{d}{2a}+\sqrt{\left(\dfrac{d}{2a}\right)^2-1}\right]$	$\dfrac{1.26a}{b}$
C (pF-m^{-1})	$\dfrac{55.6}{\ln(b/a)}$	$\dfrac{27.8}{\ln\left[\dfrac{d}{2a}+\sqrt{\left(\dfrac{d}{2a}\right)^2-1}\right]}$	$\dfrac{8.85b}{a}$
R (Ω-m^{-1})	$\dfrac{4.15\times10^{-8}(a+b)\sqrt{f}}{ab}$	$\dfrac{8.3\times10^{-8}\sqrt{f}}{a}$	$\dfrac{5.22\times10^{-7}\sqrt{f}}{b}$
G** (S-m^{-1})	$\dfrac{7.35\times10^{-4}}{\ln(b/a)}$	$\dfrac{3.67\times10^{-4}}{\ln\left[\dfrac{d}{2a}+\sqrt{\left(\dfrac{d}{2a}\right)^2-1}\right]}$	$\dfrac{1.17\times10^{-4}b}{a}$
Z_0 (Ω)	$60\ln(b/a)$	$120\ln\left[\dfrac{d}{2a}+\sqrt{\left(\dfrac{d}{2a}\right)^2-1}\right]$	$\dfrac{377a}{b}$

*Valid for $b\gg a$.
**For polyethylene at 3 GHz.

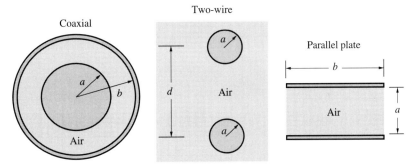

FIGURE B.1. **Cross-sectional view of three common uniform transmission lines.**
Expressions for the circuit parameters L, R, C, and G for these coaxial, two-wire, and parallel-plate lines are provided in Table B.1.

TABLE B.2. Relative permittivity and dielectric strength of selected materials

Material	Relative permittivity (ϵ_r) (at room temperature)	Dielectric strength (MV-m^{-1}) (at room temp. and 1 atm)
Air	1	~3
Alumina (Al_2O_3)	~8.8	
Amber	2.7	
Bakelite	~4.8	25
Barium titanate ($BaTiO_3$)	1200	7.5
Freon	1	~8
Fused quartz (SiO_2)	3.9	~1000
Gallium arsenide (GaAs)	13.1	~40
Germanium (Ge)	16	~10
Glass	~4–9	~30
Glycerin	50	
Ice	3.2	
Mica (ruby)	5.4	200
Nylon	~3.6–4.5	
Oil	2.3	15
Paper	1.5–4	15
Paraffin wax	2.1	30
Plexiglass	3.4	
Polyethylene	2.26	
Polystyrene	2.56	20
Porcelain	~5–9	11
Rubber	~2.4–3.0	25
Rutile (TiO_2)	100	
Silicon (Si)	11.9	~30
Silicon nitride (Si_3N_4)	7.2	~1000
Sodium chloride (NaCl)	5.9	
Styrofoam	1.03	
Sulphur	4	
Tantalum pentoxide (Ta_2O_5)	~25	
Teflon (PTFE)	2.1	
Vaseline	2.16	
Water (distilled)	81	
Wood (balsa)	1.4	

TABLE B.3. Conductivities and temperature coefficients of selected materials

Material	Conductivity σ (S-m^{-1}) (at 20°C)	Temperature coefficient α_σ [(°C)$^{-1}$]
Aluminum	3.82×10^7	0.0039
Bismuth	8.70×10^5	0.004
Brass (66 Cu, 34 Zn)	2.56×10^7	0.002
Carbon (graphite)	7.14×10^4	−0.0005
Constantan (55 Cu, 45 Ni)	2.26×10^6	0.0002
Copper (annealed)	5.80×10^7	0.0039
Dry, sandy soil	$\sim 10^{-3}$	
Distilled water	$\sim 10^{-4}$	
Fresh water	$\sim 10^{-2}$	
Germanium (intrinsic)	~ 2.13	−0.048
Glass	$\sim 10^{-12}$	−0.07
Gold	4.10×10^7	0.0034
Iron	1.03×10^7	0.0052–0.0062
Lead	4.57×10^6	0.004
Marshy soil	$\sim 10^{-2}$	
Mercury (liquid)	1.04×10^6	0.00089
Mica	$\sim 10^{-15}$	−0.07
Nichrome (65 Ni, 12 Cr, 23 Fe)	1.00×10^6	0.00017
Nickel	1.45×10^7	0.0047
Niobium	8.06×10^6	
Platinum	9.52×10^6	0.003
Polystyrene	$\sim 10^{-16}$	
Porcelain	$\sim 10^{-14}$	
Quartz (fused)	$\sim 10^{-17}$	
Rubber (hard)	$\sim 10^{-15}$	
Seawater	~ 4	
Silicon (intrinsic)	$\sim 4.35 \times 10^{-4}$	
Silver	6.17×10^7	0.0038
Sodium	2.17×10^7	
Stainless steel	1.11×10^6	
Sulfur	$\sim 10^{-15}$	
Tin	8.77×10^6	0.0042
Titanium	2.09×10^6	
Tungsten	1.82×10^7	0.0045
Y Ba$_2$Cu$_3$O$_7$ (at < 80K)	$\sim 10^{20}$	
Wood	10^{-11}–10^{-8}	
Zinc	1.67×10^7	0.0037

TABLE B.4.   Relative permeability of selected materials

Material	Relative permeability (μ_r)
Air	1.00000037
Aluminum	1.000021
Bismuth	0.999833
Cobalt	250
Copper	0.9999906
Iron (Purified: 99.96% Fe)	280,000
Iron (Motor grade: 99.6%)	5,000
Lead	0.9999831
Manganese	1.001
Manganese-zinc ferrite	1,200
Mercury	0.999968
Nickel	600
Nickel-zinc ferrite	650
Oxygen	1.000002
Palladium	1.0008
Permalloy: 78.5% Ni, 21.5% Fe	70,000
Platinum	1.0003
Silver	0.9999736
Supermalloy: 79% Ni, 15% Fe, 5% Mo, 0.5% Mn	1,000,000
Tungsten	1.00008
Water	0.9999912

Symbols and Units for Basic Quantities

Symbol	Quantity	SI Unit	Comments
$\mathscr{A}$	Magnetic vector potential	webers (Wb)-m^{-1}	
A	Magnetic vector potential	Wb-m^{-1}	Phasor
B	Magnetic field (**B** field)	Wb-m^{-2} or tesla (T)	Phasor
$\overline{\mathscr{B}}$	Magnetic field (**B** field)	Wb-m^{-2}	
C	Capacitance	farads (F)	
D	Electric flux density	C-m^{-2}	Phasor
$\overline{\mathscr{D}}$	Electric flux density	C-m^{-2}	
E	Electric field intensity	V-m^{-1}	Phasor
$\overline{\mathscr{E}}$	Electric field intensity	V-m^{-1}	
F	Force	newtons (N)	Phasor
$\overline{\mathscr{F}}$	Force	N	
f	Frequency	hertz (Hz)	
f_p	Plasma frequency	Hz	
$\mathscr{G}$	Electromagnetic momentum	N-s	
$\overline{g}$	Momentum density	N-s-m^{-3}	
g_d	Directive gain		
H	Magnetic field intensity	A-m^{-1}	Phasor
$\overline{\mathscr{H}}$	Magnetic field intensity	A-m^{-1}	
I	Current	amperes (A)	Phasor
$\mathscr{I}$	Current	A	
J	Current density	A-m^{-2}	Phasor
$\overline{\mathscr{J}}$	Current density	A-m^{-2}	
J$_s$	Surface current density	A-m^{-1}	Phasor
$\overline{\mathscr{J}}_s$	Surface current density	A-m^{-1}	
L	Inductance	henrys (H)	
$\overline{\mathscr{L}}$	Angular momentum	kg-m^2-s^{-1}	
l	Length	m	
$\overline{\mathscr{M}}$	Magnetization vector	A-m^{-1}	
M	Magnetization vector	A-m^{-1}	Phasor
m	Mass	kg	
m	Magnetic dipole moment	A-m^{-2}	
n	Index of refraction		
P	Power	watts (W)	
$\overline{\mathscr{P}}$	Electric polarization vector	C-m^{-2}	
P	Electric polarization vector	C-m^{-2}	Phasor
P$_{av}$	Time-average pressure	N-m^{-2}	
p	Electric dipole moment	C-m	
Q, q	Electric charge	coulombs (C)	
Q	Quality factor		
R	Resistance	Ω	
R_s	Surface resistance	Ω	
S	Standing wave ratio		
$\overline{\mathscr{S}}$	Poynting vector	W-m^{-2}	
S$_{av}$	Poynting vector	W-m^{-2}	Time-average quantity
S	Complex Poynting vector	W-m^{-2}	
$\overline{\mathscr{T}}$	Torque	N-m	

Symbol	Quantity	SI Unit	Comments
$\mathcal{T}$	Transmission coefficient		$\mathcal{T} = \tau e^{j\phi_{\mathcal{T}}}$
t	Time	s	
U	Radiation intensity	W-(unit solid angle)$^{-1}$	
$\mathcal{N}_e$	Electron density	m^{-3}	
N_e	Electron density	m^{-3}	
$\tilde{\mathbf{v}}$	Velocity vector	m-s^{-1}	
$\mathbf{v}$	Velocity vector	m-s^{-1}	Phasor
v_p	Phase velocity	m-s^{-1}	
$\bar{v}_p$	Phase velocity in waveguide	m-s^{-1}	
V	Voltage	volts (V)	Phasor
$\mathcal{V}$	Voltage	V	
$\mathcal{V}_{\text{emf}}$	Electromotive force	V	
$\mathcal{V}_{\text{ind}}$	Induced emf	V	
W	Work (energy)	joules (J)	
w	Energy density	J-m^{-3}	
X	Reactance	Ω	
Y	Admittance	S	
Z	Impedance	Ω	
Z_0	Characteristic impedance	Ω	$Z_0 = \sqrt{L/C}$ for lossless line
α	Attenuation constant	nepers (np)-m^{-1}	Real part of γ
$\bar{\alpha}$	Attenuation constant in waveguide	np-m^{-1}	Real part of $\bar{\gamma}$
α_c	Attenuation constant	np-m^{-1}	Due to conduction losses
α_d	Attenuation constant	np-m^{-1}	Due to dielectric losses
β	Propagation constant	radians (rad)-m^{-1}	$\beta = \omega\sqrt{\mu\epsilon}$
β	Phase constant	rad-m^{-1}	Imaginary part of γ
$\bar{\beta}$	Phase constant in waveguide	rad-m^{-1}	Imaginary part of $\bar{\gamma}$
Γ	Reflection coefficient		$\Gamma = \rho e^{j\phi_{\Gamma}}$
γ	Propagation constant	m^{-1}	$\gamma = \alpha + j\beta$
$\bar{\gamma}$	Propagation constant in waveguide	m^{-1}	$\bar{\gamma} = \bar{\alpha}$ or $\bar{\gamma} = j\bar{\beta}$
γ_{m}	Gyromagnetic ratio	rad-s^{-1}-T^{-1}	
δ	Skin depth	m	$\delta = (\pi f \mu \sigma)^{-1/2}$
$\tan \delta_c$	Loss tangent		$\tan \delta_c = \epsilon''/\epsilon'$
ϵ, ϵ_0	Permittivity	F-m^{-1}	
ϵ_r	Relative permittivity		
ϵ_c	Complex permittivity	F-m^{-1}	$\epsilon_c = \epsilon' - j\epsilon''$
$\theta_i, \theta_r, \theta_t$	Incident, reflected, transmitted angles	rad	
$\theta_{i\text{B}}$	Brewster angle	rad	
$\theta_{i\text{c}}$	Critical angle	rad	
θ_{F}	Faraday rotation angle	rad	
η	Intrinsic impedance	Ω	$\eta = \sqrt{\mu/\epsilon}$
η_c	Impedance of a lossy medium	Ω	$\eta_c = |\eta_c|e^{j\phi_{\eta}}$
λ	Wavelength	m	$\lambda = 2\pi/\beta$
$\bar{\lambda}$	Wavelength in waveguide	m	$\bar{\lambda} = 2\pi/\bar{\beta}$
μ, μ_0	Permeability	H-m^{-1}	
μ_r	Relative permeability		
μ_c	Complex permeability	H-m^{-1}	$\mu_c = \mu' - j\mu''$
ρ	Magnitude of Γ		
ρ	Volume charge density	C-m^{-3}	Phasor
$\tilde{\rho}$	Volume charge density	C-m^{-3}	
ρ_s	Surface charge density	C-m^{-2}	

Symbol	Quantity	SI Unit	Comments
ρ_l	Linear charge density	$C\text{-m}^{-1}$	
σ	Conductivity	S	
τ	Magnitude of $\mathcal{T}$		
τ	Time constant	s	
τ_r	Relaxation time	s	$\tau_r = \epsilon/\sigma$
$\tilde{\Phi}$	Electrostatic potential	volts (V)	
Φ	Electrostatic potential	V	Phasor
χ_e	Electrical susceptibility		
χ_m	Magnetic susceptibility		
ψ	True angle of refraction	rad	
ω	Angular frequency	rad-s^{-1}	
ω_p	Plasma frequency	rad-s^{-1}	
ω_c	Gyrofrequency	rad-s^{-1}	
ω_0	Larmor frequency	rad-s^{-1}	Section 6.4
ω_0	Resonance frequency	rad-s^{-1}	Section 5.3
ω_0	Characteristic frequency of bound electrons	rad-s^{-1}	Section 6.3

General Bibliography

In addition to the specific references to numerous books and articles provided in the footnotes, the following books on electromagnetic fields and waves were found to be generally useful in developing this manuscript.

Adler, R. B., L. J. Chu, and R. M. Fano, *Electromagnetic Energy Transmission and Radiation*, John Wiley & Sons, Inc., 1960.

Atwater, H. A., *Introduction to Microwave Theory*, McGraw-Hill, Inc., 1962.

Cheng, D. K., *Field and Wave Electromagnetics*, Addison Wesley, 2nd ed., 1989.

Johnson, C. C., *Field and Wave Electrodynamics*, McGraw-Hill, Inc., 1965.

Feynman, R. P., R. O. Leighton, and M. Sands, *The Feynman Lectures on Physics*, Addison Wesley, 1964.

Jordan, E. C., and K. G. Balmain, *Electromagnetic Waves and Radiating Systems*, 2nd ed., Prentice-Hall, 1968.

Kraus, J. D., *Electromagnetics*, 4th ed., McGraw-Hill, 1992.

Paris, D. T., and F. K. Hurd, *Basic Electromagnetic Theory*, McGraw-Hill, Inc., 1969.

Plonsey, R., and R. E. Collin, *Principles and Applications of Electromagnetic Fields*, 2nd ed., McGraw-Hill, 1982.

Ramo, S., J. R. Whinnery, and T. Van Duzer, *Fields and Waves in Communication Electronics*, 3rd ed., Wiley, 1994.

Skilling, H. H., *Fundamentals of Electric Waves*, 2nd ed., Wiley, 1948; reprinted by Robert E. Krieger Publishing, Co., 1974.

Answers to Odd-Numbered Problems

CHAPTER 2

2-1. (a) $\beta \simeq 17.8$ rad-m^{-1}, $\lambda \simeq 35.3$ cm. (b) $\mathscr{E}_z(0, t) = 0.15\cos(1.7\pi \times 10^9 t)$, $\mathscr{E}_z(\lambda/4, t) = 0.15\sin(1.7\pi \times 10^9 t)$. (c) $\mathscr{E}_z(x, 0) \simeq 0.15\cos(17.8x)$, $\mathscr{E}_z(x, \pi/\omega) \simeq -0.15\cos(17.8x)$.

2-3. (a) $f \simeq 444$ MHz. (b) $\overline{\mathscr{E}}(z, t) \simeq -\hat{\mathbf{y}}\sin[2\pi(444 \times 10^6)t - 9.3]$ mV-m^{-1}.

2-5. (a) $f = 100$ MHz, $\lambda = 2$ m, $v_p = 2\times 10^8$ m-s^{-1}. (b) $\epsilon_r \simeq 2.25$, $\eta \simeq 80\pi\Omega$. (c) $\overline{\mathscr{E}} \simeq \hat{\mathbf{z}}50\sin[2\pi(10^8 t + 0.5x - 0.125)]$ V-m^{-1}. (d) $|\mathbf{S}_{av}| \simeq 4.97$ W-m^{-1}.

2-7. (a) $\overline{\mathscr{H}}(y, z, t) = \hat{\mathbf{y}}[bE_0/(\omega\mu_0)]\cos(ay)\cos(\omega t - bz) - \hat{\mathbf{z}}[aE_0/(\omega\mu_0)]\sin(ay)\sin(\omega t - bz)$. (b) $a^2 + b^2 = \omega^2\epsilon_0\mu_0$. (c) $\hat{\mathbf{u}}_1 = -a\hat{\mathbf{y}} + b\hat{\mathbf{z}}/\sqrt{a^2 + b^2}$, $\hat{\mathbf{u}}_2 = a\hat{\mathbf{y}} + b\hat{\mathbf{z}}/\sqrt{a^2 + b^2}$.

2-9. (a) $\beta \simeq 70\pi$ rad-m^{-1}. (b) $\overline{\mathscr{H}} \simeq \hat{\mathbf{y}}0.126[\cos(21\pi \times 10^9 t - 70\pi z) + \cos(21\pi \times 10^9 t + 70\pi z)]$ mA-m^{-1}.

2-11. $\overline{\mathscr{E}}(x, y, t) \simeq \hat{\mathbf{x}}60\cos(2\pi \times 10^{10}t - 125.7x + 167.6y + 67.9°) - \hat{\mathbf{y}}80\cos(2\pi \times 10^{10}t - 125.7x + 167.6y + 67.9°)$ V-m^{-1}, $\overline{\mathscr{H}}(x, y, t) \simeq \hat{\mathbf{z}}(100/377)\cos(2\pi \times 10^{10}t - 125.7x + 167.6y + 67.9°)$ A-m^{-1}.

2-13. (a) ~ 0.55 np-m^{-1}, ~ 1.43 rad-m^{-1}, ~ 4.38 m, $\sim 8.76 \times 10^7$ m-s^{-1}, ~ 1.82 m, $\sim 103e^{j21.0°}\Omega$. (b) ~ 0.0109 np-m^{-1}, ~ 0.726 rad-m^{-1}, ~ 8.66 m, $\sim 1.73 \times 10^8$ m-s^{-1}, ~ 92 m, $\sim 218e^{j0.858°}\Omega$.

2-15. (a) $\epsilon_r \simeq 3.12$, $\sigma_{\text{eff}} \simeq 1.73$ mS-m^{-1}. (b) ~ 1.61 dB.

2-17. (a) $d \simeq 26.5$ cm. (b) $d \simeq 22.1$ cm.

2-19. (a) $\epsilon_r' \simeq 50$, $\epsilon_r'' \simeq 30$. (b) $\overline{\mathscr{H}}(y, t) \simeq -\hat{\mathbf{z}}10.1e^{-39y}\cos(1.83\pi \times 10^9 t - 141y - 15.5°)$ mA-m^{-1}. (c) $\mathbf{S}_{av} \simeq \hat{\mathbf{y}}2.44e^{-78y}$ mW-m^{-2}.

2-21. $\lambda_{\text{sw}}(\text{m}) \simeq 1581/\sqrt{f(\text{Hz})}$ for $f \ll \sim 1$ GHz.

2-23. (a) ~ 12.3 m. (b) ~ 38.9 m.

2-25. (a) $k_1 \simeq 1581$ m-s$^{-1/2}$. (b) $\sim 298°$.

2-27. (a) $d \simeq 3.06$ cm. (b) $d \simeq 1.71$ cm. 915 MHz signal penetrates deeper. (c) ~ 4.26 dB, ~ 7.63 dB.

2-29. (a) $\sigma \simeq 3.61 \times 10^{-4}$ S-m^{-1}, $\epsilon_r \simeq 7.18$. (b) $d \simeq 40.5$ m, $\alpha \simeq 0.214$ dB-m^{-1}. (c) ~ 21.4 dB, ~ 11.5 rad.

2-31. $\tan\delta_c \simeq 35.5$, $\eta_c \simeq 449e^{-j0.808°}\Omega$, $\alpha \simeq 8.33$ dB-(cm)$^{-1}$.

2-33. For cooked beef, $\tan \delta_c \approx 0.331$ and $d \approx 2.14$ cm. For smoked bacon, $\tan \delta_c = 0.05$ and $d \approx 40.3$ cm. Therefore, cooked beef has a much higher microwave absorption rate.

2-35. $\sigma \approx 3$ S-m^{-1}, $\epsilon_r \approx 77.4$.

2-37. (a) $\overline{\mathcal{H}}_\phi = (E_0/r)e^{r^2/\omega^2}\cos(\omega t - \beta z)$. (c) $P_{total} = E_0^2 \pi \omega^2/(4\eta)$. $|S_{av}(r=0)| \approx$ 20 kW-m^{-2}. (d) ~$0.265 R_{SE}$ where $R_{SE} \approx 1.5 \times 10^8$ km. (e) $E_0 \approx 625$ GV-m$^{-1} \gg$ 3 MV-m^{-1}, ~1.73×10^{12} N-m^{-2}, $m \approx 34.7$ kg.

2-39. (a) $\overline{\mathcal{V}}_{in} \approx 61.8 \sin(\omega t)$ mV. (b) Same as (a).

2-41. (a) All safe. (b) $P = 85$ W.

2-43. $t_{max} \approx 3.7$ hrs.

2-45. (a) ~4.67 μN-m^{-2}. (b) ~9.33 μN-m^{-2}.

2-47. ~3.33×10^{10} bars.

2-49. LHCP.

2-51. (a) $\overline{\mathcal{E}}_1(y, t) \approx \hat{x}32.5\cos[2\pi \times 10^{10}t - (200\pi/3)y]$ V-m^{-1}, $\overline{\mathcal{E}}_2(y, t) \approx \hat{x}32.5\cos[2\pi \times 10^{10}t - (200\pi/3)y] + \hat{z}32.5\sin[2\pi \times 10^{10}t - (200\pi/3)y]$ V-m^{-1}. (b) $|S_{av}|_{CP}/|S_{av}|_L = 2$.

2-53. (a) Yes, $f = 1.5$ GHz. (b) $\hat{u} = \hat{x}0.8 - \hat{y}0.6$, LHCP. (c) $\mathbf{H} \approx [-j7.96\hat{x} - j10.6\hat{y} + 13.3\hat{z}]e^{-j(8x-6y)\pi}$ μA-m^{-1}.

2-55. (a) Straight line (along the x axis). (b) Circle (on the x-y plane). (c) Ellipse (on the x-y plane).

CHAPTER 3

3-1. (a) $f \approx 2.5$ GHz, $|S_{av}| \approx 0.3$ mW-m^{-2}. (b) $\overline{\mathcal{E}}_1(y, t) = 3\sin(50\pi y/3)\sin(5\pi \times 10^9 t)$ V-m^{-1}. (c) $\overline{\mathcal{H}}_1(y, t) \approx 7.96\cos(50\pi y/3)\cos(5\pi \times 10^9 t)$ mA-m^{-1}. (d) $\overline{\mathcal{H}}_1(y = -0.75, t) = 0$.

3-3. (a) $f \approx 6$ GHz, $\lambda = 5$ cm. (b) $\mathbf{H}_i \approx (60/377)e^{-j40\pi x}(\hat{z} + j\hat{y})$ A-m^{-1}. (c) $\mathbf{E}_r = 60e^{j40\pi x}(-\hat{y} + j\hat{z})$ V-m^{-1}, $\mathbf{H}_r \approx (60/377)e^{j40\pi x}(\hat{z} + j\hat{y})$ A-m^{-1}. (d) $\mathbf{E}_1(x) = -\hat{y}120j\sin(40\pi x) - \hat{z}120\sin(40\pi x)$ V-m^{-1}.

3-5. $\epsilon_{2r} = 5$, $\mu_{2r} = 1.8$.

3-7. (a) $\mathbf{H}_r(z) \approx \hat{x}5.66e^{-j210z}$ mA-m^{-1}, $\mathbf{H}_t(z) \approx \hat{x}15.7e^{j210z}$ mA-m^{-1}. (b) $|(S_{av})_i| \approx$ 18.9 mW-m^{-1}, $|(S_{av})_r| \approx 6.03$ mW-m^{-1}, and $|(S_{av})_t| \approx 12.8$ mW-m^{-1}. (c) $\mathbf{H}_1(z) \approx \hat{x}[10e^{j210z} + 5.66e^{-j210z}]$ mA-m^{-1}.

3-9. $d_{max} \approx 10.5$ m.

3-11. (a) ~34.7%, ~12.9%. (b) $d_{wet} \approx 21.3$ cm, $d_{dry} \approx 6.75$ m.

3-13. ~66.8%.

3-15. (a) $d_{\min} \simeq 4.66$ cm. (b) ~65.43% at 1 GHz and ~65.43% at 2 GHz.

3-17. (a) $f_1 \simeq 2.04$ GHz, $f_2 \simeq 4.06$ GHz, $f_3 \simeq 5.30$ GHz. (b) $f \simeq 5.30$ GHz.
(c) Bandwidth $\simeq 10.13$ GHz. (d) See Figure 3-14; Γ_{eff} rises faster on the high frequency
end (i.e., reflections increase as λ approaches 400 nm) so that higher frequencies (blue,
violet) are reflected more effectively.

3-19. (a) $n \simeq 1.51$ (benzene). (b) $\Gamma_{\mathrm{eff}} \simeq 0$. (c) $\Gamma_{\mathrm{eff}} \simeq 0.224e^{-j136.4°}$.

3-21. ~29.38%.

3-23. $|\Gamma_{\mathrm{eff}}(\lambda_1 = 3.5\mu\mathrm{m})| \simeq 0.021$.

3-25. (a) $d_{\min} \simeq 16.8$ cm. (b) $|\mathbf{E}_r/\mathbf{E}_i| \simeq 0.807$.

3-27. (a) ~ 36.2%. (b) ~ 17.0%. (c) ~ 13.1%.

3-29. (a) $E_0 \simeq 200$ V-m^{-1}, $f \simeq 1.2$ GHz, $\theta_i \simeq 36.9°$. (b) $\mathbf{E}_r \simeq -\hat{\mathbf{y}}200e^{-j4.8\pi x}e^{+j6.4\pi z}$
V-m^{-1}. (c) $\mathbf{E}_1 \simeq \hat{\mathbf{y}}400\sin(6.4\pi z)e^{-j(4.8\pi x+\pi/2)}$ V-m^{-1}. $z_{\min} = 0$, $z_{\max} \simeq 7.81$ cm,
$z_{\min} \simeq 15.6$ cm, etc.

3-31. (a) $\mathbf{E}_{r\perp} = -\hat{\mathbf{y}}E_0 e^{-jk(x+z)/\sqrt{2}}$, $\mathbf{E}_{r\parallel} = (\hat{\mathbf{x}} - \hat{\mathbf{z}})(jE_0/\sqrt{2})e^{-jk(x+z)/\sqrt{2}}$. (b) $\mathbf{E}_i$ is LHCP,
$\mathbf{E}_r$ is RHCP.

3-33. (a) $\mathbf{E}_i \simeq \hat{\mathbf{y}}0.614e^{-j40\pi(x-z)/\sqrt{2}}$ V-m^{-1}, $\mathbf{E}_r \simeq -\hat{\mathbf{y}}0.422e^{-j40\pi(x+z)/\sqrt{2}}$ V-m^{-1}, and
$\mathbf{E}_t \simeq \hat{\mathbf{y}}0.192e^{-j40\pi(x-\sqrt{29}z)/\sqrt{2}}$ V-m^{-1}. (b) $\mathbf{E}_i \simeq 0.434(\hat{\mathbf{x}} + \hat{\mathbf{z}})e^{-j40\pi(x-z)/\sqrt{2}}$ V-m^{-1}, $\mathbf{E}_r \simeq$
$0.205(-\hat{\mathbf{x}} + \hat{\mathbf{z}})e^{-j40\pi(x+z)/\sqrt{2}}$ V-m^{-1}, and $\mathbf{E}_t \simeq (\hat{\mathbf{x}}0.0922 + \hat{\mathbf{z}}0.0171)e^{-j40\pi(x-\sqrt{29}z)/\sqrt{2}}$
V-m^{-1}.

3-35. $\Gamma_\perp \simeq -0.976$, $\mathcal{T}_\perp \simeq 0.0244$.

3-37. $h \simeq 850$ m.

3-39. (a) $n_a \simeq 1.57$. (b) $\theta_{iB} \simeq 53.1°$.

3-41. (a) $n_1 \simeq 1.41$, $n_1 \simeq 1.15$, and $n_1 = 2.0$. When in water, $n_1 \simeq 1.88$, $n_1 \simeq 1.54$, and
$n_1 \simeq 2.66$. (b) $\theta_{te} = 0$, $\theta_{te} \simeq 35.3°$, $\theta_{te} = 90°$.

3-43. (b) $n_p \simeq 1.68$.

3-45. (a) $\psi_d \simeq 76.9°$. (b) $n \simeq 1.336$.

3-47. (a) No such angle exists. (b) $\theta_i \geq$ ~48.75°.

3-49. (a) Perpendicularly polarized. (b) $A \simeq 53.021°$. (c) $n_{\mathrm{SiO}_2} \simeq 1.4472$. (d) $n_{\mathrm{SiO}_2} \simeq$
1.4444.

3-51. (b) $|\Gamma_{eff}| \approx 0.437$, (c) $|\Gamma_{eff}| \approx 0.437$.

3-53. Same as equations [3.7] and [3.8], but with the following substitutions: $\eta_1 = \eta_3 \leftrightarrow \eta_1/(\cos\theta_i)$, $\eta_2 \leftrightarrow \eta_2/(\cos\theta_2)$, $d \leftrightarrow d/(\cos\theta_2)$, where $\sin\theta_2 = \sin\theta_i\sqrt{\epsilon_0/\epsilon_2}$.

CHAPTER 4

4-1. (a) $\bar{\gamma}_0 = j100\pi/3$ rad-m^{-1}, $\bar{\gamma}_1 \approx 117$ np-m^{-1}, $\bar{\gamma}_2 \approx 296$ np-m^{-1}. (b) $\bar{\gamma}_0 = j200\pi/3$ rad-m^{-1}, $\bar{\gamma}_1 \approx j44.1\pi$ rad-m^{-1}, $\bar{\gamma}_2 \approx j234$ rad-m^{-1}. (c) $\bar{\gamma}_0 = j400\pi/3$ rad-m^{-1}, $\bar{\gamma}_1 \approx j124\pi$ rad-m^{-1}, $\bar{\gamma}_2 \approx j277$ rad-m^{-1}.

4-3. (a) TM$_0$, TE$_1$, TM$_1$. (b) TM$_0$, TE$_1$, TM$_1$, TE$_2$, TM$_2$.

4-5. (a) TEM. (b) TEM. (c) TEM, TE$_1$, TM$_1$.

4-7. $l_{min} \approx 3.71$ cm.

4-9. (a) $\lambda_{c_2} = 3$ cm, $\lambda_{c_3} = 2$ cm, $\lambda_{c_4} = 1.5$ cm. (b) $\bar{\lambda}_0 = 3.75$ cm, $\bar{\lambda}_1 \approx 4.80$ cm.

4-11. (a) $f_{c_0} = 0$, $f_{c_1} = 3$ GHz, $f_{c_2} = 6$ GHz. (b) $\bar{v}_{p_0} = c$, $\bar{v}_{p_1} \approx 1.08c$, $\bar{v}_{p_2} \approx 1.51c$, $\bar{\lambda}_0 = 3.75$ cm, $\bar{\lambda}_1 \approx 4.05$ cm, $\bar{\lambda}_2 \approx 5.67$ cm, $Z_{TM_0} \approx 377\Omega$, $Z_{TM_1} \approx 349.5\Omega$, $Z_{TM_2} \approx 249\Omega$. (c) TM$_6$.

4-13. $\theta_{i_1} \approx 75.5°$, $\theta_{i_2} = 60°$, $\theta_{i_3} \approx 41.4°$, $\theta_{i_4} = 0$.

4-15. (a) Propagating TM$_4$ mode. (b) $f = 40$ GHz.
(c) $E_x \approx j226C_1\sin(160\pi y)e^{-j640\pi x/3}$, $E_y \approx 301.3C_1\cos(160\pi y)e^{-j640\pi x/3}$. (d) TM$_7$.

4-17. $P_{av} \approx 167$ kW-(cm)$^{-1}$.

4-19. $P_{av}^{TEM} \approx 3$ GW-m^{-2}, $P_{av}^{TE_1} \approx 1.11$ GW-m^{-2}, $P_{av}^{TM_1} \approx 2$ GW-m^{-2}.

4-21. (b) $f = \sqrt{3}f_{c_m}$. (c) $\alpha_{c_{min}} = [2(3^{0.75})/(a\eta)]\sqrt{\pi\mu_0 f_{c_m}/(2\sigma)}$. (d) $\alpha_{c_{TEM}} \approx 2.82 \times 10^{-3}$ np-m^{-1}, $\alpha_{c_{TE_1}} \approx 2.30 \times 10^{-3}$ np-m^{-1}, $\alpha_{c_{TM_1}} \approx 6.91 \times 10^{-3}$ np-m^{-1}.

4-23. (a) $P_{av} = [a\cos\theta_i/\sqrt{\mu_0/\epsilon_0}][1 + \lambda\sin[(2\pi a/\lambda)\sin\theta_i]/(2a\pi\sin\theta_i)]$, (b) for $\sin\theta_i = \sqrt{3}/2$ and $a = 1$ cm, $P_{av} = (a/2)\sqrt{\epsilon_0/\mu_0} \approx 26.5$ μW.

4-25. (a) Odd TM$_1$ and TE$_1$, Even TM$_2$ and TE$_2$. (b) Odd TM$_1$ and TE$_1$, Even TM$_2$ and TE$_2$, Odd TM$_3$ and TE$_3$, and Even TM$_4$ and TE$_4$. (c) Odd TM$_1$ and TE$_1$.

4-27. $d \leq 0.289$ mm.

4-29. (a) Odd TM$_1$ and TE$_1$ and Even TM$_2$ and TE$_2$. (b) $\alpha_x^{TM_1} \approx 3.01$ np-(cm)$^{-1}$, $\beta_x^{TM_1} \approx 1.47$ rad-(cm)$^{-1}$, $\alpha_x^{TM_2} \approx 1.79$ np-(cm)$^{-1}$, $\beta_x^{TM_2} \approx 2.83$ rad-(cm)$^{-1}$, $\alpha_x^{TM_3} \approx 0.140$ np-(cm)$^{-1}$, $\beta_x^{TM_3} \approx 3.35$ rad-(cm)$^{-1}$, $\alpha_x^{TE_1} \approx 3.13$ np-(cm)$^{-1}$, $\beta_x^{TE_1} \approx 1.20$ rad-(cm)$^{-1}$, $\alpha_x^{TE_2} \approx 2.38$ np-(cm)$^{-1}$, $\beta_x^{TE_2} \approx 2.36$ rad-(cm)$^{-1}$, $\alpha_x^{TE_3} \approx 0.548$ np-(cm)$^{-1}$, and $\beta_x^{TE_3} \simeq 3.31$ rad-(cm)$^{-1}$. (c) $\theta_i^{TM_1} \simeq 66.9°$, $\theta_i^{TM_2} \simeq 40.8°$, $\theta_i^{TM_3} \simeq 26.7°$, $\theta_i^{TE_1} \simeq$ 71.3°, $\theta_i^{TE_2} \simeq 50.9°$, $\theta_i^{TE_3} \simeq 28.1°$.

4-31. $d \simeq 11.0$ μm.

4-33. (a) $f_{c_1} = 0$ (Odd TM$_1$), $f_{c_2} \simeq 1.33$ GHz (Even TE$_2$), $f_{c_3} \simeq 2.67$ GHz (Odd TM$_3$), and $f_{c_4} = 4$ GHz (Even TE$_4$). (b) No TE$_m$ mode propagates. (c) Even TE$_2$ and TE$_4$ modes.

4-35. $v_g \simeq k_1\sqrt{f}/(1 - \sqrt{f})$.

4-37. $v_g = (\overline{\beta}/\omega)[1 + (\alpha_x/\beta_x)(d\alpha/d\beta_x)]/[(\alpha_x/\beta_x)(d\alpha_x/d\beta_x)(\mu_d\epsilon_d) + \mu_0\epsilon_0]$, where $(d\alpha_x/d\beta_x) = (\beta_x/\alpha_x)(\epsilon_0/\epsilon_d)^2 \tan(\beta_xd/2)[\tan(\beta_xd/2) + (\beta_xd/2)\sec^2(\beta_xd/2)]$.

CHAPTER 5

5-1. (a) $f_{c_{10}} \simeq 59.1$ GHz. (b) $f_{c_{11}} = f_{c_{20}} \simeq 118$ GHz. (c) Bandwidth $\simeq 59.1$ GHz.

5-3. (a) $f_{c_{10}} \simeq 8.82$ MHz, $f_{c_{01}} \simeq 30.6$ MHz. (b) TE$_{92}$ and TM$_{92}$.

5-5. (a) $f_{c_{10}} \simeq 4.31$ GHz, $f_{c_{20}} \simeq 8.61$ GHz, $f_{c_{01}} \simeq 9.49$ GHz, $f_{c_{11}} \simeq 10.4$ GHz. (b) At $f = 5.85$ GHz, $\overline{v}_{p_{10}} \simeq 1.48c$ and $\overline{\lambda}_{10} \simeq 7.57$ cm. At $f = 8.2$ GHz, $\overline{v}_{p_{10}} \simeq 1.175c$ and $\overline{\lambda}_{10} \simeq 4.30$ cm.

5-7. (a) $\overline{\lambda}_{10} \simeq 0.3\{f(GHz)\sqrt{1 - [2.08/f(GHz)]^2}\}^{-1}$ for $f > \sim2.08$ GHz.
(b) $\overline{\lambda}_{10} \simeq 33.3\{f(MHz)\sqrt{1 - [231/f(MHz)]^2}\}^{-1}$ for $f > \sim231$ MHz.
(c) $\overline{\lambda}_{10} \simeq 0.3\{f(GHz)\sqrt{1 - [4.16/f(GHz)]^2}\}^{-1}$ for $f > \sim4.16$ GHz, $\overline{\lambda}_{10} \simeq 33.3\{f(MHz)\sqrt{1 - [462/f(MHz)]^2}\}^{-1}$ for $f > \sim462$ MHz.

5-9. (a) $\epsilon_r \simeq 9$. (b) No mode other than TE$_{10}$ propagates. (c) No propagation occurs.

5-11. (a) $\overline{\lambda}_{10} \simeq 76.2$ cm. (b) $\overline{v}_{p_{10}} \simeq 1.11c$, $Z_{TE_{10}} \simeq 419\Omega$. (c) Only TE$_{10}$ mode. (d) TE$_{10}$, TE$_{20}$, TE$_{01}$, TE$_{11}$, TM$_{20}$, and TM$_{11}$.

5-13. (a) $f_{c_{10}} \simeq 25$ MHz, $f_{c_{20}} \simeq 50$ MHz, $f_{c_{30}} \simeq 75$ MHz, $f_{c_{40}} \simeq 100$ MHz, $f_{c_{50}} \simeq 125$ MHz. (b) $f_{c_{10}} \simeq 2.78$ MHz, $f_{c_{20}} \simeq 5.56$ MHz, $f_{c_{30}} \simeq 8.33$ MHz, $f_{c_{40}} \simeq 11.1$ MHz, $f_{c_{50}} \simeq 13.9$ MHz. (c) $f_{c_{10}} \simeq 23.8$ MHz, $f_{c_{20}} \simeq 47.6$ MHz, $f_{c_{01}} \simeq 55.6$ MHz, $f_{c_{11}} \simeq 60.4$ MHz, $f_{c_{30}} \simeq 71.4$ MHz.

5-15. (a) TE$_{10}$. (b) $f_{c_{01}} - f_{c_{10}} \simeq 282$ MHz. (c) $f_{c_{11}} - f_{c_{10}} \simeq 541$ MHz.

5-17. (a) $a \simeq 10.833$ cm, $b \simeq 5.4165$ cm, $f \simeq 2.45$ GHz. (b) $P_{av} \simeq 964.4$ W.

5-19. (a) $a \simeq 5.814$ cm, $b \simeq 2.907$ cm. (b) $f_{c_{20}} \simeq 5.16$ GHz. (c) $P_{av} \simeq 1.93$ MW.

5-21. (a) $P_{av} \simeq 1.94$ MW. (b) $P_{av} \simeq 7.37$ MW, almost a factor of 4 increase.

5-25. (a) $E_0 \simeq 766$ kV-m^{-1}. (b) $P_{av} \simeq 383$ kW.

5-27. (a) ~4.40 m. (b) ~4.40 mm.

5-29. (a) ~58.6 cm, ~6.51 cm. (b) ~8.79 cm. (c) TE_{11} and TM_{01}.

5-31. (a) No AM propagation. (b) 21 FM modes. (c) No AM, 9 FM modes.

5-33. (a) TE_{31}, propagating. (b) $f \simeq 8.00$ GHz. (c) TM_{02} mode. (d) $E_z \simeq C_2 J_0(158r)e^{-j18\pi z}$, $H_z = 0$.

5-35. (a) TE_{11}, TM_{01}, and TE_{21}. (b) $\bar{v}_{p_{11}} \simeq 1.19c$, $\bar{\lambda}_{11} \simeq 2.23$ cm, $Z_{TE_{11}} \simeq 449\Omega$, $\bar{v}_{p_{01}} \simeq 1.42c$, $\bar{\lambda}_{01} \simeq 2.66$ cm, $Z_{TM_{01}} \simeq 266\Omega$, $\bar{v}_{p_{21}} \simeq 2.30c$, $\bar{\lambda}_{21} \simeq 4.31$ cm, $Z_{TE_{21}} \simeq 867\Omega$.

5-37. (a) $2a \simeq 3.22$ cm, $f \simeq 108$ GHz. (b) 339 modes.

5-39. (a) TE_{11} and TM_{01}. (b) TE_{11}, TM_{01}, TE_{21}, TM_{11}, TE_{01}, and TE_{31}.

5-41. (a) $a \simeq 2.599$ cm. (b) TE_{11}, TM_{01}, TE_{21}. (c) TE_{11}, TM_{01}, TE_{21}, TE_{01}, TM_{11}, TE_{31}, and TM_{21}.

5-43. At 4 GHz, TE_{11} and TM_{01}. At 6 GHz, TE_{11} to TE_{31} (6 modes). At 8 GHz, TE_{11} to TM_{02} (10 modes). At 11 GHz, TE_{11} to TE_{32} (18 modes).

5-45. TE_{11} mode only.

5-47. Circular waveguide minimizes losses.

5-49. (b) $Z_0 \simeq 76.7\Omega$. (c) No.

5-51. (a) $f \simeq 1.94$ GHz. (b) $f \simeq 1.19$ GHz. (c) Higher-order modes propagate.

5-53. Choosing $a = 2b$, $a \simeq 1.677$ cm, $b \simeq 8.385$ mm, and $d \simeq 3.354$ cm.

5-55. $f_{101} \simeq 15.8$ GHz, $f_{110} \simeq 16.8$ GHz, $f_{111} \simeq 17.5$ GHz, and $f_{102} \simeq 18.0$ GHz.

5-57. (a) TE_{104}. (b) $Q_{c_{104}} \simeq 10,692$.

5-59. (a) TE_{106}. (b) $a \simeq 1.905$ cm, $b \simeq 9.525$ mm, $d \simeq 9.525$ cm. (c) $f_{c_{101}} \simeq 8.03$ GHz, $Q_{c_{101}} \simeq 20,605$.

5-61. $f_{TE_{111}} \simeq 18.6$ GHz, $f_{TE_{112}} \simeq 19.1$ GHz, $f_{TE_{113}} \simeq 19.9$ GHz, $f_{TE_{114}} \simeq 21.0$ GHz.

5-63. ~1.451 cm $< d <$ ~2.897 cm.

5-65. (a) $d \simeq 5.00$ cm, $Q_c \simeq 1760$. (b) $d \simeq 3.45$ cm, $Q = Q_c Q_d/(Q_c + Q_d) \simeq 1372$.

CHAPTER 6

6-1. The protons move apart with nonlinearly increasing velocities, reaching to a nearly constant value of ~371 m-s^{-1} within ~10 ns.

6-3. (a) $E_{total}(x_0) = E_0 + q/[4\pi\epsilon_0[x_0 - (q/k)E_0]$, (b) $E_{total}(x_0) = E_0 + q/(4\pi\epsilon_0 x_0)$, (c) The resultant field is larger since q moves closer to the point x_0.

6-5. (a) $\sim 1.03 \times 10^{12}$ m-s^{-1}, (b) $x = -0.03$ m, $\overline{\mathcal{E}} = -\hat{x}\, 5.33 \times 10^4$ V-m^{-1}.

6-7. (a) $v_x(t) = -v_0 \cos[(qB_0/m)t]$, $v_y(t) = v_0 \sin[(qB_0/m)t]t$, $x(t) = [mv_0/(qB_0)] \sin[(qB_0/m)t]$, $y(t) = 1 - \cos[(qB_0/m)t]$. (b) Orbit is a circle described by: $[x(t)]^2 + [y(t) - mv_0/(qB_0)]^2 = [mv_0/(qB_0)]^2$.

6-13. (a) n-type. (b) $N_e \simeq 2.08 \times 10^{16}$ (cm)$^{-3}$. (c) $\mu_e \simeq 900$ cm^2-(V-s)$^{-1}$.

6-15. $f \approx 343$ GHz.

6-17. (a) $\theta_i \approx 70.2°$. (b) ~ 4401 km, (c) $f \le \sim 14.8$ MHz.

6-19. (a) $\omega_c = [\pi/(a\sqrt{\mu_0\epsilon_0})][1 - (\omega_p/\omega)^2]^{-1/2}$. (b) $Q_c = (a/3)[\pi f_{101}\mu_0\sigma]^{-1/2}$, where $f_{101} = (c/a)[2(1 - (\omega_p/\omega)^2]^{-1/2}$.

6-21. $\sim 1.17 \times 10^{-6} N\nu^2/f^2$ dB-m^{-1}.

6-23. $\sim 7.05 \times 10^5$ light years.

6-25. (a) $d \approx 10.47$ cm. (b) > 3.54 GHz.

6-27. ~ 10.05.

6-29. $\sim 88.4\%$.

6-31. (a) (i) Circular. (ii) Circular. (iii) Circular. (iv) Linear. (b) Circle. (c) 50% each. (d) [a] (i) Linear. (ii) Linear. (iii) Linear. (iv) Linear. [b] Circle. [c] 50% each.

Index

List of Tables